D1365633

BASIC COLLEGE
mathematics
A REAL-WORLD APPROACH

IGNACIO BELLO

BASIC COLLEGE mathematics

A REAL-WORLD APPROACH

THIRD EDITION

Ignacio Bello
Hillsborough Community College
Tampa, Florida

McGraw-Hill Higher Education

Boston Burr Ridge, IL Dubuque, IA New York San Francisco St. Louis
Bangkok Bogotá Caracas Kuala Lumpur Lisbon London Madrid Mexico City
Milan Montreal New Delhi Santiago Seoul Singapore Sydney Taipei Toronto

BASIC COLLEGE MATHEMATICS: A REAL-WORLD APPROACH, THIRD EDITION

Published by McGraw-Hill, a business unit of The McGraw-Hill Companies, Inc., 1221 Avenue of the Americas, New York, NY 10020.

This book is printed on acid-free paper.

4 5 6 7 8 9 0 BAN/BAN 0 9

ISBN 978–0–07–353344–5
MHID 0–07–353344–0

ISBN 978–0–07–319994–8 (Annotated Instructor's Edition)
MHID 0–07–319994–X

Editorial Director: *Stewart K. Mattson*
Senior Sponsoring Editor: *David Millage*
Developmental Editor: *Michelle Driscoll*
Marketing Manager: *Victoria Anderson*
Senior Project Manager: *Vicki Krug*
Senior Production Supervisor: *Sherry L. Kane*
Senior Media Project Manager: *Sandra M. Schnee*
Senior Designer: *David W. Hash*
Cover/Interior Designer: *Asylum Studios*
(USE) Cover Image: *©Latin Percussion*
Senior Photo Research Coordinator: *Lori Hancock*
Photo Research: *Connie Mueller*
Supplement Coordinator: *Melissa M. Leick*
Compositor: *ICC Macmillan Inc.*
Typeface: *10/12 Times Roman*
Printer: *RR Donnelley*

www.mhhe.com

About the Author

Ignacio Bello

attended the University of South Florida (USF), where he earned a B.A. and M.A. in Mathematics. He began teaching at USF in 1967, and in 1971 became a member of the Faculty at Hillsborough Community College (HCC) and Coordinator of the Math and Sciences Department. Professor Bello instituted the USF/HCC remedial program, a program that started with 17 students taking Intermediate Algebra and grew to more than 800 students with courses covering Developmental English, Reading, and Mathematics. Aside from the present series of books *(Basic College Mathematics, Introductory Algebra,* and *Intermediate Algebra),* Professor Bello is the author of more than 40 textbooks including *Topics in Contemporary Mathematics* (ninth edition), *College Algebra, Algebra and Trigonometry,* and *Business Mathematics*. Many of these textbooks have been translated into Spanish. With Professor Fran Hopf, Bello started the Algebra Hotline, the only live, college-level television help program in Florida. Professor Bello is featured in three television programs on the award-winning Education Channel. He has helped create and develop the USF Mathematics Department Website (http://mathcenter.usf.edu), which serves as support for the Finite Math, College Algebra, Intermediate Algebra, and Introductory Algebra, and CLAST classes at USF. You can see Professor Bello's presentations and streaming videos at this website, as well as at http://www.ibello.com. Professor Bello is a member of the MAA and AMATYC and has given many presentations regarding the teaching of mathematics at the local, state, and national levels.

Contents

Contents

Chapter three

3 Decimals

Chapter four

4 Ratio, Rate, and Proportion

Chapter five

5 Percent

Contents

Chapter
six

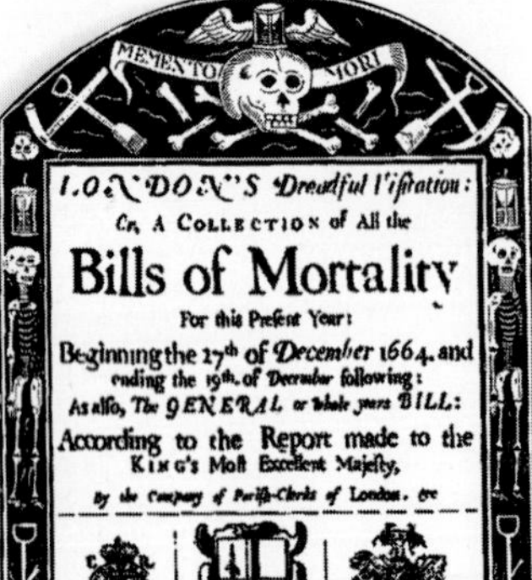

MEMENTO MORI

LONDONS Dreadful Visitation:
Or, A COLLECTION of All the
Bills of Mortality
For this Present Year:
Beginning the 27th of December 1664. and ending the 19th. of December following:
As also, The GENERAL or whole years BILL:
According to the Report made to the KING's Most Excellent Majesty,
By the Company of Parish-Clerks of London. &c

LONDON:
Printed and are to be sold by E. Cotes living in Aldersgate-street. Printer to the said Company 1665.

6 Statistics and Graphs

Chapter
seven

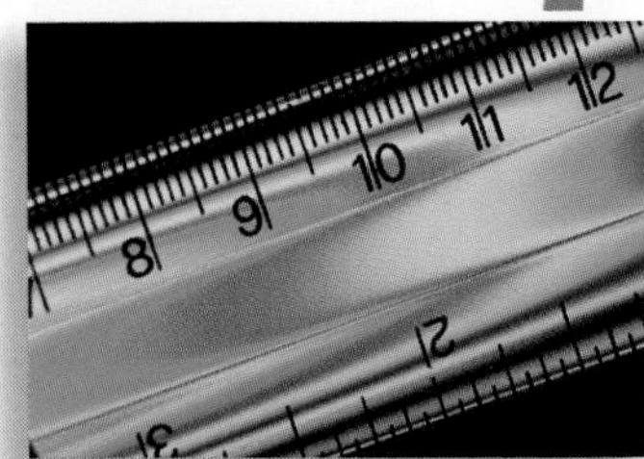

7 Measurement and the Metric System

Contents

Chapter eight

8 Geometry

Chapter nine

9 The Real Numbers

Chapter ten

10 Introduction to Algebra

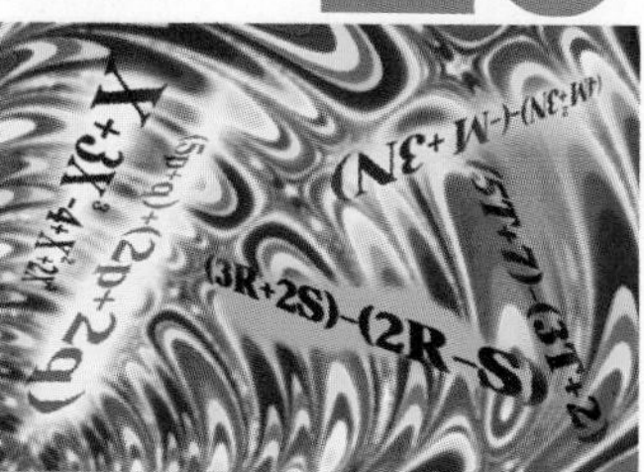

Contents

From the Author

The Inspiration for My Teaching

I was born in Havana, Cuba, and I encountered some of the same challenges in mathematics that many of my current students face, all while attempting to overcome a language barrier. In high school, I failed my freshman math course, which at the time was a complex language for me. However, perseverance being one of my traits, I scored 100% on the final exam the second time around. After working in various jobs (roofer, sheetrock installer, and dock worker), I finished high school and received a college academic scholarship. I enrolled in calculus and made a "C." Never one to be discouraged, I became a math major and learned to excel in the courses that had previously frustrated me. While a graduate student at the University of South Florida (USF), I taught at a technical school, Tampa Technical Institute, a decision that contributed to my resolve to teach math and make it come alive for my students the way brilliant instructors such as Jack Britton, Donald Rose, and Frank Cleaver had done for me. My math instructors instilled in me the motivation to be successful. I have learned a great deal about the way in which students learn and how the proper guidance through the developmental mathematics curriculum leads to student success. I believe I have accomplished a strong level of guidance in my textbook series to further explain the language of mathematics carefully to students to help them to reach success as well.

A Lively Approach to Reach Today's Students

Teaching math at the University of South Florida was a great new career for me, but I was disappointed by the materials I had to use. A rather imposing, mathematically correct but boring book was in vogue. Students hated it, professors hated it, and administrators hated it. I took the challenge to write a better book, a book that was not only mathematically correct, but **student-oriented** with **interesting applications**—many suggested by the students themselves—and even, dare we say, entertaining! That book's approach and philosophy proved an instant success and was a precursor to my current series.

Students fondly called my class "The Bello Comedy Hour," but they worked hard, and they performed well. Because my students always ranked among the highest on the common final exam at USF, I knew I had found a way to motivate them through **common-sense language** and humorous, **realistic math applications.** I also wanted to show students they could overcome the same obstacles I had in math and become successful, too. If math has been a subject that some of your students have never felt comfortable with, then they're not alone! I wrote this book with the **math-anxious** students in mind, so they'll find my tone is jovial, my explanations are patient, and instead of making math seem mysterious, I make it down-to-earth and easily digestible. For example, after I've explained the different methods for simplifying fractions, I speak directly to readers: "Which way should you simplify fractions? The way you understand!" Once students realize that math is within their grasp and not a foreign language, they'll be surprised at how much more confident they feel.

A Real-World Approach: Applications, Student Motivation, and Problem Solving

What is a "real-world approach"? I found that most textbooks put forth "real-world" applications that meant nothing to the real world of my students. How many of my students would really need to calculate the speed of a bullet (unless they are in its way) or cared to know when two trains traveling in different directions would pass by each other (disaster will certainly occur if they are on the same track)? For my students, both traditional and nontraditional, the real world consists of questions such as, "How do I find the best cell phone plan?" and "How will I pay my tuition and fees if they increase by x%?" That is why I introduce mathematical concepts through everyday applications with **real data** and give homework using similar, well-grounded situations (see the Getting Started application that introduces every section's topic and the word problems in every exercise section). Putting math in a real-world context has helped me overcome one of the problems we all face as math educators: **student motivation.** Seeing math in the real world makes students perk up in a math class in a way I have never seen before, and realism has proven to be the best motivator I've ever used. In addition, the real-world approach has enabled me to enhance students' **problem-solving skills** because they are far more likely to tackle a real-world problem that matters to them than one that seems contrived.

Diverse Students and Multiple Learning Styles

We know we live in a pluralistic society, so how do you write one textbook for everyone? The answer is to build a flexible set of teaching tools that instructors and students can adapt to their own situations. Are any of your students members of a **cultural minority?** So am I! Did they learn **English as a second language?** So did I! You'll find my book speaks directly to them in a way that no other book ever has, and fuzzy explanations in other books will be clear and comprehensible in mine.

Do your students all have the same **learning style?** Of course not! That's why I wrote a book that will help students learn mathematics regardless of their personal learning style. **Visual learners** will benefit from the text's clean page layout, careful use of color highlighting, "*Web Its,*" and the video lectures on the text's website. **Auditory learners** will profit from the audio *e-Professor lectures* on the text's website, and both **auditory** and **social learners** will be aided by the *Collaborative Learning* projects. **Applied** and **pragmatic learners** will find a bonanza of features geared to help them: *Pretests* can be found in MathZone providing practice problems by every example, and *Mastery Tests* appearing at the end of every section, to name just a few. **Spatial learners** will find the chapter *Summary* is designed especially for them, while **creative learners** will find the *Research Questions* to be a natural fit. Finally, **conceptual learners** will feel at home with features like "*The Human Side of Mathematics*" and the "*Write On*" exercises. Every student who is accustomed to opening a math book and feeling like they've run into a brick wall will find in my books that a number of doors are standing open and inviting them inside.

A Preparation for Algebra

This text offers two optional features to help students in basic college mathematics prepare themselves to take the following course in algebra: (1) "Algebra Bridge" activities occur in the margins and exercise sets where appropriate in Chapters 1 through 3 and provide side-by-side comparisons between numeric and variable expressions; (2) Chapters 1 through 3 end in sections on Equations and Problem Solving that apply the chapter topic to the context of finding unknowns, which gives students exposure to the fundamentals they will need to perform algebra at a later date.

Listening to Student and Instructor Concerns

McGraw-Hill has given me a wonderful resource for making my textbook more responsive to the immediate concerns of students and faculty. In addition to sending my manuscript out for review by instructors at many different colleges, several times a year McGraw-Hill holds symposia and focus groups with math instructors where the emphasis is *not* on selling products but instead on the **publisher listening** to the needs of faculty and their students. These encounters have provided me with a wealth of ideas on how to improve my chapter organization, make the page layout of my books more readable, and fine-tune exercises in every chapter so that students and faculty will feel comfortable using my book because it incorporates their specific suggestions and anticipates their needs.

R-I-S-E to Success in Math

Why are some students more successful in math than others? Often it is because they know how to manage their time and have a plan for action. Students can use models similar to these tables to make a weekly schedule of their time (classes, study, work, personal, etc.) and a semester calendar indicating major course events like tests, papers, and so on. Have them try to do as many of the suggestions on the **"R-I-S-E"** list as possible. (Larger, printable versions of these tables can be found in MathZone at www.mhhe.com/bello.)

Weekly Time Schedule

Time	S	M	T	W	R	F	S
8:00							
9:00							
10:00							
11:00							
12:00							
1:00							
2:00							
3:00							
4:00							
5:00							
6:00							
7:00							
8:00							
9:00							
10:00							
11:00							

Semester Calendar

Wk	M	T	W	R	F
1					
2					
3					
4					
5					
6					
7					
8					
9					
10					
11					
12					
13					
14					
15					
16					

R—Read and/or view the material before and after each class. This includes the textbook, the videos that come with the book, and any special material given to you by your instructor.

I—Interact and/or practice using the CD that comes with the book or the Web exercises suggested in the sections, or seeking tutoring from your school.

S—Study and/or discuss your homework and class notes with a study partner/ group, with your instructor, or on a discussion board if available.

E—Evaluate your progress by checking the odd numbered homework questions with the answer key in the back of the book, using the mastery questions in each section of the book as a selftest, and using the Chapter Reviews and Chapter Practice Tests as practice before taking the actual test.

As the items on this list become part of your regular study habits, you will be ready to **"R-I-S-E"** to success in math.

Improvements in the Third Edition

Based on the valuable feedback of many reviewers and users over the years, the following improvements were made to the third edition of *Basic College Mathematics.*

Organizational Changes

- Chapter 2 features one new section: 2.4 The Least Common Multiple.
- Chapter 8 has been reorganized and now starts with the basic ideas of Geometry: Points, Lines, Angles, and Triangles (Section 8.1), followed by the study of Perimeters (Section 8.2), Areas (Section 8.3), Volumes of Solids (Section 8.4), and Square Roots and the Pythagorean theorem (Section 8.5).
- Chapter 9 has a new revised section (9.1) emphasizing the uses of integers and how to compare and graph them on the number line.

Pedagogical Changes

- *Real-World Applications*—many examples, applications, and real-data problems have been added or updated to keep the book's content current.
- *Web Its*—now found in the margin of the Exercises and on MathZone (www.mhhe.com/bello) to encourage students to visit math sites while they're Web surfing and discover the many informative and creative sites that are dedicated to stimulating better education in math.
- *Calculator Corners*—found before the exercise sets, these have been updated with recent information and keystrokes relevant to currently popular calculators.
- *Concept Checkers*—these have been added to the end-of-section exercises to help students reinforce key terms and equations.
- *Pretests*—can be found in MathZone (www.mhhe.com/bello), providing practice problems for every example. These Pretests results can be compared to the Practice Tests at the end of the chapter to evaluate and analyze student success.
- *The RSTUV approach* to problem solving has been expanded and used throughout this edition as a response to positive comments from both students and users of the previous edition.
- *Translate It*—boxes appear periodically before word problem exercises to help students translate phrases into equations, reinforcing the RSTUV method.
- *Skill Checker*—now appears at the end of every exercise set, making sure the students have the necessary skills for the next section.
- *Practice Tests (Diagnostic)*—at the end of every chapter give students immediate feedback and guidance on which section, examples, and pages to review.

Acknowledgments

I would like to thank the following people associated with the third edition: David Dietz, who provided the necessary incentives and encouragement for creating this series with the cooperation of Bill Barter; Christien Shangraw, my first developmental editor, who worked many hours getting reviewers and gathering responses into concise and usable reports; Randy Welch, who continued and expanded the Christien tradition into a well-honed editing engine with many features, including humor, organization, and very hard work; Liz Haefele, my former editor and publisher, who was encouraging, always on the lookout for new markets; Lori Hancock and her many helpers (LouAnn, Emily, David, and Connie), who always get the picture; Dr. Tom Porter, of Photos at Your Place, who improved on the pictures I provided; Vicki Krug, one of the most exacting persons at McGraw-Hill, who will always give you the time of day and then solve the problem; Hal Whipple, for his help in preparing the answers manuscript; Cindy Trimble, for the accuracy of the text; Jeff Huettman, one of the best 100 producers in the United States, who learned Spanish in anticipation of this project; Marie Bova, for her detective work in tracking down permission rights; and to Professor Nancy Mills, for her expert advice on how my books address multiple learning styles. David Millage, senior sponsoring editor; Michelle Driscoll and Lisa Collette, developmental editors; Torie Anderson, marketing manager; and especially Pat Steele, our very able copy editor. Finally, thanks to our attack secretary, Beverly DeVine, who still managed to send all materials back to the publisher on time. To everyone, my many thanks.

I would also like to extend my gratitude to the following reviewers of the Bello series for their many helpful suggestions and insights. They helped me write better textbooks:

Tony Akhlaghi, *Bellevue Community College*

Theresa Allen, *University of Idaho*

John Anderson, *San Jacinto College–South Campus*

Ken Anderson, *Chemeketa Community College*

Tiffany Andrade, *Fresno City College*

Keith A. Austin, *DeVry University–Arlington*

Sohrab Bakhtyari, *St. Petersburg College–Clearwater*

Fatemah Bicksler, *Delgado Community College*

Brenda Blankenship, *Volunteer State Community College*

Rich Bogdanovich, *Community College of Aurora*

Ann Brackebusch, *Olympic College*

Margaret A. Brock, *Central New Mexico Community College*

Gail G. Burkett, *Palm Beach Community College*

Linda Burton, *Miami Dade College*

Jim Butterbach, *Joliet Junior College*

Susan Caldiero, *Cosumnes River College*

Judy Carlson, *Indiana University–Purdue University Indianapolis*

Edie Carter, *Amarillo College*

Randall Crist, *Creighton University*

Mark Crawford, *Waubonsee Community College*

Mark Czerniak, *Moraine Valley Community College*

Acknowledgments

Antonio David, *Del Mar College*
Robert Diaz, *California State University–Northridge*
Parsla Dineen, *University of Nebraska–Omaha*
Sue Duff, *Guilford Technical Community College*
Lynda Fish, *St. Louis Community College–Forest Park*
Donna Foster, *Piedmont Technical College*
Jeanne H. Gagliano, *Delgado Community College*
Debbie Garrison, *Valencia Community College*
Donald K. Gooden, *Northern Virginia Community College–Woodbridge*
William Graesser, *Ivy Tech Community College*
Edna Greenwood, *Tarrant County College–Northwest Campus*
Ken Harrelson, *Oklahoma City Community College*
Joseph Lloyd Harris, *Gulf Coast Community College*
Tony Hartman, *Texarkana College*
Susan Hitchcock, *Palm Beach Community College*
Kayana Hoagland, *South Puget Sound Community College*
Patricia Carey Horacek, *Pensacola Junior College*
Peter Intarapanich, *Southern Connecticut State University*
Judy Ann Jones, *Madison Area Technical College*
Eric Kaljumagi, *Mt. San Antonio College*
Linda Kass, *Bergen Community College*
Joe Kemble, *Lamar University*
Joanne Kendall, *Blinn College–Brenham*
Bernadette Kocyba, *J S Reynolds Community College*
Theodore Lai, *Hudson County Community College*
Marie Agnes Langston, *Palm Beach Community College*
Kathryn Lavelle, *Westchester Community College*
Angela Lawrenz, *Blinn College–Bryan*
Richard Leedy, *Polk Community College*
Edith B. Lester, *Volunteer State Community College*
Mickey Levendusky, *Pima Community College*
Sharon Louvier, *Lee College*
Judith L. Maggiore, *Holyoke Community College*
Timothy Magnavita, *Bucks Community College*
Tsun-Zee Mai, *University of Alabama*
Harold Mardones, *Community College of Denver*
Lois Martin, *Massasoit Community College*
Louise Matoax, *Miami Dade College*
Gary McCracken, *Shelton State Community College*
Tania McNutt, *Community College of Aurora*
Kathryn Merritt, *Pensacola Junior College*

Barbara Miller, *Lexington Community College*
Danielle Morgan, *San Jacinto College–South Campus*
Shauna Mullins, *Murray State University*
Ken Nickels, *Black Hawk College*
Diana Orrantia, *El Paso Community College–Transmountain Campus*
Mohammed L. Pasha, *Del Mar College*
Joanne Peeples, *El Paso Community College*
Faith Peters, *Miami Dade College–Wolfson*
Jane Pinnow, *University of Wisconsin–Parkside*
Marilyn G. Platt, *Gaston College*
Janice F. Rech, *University of Nebraska–Omaha*
Libbie Reeves, *Mitchell Community College*
Tian Ren, *Queensborough Community College*
Karen Roothaan, *Harold Washington College*
Lisa Rombes, *Washtenaw Community College*
Don Rose, *College of the Sequoias*
Pascal Roubides, *Miami Dade College–Wolfson*
Juan Saavedra, *Albuquerque Technical Vocational Institute*
Judith Salmon, *Fitchburg State College*
Mansour Samimi, *Winston–Salem State University*
Susan Santolucito, *Delgado Community College*
Ellen Sawyer, *College of DuPage*
Gretchen Syhre, *Hawkeye Community College*
Kenneth Takvorian, *Mount Wachusett Community College*
Sharon Testone, *Onondaga Community College*
Stephen Toner, *Victor Valley College*
Michael Tran, *Antelope Valley College*
Bettie Truitt, *Black Hawk College*
William L. Van Alstine, *Aiken Technical College*
Julian Viera, *University of Texas at El Paso*
Andrea Lynn Vorwark, *Metropolitan Community College–Kansas City*
Pat Widder, *William Rainey Harper College*

Features and Supplements

Motivation for a Diverse Student Audience

A number of features exist in every chapter to motivate students' interest in the topic and thereby increase their performance in the course:

The Human Side of Mathematics

To personalize the subject of mathematics, the origins of numerical notation, concepts, and methods are introduced through the lives of real people solving ordinary problems.

The Human Side of Mathematics

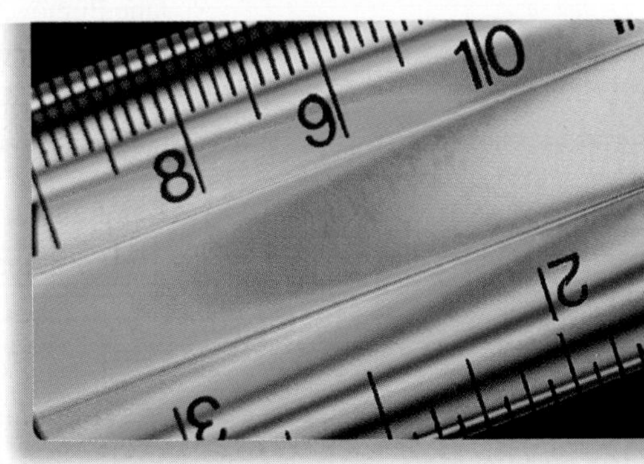

Here is a quiz for you: name one country that uses the metric system. Almost any country will do *except* the United States. Take a romantic vacation in France. It is a cool day; so you take a 30-minute walk to a nearby village, get some cheese and some wine. The temperature (in degrees Celsius, not Fahrenheit), the distance (kilometers, not miles), the cheese (kilograms, not pounds), and the wine (liters, not quarts) are all in the metric system. How did this happen?

Weights and measures are among the earliest tools invented by humans. Babylonians and Egyptians measured length with the forearm, hand, or finger, but the system used in the United States is based on the English system. The need for a single, worldwide, coordinated system of measurement was recognized over 300 years ago by Gabriel Mouton, vicar of St. Paul's Church in Lyons, France. In 1790, the National Assembly of France requested the French Academy of Sciences to "deduce an invariable standard for all measures and weights." The result was the metric system, made compulsory in France in 1840. What about the United States? As early as 1789, Thomas Jefferson submitted a report proposing a decimal-based system of measurement. Congress took no action on his proposal but later, in 1832, directed the Treasury Department to standardize the measures used by customs officials at U.S. ports. Congress allowed this report to stand without taking any formal action. Finally, in 1866, Congress legalized the use of the metric system and in 1875, the United States became one of the original signers of the Treaty of the Meter.

Getting Started

Each topic is introduced in a setting familiar to students' daily lives, making the subject personally relevant and more easily understood.

Getting Started

The sign on the left shows the price of 1 gallon of gasoline using the fraction, $\frac{9}{10}$. However, the sign on the right shows this price as the decimal 0.9. If we are given a fraction, we can sometimes find its decimal equivalent by multiplying the numerator and denominator by a number that will cause the denominator to be a power of 10 (10, 100, 1000, etc.) and then writing the decimal equivalent. For example,

$$\frac{2}{5} = \frac{2 \cdot 2}{5 \cdot 2} = \frac{4}{10} = 0.4$$

$$\frac{3}{4} = \frac{3 \cdot 25}{4 \cdot 25} = \frac{75}{100} = 0.75$$

$$\frac{3}{125} = \frac{3 \cdot 8}{125 \cdot 8} = \frac{24}{1000} = 0.024$$

Web It

Appearing in the margin of the section exercises, this URL refers students to the abundance of resources available on the Web that can show them fun, alternative explanations, and demonstrations of important topics.

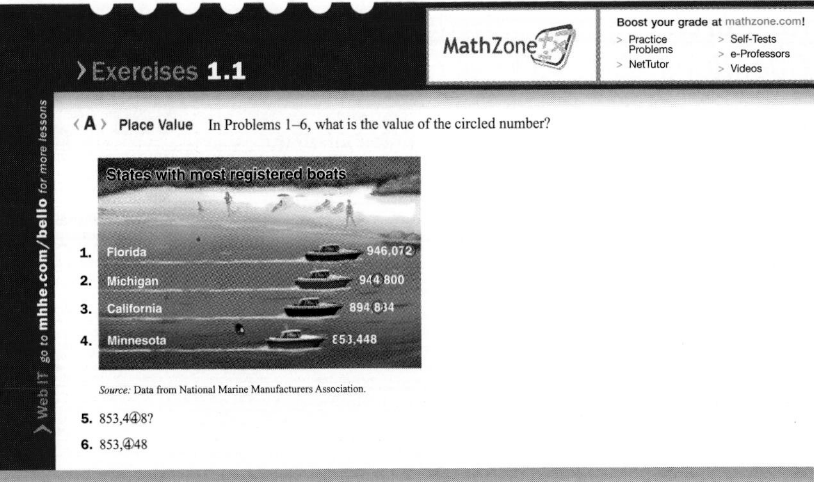

Exercises 1.1

MathZone

Boost your grade at mathzone.com!
> Practice Problems
> NetTutor
> Self-Tests
> e-Professors
> Videos

> Web IT go to mhhe.com/bello for more lessons

⟨A⟩ Place Value In Problems 1–6, what is the value of the circled number?

Source: Data from National Marine Manufacturers Association.

5. 853,4④8?

6. 853,④48

Write On

Write On

52. What are the three variables (factors) used when calculating simple interest?

53. Write in your own words which would be better for you: to take a 20% discount on an item, or to take 10% off and then 10% off the reduced price.

54. Which investment is better for you: \$10,000 invested at 5% compounded semiannually or \$10,000 invested at 4% compounded monthly. Explain why.

55. Most people give 10%, 15%, or 20% of the total bill as a tip. (See Problem 42.)

a. In your own words, give a rule to find 10% of any amount. (*Hint:* It is a matter of moving the decimal point.)

b. Based on the rule in part **a,** state a rule that can be used to find 15% of any amount.

c. Based on the rule in part **a,** state a rule that can be used to find 20% of any amount.

Writing exercises give students the opportunity to express mathematical concepts and procedures in their own words, thereby expressing and verbalizing what they have learned.

Collaborative Learning

Concluding the chapter are exercises for collaborative learning that promote teamwork by students on interesting and enjoyable exploration projects.

Collaborative Learning

Planets	Time of Revolution Around the Sun
Mercury	88 days
Venus	224.7 days
Earth	365.25 days
Mars	687 days
Jupiter	11.86 years
Saturn	29.46 years
Uranus	84.01 years
Neptune	164.8 years
Pluto*	248.5 years

* Pluto is not classified as a planet at this time.

Weight	Multiplier	Factor
Earth weight	×	0.38
Earth weight	×	0.91
Earth weight	×	1.00
Earth weight	×	0.38
Earth weight	×	2.60
Earth weight	×	1.10
Earth weight	×	0.90
Earth weight	×	1.20
Earth weight	×	0.08

Suppose you want to become younger (in age) and trimmer (in weight). We can help you with that, but it will require a little travel. Let us start with the age issue. On Earth, a year is 365 days long but the year on other planets is dependent on the distance the planet is away from the sun. The first table shows the time it takes each of the planets to go around the sun. Now, suppose you are 18 years old. What is your Mercurian age? Since Mercury goes around the sun every 88 days, you would be much older! (You have been around more!) As a matter of fact your Mercurian age M would be $M = 18 \times \frac{365}{88} \approx 75$ years old.

In which planet will you become youngest? We will let you find out for yourself. *Hint:* The denominator has to be in days, for example, 88 days.

Now, for the weight issue. Suppose you weigh 130 pounds on Earth. How much would you weigh on Mercury? Since your weight is dependent on the laws of gravity of the planet you are visiting, your weight will change by the factor shown in the second table.

On Mercury, your weight would be: $130 \times 0.38 \approx 49$ pounds.

Guided Tour

Research Questions

Research questions provide students with additional opportunities to explore interesting areas of math, where they may find the questions can lead to surprising results.

> Research Questions

1. Where did the decimal point first appear, in what year, and who used the decimal point?

2. Write a brief description of the decimal system we use, how it works, and how decimals are used.
Reference: http://www.infoplease.com

3. Write a brief description of the Dewey decimal system, how it works, and for what it is used.
Reference: http://www.mtsu.edu

4. Write a paragraph about Bartholomeus Pitiscus and the ways in which he used the decimal point.

5. Write a paragraph explaining where our decimal system comes from and its evolution throughout the years.
Reference: http://www.scit.wlv.ac.uk

Abundant Practice and Problem Solving

Bello offers students many opportunities and different skill paths for developing their problem-solving skills.

Pretest

An optional Pretest can be found in MathZone at www.mhhe.com/bello and is especially helpful for students taking the course as a review who may remember some concepts but not others. The answer grid is also found online and gives students the page number, section, and example to study in case they missed a question. The results of the Pretest can be compared with those of the Practice Test at the end of the chapter to evaluate progress and student success.

> Pretest Chapter 5

1. a. Write 87% as a decimal.
b. Write 29.8% as a decimal.

2. a. Write $6\frac{1}{4}\%$ as a decimal.
b. Write $25\frac{2}{3}\%$ as a repeating decimal.

3. a. Write 0.08 as a percent.
b. Write 6.2 as a percent.

4. a. Write 21% as a reduced fraction.
b. Write 80% as a reduced fraction.

5. Write $37\frac{1}{2}\%$ as a reduced fraction.

6. Write $\frac{5}{8}$ as a percent.

7. 30% of 50 is what number?

8. Find $10\frac{1}{2}\%$ of 60.

9. What is $66\frac{2}{3}\%$ of 39?

10. Find what percent of 50 is 20.

11. 20 is 50% of what number?

12. Set up a proportion and find 20% of 40.

13. Set up a proportion and find what percent of 80 is 40.

14. Set up a proportion and find 20% of what number is 10.

15. The sales tax rate in one city is 5%. Find the total price paid for a pair of shoes that cost $56.60.

16. Find the simple interest earned on $500 invested at 8.5% for 2 years.

> Answers to Pretest Chapter 5

Answer		If You Missed	Review		
		Question	Section	Examples	Page
1. a. 0.87	**b.** 0.298	1	5.1	1	289
2. a. 0.0625	**b.** $0.25\overline{6}$	2	5.1	2	289
3. a. 8%	**b.** 620%	3	5.1	3	290
4. a. $\frac{21}{100}$	**b.** $\frac{4}{5}$	4	5.1	4	290
5. $\frac{3}{8}$		5	5.1	5	291
6. $62\frac{1}{2}\%$		6	5.1	6, 7	291–292
7. 15		7	5.2	1	299
8. 6.3		8	5.2	2	300
9. 26		9	5.2	3	300

Paired Examples/ Problems

Examples are placed adjacent to similar problems intended for students to obtain immediate reinforcement of the skill they have just observed. These are especially effective for students who learn by doing and who benefit from frequent practice of important methods. Answers to the problems appear at the bottom of the page.

EXAMPLE 1 Credit card comparisons

Suppose you have a card with a \$25 annual fee and an 18% annual percentage rate (APR). If the average monthly balance on your card is \$500 and you can get a different card with a 14% APR and no annual fee, how much will your savings be if you change to the second card?

SOLUTION You will be saving

4% (18% − 14%) of \$500, or \$20, in interest and the \$25 annual fee.

The total savings amount to \$20 + \$25 = \$45.

PROBLEM 1

Suppose you have a card with a \$50 annual fee and a 10% APR. If your average monthly balance is \$1000 and you can get a card with a 14% APR and no annual fee, how much will your total savings be?

Answers to PROBLEMS

1. \$10

RSTUV Method

The easy-to-remember **"RSTUV"** method gives students a reliable and helpful tool in demystifying word problems so that they can more readily translate them into equations they can recognize and solve.

- **R**ead the problem and decide what is being asked.
- **S**elect a letter or □ to represent this unknown.
- **T**ranslate the problem into an equation.
- **U**se the rules you have studied to solve the resulting equation.
- **V**erify the answer.

RSTUV PROCEDURE TO SOLVE WORD PROBLEMS

1. Read the problem carefully and decide what is asked for (the unknown).
2. Select □ or a letter to represent the unknown.
3. Translate the problem into an equation.
4. Use the rules we have studied to solve the equation.
5. Verify the answer.

Translate This

These boxes appear periodically before word-problem exercises to help students translate phrases into equations, reinforcing the RSTUV method.

TRANSLATE THIS

1. The average cost for tuition and fees at a 2-year public college is \$2191. This represents a 5.4% increase from last year's cost L.

Source: www.CollegeBoard.com.

2. The average 2-year public college student receives grant aid (G) that reduces the average tuition and fees of \$2191 to a net price of \$400.

Source: www.CollegeBoard.com.

3. According to the Census Bureau a person with a bachelor's degree earns 62% more on average than those with only a high school diploma. If a person with a high school diploma makes h dollars a year, what is the pay b for a person with a bachelor's degree?

Source: www.CollegeBoard.com.

4. At 2-year public institutions, the average tuition and fees is \$2191, which is \$112 more than last year's cost L. What was the cost of tuition and fees last year?

Source: www.CollegeBoard.com.

5. At American University, the cost per credit for undergraduate courses is \$918. What is the cost C for an undergraduate taking h credits?

Source: American University.

The third step in the RSTUV procedure is to **TRANSLATE** the information into an equation. In Problems 1–10 **TRANSLATE** the sentence and match the correct translation with one of the equations A–O.

A. $T + 3.8 = 103.8$
B. $2191 - G = 400$
C. $P = 39 + 0.45m$
D. $D = 10.9391P$
E. $2191 = L + 5.4\%L$
F. $b = h + 62\%h$
G. $C = 13.75h$
H. $C = 918h$
I. $G - 2191 = 400$
J. $h = b + 62\%h$
K. $C = 51W$
L. $P = 10.9391D$
M. $W = 51C$
N. $2191 = L + 112$
O. $D = 10.9391P$

6. The monthly access fee for a cell phone plan is \$39 plus \$0.45 for each minute after a 300-minute allowance. If Joey used m minutes over 300, write an equation for the amount P Joey paid.
7. If the body temperature T of a penguin were 3.8°F warmer, it would be as warm as a goat, 103.8°F.
8. The Urban Mobility Institute reports that commuters say they would be willing to pay \$13.75 an hour to avoid traffic congestion. What would be the cost C for h hours of avoiding traffic congestion?
9. Since the price of gas is higher, commuters are willing to pay more than \$13.75 an hour to avoid traffic congestion, say W dollars per hour. South Florida travelers lose 51 hours per year due to traffic congestion. Write an equation that will give the total cost C for Florida travelers to avoid losing any time to traffic congestion.
10. As of this writing, one U.S. dollar is worth 10.9391 Mexican pesos. Write the formula to convert D dollars to P Mexican pesos.

Exercises

A wealth of exercises for each section are organized according to the learning objectives for that section, giving students a reference to study if they need extra help.

Exercises 1.6

MathZone

Boost your grade at mathzone.com!
- Practice Problems
- NetTutor
- Self-Tests
- e-Professors
- Videos

⟨A⟩ Writing Division as Multiplication In Problems 1–20, divide and check using multiplication.

1. $30 \div 5$	**2.** $63 \div 9$	**3.** $28 \div 4$	**4.** $6 \div 1$
5. $21 \div 7$	**6.** $2 \div 0$	**7.** $0 \div 2$	**8.** $54 \div 9$
9. $7 \div 7$	**10.** $56 \div 8$	**11.** $\frac{36}{9}$	**12.** $54 \div 6$
13. $\frac{3}{0}$	**14.** $\frac{0}{3}$	**15.** $\frac{32}{8}$	**16.** $\frac{13}{13}$
17. $\frac{24}{1}$	**18.** $\frac{45}{9}$	**19.** $\frac{62}{6}$	**20.** $\frac{48}{5}$

Guided Tour

Applications

Students will enjoy the exceptionally creative applications in most sections that bring math alive and demonstrate that it can even be performed with a sense of humor.

Applications

78. *Distance traveled by* Pioneer *in 10 months* The first manufactured object to leave the solar system was *Pioneer 10*, which attained a velocity of 5.1×10^4 kilometers per hour when leaving Earth on its way to Jupiter. After 10 months, it had traveled

$$(5.1 \times 10^4) \times (7.2 \times 10^3), \text{ or}$$

$$(5.1 \times 7.2) \times (10^4 \times 10^3) \text{ km.}$$

How many kilometers is that?

79. *Distance traveled by* Pioneer *in 10 years* The top speed of *Pioneer 10* was 1.31×10^5 kilometers per hour. At this rate, in 8.7×10^4 hours (about 10 years), *Pioneer 10* would go

$$(1.31 \times 10^5) \times (8.7 \times 10^4), \text{ or}$$

$$(1.31 \times 8.7) \times (10^5 \times 10^4) \text{ km.}$$

How many kilometers is that?

80. *Petroleum and gas reserves* The estimated petroleum and gas reserves for the United States are about 2.8×10^{17} kilocalories. If we consume these reserves at the rate of 1.4×10^{16} kilocalories per year, they will last $\frac{2.8 \times 10^{17}}{1.4 \times 10^{16}}$, or $\frac{2.8}{1.4} \times \frac{10^{17}}{10^{16}}$ years. How many years is that?

81. *Internet sites* During January 2006 almost 3×10^7 people visited flower, greeting, and gift sites on the Internet. On average, each visitor spent about \$9 from January 1 to February 9 on these sites. How many million dollars were spent on flowers, greetings, and gift sites during this period?

Using Your Knowledge

Optional, extended applications give students an opportunity to practice what they've learned in a multistep problem requiring reasoning skills in addition to numerical operations.

Using Your Knowledge

What is the corresponding dose (amount) of medication for children when the adult dosage is known? There are several formulas that tell us.

41. Friend's rule (for children under 2 years):

(Age in months · adult dose) ÷ 150 = child's dose

Suppose a child is 10 months old and the adult dose is a 75-milligram tablet. What is the child's dose? [*Hint:* Simplify (10 · 75) ÷ 150.]

42. Clarke's rule (for children over 2 years):

(Weight of child · adult dose) ÷ 150 = child's dose

If a 7-year-old child weighs 75 pounds and the adult dose is 4 tablets a day, what is the child's dose? [*Hint:* Simplify (75 · 4) ÷ 150.]

43. Young's rule (for children between 3 and 12):

(Age · adult dose) ÷ (age + 2) = child's dose

Suppose a child is 6 years old and the adult dose of an antibiotic is 4 tablets every 12 hours. What is the child's dose? [*Hint:* Simplify (6 · 4) ÷ (6 + 2).]

You already know how to use the order of operations to evaluate expressions. Here is a more challenging game: Use the numbers 1, 2, 5, and 6, any of the operations we have studied, and parentheses to make an expression whose value is 24. Here is one: (1 + 5) · (6 − 2).

Now, you make your own expression using the given numbers.

44. 1, 2, 7, 2 **45.** 2, 3, 3, 8 **46.** 4, 6, 9, 5 **47.** 2, 3, 8, 9

Study Aids to Make Math Accessible

Because some students confront math anxiety as soon as they sign up for the course, the Bello system provides many study aids to make their learning easier.

Objectives

The objectives for each section not only identify the specific tasks students should be able to perform, they organize the section itself with letters corresponding to each section heading, making it easy to follow.

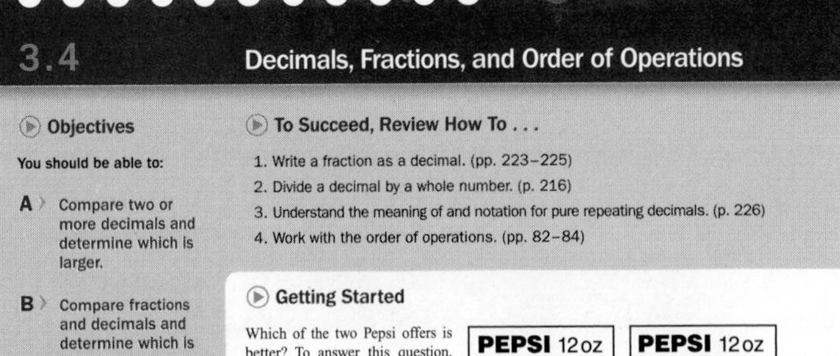

3.4 Decimals, Fractions, and Order of Operations

Objectives

You should be able to:

A › Compare two or more decimals and determine which is larger.

B › Compare fractions and decimals and determine which is larger.

C › Solve applications involving fractions, decimals, and the order of operations.

To Succeed, Review How To . . .

1. Write a fraction as a decimal. (pp. 223–225)
2. Divide a decimal by a whole number. (p. 216)
3. Understand the meaning of and notation for pure repeating decimals. (p. 226)
4. Work with the order of operations. (pp. 82–84)

Getting Started

Which of the two Pepsi offers is better? To answer this question, we can find the price of each can by dividing \$2.00 by 8 and \$2.99 by 12 and then comparing the individual price of each can.

Reviews

Every section begins with "To succeed, review how to . . . ," which directs students to specific pages to study key topics they need to understand to successfully begin that section.

Concept Checker

This feature has been added to the end-of-section exercises to help students reinforce key terms and concepts.

Concept Checker

Fill in the blank(s) with the correct word(s), phrase, or mathematical statement.

81. To convert a **percent** to a **decimal,** move the decimal point ______ places to the ______ and omit the % symbol.

82. To convert a **decimal** to a **percent,** move the decimal point ______ places to the ______ and attach the % symbol.

83. To convert a **percent** to a **fraction,** follow these steps:

a. Write the number over ______

b. ______ the resulting fraction

divide	**attach**
left	**reduce**
omit	**multiply**
two	**percent**
right	**fraction**
100	**%**

Mastery Test

85. Among taxpayers who owed the IRS money, 82% said they would use money from their savings or checking accounts to pay their taxes. Write 82% as

a. a reduced fraction. **b.** a decimal.

86. Write as a percent:

a. $\frac{3}{8}$ **b.** $\frac{3}{5}$

87. Write as a reduced fraction:

a. $6\frac{1}{2}\%$ **b.** 6.55%

88. Write as a reduced fraction:

a. 19% **b.** 80%

Mastery Tests

Brief tests in every section give students a quick checkup to make sure they're ready to go on to the next topic.

Skill Checkers

These brief exercises help students keep their math skills well honed in preparation for the next section.

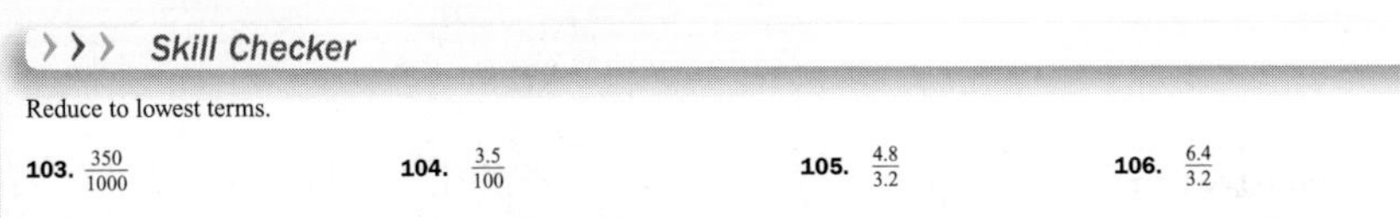

Skill Checker

Reduce to lowest terms.

103. $\frac{350}{1000}$ **104.** $\frac{3.5}{100}$ **105.** $\frac{4.8}{3.2}$ **106.** $\frac{6.4}{3.2}$

Calculator Corner

Most calculators contain a set of parentheses on the keyboard. These keys will allow you to specify the exact order in which you wish operations to be performed. Thus to find the value of $(4 \cdot 5) + 6$, key in

(4 × 5) + 6 ENTER

to obtain 26. Similarly, if you key in

4 × (5 + 6) ENTER

you obtain 44. However, in many calculators you won't be able to find the answer to the problem $4(5 + 6)$ unless you key in the multiplication sign × between the 4 and the parentheses.

It's important to note that some calculators follow the order of operations automatically and others do not. To find out whether yours does, enter 2 + 3 × 4 ENTER. If your calculator follows the order of operations, you should get 14 for an answer.

Calculator Corner

When appropriate, optional calculator exercises are included to show students how they can explore concepts through calculators and verify their manual exercises with the aid of technology.

Algebra Bridge

Algebra Bridges offer optional exercises to help students begin to work with variables in mathematics.

Algebra Bridge

Look at the numbers in the arithmetic column. Using x instead of 10, write an equivalent expression in the algebra column.

	Arithmetic	Algebra
106.	$7 \times 10^3 + 9 \times 10^2 + 3 \times 10 + 8$	
107.	$4 \times 10^2 + 8 \times 10 + 6$	
108.	$2 \times 10 + 4$	

Summary Chapter 3

Section	Item	Meaning	Example
3.1A	Word names	The word name for a number is the number written in words.	The word name for 4.7 is four and seven tenths.
3.1B	Expanded form	Numeral written as a sum indicating the value of each digit.	$78.2 = 70 + 8 + \frac{2}{10}$
3.2B	Multiplying by powers of 10	A product involving 10, 100, 1000, and so on as a factor.	$93.78 \times 100 = 9378$

Summary

An easy-to-read grid summarizes the essential chapter information by section, providing an item, its meaning, and an example to help students connect concepts with their concrete occurrences.

Review Exercises

Chapter review exercises are coded by section number and give students extra reinforcement and practice to boost their confidence.

Review Exercises Chapter 3

(If you need help with these exercises, look in the section indicated in brackets.)

1. ‹ **3.1A** › *Give the word name.*
a. 23.389
b. 22.34
c. 24.564

2. ‹ **3.1B** › *Write in expanded form.*
a. 37.4
b. 59.09
c. 145.035
d. 150.309
e. 234.003

3. ‹ **3.1C** › *Add.*
a. 8.51 + 13.43
b. 9.6457 + 15.78
c. 5.773 + 18.0026
d. 6.204 + 23.248
e. 9.24 + 14.28

Practice Test with Answers

The chapter Practice Test offers students a nonthreatening way to review the material and determine whether they are ready to take a test given by their instructor. The answers to the Practice Test give students immediate feedback on their performance, and the answer grid gives them specific guidance on which section, example, and pages to review for any answers they may have missed.

Practice Test Chapter 3

(Answers on page 252)

Visit www.mhhe.com/bello to view helpful videos that provide step-by-step solutions to several of the problems below.

1. Give the word name for 342.85.

2. Write 24.278 in expanded form.

3. 9 + 12.18 = ____

4. 46.654 + 8.69 = ____

5. 447.42 − 18.5 = ____

6. 5.34 · 0.013 = ____

Answers to Practice Test Chapter 3

Answer	If You Missed	Review		
	Question	Section	Examples	Page
1. Three hundred forty-two and eighty-five hundredths	1	3.1	1	201
2. $20 + 4 + \frac{2}{10} + \frac{7}{100} + \frac{8}{1000}$	2	3.1	2	202
3. 21.18	3	3.1	3	203
4. 55.344	4	3.1	4, 5	203–204

Cumulative Review

The Cumulative Review covers material from the present chapter and any of the chapters prior to it and can be used for extra homework or for student review to improve their retention of important skills and concepts.

Cumulative Review Chapters 1–3

1. Write 300 + 90 + 4 in standard form.

2. Write three thousand, two hundred ten in standard form.

3. Write the prime factors of 20.

4. Write 60 as a product of primes.

5. Multiply: $2^2 \times 5 \times 5^0$

6. Simplify: 49 ÷ 7 · 7 + 8 − 5

7. Classify $\frac{5}{4}$ as proper or improper.

8. Write $\frac{11}{2}$ as a mixed number.

9. Write $5\frac{3}{8}$ as an improper fraction.

10. $\frac{2}{3} = \frac{?}{27}$

11. $\frac{5}{7} = \frac{25}{?}$

12. Multiply: $\frac{1}{2} \cdot 5\frac{1}{6}$

Supplements for Instructors

Annotated Instructor's Edition

This version of the student text contains **answers** to all odd- and even-numbered exercises in addition to helpful **teaching tips.** The answers are printed on the same page as the exercises themselves so that there is no need to consult a separate appendix or answer key.

Computerized Test Bank (CTB) Online

Available through MathZone, this **computerized test bank,** utilizes Brownstone Diploma®, an algorithm-based testing software to quickly create customized exams. This user-friendly program enables instructors to search for questions by topic, format, or difficulty level; to edit existing questions or to add new ones; and to scramble questions and answer keys for multiple versions of the same test. Hundreds of text-specific open-ended and multiple-choice questions are included in the question bank. Sample chapter tests and final exams in Microsoft Word® and PDF formats are also provided.

Instructor's Solutions Manual

Available on Mathzone, the Instructor's Solutions Manual provides comprehensive, **worked-out solutions** to all exercises in the text. The methods used to solve the problems in the manual are the same as those used to solve the examples in the textbook.

www.mathzone.com

McGraw-Hill's MathZone is a complete online tutorial and homework management system for mathematics and statistics, designed for greater ease of use than any other system available. Instructors have the flexibility to create and share courses and assignments with colleagues, adjunct faculty, and teaching assistants with only a few clicks of the mouse. All algorithmic exercises, online tutoring, and a variety of video and animations are directly tied to text-specific materials. Completely customizable, MathZone suits individual instructor and student needs. Exercises can be easily edited, multimedia is assignable, importing additional content is easy, and instructors can even control the level of help available to students while doing their homework. Students have the added benefit of full access to the study tools to individually improve their success without having to be part of a MathZone course. MathZone allows for automatic grading and reporting of easy-to assign algorithmically generated homework, quizzes and tests. Grades are readily accessible through a fully integrated grade book that can be exported in one click to Microsoft Excel, WebCT, or BlackBoard.

MathZone Offers

- Practice exercises, based on the text's end-of-section material, generated in an unlimited number of variations, for as much practice as needed to master a particular topic.
- Subtitled videos demonstrating text-specific exercises and reinforcing important concepts within a given topic.
- Net Tutor™, integrating online whiteboard technology with live personalized tutoring via the Internet.

- Assessment capabilities, powered through ALEKS, which provide students and instructors with the diagnostics to offer a detailed knowledge base through advanced reporting and remediation tools.
- Faculty with the ability to create and share courses and assignments with colleagues and adjuncts, or to build a course from one of the provided course libraries.
- An Assignment Builder that provides the ability to select algorithmically generated exercises from any McGraw-Hill math textbook, edit content, as well as assign a variety of MathZone material including an ALEKS Assessment.
- Accessibility from multiple operating systems and Internet browsers.

ALEKS® (www.aleks.com)

ALEKS (**A**ssessment and **LE**arning in **K**nowledge **S**paces) is a dynamic online learning system for mathematics education, available over the Web 24/7. ALEKS assesses students, accurately determines their knowledge, and then guides them to the material that they are most ready to learn. With a variety of reports, Textbook Integration Plus, quizzes, and homework assignment capabilities, ALEKS offers flexibility and ease of use for instructors.

- ALEKS uses artificial intelligence to determine exactly what each student knows and is ready to learn. ALEKS remediates student gaps and provides highly efficient learning and improved learning outcomes.
- ALEKS is a comprehensive curriculum that aligns with syllabi or specified textbooks. Used in conjunction with McGraw-Hill texts, students also receive links to text-specific videos, multimedia tutorials, and textbook pages.
- Textbook Integration Plus allows ALEKS to be automatically aligned with syllabi or specified McGraw-Hill textbooks with instructor chosen dates, chapter goals, homework, and quizzes.
- ALEKS with AI-2 gives instructors increased control over the scope and sequence of student learning. Students using ALEKS demonstrate a steadily increasing mastery of the content of the course.
- ALEKS offers a dynamic classroom management system that enables instructors to monitor and direct student progress toward mastery of course objectives.

Supplements for Students

Student's Solutions Manual

This supplement contains complete worked-out solutions to all odd-numbered exercises and all odd- and even-numbered problems in the Review Exercises and Cumulative Reviews in the textbook. The methods used to solve the problems in the manual are the same as those used to solve the examples in the textbook. This tool can be an invaluable aid to students who want to check their work and improve their grades by comparing their own solutions to those found in the manual and finding specific areas where they can do better.

www.mathzone.com

McGraw-Hill's MathZone is a complete online tutorial and homework management system for mathematics and statistics, designed for greater ease of use than any other system available. All algorithmic exercises, online tutoring, and a variety of video and animations are directly tied to text-specific materials.

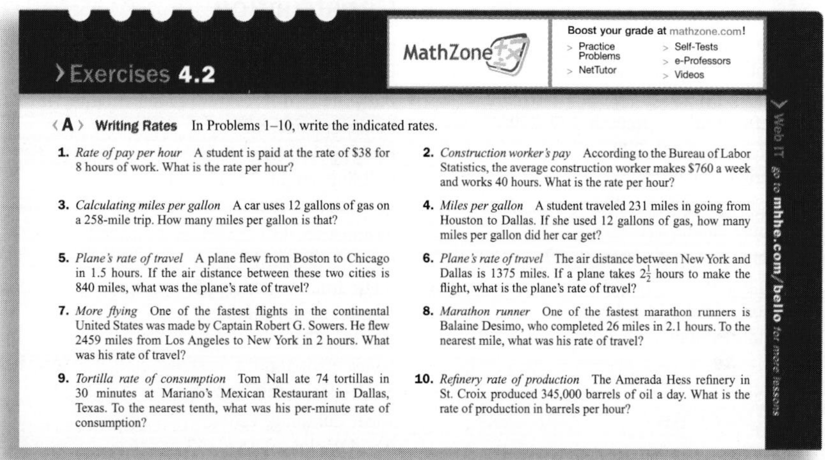

> Exercises 4.2

MathZone

Boost your grade at mathzone.com!
> Practice Problems
> NetTutor
> Self-Tests
> e-Professors
> Videos

〈A〉 **Writing Rates** In Problems 1–10, write the indicated rates.

1. *Rate of pay per hour* A student is paid at the rate of $38 for 8 hours of work. What is the rate per hour?
2. *Construction worker's pay* According to the Bureau of Labor Statistics, the average construction worker makes $760 a week and works 40 hours. What is the rate per hour?
3. *Calculating miles per gallon* A car uses 12 gallons of gas on a 258-mile trip. How many miles per gallon is that?
4. *Miles per gallon* A student traveled 231 miles in going from Houston to Dallas. If she used 12 gallons of gas, how many miles per gallon did her car get?
5. *Plane's rate of travel* A plane flew from Boston to Chicago in 1.5 hours. If the air distance between these two cities is 840 miles, what was the plane's rate of travel?
6. *Plane's rate of travel* The air distance between New York and Dallas is 1375 miles. If a plane takes $2\frac{1}{2}$ hours to make the flight, what is the plane's rate of travel?
7. *More flying* One of the fastest flights in the continental United States was made by Captain Robert G. Sowers. He flew 2459 miles from Los Angeles to New York in 2 hours. What was his rate of travel?
8. *Marathon runner* One of the fastest marathon runners is Balaine Desimo, who completed 26 miles in 2.1 hours. To the nearest mile, what was his rate of travel?
9. *Tortilla rate of consumption* Tom Nall ate 74 tortillas in 30 minutes at Mariano's Mexican Restaurant in Dallas, Texas. To the nearest tenth, what was his per-minute rate of consumption?
10. *Refinery rate of production* The Amerada Hess refinery in St. Croix produced 345,000 barrels of oil a day. What is the rate of production in barrels per hour?

> Web IT go to mhhe.com/bello for more lessons

ALEKS® (www.aleks.com)

ALEKS (**A**ssessment and **LE**arning in **K**nowledge **S**paces) is a dynamic online learning system for mathematics education, available over the Web 24/7. ALEKS assesses students, accurately determines their knowledge, and then guides them to the material that they are most ready to learn. With a variety of reports, Textbook Integration Plus, quizzes, and homework assignment capabilities, ALEKS offers flexibility and ease of use for instructors.

Bello Video Series

The video series is available on DVD and VHS tape and features the authors introducing topics and working through selected odd-numbered exercises from the text, explaining how to complete them step by step. The DVDs are **closed-captioned** for the hearing impaired and are also **subtitled in Spanish.**

Math for the Anxious: Building Basic Skills, by Rosanne Proga

Math for the Anxious: Building Basic Skills is written to provide a practical approach to the problem of math anxiety. By combining strategies for success with a pain-free introduction to basic math content, students will overcome their anxiety and find greater success in their math courses.

Applications Index

Animals

Business

Construction

Consumer Issues

Education

Entertainment

Finances

Food

Applications Index

Geometry

Government

Home and Garden

Law Enforcement

Medicine and Health

Science

Applications Index

Social Studies

Sports

Transportation

Weather

Section

Chapter

1 *one*

Whole Numbers

The Human Side of Mathematics

The development of the number system used in arithmetic has been a multicultural undertaking. More than 20,000 years ago, our ancestors needed to count their possessions, their livestock, and the passage of days. Australian aborigines counted to two, South American Indians near the Amazon counted to six, and the Bushmen of South Africa were able to count to ten using twos ($10 = 2 + 2 + 2 + 2 + 2$).

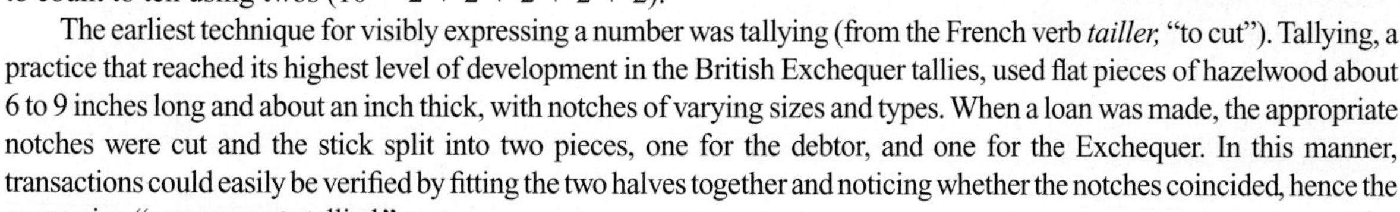

The earliest technique for visibly expressing a number was tallying (from the French verb *tailler,* "to cut"). Tallying, a practice that reached its highest level of development in the British Exchequer tallies, used flat pieces of hazelwood about 6 to 9 inches long and about an inch thick, with notches of varying sizes and types. When a loan was made, the appropriate notches were cut and the stick split into two pieces, one for the debtor, and one for the Exchequer. In this manner, transactions could easily be verified by fitting the two halves together and noticing whether the notches coincided, hence the expression "our accounts tallied."

The development of written numbers was due mainly to the Egyptians (about 3000 B.C.), the Babylonians (about 2000 B.C.), the early Greeks (about 400 B.C.), the Hindus (about 250 B.C.), and the Arabs (about 200 B.C.). The table shows the numbers some of these civilizations used:

In this chapter, we shall study operations with whole numbers and their uses in present-day society.

Egyptian, about 3000 B.C.

𓏺	𓎆	𓍢	𓆼	𓂭	𓆐	𓁨
1	10	100	1000	10,000	100,000	1,000,000

Babylonian, about 2000 B.C.

𒑱	𒁹	𒌋	𒌋𒐖	𒎙	𒁹	𒁹𒌋
0	1	10	12	20	60	600

Early Greek, about 400 B.C.

Ι	Γ	Δ	𐅄	Η	𐅅	𐅆
1	5	10	50	100	500	5000

1.1 Standard Numerals

Objectives

You should be able to:

A › Determine the place value of a digit in a numeral.

B › Write a standard numeral in expanded form.

C › Write an expanded numeral in standard form.

D › Write a standard numeral in words.

E › Write a numeral given in words in standard form.

F › Write the number corresponding to the given application.

To Succeed, Review How To . . .

Recognize the counting numbers (1, 2, 3, and so on).

Getting Started

In the following cartoon, Peter has used the **counting numbers** 1, 2, 3, and so on as mileage indicators. Unfortunately, he forgot about 4! In our number system we use the ten **digits** 0, 1, 2, 3, 4, 5, 6, 7, 8, and 9 to build **numerals** that name **whole numbers. Number** and **numeral** are closely related concepts (sometimes we use the terms interchangeably). A **number** is an abstract idea used to represent a quantity, whereas a **numeral** is a symbol that represents a number. In the cartoon, "four" is a number represented by the numeral 4. In Roman times, the number "four" would be represented by the numeral IV. Similarly, the marker on the last mileage indicator is "two hundred one," or 201, or CCI in Roman numerals. We have written 201 two ways: in the standard form **201** and in words, **two hundred one.** In this section we will learn how to write numerals three ways: in standard form, in expanded form, and in words, but before doing so we will explore how digits can have different values depending on their placement in a numeral.

By Permission of John L. Hart FLP, and Creators Syndicate, Inc.

A › Place Value

The Population of the Earth is **6,511,257,348**

To see the current population, try http://www.census.gov

The position of each digit in a number determines the digit's **place value.** Look at the world population clock. Which is the only digit that is missing? Which digits are repeated? What is the **value** of the digits that are repeated? It depends! To help you with the answer, we use a place value chart in which each group of three digits is called a **period.** We name these periods *ones, thousands, millions, billions,* and so on. Each period has three categories: *ones (units), tens,* and *hundreds* separated by commas. (Commas are usually omitted in four-digit numbers, such as 3248 and 5093.)

Now, place the **6,511,257,348** in the chart

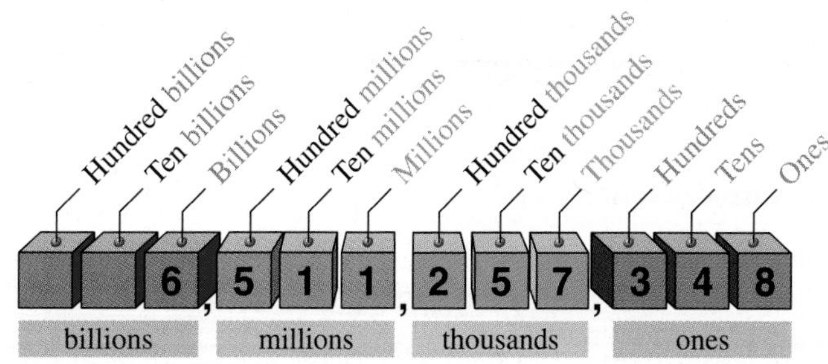

What is the value of the **1** (there are two of them)? Either **1** **million** or **1** **ten millions**. The value of **5** is either 5 **ten thousands** or 5 **hundred millions**. Get the idea? Just read the column (upward) using the category and the period in which the desired number appears.

EXAMPLE 1 **Finding the value of a digit**

Referring to the population of the earth, find the value of

a. 6 **b.** 3 **c.** 4

SOLUTION

a. The 6 appears in the **billions** column; its value is **6 billion.**

b. The 3 is in the **hundreds** column; its value is **3 hundred.**

c. The 4 is in the **tens** column; its value is **4 tens.**

PROBLEM 1

Find the value of

a. 2 **b.** 7 **c.** 8

The numeral **6,511,257,348** is an example of a *whole number.* The set of whole numbers is

0, 1, 2, 3, 4, 5, 6, 7, 8, 9, 10, 11, 12, . . .

The *smallest* whole number is 0, and the pattern continues *indefinitely,* as indicated by the three dots (. . .) called an *ellipsis*. This means that there is no *largest* whole number. If 0 is omitted from the set of whole numbers, the new set of numbers is called the **natural** or **counting** numbers. The natural numbers are

1, 2, 3, . . .

As promised, we will now learn how to write numbers in standard form, in expanded form, and in words.

B > Expanded Form of a Numeral

The **standard numeral** for the last mileage indicator in the cartoon is 201. Here are other standard numerals: 4372 and 68. Standard numerals such as 4372, 201, and 68 can be written in **expanded form** like this:

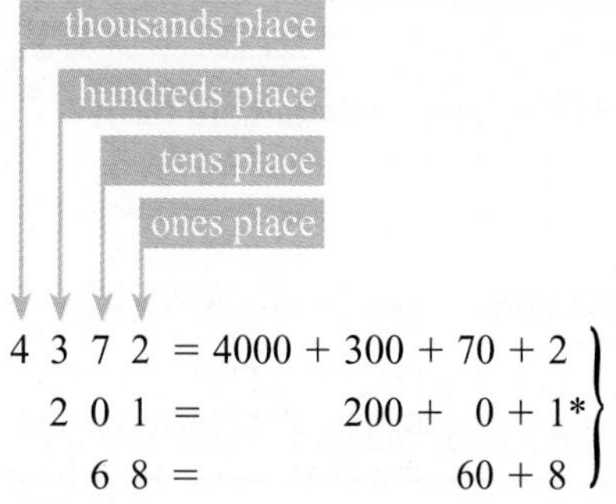

$$\left.\begin{aligned} 4372 &= 4000 + 300 + 70 + 2 \\ 201 &= 200 + 0 + 1^{*} \\ 68 &= 60 + 8 \end{aligned}\right\}$$

* 201 can also be written in expanded form as 200 + 1.

Answers to PROBLEMS

1. a. 2 hundred thousand
b. 7 thousand
c. 8 ones (units)

EXAMPLE 2 **Writing numbers in expanded form**
Write 4892 in expanded form.

SOLUTION

$$4892 = 4000 + 800 + 90 + 2$$

PROBLEM 2
Write 9241 in expanded form.

EXAMPLE 3 **Writing numbers in expanded form**
Write 765 in expanded form.

SOLUTION

$$765 = 700 + 60 + 5$$

PROBLEM 3
Write 197 in expanded form.

EXAMPLE 4 **Writing numbers in expanded form**
Write 41,205 in expanded form.

SOLUTION

$$41{,}205 = 40{,}000 + 1000 + 200 + 0 + 5 \text{ or}$$
$$= 40{,}000 + 1000 + 200 + 5$$

Note that when zeros occur in the given numeral, the expanded form is shorter.

PROBLEM 4
Write 98,703 in expanded form.

EXAMPLE 5 **Urban local public road miles**
According to the Bureau of Transportation Statistics, the number of urban local U.S. public road miles is 598,421. Write 598,421 in expanded form.

SOLUTION

$$598{,}641 = 500{,}000 + 90{,}000 + 8000 + 400 + 20 + 1$$

PROBLEM 5
The number of urban minor arterial local U.S. public road miles is 89,789. Write 89,789 in expanded form.

C › Standard Form of a Numeral

We can also do the reverse process, that is, we can write the standard form for

3000 + 200 + 80 + 9 as

3 2 8 9

EXAMPLE 6 **Writing numbers in standard form**
Write 7000 + 800 + 90 + 2 in standard form.

SOLUTION

7000 + 800 + 90 + 2 is written as

7 8 9 2

PROBLEM 6
Write 9000 + 200 + 20 + 5 in standard form.

EXAMPLE 7 **Writing numbers in standard form**
Write 70,000 + 6000 + 300 + 20 + 1 in standard form.

SOLUTION

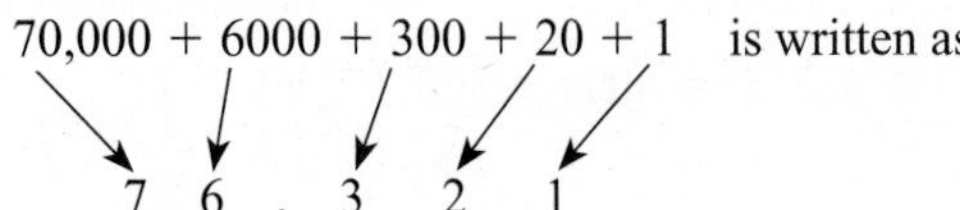

PROBLEM 7
Write 10,000 + 1000 + 100 + 10 + 1 in standard form.

Answers to PROBLEMS

2. 9000 + 200 + 40 + 1 **3.** 100 + 90 + 7 **4.** 90,000 + 8000 + 700 + 3 **5.** 80,000 + 9000 + 700 + 80 + 9 **6.** 9225 **7.** 11,111

EXAMPLE 8 **Writing numbers in standard form**

Write 90,000 + 600 + 1 in standard form.

SOLUTION

90,000 + 600 + 1 is written as

9 0 , 6 0 1

Note that a zero appears in the tens and in the thousands places, making the expanded numeral shorter.

PROBLEM 8

Write 50,000 + 200 + 6 in standard form.

D > Numerals to Words

The amount of the lottery check shown here is $294,000,000. Can you write this amount in words? Here we will show you how to do that, but we warn you that the word "and" used in writing "Two hundred and ninety four million" in the check is used incorrectly. Moreover, the amount in the check is not that large. According to the Guinness Book of Records, the greatest amount ever paid by a single check is $16,640,000,000 and was paid by the U.S. government to the Ministry of Finance in India. Do you know how to read and write 16,640,000,000 in words? This amount is written as

sixteen billion six hundred forty million*

The number 16,640,000,000 can be placed in a chart like this:

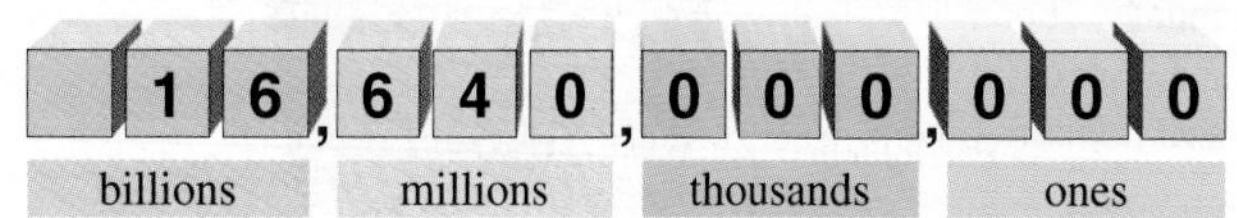

A **period** is a group of *three digits.*

In each period, digits are read in the normal way ("sixteen," "six hundred forty") and then, for each period except ones, the name of the period is added ("billion," "million"). Note that "and" is not used in writing the word names for whole numbers.

* In Europe and South America, a billion is a million millions, or 1,000,000,000,000.

Answers to PROBLEMS

8. 50,206

EXAMPLE 9 **Writing numbers in words**
Write word names for each of the numbers.

SOLUTION

a. 85 Eighty-five

b. 102,682 We first label each period for easy reference as shown.
One hundred two *thousand,* six hundred eighty-two.

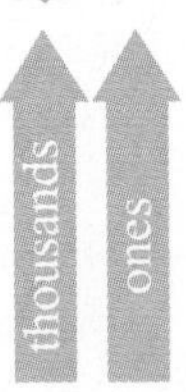

c. 13,012,825,476 We first label each period for easy reference, as shown.
The word name is thirteen *billion,* twelve *million,* eight hundred twenty-five *thousand,* four hundred seventy-six.

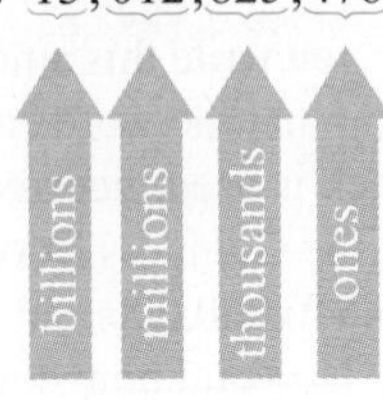

PROBLEM 9
Write word names for each of the numbers.

a. 93

b. 209,376

c. 75,142,642,893

E > Words to Numerals

We can reverse the process in Example 9 and write the word name of a numeral in standard form, as shown next.

EXAMPLE 10 **Writing numbers in standard form**
Write one hundred three million, eight hundred forty-seven thousand, six hundred eleven in standard form.

SOLUTION The standard form is 103,847,611.

PROBLEM 10
Write three hundred ten million, six hundred ninety-two thousand, seven hundred twelve in standard form.

EXAMPLE 11 **Writing numbers in standard form**
Write four billion in standard form.

Gobbling popcorn
Americans eat about 4 billion gallons of popped popcorn each year, with the average person munching about 15 gallons annually. That's enough popped corn to fill 70 four-cup-size popcorn boxes.

SOLUTION We have to include billions, millions, thousands, and ones.

4, 000, 000, 000

PROBLEM 11
Americans eat one billion, one hundred twenty-five million pounds of popcorn each year. Write 1125 million in standard form.
Source: The Popcorn Board.

Answers to PROBLEMS

9. a. Ninety-three **b.** Two hundred nine thousand, three hundred seventy-six **c.** Seventy-five billion, one hundred forty-two million, six hundred forty-two thousand, eight hundred ninety-three **10.** 310,692,712 **11.** 1,125,000,000

F > Applications Involving Standard Numerals

The ideas discussed here can be used in everyday life. For example, the amount of electricity used in your house is measured by an electric meter that records kilowatt-hours (kWh). (An electric meter actually has *six* dials.) To read the meter we must use the standard form of the number involved. Thus, the meter below is read by starting at the left and writing the figures in standard form. When the pointer is between two numbers, we use the *smaller* one. In this manner, the reading on the meter is

5 8 2 4

EXAMPLE 12 **Reading your electric meter**

Read the meter.

SOLUTION The first number is 6 (because the pointer is between 6 and 7, and we must choose the smaller number), and the next is 3, followed by 8 and 1. Thus the reading is 6381 kilowatt-hours.

PROBLEM 12

Read the meter.

> Algebra Bridge

We have now learned how to write numbers in expanded form. In algebra, we use **letters** such as x, y, and z to represent numbers. In arithmetic we write 397 as $300 + 90 + 7$ or in the equivalent form $3 \times 10^2 + 9 \times 10 + 7$, where 10^2 means 10×10. Notice the similarities between arithmetic and algebra.

For more applications to everyday life, see Using Your Knowledge on page 11.

Arithmetic	Algebra
$3 \times 10^2 + 9 \times 10 + 7$	$3x^2 + 9x + 7$
$8 \times 10^3 + 2 \times 10^2 + 5 \times 10 + 2$	$8x^3 + 2x^2 + 5x + 2$
$3 \times 10 + 9$	$3x + 9$

We will point out many more similarities and relationships between arithmetic and algebra later in the chapter.

Calculator Corner

To the student and instructor: CALCULATOR CORNER exercises give the students the opportunity to see how the material relates to the features of an inexpensive scientific calculator.

1. What is the largest number you can enter on your calculator? Write the answer as a numeral and in words.
2. Do you have to use commas to enter the number 32,456 on your calculator?

Answers to PROBLEMS

12. 1784

> Exercises 1.1

Boost your grade at mathzone.com!
> Practice Problems
> NetTutor
> Self-Tests
> e-Professors
> Videos

⟨ A ⟩ Place Value In Problems 1–6, what is the value of the circled number?

Source: Data from National Marine Manufacturers Association.

5. 853,4④8?

6. 853,④48

In Problems 7–10, what is the underlined place value?

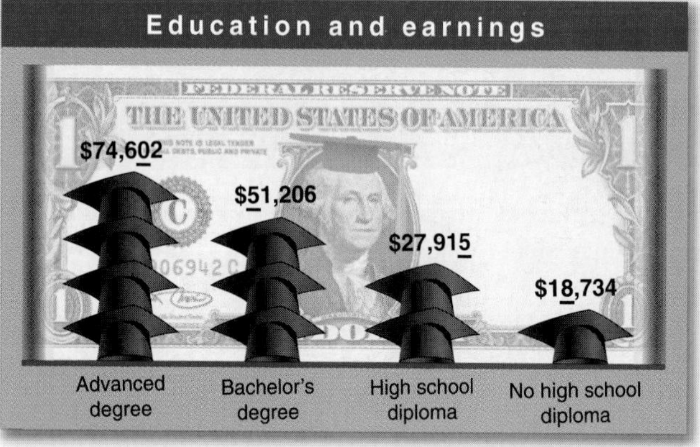

Source: Data from U.S. Census Bureau.

⟨ B ⟩ Expanded Form of a Numeral In Problems 11–30, write in expanded form.

11. 34 **12.** 27 **13.** 108

14. 375 **15.** 2500 **16.** 8030

17. 7040 **18.** 3990 **19.** 23,018

20. 30,013 **21.** 604,000 **22.** 82,000

23. 91,387 **24.** 13,058 **25.** 68,020

26. 30,050 **27.** 80,082 **28.** 50,073

29. 70,198 **30.** 90,487

> Web IT go to mhhe.com/bello for more lessons

⟨ C ⟩ Standard Form of a Numeral In Problems 31–50, write in standard form.

31. 70 + 8
32. 60 + 3
33. 300 + 8
34. 600 + 5
35. 800 + 20 + 2
36. 600 + 30 + 6
37. 700 + 1
38. 900 + 4
39. 3000 + 400 + 70 + 3
40. 1000 + 600 + 10 + 2
41. 5000 + 200 + 50
42. 7000 + 500 + 20
43. 2000 + 30
44. 5000 + 60
45. 8000 + 90
46. 6000 + 3
47. 7000 + 1
48. 1000 + 300
49. 6000 + 600
50. 8000 + 70

⟨ D ⟩ Numerals to Words In Problems 51–60, write word names for the numerals.

51. 57
52. 109
53. 3408
54. 43,682
55. 181,362
56. 6,547,210
57. 41,300,000
58. 341,310,000
59. 1,231,341,000
60. 10,431,781,000

⟨ E ⟩ Words to Numerals In Problems 61–70, write the given numeral in standard form.

61. Eight hundred nine
62. Six hundred fifty-three
63. Four thousand eight hundred ninety-seven
64. Eight thousand six hundred twenty-seven
65. Two thousand three
66. One million, two thousand
67. Two million, twenty-three thousand, forty-five
68. Seventeen million, forty-seven thousand, ninety-seven
69. Three hundred forty-five million, thirty-three thousand, eight hundred ninety-four
70. Nine billion, nine hundred ninety-nine million, nine hundred ninety

⟨ F ⟩ Applications Involving Standard Numerals

71. *Cost of raising a child* The U.S. Department of Agriculture has determined that it costs about $173,880 to raise a child from birth to age 18. Write the numeral 173,880 in words.

72. *Germs on your phone* The average phone has 25,127 germs per square inch. Write the numeral 25,127 in words.

Source: Microbiologist Charles Gerba.

73. *School attendance* On an average day in America, 13,537,000 students attend secondary school. Write the numeral 13,537,000 in words.

74. *Rainfall on an acre* A rainfall of 1 inch on an acre of ground will produce six million, two hundred seventy-two thousand, six hundred forty cubic inches of water. Write this number in standard form.

75. *College attendance* On an average year in America, fourteen million, nine hundred seventy-nine thousand students are enrolled in colleges and universities. Write this number in standard form.

76. *Middle East oil reserves* The proven oil reserves of the Middle East are about six hundred eighty-five billion barrels. Write this number in standard form.

Source: BP Statistical Review of World Energy.

Web IT go to mhhe.com/bello for more lessons

Applications

Projected college expenses The Projected College Expenses from 2006 to 2010 are as shown in the table. In Exercises 77–81, write the numerals for the specified four amounts (Annual Cost Public, Four-Year Public, Annual Cost Private, Four-Year Private) in words.

77. Corresponding to year 2006

78. Corresponding to year 2007

79. Corresponding to year 2008

80. Corresponding to year 2009

Projected College Expenses

Beginning School Year	Annual Cost Public	Four-Year Projected Public	Annual Cost Private	Four-Year Projected Private
2006	$14,872	$63,627	$32,044	$137,091
2007	$15,542	$66,491	$33,486	$143,260
2008	$16,241	$69,482	$34,993	$149,707
2009	$16,972	$72,609	$36,568	$156,444
2010	$17,736	$75,876	$38,213	$163,484

Source: College Board Annual Survey of Colleges.

81. Corresponding to year 2010

In Exercises 82–85, write the word name for the numeral.

82. $29,470: Annual median income of African-American households in 1999

83. 823,500: Number of African-American–owned businesses in the United States

84. $86,500: Average receipts of an African-American–owned firm

85. 954,000: Number of African-Americans age 25 and over who have a graduate professional degree

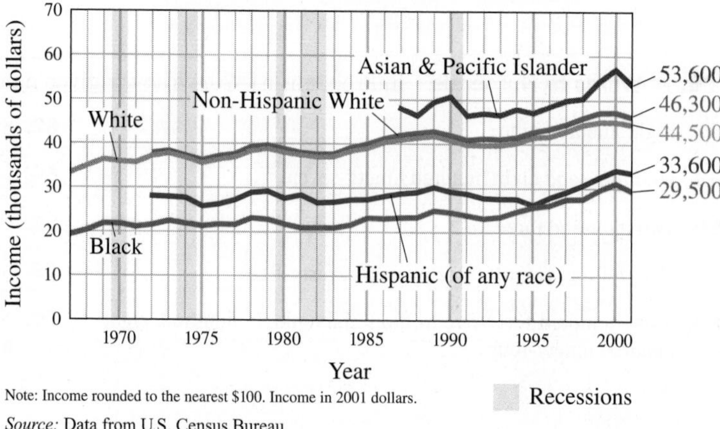

Note: Income rounded to the nearest $100. Income in 2001 dollars.

Source: Data from U.S. Census Bureau.

In Exercises 86–90, read the graph above and write the word name for the numeral (to the right of the graph, in color) that describes the situation.

86. The median Asian and Pacific Islander's household income

87. The median Non-Hispanic White household income

88. The White household income

89. The Hispanic (any race) household income

90. The Black household income

In Problems 91–94, read the meter.

91.

92.

93. 

94. 

Using Your Knowledge

Writing Checks Check this out! In problems 95–99, fill in the appropriate blank in the check.

95.

CHARLES SMITH
205 FAIR ST.
FORT BLAT, IL 64521

60–8124/7233
1000824431

1024

DATE June 20

PAY TO THE ORDER OF Major Motors $ ____

Nine thousand, seven hundred ninety-nine and 00/100 DOLLARS

CITIZENS BANK
OF EAGLE POINT
EAGLE POINT, IL 64502

MEMO New car — Chuck Smith

⑆7233602401⑆ 1000824431⑈ 1024

96.

JAMES L. SEAGOER
1004 CREEKSIDE AVE.
FORT BLAT, IL 64539

60–8124/7233
1000824628

2872

DATE August 23

PAY TO THE ORDER OF Ship in Shake Co. $ ____

Eight thousand, six hundred ninety-six and 00/100 DOLLARS

CITIZENS BANK
OF EAGLE POINT
EAGLE POINT, IL 64502

MEMO New boat — James Seagoer

⑆7233602401⑆ 1000824628⑈ 2872

97.

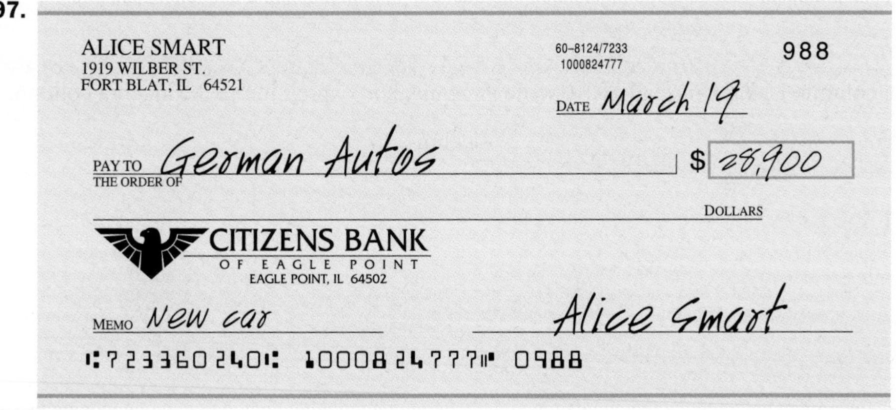

ALICE SMART
1919 WILBER ST.
FORT BLAT, IL 64521

60–8124/7233
1000824777

988

DATE March 19

PAY TO THE ORDER OF German Autos $ 28,900

____ DOLLARS

CITIZENS BANK
OF EAGLE POINT
EAGLE POINT, IL 64502

MEMO New car — Alice Smart

⑆7233602401⑆ 1000824777⑈ 0988

98.

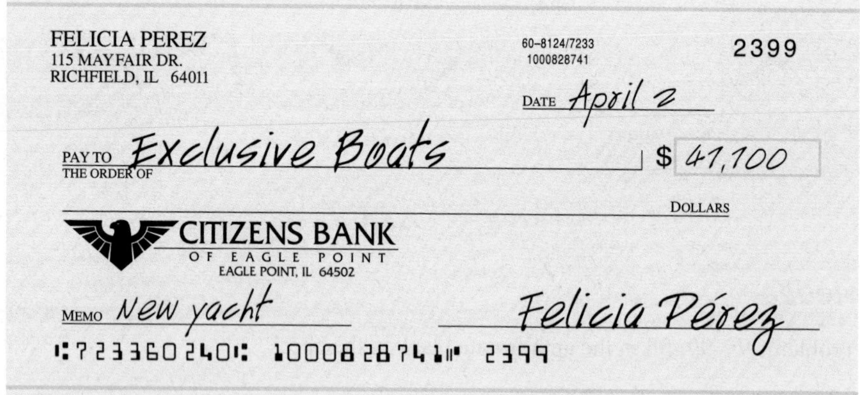

FELICIA PEREZ
115 MAYFAIR DR.
RICHFIELD, IL 64011

60–8124/7233
1000828741

2399

DATE April 2

PAY TO THE ORDER OF Exclusive Boats $ 41,100

DOLLARS

CITIZENS BANK
OF EAGLE POINT
EAGLE POINT, IL 64502

MEMO New yacht Felicia Pérez

⑆723360240⑆ 1000828741⑈ 2399

99.

JOSEPH CLEMENTE
1316 PARK ST.
FORT BLAT, IL 64539

60–8124/7233
1000824119

1381

DATE July 7

PAY TO THE ORDER OF Home Makers, Inc. $ 90,800

DOLLARS

CITIZENS BANK
OF EAGLE POINT
EAGLE POINT, IL 64502

MEMO New house Joe Clemente

⑆723360240⑆ 1000824119⑈ 1381

››› Write On

To the student and instructor: The WRITE ON exercises give you an opportunity to express your thoughts in writing. They usually can be answered in a few sentences. Most of the answers to these exercises do not appear at the back of the book.

100. What is the definition of a period?

101. Why do we use commas when writing large numbers?

102. In the numeral 5678, what is the value of the 5?

103. In the numeral 5678, what is the value of the 8?

104. In the numeral 5678, what digit tells the number of hundreds?

105. In the numeral 5678, what digit tells the number of tens?

››› Algebra Bridge

Look at the numbers in the arithmetic column. Using x instead of 10, write an equivalent expression in the algebra column.

Arithmetic **Algebra**

106. $7 \times 10^3 + 9 \times 10^2 + 3 \times 10 + 8$

107. $4 \times 10^2 + 8 \times 10 + 6$

108. $2 \times 10 + 4$

Concept Checker

Fill in the blank(s) with the correct word(s), phrase, or mathematical statement.

number	**natural**
numeral	**standard**
whole	**digits**
expanded	

109. 0, 1, 2, 3, . . . are the ________ numbers.

110. A ________ is an abstract idea used to **represent** a quantity.

111. 1, 2, 3, 4, . . . are the ________ numbers.

112. When a numeral is written as an **addition** of their place values it is in ________ form.

113. A ________ is a **symbol** that represents a number.

Mastery Test

114. Write 8000 + 600 + 90 + 3 in standard form.

115. Write the numeral 12,849 in words.

116. Write 785 in expanded form.

117. What is the place value of 6 in 689?

118. Write "fifty-six thousand, seven hundred eighty-five" in standard form.

119. Write 305 in words.

1.2 Ordering and Rounding Whole Numbers

Objectives

You should be able to:

A Determine if a given number is less than or greater than another number.

B Round whole numbers to the specified place value.

C Solve applications involving the concepts studied.

To Succeed, Review How To . . .

Find the place value of a digit in a numeral. (p. 3)

Getting Started

How long is a large paper clip? To the nearest inch, it is 2 inches.

A > Ordering Numbers

We know that 2 is greater than one because the 2 on the ruler is to the **right** of the one. Whole numbers can be compared using the number line shown:

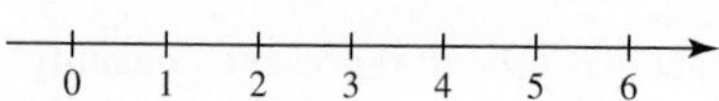

FOR ANY WHOLE NUMBERS *a* AND *b*

1. $a < b$ (read "a is less than b") if a is to the left of b on a number line.
2. $a > b$ (read "a is greater than b") if a is to the right of b on the number line.

Thus, $3 < 5$ because 3 is to the *left* of 5 on the number line. Similarly, $5 > 3$ because 5 is to the *right* of 3 on the number line. Sentences such as $3 < 5$ or $5 > 3$ are called **inequalities.** The inequality $3 < 5$ is true, but the inequality $3 > 8$ is false.

EXAMPLE 1 Creating true statements

Fill in the blank with < or > to make the resulting inequalities true.

a. 27 ____ 28 **b.** 33 ____ 25

SOLUTION We make a number line starting with 25.

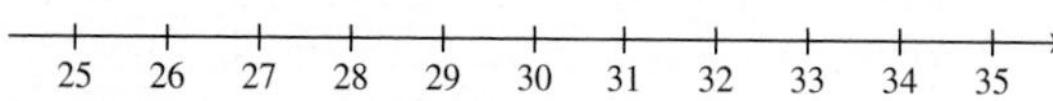

Since 27 is to the *left* of 28, $27 < 28$.
Since 33 is to the *right* of 25, $33 > 25$.

PROBLEM 1

Fill in the blank with < or > to make the resulting inequalities true.

a. 23 ____ 25 **b.** 31 ____ 27

EXAMPLE 2 Ordering numbers

The graph shows the cities with the most auto thefts per 100,000 people.

a. Order the *number* of auto thefts from smallest to largest.

b. Which city had the smallest number of auto thefts?

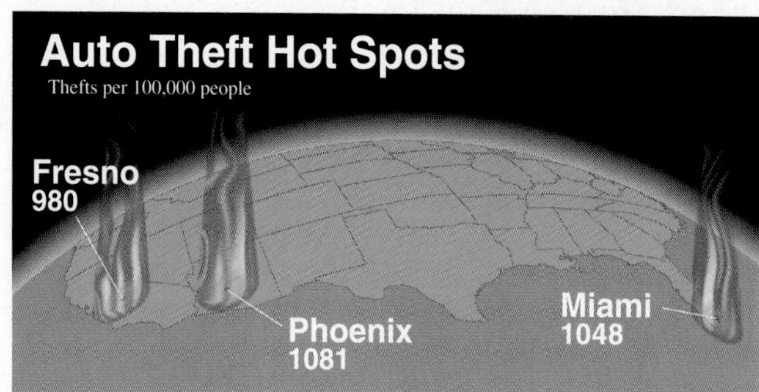

Source: Data from National Insurance Crime Bureau Study.

SOLUTION

a. To make sure we include the numbers 980 to 1081, we construct a number line starting at 950 and ending at 1100, as shown in Figure 1.1.

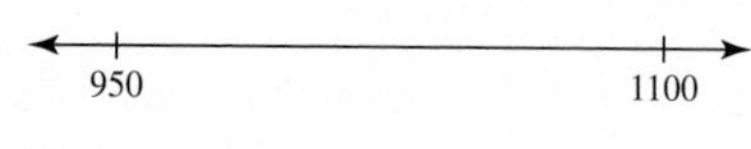

>Figure 1.1

PROBLEM 2

a. Order the number of auto thefts from largest to smallest.

b. Which city had the largest number of auto thefts?

Answers to PROBLEMS

1. a. < **b.** > **2. a.** 1081 > 1048 > 980 **b.** Phoenix

For convenience, we make 50 unit subdivisions, that is, we count by fifties. The line looks like Figure 1.2.

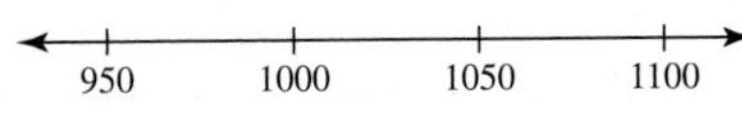

>Figure 1.2

The numbers 980, 1048, and 1081 are then placed on the line from left to right, as shown in Figure 1.3.

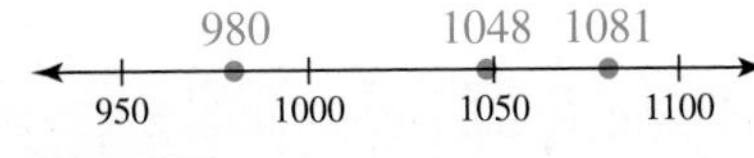

>Figure 1.3

Since 1048 is to the right of 980 and 1081 is to the right of 1048, we have

980	<	1048	<	1081
Fresno		Miami		Phoenix

b. The city with the smallest number of auto thefts is Fresno.

B > Rounding Whole Numbers

When finding the length of a paper clip, we might be told to **round** the answer to the nearest inch. In this case, the length is 2 inches. One use of **rounding numbers** is in dealing with very large numbers (such as the amount owed by the U.S. government in a recent year, $6409 billion) or where the numbers change so fast that it is not possible to give an exact figure (for example, the population of a certain city might be about 250,000), or when estimating answers in solving problems.

To **round** a number, we specify the **place** value to which we are to round and underline it. Thus, when rounding 78 to the nearest ten, we write

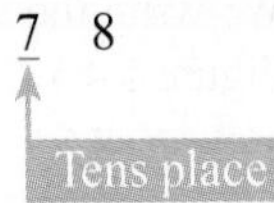

On the number line, this means we count by tens and find the group of tens closest to 78, which is 80, as shown in Figure 1.4.

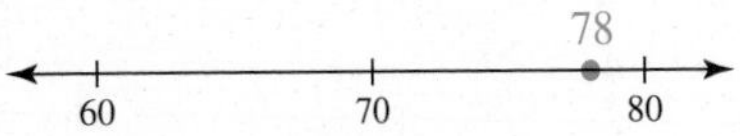

>Figure 1.4

When rounding 813 to the nearest hundred, write

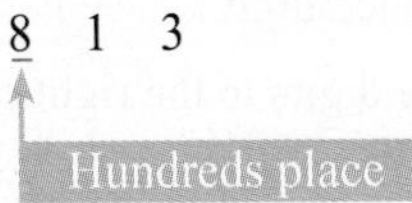

and use a number line with intervals of 100. You can see that 813 is closest to 800, as indicated in Figure 1.5.

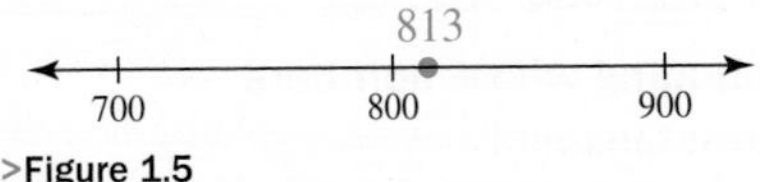

>Figure 1.5

When rounding 3500 to the nearest thousand, write

and use a number line with intervals of 1000. You can see that 3500 lies exactly in between 3000 and 4000, as shown in Figure 1.6.

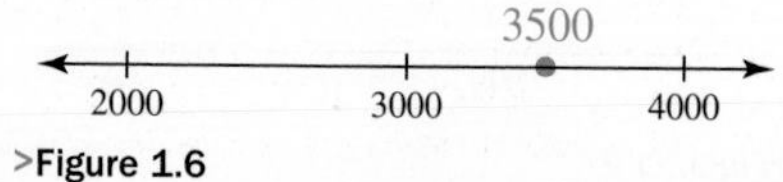

>Figure 1.6

To do the actual **rounding,** we use the following rule:

RULE FOR ROUNDING WHOLE NUMBERS*

Step 1. *Underline* the place to which you are rounding.

Step 2. If the first digit to the *right* of the underlined place is 5 *or more, add one* to the underlined digit. Otherwise *do not change* the underlined digit.

Step 3. *Change* all the digits to the *right* of the underlined digit to zeros.

Thus, to round 78 to the nearest ten, we use the three steps given.

Step 1. Underline the place to which we are rounding. 7̲ 8

Tens place

Step 2. If the first digit to the right of the underlined place (the 8) is 5 or more, add one to the underlined digit. 7̲ 8

Step 3. The digit to the right of the underlined digit becomes zero. 8 0

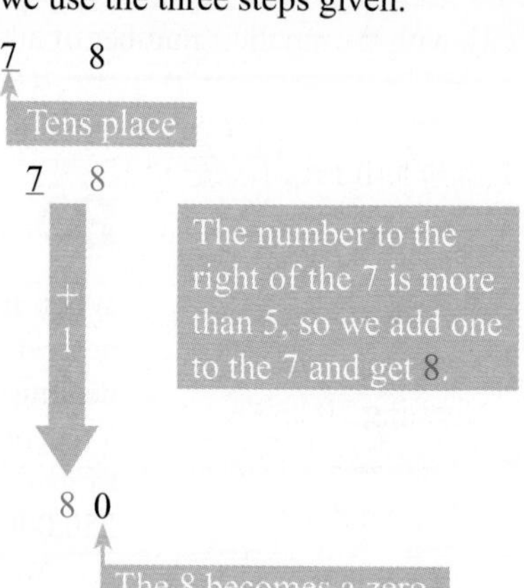

The 8 becomes a zero.

We write the answer like this: 7̲8 ≈ 80. Thus, 78 rounded to the nearest 10 is 80. (See Figure 1.4.)

EXAMPLE 3 **Rounding whole numbers**

Round 813 to the nearest hundred.

SOLUTION

Step 1. Underline the place to which we are rounding. 8̲ 1 3

Step 2. The first digit to the right of 8 is 1, so we do not change the underlined digit. 8̲ 3

Step 3. We then change all the digits to the right of the underlined digit to zeros. 8̲ 0 0

We then have 8̲13 ≈ 800. Thus, 813 rounded to the nearest hundred is 800. (See Figure 1.5.)

PROBLEM 3

Round 347 to the nearest hundred.

EXAMPLE 4 **Rounding whole numbers**

Round 3500 to the nearest thousand.

SOLUTION

Step 1. Underline the place to which we are rounding. 3̲500

PROBLEM 4

Round 6508 to the nearest thousand.

* Some textbooks round a number ending in 5 so that the last retained digit is even.

Answers to PROBLEMS

3. 3̲47 ≈ 300 **4.** 6̲508 ≈ 7000

Step 2. The first digit to the right of the underlined digit is 5, so we add one to the underlined digit (obtaining 4). 4 00

Step 3. Change all the digits to the right of the underlined digit to zeros. 4000

Thus, 3500 ≈ 4000 so that 3500 rounded to the nearest thousand is 4000. (See Figure 1.6.)

C > Applications Involving Whole Numbers

EXAMPLE 5 **Rounding whole numbers**

A planner estimated that on an average day 1,169,863 persons take a taxi. Round this number to the nearest thousand.

SOLUTION

Step 1. Underline the place to which we are rounding. 1,169,863

Step 2. The first digit to the right of the underlined digit is 8 (more than 5), so we add one to the 9, obtaining 10, and write the answer on the next line. You can also think of this as adding 1 to 69 and getting 70 on the next line. 1,1610,863 1,170,863

Step 3. Change all the digits to the right of the underlined digit to zeros. 1,170,000

We then have 1,169,863 ≈ 1,170,000, which means that 1,169,863 rounded to the nearest thousand is 1,170,000.

PROBLEM 5

On an average day 6375 couples wed in the United States. Round this number to the nearest thousand.

Source: U.S. Census Bureau.

Sometimes the same number is rounded to different places. For example, on an average day 231,232,876 eggs are laid (honest!). This number can be rounded to

The nearest *hundred* 231,232,876 ≈ 231,232,900
The nearest *thousand* 231,232,876 ≈ 231,233,000
The nearest *million* 231,232,876 ≈ 231,000,000

We use this idea in Example 6, which should be of interest to you.

EXAMPLE 6 **Rounding whole numbers**

It has been estimated that by retirement age a high school graduate will outearn a non–high school graduate by $405,648. Round this number to

a. the nearest hundred.
b. the nearest thousand.
c. the nearest ten thousand.

SOLUTION

a. Step 1. 405,648

Underline the 6 (the hundreds place).

PROBLEM 6

A college graduate will outearn a high school graduate by $1,013,088. Round this number to

a. the nearest hundred.
b. the nearest thousand.
c. the nearest ten thousand

(continued)

Answers to PROBLEMS

5. 6375 ≈ 6000 **6. a.** $1,013,100 **b.** $1,013,000 **c.** 1,010,000

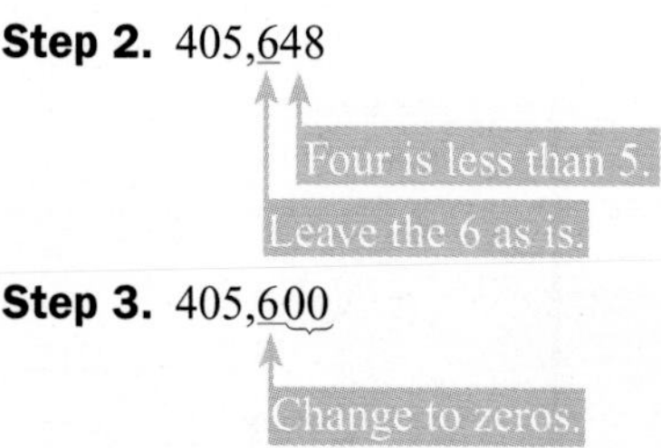

Step 3. 405,600

Change to zeros.

Thus, 405,648 rounded to the nearest hundred is 405,600.

b. Step 1. 405,648

Underline the 5 (the thousands place).

Step 2. 406,648

Six is more than 5 (add 1 to the 5).

5 + 1 = 6

Step 3. 406,000

Change to zeros.

Thus, 405,648 rounded to the nearest thousand is 406,000.

c. Step 1. 405,648

Underline the zero (the ten thousands place).

Step 2. 415,648

Five is equal to 5 (add 1 to the 0).

0 + 1 = 1

Step 3. 410,000

Change to zeros

Thus, 405,648 rounded to the nearest ten thousand is 410,000.

EXAMPLE 7 Rounding whole numbers

This year, the best-selling car in America is the Toyota Camry. Use the chart to round the specified prices. Suppose you have a $22,000 budget.

a. Round the True Market Value (TMV) Base Price to the nearest hundred.

b. Round the TMV price of the GJ package #3 to the nearest hundred.

c. Round the TMV price of the BE package to the nearest hundred.

SOLUTION

a. The base price is $20,080. To round $20,080 to the nearest hundred, underline the hundreds place, that is, the 0. Since the 8 to the right of 0 is more than 5, add one to the 0, write 1, and change the last two numbers to zeros to get the estimate $20,100 as shown.

b. The GJ package costs $1475. Underline the hundreds place, that is, the 4. Since the 7 to the right of 4 is more than 5, add one to the 4, write 5, and add two zeros at the end to get the estimate, $1500.

c. The BE package is $438. Underline the 4. Since the 3 to the right of the 4 is less than 5, leave the 4 alone and add two zeros at the end to get the estimate, $400.

PROBLEM 7

Estimate, to the nearest hundred, the TMV price of

a. the GU package.

b. the SR package.

c. the XV package.

Answers to PROBLEMS

7. a. $1000 **b.** $700 **c.** $100

PRICE WITH OPTIONS

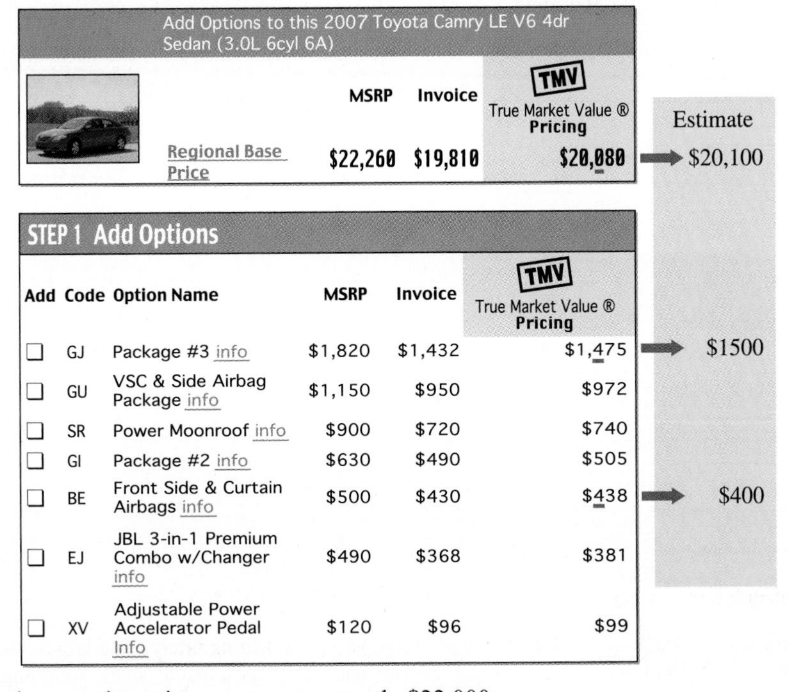

Add Options to this 2007 Toyota Camry LE V6 4dr Sedan (3.0L 6cyl 6A)

	MSRP	Invoice	TMV True Market Value ® Pricing	Estimate
Regional Base Price	$22,260	$19,810	$20,080	→ $20,100

STEP 1 Add Options

Add	Code	Option Name	MSRP	Invoice	TMV True Market Value ® Pricing	Estimate
☐	GJ	Package #3 info	$1,820	$1,432	$1,475	→ $1500
☐	GU	VSC & Side Airbag Package info	$1,150	$950	$972	
☐	SR	Power Moonroof info	$900	$720	$740	
☐	GI	Package #2 info	$630	$490	$505	
☐	BE	Front Side & Curtain Airbags info	$500	$430	$438	→ $400
☐	EJ	JBL 3-in-1 Premium Combo w/Changer info	$490	$368	$381	
☐	XV	Adjustable Power Accelerator Pedal Info	$120	$96	$99	

By the way, the estimate comes to exactly $22,000.

› Exercises 1.2

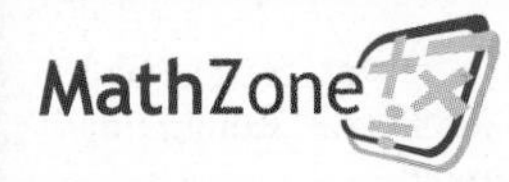

Boost your grade at mathzone.com!
- Practice Problems
- NetTutor
- Self-Tests
- e-Professors
- Videos

‹ A › Ordering Numbers

In Problems 1–10, fill in the blank with < or > to make a true inequality.

1. 8 ____ 10 **2.** 6 ____ 16 **3.** 8 ____ 0 **4.** 0 ____ 10

5. 102 ____ 120 **6.** 808 ____ 880 **7.** 999 ____ 990 **8.** 777 ____ 770

9. 1001 ____ 1010 **10.** 2002 ____ 2020

‹ B › Rounding Whole Numbers

In Problems 11–30, round to the underlined place.

11. $\underline{7}3$ **12.** $\underline{8}4$ **13.** $\underline{8}6$ **14.** $\underline{4}7$

15. $\underline{9}8$ **16.** $\underline{9}7$ **17.** $\underline{1}03$ **18.** $\underline{2}04$

19. $\underline{3}86$ **20.** $\underline{4}76$ **21.** $\underline{9}50$ **22.** $\underline{9}63$

23. $\underline{2}308$ **24.** $\underline{6}209$ **25.** $\underline{6}999$ **26.** $\underline{8}999$

27. $\underline{9}999$ **28.** $\underline{9}990$ **29.** $\underline{9}099$ **30.** $\underline{9}011$

❯ Web IT go to mhhe.com/bello for more lessons

In Problems 31–40, round the given number to the nearest ten, the nearest hundred, and the nearest thousand.

	Ten	Hundred	Thousand
31. 586	______	______	______
32. 650	______	______	______
33. 29,450	______	______	______
34. 39,990	______	______	______
35. 49,992	______	______	______
36. 349,908	______	______	______
37. 259,906	______	______	______
38. 349,904	______	______	______
39. 289,000	______	______	______
40. 999,000	______	______	______

‹ C › Applications Involving Whole Numbers

41. *Rapid typing* The record for rapid typing with a standard typewriter is held by Albert Tagora. On October 23, 1923, he typed an average of 147 words a minute. Round 147 to the nearest ten.

42. *Fishing* Have you gone fishing lately? The largest fish ever caught by rod and reel was a white shark weighing 2664 pounds. Round 2664 to the nearest ten.

43. *Weight loss* Do you have a weight problem? The heaviest man on record was Robert Earl Hughes, who tipped the scales at 1069 pounds. Round 1069 to the nearest hundred.

44. *Heavy lifting* The heaviest weight ever lifted by a human being was 6270 pounds, lifted by Paul Anderson in 1957. Round 6270 to the nearest hundred.

45. *Really smoking!* If you smoke $1\frac{1}{2}$ packs of cigarettes a day, you will smoke about 10,950 cigarettes a year. Round 10,950 to the nearest thousand.

46. *Hertz used cars* A survey conducted by Hertz shows that the typical used car purchased in a certain year showed 29,090 miles on the odometer. Round 29,090 to the nearest thousand.

47. *New York population* According to a recent census, the number of people in New York City is 7,895,563. Round this number to the nearest one hundred thousand.

48. *Population of Nevada* The number of residents of Nevada was counted and found to be 2,070,000. Round this number to the nearest one hundred thousand.

49. *Godfather money!* Did you see the movie *The Godfather?* A lot of people did! In fact, during its first three years in circulation the movie made $85,747,184. Round this number to the nearest million.

50. *Movie revenue* *The Sound of Music* is another famous movie. In its first ten years it made $83,891,000. Round this number to the nearest million.

51. *Cheap as Dell* Here are the prices for three Dell computer models.

Use an inequality to compare the prices:

a. From lowest to highest

b. From highest to lowest

c. Which is the most expensive model?

d. Which is the least expensive model?

Cutting Edge	Performance	Affordability
Dimension 8400	Dimension E310	Dimension F510
from $1019	from $689	from $968

52. *Price of Gateways* Here are the prices for three Gateway computer models.

Use an inequality to compare the prices:

a. From lowest to highest

b. From highest to lowest

c. Which is the most expensive model?

d. Which is the least expensive model?

Feature Rich	High Performance	Value
Gateway GM 5072 starting at $1299	Gateway GT 5058 starting at $899	Gateway GT 4016 starting at $449

53. The following graph shows the percentages of Internet users by race and gender.

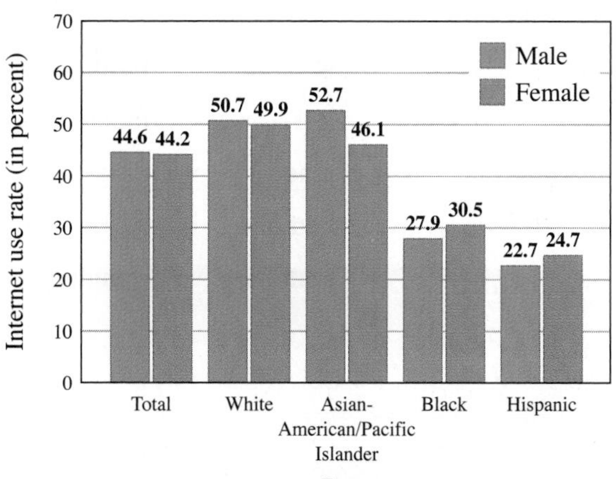

Source: NTIA and ESA, U.S. Department of Commerce, using U.S. Bureau of the Census Current Population Survey.

Fill in the blank with > or < to make a true statement.

a. The percent of black males ____ the percent of black females.

b. The percent of male Hispanics ____ the percent of female Hispanics.

c. The percent of white males ____ the percent of white females.

54. The following graph shows the percentage of Internet/e-mail use at work by gender and age.

Source: NTIA and ESA, U.S. Department of Commerce, using U.S. Census Bureau Current Population Survey Supplements.

Fill in the blank with > or < to make a true statement.

a. At age 45, the percent of female users ____ the percent of male users.

b. At age 55, the percent of female users ____ the percent of male users.

c. At age 65, the percent of female users ____ the percent of male users.

Web IT go to mhhe.com/bello for more lessons

Using Your Knowledge

Here is an activity that you cannot evade: filing your income tax return. The U.S. government has a booklet called Publication 796 to help you do it! This publication states:

> All money items appearing on your return may be rounded off to **whole dollars** on your returns and schedules, provided you do so for all entries on the return.

Use the knowledge you have obtained in this section to round off the numbers in the following form.

Income	7	Wages, salaries, tips, etc. Attach Form(s) W-2	7	23,899 | 56	**55.**
	8a	**Taxable** interest. Attach Schedule B if required	8a	39 | 06	**56.**
Attach Forms W-2 and W-2G here. Also attach Form(s) 1099-R if tax was withheld.	b	**Tax-exempt** interest. **Do not** include on line 8a . . . 8b			
	9a	Ordinary dividends. Attach Schedule B if required	9a		
	b	Qualified dividends (see page 23) . . . 9b			
	10	Taxable refunds, credits, or offsets of state and local income taxes (see page 23)	10		
	11	Alimony received	11		
	12	Business income or (loss). Attach Schedule C or C-EZ	12	349 | 48	**57.**
	13a	Capital gain or (loss). Attach Schedule D if required. If not required, check here ▶ ☐	13a		
	b	If box on 13a is checked, enter post-May 5 capital gain distributions 13b			
If you did not get a W-2, see page 22.	14	Other gains or (losses). Attach Form 4797	14		
	15a	IRA distributions . . 15a ; b Taxable amount (see page 25)	15b		
	16a	Pensions and annuities 16a ; b Taxable amount (see page 25)	16b	1,387 | 53	**58.**
Enclose, but do not attach, any payment. Also, please use **Form 1040-V.**	17	Rental real estate, royalties, partnerships, S corporations, trusts, etc. Attach Schedule E	17		
	18	Farm income or (loss). Attach Schedule F	18		
	19	Unemployment compensation	19		
	20a	Social security benefits . 20a ; b Taxable amount (see page 27)	20b		
	21	Other income. List type and amount (see page 27)	21		
	22	Add the amounts in the far right column for lines 7 through 21. This is your **total income** ▶	22	25,675 | 63	**59.**

Write On

60. Write in your own words the procedure you will use to round a number when the digit to the right of the place to which you are rounding is less than 5.

61. Write in your own words the procedure you will use to round a number when the digit to the right of the place to which you are rounding is 5 or more.

62. Can you think of three situations in which estimating can be useful?

Concept Checker

Fill in the blank(s) with the correct word(s), phrase, or mathematical statement.

$<$	**right**
left	**inequalities**
place	**delete**
add one	$>$
underline	**rounding**

63. $a < b$ means that a is to the _____ of b on the number line.

64. If a is to the right of b on the number line, a _____ b.

65. To **round** a number, we specify the _____ to which we are to round.

66. The first step in the **rule** for rounding whole numbers is to _______ the place to which we are rounding.

67. When rounding the number ▽□ to the nearest **ten,** if the number in the □ box is 5 or more, _______ to ▽.

68. $a < b$ and $b > a$ are examples of _________.

Mastery Test

69. Round to the underlined place:

a. $7\underline{6}5$ **b.** $3\underline{6}4$

c. $\underline{8}62$

70. Fill in the blank with $<$ or $>$ to make the resulting inequalities true.

a. 349 ___ 399

b. 57 ___ 27

c. 1000 ___ 999

d. 1099 ___ 1199

71. The median income of an MTV median household is \$49,773. Round \$49,773 to the nearest thousand.

Source: Nielsen Media Research.

72. The domestic gross ticket receipts (in millions) of the top five grossing movies listed alphabetically are as follows:

E.T., \$435
Spider-Man, \$404
Star Wars, \$461
Star Wars: The Phantom Menace, \$431
Titanic, \$601

a. List the *amounts* of ticket receipts from highest to lowest using inequalities.

b. List the *amounts* of ticket receipts from lowest to highest using inequalities.

Skill Checker

What kind of car do you drive? Here are the six least-expensive cars this year. Write the numerals in words.

73. Chevrolet Aveo, \$9995

74. Kia Rio, \$10,280

75. Hyundai Accent, \$10,544

76. Chevrolet Cavalier, \$10,890

77. Toyota Echo, \$10,995

78. Pontiac Sunfire, \$11,460

Source: Data for Exercises 73–78 from Edmunds.com, http://www.edmunds.com

1.3 Addition

Objectives

You should be able to:

A Add two or more whole numbers, regrouping (carrying) if necessary.

B Use addition to find perimeters of polygons.

To Succeed, Review How To . . .

Use the addition facts in the table on page 24.

Getting Started

Configuration	Invoice	MSRP
Base Model	\$19,810	\$22,260
Destination	\$485	\$485

What is the invoice price for the car? To find out, we have to *add* the Base Model Price (\$19,810) and the Destination charge (\$485). To answer the question "how many?" we use the set of *whole numbers* 0, 1, 2, 3, . . . , and the operation of **addition.**

A › Adding Two or More Whole Numbers

After you do the examples, you will see that the invoice price will be \$20,295.

Addition can be explained by counting. For example, the sum

$$6 + 2$$

can be found by using a set of 6 objects and another set of 2 objects, putting them together, and counting all the objects, as shown:

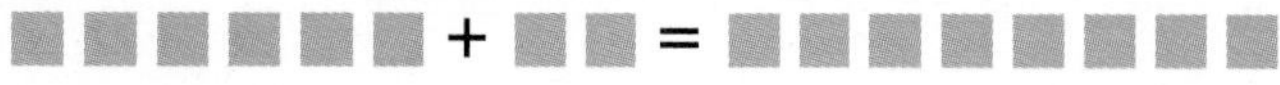

The numbers to be added, in this case 6 and 2, are called **addends,** and the result 8 is called the **sum,** or **total.** The procedure is usually written as

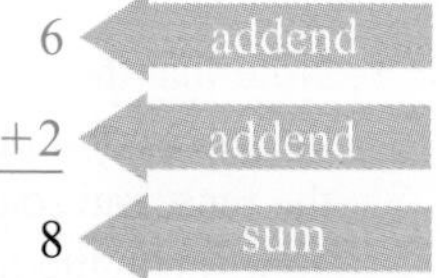

where we have used the plus (+) sign to indicate addition. All the addition facts you need are in the table, but you either know these facts already, or should take the time to memorize them now!

ADD (+)	0	1	2	3	4	5	6	7	8	9
0	0	1	2	3	4	5	6	7	8	9
1	1	2	3	4	5	6	7	8	9	10
2	2	3	4	5	6	7	8	9	10	11
3	3	4	5	6	7	8	9	10	11	12
4	4	5	6	7	8	9	10	11	12	13
5	5	6	7	8	9	10	11	12	13	14
6	6	7	8	9	10	11	12	13	14	15
7	7	8	9	10	11	12	13	14	15	16
8	8	9	10	11	12	13	14	15	16	17
9	9	10	11	12	13	14	15	16	17	18

Zero is called the **identity** for addition.

IDENTITY FOR ADDITION

Any number *a* added to zero equals the number; that is, adding 0 to a number does not change the number.

$$a + \mathbf{0} = a = \mathbf{0} + a$$

Also, the order in which two numbers a and b are added is not important. This is called the **commutative property of addition.** Thus, $3 + 2 = 2 + 3$ and $5 + 7 = 7 + 5$. In general,

COMMUTATIVE PROPERTY OF ADDITION

Changing the order of two addends does not change their sum.

$$a + b = b + a$$

Now let us try another problem:

Add: 46 + 52 = ______

Before adding, we note that the answer, to the nearest ten, should be 50 + 50, or about 100. This kind of approximation, or **estimation,** can give a valuable check on the answer. We then write the problem like this:

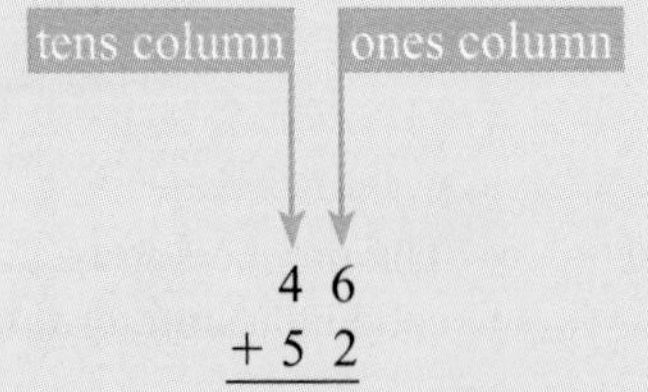

Note that the numbers have been arranged vertically in columns with the ones digits in the ones column and the tens digits in the tens column. We then add the ones first, then the tens, and so on. The short form of the addition is on the left, and the expanded form appears on the right.

Short Form	Expanded Form
$\begin{array}{r} 46 \\ +\ 52 \\ \hline 98 \end{array}$	$\begin{array}{r} 40 + 6 \\ +\ 50 + 2 \\ \hline 90 + 8 = 98 \end{array}$

(98: add tens ↑ add ones first ↑)

In this case, if the actual answer (98) is rounded to the nearest ten, we obtain our estimated answer (100).

Our estimate (100) is near the actual answer (98).

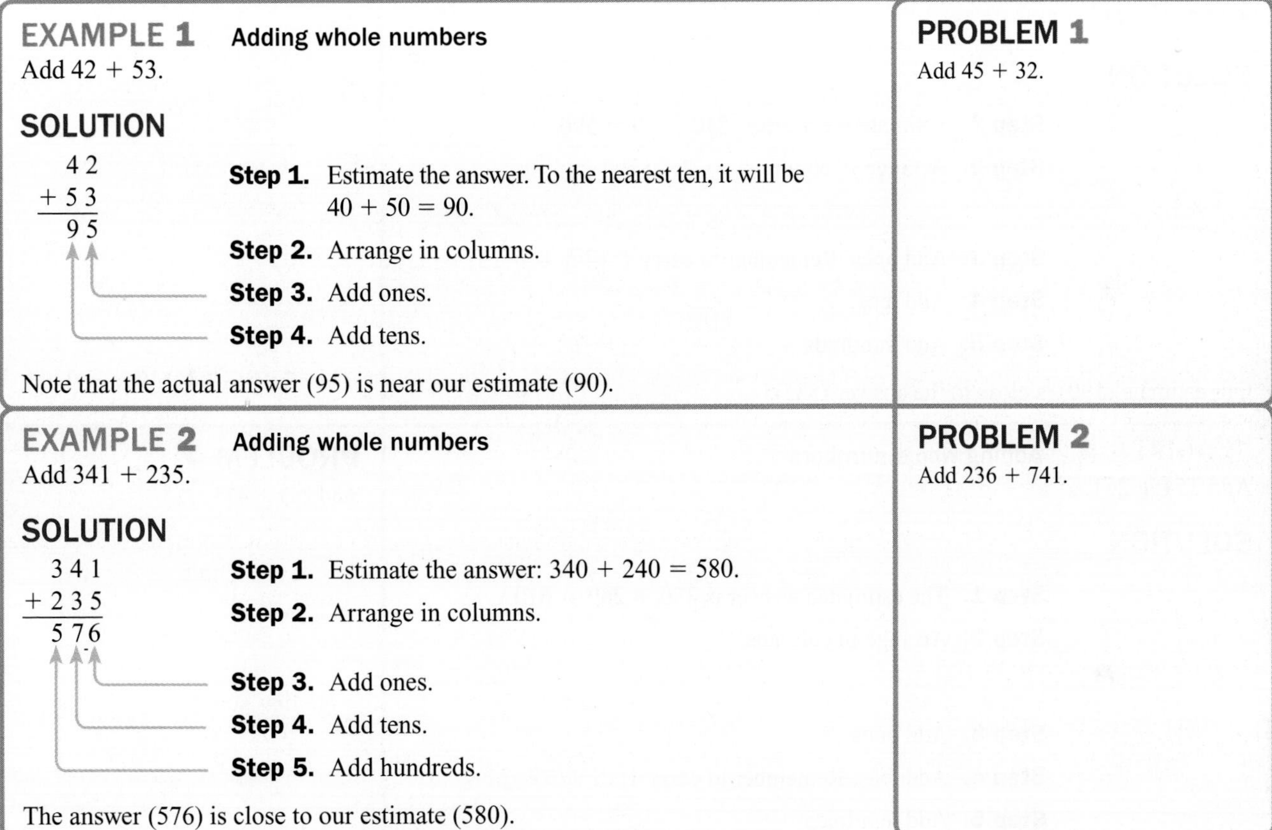

EXAMPLE 1 **Adding whole numbers**

Add 42 + 53.

SOLUTION

$$\begin{array}{r} 42 \\ +\ 53 \\ \hline 95 \end{array}$$

Step 1. Estimate the answer. To the nearest ten, it will be 40 + 50 = 90.

Step 2. Arrange in columns.

Step 3. Add ones.

Step 4. Add tens.

Note that the actual answer (95) is near our estimate (90).

PROBLEM 1

Add 45 + 32.

EXAMPLE 2 **Adding whole numbers**

Add 341 + 235.

SOLUTION

$$\begin{array}{r} 341 \\ +\ 235 \\ \hline 576 \end{array}$$

Step 1. Estimate the answer: 340 + 240 = 580.

Step 2. Arrange in columns.

Step 3. Add ones.

Step 4. Add tens.

Step 5. Add hundreds.

The answer (576) is close to our estimate (580).

PROBLEM 2

Add 236 + 741.

Here is another problem: 56 + 38 = ______. Compare the expanded form and the short form.

Short Form	Expanded Form
$\begin{array}{r} 1 \\ 56 \\ +\ 38 \\ \hline 94 \end{array}$	$\begin{array}{rl} 50 + 6 & \\ +\ 30 + 8 & \\ \hline 80 + \underbrace{14} & = 80 + \underbrace{10 + 4} \\ & = 90 + 4 \\ & = 94 \end{array}$

Note that the 1 "carried" over to the tens column is really a 10. The 14 is rewritten as $\underline{1}0$ + 4. Here is another way of showing this:

Answers to PROBLEMS

1. 77 **2.** 977

Step 1	Step 2	Step 3
56 + 38 = 14 Add ones (6 + 8 = 14).	56 + 38: 14, 80 Add tens (50 + 30 = 80).	56 + 38: 14 + 80 = 94 Add partial sums (14 + 80 = 94).

The 1 we "carry" is the 1 in 14. Of course, you should do your addition using the short form to save time.

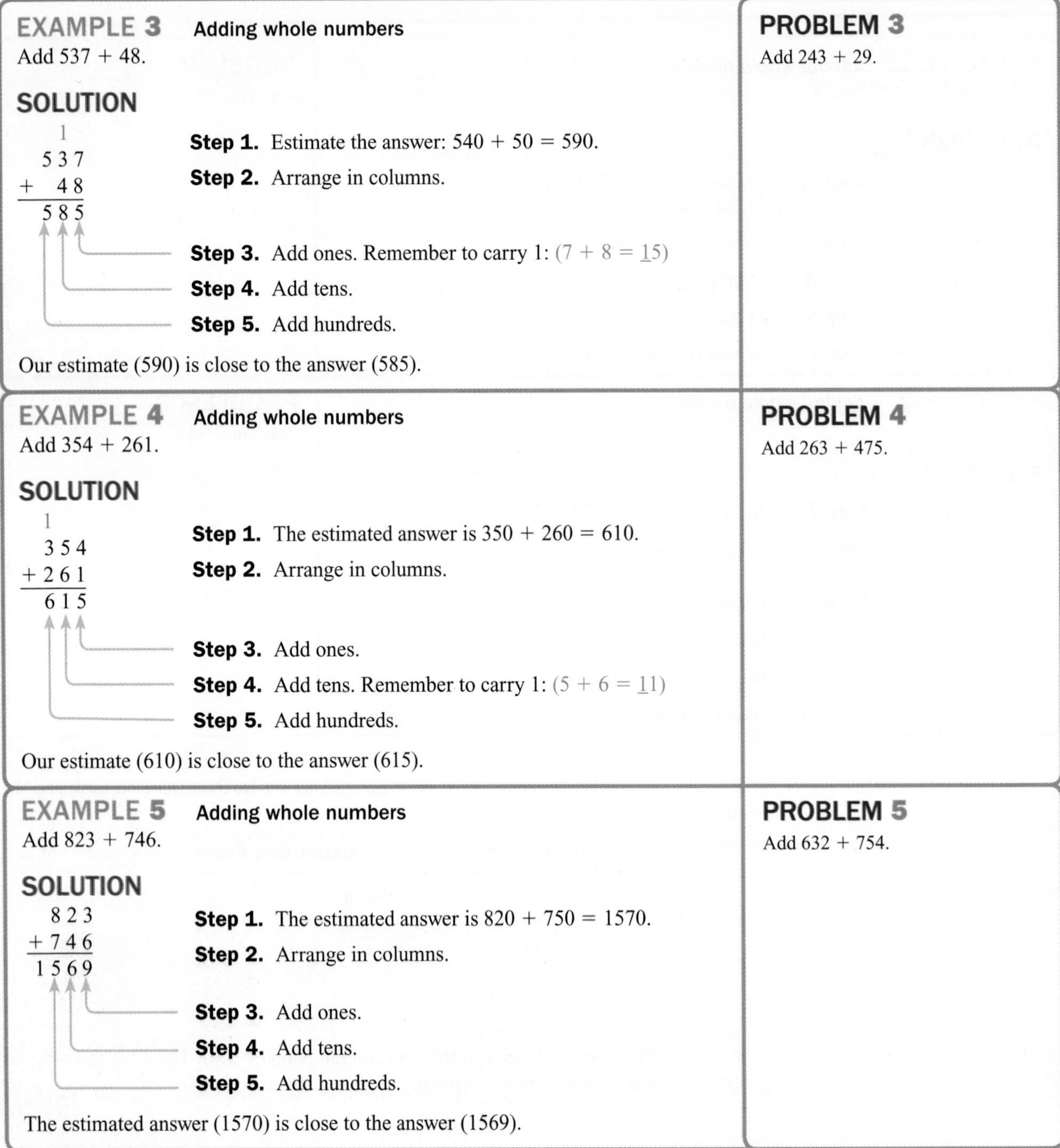

EXAMPLE 3 Adding whole numbers

Add 537 + 48.

SOLUTION

```
   1
  537
+  48
  585
```

Step 1. Estimate the answer: 540 + 50 = 590.

Step 2. Arrange in columns.

Step 3. Add ones. Remember to carry 1: (7 + 8 = 15)

Step 4. Add tens.

Step 5. Add hundreds.

Our estimate (590) is close to the answer (585).

PROBLEM 3

Add 243 + 29.

EXAMPLE 4 Adding whole numbers

Add 354 + 261.

SOLUTION

```
  1
  354
+ 261
  615
```

Step 1. The estimated answer is 350 + 260 = 610.

Step 2. Arrange in columns.

Step 3. Add ones.

Step 4. Add tens. Remember to carry 1: (5 + 6 = 11)

Step 5. Add hundreds.

Our estimate (610) is close to the answer (615).

PROBLEM 4

Add 263 + 475.

EXAMPLE 5 Adding whole numbers

Add 823 + 746.

SOLUTION

```
   823
 + 746
  1569
```

Step 1. The estimated answer is 820 + 750 = 1570.

Step 2. Arrange in columns.

Step 3. Add ones.

Step 4. Add tens.

Step 5. Add hundreds.

The estimated answer (1570) is close to the answer (1569).

PROBLEM 5

Add 632 + 754.

Answers to PROBLEMS

3. 272 **4.** 738 **5.** 1386

EXAMPLE 6 Adding whole numbers

Add 704 + 5642.

SOLUTION

```
  1
   7 0 4
+ 5 6 4 2
  6 3 4 6
```

Step 1. To the nearest hundred, the estimate is 700 + 5600 = 6300. (You could estimate the answer to the nearest ten, but such an estimate is as involved as the original problem.)

Step 2. Arrange in columns.

Step 3. Add ones.

Step 4. Add tens.

Step 5. Add hundreds. Remember to carry 1: (7 + 6 = 13)

Step 6. Add thousands.

Our estimate (6300) is close to the answer (6346).

PROBLEM 6

Add 813 + 1702.

EXAMPLE 7 Adding whole numbers

Add 5471 + 2842.

SOLUTION

```
   1 1
  5 4 7 1
+ 2 8 4 2
  8 3 1 3
```

Step 1. To the nearest thousand, the estimate is 5000 + 3000 = 8000.

Step 2. Arrange in columns.

Step 3. Add ones.

Step 4. Add tens. Remember to carry 1: (7 + 4 = 11)

Step 5. Add hundreds. Remember to carry 1: (1 + 4 + 8 = 13)

Step 6. Add thousands.

Note that our estimate (8000) is close to the actual answer (8313).

PROBLEM 7

Add 3943 + 4672.

Sometimes it is necessary to add more than two numbers. The procedure is similar to the one explained previously. For example, to do the addition

4272 + 2367 + 7489 + 1273

we proceed as follows:

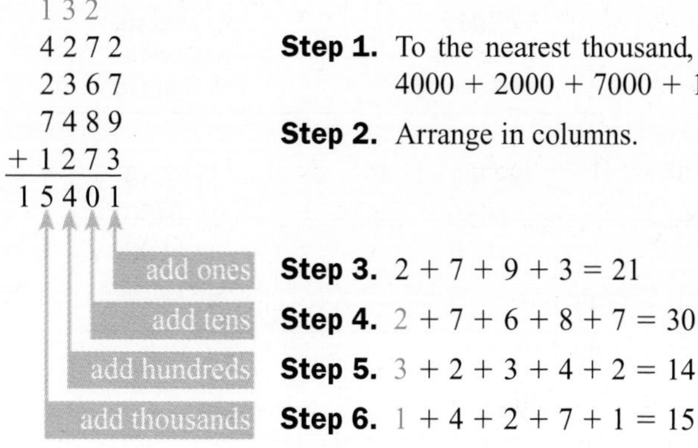

```
   1 3 2
   4 2 7 2
   2 3 6 7
   7 4 8 9
 + 1 2 7 3
 1 5 4 0 1
```

Step 1. To the nearest thousand, our estimate is 4000 + 2000 + 7000 + 1000 = 14,000.

Step 2. Arrange in columns.

add ones **Step 3.** 2 + 7 + 9 + 3 = 21

add tens **Step 4.** 2 + 7 + 6 + 8 + 7 = 30

add hundreds **Step 5.** 3 + 2 + 3 + 4 + 2 = 14

add thousands **Step 6.** 1 + 4 + 2 + 7 + 1 = 15

Answers to PROBLEMS

6. 2515 **7.** 8615

EXAMPLE 8 **Adding whole numbers**

Add 1343 + 5632 + 8789 + 7653.

SOLUTION

```
   221
  1343
  5632
  8789
+ 7653
23,417
```

PROBLEM 8

Add 2451 + 4741 + 7879 + 6563.

When adding a long list of digits, it is often easier to look for pairs of numbers whose sum is ten or a multiple of ten and add these first. Here is how you do it:

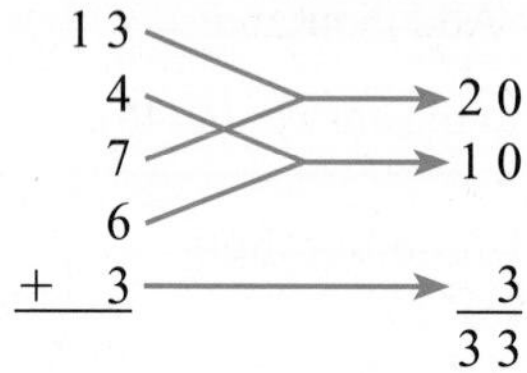

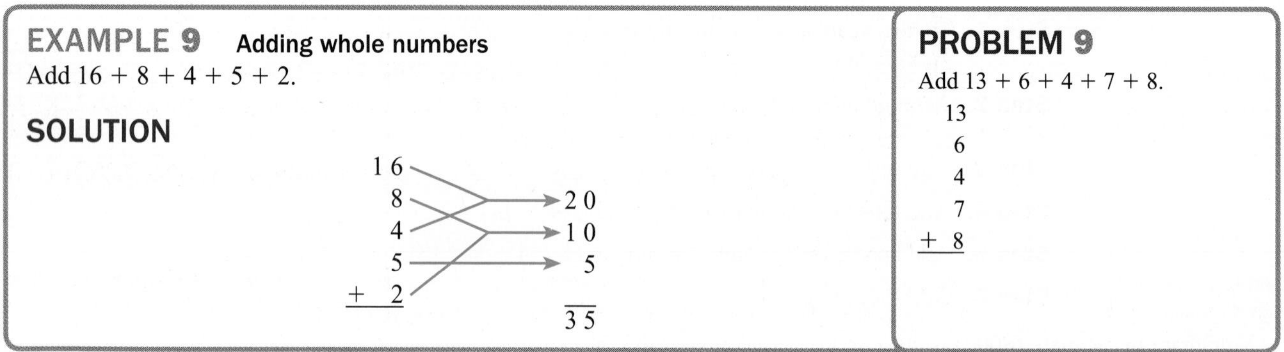

EXAMPLE 9 **Adding whole numbers**

Add 16 + 8 + 4 + 5 + 2.

SOLUTION

PROBLEM 9

Add 13 + 6 + 4 + 7 + 8.

```
 13
  6
  4
  7
+ 8
```

There are many problems that require *addition* for their solution. We give one that may be of interest to you in the next example.

EXAMPLE 10 **Tuition costs**

How many thousands of dollars will it cost a Florida resident to attend the fall and spring semesters?

Tuition Costs and Fees	Florida Residents Fall & Spring	Out-of-State Residents Fall & Spring
Undergraduate Student		
Full-Time Tuition	$2700	$12,240
Room & Board	6450	6450
Books/Supplies	800	800
SUBTOTAL	_____	_____

PROBLEM 10

How many thousands of dollars will it cost an out-of-state resident to attend the fall and spring semesters?

Answers to PROBLEMS

8. 21,634 **9.** 38
10. $19,000

SOLUTION To find the answer, we first round each of the numbers in the first column ($2700, $6450, and $800) to the nearest thousand and then add the thousands.

$$\underline{2}700 \longrightarrow 3000$$
$$\underline{6}450 \longrightarrow 6000$$
$$800 \longrightarrow \underline{1000}$$
$$\$10{,}000$$

Note that we can add 3 and 6 and then 1 like this:

$(3 + 6) + 1 = 9 + 1 = 10$ The parentheses tell us to add 3 and 6 first.

or $3 + (6 + 1) = 3 + 7 = 10$ Same answer

This is an illustration of the associative property of addition. It tells us that it does not matter how we group the numbers, the answer is the same.

ASSOCIATIVE PROPERTY OF ADDITION

The grouping of numbers (addends) does not change the final answer (sum).

$$(a + b) + c = a + (b + c)$$

B › Perimeter of Polygons

A **polygon** is a flat geometric figure with many sides. Some examples of polygons are triangles, squares, rectangles, and pentagons.

The associative law can be used to find the *perimeter* of polygons. What is the perimeter? Here is the definition:

PERIMETER

The distance around an object is its *perimeter.*

The perimeter of a **polygon** is the sum of the length of all sides.

EXAMPLE 11 **Perimeter of a triangle**

Find the perimeter of the triangle (in. means inches).

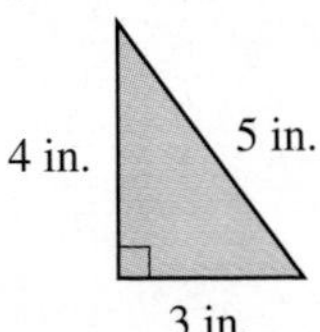

SOLUTION You can add the length of sides 4 and 3 first, and then add 5, that is,

$(4 + 3) + 5 = 7 + 5 = 12$ inches

or $4 + (3 + 5) = 4 + 8 = 12$ inches

Because of the associative property, the answer is the same.

PROBLEM 11

Find the perimeter of the triangle (mi means miles).

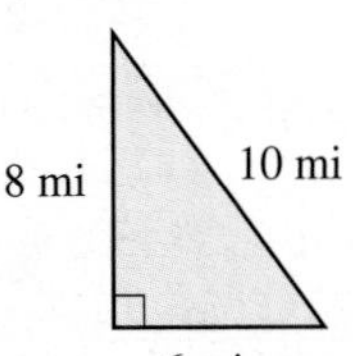

Answers to PROBLEMS

11. 24 mi

Algebra Bridge

Here is an instance in which algebra is easier than arithmetic. We mentioned that in algebra we work with expressions such as $3x^2 + 5x + 7$. Suppose we wish to add $4x^2 + 8x + 9$ to this expression. As in ordinary addition, we place both expressions in columns.

$$\begin{array}{r} 3x^2 + 5x + 7 \\ + \; 4x^2 + 8x + 9 \\ \hline 7x^2 + 13x + 16 \end{array}$$

Step 1. Add the numbers.

Step 2. Add the x's.

Step 3. Add the x^2's.

There is no carrying process in algebra.

Calculator Corner

To add numbers on a calculator with algebraic logic, you can **key** in the problem by pressing the appropriate keys. For example, to do the problem

$$\begin{array}{r} 6 \\ +8 \\ \hline \end{array}$$

you simply press 6 + 8 ENTER. The display will give the correct answer, 14. If more than two numbers are to be added, the procedure is similar. For example, to obtain the answer for Problem 51, we proceed as follows:

1 1 + 6 5 + 2 0 1 + 3 0 5 ENTER

The display will give the correct answer, 582.

Exercises 1.3

Web IT go to mhhe.com/bello for more lessons

⟨ A ⟩ Adding Two or More Whole Numbers In Problems 1–50, add.

1. $\begin{array}{r} 8 \\ +3 \\ \hline \end{array}$	**2.** $\begin{array}{r} 6 \\ +4 \\ \hline \end{array}$	**3.** $\begin{array}{r} 4 \\ +7 \\ \hline \end{array}$	**4.** $\begin{array}{r} 0 \\ +6 \\ \hline \end{array}$	**5.** $\begin{array}{r} 4 \\ 6 \\ +6 \\ \hline \end{array}$	**6.** $\begin{array}{r} 5 \\ 5 \\ +0 \\ \hline \end{array}$
7. $\begin{array}{r} 10 \\ +20 \\ \hline \end{array}$	**8.** $\begin{array}{r} 30 \\ +30 \\ \hline \end{array}$	**9.** $\begin{array}{r} 82 \\ +83 \\ \hline \end{array}$	**10.** $\begin{array}{r} 71 \\ +65 \\ \hline \end{array}$	**11.** $\begin{array}{r} 4 \\ 53 \\ +72 \\ \hline \end{array}$	**12.** $\begin{array}{r} 5 \\ 93 \\ +91 \\ \hline \end{array}$
13. 9 + 51	**14.** 7 + 33	**15.** 0 + 11	**16.** 7 + 71	**17.** 28 + 6	**18.** 32+ 0
19. 26 + 9	**20.** 39 + 7	**21.** $\begin{array}{r} 4 \\ 13 \\ +\;6 \\ \hline \end{array}$	**22.** $\begin{array}{r} 5 \\ 12 \\ +\;5 \\ \hline \end{array}$	**23.** $\begin{array}{r} 3 \\ 31 \\ +47 \\ \hline \end{array}$	**24.** $\begin{array}{r} 2 \\ 38 \\ +61 \\ \hline \end{array}$
25. $\begin{array}{r} 21 \\ 19 \\ +87 \\ \hline \end{array}$	**26.** $\begin{array}{r} 51 \\ 39 \\ +28 \\ \hline \end{array}$	**27.** $\begin{array}{r} 67 \\ +58 \\ \hline \end{array}$	**28.** $\begin{array}{r} 65 \\ +25 \\ \hline \end{array}$	**29.** $\begin{array}{r} 97 \\ +35 \\ \hline \end{array}$	**30.** $\begin{array}{r} 86 \\ +24 \\ \hline \end{array}$

31. $\begin{array}{r} 85 \\ +67 \\ \hline \end{array}$ **32.** $\begin{array}{r} 36 \\ +98 \\ \hline \end{array}$ **33.** $\begin{array}{r} 386 \\ +\ 14 \\ \hline \end{array}$ **34.** $\begin{array}{r} 466 \\ +\ 89 \\ \hline \end{array}$ **35.** $\begin{array}{r} 6347 \\ +\ 426 \\ \hline \end{array}$ **36.** $\begin{array}{r} 7479 \\ +\ 521 \\ \hline \end{array}$

37. $\begin{array}{r} 432 \\ +1381 \\ \hline \end{array}$ **38.** $\begin{array}{r} 795 \\ +2160 \\ \hline \end{array}$ **39.** $\begin{array}{r} 136 \\ +3587 \\ \hline \end{array}$ **40.** $\begin{array}{r} 384 \\ +4439 \\ \hline \end{array}$ **41.** $\begin{array}{r} 4605 \\ +\ 39 \\ \hline \end{array}$ **42.** $\begin{array}{r} 7870 \\ +\ 46 \\ \hline \end{array}$

43. $\begin{array}{r} 108 \\ 2134 \\ +\ 98 \\ \hline \end{array}$ **44.** $\begin{array}{r} 706 \\ 3629 \\ +\ 83 \\ \hline \end{array}$ **45.** $\begin{array}{r} 305 \\ 6312 \\ +8573 \\ \hline \end{array}$ **46.** $\begin{array}{r} 609 \\ 6205 \\ +9503 \\ \hline \end{array}$ **47.** $\begin{array}{r} 82{,}583 \\ +\ 8692 \\ \hline \end{array}$ **48.** $\begin{array}{r} 99{,}989 \\ +\ 3454 \\ \hline \end{array}$

49. $\begin{array}{r} 63{,}126 \\ 77{,}684 \\ +80{,}000 \\ \hline \end{array}$ **50.** $\begin{array}{r} 89{,}137 \\ 60{,}470 \\ +36{,}409 \\ \hline \end{array}$

››› Applications

51. *A kettle of descendants* Captain Wilson Kettle, who died on January 25, 1963, left 11 children, 65 grandchildren, 201 great-grandchildren, and 305 great-great-grandchildren. How many descendants did he leave behind at the time of his death?

52. *Sugar fights* Sugar Ray Robinson had a record that included 175 victories, 6 draws, and 21 losses. How many total fights did he have?

53. *Recovery efforts* The amounts of materials recovered (in millions of tons) in a recent year are as follows: paper and paperboard, 37; ferrous metals, 5; glass and plastics, 5; yard and other waste, 17. How many millions of tons is that?

54. *People in plane* A jumbo jet carries 383 passengers and a crew of 9. How many people are there in the plane?

55. *Milkshake gallons* One of the largest milkshakes ever made started out with 118 gallons of vanilla ice cream, 60 gallons of milk, and 17 gallons of strawberry flavoring. How many gallons of milkshake were mixed in order to make the milkshake?

56. *Bible chapters* In the Judeo-Christian Bible, the Old Testament has 929 chapters and the New Testament has 270. How many total chapters are there in the Judeo-Christian Bible?

57. *Empire State total height* The Empire State Building is 1250 feet tall. The antenna on top of the building is 222 feet tall. How tall is the building with its antenna?

58. *Total calories* A lunch at McDonald's consisted of a Big Mac (570 calories), regular fries (220 calories), a 12-oz Sprite (144 calories), and a hot fudge sundae (357 calories). What is the total number of calories in this lunch?

59. *Gold in a nugget* The purest gold nugget ever found yielded 2248 ounces of pure gold and 32 ounces of impurities. How many ounces did the nugget weigh?

60. *Lots of paella!* One of the largest paellas made included 8140 pounds of rice, 6600 pounds of meat, 3300 pounds of mussels, 3080 pounds of beans and peppers, and 440 pounds of garlic. How many pounds of ingredients is that?

‹B› Perimeter of Polygons

61. *Perimeter* How far does a batter have to travel when circling the bases in the major league baseball stadium shown?

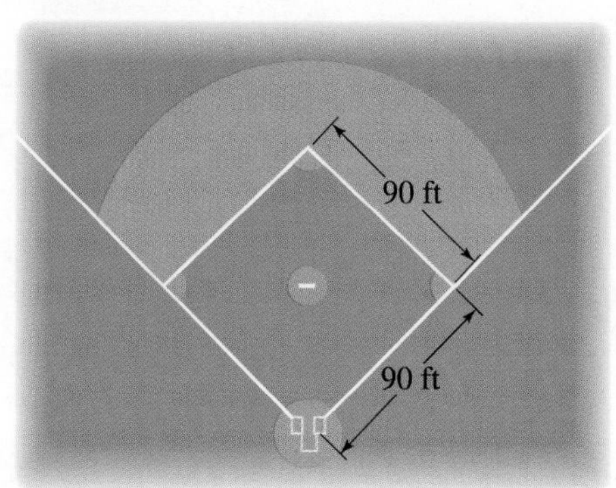

62. *Perimeter of basketball court* What is the perimeter of the standard basketball court shown?

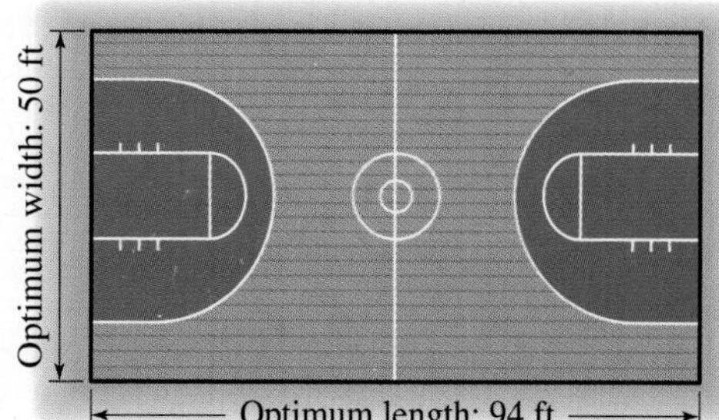

63. *Perimeter of football field* What is the perimeter of the football field?

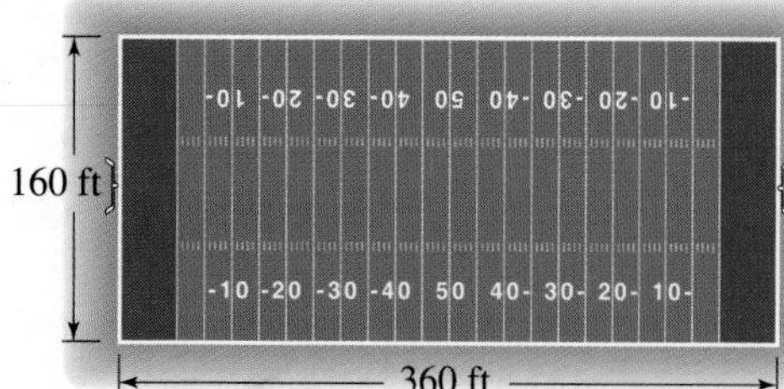

64. *Perimeter of hockey rink* The diagram shows half of a hockey rink. What is the perimeter of the whole rink?

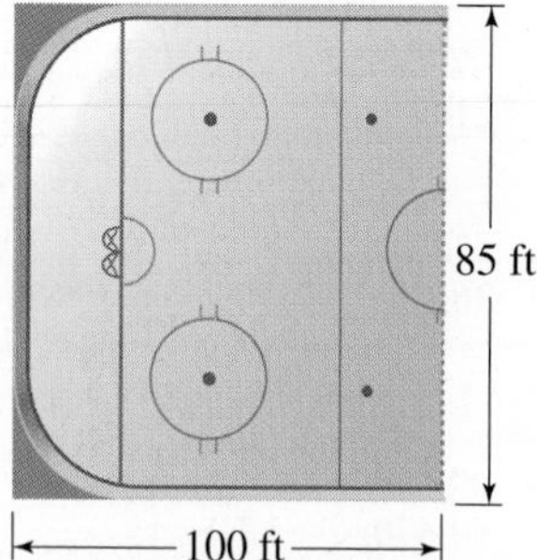

65. Do you know what a Hogan is? It is a traditional Navajo home usually built with the entrance facing east. There are two kinds of Hogans: male and female. The modern female Hogan has eight sides and each of the walls is about 8 feet wide. What is the perimeter of a female Hogan?

66. The Pentagon is a five-sided building housing the Defense Department. If each of the outside walls is 921 feet long, what is the perimeter of the building?

Using Your Knowledge

67. Individuals and businesses do banking transactions. A common transaction is to deposit money in a savings account. How much money did this person deposit?

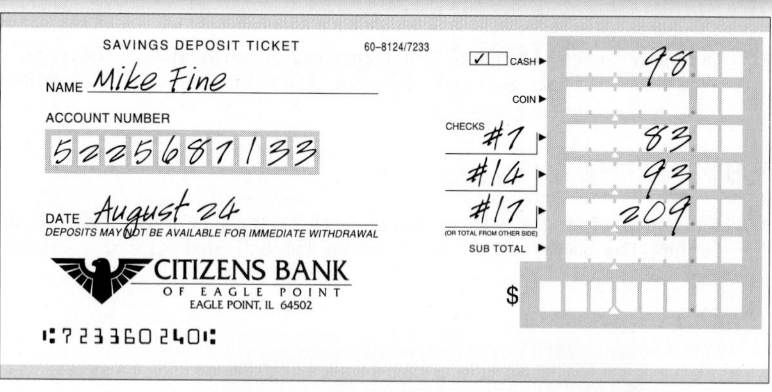

SAVINGS DEPOSIT TICKET 60–8124/7233

NAME Mike Fine

ACCOUNT NUMBER 5225687133

DATE August 24

DEPOSITS MAY NOT BE AVAILABLE FOR IMMEDIATE WITHDRAWAL

CITIZENS BANK OF EAGLE POINT EAGLE POINT, IL 64502

723360240

✓ CASH	98
COIN	
CHECKS #7	83
#14	93
#17	209
(OR TOTAL FROM OTHER SIDE) SUB TOTAL	
$	

68. Another transaction with a bank is to deposit money in a checking account. How much money did this person deposit?

DEPOSIT TICKET 60–8124/7233 1000821897

DAVE ROSE
124 FLOWERY ROAD
HILL CITY, IL 64715

DATE August 22

DEPOSITS MAY NOT BE AVAILABLE FOR IMMEDIATE WITHDRAWAL

SIGN HERE FOR CASH RECEIVED (IF REQUIRED) ★

CITIZENS BANK OF EAGLE POINT EAGLE POINT, IL 64502

723360240 1000821897 02

✓ CASH	44
COIN	14
CHECKS #2	138
#8	204
#16	77
(OR TOTAL FROM OTHER SIDE) SUB TOTAL	
$	

69. A budget is a plan for estimating how your income is to be spent (or saved). Find the estimated weekly expenditures of Family A and Family B:

	Family A	Family B
Housing	$107	$258
Food	98	175
Clothing	75	130
Transportation	150	190
Recreation	90	105
Savings	100	150

70. One of the items in the budget is the amount of money spent on transportation. In a recent year the total cost (in cents per mile) of owning and operating different automobiles and vans was as follows:

	Standard	Compact	Passenger Van
Maintenance	6	5	6
Gas and oil	9	8	11
Garage, parking	3	3	3
Insurance	7	4	8
Taxes, license, registration	2	1	2

How many cents per mile does it cost to operate

a. A standard car?

b. A compact car?

c. A passenger van?

71. Heart disease is a major cause of death in America. There is a simple test based on a study made in Framingham, Mass., that can tell you your risk of suffering a heart attack or stroke. You can find out by *addition*. The chart giving the risk factors, along with the scores for a certain individual, is shown here. (The person scored 14 points, which means that the total risk of having a heart attack or stroke is average.) Add the scores for A, B, and C to determine the total risk for each.

Risk Factor	Scores
Smoking	0 Nonsmoker 2 Less than 20 cigarettes a day 4 20 cigarettes or more a day
Weight	0 Desirable 2 Up to 10 percent over 4 More than 10 percent over
Systolic blood pressure	0 Less than 120 2 120 to 140 4 Over 140
Cholesterol level	0 Less than 150 2 150 to 250 4 Over 250
Physical activity	0 Regular vigorous exercise 2 Moderate exercise 4 Sedentary
Stress and tension	0 Rarely tense or anxious 2 Feel tense two or three times a day 4 Extremely tense
Total risk	0–4 Low 5–9 Below average 10–14 Average 15–20 High 21–24 Very high

Certain Individual	A	B	C
0	2	4	0
2	0	2	2
2	0	4	2
4	2	2	0
2	4	4	0
4	2	2	2
14			

72. Suppose you want to buy a car. You should start by learning how to read a car invoice, finding out how much the dealer pays for the car, and estimating how much you should offer the dealer. (For a lesson on how to do this and many other tips, go to link 1-3-3 on the Bello website at mhhe.com/bello.)

Using the car dealer example shown here:

a. Find the suggested list price.

b. Find the total of fees.

c. Find the total list price (including options).

Car Dealer Example

BASE PRICE	$18,580
Destination Charge	$435
Floor Mats	$50
Preferred Value Package	$2571
SUGGESTED LIST PRICE	☐
EXTRA FEES:	
Sales Promotion Fund	$100
Dealer Advertising Association	$484
Holdback	$371
Dealer Flooring Assistance	$185
TOTAL OF FEES:	☐
TOTAL LIST PRICE (including options)	☐

73. Do the fees seem reasonable to you? Here is the story (reprinted with permission of Jeff Ostroff and Consumer.Net).

A farmer had been ripped off before by a local car dealer. One day, the car dealer told the farmer he was coming over to buy a cow. The farmer priced that cow with the invoice on the right:

a. Round all prices to the nearest dollar.

b. Find the farmer's actual suggested list price and the list price you get by adding the answers in part **a.**

c. Find the actual total list price.

d. What is the highest-priced item for the basic cow?

e. What is the least expensive item for the basic cow?

Cow Dealer Example

BASIC COW	**$499.95**
Shipping and handling	35.75
Extra stomach	79.25
Two-tone exterior	142.10
Produce storage compartment	126.50
Heavy-duty straw chopper	189.60
4 spigot/high output drain system	149.20
Automatic fly swatter	88.50
Genuine cowhide upholstery	179.90
Deluxe dual horns	59.25
Automatic fertilizer attachment	339.40
4 × 4 traction drive assembly	884.16
Pre-delivery wash and comb (Farmer Prep)	69.80
FARMER'S SUGGESTED LIST PRICE	☐
Additional Farmer Markup and hay fees	300.00
TOTAL LIST PRICE (including options)	☐

For a cow that's worth maybe $2500.

74. Now that you know how to buy a car, let us see how far we can go with it. The map shows distances between cities (red numbers) in the state of Florida. For instance, the driving distance from Tampa to Orlando is 82 miles and from Orlando to Cocoa is 49 miles.

a. Find the driving distance from Tampa to Cocoa via Orlando.

b. Find the driving distance from Tampa to Ft. Pierce via Orlando and Cocoa.

c. Find the driving distance from Bradenton to Miami via West Palm Beach.

d. Find the driving distance from Bradenton to Miami via Ft. Myers and Naples. Is this driving distance shorter than the one obtained in part **c**?

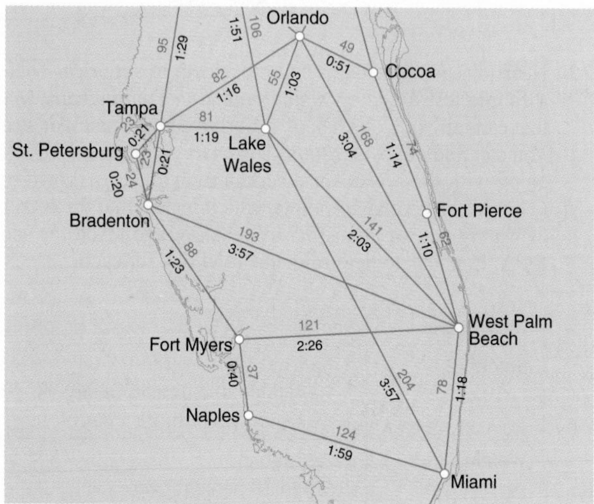

75. Next, let us travel in California.

a. Find the driving distance from San Francisco to Manteca, via Oakland.

b. Find the driving distance from Monterey to Los Angeles via San Luis Obispo and Santa Barbara.

c. Find another route that takes the same number of miles as the one in part **b.**

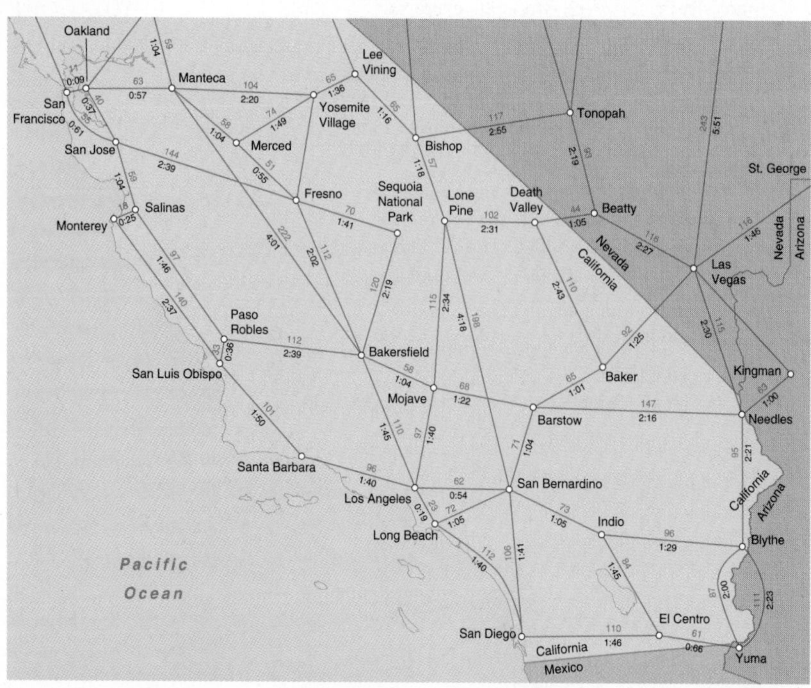

76. **a.** The roads between Los Angeles, Long Beach, and San Bernardino form a triangle. The perimeter of this triangle is the driving distance for a round-trip from Los Angeles via Long Beach and San Bernardino. What is this perimeter?

b. What is the round-trip distance between Long Beach, San Diego, and San Bernardino?

77. The roads connecting San Bernardino, Indio, Blythe, Needles, Barstow, and back to San Bernardino form a pentagon (a five-sided figure). What is the perimeter of this pentagon?

78. Find a pentagon and its perimeter if you start at San Luis Obispo.

Write On

79. Write in your own words what the identity for addition means.

80. Write in your own words what the commutative law of addition means.

81. Write in your own words what the associative law of addition means.

82. The driving distance from Tampa to Cocoa is 82 + 49 miles. The return trip is the same distance, 49 + 82 miles. Which property of addition tells us that the driving distances are the same?

83. Tyrone drove from San Luis Obispo to Los Angeles via Santa Barbara, (101 + 96) miles. He then drove 62 more miles to San Bernardino.

a. Write an expression that illustrates this situation and then find the distance from San Luis Obispo to San Bernardino.

b. Maria started from San Luis Obispo but first stopped at Santa Barbara, 101 miles. She then drove to San Bernardino via Los Angeles, (96 + 62) miles. Write an expression that illustrates Maria's driving distance.

c. Which property of addition justifies that the driving distances for Tyrone and Maria are the same?

84. The price P of a car is its base price (B) plus destination charges D, that is, $P = B + D$. Tran bought a Nissan in Smyrna, Tennessee, and there was no destination charge.

a. What is D?

b. Fill in the blank in the equation $P = B +$ ______.

c. What property of addition justifies the equation in part **b** is correct?

Algebra Bridge

85. Add $3x + 9$ and $8x + 7$.

86. Add $5x^2 + 7x + 8$ and $9x^2 + 8x + 17$.

87. Add $3x^3 + 8x^2 + 7x + 9$ and $9x^3 + 8x^2 + 4x + 5$.

Concept Checker

Fill in the blank(s) with the correct word(s), phrase, or mathematical statement.

88. The **result** of addition is called the ________.

89. In an addition, the numbers to be **added** are called ________.

90. The fact that $a + b = b + a$ is called the ____________ Property of Addition.

91. The **identity** for addition is ________.

92. The Associative Property of Addition states that $a + (b + c) =$ __________.

93. The distance around an object is called its ________.

0	$a + (b + c)$
total	**Perimeter**
$(a + b) + c$	**1**
addends	**area**
sum	**Associative**
Commutative	

Mastery Test

94. Find the perimeter of the triangle.

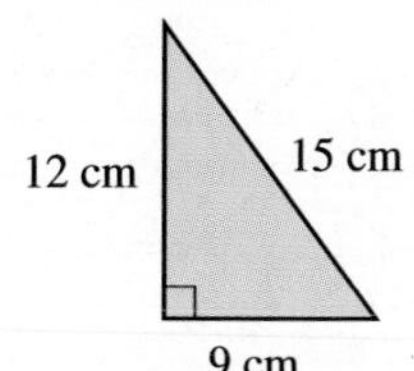

95. Add: 2454 + 6743 + 8789 + 7563

96. Add: 5374 + 3478 + 8598 + 2382

97. Add: 712 + 635

98. Add: 32 + 54

99. How many thousands of dollars (to the nearest thousand) will it cost a Florida resident to attend the fall and spring semesters?

Fall and Spring Tuition

Florida Residents	
Tuition/Fees	$2538
Room	$3120
Board	$3330

Skill Checker

100. A Wendy's triple cheeseburger contains 1040 calories, 225 grams of cholesterol, and 72 grams of protein.

a. Round 1040 to the nearest thousand.

b. Round 225 to the nearest hundred.

c. Round 72 to the nearest ten.

101. Write 1040, 225, and 72 in words.

1.4 Subtraction

Objectives

You should be able to:

A Subtract one whole number from another, regrouping (borrowing) if necessary.

B Solve applications involving the concepts studied.

To Succeed, Review How To . . .

1. Add two numbers. (p. 25)
2. Write a number in expanded form. (p. 4)

Getting Started

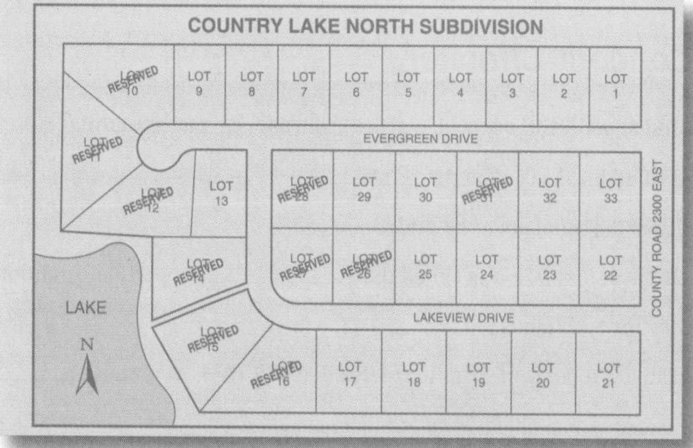

The subdivision has 33 lots and 10 are reserved. How many lots are available? To find the answer, we use the operation of subtraction, indicated by a minus sign (−), and write

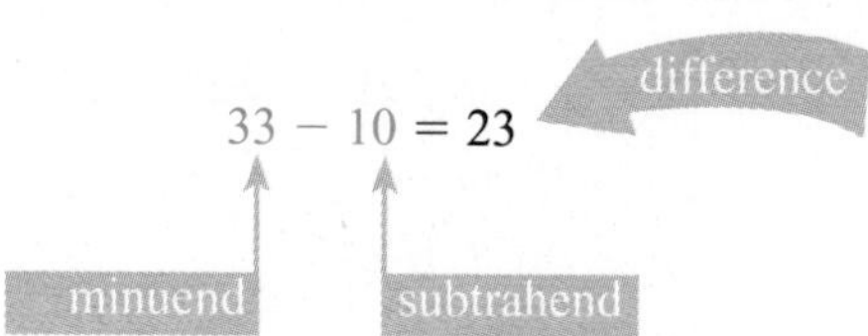

The procedure is usually written as

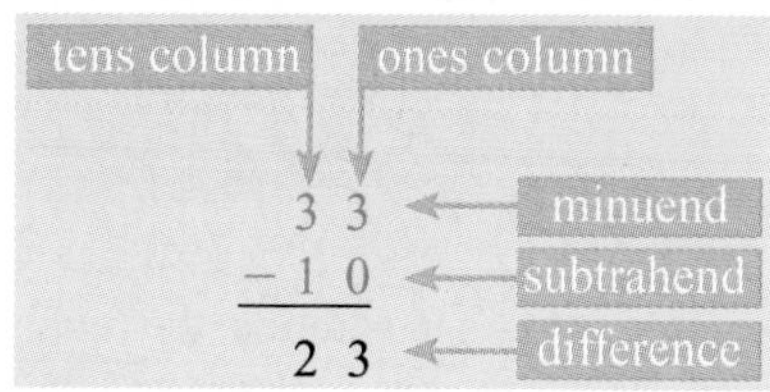

To find how many lots are available you could also think this way: the number reserved is now 10; how many lots have to be reserved before all 33 are reserved? We write

$$10 + \square = 33$$

which means that there is a number $\square$ that added to 10 gives 33. The answer is $\square = 23$.

A › Subtracting Whole Numbers

Here is the definition of subtraction.

SUBTRACTION

The difference $a - b$ is a unique number c so that $a = c + b$

From the discussion in the *Getting Started,* you can see that to subtract whole numbers, we start with two numbers, the **minuend,** which is the larger of the two,* and the **subtrahend,** which is the one being subtracted, and get a third number called the **difference.** As we mentioned, addition and subtraction are related. Thus,

$$7 - 3 = 4 \quad \text{because} \quad 7 = 3 + 4$$
$$8 - 5 = 3 \quad \text{because} \quad 8 = 5 + 3$$

Because addition and subtraction are related (they are *inverse* operations), you can use addition to check your answers in subtraction. For example, you can write

$$\left.\begin{array}{r} 7 \\ -3 \\ \hline 4 \end{array}\right\} \text{and check by adding } 3 + 4 = 7$$

EXAMPLE 1 **Subtracting whole numbers**

Subtract 13 − 8.

SOLUTION

$$\begin{array}{r} 13 \\ -8 \\ \hline 5 \end{array}$$

CHECK

$$\left.\begin{array}{r} 13 \\ -8 \\ \hline 5 \end{array}\right\} 8 + 5 = 13$$

PROBLEM 1

Subtract 15 − 9.

Here is another problem:

$$59 - 36 = \underline{\qquad}$$

As we did before, we estimate that the answer to the nearest ten should be $60 - 40 = 20$. We then write the problem with the ones and the tens arranged in a column as shown

* Subtractions such as 3 − 7 and 5 − 8 are discussed later.

Answers to PROBLEMS

1. 6

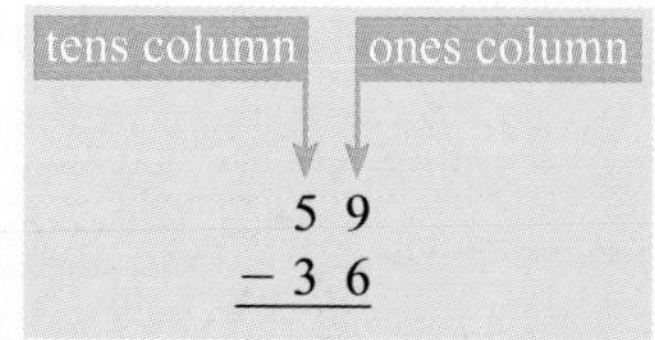

As in addition, the numbers were arranged vertically in columns, with the ones digits in the ones column and the tens digits in the tens column. The short form of the subtraction is on the left, and the expanded form is on the right.

Short Form	**Expanded Form**
59	50 + 9
−36	(−)30 + 6
23	20 + 3 = 23

We can check our work by adding 36 + 23. Since 36 + 23 = 59, our answer is correct. Note that our estimate (20) is close to the actual answer (23).

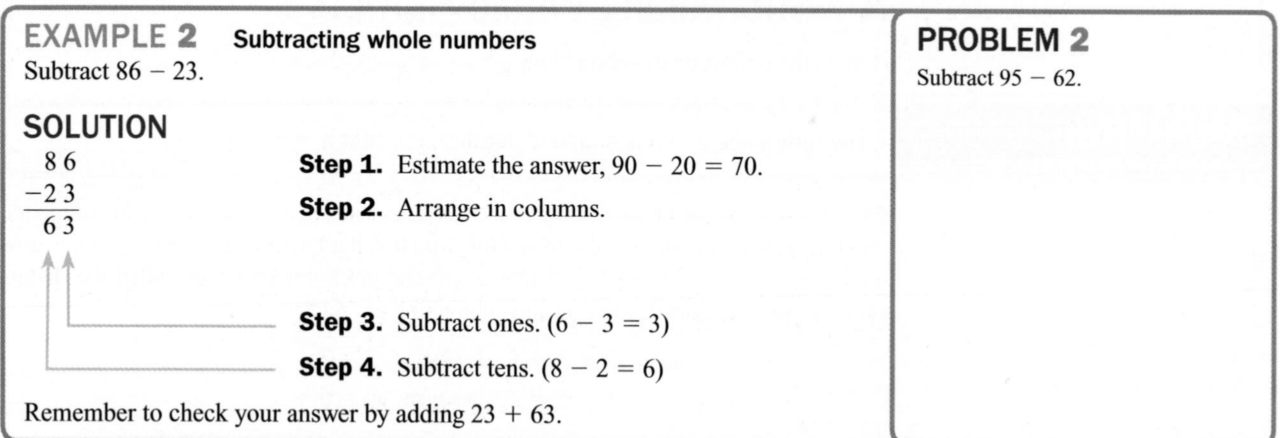

EXAMPLE 2 **Subtracting whole numbers**

Subtract 86 − 23.

SOLUTION

$$\begin{array}{r} 86 \\ -23 \\ \hline 63 \end{array}$$

Step 1. Estimate the answer, 90 − 20 = 70.

Step 2. Arrange in columns.

Step 3. Subtract ones. (6 − 3 = 3)

Step 4. Subtract tens. (8 − 2 = 6)

Remember to check your answer by adding 23 + 63.

PROBLEM 2

Subtract 95 − 62.

Now consider the problem

$$\begin{array}{r} 46 \\ -28 \\ \hline \end{array}$$

We cannot subtract 8 from 6 and obtain a whole number, so we have to "borrow" 10 from the 4 (which represents 4 tens, or 40) and think of the 46 (regroup or rewrite) as 30 + 16. We write

(40 − 10 = 30) ⟶ 3 16 ⟵ (10 + 6 = 16)

$$\begin{array}{r} \not4\,\not6 \\ -2\,8 \\ \hline \end{array}$$

We then subtract 8 from 16 and 2 from 3, as shown. Again, compare the short form and the expanded form.

Short Form	**Expanded Form**
3 16	
$\not4$ $\not6$	30 + 16
−2 8	(−)20 + 8
1 8	10 + 8 = 18

CHECK 28 + 18 = 46

Answers to PROBLEMS

2. 33

EXAMPLE 3 Subtracting whole numbers

Subtract 742 − 327.

SOLUTION To the nearest hundred, the answer should be about 700 − 300 = 400.

Short Form		Expanded Form
3 12 7 ~~4~~ ~~2~~ −3 2 7 4 1 5	**Step 1.** Estimate the answer. **Step 2.** Arrange in columns. **Step 3.** Write the 42 in 742 as 30 + 12 ("borrow 1"), as shown. **Step 4.** Subtract ones. **Step 5.** Subtract tens. **Step 6.** Subtract hundreds.	700 + 30 + 12 (−)300 + 20 + 7 400 + 10 + 5 = 415

CHECK

$$\begin{array}{r} 327 \\ +415 \\ \hline 742 \end{array}$$

So the result is correct.

PROBLEM 3

Subtract 654 − 239.

Sometimes, we have to rewrite the number in the hundreds place. For example, to subtract 81 from 346 we proceed as follows:

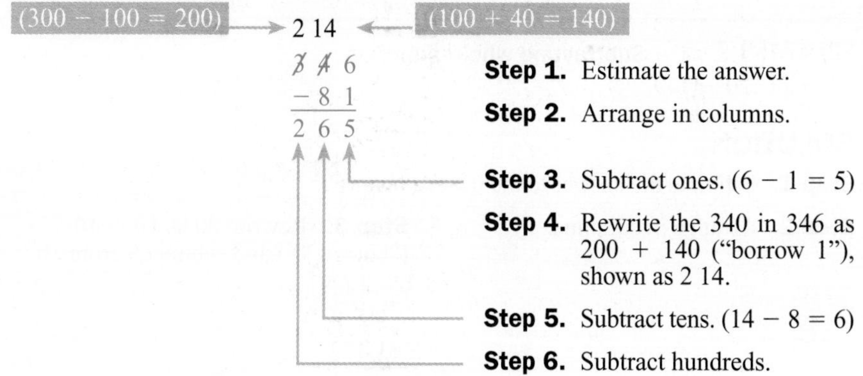

CHECK

$$\begin{array}{r} {}^{1} \\ 81 \\ +265 \\ \hline 346 \end{array}$$

The expanded form for this procedure is

$$\begin{array}{rrl} 346 & & 200 + 140 + 6 \\ -\ 81 & (-) & 80 + 1 \\ \hline & & 200 + 60 + 5 = 265 \end{array}$$

Answers to PROBLEMS

3. 415

EXAMPLE 4 **Subtracting whole numbers**

Subtract 937 − 53.

SOLUTION

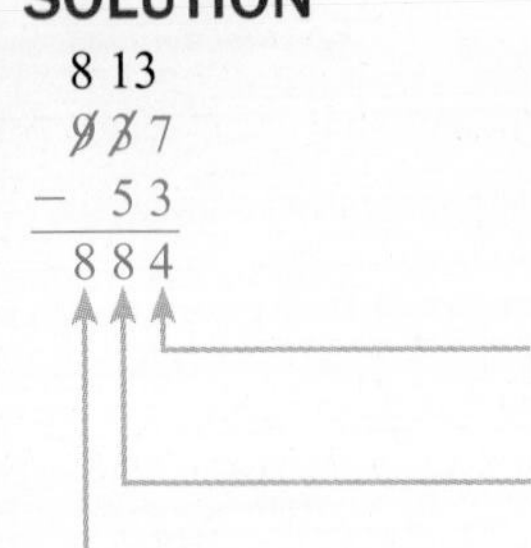

Step 1. Estimate the answer.

Step 2. Arrange in columns.

Step 3. Subtract ones. (7 − 3 = 4)

Step 4. Rewrite the 93 in 937 as 80 + 13 ("borrow 1"), shown as 8 13.

Step 5. Subtract tens.

Step 6. Subtract hundreds.

PROBLEM 4

Subtract 846 − 72.

There are some problems in which we have to rewrite the numbe rs involved more than once. Examine the subtraction problem 732 − 453 = ________.

Expanded Form

$$\begin{array}{r} 700 + 30 + 2 \\ (-)\ 400 + 50 + 3 \\ \hline \end{array} \quad \text{or} \quad \begin{array}{r} 700 + 20 + 12 \\ (-)\ 400 + 50 + 3 \\ \hline 9 \end{array} \quad \text{or} \quad \begin{array}{r} 600 + 120 + 12 \\ (-)\ 400 + 50 + 3 \\ \hline 200 + 70 + 9 = 279 \end{array}$$

We cannot subtract 3 from 2, so we *rewrite* 732 as shown on the right.

We cannot subtract 50 from 20, so we *rewrite* 732 again.

The short form is

$$\begin{array}{r} \scriptstyle 12 \\ \scriptstyle 6\ \not{2}\ 12 \\ \not{7}\ \not{3}\ \not{2} \\ -4\ 5\ 3 \\ \hline 2\ 7\ 9 \end{array}$$

EXAMPLE 5 **Subtracting whole numbers**

Subtract 520 − 149.

SOLUTION

Step 1. Estimate the answer to the nearest hundred.

Step 2. Arrange in columns.

$$\begin{array}{r} 520 \\ -149 \\ \hline \end{array}$$

Step 3. Rewrite 20 as 10 + 10 ("borrow 1") and subtract 9 from 10.

$$\begin{array}{r} \scriptstyle 1\ 10 \\ 5\ \not{2}\ \not{0} \\ -1\ 4\ 9 \\ \hline 1 \end{array}$$

Step 4. Rewrite 51 as 40 + 11 ("borrow 1") and subtract 4 from 11.

$$\begin{array}{r} \scriptstyle 4\ 11\ 10 \\ \not{5}\ \not{2}\ \not{0} \\ -1\ 4\ 9 \\ \hline 7\ 1 \end{array}$$

Step 5. Subtract 1 from 4.

$$\begin{array}{r} \scriptstyle 4\ 11\ 10 \\ \not{5}\ \not{2}\ \not{0} \\ -1\ 4\ 9 \\ \hline 3\ 7\ 1 \end{array}$$

CHECK

$$\begin{array}{r} \scriptstyle 11 \\ 149 \\ +371 \\ \hline 520 \end{array}$$

PROBLEM 5

Subtract 680 – 295.

Answers to PROBLEMS

4. 774 **5.** 385

The complete procedure is usually written as

$$\begin{array}{r} {\scriptstyle 4\ 11\ 10} \\ \not{5}\ \not{2}\ \not{0} \\ -1\ 4\ 9 \\ \hline 3\ 7\ 1 \end{array}$$

EXAMPLE 6 Subtracting whole numbers

Subtract 8340 − 2459.

SOLUTION

Step 1. Estimate the answer to the nearest thousand.

Step 2. Arrange in columns.

$$\begin{array}{r} 8340 \\ -2459 \\ \hline \end{array}$$

Step 3. Rewrite 40 as 30 + 10 ("borrow 1") and subtract 9 from 10.

$$\begin{array}{r} {\scriptstyle 3\ 10} \\ 8\ 3\ \not{4}\ \not{0} \\ -2\ 4\ 5\ 9 \\ \hline 1 \end{array}$$

Step 4. Rewrite 33 as **20** + 13 ("borrow 1") and subtract 5 from 13.

$$\begin{array}{r} {\scriptstyle 2\ 13\ 10} \\ 8\ \not{3}\ \not{4}\ \not{0} \\ -2\ 4\ 5\ 9 \\ \hline 8\ 1 \end{array}$$

Step 5. Rewrite 82 as **70** + 12 ("borrow 1") and subtract 4 from 12.

$$\begin{array}{r} {\scriptstyle 12} \\ {\scriptstyle 7\ 2\ 13\ 10} \\ \not{8}\ \not{3}\ \not{4}\ \not{0} \\ -2\ 4\ 5\ 9 \\ \hline 8\ 8\ 1 \end{array}$$

Step 6. Subtract 2 from 7.

$$\begin{array}{r} {\scriptstyle 12} \\ {\scriptstyle 7\ \not{2}\ 13\ 10} \\ \not{8}\ \not{3}\ \not{4}\ \not{0} \\ -2\ 4\ 5\ 9 \\ \hline 5\ 8\ 8\ 1 \end{array}$$

CHECK

$$\begin{array}{r} {\scriptstyle 111} \\ 2459 \\ +5881 \\ \hline 8340 \end{array}$$

PROBLEM 6

Subtract 5250 − 1478.

Now let us find the difference 705 − 238 = ________. As usual, we write

$$\begin{array}{r} 705 \\ -238 \\ \hline \end{array}$$

Since we cannot subtract 8 from 5 and obtain a whole number answer, we rewrite 705 as 690 + 15 and then subtract as shown:

$$\begin{array}{r} {\scriptstyle 6\ 9\ 15} \\ \not{7}\ \not{0}\ \not{5} \\ -2\ 3\ 8 \\ \hline 4\ 6\ 7 \end{array}$$

What we have done is rewritten

$$\begin{aligned} 705 &= 70 \text{ tens} + 5 \\ &= 69 \text{ tens} + 15 \end{aligned}$$

Answers to PROBLEMS

6. 3772

EXAMPLE 7 **Subtracting whole numbers**

Subtract 803 − 346.

SOLUTION

```
  7 9 13
  8 0 3  (crossed out)
 −3 4 6
  4 5 7
```

Step 1. Estimate the answer.

Step 2. Arrange in columns.

Step 3. Write 803 as 79 tens + 13.

Step 4. Subtract ones.

Step 5. Subtract tens.

Step 6. Subtract hundreds.

PROBLEM 7

Subtract 605 − 247.

EXAMPLE 8 **Subtracting whole numbers**

Subtract 7006 − 6849.

SOLUTION

```
  6 9 9 16
 −7 0 0 6  (crossed out)
 −6 8 4 9
  0 1 5 7
```

Step 1. Estimate the answer.

Step 2. Arrange in columns.

Step 3. Rewrite 7006 as 6990 + 16.

Step 4. Subtract ones.

Step 5. Subtract tens.

Step 6. Subtract hundreds.

Step 7. Subtract thousands.

The answer is written as 157.

CHECK

```
   111
  6849
 + 157
  7006
```

PROBLEM 8

Subtract 6002 − 5843.

B › Applications Involving Subtraction

Subtraction is used in everyday life. For example, if you have a checking account, the **stub** gives you a record of your balance, the amount you paid, and to whom you paid it. To find the balance carried forward, we need to subtract 38 from 213. This is done in Example 9.

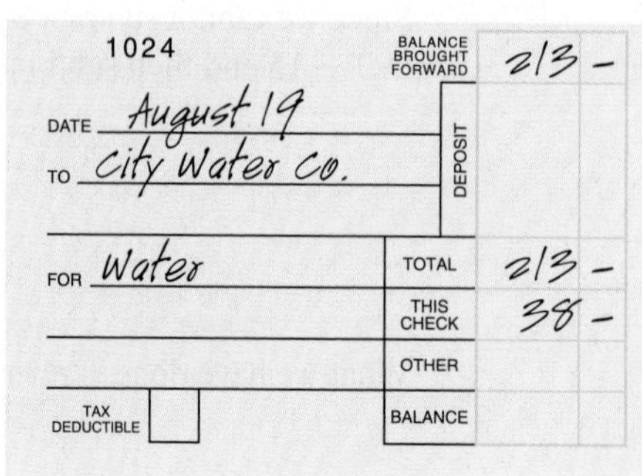

Answers to PROBLEMS

7. 358 **8.** 159

EXAMPLE 9 Bank stub

Find the balance carried forward on the stub shown by subtracting 38 from 213.

SOLUTION

New balance
Amount of this check
Balance carried forward

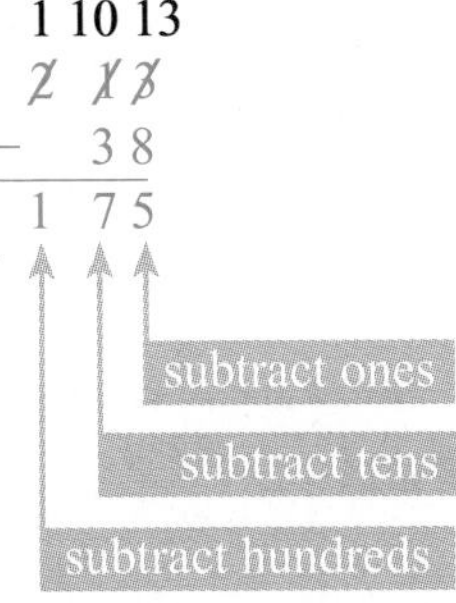

Thus, the balance carried forward is $175.

PROBLEM 9

The new balance in an account is $317. A check for $59 is written. What is the balance carried forward?

EXAMPLE 10 Subtracting discount and rebate

The ad says that if you subtract the discount and rebate ($4043) from the factory MSRP ($19,266) you obtain the clearance price $15,233. Use addition to check this fact.

Source: Tampa Tribune, 2/23/03, Money, 12.

SOLUTION

Remember that $a - b = c$ means that $a = c + b$, that is, $19,266 − $4043 = $15,233 means that $19,266 = $15,233 + $4043, which is not a true statement. (Add 15,233 and 4043 to check it out!) The ad was incorrect!

PROBLEM 10

Check the clearance price in the following ad:

Algebra Bridge

In algebra, we can subtract $2x^2 + 3x + 6$ from $5x^2 + 7x + 9$. As usual, we align the expressions in columns and subtract like this:

$$\begin{array}{r} 5x^2 + 7x + 9 \\ (-)\ 2x^2 + 3x + 6 \\ \hline 3x^2 + 4x + 3 \end{array}$$

Step 1. Subtract the numbers.

Step 2. Subtract the x's.

Step 3. Subtract the x^2's.

Answers to PROBLEMS

9. $258 **10.** Not a true statement; should be $10,998

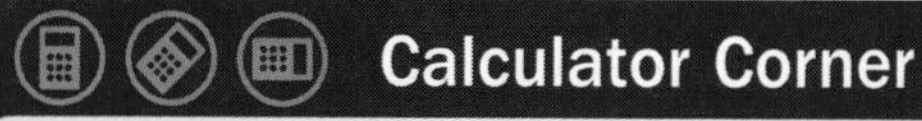

Calculator Corner

Subtraction with a calculator is easy. As with addition, you simply press the appropriate keys. Thus, to do the problem

$$\begin{array}{r} 35 \\ -15 \\ \hline \end{array}$$

you press 3 5 − 1 5 ENTER. The display shows the answer, 20.

› Exercises 1.4

› Web IT go to mhhe.com/bello for more lessons

‹A› Subtracting Whole Numbers In Problems 1–50, subtract.

1. $14 - 6$	**2.** $8 - 5$	**3.** $13 - 6$	**4.** $16 - 9$	**5.** $17 - 0$	**6.** $19 - 9$
7. $17 - 9$	**8.** $15 - 7$	**9.** $66 - 32$	**10.** $53 - 21$	**11.** $83 - 20$	**12.** $55 - 30$
13. $90 - 56$	**14.** $58 - 40$	**15.** $73 - 58$	**16.** $96 - 39$	**17.** $503 - 291$	**18.** $605 - 332$
19. $578 - 499$	**20.** $937 - 888$	**21.** 725 − 318	**22.** 835 − 608		
23. 872 − 657	**24.** 742 − 345	**25.** 560 − 278	**26.** 590 − 445		
27. 905 − 726	**28.** 308 − 199	**29.** 607 − 398	**30.** 804 − 297		
31. $5837 - 3216$	**32.** $6784 - 3271$	**33.** $4783 - 1278$	**34.** $6892 - 4583$	**35.** $6853 - 2765$	**36.** $4721 - 1242$
37. $5325 - 2432$	**38.** $9423 - 2540$	**39.** $3860 - 2971$	**40.** $4520 - 3362$	**41.** $7634 - 388$	**42.** $3425 - 137$
43. $5073 - 782$	**44.** $6068 - 697$	**45.** $6003 - 289$	**46.** $9001 - 539$	**47.** $13{,}456 - 7{,}576$	**48.** $27{,}333 - 9666$
49. $50{,}000 - 23{,}569$	**50.** $80{,}000 - 65{,}687$				

‹ B › Applications Involving Subtraction

51. *Tallest structure in the world* The tallest structure in the world is the Warszawa radiomast, 2117 feet tall. If the Empire State Building is 1250 feet high, what is the difference between the two heights?

52. *Elevation difference* The elevation of Lhasa, Tibet, is 12,087 feet above sea level. La Paz, Bolivia, is 11,916 feet above sea level. What is the difference in altitudes?

53. *Young authors* The youngest recorded commercially published author is Dorothy Straight, born in 1958. She published her book in 1962. How old was she then?

54. *Trip length* At the beginning of a trip the odometer on a car read 37,742 miles. At the end it read 43,224. How many miles was the trip?

55. *Fall difference* Niagara Falls is 193 feet tall. Angel Falls is 2212 feet. What is the difference in heights?

56. *Passing and rushing in Miami* The professional football team that gained the most yards in a season is Miami, with 6936 yards gained in 1984. If 5018 yards were gained passing, how many did they gain rushing?

57. *Motorcycle sale* A $3200 motorcycle is on sale for $2999. How much can be saved by buying it at the sale price?

58. *Bank balance* Your bank balance is $347. You write checks for $59 and $39. What is your new balance?

59. *Making money* A woman bought a car for $4500. She spent $787 on repairs and then sold it for $6300. How much profit did she make on the car?

61. *Real buying power* The *real buying power* of your dollar is going down fast! How can you figure it out? Here are the steps:

		Jane Dough
Adjusted gross income	$19,205	$20,000
Subtract federal tax	− 3,574	− 4,672
Income less federal tax	15,631	☐
Subtract Social Security	− 1,440	− 1,500
Net income	14,191	☐
Subtract 10% of net income for inflation	− 1,419	− 1,383
Real buying power	$12,772	☐

This means that your net income of $14,191 will buy only $12,772 worth of goods this year. Find the real buying power for Jane Dough using the given figures.

60. *Income after deductions* A man makes $682 a week. The deductions are:

Withholding tax	$87
Social Security	$46
Union dues	$13

What is his net income after these deductions?

62. *Real buying power* Joe Worker is single and has an adjusted gross income of $19,000. His federal tax amounts to $4264, his Social Security contribution is $1273. His 10% loss due to inflation amounts to $1346. What is his real buying power?

According to Jeff Ostroff and ConsumerNet, Inc, when you buy a new car, the *dealer's actual cost* is given by:

Invoice Price − Factory To Dealer Incentive − Factory Holdback

Source: http://carbuyingtips.com

63. You want to buy a Toyota Camry LE V6 with an invoice price of $19,922. You discovered a $500 factory to dealer incentive exists, and a 2% holdback of the MSRP ($447).

a. What is the dealer's actual cost?

b. If the dealer offers to sell you the Camry for $22,000, how much is the dealer making?

64. You want a used Toyota Corolla CE. Surf over to Edmund's; the invoice price is $13,853. Edmund's also shows a $900 dealer incentive and a 2% holdback of $302.

a. What is the dealer's actual cost?

b. If you are letting the dealer make $500 on the deal, how much should you offer for the car?

These computer prices will be used in Exercises 65–70.

Cutting Edge	Performance	Affordability
Dimension 8400 from $1019	Dimension G310 from $689	Dimension F310 from $968

65. What is the difference in price between a Dimension 8400 and the Dimension F310?

66. What is the difference in price between a Dimension 8400 and Dimension G310?

67. What is the difference in price between a Dimension G310 and a Dimension F310?

68. If you have enough money to buy a Dimension F310, how much more money do you need to upgrade to the 8400?

69. If you can afford a Dimension G310, how much more money do you need to upgrade to the 8400?

70. What is the difference in price between the most expensive and the least expensive of the computers?

Web IT go to mhhe.com/bello for more lessons

The following chart shows the median household income by race and will be used in Exercises 71–76.

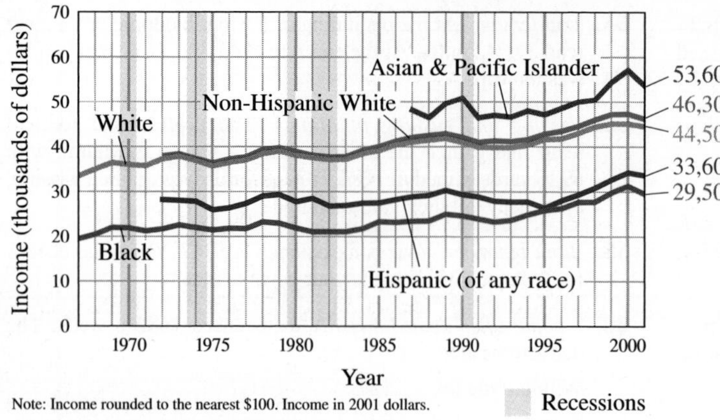

Note: Income rounded to the nearest $100. Income in 2001 dollars.

Source: Data from U.S. Census Bureau.

71. Find the difference in household income between Blacks and Hispanics in 2001.

72. Find the difference in household income between Hispanics and Whites in 2001.

73. Find the difference in household income between Non-Hispanic Whites and Blacks in 2001.

74. How much does the household income of Blacks have to increase to reach the level of Asian and Pacific Islanders?

75. Which two races are closest in their household incomes?

76. Which two races are farthest apart in their household incomes?

The table will be used in Problems 77–80.

Trends in College and University Pricing

	4-Year Public	2-Year Public	4-Year Private
Total tuition and fees	$5491	$2191	$21,235

77. What is the price difference between a 4-year public college and a 2-year public college?

78. What is the price difference between a 4-year public college and a 4-year private college?

79. What is the price difference between a 4-year private college and a 2-year public college?

80. If you have saved $3000 and want to go to a 4-year public college, how much more money do you need?

Source: Trends in College Pricing, The College Board, http://www.ed.gov/about/bdscomm/list/hiedfuture/2nd-meeting/trends.pdf

Using Your Knowledge

One of the most important applications of addition and subtraction occurs in banking. To help customers keep a record of the money in their checking accounts, banks provide a **check stub** like the one shown on the right.

The first line of the stub shows that the check number (NO.) is 70. "This check" indicates that the person is paying $120. Since the balance brought forward (the amount of money in the account) is $380 and the amount of the check is $120, we *subtract* $120 from $380, obtaining $260, the new balance in the account. This new balance is then entered in the stub number 71, which is shown on the left below. Follow this procedure and find the new balance on each of the given stubs.

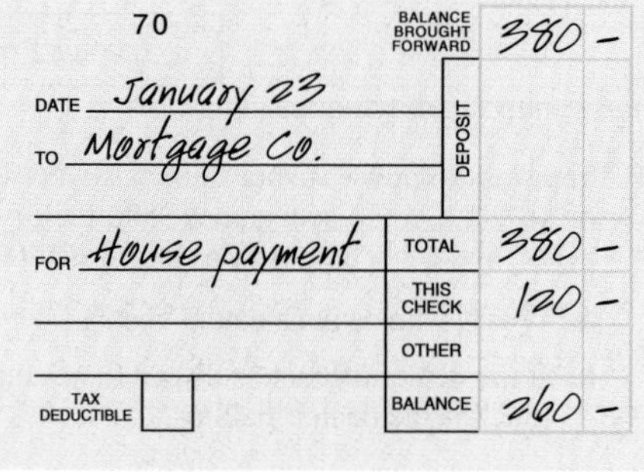

81.

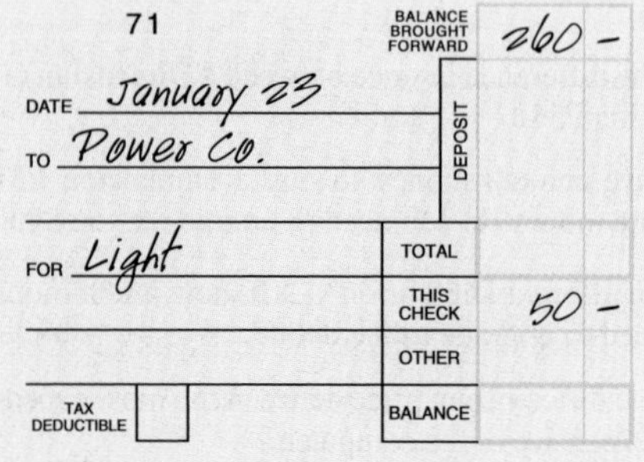

82.

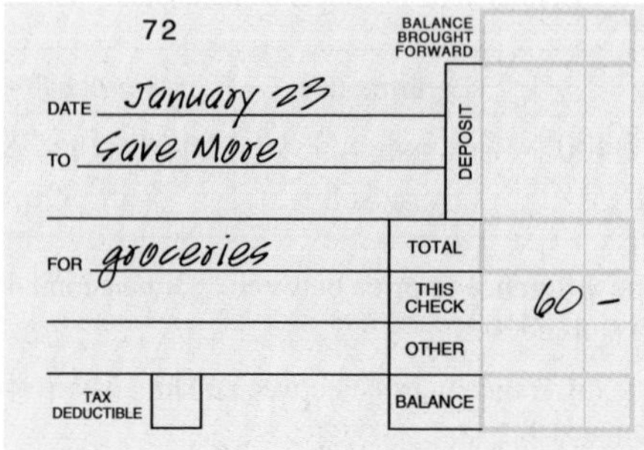

83.

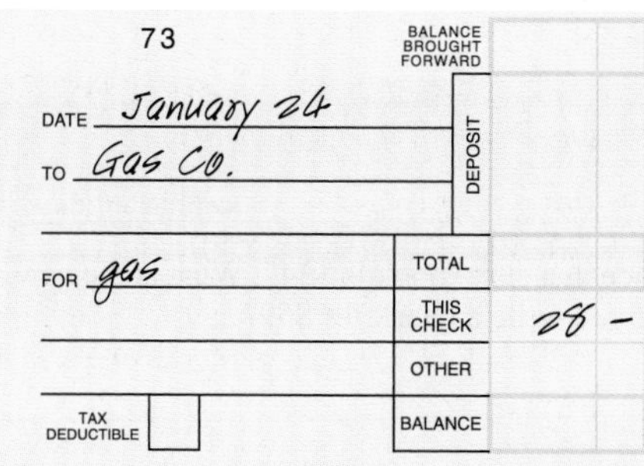

73

BALANCE BROUGHT FORWARD

DATE January 24

TO Gas Co.

DEPOSIT

FOR gas

TOTAL

THIS CHECK 28 –

OTHER

TAX DEDUCTIBLE

BALANCE

84.

74

BALANCE BROUGHT FORWARD

DATE January 25

TO Footsie Shoes

DEPOSIT

FOR shoes

TOTAL

THIS CHECK 23 –

OTHER

TAX DEDUCTIBLE

BALANCE

A Burning Issue The kindling point of a substance is the amount of heat needed to make the substance burst into flames. For example, the kindling point of natural gas is 1225°F. If the kindling point of a certain piece of wood is 500°F, the difference between the kindling point of natural gas and that of wood is

$$\begin{array}{r} 1225 \\ -\ 500 \\ \hline 725°\text{F} \end{array}$$

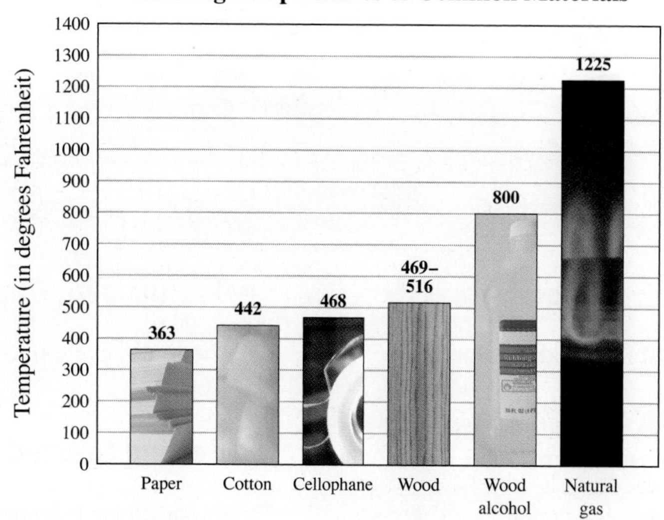

85. Find the difference (°F) between the kindling point of paper and cotton (look at the graph).

86. Find the difference (°F) between the kindling point of cotton and cellophane.

87. Find the difference (°F) between the kindling point of cellophane and wood alcohol.

88. Find the difference (°F) between the kindling point of paper and wood alcohol.

89. Find the difference (°F) between the kindling point of cotton and wood alcohol.

90. Find the difference (°F) between the kindling point of paper and cellophane.

Write On

91. A subtraction problem can be changed to an addition problem; for example $20 - 15 = 5$ means that $20 = 15 + 5$. Write two more examples to demonstrate this.

92. Is the operation of subtraction commutative? Explain and give examples of why or why not.

93. Is the operation of subtraction associative? Explain and give examples of why or why not.

94. Is there an identity for subtraction? Explain.

Algebra Bridge

95. Subtract $3x^2 + 6x + 2$ from $5x^2 + 8x + 7$.

96. Subtract $5x^2 + 3x + 6$ from $8x^2 + 7x + 9$.

97. Subtract $2x^2 + 5x + 3$ from $3x^3 + 5x^2 + 7x + 9$.

Concept Checker

Fill in the blank(s) with the correct word(s), phrase, or mathematical statement.

98. The **difference** of a and b is __________.

99. $a - b$ is a unique number c so that __________.

100. In the subtraction problem **106 − 79 = 27,** the **106** is called the __________, **79** is the __________, and **27** is the __________.

101. Addition and subtraction are __________ **operations.**

inverse	**associative**
$a - b$	**subtrahend**
minuend	$c = b - a$
commutative	**difference**
$a = c + b$	$b - a$

Mastery Test

102. 632 − 216

103. 857 − 62

104. 720 − 169

105. 703 − 257

106. 8006 − 7859

107. The balance in a checkbook is $347. What is the new balance if you write a check for $59?

Skill Checker

108. Add 28 + 120.

109. Add 345 + 5400.

110. Add 168 + 2349.

111. Write 2348 in expanded form.

112. Write 5250 in expanded form.

1.5 Multiplication

Objectives

You should be able to:

A Multiply two whole numbers.

B Multiply a whole number by any multiple of 10 (10, 100, 1000, etc.).

C Solve applications involving the concepts studied.

D Find areas using multiplication.

To Succeed, Review How To . . .

Use the multiplication facts in the table on page 49.

Getting Started

There's something fishy here! How many fishbowls do you see in the photo? To find the answer you may do one of the following:

1. Count the fishbowls.
2. Note that there are three bowls in each row, then add the rows like this:

$$3 + 3 + 3 + 3 = 12$$

3. Multiply.

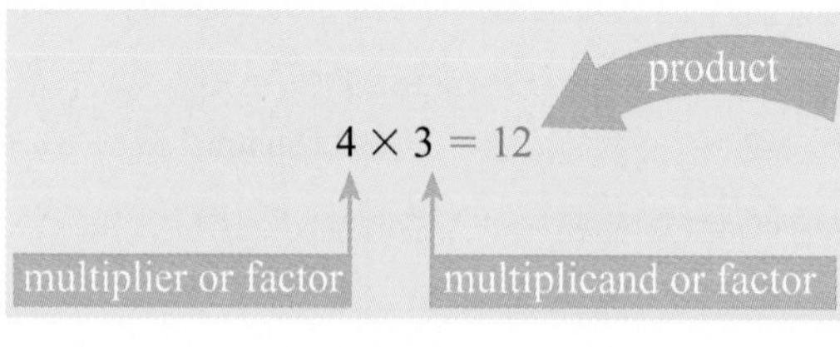

We used the product 4×3 to indicate the repeated addition $3 + 3 + 3 + 3$. We use a multiplication sign (×) to indicate multiplication. A German mathematician by the name of Leibniz, who thought that × could be confused with the letter x, started using the dot (·) to indicate multiplication; thus 4×3 can be written as $4 \cdot 3$ or using parentheses as (4)(3).

A > Multiplication of Whole Numbers

The procedure discussed is usually written as

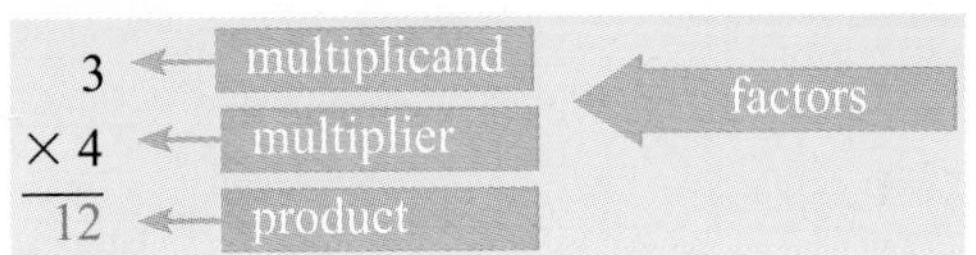

The multiplicand and the multiplier—that is, the numbers to be multiplied—are sometimes simply called **factors.** The **product** is the result of multiplication. All the multiplication facts you should know appear in the following table. As with the addition facts, you should know these facts already or take the time to memorize them now.

×	0	1	2	3	4	5	6	7	8	9
0	0	0	0	0	0	0	0	0	0	0
1	0	1	2	3	4	5	6	7	8	9
2	0	2	4	6	8	10	12	14	16	18
3	0	3	6	9	12	15	18	21	24	27
4	0	4	8	12	16	20	24	28	32	36
5	0	5	10	15	20	25	30	35	40	45
6	0	6	12	18	24	30	36	42	48	54
7	0	7	14	21	28	35	42	49	56	63
8	0	8	16	24	32	40	48	56	64	72
9	0	9	18	27	36	45	54	63	72	81

As with addition, multiplication has many useful properties. Notice in the table that when any number is multiplied by zero, the product is zero. This is called the multiplication property of zero.

MULTIPLICATION PROPERTY OF ZERO

The product of a number and zero is zero.

$$a \cdot 0 = 0$$

Also notice that when any number is multiplied by 1, the product is the same number. This is called the **Identity for Multiplication.**

IDENTITY FOR MULTIPLICATION

The product of any number and one is the number.

$$a \cdot 1 = a$$

Like addition, multiplication is both commutative and associative.

COMMUTATIVE PROPERTY OF MULTIPLICATION

Changing the *order* of two factors does not change their product.

$$a \cdot b = b \cdot a$$

Now let us try another problem.

$$3 \times 32 = \_\_\_\_\_\_$$

First, we write the problem like this:

$$\begin{array}{r} 32 \\ \times\ 3 \\ \hline \end{array}$$

We can estimate that the answer is about $30 \times 3 = 90$. We then proceed by steps.

Step 1	Step 2	Step 3
$\begin{array}{r} 32 \\ \times\ 3 \\ \hline 6 \end{array}$	$\begin{array}{r} 32 \\ \times\ 3 \\ \hline 6 \\ 90 \\ \hline \end{array}$	$\begin{array}{r} 32 \\ \times\ 3 \\ \hline 6 \\ +90 \\ \hline 96 \end{array}$
Multiply 3 by 2.	Multiply 3 by 30.	Add 6 + 90.
$3 \times 2 = 6$	$3 \times 30 = 90$	$6 + 90 = 96$

The expanded form is

$$\begin{array}{r} 30 + 2 \\ \times\ 3 \\ \hline 90 + 6 = 96 \end{array}$$

The fact that we can do this multiplication is based on a property called the **Distributive Property** which says that to multiply a number by a sum we can multiply each addend by the number and then add. Thus,

DISTRIBUTIVE PROPERTY OF MULTIPLICATION

$$a \times (b + c) = (a \times b) + (a \times c)$$

$$\text{or} \quad a \cdot (b + c) = \underbrace{a \cdot b} + \underbrace{a \cdot c}$$

In our case, $3 \times (30 + 2)$

$$3 \times (30 + 2) = (3 \times 30) + (3 \times 2)$$

Note that the actual answer (96) is roughly equal to our estimate (90).

EXAMPLE 1 Multiplying whole numbers

Multiply 4×37.

SOLUTION The answer should be about $4 \times 40 = 160$. We then proceed by steps.

Step 1	Step 2	Step 3
$\begin{array}{r} 37 \\ \times\ 4 \\ \hline 28 \end{array}$	$\begin{array}{r} 37 \\ \times\ 4 \\ \hline 28 \\ 120 \\ \hline \end{array}$	$\begin{array}{r} 37 \\ \times\ 4 \\ \hline 28 \\ 120 \\ \hline 148 \end{array}$
Multiply 4 by 7.	Multiply 4 by 30.	Add 28 and 120.
$4 \times 7 = 28$	$4 \times 30 = 120$	$28 + 120 = 148$

PROBLEM 1

Multiply 6×23.

Answers to PROBLEMS

1. 138

The procedure used to solve Example 1 can be shortened after some practice. Here is how.

Step 1	Step 2
$\begin{array}{r} ^{2}\\ 37 \\ \times\ 4 \\ \hline 8 \end{array}$	$\begin{array}{r} ^{2}\\ 37 \\ \times\ 4 \\ \hline 148 \end{array}$
$4 \times 7 = 28$, write 8 and carry 20.	$4 \times 30 = 120$, plus the 20 carried, is 140; write 14.

The expanded form illustrating the procedure is

$$\begin{array}{r} 30 + 7 \\ \times 4 \\ \hline 120 + 28 \end{array} \begin{array}{l} \\ \\ = 120 + 20 + 8 \\ = 140 + 8 \\ = 148 \end{array}$$

Remember, we can indicate a multiplication by using the dot ·, as is done in Example 2.

EXAMPLE 2 Multiplying whole numbers

Multiply $7 \cdot 46$.

SOLUTION The answer is about $7 \cdot 50 = 350$. Here are the steps.

Step 1	Step 2	Step 3
$\begin{array}{r} 46 \\ \times\ 7 \\ \hline 42 \end{array}$	$\begin{array}{r} 46 \\ \times\ 7 \\ \hline 42 \\ 280 \\ \hline \end{array}$	$\begin{array}{r} 46 \\ \times\ 7 \\ \hline 42 \\ 280 \\ \hline 322 \end{array}$
Multiply 7 by 6.	Multiply 7 by 40.	Add 42 and 280.
$7 \times 6 = 42$	$7 \times 40 = 280$	$42 + 280 = 322$

PROBLEM 2

Multiply $6 \cdot 53$.

Multiplication involving two-digit factors is done in the same manner. Thus, to multiply $32 \cdot 53$, we proceed as follows:

Step 1	Step 2	Step 3
$\begin{array}{r} 53 \\ \times 32 \\ \hline 106 \end{array}$	$\begin{array}{r} 53 \\ \times 32 \\ \hline 106 \\ 1590 \\ \hline \end{array}$	$\begin{array}{r} 53 \\ \times 32 \\ \hline 106 \\ 159 \\ \hline 1696 \end{array}$
Multiply 2 by 53.	Multiply 30 by 53.	Add 106 and 1590.
$2 \times 53 = 106$	$30 \times 53 = 1590$ (The 0 in 1590 is usually omitted.)	$106 + 1590 = 1696$ (Note that the 0 in 1590 is omitted.)

Answers to PROBLEMS

2. 318

EXAMPLE 3 **Multiplying whole numbers**

Multiply 43 · 56.

SOLUTION The answer should be near 40 × 60 = 2400.

Step 1	Step 2	Step 3
$\begin{array}{r} ^{1} \\ 56 \\ \times 43 \\ \hline 168 \end{array}$	$\begin{array}{r} ^{2} \\ 56 \\ \times 43 \\ \hline 168 \\ 224 \\ \hline \end{array}$	$\begin{array}{r} 56 \\ \times 43 \\ \hline 168 \\ 224 \\ \hline 2408 \end{array}$
Multiply 3 by 56.	Multiply 40 by 56.	Add 168 and 2240.
3 × 56 = 168	40 × 56 = 2240 (But the 0 is omitted.)	168 + 2240 = 2408

PROBLEM 3

Multiply 52 · 38.

EXAMPLE 4 **Multiplying whole numbers**

Multiply 132 × 418.

SOLUTION The answer should be about 100 × 400 = 40,000.

Step 1	Step 2	Step 3	Step 4
$\begin{array}{r} ^{1} \\ 418 \\ \times 132 \\ \hline 836 \end{array}$	$\begin{array}{r} ^{2} \\ 418 \\ \times 132 \\ \hline 836 \\ 1254 \\ \hline \end{array}$	$\begin{array}{r} 418 \\ \times 132 \\ \hline 836 \\ 1254 \\ 418 \\ \hline \end{array}$	$\begin{array}{r} 418 \\ \times 132 \\ \hline 836 \\ 1254 \\ 418 \\ \hline 55176 \end{array}$
Multiply 2 by 418.	Multiply 30 by 418.	Multiply 100 by 418.	Add.
2 × 418 = 836	30 × 418 = 12,540	100 × 418 = 41,800	836 + 12,540 + 41,800 = 55,176

PROBLEM 4

Multiply 213 × 514.

EXAMPLE 5 **Multiplying whole numbers**

Multiply 203 × 417.

SOLUTION Here we should get about 200 × 400 = 80,000.

Step 1	Step 2	Step 3	Step 4
$\begin{array}{r} ^{2} \\ 417 \\ \times 203 \\ \hline 1251 \end{array}$	$\begin{array}{r} 417 \\ \times 203 \\ \hline 1251 \\ 000 \end{array}$	$\begin{array}{r} ^{1} \\ 417 \\ \times 203 \\ \hline 1251 \\ 000 \\ 834 \\ \hline \end{array}$	$\begin{array}{r} 417 \\ \times 203 \\ \hline 1251 \\ 000 \\ 834 \\ \hline 84651 \end{array}$
Multiply 3 by 417.	Multiply 0 by 417.	Multiply 200 by 417.	Add.
3 × 417 = 1251	0 × 417 = 0	200 × 417 = 83,400	1251 + 000 + 83,400 = 84,651

PROBLEM 5

Multiply 304 × 512.

Answers to PROBLEMS

3. 1976 **4.** 109,482 **5.** 155,648

Of course, you can skip the second step and simply multiply 417 by 2, writing the 4 in 834 under the 2 in 1251 (instead of under the 5), as shown in the following short form.

```
  417
× 203
 1251   We attach a 0 to 834 to keep the
 8340   columns lined up correctly.
84651
```

EXAMPLE 6 Multiplying whole numbers

Multiply 350 × 429.

SOLUTION

Step 1	Step 2	Step 3	Step 4
429 ×350 000	14 429 ×350 000 2145	2 429 ×350 000 2145 1287	429 ×350 000 2145 1287 150150
Multiply 0 by 429.	Multiply 50 by 429.	Multiply 300 by 429.	Add.
0 × 429 = 0	50 × 429 = 21,450	300 × 429 = 128,700	000 + 21,450 + 128,700 = 150,150

Note that you can save time by multiplying 35 × 429 and then attaching a zero to the result.

PROBLEM 6

Multiply 290 × 134.

EXAMPLE 7 Multiplying whole numbers

Multiply 430 × 219.

SOLUTION We first multiply 43 × 219, then attach zero to the result to obtain the answer.

Step 1	Step 2	Step 3
2 219 × 43 657	3 219 × 43 657 876	219 × 43 657 876 9417
Multiply 3 by 219.	Multiply 40 by 219.	Add.
3 × 219 = 657	40 × 219 = 8760	657 + 8760 = 9417

Attaching a zero to the result 9417, we have 94,170. Thus 430 × 219 = 94,170.

PROBLEM 7

Multiply 620 × 318.

B › Multiplying by Multiples of 10

We are able to shorten the labor in Examples 6 and 7 because one of the numbers involved was a **multiple** of 10, that is, 10 multiplied by another number. Multiplication involving multiples of 10 can always be shortened. If you have been estimating the answers in Examples 1–5, the following pattern will be easy to follow.

Answers to PROBLEMS

6. 38,860 **7.** 197,160

$$\left.\begin{aligned}10 \times 3 &= 30\\ 10 \times 6 &= 60\end{aligned}\right\}\quad \text{Add 1 zero.}$$

$$\left.\begin{aligned}100 \times 3 &= 300\\ 100 \times 6 &= 600\end{aligned}\right\}\quad \text{Add 2 zeros.}$$

$$\left.\begin{aligned}1000 \times 3 &= 3000\\ 1000 \times 6 &= 6000\end{aligned}\right\}\quad \text{Add 3 zeros.}$$

As you can see, to multiply a number by 10, 100, 1000, we simply write the number followed by as many zeros as there are in the multiplicand. Here are some more examples:

$$20 \times 3 = 10 \times 2 \times 3 = 60$$
$$30 \times 40 = 10 \times 3 \times 10 \times 4 = 1200$$
$$90 \times 70 = 10 \times 9 \times 10 \times 7 = 6300$$
$$100 \times 80 = 10 \times 10 \times 10 \times 8 = 8000$$
$$200 \times 70 = 10 \times 10 \times 2 \times 10 \times 7 = 14{,}000$$

The pattern here is found by multiplying the nonzero digits and adding as many zeros as appear in both factors.

EXAMPLE 8 **Multiplication involving multiples of 10**

a. Multiply: 1000×7

b. Multiply: 30×50

c. Multiply: 300×70

SOLUTION

a. $1000 \times 7 = 7000$ **b.** $30 \times 50 = 1500$ **c.** $300 \times 70 = 21{,}000$

PROBLEM 8

Multiply:

a. 1000×5

b. 40×90

c. 700×80

C > Applications Involving Multiplication

Sometimes we must multiply more than two factors. For example, suppose you wish to buy a $10,000 car. You put $1000 down as a down payment and need to borrow the rest (say at about 6% interest) for 4 years. Under these terms your payment is $210 per month. How much do you end up paying for the $9000 you borrowed? The answer is

$$\underbrace{\$210}_{\text{Monthly payment}} \times \underbrace{4 \times 12}_{\text{Number of months in four years}}$$

We find this answer in Example 9.

EXAMPLE 9 **Multiplying whole numbers**

Multiply $210 \times 4 \times 12$.

SOLUTION We first multiply 210 by 4, obtaining

$$\begin{array}{r} 210 \\ \times \quad 4 \\ \hline 840 \end{array}$$

Then, multiply the product by 12.

$$\begin{array}{r} 840 \\ \times \quad 12 \\ \hline 1680 \\ 840 \\ \hline 10080 \end{array}$$

Thus you end up paying $10,080 for the $9000 you borrowed.

PROBLEM 9

Multiply $120 \times 4 \times 12$.

Answers to PROBLEMS

8. a. 5000 **b.** 3600 **c.** 56,000

9. 5760

EXAMPLE 10 Strawberry fields

Here is some information about strawberries:

One pint = 15 large berries One quart = 2 pints
One flat = 8 quarts One flat = 12 pounds

a. How many pints in a flat?

b. How many berries in a flat?

c. In a recent year, 6000 acres of berries were harvested. Each acre yielded about 2000 flats valued at \$10 a flat. What was the value of the total crop?

SOLUTION

a. One flat = 8 quarts and one quart = 2 pints. So, we use substitution by exchanging 2 pints for one quart. Thus, one flat = 8 quarts = 8(2 pints) = 16 pints.

This means that one flat = 16 pints.

b. Again, we use substitution by using 8 quarts for one flat, 2 pints for one quart, and 15 berries for one pint.

One flat = 8 quarts = 8(2 pints) = 8(2)(15 berries)

Thus, one flat = 8(2)(15) berries = 240 berries. Note that we use parentheses to indicate the multiplication: 8(2)(15) berries means $8 \times 2 \times 15$ berries.

c. The value of the crop is the product of 6000 (acres), 2000 (flats), and \$10, that is,

$$6000 \times 2000 \times \$10 = \$120{,}000{,}000 \ (\$120 \text{ million})$$

PROBLEM 10

a. How many berries in a quart?

b. What is the weight of 16 quarts?

c. If 8000 acres of berries are harvested, what is the value of the crop?

In part **b** of Example 10 we can multiply 8 by 2 first and then multiply the result by 15, that is, $(8 \times 2) \times 15$. It is actually easier to multiply $8 \times (2 \times 15) = 8 \times 30 = 240$. It does not matter how we group the numbers because *multiplication is associative.*

ASSOCIATIVE PROPERTY OF MULTIPLICATION

Changing the *grouping* of two factors does not change the product.

$$a \times (b \times c) = (a \times b) \times c$$
$$a \cdot (b \cdot c) = (a \cdot b) \cdot c$$

D › Finding Areas

An application of multiplication is finding the area of a region. The **area** of a region measures the amount of **surface** in the region. To find the area of a figure, we use multiplication and find the number of square units contained in the figure. Unit squares are shown on the left below. The area of a rectangle, for example, consists of the number of square units it contains. If the rectangle measures 3 centimeters by 4 centimeters, then its area measures 3 cm × 4 cm = 12 cm² (read "12 square centimeters"), as shown in the diagram.

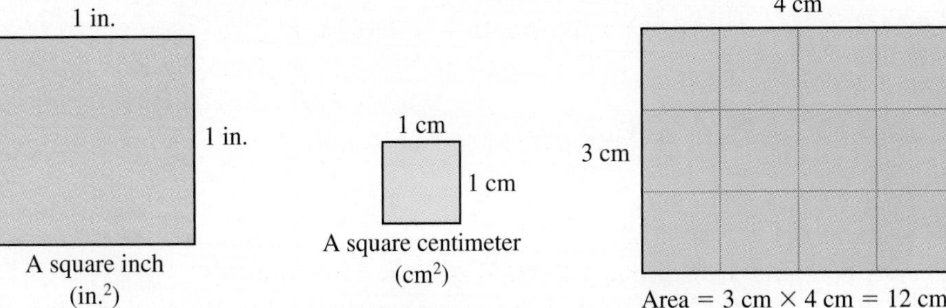

Answers to PROBLEMS

10. a. 30 berries **b.** 24 pounds **c.** \$160,000,000 or \$160 million

In general, we can find the area A of a rectangle by multiplying its length L by its width W, as given here.

AREA OF A RECTANGLE

The area A of a rectangle is found by multiplying its length L by its width W.

$$A = L \cdot W$$

EXAMPLE 11 Calculating area

One of the largest strawberry shortcakes ever created was made in Plant City, Florida. The table holding the shortcake measured 104 feet by 8 feet. What was the area of the table?

SOLUTION The area of the table is 104 ft × 8 ft = 832 ft^2. (Actually, the cake was 827 square feet, used 162,179 strawberries, 450 pounds of sugar, and 600 pounds of whipped cream.)

PROBLEM 11

The *St. Petersburg Times* of September 27, 2002, claims that Danny Julian made a cake measuring 40 feet by 38 feet. What is the area of this cake?

The 40-foot by 38-foot cake was assembled with two 40-foot tractor-trailer loads of 18-inch by 25-inch cakes, frozen and stacked together like bricks, with butter cream serving as mortar.

Algebra Bridge

Let us do multiplication in algebra. For example, let us multiply 3 by $2x + 1$. In algebra, we use parentheses ( ) and write $3 \times (2x + 1)$, or $3(2x + 1)$. Note how the arithmetic column works. The algebra column works in the same manner!

Arithmetic **Algebra**

Step 1. Write in expanded notation.

$$\begin{array}{rr} 21 = & 2(10) + 1 \\ \times\ 3 & \times \qquad\quad 3 \\ \hline \end{array} \qquad \begin{array}{r} 2x + 1 \\ \times \qquad 3 \\ \hline \end{array}$$

Step 2. Multiply 3 × 1.

$$\begin{array}{rr} 21 = & 2(10) + \mathbf{1} \\ \times\ \mathbf{3} & \times \qquad\quad \mathbf{3} \\ \hline \mathbf{3} & \mathbf{3} \end{array} \qquad \begin{array}{r} 2x + \mathbf{1} \\ \times \qquad \mathbf{3} \\ \hline \mathbf{3} \end{array}$$

Step 3. Multiply 3 × 2(10).

$$\begin{array}{rr} 21 = & \mathbf{2(10)} + 1 \\ \times\ \mathbf{3} & \times \qquad\quad \mathbf{3} \\ \hline \mathbf{63} & \mathbf{6(10)} + 3 \end{array} \qquad \begin{array}{r} \mathbf{2x} + 1 \\ \times \qquad \mathbf{3} \\ \hline \mathbf{6x} + 3 \end{array}$$

Answers to PROBLEMS

11. 1520 ft^2

Now suppose you wish to multiply $(2x + 1)$ by $(3x + 2)$. Look at the multiplication of 21 by 32 and see if you spot the similarities.

Arithmetic | **Algebra**

Step 1. Write in expanded notation.

$$\begin{array}{rr} 21 = & 2(10) + 1 \\ \times\ 32 = & \times 3(10) + 2 \\ \hline \end{array} \qquad \begin{array}{r} 2x + 1 \\ \times 3x + 2 \\ \hline \end{array}$$

Step 2. Multiply 2×1.

$$\begin{array}{rr} 21 = & 2(10) + \mathbf{1} \\ \times\ 3\mathbf{2} = & \times 3(10) + \mathbf{2} \\ \hline \mathbf{2} & \mathbf{2} \end{array} \qquad \begin{array}{r} 2x + \mathbf{1} \\ \times 3x + \mathbf{2} \\ \hline \mathbf{2} \end{array}$$

Step 3. Multiply $2 \times 2(10)$.

$$\begin{array}{rr} \mathbf{2}1 = & \mathbf{2(10)} + 1 \\ \times\ 3\mathbf{2} = & \times 3(10) + \mathbf{2} \\ \hline \mathbf{4}2 & \mathbf{4(10)} + 2 \end{array} \qquad \begin{array}{r} \mathbf{2x} + 1 \\ \times 3x + \mathbf{2} \\ \hline \mathbf{4x} + 2 \end{array}$$

Step 4. Multiply $3(10) \times 1$.

$$\begin{array}{rr} 2\mathbf{1} = & 2(10) + \mathbf{1} \\ \times\ \mathbf{3}2 = & \times \mathbf{3(10)} + 2 \\ \hline 42 & 4(10) + 2 \\ \mathbf{3}\ \ & \mathbf{3(10)}\ \ \ \ \\ \hline \end{array} \qquad \begin{array}{r} 2x + \mathbf{1} \\ \times \mathbf{3x} + 2 \\ \hline 4x + 2 \\ \mathbf{3x}\ \ \ \ \\ \hline \end{array}$$

Step 5. Multiply $3(10) \times 2(10)$.

Note that $3(10) \times 2(10) =$

$6 \times 10 \times 10 = 6 \times 10^2$.

That is, $10 \times 10 = 10^2$. | Note that $(2x)(3x) = 6x^2$.

$$\begin{array}{rr} \mathbf{2}1 = & \mathbf{2(10)} + 1 \\ \times\ \mathbf{3}2 = & \times \mathbf{3(10)} + 2 \\ \hline 42 & 4(10) + 2 \\ \mathbf{6}3\ \ & \mathbf{6(10^2)} + 3(10)\ \ \ \ \\ \hline \end{array} \qquad \begin{array}{r} \mathbf{2x} + 1 \\ \times \mathbf{3x} + 2 \\ \hline 4x + 2 \\ \mathbf{6x^2} + 3x\ \ \ \ \\ \hline \end{array}$$

Step 6. Add the columns.

$$\begin{array}{rr} 21 = & 2(10) + 1 \\ \times\ 32 = & \times 3(10) + 2 \\ \hline 42 & 4(10) + 2 \\ 63\ \ & 6(10^2) + 3(10)\ \ \ \ \\ \hline \mathbf{672} & \mathbf{6(10^2) + 7(10) + 2} \end{array} \qquad \begin{array}{r} 2x + 1 \\ \times 3x + 2 \\ \hline 4x + 2 \\ +6x^2 + 3x\ \ \ \ \\ \hline \mathbf{6x^2 + 7x + 2} \end{array}$$

Calculator Corner

As with addition and subtraction, multiplication can be done by simply pressing the appropriate keys on the calculator. Thus, to do the problem 4×3, we simply press 4 × 3 ENTER and the answer, 12, appears in the display. If more than two numbers are to be multiplied, the procedure is similar. Thus, to obtain the answer in Example 9, that is, to find $210 \times 4 \times 12$, we proceed as follows: 2 1 0 × 4 × 1 2 ENTER, obtaining the answer, 10080, in the display. Note that **you** must insert the comma in this answer to obtain the final result, 10,080.

Exercises 1.5

⟨A⟩ Multiplication of Whole Numbers

In Problems 1–34, multiply.

1. a. 3×4 b. 3×7

2. a. 5×2 b. 5×9

3. a. 9×1 b. 8×1

4. a. 0×6 b. 0×9

5. a. 5×8 b. 9×8

6. a. 9×6 b. 7×6

7. a. 0×0 b. 0×3

8. a. 7×0 b. 4×0

9. a. 1×8 b. 1×4

10. a. 1×0 b. 1×4

11. $6 \cdot 8$

12. $0 \cdot 9$

13. $1 \cdot 5$

14. $6 \cdot 1$

15. $9 \cdot 9$

16. $1 \cdot 1$

17. $0 \cdot 1$

18. $9 \cdot 8$

19. $4 \cdot 4$

20. $8 \cdot 8$

21. 10×9

22. 10×5

23. 20×8

24. 90×5

25. 53×6

26. 39×4

27. 48×17

28. 98×15

29. 608×32

30. 508×23

31. 1234×3

32. 4321×4

33. $35,209 \times 16$

34. $43,802 \times 15$

⟨B⟩ Multiplying by Multiples of 10

In Problems 35–50, multiply.

35. 63×40

36. 83×30

37. 249×50

38. 296×60

39. 346×420

40. 671×350

41. 2260×200

42. 3160×300

43. 3020×405

44. 6050×802

45. $20 \cdot 5$

46. $(30)(90)$

47. $(700)(80)$

48. $120 \cdot 30$

49. $300 \cdot 200$

50. 600×900

⟨C⟩ Applications Involving Multiplication

51. *Ostrich strides* An ostrich covers 25 feet in one stride. How far does it go in 8 strides?

52. *Patterns* Copy the pattern and write the missing numbers. Verify your answer by multiplying.

a.
$1 \times 1 = 1$
$2 \times 2 = 1 + 2 + 1$
$3 \times 3 = 1 + 2 + 3 + 2 + 1$
$4 \times 4 = 1 + 2 + 3 + 4 + 3 + 2 + 1$
$= 1 + 2 + 3 + 4 + 5 + 4 + 3 + 2 + 1$

b.
$999,999 \times 2 = 1,999,998$
$999,999 \times 3 = 2,999,997$
$999,999 \times 4 =$

c.
$1 \times 9 + 2 = 11$
$12 \times 9 + 3 = 111$
$123 \times 9 + 4 = 1111$
$1234 \times 9 + 5 =$

53. *Distance traveled by light* Light travels 300,000 kilometers per second. How far would light travel in

a. 20 seconds?

b. 25 seconds?

c. 30 seconds?

54. *Hours in day, week, month* How many hours are there in

a. 3 days?

b. A week?

c. A 30-day month?

d. A year (365 days)?

55. *Borrowing and interest* You borrow $8000 (at 11% interest) for a period of 4 years. Your monthly payments amount to $206. How much will you end up paying for the $8000 borrowed? (See Example 9 in this section.)

56. *How far can you go?* A car can go about 28 miles on one gallon of gas. How far could it go on

a. 2 gallons?

b. 5 gallons?

c. 10 gallons?

57. *Weeks* There are 52 weeks in a year. How many weeks are there in

a. 5 years?

b. 10 years?

58. *Earth weight* An object weighs six times as much on earth as on the moon. Find the earth weights of these:

a. An astronaut weighing 27 pounds on the moon.

b. The Lunar Rover used by the astronauts of *Apollo 15,* which weighs 75 pounds on the moon.

c. The rocks brought back by the *Apollo 16* crew, which weighed 35 pounds on the moon.

59. *Oil consumption* In 2010, the estimated annual oil consumption of the United States is 8 billion barrels. At this rate, how much oil will be used in the next 8 years?

60. *Caloric intake* The caloric intake needed to maintain your weight is found by multiplying your weight by 15. For example, a 200-pound man needs

$$15 \times 200 = 3000 \text{ calories a day}$$

Find the caloric intake for:

a. A 150-pound man.

b. A 120-pound man.

c. A 170-pound man.

‹ D › Finding Areas

61. *Lots of eggs!* One of the largest omelets ever cooked was 30 feet long by 10 feet wide. What was its area?

62. *Swimming, anyone?* One of the largest rectangular pools in the world is in Casablanca, Morocco, and it is 480 meters long and 75 meters wide. What is the area covered by the pool?

63. *Area of baseball diamond* Find the area of the region enclosed by the base lines on the baseball diamond shown.

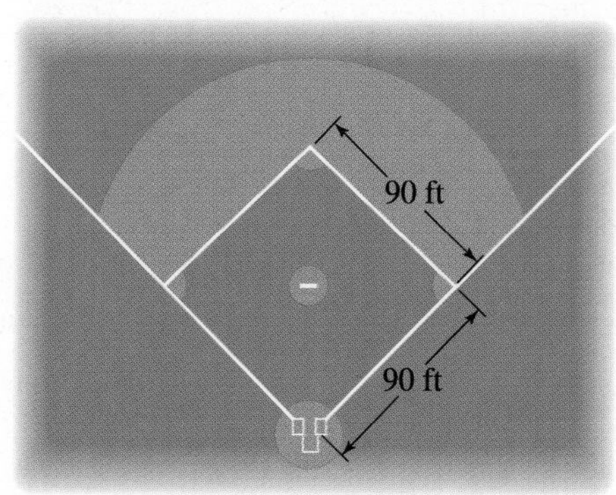

64. *Area of basketball court* Find the area of the basketball court shown.

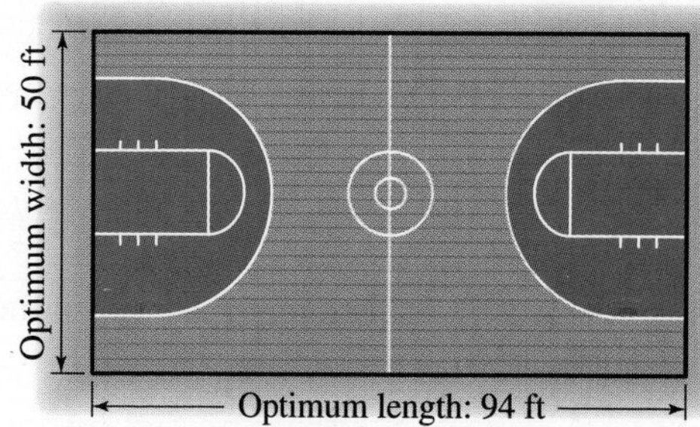

65. *Area of football field* Find the area of the football field.

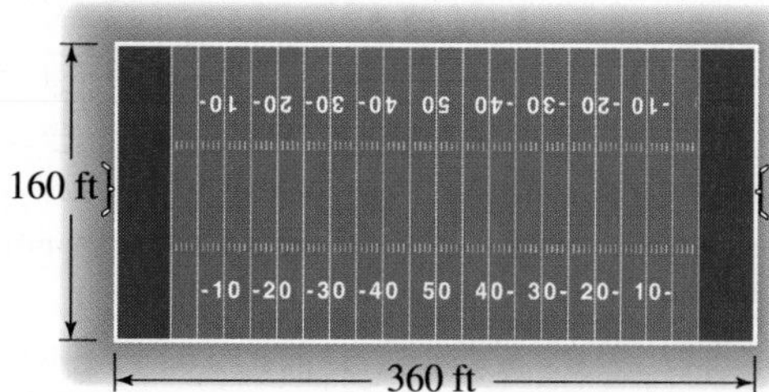

66. *Area* What is the area of the hockey rink?

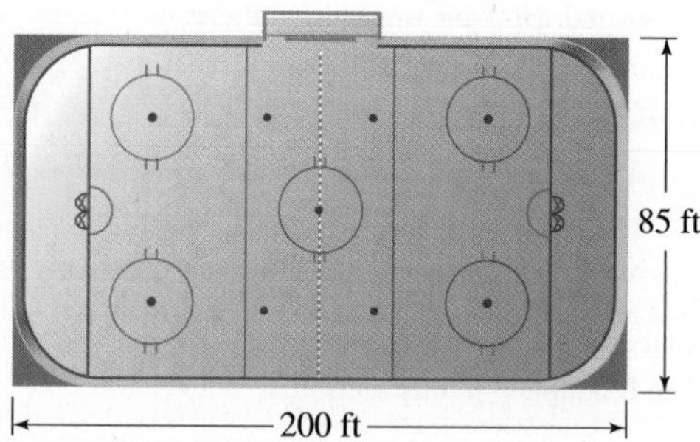

67. *Trees* The average American uses the equivalent of seven trees a year in paper, wood, and other products. If there are about 300 million Americans, how many trees are used in a year?
Source: http://members.aol.com

68. *Water flow* A standard shower head has a flow rate of 5 to 10 gallons of water per minute. Most people take 5-minute showers. How many gallons of water are used for a 5-minute shower? How many gallons of water does a family of four use in a week (7 days) assuming they each take a daily 5-minute shower?

Source: http://www.asheonline.com

69. *Water flow* Assuming a 5 gallon per minute flow, how much water could Tran save in a year (365 days) by taking a daily 4-minute shower instead of a daily 5-minute shower?

Source: http://www.asheonline.com

70. *Gallons of gallons* A bath uses 50 gallons of water. Assuming a 5 gallon per minute flow, how much water could a person save in a week (7 days) by taking a daily 5-minute shower instead of a bath?
Source: http://www.asheonline.com

71. *Don't dump it!* A gallon of paint or a quart of motor oil can seep into the earth and pollute 250,000 gallons of drinking water. When changing the oil in your car, 5 quarts of motor oil are accidentally spilled and seep into the earth. How many gallons of drinking water can be polluted by the spill?
Source: http://www.leeric.lsu.edu

72. *Watch that tank!* A spilled gallon of gasoline can pollute 750,000 gallons of water. How many gallons of water can be polluted if your 13 gallon tank of gas leaks into the ground?
Source: http://www.leeric.lsu.edu

Using Your Knowledge

A Weighty Matter How much should you weigh? There is a formula to estimate your ideal weight. For an average man the formula is like this:

$$\text{Ideal weight} = \text{height (inches)} \times 4 - 130$$

For women,

$$\text{Ideal weight} = \text{height (inches)} \times 4 - 140$$

For example, the ideal weight for a 70-inch woman is calculated as follows:

$$\begin{aligned}\text{Ideal weight} &= 70 \times 4 - 140\\ &= 280 - 140\\ &= 140\end{aligned}$$

73. Use the formula to find the ideal weight of a man 6 feet tall (72 inches).

74. Use the formula to find the ideal weight of a woman 5 feet 8 inches tall (68 inches).

The amount of food (calories) needed to maintain your ideal weight depends on your activities. Rate yourself on the following scale.

13	Very inactive
14	Slightly inactive
15	Moderately active
16	Relatively active
17	Frequently active

To find the daily calories needed to maintain your ideal weight, use the following formula:

$$\text{Calories needed} = \text{ideal weight} \times \text{activity level (from table)}$$

For example, a 200-pound office worker who is 13 on the scale calculated his need like this:

$$\text{Calories needed} = 200 \times 13 = 2600$$

75. Find the calories needed by a 150-pound man who is:

a. Moderately active. **b.** Frequently active.

Write On

76. Write in your own words the procedure you use to multiply a whole number by 10, 100, or 1000.

77. Write in your own words the meaning of the commutative property of multiplication. How does this property compare to the commutative property of addition?

78. Write in your own words what it means when we say that 1 is the identity for multiplication. Why do you think they use the word identity to describe the number 1? What is the identity for addition?

79. Write in your own words the meaning of the associative property of multiplication. How does this property compare to the associative property of addition?

80. Which of the properties mentioned (commutative, associative, identity), if any, apply to the operation of subtraction? Explain and give examples.

Algebra Bridge

Multiply.

81. $5 \times (2x + 7)$

82. $3(4x + 2)$

83. $3x + 1$ by $2x + 2$

84. $(2x + 3)(3x + 1)$

Concept Checker

Fill in the blank(s) with the correct word(s), phrase, or mathematical statement.

85. The **product** of a and b can be written three ways: ______, ______, and ______.

86. When we write **5 × 4 = 20,** the 5 is the ______, the 4 is the ______, and the 20 is the ______.

87. The **multiplicand** and the **multiplier** are also called ______.

88. The **Commutative Property** of multiplication states that ______.

89. The **identity** for multiplication is ____.

90. $a \cdot (b + c) =$ ______.

91. The **Associative Property** of multiplication states that $(a \cdot b) \cdot c =$ ______.

92. The **area** of a rectangle of length L and width W is ______.

multiplicand	**1**
$L \cdot W$	$a \cdot (b \cdot c)$
$(a \cdot b) \cdot c$	$a \times b$
multiplier	$a \cdot b$
0	$(a)(b)$
product	$a \cdot b + a \cdot c$
factors	
$a \times b = b \times a$	

Mastery Test

93. Multiply $210 \times 5 \times 12$.

94. Multiply 4000×80.

95. Multiply 450×319.

96. Multiply 403×319.

97. Multiply 129×318.

98. Multiply 53×65.

99. Multiply 7×56.

100. Multiply 1000×9.

101. One pint contains about 15 large berries, and one quart is equivalent to two pints. How many berries in a quart?

102. A volleyball playing court is 30 feet by 60 feet. What is the area of the court?

Skill Checker

103. Write 234 in expanded notation.

104. Write 758 in expanded notation

105. Add 349 and 786.

106. Add 1289 and 7893.

107. Subtract 728 from 3500.

108. Subtract 999 from 2300.

1.6 Division

Objectives

You should be able to:

A Write a division problem as an equivalent multiplication problem.

B Divide a counting number by another using long division.

C Solve applications using the concepts studied.

To Succeed, Review How To . . .

1. Use the definition of a counting number. (p. 3)
2. Use the multiplication facts. (p. 49)

Getting Started

Suppose you have 12 cents to spend on 3-cent stamps. How many stamps can you buy? If you have 12 pennies, you can separate them into equal groups of 3. Since there are four groups, you can buy four 3-cent stamps with 12 cents, as shown in the photo. We have separated a quantity into 4 equal groups. This process is called **division.** Of course, the problem could have been solved by dividing the money available (12 cents) by the cost of each stamp (3 cents).

A > Writing Division as Multiplication

In mathematics the phrase "twelve divided by 3" is written like this:

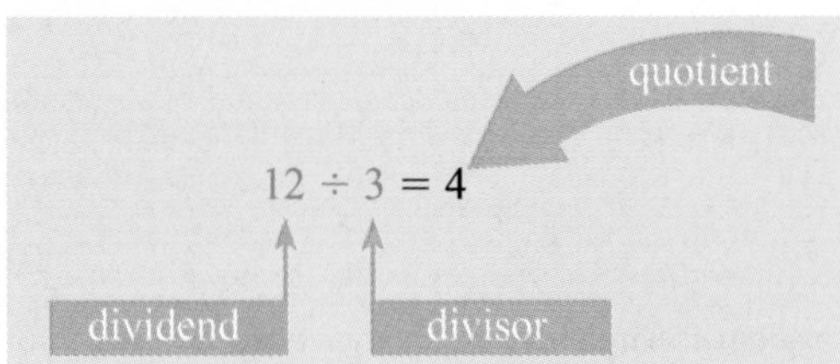

where ÷ is the sign used to indicate division. In this problem, 12, the number being divided, is the **dividend;** 3, the number used to divide, is the **divisor;** and 4, the result of the division, is called the **quotient.** The mathematical statement 12 ÷ 3 = 4 (twelve divided by three equals four) can also be written as

$$\frac{12}{3} = 4 \qquad 12/3 = 4 \qquad 3\overline{)12}^{\,4}$$

Here is the definition.

DIVISION

The quotient $a \div b$, where $b \neq 0$, is the unique whole number c so that $a = b \times c$.

How can you check the fact that $12/3 = 4$? One way is to use multiplication. By definition,

$$12 \div 3 = \square \quad \text{means that} \quad 12 = 3 \times \square$$

This makes multiplication the inverse process of division. By placing 4 in the $\square$ you make the statement $12 = 3 \times \square$ true. Similarly, to find $42 \div 6 = \square$ you think "$42 = 6 \times \square$." Since $42 = 6 \times 7$, $42 \div 6 = 7$.

EXAMPLE 1 Rewriting division as multiplication

Write each division problem as an equivalent multiplication problem, then find the answer.

a. $36 \div 9 = \square$ **b.** $54 \div 6 = \square$

SOLUTION

a. $36 \div 9 = \square$ can be written as $36 = 9 \times \square$. Since $36 = 9 \times \mathbf{4}$, $36 \div 9 = \mathbf{4}$.

b. $54 \div 6 = \square$ can be written as $54 = 6 \times \square$. Since $54 = 6 \times \mathbf{9}$, $54 \div 6 = \mathbf{9}$.

PROBLEM 1

Write each division problem as an equivalent multiplication problem, then find the answer.

a. $63 \div 9 = \square$

b. $56 \div 8 = \square$

As with addition and multiplication, the operation of division has some important properties.

DIVISION PROPERTIES OF 1

1. **For any nonzero number a, $a \div a = 1$. A nonzero number divided by itself is 1.**
2. **For any number a, $a \div 1 = a$. Any number divided by 1 is the number.**

DIVISION PROPERTIES OF 0

1. **For any nonzero number a, $0 \div a = 0$. Zero divided by any nonzero number is 0.**
2. **For any number a, $a \div 0$ is undefined. Division by zero is undefined.**

EXAMPLE 2 Rewriting division as multiplication

Write each division problem as an equivalent multiplication problem, then find the answer if possible.

a. $8 \div 8 = \square$ **b.** $8 \div 1 = \square$

c. $0 \div 8 = \square$ **d.** $8 \div 0 = \square$

SOLUTION

a. $8 \div 8 = \square$ can be written as $8 = 8 \times \square$. Since $8 \times \mathbf{1} = 8$, $8 \div 8 = \mathbf{1}$.

b. $8 \div 1 = \square$ can be written as $8 = 1 \times \square$. Since $1 \times \mathbf{8} = 8$, $8 \div 1 = \mathbf{8}$.

c. $0 \div 8 = \square$ can be written as $0 = 8 \times \square$. Since $8 \times \mathbf{0} = 0$, $0 \div 8 = \mathbf{0}$.

d. $8 \div 0 = \square$ can be written as $8 = 0 \times \square$. Since there is *no* number such that $8 = 0 \times \square$, $8 \div 0$ is not defined.

PROBLEM 2

Write each division problem as an equivalent multiplication problem, then find the answer if possible.

a. $9 \div 9 = \square$

b. $9 \div 1 = \square$

c. $0 \div 9 = \square$

d. $9 \div 0 = \square$

Answers to PROBLEMS

1. a. $63 = 9 \times \square$; 7 **b.** $56 = 8 \times \square$; 7

2. a. $9 = 9 \times \square$; 1 **b.** $9 = 1 \times \square$; 9 **c.** $0 = 9 \times \square$; 0 **d.** $9 \div 0 = \square$; means $9 = 0 \times \square$, which is undefined. You cannot divide by 0.

How do we divide 67 by 13? Since we wish to find out how many groups of 13 there are in 67, we can do it by repeated subtractions, like this.

$$\begin{array}{r l}
67 & \\
\underline{-13} & \rightarrow \text{First subtraction} \\
54 & \\
\underline{-13} & \rightarrow \text{Second subtraction} \\
41 & \\
\underline{-13} & \rightarrow \text{Third subtraction} \\
28 & \\
\underline{-13} & \rightarrow \text{Fourth subtraction} \\
15 & \\
\underline{-13} & \rightarrow \text{Fifth subtraction} \\
\text{Remainder} \rightarrow \quad 2 &
\end{array}$$

Thus, there are 5 thirteens in 67, with a **remainder** of 2. The remainder is what is left over. We then write

$$67 \div 13 = 5 \textbf{ r } 2$$

Dividend Divisor Quotient Remainder

As before, we can check this by multiplying $5 \times 13 = 65$ *and then adding the remainder,* 2, to the 65, obtaining the required 67.

Here is a better procedure!

B › Long Division

When doing long division, we still use repeated subtractions, but we write the procedure in a different way. We use the symbol $\overline{)\ }$ to indicate division, placing the **divisor** to the left of the division sign $\overline{)\ }$ and the **dividend** inside. Here is the set up:

$$\begin{array}{r}\text{quotient}\\ \text{divisor}\overline{)\text{dividend}}\end{array}$$

Next, we want to find the *first digit* in the quotient. Here is the way to do it:

1. If the first digit in the dividend is greater than or equal to the divisor, divide it by the divisor and write the quotient of that division directly above the *first digit* of the dividend. If this is not the case, then go to step 2.
2. Divide the first *two digits* of the dividend by the divisor and write the quotient of that division directly above the *second digit* of the dividend.

We illustrate this procedure in Example 3, where we are dividing 786 by 6.

EXAMPLE 3 Using long division

Divide $786 \div 6$.

SOLUTION

$6\overline{)786}$ **Step 1.** Arrange the divisor and dividend horizontally, as shown.

$$\begin{array}{r}
131\\
6\overline{)786}\\
\underline{-6}\\
18\\
\underline{-18}\\
06\\
\underline{-6}\\
0
\end{array}$$

Step 2. 6 goes into 7 once. Write 1 above the 7.

Step 3. $6 \times 1 = 6$; subtract $7 - 6 = 1$.

Step 4. Bring down the 8. Now, 6 goes into 18 three times. Write 3 above the 8.

Step 5. $6 \times 3 = 18$; subtract $18 - 18 = 0$.

Step 6. Bring down the 6. The 6 goes into 6 once. Write 1 above the 6.

Step 7. $6 \times 1 = 6$; subtract $6 - 6 = 0$.

Thus, $786 \div 6 = 131$.

PROBLEM 3

Divide $917 \div 7$.

Answers to PROBLEMS

3. 131

In Examples 1, 2, and 3, the remainders have been zero. This means that the dividend is exactly divisible by the divisor. Of course, not all problems are like this. Here is one in which a nonzero remainder occurs. In such cases we continue the long division process until the *remainder is less than the divisor.*

EXAMPLE 4 Using long division

Divide $1729 \div 9$.

SOLUTION

$9\overline{)1729}$ **Step 1.** Arrange the divisor and dividend horizontally, as shown.

$$\begin{array}{r} 192 \\ 9\overline{)1729} \\ \underline{-9} \\ 82 \\ \underline{-81} \\ 19 \\ \underline{-18} \\ 1 \end{array}$$

Step 2. You cannot divide 1 by 9, so divide 17 by 9. 9 goes into 17 once. Write 1 above the 7.

Step 3. $9 \times 1 = 9$; subtract $17 - 9 = 8$.

Step 4. Bring down the 2. Now, 9 goes into 82 nine times. Write 9 above the 2.

Step 5. $9 \times 9 = 81$; subtract $82 - 81 = 1$.

Step 6. Bring down the 9. The 9 goes into 19 twice. Write 2 above the 9.

Step 7. $9 \times 2 = 18$; subtract $19 - 18 = 1$.

The remainder 1 is less than the divisor 9, so we can stop. The answer is 192 with a remainder of 1 and we write $1729 \div 9 = 192$ r 1. This is true because $1729 = 9 \times 192 + 1$, as shown in the diagram:

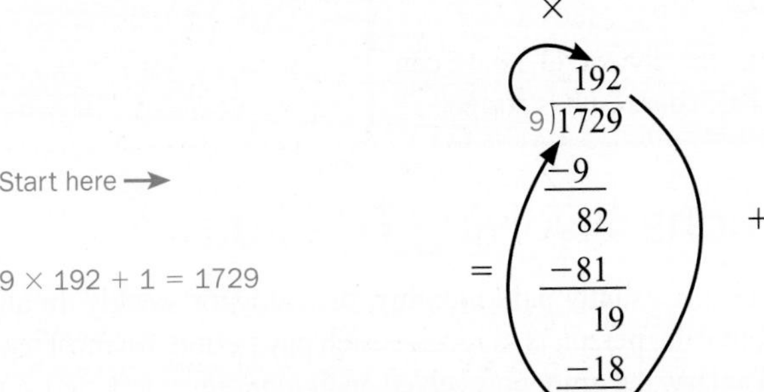

PROBLEM 4

Divide $1367 \div 4$.

Here is a slightly different example.

EXAMPLE 5 Using long division

Divide $809 \div 8$.

SOLUTION

$8\overline{)809}$ **Step 1.** Arrange the divisor and dividend horizontally, as shown.

$$\begin{array}{r} 101 \\ 8\overline{)809} \\ \underline{-8} \\ 009 \\ \underline{-8} \\ 1 \end{array}$$

Step 2. 8 goes into 8 once. Write 1 above the 8.

Step 3. $8 \times 1 = 8$; subtract $8 - 8 = 0$.

Step 4. Bring down the 0. The 0 divided by 8 is 0. Write 0 above the 0.

Step 5. Bring down the 9. Now, 8 goes into 9 once. Write 1 above the 9.

Step 6. $8 \times 1 = 8$; subtract $9 - 8 = 1$.

(continued)

PROBLEM 5

Divide $709 \div 7$.

Answers to PROBLEMS

4. 341 r 3 **5.** 101 r 2

The remainder 1 is less than the divisor 8, so we can stop. The answer is 101 with a remainder of 1 and we write 809 ÷ 8 = 101 r 1. This can be checked by following the diagram:

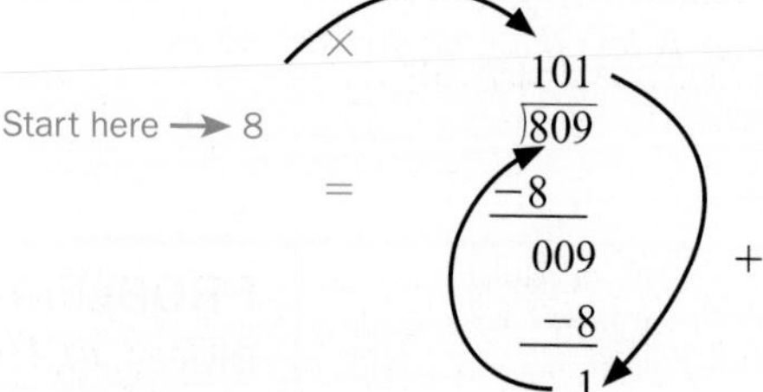

That is, 8 × 101 + 1 = 809.

We now give an example involving division by a two-digit number.

EXAMPLE 6 **Using long division**

Find the quotient and the remainder: 1035 ÷ 43.

SOLUTION

```
     24
43)1035
   -86
    175
   -172
      3
```

Step 1. Arrange the divisor and dividend horizontally, as shown.

Step 2. 43 does not go into 10. Try to divide 103 by 43. 103 divided by 43 should be about 2 since 100 divided by 50 is 2. Since 43 × 2 = 86, it goes twice. Write 2 above the 3.

Step 3. 43 × 2 = 86; subtract 103 − 86 = 17.

Step 4. Bring down the 5. The 43 goes into 175 four times. Write 4 above the 5.

Step 5. 43 × 4 = 172; subtract 175 − 172 = 3.

The quotient is 24 with a remainder of 3. Now, 3 is less than the divisor 43, so we can stop and write 1035 ÷ 43 = 24 r 3. Since 43 × 24 + 3 is 1035, our result is correct.

PROBLEM 6

Find the quotient and the remainder: 1029 ÷ 45.

C › Applications Involving Division

A person's annual salary is usually paid monthly, biweekly, or weekly. In any case, to find the amount of money the person is to receive each pay period, we must use **division.** For a $19,500 annual salary, the amount received each *monthly* pay period should be

$$\frac{19{,}500}{12} \begin{array}{l}\leftarrow \text{Annual salary}\\ \leftarrow \text{Months in a year}\end{array}$$

In Example 7 we find out how much this is.

EXAMPLE 7 **Using long division**

Divide 19,500 by 12.

SOLUTION

```
12)19500
```

Step 1. Arrange the divisor and dividend horizontally, as shown.

```
   1625
12)19500
  -12
    75
   -72
    30
   -24
    60
   -60
     0
```

Step 2. 12 goes into 19 once. Write 1 above the 9.

Step 3. 12 × 1 = 12; subtract 19 − 12 = 7.

Step 4. Bring down the 5. The 12 goes into 75 six times. Write 6 above the 5.

Step 5. 12 × 6 = 72; subtract 75 − 72 = 3.

Step 6. Bring down the 0. Now, 12 goes into 30 twice. Write 2 above the first 0.

Step 7. 12 × 2 = 24; subtract 30 − 24 = 6.

PROBLEM 7

If a person earns $19,500 annually, the biweekly salary will be

$$\frac{\$19{,}500}{26}$$

Divide $19,500 by 26.

Answers to PROBLEMS

6. 22 r 39 **7.** $750

Step 8. Bring down the other 0. The 12 goes into 60 five times. Write 5 above the second 0.

Step 9. $12 \times 5 = 60$; subtract $60 - 60 = 0$.

The answer is 1625 with no remainder. Thus, if a person earns $19,500 annually, the monthly salary will be $1625.

Algebra Bridge

If you understand the division problems we have discussed, you can apply this information to division problems in algebra. Look at the two columns. On the left we are dividing 275 by 13, which is $275 = 2 \times 10^2 + 7 \times 10 + 5$ divided by $13 = 10 + 3$. On the right we divide $2x^2 + 7x + 5$ by $x + 3$.

Arithmetic	Algebra
Step 1. Arrange the dividend and divisor in the usual way. $13\overline{)275}$	$x + 3\overline{)2x^2 + 7x + 5}$
Step 2. You cannot divide 2 by 13, so use the first two digits in the dividend and divide 27 by 13. Write the answer, 2, above the 7. $\begin{array}{r} 2 \\ 13\overline{)275} \end{array}$	You cannot divide 2 by $x + 3$, so divide $2x^2$ by x. $\frac{2x^2}{x} = \frac{2 \cdot x \cdot x}{x} = 2x$ Write the $2x$ above the $7x$. $\begin{array}{r} 2x \\ x + 3\overline{)2x^2 + 7x + 5} \end{array}$
Step 3. Multiply 2 by 13 and subtract the result (26) from 27, obtaining 1. Bring down the 5. $\begin{array}{r} 2 \\ 13\overline{)275} \\ \underline{26} \\ 15 \end{array}$	Multiply $2x$ by $x + 3$ and subtract the result, $2x^2 + 6x$, from the dividend, obtaining x. Bring down the 5. $\begin{array}{r} 2x \\ x + 3\overline{)2x^2 + 7x + 5} \\ \underline{2x^2 + 6x} \\ x + 5 \end{array}$
Step 4. Divide 15 by 13, and write the answer, 1, over the 5. Multiply 1 by 13, write the 13 under 15, and subtract. The remainder is 2. $\begin{array}{r} 21 \\ 13\overline{)275} \\ \underline{26} \\ 15 \\ \underline{13} \\ 2 \end{array}$	Divide x by x, and write the answer, 1, over the 5. Multiply 1 by $x + 3$, write the $x + 3$ under the $x + 5$, and subtract. The remainder is 2. $\begin{array}{r} 2x + 1 \\ x + 3\overline{)2x^2 + 7x + 5} \\ \underline{2x^2 + 6x} \\ x + 5 \\ \underline{x + 3} \\ 2 \end{array}$

Calculator Corner

Division is faster with a calculator, especially if there is *no* remainder. Thus, to divide 12 by 3, we press 1 2 ÷ 3 ENTER and the result, 4, appears in the display. If there is a remainder, the calculator gives that remainder in the form of a **decimal.**

Boost your grade at mathzone.com!
> Practice Problems
> NetTutor
> Self-Tests
> e-Professors
> Videos

Exercises 1.6

Web IT go to mhhe.com/bello for more lessons

⟨A⟩ Writing Division as Multiplication

In Problems 1–20, divide and check using multiplication.

1. $30 \div 5$ **2.** $63 \div 9$ **3.** $28 \div 4$ **4.** $6 \div 1$

5. $21 \div 7$ **6.** $2 \div 0$ **7.** $0 \div 2$ **8.** $54 \div 9$

9. $7 \div 7$ **10.** $56 \div 8$ **11.** $\frac{36}{9}$ **12.** $54 \div 6$

13. $\frac{3}{0}$ **14.** $\frac{0}{3}$ **15.** $\frac{32}{8}$ **16.** $\frac{13}{13}$

17. $\frac{24}{1}$ **18.** $\frac{45}{9}$ **19.** $\frac{62}{6}$ **20.** $\frac{48}{5}$

⟨B⟩ Long Division

In Problems 21–50, use long division to divide.

21. $6\overline{)366}$ **22.** $7\overline{)371}$ **23.** $8\overline{)5048}$ **24.** $7\overline{)6097}$

25. $4\overline{)2055}$ **26.** $9\overline{)6013}$ **27.** $336 \div 14$ **28.** $340 \div 17$

29. $399 \div 19$ **30.** $406 \div 13$ **31.** $605 \div 10$ **32.** $600 \div 27$

33. $\frac{704}{16}$ **34.** $\frac{903}{17}$ **35.** $\frac{805}{81}$ **36.** $11\overline{)341}$

37. $12\overline{)505}$ **38.** $46\overline{)508}$ **39.** $22\overline{)1305}$ **40.** $53\overline{)1325}$

41. $42\overline{)9013}$ **42.** $111\overline{)3414}$ **43.** $123\overline{)5583}$ **44.** $253\overline{)8096}$

45. $417\overline{)36,279}$ **46.** $505\overline{)31,815}$ **47.** $50\overline{)31,500}$ **48.** $600\overline{)188,400}$

49. $654\overline{)611,302}$ **50.** $703\overline{)668,553}$

⟨C⟩ Applications Involving Division

51. *Shares sold* A stockbroker sold $12,600 worth of stocks costing $25 per share. How many shares were sold?

52. *Tickets purchased* Receipts at a football game were $52,640. If tickets sold for $7, how many tickets were purchased for the game?

53. *Loads in the dishwasher* A dishwasher uses 14 gallons of hot water per load. If 42 gallons were used, how many loads were washed?

54. *Figuring miles per gallon* A car is driven 348 miles and uses 12 gallons of gas. How many miles per gallon is that?

55. *Finding the weekly salary* A teacher makes $31,200 annually. What is the weekly salary? (There are 52 weeks in a year.)

56. *Cost per credit hour* A part-time student in a community college paid $156 for tuition. The student was taking 3 credit hours. What was the cost of each credit hour?

57. *BTUs for 100-watt bulb* One 100-watt bulb burning for 10 hours uses 11,600 Btu. How many Btu per hour does the bulb burn?

58. *Running out of gas* The United States uses 23 trillion cubic feet of natural gas each year. The reserves of natural gas are 253 trillion cubic feet. If no more gas is discovered, how many more years will it take before we run out of natural gas?

59. *Words per minute in shorthand* The fastest recorded shorthand writing speed under championship conditions is 1500 words in 5 minutes. How many words per minute is that?

60. *Pay per word* One of the highest rates ever offered to a writer was $30,000 to Ernest Hemingway for a 2000-word article on bullfighting. How much was he paid per word?

Monthly expenses In Problems 61–64 find the **monthly** expense for each of the categories.

	Category	Annual (12-Month) Expenditures
61.	Apparel and services	$1368
62.	Entertainment	$1164
63.	Housing	$7656
64.	Personal Care Products	$336

Source: Bureau of Labor Statistics, Consumer Expenditure Survey for persons under 25 http://data.bls.gov/PDQ/outside.jsp?survey = cx.

Using Your Knowledge

An Average Problem The idea of an **average** is used in many situations. The average (mean) is the result obtained by dividing the sum of two or more quantities by the number of quantities. For example, if you have taken three tests and your scores were 90, 72, and 84, your average for the three tests is:

$$\begin{array}{l} \text{Sum of the scores} \rightarrow \\ \text{Number of tests} \rightarrow \end{array} \frac{90 + 72 + 84}{3} = \frac{246}{3} = 82$$

Thus the average for the three tests is 82.

The accompanying table shows the average income for a person with different numbers of years of schooling. For example, the average income for a person with a master's degree is obtained by dividing the *life income,* \$2,500,000, by *the number of years the average person receives income,* 40.

$$\begin{array}{l} \text{Income} \rightarrow \\ \text{Years} \rightarrow \end{array} \frac{2,500,000}{40} = \$62,500$$

Thus, the average income is \$62,500

Years of School	Average Income	Life (40-yr) Income
Master's	\$62,500	\$2,500,000
Bachelor's	☐	2,100,000
Associate	☐	☐
Some College	☐	☐
High School	☐	☐
No High School	☐	☐

Work Life (40 years) Earnings Estimates (in Millions) for Full-Time Year-Round Workers by Educational Attainment

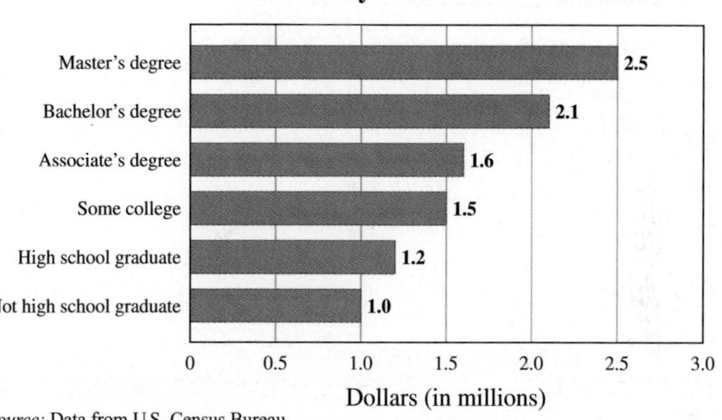

Source: Data from U.S. Census Bureau.

65. Find the average income for a person with a bachelor's degree.

66. Find the average income for a person with an associate's degree.

67. Find the average income for a person with some college education.

68. Find the average income for a high school graduate.

69. Find the average income for a person who is not a high school graduate.

70. What is the average number of calories in a hamburger? Estimate the answer using the following information:

Burger King:	275 calories
Dairy Queen:	360 calories
Hardee's	275 calories
McDonald's	265 calories
Wendy's	350 calories

71. A person earned \$250, \$210, \$200, and \$240 during the four weeks of a certain month. What was the average weekly salary?

72. The weight of the players in the defensive line of a football team is as follows: 240, 237, 252, and 263. What is the average weight of these players?

73. The price of a certain stock during each day of last week was as follows: \$24, \$25, \$24, \$26, \$26. What was the average price of the stock?

74. The ages of the players on a basketball team are as follows: 18, 17, 20, 18, 17. What is the average age of the players on this team?

Write On

75. Look at the definition of division and write the procedure you use to check a division problem using multiplication. Make sure the procedure includes the instances in which there is a remainder.

76. If you divide a number by itself, the answer is 1. Write in your own words why this rule does not work when you divide 0 by 0.

77. Explain in your own words why division by 0 is not defined.

78. Is division commutative? Why or why not? Give examples.

Algebra Bridge

Divide.

79. $x^2 + 4x + 5$ by $x + 1$

80. $x^2 + 5x + 7$ by $x + 2$

81. $x^2 + 4x + 6$ by $x + 3$

82. $x^2 + 6x + 11$ by $x + 4$

Concept Checker

Fill in the blank(s) with the correct word(s), phrase, or mathematical statement.

83. The **quotient** $a \div b$ $(b \neq 0)$ is the unique whole number c so that ____________.

84. When we write **20 ÷ 5 = 4,** 20 is called the ____________, 5 is the ____________, and 4 is the ____________.

85. **Multiplication** is the ____________ process of division.

86. $a \div a =$ ____________ $(a \neq 0)$.

87. $a \div 1 =$ ____________.

88. $0 \div a =$ ____________ $(a \neq 0)$.

divisor	**0**
1	**inverse**
$a = b \times c$	**quotient**
$c = a \times b$	a
dividend	$b = a \times c$

Mastery Test

89. Write as a multiplication problem and find the answer if possible.

a. 48 ÷ 6 = □

b. 37 ÷ 1 = □

90. Write as a multiplication problem and find the answer if possible.

a. 0 ÷ 8 = □

b. 8 ÷ 0 = □

91. Write as a multiplication problem and find the answer if possible.

a. 9 ÷ 9 = □

b. 7 ÷ 1 = □

92. Write as a multiplication problem and find the answer if possible.

a. 99 ÷ 9 = □

b. 9 ÷ 3 = □

93. Divide using long division. (Show the remainder if there is one.)

a. 792 ÷ 6 = ___ **b.** 1728 ÷ 9 = ___

94. A person earns $27,600 annually. How much does the person earn each month?

Skill Checker

Fill in with $<$ or $>$ to make the resulting inequality true.

95. 345 ____________ 354

96. 908 ____________ 809

Multiply:

97. 305 × 1003

98. 908 × 1203

1.7 Primes, Factors, and Exponents

Objectives

You should be able to:

A Determine if a number is prime or composite.

B Find the prime factors of a number.

C Write a number as a product of primes, using exponential notation if necessary.

D Write two or more numbers containing exponents as a product and then find the product.

To Succeed, Review How To . . .

1. Use the multiplication facts. (p. 49)
2. Apply the meaning of the word *factor*. (p. 49)

Getting Started

Look at the numbers in the blue boxes (2, 3, 5, and so on). They are **prime numbers.** The spiral was constructed by arranging the counting numbers in a counterclockwise fashion and placing the numbers with exactly two factors (themselves and 1) in a blue box. As you can see, the prime numbers tend to form diagonal lines. Why? Nobody knows!

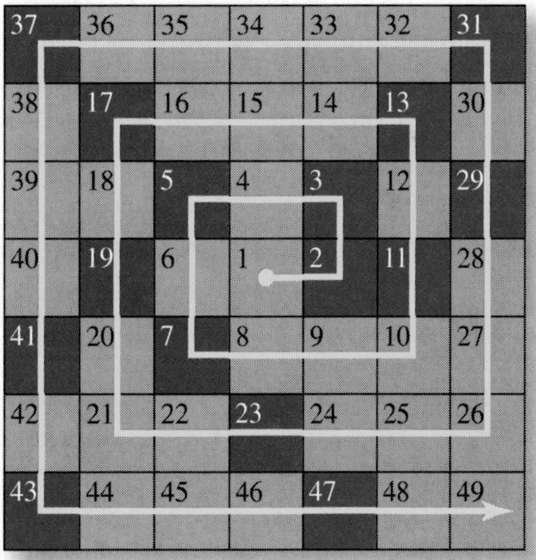

A › Primes and Composites

As you recall from Section 1.5, when a number is written as the product of other numbers, these other numbers are called **factors.** Thus, 2 and 3 are factors of 6 because $6 = 2 \times 3$. Note that 2×3 is an example of a **factorization** of 6. Similarly, 1 and 6 are factors of 6 because $6 = 1 \times 6$.

PRIME NUMBER

A **prime number** is a counting number having exactly two *different* factors, itself and 1.

Here are the first few counting numbers written as products of factors:

$1 = 1 \times 1 \qquad 4 = 1 \times 4 \text{ or } 2 \times 2$

$2 = 1 \times 2 \qquad 5 = 1 \times 5$

$3 = 1 \times 3 \qquad 6 = 1 \times 6 \text{ or } 2 \times 3$

As you can see:

1 has only one factor:	1
2 has two factors:	1 and 2
3 has two factors:	1 and 3
4 has three factors:	1, 2, and 4
5 has two factors:	1 and 5
6 has four factors:	1, 2, 3, and 6

Recall that $4 = 1 \times 4$ and 2×2, $6 = 1 \times 6$ and 2×3.

Since a number is prime when it has exactly two *distinct* factors, itself and 1, the prime numbers in the list above are 2, 3, and 5. The other numbers (1, 4, 6) are **not prime.** Note that 1 is *not* prime because it does not have two *distinct* factors.

COMPOSITE NUMBER

A counting (natural) number greater than 1 that is not prime is said to be composite.

Thus in the preceding list, 4 and 6 are composite.

EXAMPLE 1 Finding if a number is prime or composite

Determine if the given numbers are prime or composite.

a. 14 **b.** 17

SOLUTION

a. 14 has more than two factors since $14 = 1 \times 14$ or 2×7. Thus 14 is composite.

b. $17 = 1 \times 17$. Since 17 has *exactly* two factors, itself and 1, 17 is a prime number.

PROBLEM 1

Determine if the given numbers are prime or composite.

a. 19 **b.** 15

B > Finding Prime Factors

In Example 1 we noticed that 14 is a composite number with four factors (1, 2, 7, 14), two of which (2 and 7) are prime. The *prime* factors of a number are those factors that are prime. For example, the factors of 18 are 1, 2, 3, 6, 9, and 18, but the **prime factors** of 18 are 2 and 3. The numbers 1, 6, 9, and 18 are *not* prime and thus cannot be prime factors of 18.

EXAMPLE 2 Finding prime factors

Find the prime factors of these numbers.

a. 10 **b.** 11

SOLUTION

a. $10 = 1 \times 10$ or 2×5. Thus the factors of 10 are 1, 2, 5, and 10. Of these, 2 and 5 are prime factors.

b. $11 = 1 \times 11$. Thus 11 has only one prime factor, 11. Any prime number has only one prime factor, itself.

PROBLEM 2

Find the prime factors of these numbers.

a. 13 **b.** 12

It is easy to find the prime factors of 10 because 10 is a small number. How can we find the prime factors of larger numbers? To do this we first need to know if the given number is prime. Unfortunately, nobody has discovered a simple formula for finding all prime numbers. However, a Greek mathematician named Eratosthenes invented a

Answers to PROBLEMS

1. a. prime **b.** composite
2. a. 13 **b.** 2 and 3

procedure to find all prime numbers smaller than a given number—say 50. He listed all the numbers 1 through 50 in a table, as shown here:

Sieve of Eratosthenes

~~1~~	2	3	~~4~~	5	~~6~~	7	~~8~~	~~9~~	~~10~~
11	~~12~~	13	~~14~~	~~15~~	~~16~~	17	~~18~~	19	~~20~~
~~21~~	~~22~~	23	~~24~~	~~25~~	~~26~~	~~27~~	~~28~~	29	~~30~~
31	~~32~~	~~33~~	~~34~~	~~35~~	~~36~~	37	~~38~~	~~39~~	~~40~~
41	~~42~~	43	~~44~~	~~45~~	~~46~~	47	~~48~~	~~49~~	~~50~~

He then reasoned as follows: The number 1 is not a prime (it has only one factor), so he marked it out. He then concluded that 2 is a prime, but any number with 2 as a factor (4, 6, 8, etc.) is composite. He then circled 2 as a prime and crossed out 4, 6, 8, and so forth. Next, since 3 is a prime, he circled it and crossed out all numbers having 3 as a factor, starting with 9 (6 was already out). Similarly, he circled 5 (the next prime) and crossed out the numbers having 5 as a factor, starting with 25 (10, 15, and 20 were already out). He continued this process until he reached 11, noting that all the numbers having 11 as a factor (2×11, 3×11, and 4×11) had been eliminated when he had crossed out the numbers having factors of 2 and 3. Hence all the numbers that are left are prime. They are circled and set in color in the table.

C > Writing Composite Numbers as Products of Primes

Now we are ready to write a composite number as a product of primes; that is, to find the prime factorization of a number (a procedure that is necessary to reduce and add fractions). This is done by dividing by the first few primes, 2, 3, 5, 7, 11, and so on. To help you in this process, we give the rules that will tell you if a number is divisible by 2, 3, and 5. (There are rules to tell if a number is divisible by 7 or 11, but they are so complicated that you may as well try dividing by 7 or 11. The Using Your Knowledge section has additional divisibility rules.)

RULE TO TELL IF A NUMBER IS DIVISIBLE BY 2

A number is divisible by 2 *if it ends in an even number* (2, 4, 6, 8, or 0).

For example, 42, 68, and 90 are divisible by 2. When a number is divisible by 2, it is called an **even** number, so 42, 68, and 90 are even. If a whole number is *not* even (not divisible by 2) it is called an **odd** number. Thus, 43, 71, and 95 are odd.

RULE TO TELL IF A NUMBER IS DIVISIBLE BY 3

A number is divisible by 3 *if the sum of its digits is divisible by 3.*

Thus, 273 is divisible by 3 because the sum of its digits (12) is divisible by 3 as shown:

$$\begin{array}{ccccc} 2 & & 7 & & 3 \\ \downarrow & & \downarrow & & \downarrow \\ 2 & + & 7 & + & 3 = 12 \end{array} \qquad \text{(divisible by 3)}$$

RULE TO TELL IF A NUMBER IS DIVISIBLE BY 5

A number is divisible by 5 if it *ends in 0 or 5.*

For example, 300 and 95 are divisible by 5.

We will use these divisibility tests to help us write a composite number as a product of primes. This is called prime factorization. Let us try to write 24 as a product of primes; that is, find the prime factorization of 24. To do so, we divide by successive primes (2, 3, 5, and so on).

Divide 24 by 2 (the first prime).	$\rightarrow 2 \lfloor 24$
Divide 12 by 2.	$\rightarrow 2 \lfloor 12$
Divide 6 by 2.	$\rightarrow 2 \lfloor 6$
3 cannot be divided by 2, so we divide by the next prime, 3.	$\rightarrow 3 \lfloor 3$
	1

Thus,

$$24 = 2 \times 2 \times 2 \times 3$$

You can also use a factor tree, as follows:

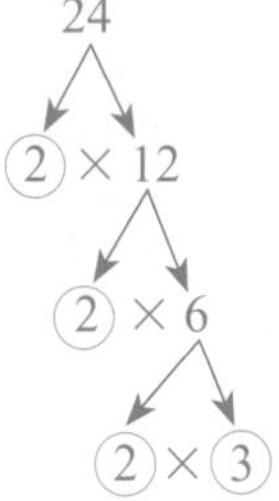

Write 24 as a prime (2) times a composite number (12).

Write 12 as a prime (2) times a composite.

Write 6 as a prime (2) times **another** prime.

Now we are finished, $24 = 2 \times 2 \times 2 \times 3$.

Note: You can also use a dot (·) and write $24 = 2 \cdot 2 \cdot 2 \cdot 3$. In this section we will use the multiplication sign (×) to indicate a multiplication.

What about trying to write 79 as a product of primes? We do this by dividing by successive primes (2, 3, 5, etc.) if possible.

79 is not divisible by 2 (it ends in 9).

79 is not divisible by 3 ($7 + 9 = 16$, which is not divisible by 3).

79 is not divisible by 5.

79 is not divisible by 7. If we divide 79 by 7 we have a remainder of 2.

$$\begin{array}{r} 11 \\ 7\overline{)79} \\ \underline{-7} \\ 09 \\ \underline{-7} \\ 2 \end{array} \leftarrow \text{Remainder}$$

79 is not divisible by 11. If we divide 79 by 11 we have a remainder of 2.

$$\begin{array}{r} 7 \\ 11\overline{)79} \\ \underline{-77} \\ 2 \end{array} \leftarrow \text{Remainder}$$

Note that we can stop dividing here because dividing by 11 gives a quotient (7) less than the divisor (11). In general, when testing to see if a number is prime, you can stop dividing *when the quotient is less than the divisor*. Since 79 is *not* divisible by any of our prime divisors, 79 is a *prime* number.

EXAMPLE 3 Prime factorization: Writing as product of primes

Write (if possible) the given number as a product of primes.

a. 45 **b.** 89 **c.** 32

SOLUTION

a. We must divide by the first few primes, 2, 3, 5, 7, and so on. Since 45 is not divisible by 2, we start by dividing by 3 as shown:

Divide by 3. $\rightarrow 3\,\underline{|\,45}$

Divide by 3. $\rightarrow 3\,\underline{|\,15}$

Divide by 5. $\rightarrow 5\,\underline{|\,5}$

1

Thus,

$$45 = 3 \times 3 \times 5$$

The factor tree is

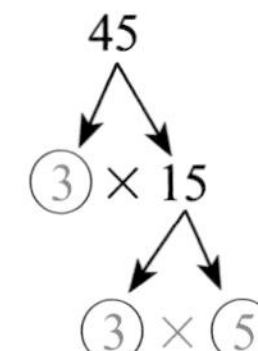

Write 45 as a prime (3) times a composite number.

Write 15 as a prime (3) times another prime. As before, 45 = 3 × 3 × 5

b. We try to divide by successive primes.

89 is not divisible by 2 (it ends in 9).

89 is not divisible by 3 (8 + 9 = 17, which is not divisible by 3).

89 is not divisible by 5.

89 is not divisible by 7.

$$\begin{array}{r} 12 \\ 7\overline{)89} \\ -7 \\ \hline 19 \\ -14 \\ \hline 5 \end{array} \leftarrow \text{Remainder}$$

89 is not divisible by 11.

$$\begin{array}{r} 8 \\ 11\overline{)89} \\ -88 \\ \hline 1 \end{array} \leftarrow \text{Remainder}$$

Since the quotient (8) is smaller than the divisor (11), we can stop here. Because 89 is not divisible by any of our prime divisors, we can say that 89 is a **prime number.**

c. Divide by 2. $\rightarrow 2\,\underline{|\,32}$

Divide by 2. $\rightarrow 2\,\underline{|\,16}$

Divide by 2. $\rightarrow 2\,\underline{|\,8}$ or

Divide by 2. $\rightarrow 2\,\underline{|\,4}$

Divide by 2. $\rightarrow 2\,\underline{|\,2}$

1

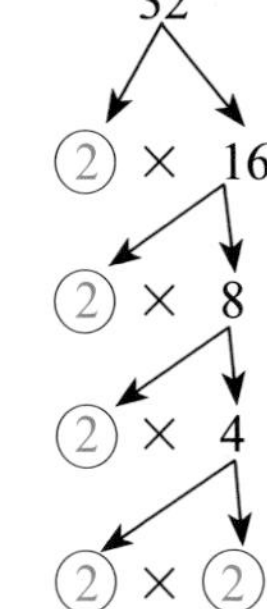

Thus,

$$32 = 2 \times 2 \times 2 \times 2 \times 2$$

PROBLEM 3

Prime factorization: Write (if possible) the given number as a product of primes.

a. 35 **b.** 16 **c.** 97

Answers to PROBLEMS

3. a. 5 × 7 **b.** 2 × 2 × 2 × 2
c. Prime

Products like $2 \times 2 \times 2 \times 2 \times 2$ are easier to write using **exponential notation.** Using this notation, we write

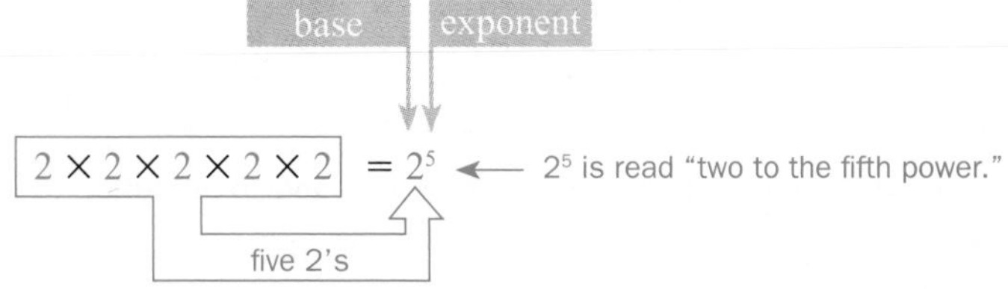

Here 2 is called the **base** and 5 is called the **exponent.** The exponent 5 tells us how many times the base 2 must be used as a factor. Similarly.

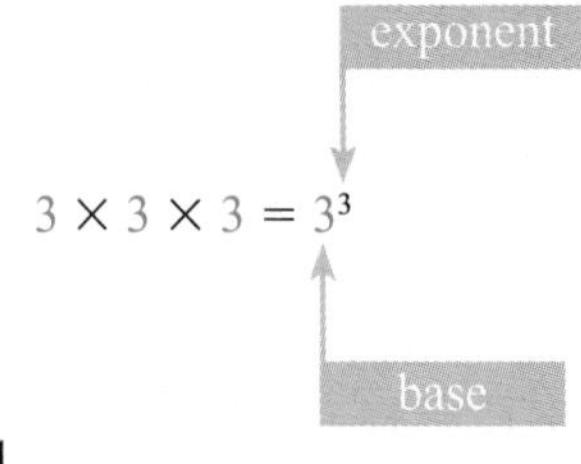

3^3 is read "three to the third power" or "three cubed."

and

exponent

$6 \times 6 = 6^2$

base

6^2 is read "six to the second power" or "six squared."

EXAMPLE 4 Prime factorization: Writing numbers with exponential notation

Use exponential notation to write the given number as a product of primes.

a. 18 **b.** 72

SOLUTION

a. Divide by 2 → 2|18

Divide by 3 → 3|9 $\frac{18}{2} = 9$

Divide by 3 → 3|3 $\frac{9}{3} = 3$

1 $\frac{3}{3} = 1$

Thus,

$$18 = 2 \times 3 \times 3 = 2 \times 3^2$$

b. Divide by 2 → 2|72

Divide by 2 → 2|36 $\frac{72}{2} = 36$

Divide by 2 → 2|18 $\frac{36}{2} = 18$

Divide by 3 → 3|9 $\frac{18}{2} = 9$

Divide by 3 → 3|3 $\frac{9}{3} = 3$ and $\frac{3}{3} = 1$

1

Thus,

$$72 = 2 \times 2 \times 2 \times 3 \times 3 = 2^3 \times 3^2$$

$2^3 \times 3^2$ is read "2 cubed times 3 squared."

PROBLEM 4

Prime factorization: Use exponential notation to write the given number as a product of primes.

a. 27 **b.** 98

Answers to PROBLEMS

4. a. 3^3 **b.** 2×7^2

D > Products Involving Exponents

Sometimes we are given a number such as $2^2 \times 3^2$ and we want to write the answer in standard notation. Thus,

$$2^2 \times 3^2 = 2 \times 2 \times 3 \times 3 = 36$$

EXAMPLE 5 Prime factorization: Finding products involving exponents

Write the given numbers as a product of factors; then find the product.

a. $2^3 \times 3^2$ **b.** $2^2 \times 5^3$

SOLUTION

a. $2^3 \times 3^2 = 2 \times 2 \times 2 \times 3 \times 3 = 8 \times 9 = 72$

b. $2^2 \times 5^3 = 2 \times 2 \times 5 \times 5 \times 5 = 4 \times 125 = 500$

PROBLEM 5

Prime factorization: Find the products by writing as a product of factors.

a. $2^2 \times 3^3$ **b.** $2^3 \times 5^2$

Now look at this pattern.

$$10^5 = 100{,}000$$
$$10^4 = 10{,}000$$
$$10^3 = 1000$$
$$10^2 = 100$$
$$10^1 = \underline{\quad ? \quad}$$
$$10^0 = \underline{\quad ? \quad}$$

Here, the exponent of a number whose base is 10 tells us how many zeros there are in the final product. Can you now find the values of 10^1 and 10^0? Since the exponent tells us the number of zeros, then we have

$$10^1 = 10 \quad \text{and} \quad 10^0 = 1$$

In general, for any number a, we define

$$a^1 = a \qquad a^0 = 1, a \neq 0$$

Thus, $2^1 = 2$, $3^1 = 3$, $2^0 = 1$, and $3^0 = 1$.

EXAMPLE 6 Prime factorization: Finding products involving exponents

Write these as a product of factors, then find the product.

a. $3^2 \times 2^1$ **b.** $5^2 \times 2^3 \times 7^0$

SOLUTION

a. $3^2 \times 2^1 = 9 \times 2 = 18$

b. $5^2 \times 2^3 \times 7^0 = 25 \times 8 \times 1 = 200$

PROBLEM 6

Prime factorization: Find the products by writing as a product of factors.

a. $3^1 \times 4^2$ **b.** $2^2 \times 3^0 \times 5^3$

Answers to PROBLEMS

5. a. 108 **b.** 200

6. a. 48 **b.** 500

Calculator Corner

We can easily determine if a number is prime using a calculator. We simply divide the number by the first few primes. Thus, to determine if 79 is prime, we divide 79 by 2, 3, 5, 7, and 11. Since none of the divisions is exact, 79 is a prime number.

Regarding multiplying factors involving exponents, we can proceed two ways:

Method 1. By repeated multiplications.
Method 2. By using the y^x key, if your calculator has one.

Thus, to multiply $2^3 \times 3^2$ (as in Example 5), we can proceed as follows:

Method 1. $2^3 \times 3^2 =$ 2 × 2 × 2 × 3 × 3 ENTER 72

Method 2. If you have a y^x key, you can find 2^3 first by pressing 2 y^x 3 ENTER. The display will show 8. To multiply this answer by 3^2, key in × 3 y^x 2 ENTER. As before, the answer will be 72. The complete sequence is 2 y^x 3 ENTER × 3 y^x 2 ENTER.

Exercises 1.7

⟨A⟩ Primes and Composites In Problems 1–10, determine if the number is prime or composite. If composite, list *all* its **factors.**

1. 7 **2.** 28 **3.** 6 **4.** 17
5. 24 **6.** 8 **7.** 25 **8.** 26
9. 23 **10.** 30

⟨B⟩ Finding Prime Factors In Problems 11–20, find the *prime* factors of each number.

11. 14 **12.** 16 **13.** 18 **14.** 23 **15.** 29
16. 30 **17.** 22 **18.** 20 **19.** 21 **20.** 31

⟨C⟩ Writing Composite Numbers as Products of Primes In Problems 21–30, write each number as a product of primes using exponents. If the number is prime, indicate so.

21. 34 **22.** 31 **23.** 41 **24.** 48 **25.** 64
26. 81 **27.** 91 **28.** 110 **29.** 190 **30.** 200

⟨D⟩ Products Involving Exponents In Problems 31–40, find the product by writing as a product of factors.

31. $3^0 \times 2^2$ **32.** $10^3 \times 2^2$ **33.** $2^0 \times 10^0$ **34.** $4^2 \times 3^1$ **35.** $5^2 \times 2^2$
36. $2^2 \times 5^0 \times 3^2$ **37.** $4^3 \times 2^1 \times 4^0$ **38.** $10^0 \times 3^2 \times 10^3$ **39.** $5^2 \times 2^3 \times 11^0$ **40.** $2^3 \times 5^2 \times 6^1$

Applications

41. *Prime numbers less than 50* How many prime numbers are less than 50?

42. *Prime number between 50 and 100* How many prime numbers are between 50 and 100? (*Hint:* Use a sieve of Eratosthenes.)

43. *Cell phones and exponents* At the end of 2005, the number of worldwide cell phone subscribers was 2,000,000,000. Write this number as a product using exponents.

44. *Alaskan oil reserves* The proved Alaskan oil reserves are 5,000,000,000 barrels. Write this number as a product by using exponents.

45. *Folding paper* Take a sheet of notebook paper. If you fold it in half once, it is 2 layers thick. If you fold it in half again, it is 4 layers thick. If you fold it in half another time, it is 8 layers thick.

Here is the table relating the number of folds and the number of layers thick. Complete this table.

Number of Folds	Number of Layers Thick
0	$2^0 = 1$
1	$2^1 = 2$
2	$2^2 = 4$
3	$2^3 = 8$
4	$2 =$
5	$2 =$

46. *Patterns in paper folding* Is there a pattern for finding the number of layers thick by using exponents for 2? If so, how many layers thick will it be after:

a. 6 folds?

b. 10 folds?

47. *Pioneer journey* *Pioneer 10*, a crewless U.S. spacecraft, was launched March 2, 1972, on a 639-day, 1,000,000,000-kilometer journey past Jupiter. Write this distance using exponents.

48. *Hemoglobin in red blood cells* A red blood cell contains about 3×10^8 hemoglobin molecules. Write this number as a product of factors and multiply out.

49. *Distance to Milky Way* Our solar system is 3×10^4 light-years from the center of the Milky Way galaxy. Write 3×10^4 as a product of factors and multiply out.

50. *Light-years to Milky Way* A light-year (the distance light travels in 1 year at 186,000 miles per second) is 6×10^{12} miles. Since we are 3×10^4 light-years from the center of the Milky Way, our distance from this center is $(3 \times 10^4) \times (6 \times 10^{12})$ miles. Write this number as a product of factors and multiply out.

51. *Letters in a manuscript* A typical typed page of text contains about $3000 = 3 \times 10^3$ letters. Thus, a 500-page manuscript will have $\mathbf{(5 \times 10^2) \times (3 \times 10^3)}$ letters. Write this number as a product of factors and multiply.

52. *Hamburger revenue* A large hamburger costs ($\$3 \times 10^0$). If 2 million = (2×10^6) of these burgers are sold, the revenue would be $\mathbf{(\$3 \times 10^0) \times (2 \times 10^6)}$. Write this number as a product of factors and multiply.

53. *0's in a googol* The definition of a googol is: **1 googol = 10^{100}**. How many 0's are in a googol? *Hint:* 10^2 has 2 zeros, 10^3 has 3 zeros.

54. *Exponents* If a **googolplex = 10^{googol}**, write a googolplex using powers of 10.

Using Your Knowledge

Divide and Conquer (Prime Factorization) To write a number as a product of primes, you have to divide the number by 2, 3, 5, 7, and so on. We restate the rules to tell, without dividing, if a number is divisible by 2, 3, or 5.

55. A number is divisible by 2 if its last digit is 2, 4, 6, 8, or 0. Are these numbers divisible by 2?

a. 12 **b.** 13 **c.** 20

56. A number is divisible by 3 if the sum of its digits is divisible by 3. For example, 462 is divisible by 3 because $4 + 6 + 2 = 12$, which is divisible by 3. Are these numbers divisible by 3?

a. 493 **b.** 112 **c.** 111

57. A number is divisible by 5 if its last digit is 5 or 0. Are these numbers divisible by 5?

a. 125 **b.** 301 **c.** 240

58. Use the information in problems 55, 56, and 57 to determine if the given numbers are divisible by 2, 3, or 5. (For example, 42 is divisible by 2 and 3, as indicated by the check marks.)

	Number	Divisible by		
		2	3	5
	42	✓	✓	
a.	24			
b.	50			
c.	19			
d.	30			
e.	91			

59. A number is divisible by 4 if the number created by the last two digits in the number is divisible by 4. For example, 384 is divisible by 4, since the last two digits in the number form the number 84, which is divisible by 4. On the other hand, the number 319 is *not* divisible by 4, since the last two digits form the number 19, which is not divisible by 4. Are these numbers divisible by 4?

a. 420 **b.** 308

c. 1234 **d.** 1236

60. A number is divisible by 6 if it is divisible by 2 and 3. Thus, 234 is divisible by 6 because 234 is divisible by 2 and by 3 $(2 + 3 + 4 = 9$, which is divisible by 3, so 234 is divisible by 3). On the other hand, 368 is *not* divisible by 6 because 368 is divisible by 2 but not by 3 $(3 + 6 + 8 = 17$, which is not divisible by 3). Are these numbers divisible by 6?

a. 432 **b.** 315

c. 3126 **d.** 4123

61. A number is divisible by 8 if the number formed by the last three digits in the number is divisible by 8. For example, 3416 is divisible by 8 because the last three digits in the number form the number 416, which is divisible by 8. On the other hand, the number 1319 is *not* divisible by 8, since the last three digits form the number 319, which is not divisible by 8. Are these numbers divisible by 8?

a. 1424 **b.** 1630

c. 2360 **d.** 2148

62. A number is divisible by 9 if the sum of the digits of the number is divisible by 9. This means that, 342 is divisible by 9, since $3 + 4 + 2 = 9$, which is divisible by 9. However, 1352 is *not* divisible by 9, since $1 + 3 + 5 + 2 = 11$, which is not divisible by 9. Are these numbers divisible by 9?

a. 348 **b.** 564

c. 2386 **d.** 6570

63. A number is divisible by 10 if it ends with a 0. Thus, 980 and 340 are divisible by 10, but 342 and 786 are not. Are these numbers divisible by 10?

a. 450 **b.** 432

c. 567 **d.** 980

64. Christian Goldbach made a famous conjecture (guess) that has not yet been proved or disproved. He said that every even number greater than 2 can be written as the sum of two primes. For example,

$$4 = 2 + 2$$
$$6 = 3 + 3$$
$$8 = 3 + 5$$

Write each even number from 10 through 20 as the sum of two primes.

Write On

65. In 1771 Leonhard Euler, a Swiss mathematician, discovered the prime number $2^{31} - 1$. This is an example of a Mersenne prime, a prime of the form $2^p - 1$. The largest prime at this time is $2^{13466917} - 1$.

a. Do you think this number is odd or even? Write the reasons.

b. How many digits do you think this prime has?

c. How many lines do you think it would take to type it?

66. Do you think the number of primes is finite or infinite? Write your reasons.

67. What is the difference between "the number of factors" of a number and "the number of prime factors" of a number? Explain and give examples.

Concept Checker

Fill in the blank(s) with the correct word(s), phrase, or mathematical statement.

68. A number is **divisible by 2** if it ends in an ______________ number.

69. A number is **divisible by 3** if the ______________ of its digits is divisible by 3.

70. A number is **divisible by 5** if it ends in ______________ or in ______________.

71. A number is ______________ if it is **divisible by 2.**

72. $a^1 =$ ______________ $(a \neq 0)$.

73. $a^0 =$ ______________ $(a \neq 0)$.

1	**sum**
a	**product**
factors	**even**
5	**0**
odd	

Mastery Test

74. Find the prime factors of 40.

75. Determine if the following numbers are prime or composite:

a. 41 **b.** 39

76. Write 84 as a product of primes.

77. Use exponents to write the number as a product of primes.

a. 3600 **b.** 360

78. Find the product: $2^4 \times 3^2$

79. Find the product: $5^2 \times 2^2 \times 7^0$

80. Find the product: $5^2 \times 2^2 \times 3^3$

Skill Checker

In Section 1.1 we discussed how to write a number in expanded form. Thus, $387 = 300 + 80 + 7$. We can now write the expanded form using exponents. To do this, recall that $10^4 = 10{,}000$, $10^3 = 1000$, $10^2 = 100$, $10^1 = 10$, and $10^0 = 1$. Thus,

$$387 = 3 \times 10^2 + 8 \times 10^1 + 7 \times 10^0$$

Write each number in expanded form using exponents.

81. 138

82. 345

83. 1208

84. 3046

1.8 Order of Operations and Grouping Symbols

Objectives

You should be able to:

A Simplify expressions using the order of operations.

B Remove grouping symbols within grouping symbols.

C Solve applications using the concepts studied.

To Succeed, Review How To . . .

1. Use arithmetic facts (+, −, ×, ÷). (pp. 24, 37, 49, 62)
2. Write a number containing exponents as a product and how to find this product. (pp. 77–78)

Getting Started

Suppose all 239 rooms in the motel are taken (for simplicity, do not include extra persons in the rooms). How can we figure out how much money we are going to collect? To do this, we *first* multiply the price of each room ($24, $28, and $38) times the number of rooms (44, 150, 45). *Next,* we add all these figures and get the result. Note that we have *multiplied* before *adding*. (Did you get $6966 for the answer?)

A › Order of Operations

If we want to find the answer to $3 \cdot 4 + 5$, do we (1) add 4 and 5 first and then multiply by 3, or (2) multiply 3 by 4 first and then add 5? In (1) the answer is 27. In (2) the answer is 17. What if we write $3 \cdot (4 + 5)$ or $(3 \cdot 4) + 5$? What do the parentheses mean? By the way, $3 \cdot (4 + 5)$ can also be written as $3(4 + 5)$ without the multiplication dot $\cdot$. To obtain an answer we can agree upon, we must make rules regarding the order in which we perform operations. The rules are as follows:

ORDER OF OPERATIONS (PEMDAS)

1. Do all calculations *inside* grouping symbols like parentheses ( ), brackets [ ], or braces { } first.
2. Evaluate all *exponential* expressions.
3. Do *multiplications* and *divisions* as they occur in order from left to right.
4. Do *additions* and *subtractions* as they occur in order from left to right.

You can remember this order if you remember the phrase

Please	(**p**arentheses)
Excuse	(**e**xponential expressions)
My	(**m**ultiplication)
Dear	(**d**ivision)
Aunt	(**a**ddition)
Sally	(**s**ubtraction)

Note that $8 \cdot 4 \div 2 = 32 \div 2 = 16$. Done *in order* from left to right multiplication was done *before* division.

With these conventions:

$3 \cdot 2^2 + 5 = 3 \cdot 4 + 5$ Do exponents ($2^2 = 2 \cdot 2 = 4$).
$= 12 + 5$ Multiply ($3 \cdot 4 = 12$).
$= 17$ Add ($12 + 5 = 17$).

Similarly,

$3 \cdot (2^2 + 5) = 3(4 + 5)$ To add inside parentheses, first do exponents ($2^2 = 2 \cdot 2 = 4$).
$= 3 \cdot 9$ Add inside parentheses ($4 + 5 = 9$).
$= 27$ Multiply 3 by 9.

But

$(3 \cdot 2^2) + 5 = (3 \cdot 4) + 5$ To multiply inside parentheses, first do exponents ($2^2 = 2 \cdot 2 = 4$).
$= 12 + 5$ Multiply inside parentheses ($3 \cdot 4 = 12$).
$= 17$ Add ($12 + 5 = 17$).

EXAMPLE 1 Simplifying numerical expressions

Simplify.

a. $8 \cdot 3^2 - 3$ **b.** $3^3 + 3 \cdot 5$

SOLUTION

a. $8 \cdot 3^2 - 3 = 8 \cdot 9 - 3$ Do exponents ($3^2 = 3 \cdot 3 = 9$).
$= 72 - 3$ Do multiplications and divisions in order from left to right ($8 \cdot 9 = 72$).
$= 69$ Do additions and subtractions in order from left to right ($72 - 3 = 69$).

PROBLEM 1

Simplify.

a. $7 \cdot 2^3 - 7$ **b.** $23 + 2^2 \cdot 5$

Answers to PROBLEMS

1. a. 49 **b.** 43

b. $3^3 + 3 \cdot 5 = 27 + 3 \cdot 5$ Do exponents ($3^3 = 3 \cdot 3 \cdot 3 = 27$).
$= 27 + 15$ Do multiplications and divisions in order from left to right ($3 \cdot 5 = 15$).
$= 42$ Do additions and subtractions in order from left to right ($27 + 15 = 42$).

EXAMPLE 2 Simplifying numerical expressions

Simplify.

a. $63 \div 7 - (2 + 3)$ **b.** $8 \div 2 \cdot 2 \cdot 2 + 3 - 1$

SOLUTION

a. $63 \div 7 - (2 + 3) = 63 \div 7 - 5$ Do the operations inside parentheses first ($2 + 3 = 5$).
$= 9 - 5$ Do multiplications and divisions ($63 \div 7 = 9$).
$= 4$ Do additions and subtractions ($9 - 5 = 4$).

b. $8 \div 2 \cdot 2 \cdot 2 + 3 - 1 = 4 \cdot 2 \cdot 2 + 3 - 1$ Divide 8 by 2.
$= 16 + 3 - 1$ Do multiplication ($4 \cdot 2 \cdot 2 = 16$).
$= 19 - 1$ Do addition ($16 + 3 = 19$).
$= 18$ Do subtraction ($19 - 1 = 18$).

PROBLEM 2

Simplify.

a. $48 \div 6 - (3 + 1)$

b. $10 \div 2 \cdot 2 \cdot 2 + 2 - 1$

EXAMPLE 3 Simplifying numerical expressions

Simplify $2^3 \div 4 \cdot 2 + 3(5 - 2) - 3 \cdot 2$.

SOLUTION

$2^3 \div 4 \cdot 2 + 3(5 - 2) - 3 \cdot 2$
$= 2^3 \div 4 \cdot 2 + 3(3) - 3 \cdot 2$ Do calculations inside parentheses ($5 - 2 = 3$).
$= 8 \div 4 \cdot 2 + 3(3) - 3 \cdot 2$ Do exponents ($2^3 = 8$).
$= 2 \cdot 2 + 3(3) - 3 \cdot 2$ Divide 8 by 4.
$= 4 + 3(3) - 3 \cdot 2$ Multiply 2 by 2.
$= 4 + 9 - 3 \cdot 2$ Multiply 3 by 3.
$= 4 + 9 - 6$ Multiply 3 by 2.
$= 13 - 6$ Add 4 and 9.
$= 7$ Subtract 6 from 13.

PROBLEM 3

Simplify.

$6 \div 3 \cdot 2 + 2(5 - 3) - 2^2 \cdot 1$

B > More Than One Set of Grouping Symbols

Suppose you wish to buy 2 bedspreads and 2 mattresses. The price of *one* bedspread and *one* mattress is \$14 + \$88. Thus, the price of *two* of each is $2 \cdot (14 + 88)$. If, in addition, you wish to buy a lamp, the total price is

$$[2 \cdot (14 + 88)] + 12$$

Answers to PROBLEMS

2. a. 4 **b.** 21 **3.** 4

We have used two types of grouping symbols in the expression, parentheses ( ) and brackets [ ]. There is one more grouping symbol, braces { }. When grouping symbols occur within other grouping symbols, computations in the innermost ones are done first. Thus, to simplify $[2 \cdot (14 + 88)] + 12$, we *first* add 14 and 88 (the innermost grouping symbols), then multiply by 2, and finally add the 12. Here is the procedure:

$$[2 \cdot (14 + 88)] + 12 = [2 \cdot (102)] + 12 \quad \text{Add 14 and 88.}$$
$$= 204 + 12 \quad \text{Multiply 2 by 102.}$$
$$= 216 \quad \text{Add 204 and 12.}$$

EXAMPLE 4 Simplifying numerical expressions

Simplify $20 \div 4 + \{2 \cdot 3^2 - [3 + (6 - 2)]\}$.

SOLUTION

$20 \div 4 + \{2 \cdot 3^2 - [3 + (6 - 2)]\}$

$= 20 \div 4 + \{2 \cdot 3^2 - [3 + 4]\}$ Subtract inside parentheses ($6 - 2 = 4$).
$= 20 \div 4 + \{2 \cdot 3^2 - 7\}$ Add inside brackets ($3 + 4 = 7$).
$= 20 \div 4 + \{2 \cdot 9 - 7\}$ Do exponents ($3^2 = 9$).
$= 20 \div 4 + \{18 - 7\}$ Multiply inside braces ($2 \cdot 9 = 18$).
$= 20 \div 4 + 11$ Subtract inside braces ($18 - 7 = 11$).
$= 5 + 11$ Divide 20 by 4 ($20 \div 4 = 5$).
$= 16$ Add ($5 + 11 = 16$).

PROBLEM 4

Simplify.

$$25 \div 5 + \{3 \cdot 2^2 - [5 + (4 - 1)]\}$$

C › Applications Involving Order of Operations

EXAMPLE 5 Finding the ideal heart rate

When swimming, you expend less energy than when running to get the same benefit. To calculate your ideal heart rate while swimming, subtract your age A from 205, multiply the result by 7, and divide the answer by 10. In symbols,

$$\text{Ideal rate} = [(205 - A) \cdot 7] \div 10$$

Suppose you are 25 years old ($A = 25$). What is your ideal heart rate?

SOLUTION

Ideal rate $= [(205 - A) \cdot 7] \div 10$
$= [(205 - 25) \cdot 7] \div 10$ Let $A = 25$.
$= [180 \cdot 7] \div 10$ Subtract inside the parentheses.
$= 1260 \div 10$ Multiply inside the brackets.
$= 126$ Divide.

PROBLEM 5

Find your ideal heart rate while swimming if you are 35 years old.

EXAMPLE 6 Estimating vacation costs

You are ready for your spring break and want to go to the Bahamas with 19 friends; a total of 20 people. A travel agent charges \$50 for setting up the trip and \$400 per person. You have a coupon for \$100 off the total.

a. Write an expression for the total cost (using the coupon).

b. Evaluate the expression and find the total cost.

SOLUTION

a. The expression will consist of

\$50 setup + cost for 20 people − \$100 coupon
50 + 20(400) − 100

PROBLEM 6

You are having a party for 40 of your friends. The rental fee for the recreation hall is \$150, and the caterer charges \$25 per person but will give you a \$100 discount if you pay up front.

a. Write an expression for the total cost assuming you pay up front.

b. Evaluate the expression and find the total cost.

Answers to PROBLEMS

4. 9 **5.** 119 **6. a.** 150 + [25(40) − 100] **b.** \$1050

b. We use the order of operations to simplify the expression.

$$50 + 20(400) - 100$$
$$50 + 8000 - 100 \quad \text{Multiply.}$$
$$8050 - 100 \quad \text{Add.}$$
$$7950 \quad \text{Subtract.}$$

Thus, the total cost of the trip is $7950 after the discount.

Algebra Bridge

In algebra, the order of operations is the same as in arithmetic. The only difference is the instructions. Thus, if $x = 3$, $y = 4$, and $z = 2$, we may be asked to **evaluate** $x + y \div z$. This means to substitute for x, y, and z in the expression and then find the answer. Thus,

$$x + y \div z = 3 + 4 \div 2$$
$$= 3 + 2 \quad \text{Divide first.}$$
$$= 5 \quad \text{Add next.}$$

Calculator Corner

Does your calculator know the order of operations? We can give your calculator a test to see if it does! Enter 3 + 4 × 5 ENTER. What answer should the calculator give you? It should be 23 (multiply 4 by 5 first and then add 3). But what if we get 35 for an answer? This means that the calculator performs the operations as they are entered rather than following the order of operations. In this case, the calculator added 3 and 4 and then multiplied the result, 7, by 5. If you have a calculator like this, you have to enter the operations in the order in which you want them performed. Thus, you should enter 4 × 5 + 3 ENTER. Sometimes, you can remedy this problem by using parentheses to indicate which operation you want performed first, that is, you would enter 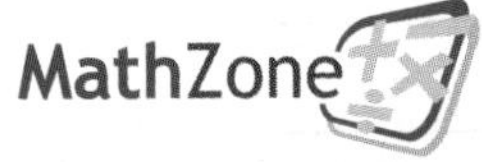 3 + (4 × 5) ENTER and obtain 23.

Exercises 1.8

⟨A⟩ Order of operations In Problems 1–20, simplify.

1. $4 \cdot 5 + 6$

2. $3 \cdot 4 + 6$

3. $7 + 3 \cdot 2$

4. $6 + 9 \cdot 2$

5. $7 \cdot 8 - 3$

6. $6 \cdot 4 - 9$

7. $20 - 3 \cdot 5$

8. $30 - 6 \cdot 5$

9. $48 \div 6 - (3 + 2)$

10. $81 \div 9 - (4 + 5)$

11. $3 \cdot 4 \div 2 + (6 - 2)$

12. $3 \cdot 6 \div 2 + (5 - 2)$

13. $6 \div 3 \cdot 3 \cdot 3 + 4 - 1$

14. $10 \div 2 \cdot 2 \cdot 2 + 3 - 2$

15. $8 \div 2 \cdot 2 \cdot 2 - 3 + 5$

16. $9 \div 3 \cdot 3 \cdot 3 - 8 + 5$

17. $10 \div 5 \cdot 2 + 8 \cdot (6 - 4) - 3 \cdot 4$

18. $15 \div 3 \cdot 3 + 2 \cdot (5 - 2) + 8 \div 4$

19. $4 \cdot 8 \div 2 - 3(4 - 1) + 9 \div 3$

20. $6 \cdot 3 \div 3 - 2(3 - 2) - 8 \div 2$

⟨B⟩ More than one set of grouping symbols In Problems 21–25, simplify.

21. $20 \div 5 + \{3 \cdot 4 - [4 + (5 - 3)]\}$

22. $30 \div 6 + \{4 \div 2 \cdot 3 - [3 + (5 - 4)]\}$

23. $(20 - 15) \cdot [20 \div 2 - (2 \cdot 2 + 2)]$

24. $(30 - 10) \cdot [52 \div 4 - (3 \cdot 3 + 3)]$

25. $\{4 \div 2 \cdot 6 - (3 + 2 \cdot 3) + [5(3 + 2) - 1]\}$

Web IT go to mhhe.com/bello for more lessons

Web IT go to mhhe.com/bello for more lessons

⟨C⟩ Applications Involving Order of Operations

26. *Babysitting pay* You have a babysitting job paying \$5 an hour. You work for 4 hours and they give you a \$10 tip.

a. Write an expression for your total earnings.

b. Evaluate the expression to find the total earnings.

27. *Mowing lawns* You are mowing lawns, making \$10 an hour. You mow a lawn in 3 hours, but they must also pay you for \$2 worth of gas and oil.

a. Write an expression for your total earnings.

b. Evaluate the expression to find the total earnings.

Is there a difference between fat calories and carbohydrate calories? The answer is yes. Each gram of fat contains 9 calories, but each gram of carbohydrate has only 4. To find the percent of fat in a dish, use this procedure:

1. Multiply the number of grams of fat by 9.
2. Divide by the total calories and multiply by 100.
3. Round to the nearest whole number.

For example, salmon smothered in cream sauce has 1024 total calories and 76 grams of fat. The percent of fat calories in the salmon is:

1. $9 \cdot 76 = 684$
2. $684/1024 \cdot 100 = 66.797$
3. The nearest whole number is 67.

Thus, 67% of the calories in this dish come from fat.

28. *Fat calories* A quarter-pound cheeseburger, large fries, and a 16-ounce soda from McDonald's have about 1200 calories and 50 grams of fat. What percent of the calories come from fat?

29. *Fat calories* Three slices of cheese pizza and a 16-ounce diet soda from Domino's have about 510 calories and 15 grams of fat. What percent of the calories come from fat?

30. *Fat calories* A taco salad and 16-ounce soda from Taco Bell have about 1045 calories and 55 grams of fat. What percent of the calories come from fat?

31. *Fat calories* One piece of fried chicken (wing), mashed potatoes and gravy, coleslaw, and a 16-ounce diet soda from KFC contain about 380 calories and 19 grams of fat. What percent of the calories come from fat?

Follow the same three steps as for calculating the percent of fat calories in food, but in step 1 multiply by 4 instead of 9.

32. *Carbohydrate calories* A half-cup serving of vanilla ice cream has about 180 calories and 15 grams of carbohydrates. What percent of the calories come from carbohydrates?

33. *Carbohydrate calories* On the other hand, one-half cup of cooked carrots has about 36 calories and 8 grams of carbohydrates. What percent of the calories come from carbohydrates?

34. *Carbohydrate calories* An Arby's Bacon Cheddar Deluxe, regular fries, and a soft drink have about 1000 calories and 100 grams of carbohydrates. What percent of the calories come from carbohydrates?

35. *Carbohydrate calories* A Burger King Whopper with cheese, regular fries, and a medium drink have about 1380 calories and 150 grams of carbohydrates. What percent of the calories come from carbohydrates?

Did you find any proteins in Problems 28–35? The percent of protein calories in food can be found by following similar steps to those in Problems 28–35. As it turns out, a gram of protein contains the **same** amount of calories as one gram of carbohydrates: **4 calories.**

Calculating the percent of protein calories in food Follow the same three steps as for calculating the percent of carbohydrate calories in food to find the *percent* of protein calories in:

36. *Protein calories* A McDonald's hamburger (105 grams) with 280 calories and 12 grams of protein.

37. *Protein calories* A Carl's Jr. Hamburger (119 grams) with 284 calories and 14 grams of protein.

38. *Protein calories* A Sonic Jr. Burger (135 grams) with 353 calories and 14 grams of protein.

39. *Protein calories* A McDonald's McChicken (147 grams) with 430 calories and 14 grams of protein.

40. *Protein calories* A Jack in the Box Chicken Sandwich (145 grams) with 390 calories and 15 grams of protein.

Source: http://www.foodfacts.info/

Using Your Knowledge

What is the corresponding dose (amount) of medication for children when the adult dosage is known? There are several formulas that tell us.

41. **Friend's rule** (for children under 2 years):

(Age in months · adult dose) ÷ 150 = child's dose

Suppose a child is 10 months old and the adult dose is a 75-milligram tablet. What is the child's dose? [*Hint:* Simplify $(10 \cdot 75) \div 150$.]

42. **Clarke's rule** (for children over 2 years):

(Weight of child · adult dose) ÷ 150 = child's dose

If a 7-year-old child weighs 75 pounds and the adult dose is 4 tablets a day, what is the child's dose? [*Hint:* Simplify $(75 \cdot 4) \div 150$.]

43. **Young's rule** (for children between 3 and 12):

(Age · adult dose) ÷ (age + 2) = child's dose

Suppose a child is 6 years old and the adult dose of an antibiotic is 4 tablets every 12 hours. What is the child's dose? [*Hint:* Simplify $(6 \cdot 4) \div (6 + 2)$.]

You already know how to use the order of operations to evaluate expressions. Here is a more challenging game: Use the numbers 1, 2, 5, and 6, any of the operations we have studied, and parentheses to make an expression whose value is 24. Here is one: $(1 + 5) \cdot (6 - 2)$.

Now, you make your own expression using the given numbers.

44. 1, 2, 7, 2

45. 2, 3, 3, 8

46. 4, 6, 9, 5

47. 2, 3, 8, 9

Write On

48. A student claimed that the answer to $15 + 5 \times 10$ is 200. What did the student do wrong?

49. The order of operations is PEMDAS, which lists multiplication before division. Using multiplication before division, what would be the value of $4/2 \cdot 2 + 3$?

The order of operations also states that the operations must be performed from left to right. Evaluate $4/2 \cdot 2 + 3$ if we do the operations as they occur from left to right.

50. In Problem 49, which answer is correct, 4 or 7? How can we make the rule more precise?

Algebra Bridge

Evaluate the following expressions when $x = 3$, $y = 2$, and $z = 4$.

51. $(x + y)(x + z)$

52. $(x + z)(y + z)$

53. $x \cdot z \div y - z$

54. $x \cdot z \div y - x$

55. $8 \cdot (z \div y + x - z)$

56. $9 \cdot (z \div y + y - x)$

Concept Checker

Fill in the blank(s) with the correct word(s), phrase, or mathematical statement.

We can remember the order of operations when evaluating an expression by remembering **PEMDAS.**

57. **P** means to do the calculations inside the grouping symbols such as __________.

58. **E** means to evaluate all __________ expressions.

59. **M** means to do all __________ in order from left to right.

60. **D** means to do all __________ in order from left to right.

61. **A** means to do all __________ in order from left to right.

62. **S** means to do all __________ in order from left to right.

exponential	**multiplication**
division	**parentheses**
addition	**powers**
subtraction	

Mastery Test

63. Simplify: $2^3 \div 8 \cdot 2 + 4(6 - 1) - 2 \cdot 3$

64. Simplify: $15 \div 3 + \{3 \cdot 2^2 - [4 + (5 - 3)]\}$

65. Simplify: $81 \div 9 - (3 + 4)$

66. A travel agent charges \$50 for setting up a cruise for 10 people, plus \$500 per person. If you get a \$100 discount:

a. Write an expression for the total cost.

b. Evaluate the expression and find the total cost.

67. Simplify: $27 \div 3 \cdot 3 \cdot 3 + 4 - 1$

68. Simplify: $7 \cdot 3^2 - 5$

69. Simplify: $2^3 + 4 \cdot 5$

Skill Checker

70. Find the product $3^4 \cdot 2^2$.

71. Find the product $2^4 \cdot 3^2$.

72. Write the product $3^2 \cdot 10^2$ in words.

73. Write the product $2^3 \cdot 3^2 \cdot 10^2$ in words.

74. Write $3^0 \cdot 10^2 \cdot 2^2$ in expanded form.

1.9 Equations and Problem Solving

Objectives

You should be able to:

A Solve equations by using number facts.

B Solve equations using the addition, subtraction, or division principle given in the text.

C Solve applications using the concepts studied.

To Succeed, Review How To . . .

1. Use the arithmetic facts (+, −, ×, ÷). (pp. 24, 37, 49, 62)
2. Check subtraction using addition. (p. 37)
3. Write a division problem as an equivalent multiplication problem. (p. 63)

Getting Started

The ad says that you can save \$30 when you buy an electric cooktop. If the cooktop costs \$280 now, how much did it cost before? Since it costs \$280 now and you are saving \$30, it used to cost \$30 *more*, that is, \$280 + 30 = \$310. If you think of the old cost as c, you can write the problem like this:

The new cost is the old cost reduced by \$30:

$$\$280 \quad = \quad c \quad - \quad \$30$$

With what number can you replace c to make the statement true? The answer is again \$310.

Electric cooktop **Cut \$30**

A > Solving Equations

Sentences such as $280 = c - 30$ or $27 = 7 + x$ that contain the equals sign (=) are called **equations.**

SOLUTIONS

The **solution** of an equation is the replacement that makes the equation a *true* statement. When we find the solution of an equation we say that we have **solved the equation.**

Thus, the *solution* of	$280 = c - 30$ is **310**
because	$280 = \mathbf{310} - 30$
Similarly, the solution of	$27 = 7 + x$ is **20**
because	$27 = 7 + \mathbf{20}$

EXAMPLE 1 Finding solutions to equations by substitution

Find the solution of each equation.

a. $x + 7 = 13$ **b.** $10 - x = 3$
c. $15 = 5x$ **d.** $24 \div x = 6$

SOLUTION

a. We have to find a number x so that when we add 7 to it, we get 13. You can do this mentally or by substitution.

If we replace x with 5, we get $5 + 7 = 13$. False!
If we replace x with 6, we get $6 + 7 = 13$. True.
Thus, the solution of $x + 7 = 13$ is $x = 6$.*

b. We need to find a number x so that when we subtract it from 10, the answer is 3. You can do this mentally or by substitution.

If we replace x with 5, we get $10 - 5 = 3$. False!
Since $10 - 5 = 5$, we need a larger number for x.
If we replace x with 6, we get $10 - 6 = 3$. False!
We need a bigger number. Try 7.
If we replace x with 7, we get $10 - 7 = 3$. True.
Thus, the solution of $10 - x = 3$ is $\mathbf{x = 7}$.

c. To solve $15 = 5x$, we need a number that multiplied by 5 would give us 15. You can do this mentally. The number is 3. Thus, the solution of $15 = 5x$ is $\mathbf{x = 3}$.

d. Here we need a number x so 24 divided by this number would yield 6. You can do this mentally or by substitution. We can try **3.** But $24 \div \mathbf{3} = 8$ (not 6). If we try **4,** we get $24 \div \mathbf{4} = 6$, which is the desired result. The solution is $\mathbf{x = 4}$.

PROBLEM 1

Find the solution of each equation.

a. $x + 6 = 15$ **b.** $13 - x = 4$
c. $24 = 8x$ **d.** $36 \div x = 9$

B > Rules for Solving Equations

So far, we have solved our equations by trial and error. We need some rules to follow. These rules are based on the idea of equivalent equations, equations that have the same solution.

EQUIVALENCE

Two equations are **equivalent** if their solutions are the same.

Answers to PROBLEMS

1. a. $x = 9$ **b.** $x = 9$
c. $x = 3$ **d.** $x = 4$

* Technically, the solution is 6, but some people write $x = 6$ instead.

For example, $x + 7 = 10$ is equivalent to $x = 3$, but $x = 3$ has an obvious solution. Let us go back to the equation at the beginning of this section.

$$\underbrace{\text{The new cost}}_{\$280}\ \underbrace{\text{is}}_{=}\ \text{the}\ \underbrace{\text{old cost}}_{c}\ \underbrace{\text{reduced by \$30}}_{-\ \$30}$$

To find the cost c we restore the \$30 cut; that is,

$$\$280 + 30 = c - 30 + 30$$

or

$$310 = c$$

This example illustrates the fact that we can *add* the same number on both sides of the equation and produce an *equivalent* equation. Here is the idea:

ADDITION PRINCIPLE

The equation $a = b$ is equivalent to $a + c = b + c$.

Thus, we can add the same number c to both sides of an equation and obtain an equivalent equation.

EXAMPLE 2 Using the addition principle

Solve.

a. $n - 17 = 20$ **b.** $30 = m - 18$

SOLUTION

a. The idea is to have n by itself on one side of the equation. Thus, we want to "restore" the 17. We do this by adding 17 to both sides.

We get:

$$\begin{aligned} n - 17 &= 20 \\ n - 17 + 17 &= 20 + 17 && \text{Add 17 on both sides.} \\ n &= 20 + 17 && n + 0 = n \\ n &= 37 && \text{Add.} \end{aligned}$$

The solution is $\boldsymbol{n = 37}$. (*Check:* Substitute 37 in the original equation to get $\mathbf{37} - 17 = 20$, a true statement.)

b. This time we add **18** to both sides, obtaining

$$\begin{aligned} 30 + 18 &= m - 18 + 18 && \text{Add 18.} \\ 48 &= m && m + 0 = m \end{aligned}$$

The solution is $48 = m$. (*Check:* $30 = 48 - 18$.)

PROBLEM 2

Solve.

a. $n - 13 = 17$ **b.** $20 = m - 3$

Now, suppose the cost of the cooktop has gone *up* \$30 and the new cost is \$280. What was the old cost c? The equation here is

$$c + 30 = 280$$

Old cost (c), up \$30 (30), new cost (280)

This time, we bring the price *down* by *subtracting* \$30 from both sides of the equation. We get

$$\begin{aligned} c + 30 &= 280 \\ c + 30 - 30 &= 280 - 30 \\ c &= 250 \end{aligned}$$

Thus, the old cost was \$250. We have used the following principle.

SUBTRACTION PRINCIPLE

The equation $a = b$ is equivalent to $a - c = b - c$.

Answers to PROBLEMS

2. a. $n = 30$ **b.** $m = 23$

EXAMPLE 3 **Using the subtraction principle**

Solve.

a. $n + 15 = 48$ **b.** $43 = 18 + m$

SOLUTION

a.

$$n + 15 = 48$$
$$n + 15 - 15 = 48 - 15 \quad \text{Subtract 15 from both sides.}$$
$$n = 48 - 15$$
$$n = 33 \quad 48 - 15 = 33$$

The solution is $\boldsymbol{n = 33}$. (You can check this in the original equation, since $\mathbf{33} + 15 = 48$.)

b.

$$43 = 18 + m$$
$$43 - 18 = 18 + m - 18 \quad \text{Subtract 18 from both sides.}$$
$$43 - 18 = m \quad 18 + m - 18 = m$$
$$25 = m \quad 43 - 18 = 25$$

The solution is $\boldsymbol{m = 25}$. (*Check:* $43 = 18 + \mathbf{25}$.)

PROBLEM 3

Solve.

a. $n + 10 = 13$

b. $39 = 18 + m$

Now, suppose the cost of the cooktop has *doubled;* it now costs \$280. What was the old cost c? Since the old cost c has doubled and it is now \$280, the equation is $2c = 280$. To find the answer, we must cut the cost in *half* by dividing by 2. Thus, the old cost c was \$140. This example suggests that we can *divide* both sides of an equation by a (*nonzero*) number and obtain an equivalent equation. Here is the principle.

DIVISION PRINCIPLE

The equation $a = b$ is equivalent to $a \div c = b \div c$ ($c \neq 0$).

EXAMPLE 4 **Using the division principle**

Solve.

a. $3x = 33$ **b.** $48 = 6x$

SOLUTION

a. We use the division principle and divide on both sides by **3,** obtaining:

$$3x = 33$$
$$3x \div 3 = 33 \div 3 \quad \text{Divide both sides by 3.}$$
$$x = 33 \div 3 \quad \text{Divide } 3x \text{ by 3.}$$
$$x = 11 \quad \text{Divide 33 by 3.}$$

b. We divide both sides by **6,** obtaining:

$$48 = 6x$$
$$48 \div 6 = 6x \div 6 \quad \text{Divide both sides by 6.}$$
$$48 \div 6 = x \quad \text{Divide } 6x \text{ by 6.}$$
$$8 = x \quad \text{Divide 48 by 6.}$$

PROBLEM 4

Solve.

a. $4x = 36$

b. $42 = 7x$

C > Applications: The RSTUV Method

Now that you've learned how to solve equations, you need to know how to apply this knowledge to solve real-world problems. These problems are usually stated in words and consequently are called **word** or **story** problems. This is an area in which many students encounter difficulties, but don't panic; we are about to give you a sure-fire method of tackling word problems. To start, let us look at this problem.

Answers to PROBLEMS

3. a. $n = 3$ **b.** $m = 21$

4. a. $x = 9$ **b.** $x = 6$

The Sears Tower is 1454 feet tall. The addition of two antennas brought the height to 1559 feet. How tall are the antennas?

Before we attempt to solve this problem, we shall discuss an effective way to solve such a problem. The procedure is as easy as 1-2-3-4-5.

RSTUV PROCEDURE TO SOLVE WORD PROBLEMS

1. Read the problem carefully and decide what is asked for (the unknown).
2. Select □ or a letter to represent the unknown.
3. Translate the problem into an equation.
4. Use the rules we have studied to solve the equation.
5. Verify the answer.

This is the method you really want to learn to master word problems.

How do we remember all of these steps? Easy. Look at the first letter in each sentence (in bold). Do you see now why we call this the RSTUV method? To help you even more, here are some hints and tips for using the method.

1. Read the problem. Mathematics is a language. You have to learn how to read it. You may not understand or even get through reading the problem the first time. That's OK. Read it again, and as you do, pay attention to key words or instructions such as *compute, draw, write, construct, make, show, identify, state, simplify, solve,* and *graph*. (Can you think of others?)

2. Select the unknown. How can you answer a question if you don't know what the question is? One good way to look for the unknown (variable) is to look for the question mark and read the material to its left. Try to determine what is given and what is missing. Remember, this is your word problem, so you can use any letter you wish for the unknown. Here is a suggestion: Use h for height, s for speed, p for population, d for distance, and so on.

3. Translate the problem into an equation or inequality. Problem solving requires many skills and strategies. Some of them are *look for a pattern; examine a related problem; use a formula; make tables, pictures, or diagrams; write an equation; work backward;* and *make a guess*. When you solve problems, your plan should lead to writing an equation or an inequality.

4. Use the rules we have studied to solve the equation. If you are studying a mathematical technique, it's almost certain that you will have to use it in solving the given problem. Look for ways the technique you're studying (addition, subtraction, and division principles, for example) could be used to solve the problem.

5. Verify the answer. Look back and check the results of the original problem. Is the answer reasonable? Can you find it some other way?

Now, let us use the RSTUV procedure to solve Example 5.

EXAMPLE 5 Problem solving: RSTUV method

The Sears Tower, in Chicago, is 1454 feet tall. The addition of two antennas brought the height to 1559 feet. How tall are the antennas?

SOLUTION

1. Read the problem. Read the problem slowly, not once but two or three times. (Reading mathematics is not like reading a magazine; mathematics problems may have to be read several times before you understand them. It is OK, as long as you do understand them.)

PROBLEM 5

Suppose the Sears Tower is 1559 feet tall. The addition of an antenna brought the height to 1710 feet. How tall is the antenna?

Answers to PROBLEMS

5. 151 feet

2. Select the unknown. The problem asks for the height of the antennas. What would you call this height? We will use h for height. h is what we are looking for and is our unknown.

3. Translate the problem into an equation or inequality. The problem says that the Sears Tower is 1454 feet high. The *addition* of two antennas brought the height to 1559 feet. Here is the translation:

Sears Tower height + two antennas height reached 1559

$$1454 \quad + \quad h \quad = \quad 1559$$

4. Use the rules we have studied to solve the equation. To solve this equation, we use the subtraction principle and subtract 1454 from both sides of the equation (so that we can have h by itself on the left side).

$$1454 - 1454 + h = 1559 - 1454$$

Simplify $\quad h = 105$

Thus, the height h of the antennas is 105 feet.

5. Verify the answer. To verify the answer, recall that: The addition of the antennas (105) brought the height to 1559. Is it true that the original height (1454) plus 105 is 1559? Yes, $1454 + 105 = 1559$, so our answer is correct! Congratulations, you solved your first word problem using the RSTUV method! Now, use the same procedure and do Problem 5.

EXAMPLE 6 Problem solving: tuition and fees

In a recent year the charges for tuition and fees T plus room and board R, for one year, at a public 4-year college amounted to \$12,127. The room and board R cost \$1145 more than the tuition and fees T. What are the costs of tuition and fees T and the costs of room and board R?

SOLUTION

1. Read the problem. This time there are **two** unknowns.

2. Select the unknowns. The two unknowns are: tuition and fees T and room and board R.

3. Translate the problem into an equation or inequality. There are two sentences to translate:

"tuition and fees T plus room and board R amounted to \$12,127"

This means that: (1) $T + R = \$12,127$

"The room and board R cost \$1145 more than tuition and fees T"

This means that: (2) $R = T + \$1145$

Since we need one unknown only, substitute $T + 1145$ for R in (1)

Like this: (1) $T + R = \$12,127$

Substituting $\quad T + (T + \$1145) = \$12,127$

4. Use the rules we have studied to solve the equation.

To solve $\quad T + (T + \$1145) = \$12,127$

Add like terms ($T + T = 2T$) $\quad 2T + \$1145 = \$12,127$

Subtract **\$1145** to isolate T $\quad 2T + \$1145 - \mathbf{\$1145} = \$12,127 - \mathbf{\$1145}$

$\quad 2T = \$10,982$

Divide both sides by 2 $\quad T = \$5491$

This means: (2) $R = \$5491 + \$1145 = \$6636$

5. Verify the answer. Is it true that tuition and fees (\$5491) plus room and board (\$6636) amount to \$12,127? Yes, $\$5491 + \$6636 = \$12,127$. Our answer is correct!

Source: http://www.ed.gov/about/bdscomm/list/hiedfuture/2ndmeeting/trends.pdf

PROBLEM 6

College charges for tuition and fees T plus room and board R, for one year, at a private 4-year college amounted to \$29,026. The tuition and fees T cost \$13,444 more than the room and board R. Use the RSTUV procedure to find the cost of tuition and fees T and the cost of room and board R.

Answers to PROBLEMS

6. $R = \$7791$, $T = \$21,235$

TRANSLATE THIS

The third step in the RSTUV procedure is to **TRANSLATE** the information into an equation. In Problems 1–10 **TRANSLATE** the sentence and match the correct translation with one of the equations A–O.

1. Of the 995 candy products introduced in a recent year, 398 were chocolate. Write an equation for N, the number of candies that were *not* chocolate.

   Source: National Confectioners Assoc.

2. Who spends the most on women's clothes? According to MapInfo, Connecticut women (C) do. As a matter of fact, they spend \$27 dollars more per household annually than women (N) in New Jersey. Write an equation for C.
3. If you are under 25 years old, your probability P of completing college is 69 percent, an increase of 12% over the probability 10 years ago.
4. You can buy a computer with a \$100 discount. If the original price of the computer was P, what is the discounted price D?
5. Do women want to be thin or smarter? In a recent survey conducted by eDiets.com the percent of women T who wanted to be thinner exceeded the percent of women S who wanted to be smarter by 12%. Write an equation for T.
6. When jogging you will burn 675 calories each hour. What is the number of calories C burned in h hours of jogging?
7. How many calories C do you need in a 24-hour day? It depends on your weight W in pounds and your daily activities.

   Here are the steps:

   1. Divide W by 2.2.
   2. Multiply by 24.

   These are the calories you need for basic function.

   3. If your daily activities are moderate, multiply the result in **2** by 0.70.
   4. Add the results of steps 2 and 3. This is C.
8. The basal metabolic rate (BMR) for an adult male is obtained by multiplying his body weight W by 10 and adding double his body weight to this value. What is the formula for the BMR of an adult male?
9. The BMR for an adult female is obtained by multiplying her body weight W by 10 and adding her body weight to this value. What is the formula for the BMR of an adult female?
10. The body mass index (BMI) is a reliable indicator of body fatness for most people. BMI is obtained by multiplying your weight P by 705 and dividing by the square of your height H.

    Note: If your BMI is between 18.5 and 24.9 you are normal!

A. $995 = N - 398$

B. $69 = P + 12$

C. $T = S + 12$

D. $C = \frac{24W}{2.2} + \frac{0.70 \cdot 24W}{2.2}$

E. $S = T + 12$

F. $D = P - 100$

G. $\text{BMR} = 2W + W$

H. $\text{BMI} = \frac{P}{705H^2}$

I. $995 - 398 = N$

J. $\text{BMR} = \frac{705P}{H^2}$

K. $C = 675h$

L. $\$100 = D - P$

M. $\text{BMR} = 10W + W$

N. $C = N + 27$

O. $\text{BMR} = 10W + 2W$

Exercises 1.9

Boost your grade at mathzone.com!
- Practice Problems
- NetTutor
- Self-Tests
- e-Professors
- Videos

⟨A⟩ Solving Equations

In Problems 1–12, find the solution of the equation, mentally or by substitution.

1. $m + 9 = 17$
2. $n + 18 = 29$
3. $13 - x = 9$
4. $19 - x = 8$
5. $20 = 4x$
6. $27 = 3x$
7. $9x = 54$
8. $8x = 88$
9. $30 \div y = 6$
10. $45 \div y = 5$
11. $9 = 63 \div t$
12. $7 = 35 \div t$

⟨B⟩ Rules for Solving Equations

In Problems 13–24, solve.

13. $z - 18 = 30$
14. $z - 13 = 41$
15. $40 = p - 12$
16. $30 = p - 9$
17. $x + 17 = 31$
18. $x + 12 = 37$
19. $30 = 17 + m$
20. $21 = 18 + m$
21. $4x = 28$
22. $7x = 49$
23. $9x = 36$
24. $11x = 121$

‹ C › Applications Using The RSTUV Method

In Problems 25–50, use the RSTUV procedure to solve the problem.

25. *Finding the speed of Thrust 2* The fastest car is the jet-engined Thrust SSC, which can travel 763 miles per hour. This is 130 miles per hour more than the Thrust 2. If s is the speed of the Thrust 2, find s.

26. *Population predictions* It is predicted that by the year 2010 the world population will be 6823 million, 743 million more than in the year 2000. If p was the world population in the year 2000, find p.

27. *Human-powered flight record* Glenn Tremml holds the record for human-powered flight. The previous record was held by Bryan Allen, who flew 22 miles across the English channel. If d is the distance flown by Glenn, and Bryan flew 15 miles less than that, find d.

28. *Calories needed for weight maintenance* The number of calories you must eat to maintain your weight w (in pounds) is given by the formula:

$$15w = \text{number of calories}$$

A man is eating 2700 calories a day and maintaining his weight. What is his weight?

29. *Weight of a brontosaurus* A brontosaurus, a prehistoric animal, weighed about 60,000 pounds. This is four times as much as an average African elephant. If w is the weight of an African elephant, find w.

30. *English Smiths* There are three times as many persons named Smith in the United States as there are in England. If the United States has 2,400,000 people named Smith, how many people named Smith are there in England?

31. *Calories in Coke* A McDonald's Quarter Pounder has 420 calories. If you have a small Coca-Cola with that, the total number of calories reaches 570. How many calories are in the Coke?

32. *Sodium in Coke* A Quarter Pounder contains 580 mg (milligrams) of sodium. If you have a small Coca-Cola with that, the total number of milligrams of sodium reaches 620. How many milligrams of sodium are in the Coke?

33. *Coke calories* A Whopper sandwich contains 640 calories. If you have a medium Coca-Cola Classic with that, the total number of calories reaches 920. How many calories are in the Coke?

34. *Coke sodium* A Whopper sandwich contains 870 mg (milligrams) of sodium. If you have a Coca-Cola Classic with that, the total number of milligrams of sodium reaches 920. How many milligrams of sodium are in the Coke?

35. *Individual calories* A McDonald's cheeseburger and small fries contain 540 calories. The cheeseburger has 120 more calories than the fries. How many calories are in each food item?

36. *Individual calories* A Whopper sandwich and medium fries contain 940 calories. The Whopper has 340 more calories than the fries. How many calories are in each food item?

Source: Olen Publishing.

37. *Tuition cost* At 2-year colleges, books and supplies cost about \$700. If you add to that tuition costs, the total is \$2272. What is the cost of tuition?

Source: Data from The Chronicle of Higher Education.

38. *Tuition, board, and book costs* If you are a student in a private four-year college, you should expect to pay about \$16,080 a year for tuition and board plus books. If the amount for tuition and board is \$14,680 more than the books, what is the cost of tuition and board and what is the cost of books?

Source: Data from The Chronicle of Higher Education.

39. *Scholarship and grants awards* The combined average financial aid award for scholarship and grants is about \$3600. If the scholarship money is \$400 less than the grant money, what are the average awards for scholarships and for grants?

Source: Data from Office of Student Financial Services.

40. *Loans and grants awards* The combined average financial aid award for loans and grants is about \$5350. If the grant money is \$1150 less than the loan money, what are the average awards for loans and for grants?

Source: Data from Office of Student Financial Services.

41. *Banking* Tran deposited his \$1000 summer earnings at a bank. He also had 5 direct deposit checks for p dollars each deposited in the account.

Write an expression for his balance.

His bank statement indicated a \$2115 balance. What was the amount p of each of the direct deposit checks?

42. *Banking* Tran had his monthly car payment of c dollars each deducted from his account. After three monthly payments, his \$2115 balance was down to \$1212. How much was his monthly car payment c?

Web IT go to mhhe.com/bello for more lessons

43. *Banking* Tran wrote four checks for \$50, \$120, \$70, and \$65. If his new balance n was \$907, what was his old (starting) balance?

44. *Transportation* Andrea needed some transportation to go to school. She decided to buy a PT Cruiser and pay for it in 5 years (60 monthly payments) at 0% interest. If her total payments amounted to \$24,000, how much was her monthly payment?

45. *Transportation* Andrea also needed some insurance for her Cruiser. Allstate told her that they would sell her a policy with the coverage she wanted and she could pay in three installments of t dollars each. If the price of the policy was \$1350, what was the amount t of each installment?

46. *Transportation* Andrea was told that the Cruiser would get about 22 miles per gallon (mpg). She wanted to travel from Tampa to Miami, a distance of about 264 miles. How many gallons of gas would Andrea need?

47. *Transportation* Andrea is driving from Tampa to Miami and will pass through Yeehaw Junction on the way. If the distance from Tampa to Yeehaw Junction is 106 miles and from Tampa to Miami is 264 miles, how many miles is it from Yeehaw Junction to Miami?

48. *Transportation* A Mapquest map told Andrea the 264-mile trip would take about 4 hours. What was Andrea's average speed? (*Hint:* Distance = Average Speed × Time.)

49. *Nutrition* Did you know that you must burn about 3500 calories to lose one pound of body fat?

a. How many calories does it take to lose 15 pounds?

b. Suppose you need 1800 calories to maintain your daily weight and you cut your daily calories to 1300. What is the number of excess calories burned each day?

c. How many days would it take you to lose 15 pounds?

50. *Nutrition* You can burn calories by exercising. Here are the calories burned in one minute of the listed activity.

Type of Exercise	Calories Burned in One Minute
Walking (5 mi/hr)	3
Cycling (12 mi/hr)	10
Weightlifting	12

To burn 3500 calories (the equivalent of one pound loss), how long do you have to (answer to the nearest minute):

a. Walk at 5 mi/hr?

b. Ride your bicycle at 12 mi/hr?

c. Lift weights?

In Problems 51 and 52, follow the procedure of Example 6 and solve.

51. *Tips* The tip T plus the cost M of a meal at a restaurant amounted to \$96. If the cost M of the meal was \$64 more than the tip T, how much was the tip and how much was the meal?

52. *Tuition and fees* The tuition and fees T for courses at a college, for one semester, was \$290 more than the cost B of the books. If the total bill (tuition and fees plus books) came to \$1150, how much was the tuition and fees and how much were the books?

››› Using Your Knowledge

53. The formula for the distance d at which a car travels in time t hours at 50 miles an hour is

$$d = 50t$$

How long would it take for the car to travel 300 miles?

54. The formula for the velocity V (in feet per second) that an object travels after falling for t seconds is given by

$$V = 32t$$

How long would it take an object to reach a velocity of 96 feet per second?

Write On

55. We have an addition, a subtraction, and a division principle. Write in your own words the reasons why we have no multiplication principle.

56. Write in your own words what the multiplication principle should be.

57. The solution of an equation is the replacement that makes the equation a true statement. What is the solution of $x + 1 = 2$? Does your answer satisfy the definition of a solution?

58. What is the solution of $5 + x = 5$? What about the solution of $5 + x = 5 + x$?

Concept Checker

Fill in the blank(s) with the correct word(s), phrase, or mathematical statement.

59. The ______________ of an equation is the **replacement** that makes the equation a true statement.

60. If the solutions of two different equations are the **same,** the equations are ______________.

61. Using the ______________ Principle, the equation $a = b$ is **equivalent to** $\boldsymbol{a + c = b + c}$.

62. The equation $\boldsymbol{a = b}$ is equivalent to the equation ______________ using the Subtraction Principle.

63. By the **Division Principle,** the equation $\boldsymbol{a = b}$ is **equivalent** to the equation ______________.

64. In the **RSTUV** procedure, the **T** stands for ______________.

$a - c = b - c$	**equivalent**
Addition	**solution**
$a \div c = b \div c;\ c \neq 0$	**same**
translate	**equal**

Mastery Test

Solve:

65. $x + 8 = 17$

66. $12 - y = 7$

67. $20 = 4x$

68. $28 \div x = 7$

69. $40 = 38 + m$

70. $49 = 7p$

71. $x - 5 = 10$

72. $10 = n - 19$

73. $20 = x + 5$

74. The CN Tower in Canada is said to be 362 feet taller than the Sears Tower I, which is 1454 feet high. What is the height of the CN Tower?

Note: Many people do not recognize the CN Tower as the world's tallest building because, so they say, the CN Tower is not a building. Most of the structure is no more than a concrete shaft housing elevators, and therefore it is not a building, one could argue.

Skill Checker

75. Use long division to divide $\frac{47}{5}$.

76. $0 \div 9 =$ ______________

Collaborative Learning

This section will involve three groups of students: Carpeters, Gardeners, and Painters. At least one of the students in each group should have access to a telephone directory or the Internet and be willing to make some phone calls or get information.

Here are the dimensions of the rooms we will work with:

		Area
Family Room:	**13′ 0″ × 15′ 0″**	
Living Room:	**14′ 0″ × 15′ 0″**	
Dining Room:	**10′ 0″ × 14′ 0″**	

Carpeters

In charge of estimating the cost of carpet for the three rooms: family room, living room, and dining room.

1. Find the area of each room.
2. Find the total area for the three rooms.
3. Find the cost of a square foot of carpeting (you may have to call several carpet places or find the prices online).
4. Find the cost of padding (the cushion under the carpet).
5. Find the cost of all materials (carpet plus padding).
6. Give the total area to a retailer or dealer and request an estimate for installing the carpet and padding. How much do you save if you install the carpet yourself?

Note: When estimating area you can check your work by searching the Web for a "carpet calculator." It will calculate the area in square feet and square yards.

Painters

In charge of estimating the cost of the paint for the three rooms. *Hint:* Each room has four walls (ignore doors, windows, and ceilings), and we assume that the ceiling (and consequently each wall) is 8 feet tall.

1. Find the area of the four walls in each room.
2. Find the area of all walls in the three rooms.
3. Find the cost of a gallon of wall paint (call several paint stores or find the prices online).
4. Find how many gallons of paint you need. The industry standard for coverage is about 400 square feet per gallon, that is, you can paint 400 square feet of wall with one gallon of paint. *Note:* You can check with a paint calculator by doing a Web search.
5. Find the cost of the paint.

Of course, you need brushes, cleaning materials, and paper or a tarp to cover the floor. For more information on measuring rooms and paint coverage, try a Web search.

Gardeners

We shall not neglect the outside of the house, so we shall install sod in our lawn.

Sod is usually sold in rectangles measuring 5 square feet (15 inches by 48 inches) or by pallets bringing 500 square feet per pallet. Dimensions and pallet sizes vary!

Assume our lawn is 40 feet by 50 feet.

1. Find the area of the lawn.
2. How many pallets of sod do we need?

3. What is the cost of each pallet (delivered)? Find the price for the sod by calling sod farms or searching online.
4. After you find the area of the lawn, call a sod farm and ask for the installed price of sod (per square foot).
5. Find the installed price of the sod for our lawn.
6. Compare the installed price and the do-it-yourself price. How much do you save if you install the sod yourself?

To find more information about sod, try a Web search for "sod installation."

Research Questions

1. There is a charming story about the long-accumulated used wooden tally sticks mentioned in *The Human Side of Mathematics* at the beginning of this chapter. Find out how their disposal literally resulted in the destruction of the old Houses of Parliament in England.
2. Write a paper detailing the Egyptian number system and the base and symbols used, and enumerate the similarities and differences between the Egyptian and our (Hindu-Arabic) system of numeration.
3. Write a paper detailing the Greek number system and the base and symbols used, and enumerate the similarities and differences between the Greek and our system of numeration.
4. Find out about the development of the symbols we use in our present numeration system. Where was the symbol for zero invented and by whom?
5. Write a short paragraph discussing the development of the grouping symbols we have studied in this chapter.

Summary Chapter 1

Section	Item	Meaning	Example
1.1	Natural (Counting) numbers	1, 2, 3, and so on	19 and 23 are counting numbers.
	Digits	0, 1, 2, 3, 4, 5, 6, 7, 8, 9	
	Whole numbers	0, 1, 2, 3, and so on	0 and 17 are whole numbers.
1.1B	Expanded form	Numeral written with the ones, tens, hundreds, and so on displayed	$278 = 200 + 70 + 8$
1.2A	$<$	$a < b$ means that a is to the *left* of b on a number line.	$5 < 8$ and $0 < 9$

(continued)

Section	Item	Meaning	Example
1.2A	$>$	$a > b$ means that a is to the *right* of b on a number line.	$8 > 5$ and $9 > 0$
	Inequalities	Sentences using $<$ or $>$	$3 < 5$ and $8 > 2$
1.2B	Rounding	Approximating a given number to a specified number of digits	637 rounded to the nearest ten is 640; 637 rounded to the nearest hundred is 600.
1.3A	Addends	The numbers to be added	In the sum $6 + 3 = 9$, 6 and 3 are the addends.
	Sum or total	The result of an addition	The sum of 3 and 6 is 9.
	Identity for Addition	0 is the identity for addition.	$0 + 5 = 5, 9 + 0 = 9$.
	Commutative Property of Addition	$a + b = b + a$	$5 + 4 = 4 + 5$
	Associative Property of Addition	$(a + b) + c = a + (b + c)$	$(1 + 2) + 3 = 1 + (2 + 3)$
1.3B	Polygon	A flat geometric region with many sides	
	Perimeter	The distance around an object In the case of a polygon, the perimeter is the sum of the length of all sides.	The perimeter of the rectangle 3 ft 2 ft is $(3 + 2 + 3 + 2)$ ft = 10 ft.
1.4A	Minuend	In $a - b = c$, a is the minuend.	In $5 - 3 = 2$, 5 is the minuend.
	Subtrahend	In $a - b = c$, b is the subtrahend.	In $5 - 3 = 2$, 3 is the subtrahend.
	Difference	In $a - b = c$, c is the difference.	In $5 - 3 = 2$, 2 is the difference.
1.5A	Multiplicand	In $a \times b = c$, a is the multiplicand.	In $3 \times 5 = 15$, 3 is the multiplicand.
	Multiplier	In $a \times b = c$, b is the multiplier.	In $3 \times 5 = 15$, 5 is the multiplier.
	Product	In $a \times b = c$, c is the product.	In $3 \times 5 = 15$, 15 is the product.
	Factors	In $a \times b = c$, a and b are factors.	In $3 \times 5 = 15$, 3 and 5 are factors.
1.5A	Multiplication Property of zero	$0 \times a = 0 = a \times 0$	$0 \times 3 = 0$ and $5 \times 0 = 0$
	Commutative Property of Multiplication	$a \times b = b \times \mathrm{a}$	$4 \times 7 = 7 \times 4$
	Identity for Multiplication	$1 \times a = a \times 1 = a$	$1 \times 4 = 4 \times 1 = 4$
	Distributive Property	$a \times (b + c) = (a \times b) + (a \times c)$	$3 \times (20 + 3) = (3 \times 20) + (3 \times 3)$
1.5B	Multiples of 10	10, 100, 1000, etc.	
1.5C	Associative Property of Multiplication	$a \times (b \times c) = (a \times b) \times c$	$4 \times (2 \times 3) = (4 \times 2) \times 3$
1.5D	Area of rectangle	Multiply the length L by the width W.	The area of a rectangle 10 in. by 6 in. is 60 in.^{2}

Section	Item	Meaning	Example
1.6A	Dividend	In $a \div b = c$, a is the dividend.	In $15 \div 3 = 5$, 15 is the dividend.
	Divisor	In $a \div b = c$, b is the divisor.	In $15 \div 3 = 5$, 3 is the divisor.
	Quotient	In $a \div b = c$, c is the quotient.	In $15 \div 3 = 5$, 5 is the quotient.
	Division Properties of One	1. $a \div a = 1$ $(a \neq 0)$ 2. $a \div 1 = a$	$5 \div 5 = 1$ and $9 \div 9 = 1$ $7 \div 1 = 7$ and $10 \div 1 = 10$
	Division Properties of Zero	1. $0 \div a = 0$ $(a \neq 0)$ 2. $a \div 0$ is undefined.	$0 \div 17 = 0$ and $0 \div 3 = 0$ $7 \div 0$ is not defined.
1.7A	Prime number	A counting number having exactly two different factors, itself and 1	17 and 41 are prime.
	Composite number	A counting number greater than 1 that is not prime	22 and 48 are composite.
1.7B	Prime factors	The factors of a number that are prime numbers	The prime factors of 22 are 2 and 11.
1.7C	Base	In the expression b^n, b is the base.	In the expression 2^3, 2 is the base.
	Exponent	In the expression b^n, n is the exponent.	In the expression 2^3, 3 is the exponent.
1.7D	1 as an exponent	$a^1 = a$	$9^1 = 9$ and $3^1 = 3$
	0 as an exponent	$a^0 = 1$ $(a \neq 0)$	$9^0 = 1$ and $3^0 = 1$
1.8A	Order of operations	PEMDAS (Parentheses, Exponents, Multiplication, Division, Addition, and Subtraction)	$36 \div 3 \times 6 - (3 + 4) + 5$ $= 36 \div 3 \times 6 - 7 + 5$ $= 12 \times 6 - 7 + 5$ $= 72 - 7 + 5$ $= 65 + 5$ $= 70$
1.9A	Equation	A sentence using an = sign	$10 = 2x$ and $x + 2 = 5$ are equations.
	Solution	The solution of an equation is (are) the numbers that makes the equation a true statement.	The solution of $10 = 2x$ is 5.
1.9B	Equivalent equations	Two equations are equivalent if their solutions are the same.	$2x + 1 = 3$ and $2x = 2$ are equivalent equations.
	The Addition Principle	The equation $a = b$ is equivalent to $a + c = b + c$.	The equation $x - 3 = 5$ is equivalent to $x - 3 + 3 = 5 + 3$.
	The Subtraction Principle	The equation $a = b$ is equivalent to $a - c = b - c$.	The equation $x + 3 = 5$ is equivalent to $x + 3 - 3 = 5 - 3$.
	The Division Principle	The equation $a = b$ is equivalent to $a \div c = b \div c$ $(c \neq 0)$.	The equation $2x = 6$ is equivalent to $2x \div 2 = 6 \div 2$.
1.9C	RSTUV procedure	**R**ead the problem. **S**elect the unknown. **T**ranslate the problem. **U**se the rules studied to solve the problem. **V**erify the answer.	

› Review Exercises Chapter 1

If you need help with these exercises, look in the section indicated in brackets.

1. ‹ **1.1A, B** › *Write in expanded form and find the value of the underlined digit.*

a. $12\underline{7}$

b. $1\underline{8}9$

c. $\underline{3}80$

d. $\underline{1}490$

e. $\underline{2}559$

2. ‹ **1.1C** › *Write in standard form.*

a. 40 + 9

b. 500 + 80 + 6

c. 500 + 3

d. 800 + 10

e. 1000 + 4

3. ‹ **1.1D** › *Write the word name for these numbers.*

a. 79

b. 143

c. 1249

d. 5659

e. 12,347

4. ‹ **1.1E** › *Write in standard form.*

a. Twenty-six

b. One hundred ninety-two

c. Four hundred sixty-eight

d. One thousand, six hundred forty-four

e. Forty-two thousand, eight hundred one

5. ‹ **1.2A** › *Fill in the blank with < or > to make a true inequality.*

a. 27 _____ 29 **b.** 30 _____ 28

c. 23 _____ 25 **d.** 19 _____ 39

e. 39 _____ 19

6. ‹ **1.2B** › *Round to the nearest hundred.*

a. 2848 **b.** 9746

c. 3550 **d.** 4444

e. 5555

7. ‹ **1.2C** › *Round the price of the car to the nearest hundred dollars.*

a. $21,090 **b.** $27,270

c. $35,540 **d.** $26,460

e. $22,990

8. ‹ **1.3A** › *Add.*

a. 3402 + 8576 **b.** 2098 + 2383

c. 3099 + 6547 **d.** 4563 + 8603

e. 3480 + 9769

9. ‹ **1.3B** › *Find the perimeter of the triangle.*

a. 3 ft, 3 ft, 1 ft

b. 4 ft, 4 ft, 3 ft

c.

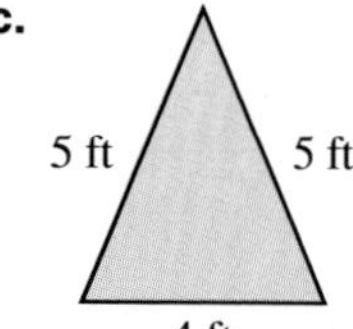

d.

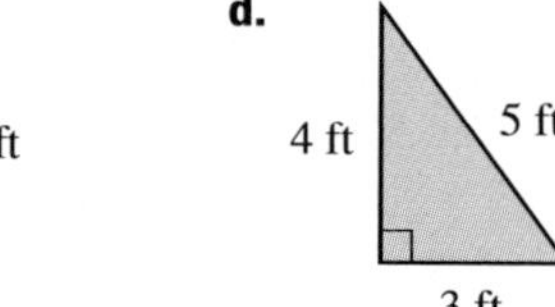

e.

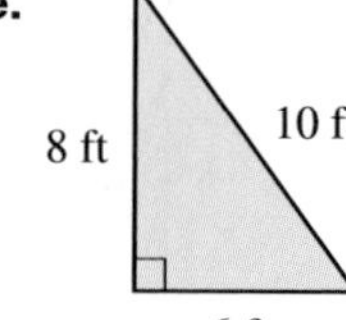

10. ⟨ **1.4A** ⟩ *Subtract.*

a. 47 − 18

b. 36 − 19

c. 55 − 26

d. 46 − 37

e. 93 − 44

11. ⟨ **1.4A** ⟩ *Subtract.*

a. 654 − 467

b. 547 − 458

c. 952 − 863

d. 851 − 673

e. 432 − 246

12. ⟨ **1.4B** ⟩ *The balance in a savings account is $5403. How much money is left in the account if*

a. $869 is withdrawn?

b. $778 is withdrawn?

c. $989 is withdrawn?

d. $676 is withdrawn?

e. $765 is withdrawn?

13. ⟨ **1.5A** ⟩ *Multiply.*

a. 36 × 45

b. 28 × 49

c. 47 × 39

d. 56 × 24

e. 48 × 92

14. ⟨ **1.5A** ⟩ *Multiply.*

a. 123 × 216

b. 231 × 413

c. 345 × 654

d. 231 × 843

e. 879 × 569

15. ⟨ **1.5B** ⟩ *Multiply.*

a. 330 × 234

b. 220 × 546

c. 550 × 324

d. 450 × 124

e. 490 × 892

16. ⟨ **1.5C** ⟩ *A person has a $220 monthly payment. What is the total amount of money paid if*

a. 36 payments are required?

b. 24 payments are required?

c. 48 payments are required?

d. 30 payments are required?

e. 60 payments are required?

17. ⟨ **1.5D** ⟩ *Find the area of a shortcake tray measuring:*

a. 36 in. by 10 in.

b. 24 in. by 10 in.

c. 30 in. by 12 in.

d. 18 in. by 10 in.

e. 24 in. by 12 in.

18. ⟨ **1.6A** ⟩ *Divide (if possible).*

a. $\frac{0}{2}$

b. $\frac{0}{5}$

c. $\frac{0}{12}$

19. ⟨ **1.6A** ⟩ *Divide (if possible).*

a. $\frac{2}{0}$

b. $\frac{5}{0}$

c. $\frac{12}{0}$

20. ⟨ **1.6A** ⟩ *Divide.*

a. 75 ÷ 5

b. 84 ÷ 7

c. 90 ÷ 6

d. 88 ÷ 8

e. 68 ÷ 4

21. ⟨ **1.6B** ⟩ *Divide.*

a. 279 ÷ 9

b. 378 ÷ 9

c. 824 ÷ 8

d. 126 ÷ 6

e. 455 ÷ 7

22. ⟨ **1.6B** ⟩ *Divide.*

a. $21\overline{)967}$

b. $24\overline{)1009}$

c. $35\overline{)876}$

d. $29\overline{)1074}$

e. $51\overline{)2450}$

23. ⟨ **1.6C** ⟩ *A person's salary is $11,232. How much does the person receive each pay period if the money is paid in*

a. 9 equal payments?

b. 12 equal payments?

c. 24 equal payments?

d. 26 equal payments?

e. 52 equal payments?

24. ⟨ **1.7A** ⟩ *Classify as prime or composite (not prime).*

a. 41

b. 26

c. 37

d. 81

e. 2

25. ⟨ **1.7B** ⟩ *Write the prime factors of these numbers.*

a. 40

b. 25

c. 75

d. 128

e. 68

26. ⟨ **1.7C** ⟩ *Write the number as a product of primes.*

a. 50

b. 34

c. 76

d. 39

e. 81

27. ⟨ **1.7D** ⟩ *Find the product by writing as a product of factors.*

a. 2^2

b. 3^2

c. 5^3

d. 2^7

e. 3^5

28. ⟨ **1.7D** ⟩ *Find the product by writing as a product of factors.*

a. $3^2 \times 5^3$

b. $3^3 \times 8^0$

c. $3^2 \times 5^2 \times 2^0$

d. $5^0 \times 2^3 \times 5^2$

e. $4^2 \times 9^0 \times 5^1$

29. ⟨ **1.7D** ⟩ *Find the product by writing as a product of factors.*

a. $2^2 \times 3 \times 8^0$

b. $5^2 \times 7^0 \times 2^1$

c. $3^3 \times 5^2 \times 6^0$

d. $5^2 \times 3^0 \times 2^0$

e. $5^0 \times 3^0 \times 2^0$

30. ⟨ **1.8A** ⟩ *Simplify.*

a. $7 \cdot 8 - 2$

b. $6 \cdot 8 - 3$

c. $5 \cdot 8 - 4$

d. $4 \cdot 8 - 5$

e. $3 \cdot 8 - 5$

31. ⟨ **1.8A** ⟩ *Simplify.*

a. $30 + 4 \cdot 5$

b. $31 + 5 \cdot 5$

c. $32 + 6 \cdot 5$

d. $33 + 7 \cdot 5$

e. $34 + 8 \cdot 5$

32. ⟨ **1.8A** ⟩ *Simplify.*

a. $48 \div 6 - (1 + 2)$

b. $48 \div 8 - (2 + 2)$

c. $48 \div 4 - (2 + 3)$

d. $48 \div 3 - (2 + 4)$

e. $48 \div 2 - (2 + 5)$

33. ⟨ **1.8A** ⟩ *Simplify.*

a. $9 \div 3 \cdot 3 \cdot 3 + 3 - 1$

b. $9 \div 3 \cdot 3 + 3 - 1$

c. $8 \div 2 \cdot 2 \cdot 2 + 2 - 1$

d. $8 \div 2 \cdot 2 + 2 - 1$

e. $8 \div 4 \cdot 4 + 4 - 1$

34. ⟨ **1.8B** ⟩ *Simplify.*

a. $20 \div 5 + \{3 \cdot 9 - [3 + (5 - 2)]\}$

b. $20 \div 5 + \{4 \cdot 9 - [3 + (5 - 3)]\}$

c. $24 \div 6 + \{5 \cdot 9 - [3 + (5 - 4)]\}$

d. $24 \div 4 + \{6 \cdot 9 - [3 + (5 - 5)]\}$

e. $24 \div 3 + \{7 \cdot 9 - [3 + (5 - 1)]\}$

35. ⟨ **1.8C** ⟩ *A mechanic charges \$30 an hour plus parts to repair a car. If the parts cost \$80, find the total cost for a job that takes:*

a. 3 hours.

b. 5 hours.

c. 2 hours.

d. 4 hours.

e. 6 hours.

36. ⟨ **1.9A** ⟩ *Find the solution.*

a. $x + 6 = 18$

b. $x + 7 = 18$

c. $x + 8 = 18$

d. $x + 9 = 18$

e. $x + 10 = 18$

37. ⟨ **1.9A** ⟩ *Find the solution.*

a. $10 - x = 3$

b. $10 - x = 4$

c. $10 - x = 5$

d. $10 - x = 6$

e. $10 - x = 7$

38. ⟨ **1.9A** ⟩ *Find the solution.*

a. $20 = 4x$

b. $20 = 5x$

c. $20 = 10x$

d. $20 = 20x$

e. $20 = 2x$

39. ⟨ **1.9A** ⟩ *Find the solution.*

a. $28 \div x = 4$

b. $24 \div x = 4$

c. $20 \div x = 4$

d. $16 \div x = 4$

e. $12 \div x = 4$

40. ⟨ **1.9B** ⟩ *Solve.*

a. $n - 10 = 11$

b. $n - 14 = 12$

c. $n - 27 = 13$

d. $n - 48 = 14$

e. $n - 18 = 15$

41. ⟨ **1.9B** ⟩ *Solve.*

a. $20 = m - 12$

b. $20 = m - 38$

c. $11 = m - 14$

d. $42 = m - 15$

e. $49 = m - 16$

42. ⟨ **1.9B** ⟩ *Solve.*

a. $33 = 18 + m$

b. $32 = 19 + m$

c. $37 = 19 + m$

d. $39 = 17 + m$

e. $46 = 17 + m$

43. ⟨ **1.9B** ⟩ *Solve.*

a. $3x = 36$

b. $4x = 52$

c. $6x = 72$

d. $7x = 63$

e. $9x = 108$

44. ⟨ **1.9B** ⟩ *Solve.*

a. $10 = 2x$

b. $16 = 4x$

c. $20 = 5x$

d. $36 = 6x$

e. $48 = 8x$

45. ⟨ **1.9C** ⟩ *A building is 1430 feet tall. An antenna is added on the roof. Find the height of the antenna if the building is now:*

a. 1520 ft tall.

b. 1530 ft tall.

c. 1540 ft tall.

d. 1515 ft tall.

e. 1505 ft tall.

› Practice Test Chapter 1

(Answers on page 107)

Visit www.mhhe.com/bello to view helpful videos that provide step-by-step solutions to several of the problems below.

1. Write 348 in expanded form and find the value of the 3.

2. Write $600 + 50 + 2$ in standard form.

3. Write 76,008 in words.

4. Write eight thousand, five hundred ten in standard form.

5. Fill in the blank with $<$ or $>$ to make a true inequality: 18 ______ 13

6. Round 3749 to the nearest hundred.

7. The price of a car is \$24,795. Round \$24,795 to the nearest hundred.

8. $501 + 9786 =$

9. Find the perimeter of a triangle with sides 24 inches, 18 inches, and 30 inches.

10. $643 - 465 =$

11. The balance in a savings account is \$4302. If \$978 is withdrawn, how much money is left in the account?

12. $420 \times 381 =$

13. A person must pay \$210 each month for 36 months. What is the total amount of money paid?

14. Find the area of a rectangular frame measuring 20 inches by 8 inches.

15. Divide: **a.** $\frac{0}{9}$ **b.** $\frac{9}{0}$

16. Divide: $328 \div 8$

17. Divide: $26\overline{)885}$

18. A person's annual salary of \$15,600 is paid in 12 equal monthly payments. How much does the person receive each month?

19. Is 49 a prime or a composite number?

20. Write the prime factors of 28.

21. Write 60 as a product of primes.

22. $2^2 \times 3^2 =$

23. $3^2 \times 5 \times 8^0 =$

24. Simplify $3 \cdot 2^2 - 5$.

25. Simplify $16 \div 4 \cdot 2^2 + 5 - 1$.

26. Simplify $15 \div 3 + \{2^2 \cdot 3 - [2 + (3 + 1)]\}$.

27. A mechanic charges \$35 an hour plus parts to fix a car. If the parts cost \$90 and it took 3 hours to do the job, what is the total cost of the repair?

28. Solve.

a. $10 = m - 6$ **b.** $30 = 20 + m$

29. Solve.

a. $4x = 24$ **b.** $35 = 7x$

30. A building is 1380 feet tall. An antenna is added on the roof of the building, which makes the building and antenna 1425 feet tall. How tall is the antenna?

Answers to Practice Test Chapter 1

Answer	If You Missed	Review		
	Question	Section	Examples	Page
1. 348 = 300 + 40 + 8; 300	1	1.1	1, 2, 3, 4, 5	3, 4
2. 652	2	1.1	6, 7, 8	4–5
3. seventy-six thousand, eight	3	1.1	9	6
4. 8510	4	1.1	10, 11	6
5. $>$	5	1.2	1, 2	14
6. 3700	6	1.2	3, 4	16–17
7. $24,800	7	1.2	5, 6, 7	17–19
8. 10,287	8	1.3	1–7	25–27
9. 72 in.	9	1.3	11	29
10. 178	10	1.4	1–8	37–42
11. $3324	11	1.4	9, 10	43
12. 160,020	12	1.5	5, 6, 7	52–53
13. $7560	13	1.5	9, 10	54–55
14. 160 in.^{2}	14	1.5	11	56
15. **a.** 0 **b.** not defined	15	1.6	1, 2	63
16. 41	16	1.6	3, 4, 5	64–66
17. 34 r 1	17	1.6	6	66
18. $1300	18	1.6	7	66–67
19. composite	19	1.7	1	72
20. 2, 7	20	1.7	2	72
21. $2 \times 2 \times 3 \times 5 = 2^2 \times 3 \times 5$	21	1.7	3, 4	75–76
22. 36	22	1.7	5, 6	77
23. 45	23	1.7	5, 6	77
24. 7	24	1.8	1	82–83
25. 20	25	1.8	2, 3	83
26. 11	26	1.8	4	84
27. $195	27	1.8	5, 6	84–85
28. **a.** $m = 16$ **b.** $m = 10$	28	1.9	2, 3	90–91
29. **a.** $x = 6$ **b.** $x = 5$	29	1.9	4	91
30. 45 ft	30	1.9	5	92–93

Section

Chapter

2 two

Fractions and Mixed Numbers

The Human Side of Mathematics

As we mentioned in Chapter 1, the concept of a whole number is one of the oldest in mathematics. On the other hand, the concept of rational numbers or fractions (so named because they are ratios of whole numbers), developed much later because nonliterate tribes had no need for such a concept. Fractions evolved over a long period of time, stimulated by the need for certain types of measurement. For example, take a rod of length 1 unit and cut it into two equal pieces. What is the length of each piece? One-half, of course. If the same rod is cut into four equal pieces, then each piece is of length $\frac{1}{4}$. Two of these pieces will have length $\frac{2}{4}$, which tells us that $\frac{2}{4} = \frac{1}{2}$, as you will see later in the chapter. It was ideas such as these that led to the development of the arithmetic of fractions.

How were fractions written? During the Bronze Age, Egyptian hieroglyphic inscriptions show the reciprocals of whole numbers by using an elongated oval sign. Thus, $\frac{1}{8}$ and $\frac{1}{20}$ were written as

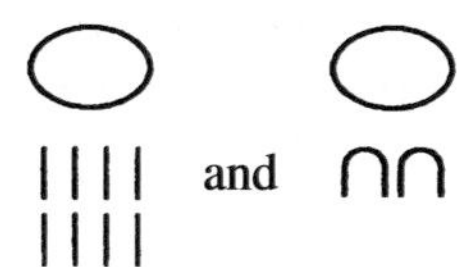

In this chapter we study the operations with fractions that are the quotients of two whole numbers and their uses today.

2.1 Fractions and Mixed Numbers

Objectives

You should be able to:

A Write a fraction corresponding to a given diagram.

B Classify a fraction as proper or improper.

C Write an improper fraction as a mixed number.

D Write a mixed number as an improper fraction.

E Solve applications using the concepts studied.

To Succeed, Review How To . . .

1. Understand the basic arithmetic (+, −, ×, ÷) facts. (pp. 24, 37, 49, 62)
2. Use the definition of a counting number. (p. 3)

Getting Started

In the cartoon, Sally is very upset with the idea of learning about **fractions.** The word *fraction* comes from the Latin word *fractio,* which means "to break" or "to divide."

PEANUTS reprinted by permission of United Feature Syndicate, Inc.

A fraction is a number (usually written as $\frac{a}{b}$, where a and b are whole numbers and b is not 0) equal to the quotient of a divided by b. Fractions are used in everyday life. For example,

$\frac{1}{3}$ of Americans do their grocery shopping at night.

$\frac{7}{10}$ of all popcorn consumed is consumed at home.

$\frac{3}{4}$ of all U.S. tomatoes are grown in California.

$\frac{3}{10}$ of M&M's Plain Chocolate Candies are brown.

There are infinitely many fractions and some examples are $\frac{1}{3}$, $\frac{7}{10}$, $\frac{3}{4}$, and $\frac{3}{10}$. Look at the symbol $\frac{2}{3}$ on the measuring cup. The number above the bar (the top number) is called the **numerator** and the number below the bar (the bottom number) is called the **denominator.**

$$\frac{2}{3} \begin{array}{l} \leftarrow \text{numerator} \\ \leftarrow \text{denominator} \end{array}$$

The denominator of a fraction tells us the number of equal parts into which a whole has been divided and the numerator tells us how many of these parts are being considered. Thus $\frac{2}{3}$ tells us that the whole (a cup) has been divided into 3 equal parts and that 2 parts are being used.

General inflation (red) is about $\frac{1}{2}$ of college cost inflation (green).

For the ten years ended 2002 college costs rose much faster than general costs

	General inflation	College cost inflation
	2.52%	5.09%

General inflation figure based on the Consumer Price Index produced by the Bureau of Labor Statistics. College inflation figure based on the Independent College 500® Index produced by the College Entrance Examination Board.

A > Diagramming Fractions

To help us understand fractions, we can represent them by using a diagram such as the following:

1 part shaded out of 2. Thus $\frac{1}{2}$ is shaded.

2 parts shaded out of 3. Thus $\frac{2}{3}$ is shaded.

3 parts shaded out of 4. Thus $\frac{3}{4}$ is shaded.

EXAMPLE 1 Using a diagram to represent fractions

Represent $\frac{3}{5}$ and $\frac{1}{4}$ by using a diagram similar to the ones shown above.

SOLUTION

$\frac{3}{5} = \frac{\text{3 shaded parts}}{\text{5 parts total}}$

$\frac{1}{4} = \frac{\text{1 shaded part}}{\text{4 parts total}}$

PROBLEM 1

Represent $\frac{2}{5}$ and $\frac{1}{7}$ in the diagrams below.

B > Proper and Improper Fractions

In Example 1, the fractions $\frac{3}{5}$ and $\frac{1}{4}$ are *less* than the whole that is used. Such fractions are called *proper fractions*.

DEFINITION OF A PROPER FRACTION

A **proper fraction** is a fraction in which the numerator is *less* (smaller) than the denominator.

Thus, $\frac{3}{5}$, $\frac{5}{16}$, $\frac{7}{32}$, and $\frac{99}{100}$ are proper fractions. On the other hand, some fractions such as $\frac{7}{7}$, $\frac{5}{1}$, or $\frac{14}{7}$ are *equal to* or **greater** (larger) than a whole. Such fractions are called *improper fractions*.

DEFINITION OF AN IMPROPER FRACTION

An **improper fraction** is a fraction in which the numerator is *equal to or greater than* the denominator.

Thus, $\frac{5}{5}$, $\frac{5}{2}$, $\frac{3}{3}$, $\frac{5}{1}$, and $\frac{17}{2}$ are improper fractions. Note that $\frac{5}{5}$ and $\frac{3}{3}$ both have the value 1. In general,

Answers to PROBLEMS

1.

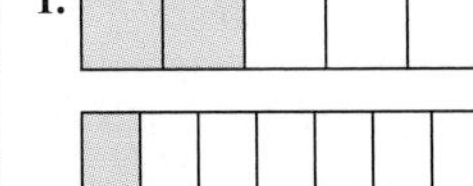

WRITING 1 AS A FRACTION

$\frac{n}{n} = 1$ for any number $n \neq 0$

On the other hand, $\frac{5}{1} = 5$, $\frac{3}{1} = 3$, and $\frac{49}{1} = 49$. In general,

WRITING n AS A FRACTION

$\frac{n}{1} = n$ for any number n

WRITING 0 AS A FRACTION

$\frac{0}{n} = 0$ for any number $n \neq 0$

Caution! $\frac{n}{0}$ is undefined.

EXAMPLE 2 Classifying fractions

Classify the given fraction as proper or improper.

a. $\frac{15}{16}$ **b.** $\frac{17}{5}$ **c.** $\frac{8}{8}$ **d.** $\frac{0}{8}$ **e.** $\frac{8}{1}$

SOLUTION

a. 15 is less than 16; thus $\frac{15}{16}$ is a proper fraction (the numerator is less than the denominator).

b. 17 is greater than 5; thus $\frac{17}{5}$ is an improper fraction.

c. 8 is equal to 8; thus, $\frac{8}{8}$ is an improper fraction.

d. 0 is less than 8. Thus, $\frac{0}{8}$ is a proper fraction.

e. 8 is greater than 1. Thus, $\frac{8}{1}$ is an improper fraction.

PROBLEM 2

Classify the given fraction as proper or improper.

a. $\frac{6}{6}$ **b.** $\frac{3}{19}$ **c.** $\frac{19}{3}$

d. $\frac{0}{3}$ **e.** $\frac{7}{1}$

C > Improper Fractions as Mixed Numbers

In Example 2 we mentioned that $\frac{17}{5}$ is an improper fraction. This fraction can also be written as a *mixed number*.

DEFINITION OF A MIXED NUMBER

A mixed number is a number representing the sum of a whole number and a proper fraction.

To write an improper fraction as a mixed number, divide the numerator by the denominator, obtaining the whole number part of the mixed number. The fractional part uses the remainder as numerator and the same denominator as the original fraction. Here is a diagram illustrating the procedure.

Note: Here is the diagram for $\frac{17}{5}$:

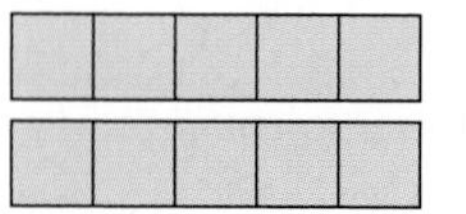

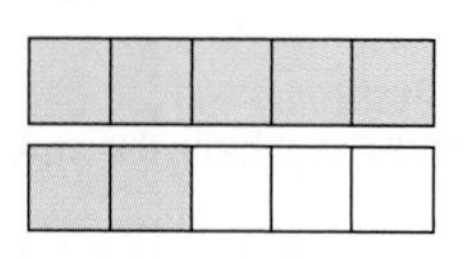

Do you see that this is $3 + \frac{2}{5}$?

Divide the numerator (17) by the denominator (5).

The answer (3) is the whole number part of the mixed number.

The remainder is the numerator of the fractional part.

improper fraction $\left\{\frac{17}{5}\right. = 17 \div 5 = 3$ with remainder $2 = \left.3\frac{2}{5}\right\}$ mixed number

remainder

whole number part

same denominator

(Note that $3\frac{2}{5} = 3 + \frac{2}{5}$, the sum of a whole number and a proper fraction.) Similarly, $\frac{5}{3} = 1\frac{2}{3}$ and $\frac{8}{5} = 1\frac{3}{5}$.

Answers to PROBLEMS

2. a. Improper **b.** Proper **c.** Improper **d.** Proper **e.** Improper

EXAMPLE 3 **Writing improper fractions as mixed numbers**

Write as mixed numbers.

a. $\frac{23}{6}$ **b.** $\frac{47}{5}$

SOLUTION

a. $\frac{23}{6} = 3$ with a remainder of 5. Thus, $\frac{23}{6} = 3\frac{5}{6}$.

b. $\frac{47}{5} = 9$ with a remainder of 2. Thus, $\frac{47}{5} = 9\frac{2}{5}$.

PROBLEM 3

Write as mixed numbers.

a. $\frac{26}{5}$ **b.** $\frac{47}{6}$

D › Mixed Numbers as Improper Fractions

We can also rewrite a mixed number as an improper fraction. For example, to write $3\frac{1}{4}$ as an improper fraction, we think of $3\frac{1}{4}$ as a sum, where the 3 is expressed as $\frac{12}{4}$, as shown in the diagram.

$$3\frac{1}{4} = 3 + \frac{1}{4} = \frac{13}{4}$$

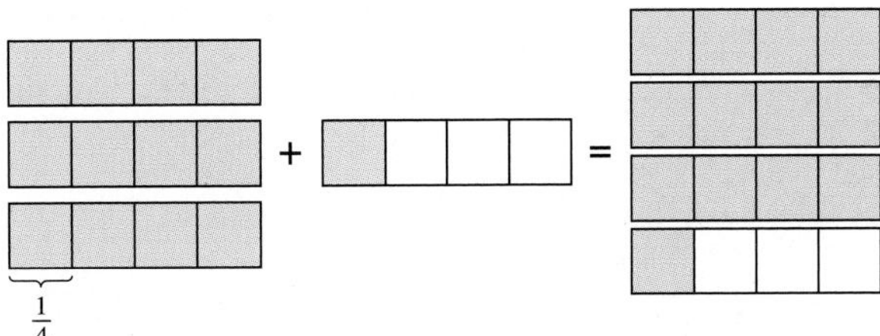

Here is a shortened form of this procedure (work clockwise from the denominator.):

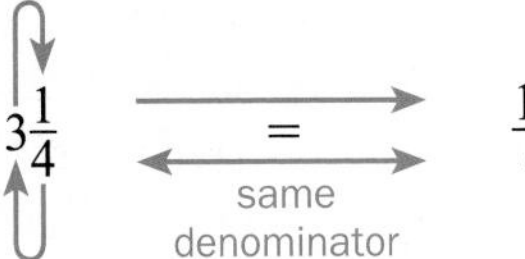

1. Multiply the denominator (4) by the whole number part (3).
2. Add the numerator (1). This is the new numerator.
3. Use the same denominator.

Here is the procedure. To write a mixed number as an improper fraction:

1. Multiply the denominator by the whole number part and add the numerator.
2. Use the number obtained in part 1 as the numerator of the improper fraction.
3. Use the same denominator.

EXAMPLE 4 **Writing mixed numbers as improper fractions**

Write as an improper fraction.

a. $6\frac{2}{7}$ **b.** $3\frac{1}{9}$

SOLUTION

a. $6\frac{2}{7} = \frac{7 \times 6 + 2}{7} = \frac{44}{7}$ **b.** $3\frac{1}{9} = \frac{9 \times 3 + 1}{9} = \frac{28}{9}$

PROBLEM 4

Write as an improper fraction.

a. $5\frac{3}{4}$ **b.** $8\frac{2}{7}$

Answers to PROBLEMS

3. a. $5\frac{1}{5}$ **b.** $7\frac{5}{6}$

4. a. $\frac{23}{4}$ **b.** $\frac{58}{7}$

E > Applications Involving Writing Fractions

Suppose you want to know what fraction of a pound (16 ounces) is in the 5-ounce can. Since the can contains 5 ounces and the whole pound is 16 ounces, the answer is $\frac{5}{16}$. Similarly, a week (7days) is $\frac{7}{31}$ of the month of January (which has 31 days), and a weekend (2 days) is $\frac{2}{7}$ of a week. Note that in all these fractions, the numerator and denominator are expressed in the same type of unit.

EXAMPLE 5 **Finding fractions of a week**

Find what fraction of a week (7 days) each amount of days represents.

a. 4 days **b.** 7 days **c.** 14 days

SOLUTION

a. 4 days $= \frac{4}{7}$ week **b.** 7 days $= \frac{7}{7}$, or 1, week

c. 14 days $= \frac{14}{7}$, or 2, weeks

PROBLEM 5

Find what fraction of a month (30 days) each amount of days represents.

a. A week **b.** 30 days

c. 60 days

What type of oil do you use for cooking? Here are the fat contents of nine different oils.

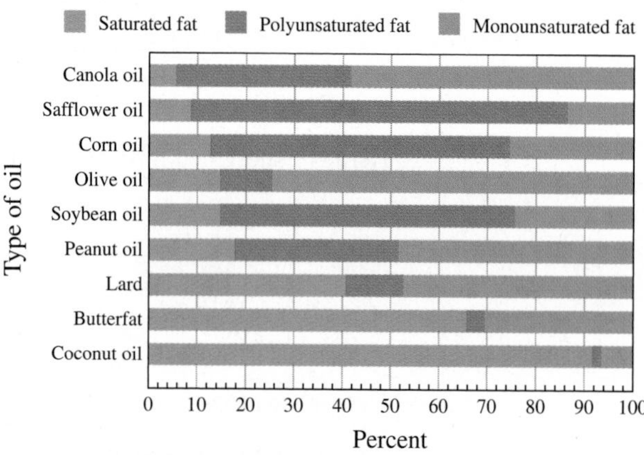

Source: NC State University.

Which oil has the most monounsaturated fat? Approximately what fraction of the oil is monounsaturated? Since monounsaturated fats are shown in blue, the oil with the most monounsaturated fat is olive oil, which is about $\frac{7}{10}$ monounsaturated fat. Why? The bar representing the olive oil is 10 units long and about 7 of them are blue, representing monounsaturated fats; thus

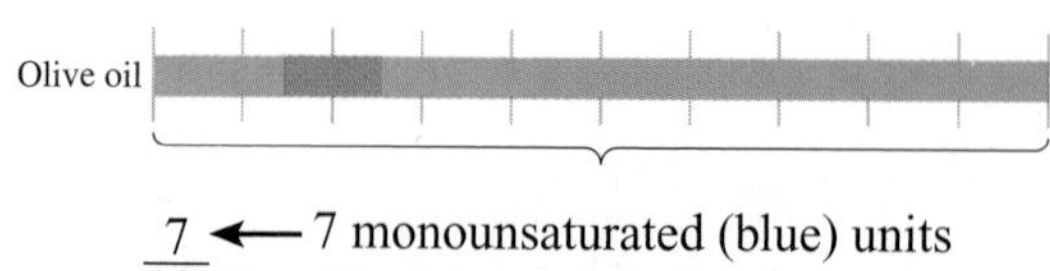

$\frac{7}{10}$ ← 7 monounsaturated (blue) units
← 10 total units of length

Answers to PROBLEMS

5. a. $\frac{7}{30}$ **b.** $\frac{30}{30} = 1$ **c.** $\frac{60}{30} = 2$

EXAMPLE 6 Looking for saturated fats

Referring to the preceding figure,

a. Which oil has the most saturated fat?

b. About what fraction of the fat in this oil is saturated?

SOLUTION

a. Coconut oil has the most saturated fat.

b. About nine (9) of the ten (10) units in the coconut oil bar are green. Thus, $\frac{9}{10}$ of the coconut oil is saturated fat.

PROBLEM 6

Referring to the figure,

a. Which oil has the most polyunsaturated fat?

b. About what fraction of the fat in this oil is polyunsaturated?

Many applications of fractions involve the idea of a ratio. A **ratio** is a quotient of two numbers.

RATIO

The ratio of a to b is written as the fraction $\frac{a}{b}$ ($b \neq 0$).

For example, the ratio of men to women in your class may be $\frac{19}{21}$ or $\frac{22}{31}$.

EXAMPLE 7 Price-to-earnings ratio of stock

The price-to-earnings (P/E) ratio of a stock is the price of the stock divided by its earnings per share. If the price of a stock is \$30 and its earnings per share are \$3, what is the P/E ratio of the stock?

SOLUTION The P/E ratio is $\frac{30}{3} = 10$

PROBLEM 7

What is the P/E ratio of a stock whose price is \$28 and whose earnings are \$4 per share?

Exercises 2.1

Boost your grade at mathzone.com!

- Practice Problems
- NetTutor
- Self-Tests
- e-Professors
- Videos

⟨A⟩ Diagramming Fractions In Problems 1–10, what part of each object is shaded?

1.

2.

3.

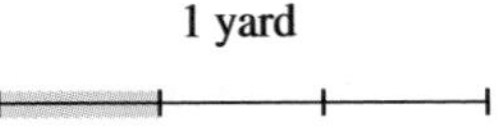

4.

5.

6.

7.

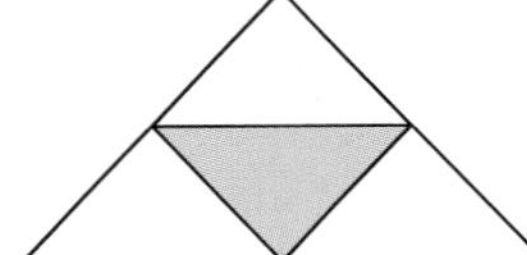

8.

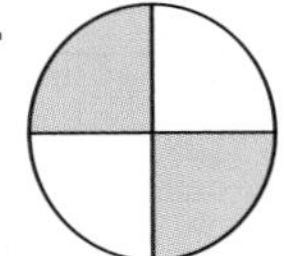

9.

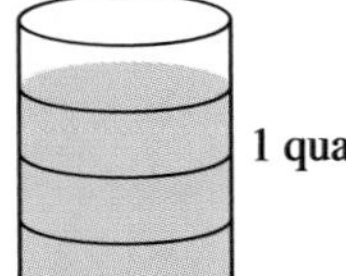

10.

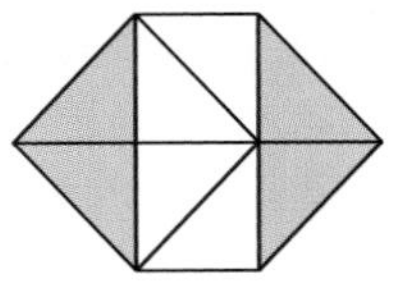

Answers to PROBLEMS

6. a. safflower **b.** $\frac{8}{10}$

7. $\frac{28}{4} = 7$

Web IT go to mhhe.com/bello for more lessons

In Problems 11–20, what fraction of the dollar bill is green (not faded)?

11.

12.

13.

14.

15.

16.

17.

18.

19.

20.

⟨ B ⟩ Proper and Improper Fractions In Problems 21–30, classify the fraction as proper or improper.

21. $\frac{9}{61}$ **22.** $\frac{61}{9}$ **23.** $\frac{4}{17}$ **24.** $\frac{17}{4}$ **25.** $\frac{8}{41}$

26. $\frac{9}{47}$ **27.** $\frac{8}{16}$ **28.** $\frac{14}{1}$ **29.** $\frac{3}{100}$ **30.** $\frac{100}{10}$

⟨ C ⟩ Improper Fractions as Mixed Numbers In Problems 31–40, write the fraction as a mixed number.

31. $\frac{31}{10}$ **32.** $\frac{46}{5}$ **33.** $\frac{8}{7}$ **34.** $\frac{59}{8}$ **35.** $\frac{29}{8}$

36. $\frac{19}{2}$ **37.** $\frac{69}{9}$ **38.** $\frac{83}{3}$ **39.** $\frac{101}{10}$ **40.** $\frac{97}{3}$

Web IT go to **mhhe.com/bello** for more lessons

⟨D⟩ Mixed Numbers as Improper Fractions

In Problems 41–50, write the mixed number as an improper fraction.

41. $5\frac{1}{7}$ **42.** $6\frac{1}{9}$ **43.** $4\frac{1}{10}$ **44.** $5\frac{3}{11}$ **45.** $1\frac{2}{11}$

46. $3\frac{2}{13}$ **47.** $8\frac{3}{10}$ **48.** $7\frac{2}{11}$ **49.** $2\frac{1}{6}$ **50.** $9\frac{7}{8}$

⟨E⟩ Applications Involving Writing Fractions

51. *Sleepy time* A person slept 7 hours. What fraction of the day (24 hours) is that?

52. *Fraction of an hour* Forty-five minutes is what fraction of an hour (60 minutes)?

53. *Fraction of a pound* A box of cereal weighs 7 ounces. What fraction of a pound (16 ounces) is that?

54. *Pizza fractions* A pizza was cut into 8 equal parts. Five pieces were eaten.

a. What fraction of the pizza was eaten?

b. What fraction of the pizza was left?

55. *Hours worked as fractions* A woman has worked for 5 hours. If her workday is 8 hours long, what fraction of the day has she worked?

56. *Fraction of reading* Sam Smart has to read 41 pages. He has already read 31. What fraction of the reading has he finished?

57. *Taxpayers' revenue* In a recent year the Internal Revenue Service collected 51 cents of every dollar (100 cents) of revenue from individual taxpayers. What fraction of the revenue came from individual taxpayers?

58. *Taxpayers' revenue* In a recent year the Internal Revenue Service collected 7 cents of every dollar (100 cents) of revenue from corporate taxes. What fraction of the revenue came from corporate taxes?

59. *Minutes in a commercial* A viewer made the following table showing the number of seconds various commercials lasted:

Complete the table by filling in the number of minutes each commercial lasted. For example, the shampoo commercial lasted $\frac{30}{60}$ or $\frac{1}{2}$ minute.

Time of Commercials

Commercial	Seconds	Minutes
Shampoo	30	$\frac{30}{60}$ or $\frac{1}{2}$
Dog food	60	**a.**
Toothpaste	90	**b.**
Soap	45	**c.**
Cereal	15	**d.**

60. In Phoenix it rained 3 out of 31 days. What fraction of the 31 days is that?

The following information will be used in Problems 61–65. Simplify your answers.

Household overcrowding Is your household overcrowded? In a recent survey it was discovered that of every 98 households in America:

25 are single-person
33 are two-person
16 are three-person
15 are four-person
6 are five-person
2 are six-person
1 are seven-person or more

61. What fraction of the households consisted of a single person?

62. What fraction of the households consisted of three persons?

63. What fraction of the households consisted of five persons?

64. What fraction of the households consisted of five persons or more?

65. What fraction of the households consisted of six persons or more?

Source: U.S. Census Bureau.

Web IT go to mhhe.com/bello for more lessons

The following information will be used in Problems 66–70.

Traveling abroad Are you taking a trip soon? In a recent survey, it was discovered that of every 99 Americans who traveled abroad, their destinations were as shown.

1.	Mexico	37	**6.**	Italy	5
2.	Canada	30	**7.**	Japan	2
3.	United Kingdom	8	**8.**	Spain	2
4.	France	6	**9.**	Netherlands	2
5.	Germany	5	**10.**	Switzerland	2

Source: Infoplease.

66. What fraction of the travelers went to Germany?

67. What fraction of the travelers went to Mexico?

68. What fraction of the travelers went to Italy?

69. What fraction of the travelers went to Spain?

70. What fraction of the travelers went to Switzerland?

FICO scores When you apply for credit—whether for a credit card, a car loan, or a mortgage—lenders want to know what risk they'd take by loaning money to you. Your FICO scores are the credit scores most lenders use to determine your credit risk. What are these scores based on? Look at the five categories in the pie chart!

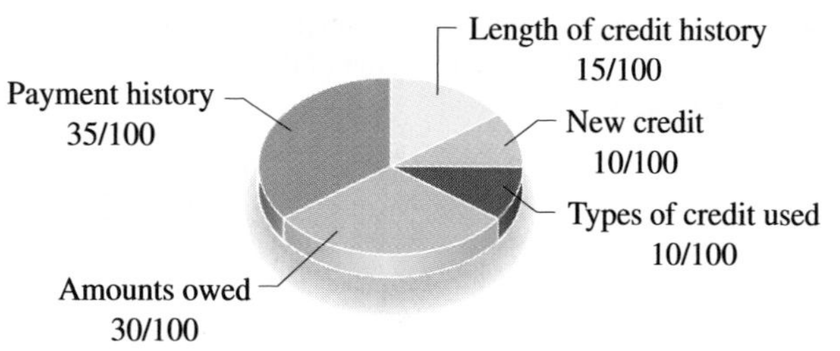

Source: http://www.myfico.com

71. What reduced fraction of the score is based on your payment history?

72. What reduced fraction of the score is based on the amounts you owe?

73. What reduced fraction of the score is based on the length of your credit history?

74. What reduced fraction of the score is based on new credit?

75. What reduced fraction of the score is based on the types of credit used?

76. Which of the categories is the most important for your score?

Using Your Knowledge

In Problems 77–80, indicate what fraction of the gas in the tank has been used and what fraction remains when:

77. The gas gauge is on ❶

78. The gas gauge is on ❷

79. The gas gauge is on ❸

80. The gas gauge is on ❹

How do you know how many miles per gallon your car gets? You look at the ratio of miles traveled to gallons of gas used. Thus, if your car travels 200 miles on 10 gallons of gas, your car gets $\frac{200}{10} = 20$ miles per gallon (mpg).

81. Your car travels 180 miles on 10 gallons of gas. How many miles per gallon does your car get?

82. You take a trip from Tampa to Miami, 260 miles. If you use 13 gallons of gas, how many miles per gallon does your car get?

83. You take a trip from Santa Rosa to Eureka, a distance of 210 miles. You use 10 gallons of gas. How many miles per gallon does your car get?

84. You know you can make it from Los Angeles to Long Beach, a distance of 23 miles, on one gallon of gas. How many miles per gallon does your car get?

85. Your car gets 25 mpg. Your gas tank takes 14 gallons of gas. How far can you go on one tank of gas?

86. You want to travel from Los Angeles to Lone Pine, a distance of 220 miles. Your car gets 20 mpg. How many gallons of gas do you need?

87. The distance from Monterey to Los Angeles is about 340 miles. Your car gets about 20 miles per gallon and the tank holds 14 gallons of gas. Can you make the trip on one tankful of gas? Explain.

88. You want to travel from Eureka to San Francisco, a distance of 260 miles. If your car gets 20 mpg and you use exactly one tank of gas, what is the capacity of your gas tank?

89. Your car gets 20 mpg. You want to travel from Las Vegas to Ely, a distance of 240 miles. What is the capacity of the smallest gas tank that would get you to Ely?

90. Your gas tank holds 14 gallons of gas. If you want to go to San Jose from Eureka, a distance of 322 miles, what is the minimum mpg your car can get so you reach your destination using one tank of gas?

Write On

91. Write in your own words why $\frac{n}{0}$ is not defined. Can $n = 0$? Explain.

92. Write in your own words why $\frac{0}{n} = 0$. Can $n = 0$?

93. Write in your own words why $\frac{n}{1} = n$. Can $n = 0$?

Concept Checker

Fill in the blank(s) with the correct word(s), phrase, or mathematical statement.

94. In the **fraction** $\frac{a}{b}$ the a is the ______________.

95. In the **fraction** $\frac{a}{b}$ the b is the ______________.

96. A **proper fraction** is a fraction in which the **numerator** is ______________ than the **denominator.**

97. An **improper fraction** is a fraction in which the **numerator** is **equal to** or ______________ than the **denominator.**

98. $\frac{0}{n} =$ ______________ for any nonzero number n.

99. $\frac{n}{0}$ is ______________ for any n.

100. A **mixed number** is a number representing the ______________ of a **whole number** and a **proper fraction.**

101. The **ratio of a to b** ($b \neq 0$) is written as the **fraction** ______________.

n	**sum**
numerator	**denominator**
less	**0**
$\frac{b}{a}$	**greater**
difference	$\frac{a}{b}$
well	**undefined**

Mastery Test

102. If the price of a stock is \$48 and its earnings are \$12 per share, what is the P/E ratio of the stock?

103. What fraction of a year is 5 months?

104. Write $7\frac{2}{3}$ as a proper fraction.

105. Write $\frac{25}{3}$ as a mixed number.

106. Is $\frac{17}{18}$ a proper or an improper fraction?

107. Is $\frac{17}{17}$ a proper or an improper fraction?

108. Represent $\frac{2}{5}$ by drawing and shading a diagram.

Skill Checker

Write as a product of primes using exponents.

109. 36

110. 90

111. 28

112. 72

113. 180

114. 200

2.2 Equivalent Fractions: Building and Reducing

Objectives

You should be able to:

A Write a fraction equivalent to a given one and with a specified numerator or denominator.

B Reduce a fraction.

C Solve applications involving the concepts studied.

To Succeed, Review How To . . .

1. Distinguish the meaning of the symbols > and <. (p. 14)
2. Write a composite number as a product of primes. (pp. 73–76)

Getting Started

The picture above illustrates the fact that one-half (dollar) is *equivalent to* two quarters. In symbols we write

$$\frac{1}{2} = \frac{2}{4}$$

When two fractions represent numerals or names for the same number, the fractions are said to be *equivalent.*

The following diagram shows that $\frac{1}{2}$, $\frac{2}{4}$, $\frac{3}{6}$, and $\frac{4}{8}$ are equivalent fractions.

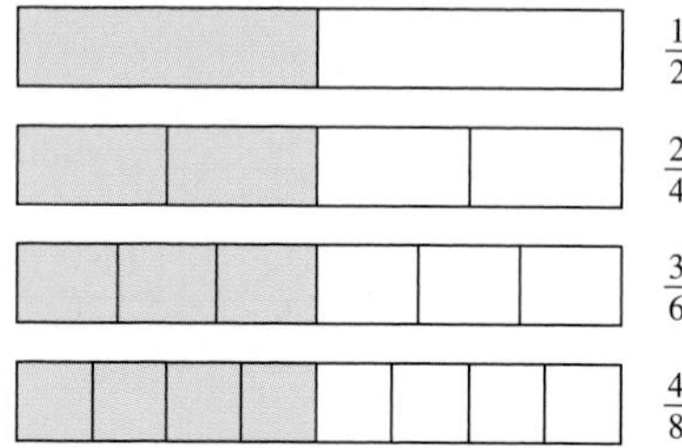

Thus, $\frac{1}{2} = \frac{2}{4} = \frac{3}{6} = \frac{4}{8}$ and so on.

A Equivalent Fractions

EQUIVALENT FRACTIONS

Two fractions are equivalent if they represent numerals or names for the same number; that is, both fractions have the same value.

The fraction $\frac{1}{2}$ is *equivalent* to many other fractions. We can always obtain fractions equivalent to any given fraction by **multiplying** both the **numerator** and **denominator** of the original fraction by the same nonzero number. In symbols,

FUNDAMENTAL PROPERTY OF FRACTIONS

If *a*, *b*, and *c* are any numbers, then

$$\frac{a}{b} = \frac{a \cdot c}{b \cdot c} \qquad (b \neq 0 \text{ and } c \neq 0)$$

and

$$\frac{a}{b} = \frac{a \div c}{b \div c} \qquad (b \neq 0 \text{ and } c \neq 0)$$

This property means that you can multiply or divide the numerator and denominator of the fraction $\frac{a}{b}$ by the same nonzero number c and obtain the equivalent fraction $\frac{a \cdot c}{b \cdot c}$ or $\frac{a \div c}{b \div c}$. In fact, this is the same as multiplying (or dividing) the fraction by 1. For example,

$$\frac{1}{3} = \frac{1 \times 2}{3 \times 2} = \frac{2}{6}$$

$$\frac{1}{3} = \frac{1 \times 3}{3 \times 3} = \frac{3}{9}$$

$$\frac{1}{3} = \frac{1 \times 4}{3 \times 4} = \frac{4}{12}$$

Thus, $\frac{1}{3} = \frac{2}{6} = \frac{3}{9} = \frac{4}{12}$. Can we find a fraction equivalent to $\frac{3}{5}$ with a denominator of 20? To do this we have to solve this problem:

$$\frac{3}{5} = \frac{?}{20}$$

5 was multiplied by 4 to get 20. (Remember that $20 \div 5 = 4$.)

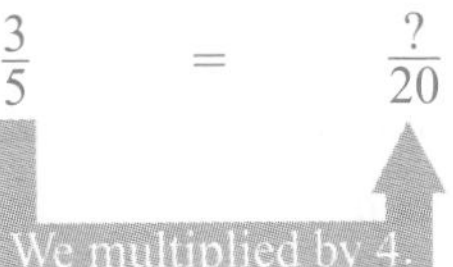

To have an equivalent fraction, also multiply the numerator 3 by 4.

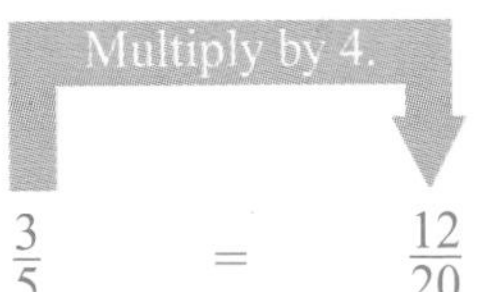

$$\frac{3}{5} = \frac{12}{20}$$

EXAMPLE 1 Finding equivalent fractions

Find the equivalent fractions.

a. $\frac{2}{3} = \frac{?}{9}$ **b.** $\frac{3}{8} = \frac{6}{?}$

SOLUTION

a. The denominator, 3, has to be multiplied by 3 to get a new denominator of 9. So, we multiply the numerator, 2, by 3. Here is the diagram.

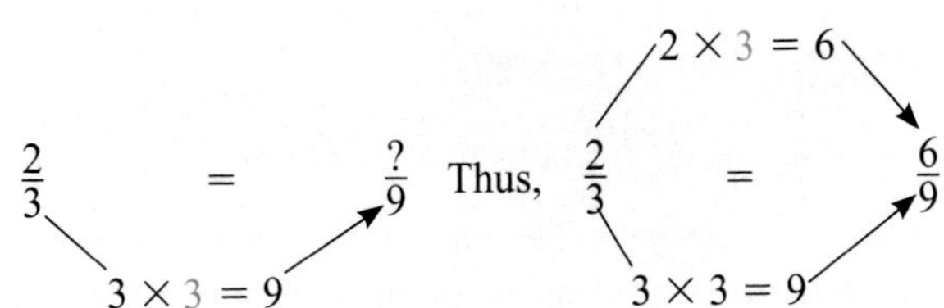

(continued)

PROBLEM 1

Find the equivalent fractions.

a. $\frac{2}{7} = \frac{?}{28}$ **b.** $\frac{5}{6} = \frac{20}{?}$

Answers to PROBLEMS

1. a. $\frac{2}{7} = \frac{8}{28}$ **b.** $\frac{5}{6} = \frac{20}{24}$

b. The numerator, 3, has to be multiplied by 2 to get the new numerator, 6. Thus we have to multiply the denominator, 8, by 2. (See the diagram.)

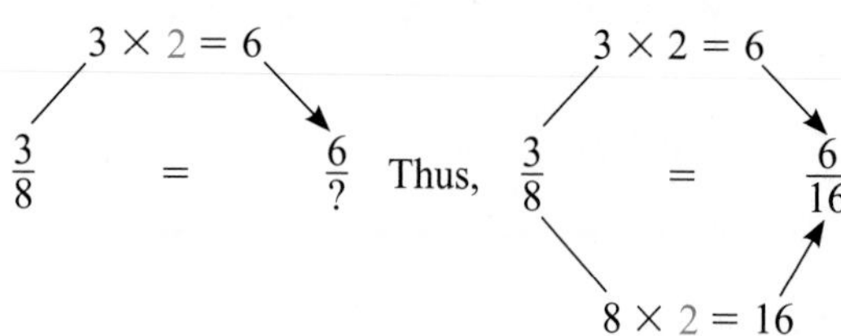

$\frac{3}{8} = \frac{6}{?}$ Thus, $\frac{3}{8} = \frac{6}{16}$

Now here is a slightly different problem. Can we find a fraction equivalent to $\frac{15}{20}$ with a denominator of 4? To do this we have to solve this problem:

$\frac{15}{20} = \frac{?}{4}$ 20 was divided by 5 to get 4.

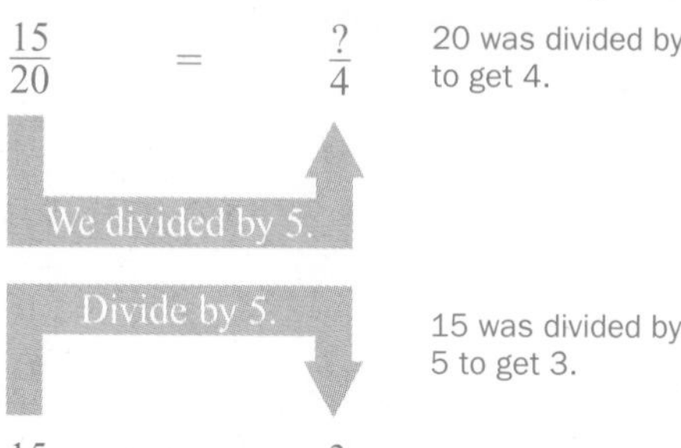

15 was divided by 5 to get 3.

$\frac{15}{20} = \frac{3}{4}$

EXAMPLE 2 Finding equivalent fractions

Find the equivalent fractions.

a. $\frac{18}{24} = \frac{?}{8}$ **b.** $\frac{20}{30} = \frac{4}{?}$

SOLUTION

a. $\frac{18}{24} = \frac{?}{8}$ ($24 \div 3 = 8$) Thus, $\frac{18}{24} = \frac{6}{8}$ ($18 \div 3 = 6$; $24 \div 3 = 8$)

b. $\frac{20}{30} = \frac{4}{?}$ ($20 \div 5 = 4$) Thus, $\frac{20}{30} = \frac{4}{6}$ ($20 \div 5 = 4$; $30 \div 5 = 6$)

PROBLEM 2

Find the equivalent fractions.

a. $\frac{42}{54} = \frac{?}{18}$ **b.** $\frac{6}{20} = \frac{3}{?}$

B › Reducing Fractions

This rule can be used to **reduce** fractions to lowest terms:

REDUCING TO LOWEST TERMS

A fraction is *reduced to lowest terms* (simplified) when there are no common factors (except 1) in the numerator and denominator.

Answers to PROBLEMS

2. a. $\frac{42}{54} = \frac{14}{18}$ **b.** $\frac{6}{20} = \frac{3}{10}$

For example, to reduce $\frac{140}{154}$ to lowest terms, we proceed by steps:

Divide 140 by 2.
Divide 70 by 2.
Divide 35 by 5.
Divide 7 by 7.

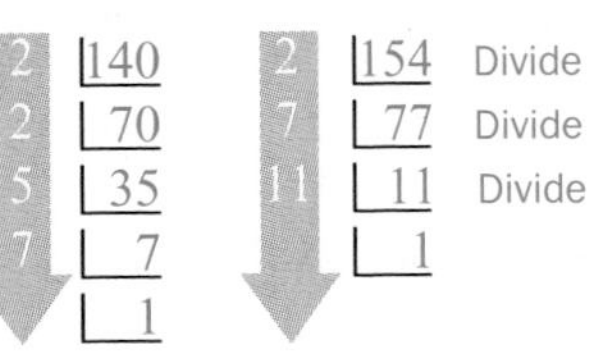

Divide 154 by 2.
Divide 77 by 7.
Divide 11 by 11.

Step 1. Write the numerator and denominator as products of primes.

$140 = 2 \times 2 \times 5 \times 7 \qquad 154 = 2 \times 7 \times 11$

$$\frac{140}{154} = \frac{2 \times 2 \times 5 \times 7}{2 \times 7 \times 11}$$

Step 2. Rewrite the fraction using the factored numerator and denominator.

$$\frac{140}{154} = \frac{2 \times \overset{1}{\cancel{2}} \times 5 \times \overset{1}{\cancel{7}}}{\underset{1}{\cancel{2}} \times \underset{1}{\cancel{7}} \times 11} = \frac{2 \times 5}{11}$$

$$= \frac{10}{11}$$

Step 3. Divide the numerator and denominator by the common factors (2 and 7).

The division of 2 by 2 and 7 by 7 is indicated by crossing out 2 and 7 and writing the quotient, 1. Note that we actually divided numerator and denominator by $2 \times 7 = 14$, which is the **greatest common factor** (GCF) of 140 and 154. Thus

$$\frac{140}{154} = \frac{140 \div 14}{154 \div 14} = \frac{10}{11}$$

The reduction of $\frac{140}{154}$ is sometimes shown like this:

The result is $\frac{10}{11}$. This method saves time. Use it!

Similarly, to reduce $\frac{60}{105}$ to lowest terms we proceed as before.

$$\frac{60}{105} = \frac{2 \times 2 \times \overset{1}{\cancel{3}} \times \overset{1}{\cancel{5}}}{\underset{1}{\cancel{3}} \times \underset{1}{\cancel{5}} \times 7} = \frac{4}{7}$$

We divide numerator and denominator by $3 \times 5 = 15$, the GCF of 60 and 105.

or, in the shortened version

$$\frac{\overset{\overset{4}{\cancel{20}}}{\cancel{60}}}{\underset{\underset{7}{\cancel{35}}}{\cancel{105}}}$$

Divide by 3 first. Divide by 5 next.

EXAMPLE 3 Simplifying fractions

Reduce (simplify) each fraction.

a. $\frac{15}{105}$ **b.** $\frac{36}{90}$

PROBLEM 3

Reduce (simplify) these.

a. $\frac{16}{80}$ **b.** $\frac{70}{155}$

(continued)

Answers to PROBLEMS

3. a. $\frac{1}{5}$ **b.** $\frac{14}{31}$

SOLUTION

a. $\frac{15}{105} = \frac{\overset{1}{\cancel{3}} \times \overset{1}{\cancel{5}}}{\underset{1}{\cancel{3}} \times \underset{1}{\cancel{5}} \times 7} = \frac{1}{7}$ or $\frac{\overset{\overset{1}{\cancel{5}}}{\cancel{15}}}{\underset{\underset{7}{\cancel{35}}}{\cancel{105}}}$ Divide by 3 first. Divide by 5 next.

Note that we divided 3 by 3, and 5 by 5, obtaining an answer of 1, which appears in the numerator and denominator of the fraction. You can also do the problem by dividing numerator and denominator by 15, the GCF of 15 and 105, obtaining

$$\frac{15}{105} = \frac{15 \div 15}{105 \div 15} = \frac{1}{7}$$

b. $\frac{36}{90} = \frac{2 \times \overset{1}{\cancel{2}} \times \overset{1}{\cancel{3}} \times \overset{1}{\cancel{3}}}{\underset{1}{\cancel{2}} \times \underset{1}{\cancel{3}} \times \underset{1}{\cancel{3}} \times 5} = \frac{2}{5}$ or $\frac{\overset{\overset{\overset{2}{\cancel{6}}}{\cancel{18}}}{\cancel{36}}}{\underset{\underset{\underset{5}{\cancel{15}}}{\cancel{45}}}{\cancel{90}}}$ Divide by 2 first. Divide by 3 next. Divide by 3 again.

The GCF of 36 and 90 is $2 \times 3 \times 3 = 18$.

You can also do the problem by dividing the numerator and denominator by 18 like this:

$$\frac{36}{90} = \frac{36 \div 18}{90 \div 18} = \frac{2}{5}$$

C › Applications Involving Reducing Fractions

EXAMPLE 4 **Origin of the 24-second shot clock**

Do you think professional basketball is boring? Danny Biasone (the owner of the Syracuse Nationals) thought the sport was boring. Biasone thought that a clock was necessary to force players to shoot at regular intervals and speed up the game. But how many seconds should be allowed between shots? Here is his reasoning: There are 48 minutes in a game, and $48 \times 60 = 2880$ seconds. An average game contained 60 shots per team, or $60 \times 2 = 120$ total shots. Thus, Biasone reasoned it should take $\frac{2880}{120}$ seconds per shot. Reduce $\frac{2880}{120}$.

SOLUTION We first note that 2880 and 120 both end in 0, that is, they both have a factor of 10. We write: $\frac{288 \times 10}{12 \times 10} = \frac{288}{12}$. If you notice that 12 goes into 288 exactly 24 times, we are finished. Otherwise, write 288 and 12 as products of primes.

```
2|288    2|12
2|144    2| 6
2| 72    3| 3
2| 36       1
2| 18
3|  9
3|  3
    1
```

$$\frac{\overset{1}{\cancel{2}} \times \overset{1}{\cancel{2}} \times 2 \times 2 \times 2 \times \overset{1}{\cancel{3}} \times 3}{\underset{1}{\cancel{2}} \times \underset{1}{\cancel{2}} \times \underset{1}{\cancel{3}}} = 24$$

And that is the way the 24-second shot clock was invented!

PROBLEM 4

Reduce $\frac{280}{120}$.

Answers to PROBLEMS

4. $\frac{7}{3}$

Algebra Bridge

In algebra, fractions can be reduced as in arithmetic. To do this, you must remember what an *exponent* means. Thus, $x^2 = x \cdot x, x^3 = x \cdot x \cdot x, x^4 = x \cdot x \cdot x \cdot x$, and so on. To reduce $\frac{6x^2}{9x^3}$, write 6 and 9 as products of primes and use the meaning of exponents to reduce x^2 and x^3. Here is the procedure:

$$\frac{6x^2}{9x^3} = \frac{2 \cdot \overset{1}{\cancel{3}} \cdot \overset{1}{\cancel{x}} \cdot \overset{1}{\cancel{x}}}{3 \cdot \underset{1}{\cancel{3}} \cdot \underset{1}{\cancel{x}} \cdot \underset{1}{\cancel{x}} \cdot x} = \frac{2}{3x}$$

Calculator Corner

Some calculators are designed to work with fractions. If you have such a calculator and you want to find a fraction equivalent to $\frac{12}{30}$ with a denominator of 10, the calculator has a key often labeled [x/y] that allows you to reduce fractions. To simplify (reduce) $\frac{15}{20}$ enter 15 [x/y] 20 ENTER and tell the calculator to **simplify** the result by pressing [SIMP], the calculator asks you what factor you want to divide numerator and denominator by. You have to know that you want to divide numerator and denominator by 5, so enter 5 ENTER and the answer 3/4 appears. To reduce 12/30 to lowest terms, enter 12 [x/y] 30 ENTER [SIMP]. The calculator asks FACTOR? Enter 6 and press ENTER and you get the answer 2/5. Here, you are dividing the numerator and denominator of 12/30 by 6.

Exercises 2.2

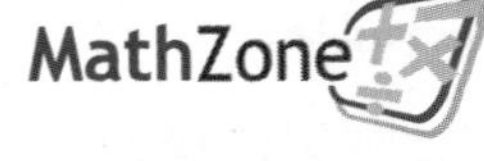

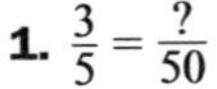

⟨A⟩ Equivalent Fractions In Problems 1–16, find the missing number.

1. $\frac{3}{5} = \frac{?}{50}$ **2.** $\frac{1}{8} = \frac{4}{?}$ **3.** $\frac{1}{6} = \frac{5}{?}$ **4.** $\frac{7}{9} = \frac{?}{27}$

5. $\frac{3}{5} = \frac{27}{?}$ **6.** $\frac{7}{12} = \frac{?}{60}$ **7.** $1\frac{2}{3} = \frac{?}{9}$ **8.** $2\frac{1}{5} = \frac{?}{15}$

9. $4\frac{1}{2} = \frac{?}{16}$ **10.** $5\frac{1}{10} = \frac{?}{90}$ **11.** $\frac{12}{15} = \frac{?}{5}$ **12.** $\frac{14}{42} = \frac{?}{6}$

13. $\frac{8}{24} = \frac{4}{?}$ **14.** $\frac{12}{18} = \frac{4}{?}$ **15.** $\frac{21}{56} = \frac{?}{8}$ **16.** $\frac{36}{180} = \frac{?}{5}$

⟨B⟩ Reducing Fractions In Problems 17–30, reduce each fraction to lowest terms.

17. $\frac{28}{30}$ **18.** $\frac{15}{12}$ **19.** $\frac{13}{52}$ **20.** $\frac{27}{54}$ **21.** $\frac{56}{24}$

22. $\frac{56}{21}$ **23.** $\frac{21}{28}$ **24.** $\frac{18}{24}$ **25.** $\frac{22}{33}$ **26.** $\frac{100}{25}$

27. $\frac{45}{210}$ **28.** $\frac{180}{160}$ **29.** $\frac{231}{1001}$ **30.** $\frac{91}{455}$

Web IT go to mhhe.com/bello for more lessons

Web IT go to mhhe.com/bello for more lessons

‹ C › Applications Involving Reducing Fractions

31. *Personal income taxes* In a recent year, 46 cents out of every dollar (100 cents) of revenue came from personal income taxes. What (reduced) fraction of the revenues is that?

32. *Corporate income taxes* In a recent year, 8 cents out of every dollar (100 cents) of revenue came from corporate income taxes. What (reduced) fraction of the revenues is that?

33. *Defense* The predicted defense budget for 2008 is \$460 billion. The predicted total budget is \$2760 billion. What (reduced) fraction of the budget will be spent on defense?

Source: The White House.

34. *Temperature* In Alabama, about 20 days out of the year (365 days) have temperatures less than 32 degrees Fahrenheit. What (reduced) fraction of the days is that?

35. *Temperature* In West Virginia, 100 days of the year (365 days) have temperatures less than 32 degrees Fahrenheit. What (reduced) fraction of the days is that?

36. *Water usage* A recent survey determined that the average household uses 80 gallons of water a day. If you fill your tub, 36 gallons are used. What (reduced) fraction of the water used per day is the 36 gallons?

Source: Center for Innovation in Engineering and Science Education.

37. *Cards* A standard deck of cards consists of 52 cards. Which (reduced) fraction of the deck is each of the following?

a. Red (26 cards are red)

b. Hearts (13 cards are hearts)

c. Kings (4 cards are kings)

38. *Baseball* A baseball player collected 210 hits in 630 times at bat. What (reduced) fraction of the time did he get a hit?

39. *Recipes* A recipe calls for $\frac{3}{5}$ cup of sugar. Another recipe calls for $\frac{1}{2}$ cup. Which recipe takes more sugar?

40. *Waste* In a recent year, about 10,000 tons of waste materials were dumped into the ocean. Of these 4500 tons were sewage.

a. What (reduced) fraction of the waste materials was sewage?

b. In the following year about $\frac{1}{5}$ of the waste materials dumped into the ocean were sewage. Proportionately, in which year was the dumping of sewage greater?

The following information will be used in Problems 41–44.

Housework study The New York State College of Human Ecology at Cornell University conducted a study of housework. The study found that the average housewife puts 8 hours a day into housework. What about the husband? He does only 96 minutes a day! Here is what the husband does in his 96 minutes of chores:

Cleaning	36 minutes
Kitchen work	12 minutes
Caring for the family	24 minutes
Shopping and paperwork	24 minutes

41. What (reduced) fraction of his time does the husband use in shopping and paperwork?

42. What (reduced) fraction of his time does the husband use in cleaning?

43. What (reduced) fraction of his time does the husband use in doing kitchen work?

44. What (reduced) fraction of his time does the husband use in caring for the family?

The following information will be used in Problems 45–47.

Big Mac A Big Mac weighs 200 grams and provides

25 grams of protein
40 grams of carbohydrates
35 grams of fat

45. What (reduced) fraction of a Big Mac is protein?

46. What (reduced) fraction of a Big Mac is carbohydrates?

47. What (reduced) fraction of a Big Mac is fat?

The following information will be used in Problems 48–50.

Two slices of a 12-inch Domino's cheese pizza weigh 140 grams and provide:

18 grams of protein
52 grams of carbohydrates
6 grams of fat

48. What (reduced) fraction of the 2 slices is protein?

49. What (reduced) fraction of the 2 slices is carbohydrates?

50. What (reduced) fraction of the 2 slices is fat?

	Serving Size (g)	Calories	Calories from Fat	Total Fat (g)	Saturated Fat (g)	Trans Fats (g)	Cholesterol (mg)	Sodium (mg)	Carbohydrates (g)	Dietary Fiber (g)	Sugars (g)	Protein (g)
6" Bourbon Chicken (Subway)	258	350	45	5.0	1.5	0	50	1020	54	4	16	25
Tendergrill™ Chicken Sandwich w/ Honey Mustard (Burger King)	258	450	90	10	2	0	75	1210	53	4	9	37

Nutritional information The chart gives the nutritional information for a 6-inch Bourbon Chicken Sandwich from Subway and a Tendergrill Chicken Sandwich from Burger King. Each of the sandwiches weighs **258** grams.

51. What reduced fraction of the 258-gram serving size grams are total fat for each sandwich?

a. Subway sandwich
b. Burger King sandwich

52. What reduced fraction of the 258-gram serving size grams are trans fats for each sandwich?

a. Subway sandwich
b. Burger King sandwich

53. What reduced fraction of the 258-gram serving size grams are carbohydrates for each sandwich?

a. Subway sandwich
b. Burger King sandwich

54. What reduced fraction of the 258-gram serving size grams are dietary fibers for each sandwich?

a. Subway sandwich
b. Burger King sandwich

55. What reduced fraction of the 258-gram serving size grams are sugars for each sandwich?

a. Subway sandwich
b. Burger King sandwich

56. What reduced fraction of the 258-gram serving size grams are proteins for each sandwich?

a. Subway sandwich
b. Burger King sandwich

››› Using Your Knowledge

A **ratio** is not only a fraction but is also a way of comparing two or more quantities. For example, if in a group of 10 people, there are 3 women and 7 men, the ratio of women to men is

$$\frac{3}{7} \begin{array}{l}\leftarrow \text{Number of women}\\ \leftarrow \text{Number of men}\end{array}$$

On the other hand, if there are 6 women and 4 men in the group, the *reduced* ratio of men to women is

$$\frac{4}{6} = \frac{2}{3} \begin{array}{l}\leftarrow \text{Number of men}\\ \leftarrow \text{Number of women}\end{array}$$

Another way to write this ratio is "2 to 3."

57. A class is composed of 25 girls and 30 boys. Find the reduced ratio of girls to boys.

58. Stock analysts consider the price-to-earnings (P/E) ratio when buying or selling stock. If the price of a certain stock is $20 and its earnings are $5, what is the P/E ratio of the stock?

Web IT go to mhhe.com/bello for more lessons

59. Do you know what the teacher–student ratio is in your school? For example, if your school has 5000 students and 250 teachers, the teacher–student ratio is

$$\frac{250}{5000} = \frac{1}{20}$$

This means that for every teacher there are 20 students.

a. Find the teacher-to-student ratio in a school with 200 teachers and 5600 students.

b. If a school wishes to maintain a $\frac{1}{20}$ ratio, and the enrollment goes up to 8000 students, you can find how many teachers you need by writing

Ratio you want $\frac{1}{20} = \frac{?}{8000}$ ← Teachers you need / ← Students you have

How many teachers are needed?

››› Write On

60. Write in your own words the meaning of the fundamental law of fractions.

61. What does it mean when we say that two fractions are equivalent? Give examples of equivalent fractions.

62. What does it mean when we say that a fraction is reduced to lowest terms? Give examples of fractions that are not reduced to lowest terms and the procedure that must be used to reduce them.

63. Is there a difference between "simplify" and "reduce to lowest terms"? Explain.

64. Write the procedure you use to find the GCF of two numbers.

››› Algebra Bridge

Reduce to lowest terms.

65. $\frac{6x^3}{8x^2}$ **66.** $\frac{8x^3}{6x^2}$ **67.** $\frac{12x^4}{18x^3}$ **68.** $\frac{16x^5}{32x^7}$

››› Concept Checker

Fill in the blank(s) with the correct word(s), phrase, or mathematical statement.

69. Two fractions are ________ if they **represent numerals** or **names** for the **same** number.

70. The **Fundamental Property of Fractions** states that for any nonzero numbers a, b, and c, $\frac{a}{b}$ = ______.

71. A **fraction** is ________ when **there are no common factors** (except 1) in the **numerator** and the **denominator.**

72. To **compare** two fractions with **different** denominators, write both fractions with a **denominator** equal to the ________ of the **original** ones.

sum
reduced
product
equivalent
$\frac{a \cdot c}{b \cdot c}$
$\frac{b \cdot c}{a \cdot c}$

››› Mastery Test

73. Find the fraction equivalent to $\frac{3}{5} = \frac{?}{25}$.

74. Find the fraction equivalent to $\frac{4}{9} = \frac{24}{?}$.

75. Find the fraction equivalent to $\frac{9}{75} = \frac{?}{25}$.

76. Find the fraction equivalent to $\frac{54}{90} = \frac{6}{?}$.

77. Reduce $\frac{20}{115}$.

78. Reduce $\frac{54}{90}$.

››› Skill Checker

Write as an improper fraction.

79. $3\frac{3}{8}$ **80.** $6\frac{5}{7}$ **81.** $7\frac{9}{10}$

82. $9\frac{2}{11}$ **83.** $10\frac{2}{13}$ **84.** $11\frac{1}{5}$

2.3 Multiplication and Division of Fractions and Mixed Numbers

Objectives

You should be able to:

A Multiply two fractions.

B Multiply a mixed number by a fraction and vice versa.

C Find the square or cube of a fraction or mixed number.

D Divide a fraction by another fraction.

E Divide a fraction by a mixed number.

F Solve applications involving the concepts studied.

G Find areas using multiplication of fractions.

To Succeed, Review How To . . .

Write a multiplication problem as a division problem (1.6). (pp. 62–63)

Getting Started

The photo shows 3 cups, each containing $\frac{1}{4}$ cup of sugar. How much sugar do they contain altogether? To find the answer we must multiply 3 by $\frac{1}{4}$, that is, we must find

$$3 \cdot \frac{1}{4}$$

We have 3 one-quarter cups of sugar, which make $\frac{3}{4}$ cup. Thus, to find the answer, we multiply 3 by $\frac{1}{4}$, obtaining

$$3 \cdot \frac{1}{4} = \frac{3}{4}$$

We can show the idea pictorially like this:

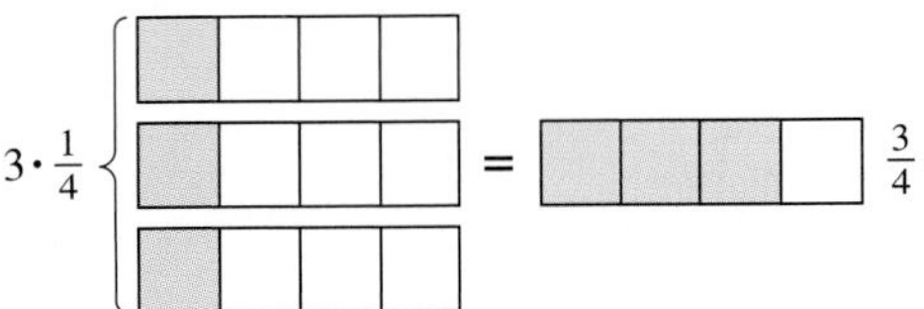

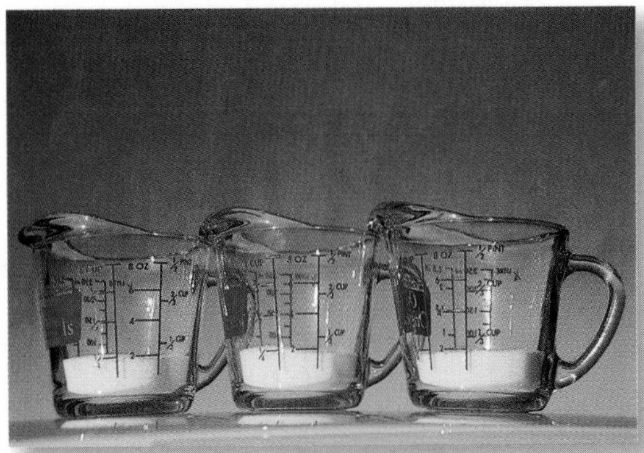

The diagram also suggests that multiplication is repeated addition; that is, $3 \times \frac{1}{4} = \frac{1}{4} + \frac{1}{4} + \frac{1}{4}$. Similarly, if a recipe calls for $\frac{1}{3}$ cup of flour and we wish to make only $\frac{1}{2}$ of the recipe, we have to find $\frac{1}{2}$ of $\frac{1}{3}$ (which means $\frac{1}{2} \times \frac{1}{3}$ because "of" is translated as "times"), that is, $\frac{1}{2} \cdot \frac{1}{3}$. Here is a diagram to help you do it.

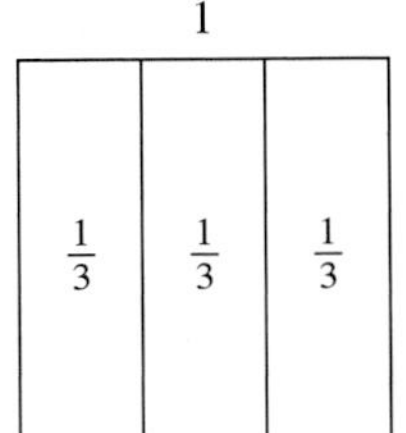

Each rectangle on the diagram represents $\frac{1}{3}$.

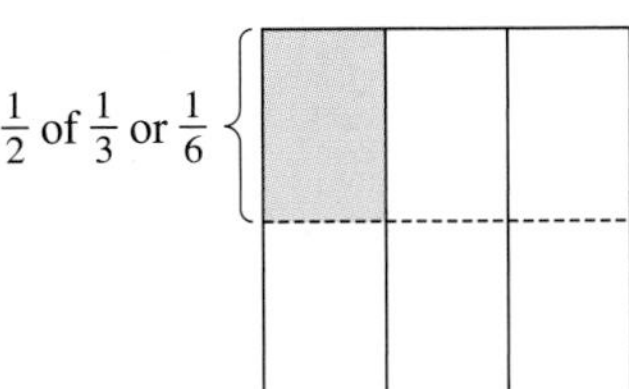

$$\frac{1}{2} \cdot \frac{1}{3} = \frac{1}{6}$$

A › Multiplying Fractions

Note that we can also find the product of $\frac{1}{2}$ and $\frac{1}{3}$ like this:

$$\frac{1}{2} \cdot \frac{1}{3} = \frac{1 \cdot 1}{2 \cdot 3} = \frac{1}{6}$$

Similarly,

$$\frac{2}{5} \cdot \frac{3}{7} = \frac{2 \cdot 3}{5 \cdot 7} = \frac{6}{35}$$

and

$$\frac{3}{2} \cdot \frac{2}{5} = \frac{3 \cdot 2}{2 \cdot 5} = \frac{6}{10} = \frac{3}{5}$$

Notice that we reduced $\frac{6}{10}$, the answer obtained when multiplying $\frac{3}{2} \cdot \frac{2}{5}$.

In general, to multiply two fractions together, we multiply their numerators and their denominators.

RULE FOR MULTIPLYING FRACTIONS

The **product** of two fractions is a fraction whose *numerator* is *the product of the numerators* of the given fractions and whose *denominator* is *the product of their denominators*. In symbols,

$$\frac{a}{b} \cdot \frac{c}{d} = \frac{a \cdot c}{b \cdot d}$$

EXAMPLE 1 Multiplying fractions

Multiply.

a. $\frac{2}{7} \cdot \frac{3}{11}$ b. $\frac{3}{5} \cdot \frac{2}{3}$

SOLUTION

a. $\frac{2}{7} \cdot \frac{3}{11} = \frac{2 \cdot 3}{7 \cdot 11} = \frac{6}{77}$ b. $\frac{3}{5} \cdot \frac{2}{3} = \frac{3 \cdot 2}{5 \cdot 3} = \frac{6}{15} = \frac{2}{5}$

PROBLEM 1

Multiply.

a. $\frac{2}{5} \cdot \frac{4}{7}$

b. $\frac{3}{4} \cdot \frac{2}{3}$

When multiplying fractions it saves time if common factors are eliminated before you multiply. Thus, to multiply $\frac{1}{3}$ by $\frac{9}{10}$, we write

$$\frac{1}{\underset{1}{\cancel{3}}} \cdot \frac{\overset{3}{\cancel{9}}}{10} = \frac{3}{10}$$

Instead of writing

$$\frac{1}{3} \cdot \frac{9}{10} = \frac{1 \cdot 9}{3 \cdot 10} = \frac{9}{30} = \frac{\cancel{3} \cdot 3}{\cancel{3} \cdot 10} = \frac{3}{10}$$

we just divided numerator and denominator by the common factor 3 before multiplying.

Answers to PROBLEMS

1. a. $\frac{8}{35}$ b. $\frac{1}{2}$

B > Multiplication with Mixed Numbers

To multiply by a mixed number, write the mixed number as a fraction *first,* since our rule for multiplying states only how to multiply *fractions*. We illustrate this in Example 2.

EXAMPLE 2 Multiplying mixed numbers by fractions

Here is a recipe for hamburger. We wish to cut it in half (take $\frac{1}{2}$ of each ingredient).

a. Find the amount of ground beef needed.
b. Find the amount of veal needed.
c. Find the amount of onion needed.
d. Find the amount of salt needed.
e. Find the amount of garlic salt needed.
f. Find the amount of pepper needed.

Hamburger
$1\frac{1}{2}$ pounds ground beef
1 pound ground veal
$\frac{1}{4}$ cup minced onion
2 teaspoons salt
$\frac{1}{2}$ teaspoon garlic salt
$\frac{1}{3}$ teaspoon pepper

SOLUTION

a. We need $\frac{1}{2} \cdot 1\frac{1}{2}$ pounds of ground beef. We first write $1\frac{1}{2}$ as $\frac{3}{2}$, then we multiply.

$$\frac{1}{2} \cdot 1\frac{1}{2} = \frac{1}{2} \cdot \frac{3}{2} = \frac{1 \cdot 3}{2 \cdot 2} = \frac{3}{4} \text{ pound of ground beef}$$

b. We need $\frac{1}{2} \cdot 1$ pound of veal.

$$\frac{1}{2} \cdot 1 = \frac{1}{2}$$

Thus, we need $\frac{1}{2}$ pound of ground veal. Remember, any number multiplied by 1 equals the original number.

c. Here we need $\frac{1}{2} \cdot \frac{1}{4}$ cup of onion.

$$\frac{1}{2} \cdot \frac{1}{4} = \frac{1 \cdot 1}{2 \cdot 4} = \frac{1}{8}$$

Thus, we need $\frac{1}{8}$ cup of minced onion.

d. Now, we need $\frac{1}{2} \cdot 2$ teaspoons of salt. To multiply by a whole number such as 2, write the whole number as a fraction with denominator 1, that is, as $\frac{2}{1}$. Then multiply.

$$\frac{1}{2} \cdot 2 = \frac{1}{2} \cdot \frac{\overset{1}{\cancel{2}}}{1} = 1$$

Thus, we need 1 teaspoon of salt.

e. This time we need $\frac{1}{2} \cdot \frac{1}{2}$ teaspoon of garlic salt.

$$\frac{1}{2} \cdot \frac{1}{2} = \frac{1 \cdot 1}{2 \cdot 2} = \frac{1}{4}$$

Thus, we need $\frac{1}{4}$ teaspoon of garlic salt.

f. Finally, we need $\frac{1}{2} \cdot \frac{1}{3}$ teaspoon of pepper.

$$\frac{1}{2} \cdot \frac{1}{3} = \frac{1 \cdot 1}{2 \cdot 3} = \frac{1}{6}$$

Thus, we need $\frac{1}{6}$ teaspoon of pepper.

PROBLEM 2

We wish to make $\frac{1}{3}$ as much hamburger. Find the needed amount of each ingredient.

a. Ground beef ________
b. Veal ________
c. Onion ________
d. Salt ________
e. Garlic salt ________
f. Pepper ________

If you multiply a fraction by 1, you get the same fraction.

Answers to PROBLEMS

2. a. $\frac{1}{2}$ pound **b.** $\frac{1}{3}$ pound **c.** $\frac{1}{12}$ cup **d.** $\frac{2}{3}$ teaspoon **e.** $\frac{1}{6}$ teaspoon **f.** $\frac{1}{9}$ teaspoon

What did we learn from Example 2?

RULE TO MULTIPLY A FRACTION BY A MIXED NUMBER

1. To multiply a fraction by a mixed number, *first* convert the mixed number to a fraction. Thus,

Convert to a fraction

$$3\frac{1}{7} \cdot \frac{2}{5} = \frac{22}{7} \cdot \frac{2}{5} = \frac{44}{35}$$

2. Any fraction multiplied by 1 equals the fraction. Thus,

$$\frac{3}{19} \cdot 1 = \frac{3}{19}$$

3. To multiply a fraction by a whole number (such as 5), write the number as a fraction with a denominator of 1. Thus, write 5 as $\frac{5}{1}$ and multiply as shown.

$$5 \cdot \frac{3}{10} = \frac{\overset{1}{\cancel{5}}}{1} \cdot \frac{3}{\underset{2}{\cancel{10}}} = \frac{3}{2}$$

$5 = \frac{5}{1}$

What about multiplying more than two fractions? See Example 3!

EXAMPLE 3 Multiplying mixed numbers by fractions

Find $2\frac{1}{3} \cdot \frac{3}{5} \cdot \frac{5}{7}$.

SOLUTION We first write $2\frac{1}{3}$ as $\frac{7}{3}$, and then we proceed as shown:

$$\frac{\overset{1}{\cancel{7}}}{\underset{1}{\cancel{3}}} \cdot \frac{\overset{1}{\cancel{3}}}{\underset{1}{\cancel{5}}} \cdot \frac{\overset{1}{\cancel{5}}}{\underset{1}{\cancel{7}}} = 1$$

Recall that $2\frac{1}{3} = \frac{2 \cdot 3 + 1}{3} = \frac{7}{3}$

The answer is 1.

PROBLEM 3

Find $3\frac{1}{4} \cdot \frac{4}{3} \cdot \frac{3}{13}$.

C › Exponents and Fractions

Do you remember what 3^2 means?

$$3^2 = 3 \cdot 3 = 9$$

What about $\left(\frac{2}{3}\right)^2$?

$$\left(\frac{2}{3}\right)^2 = \frac{2}{3} \cdot \frac{2}{3} = \frac{4}{9}$$

$\left(\frac{2}{3}\right)^2$ means raising $\frac{2}{3}$ to the **second** power.

$$\left(\frac{2}{3}\right)^3 = \frac{2}{3} \cdot \frac{2}{3} \cdot \frac{2}{3} = \frac{8}{27}$$

Here we raise $\frac{2}{3}$ to the **third** power.

We use these ideas in Example 4.

Answers to PROBLEMS

3. 1

EXAMPLE 4 **Raising a fraction or mixed number to a power**

Evaluate.

a. $\left(\frac{3}{4}\right)^3$ **b.** $\left(1\frac{1}{5}\right)^2$ **c.** $\left(\frac{3}{4}\right)^2 \cdot \left(1\frac{1}{3}\right)$

SOLUTION

a. $\left(\frac{3}{4}\right)^3 = \frac{3}{4} \cdot \frac{3}{4} \cdot \frac{3}{4} = \frac{27}{64}$

b. We first write $1\frac{1}{5}$ as $\frac{6}{5}$. Thus

$$\left(1\frac{1}{5}\right)^2 = \left(\frac{6}{5}\right)^2 = \frac{6}{5} \cdot \frac{6}{5} = \frac{36}{25}$$

c. $\left(\frac{3}{4}\right)^2 \cdot \left(1\frac{1}{3}\right) = \frac{3}{4} \cdot \frac{\overset{1}{\cancel{3}}}{\underset{1}{\cancel{4}}} \cdot \frac{\overset{1}{\cancel{4}}}{\underset{1}{\cancel{3}}} = \frac{3}{4}$

PROBLEM 4

Evaluate.

a. $\left(\frac{2}{5}\right)^3$

b. $\left(1\frac{1}{4}\right)^2$

c. $\left(\frac{2}{3}\right)^2 \cdot \left(2\frac{1}{4}\right)$

D › Division of Fractions

Now we are ready to do division!

Henry Serrano, Tampa, Florida

In the cartoon it is claimed that 5 will divide into 2 if you push it! What this means is that $2 \div 5$ is *not* a counting number but a fraction. Thus

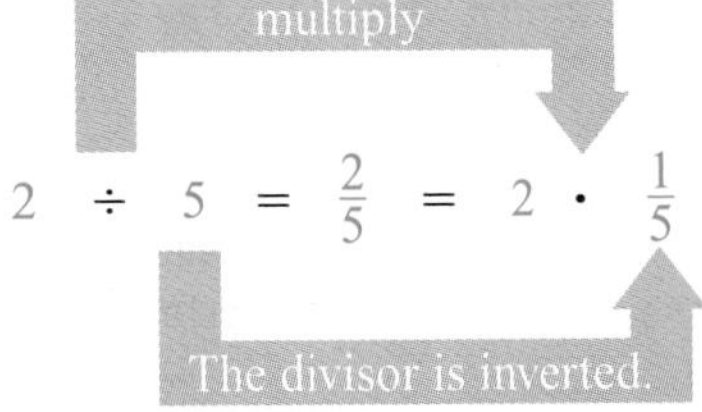

$$2 \div 5 = \frac{2}{5} = 2 \cdot \frac{1}{5}$$

Note that to divide 2 by 5 we *multiplied* 2 by $\frac{1}{5}$, where the fraction $\frac{1}{5}$ was obtained by *inverting* $5 = \frac{5}{1}$ to obtain $\frac{1}{5}$. (In mathematics, 5 and $\frac{1}{5}$ are called *reciprocals* of each other.) Here is the rule we need:

RULE FOR FINDING THE RECIPROCAL OF A FRACTION

To find the reciprocal of $\frac{a}{b}$, **invert** the fraction (interchange the numerator and denominator) to obtain $\frac{b}{a}$. This means that $\frac{a}{b}$ and $\frac{b}{a}$ are **reciprocals** of each other.

Answers to PROBLEMS

4. a. $\frac{8}{125}$ **b.** $\frac{25}{16}$ **c.** 1

Thus, the reciprocal of 5 is $\frac{1}{5}$, the reciprocal of $\frac{2}{3}$ is $\frac{3}{2}$, and the reciprocal of $\frac{5}{4}$ is $\frac{4}{5}$. Now let us try the problem $5 \div \frac{5}{7}$. If we try to do it like the previous problem we must multiply 5 by the reciprocal of $\frac{5}{7}$, that is,

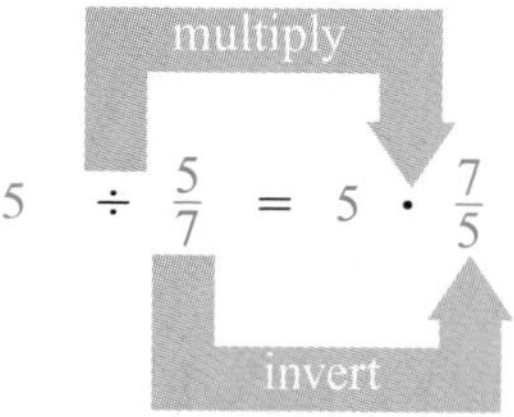

Since $5 = \frac{5}{1}$, $5 \cdot \frac{7}{5} = \frac{5}{1} \cdot \frac{7}{5} = 7$. Thus, $5 \div \frac{5}{7} = 7$. To check this answer we recall that any division problem can be written as an equivalent multiplication problem. Thus, the problem $12 \div 4 = \square$ can be written as $12 = 4 \cdot \square$. Similarly,

$$5 \div \frac{5}{7} = \square \quad \text{means} \quad 5 = \frac{5}{7} \cdot \square$$

What number can we place in the box so that $5 = \frac{5}{7} \times \square$? The answer is 7, since $\frac{5}{7} \times 7 = 5$.

We now have a simple rule for dividing fractions:

RULE FOR DIVIDING FRACTIONS: INVERT THE DIVISOR

To divide $\frac{a}{b}$ by $\frac{c}{d}$, *invert* the divisor $\frac{c}{d}$ and multiply.

invert

In symbols, $\frac{a}{b} \div \frac{c}{d} = \frac{a}{b} \cdot \frac{d}{c}$.

Note : You can also state this rule like this:

To divide $\frac{a}{b}$ by $\frac{c}{d}$, multiply $\frac{a}{b}$ by the *reciprocal* of $\frac{c}{d}$.

For example;

multiply

$$7 \div \frac{3}{5} = \frac{7}{1} \cdot \frac{5}{3} = \frac{35}{3}$$

invert

Caution To divide by a fraction, we *invert the divisor only* and then multiply.

Recall that to *invert* means to switch the numerator and denominator. Thus inverting $\frac{3}{5}$ gives $\frac{5}{3}$, inverting $\frac{1}{6}$ gives $\frac{6}{1}$, or 6, and inverting 2 gives $\frac{1}{2}$. Here are some more examples:

$$\frac{3}{4} \div \frac{5}{7} = \frac{3}{4} \cdot \frac{7}{5} = \frac{21}{20}$$

Check

$$\frac{3}{4} \div \frac{5}{7} = \frac{21}{20} \quad \text{because} \quad \frac{3}{4} = \frac{\overset{1}{\cancel{5}}}{\underset{1}{\cancel{7}}} \cdot \frac{\overset{3}{\cancel{21}}}{\underset{4}{\cancel{20}}}$$

invert

$$\frac{5}{11} \div \frac{5}{7} = \frac{5}{11} \cdot \frac{7}{5} = \frac{7}{11}$$

Check

$$\frac{5}{11} \div \frac{5}{7} = \frac{7}{11} \quad \text{because} \quad \frac{5}{11} = \frac{5}{\cancel{7}_1} \cdot \frac{\cancel{7}^1}{11}$$

EXAMPLE 5 Dividing by a fraction and by a whole number

Divide.

a. $\frac{3}{5} \div \frac{2}{7}$

b. $\frac{4}{9} \div 5$

SOLUTION

a. $\frac{3}{5} \div \frac{2}{7} = \frac{3}{5} \cdot \frac{7}{2} = \frac{21}{10}$ (invert)

b. $\frac{4}{9} \div 5 = \frac{4}{9} \cdot \frac{1}{5} = \frac{4}{45}$

PROBLEM 5

Divide.

a. $\frac{5}{7} \div \frac{3}{8}$

b. $\frac{3}{4} \div 7$

E › Division with Mixed Numbers

As in the case of multiplication, if mixed numbers are involved, they are changed to improper fractions first, like this:

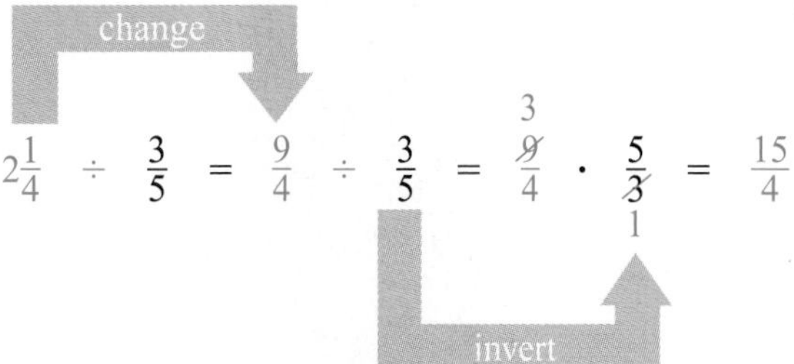

EXAMPLE 6 Division involving mixed numbers

Divide.

a. $3\frac{1}{4} \div \frac{7}{8}$

b. $\frac{11}{12} \div 7\frac{1}{3}$

SOLUTION

a. $3\frac{1}{4} \div \frac{7}{8} = \frac{13}{4} \div \frac{7}{8} = \frac{13}{\cancel{4}_1} \cdot \frac{\cancel{8}^2}{7} = \frac{26}{7}$

b. $\frac{11}{12} \div 7\frac{1}{3} = \frac{11}{12} \div \frac{22}{3} = \frac{\cancel{11}^1}{\cancel{12}_4} \cdot \frac{\cancel{3}^1}{\cancel{22}_2} = \frac{1}{8}$

PROBLEM 6

Divide.

a. $5\frac{1}{2} \div \frac{3}{4}$

b. $\frac{6}{7} \div 3\frac{1}{7}$

F › Applications Involving Multiplying and Dividing Fractions

If you are considering purchasing a home, you need to figure out how much you can afford. Home and Garden Bulletin No. 182 (U.S. Dept. of Agriculture) suggests that a family spend no more than $2\frac{1}{2}$ times its total annual income for a home.

Answers to PROBLEMS

5. a. $\frac{40}{21}$ b. $\frac{3}{28}$ 6. a. $\frac{22}{3}$ b. $\frac{3}{11}$

EXAMPLE 7 **Purchasing a home**

Supposing your family has \$24,000 in total annual income, how much can you afford to pay for a home?

SOLUTION

$2\frac{1}{2}$ times 24,000

or

$$\frac{5}{2} \cdot 24{,}000 = \frac{5}{\cancel{2}} \cdot \frac{\overset{12{,}000}{\cancel{24{,}000}}}{1} = 60{,}000$$

Thus, your family can afford a \$60,000 home.

PROBLEM 7

A family has \$41,000 in total annual income. How much can it afford for a home?

EXAMPLE 8 **Capacity of a bottle**

The 2-in-1 bottle contains 130 grams of gel. If you use $1\frac{1}{5} = \frac{6}{5}$ of a gram each time you brush your teeth, how many times can you brush with the 130 grams in the bottle?

SOLUTION Since you have 130 grams of gel and you use $1\frac{1}{5}$ grams per application, you should be able to have $130 \div 1\frac{1}{5}$ brushings. Now, $130 \div 1\frac{1}{5} = 130 \div \frac{6}{5} = 130 \cdot \frac{5}{6} = \frac{650}{6} = 108\frac{1}{3}$ or about 108 brushings.

PROBLEM 8

How many brushings are possible if you use $1\frac{2}{5}$ grams each time you brush?

G > Applications: Finding Area

As you recall from Section 1.5,

AREA OF A RECTANGLE

The area *A* of a rectangle is found by multiplying its length *L* by its width *W*.

$$A = L \cdot W$$

Apply this formula in Example 9.

EXAMPLE 9 **Find the area of a room**

A room measures $3\frac{1}{3}$ yd by $5\frac{2}{3}$ yd. How many square yards of carpet would it take to cover the floor?

SOLUTION The area of the floor is: $3\frac{1}{3} \cdot 5\frac{2}{3} = \frac{10}{3} \cdot \frac{17}{3} = \frac{170}{9} = 18\frac{8}{9}$ yd² or almost 19 yd², so it takes about 19 square yards of carpet to cover the floor.

PROBLEM 9

Find the area of a room measuring $4\frac{1}{3}$ yards by $5\frac{2}{3}$ yards.

Answers to PROBLEMS

7. \$102,500 **8.** About 93 ($92\frac{6}{7}$)
9. $24\frac{5}{9}$ yd²

Calculator Corner

You can multiply and divide fractions with a scientific calculator. The difficulty is that your answer will be a decimal. Thus, to multiply $\frac{3}{5} \times \frac{5}{8}$ you enter 3 [÷] 5 [×] 5 [÷] 8 [ENTER]. The display will show 0.375, which is the correct answer but in decimal form. With a fraction calculator with an [x/y] key, you enter 3 [x/y] 5 [×] 5 [x/y] 8 [ENTER]. The result is given as 15/40. But the answer is not in reduced form! To do that press [SIMP]. The calculator asks: FACTOR? The GCF here is 5, so enter 5 and press [ENTER]. The answer 3/8 is displayed.

You can also do multiplication involving mixed numbers if you know how to enter the mixed numbers. To do so you enter the whole number part first and then the fraction part. Thus, to enter $3\frac{4}{5}$, enter 3[UNIT] 4 [x/y] 5. If you now decide to multiply by 2, just enter [×] 2 [ENTER]. If you want the answer 38/5 displayed as a mixed number, press [A B/C] and you get $7_U3/5$, which means $7\frac{3}{5}$.

› Exercises 2.3

Boost your grade at mathzone.com!
- > Practice Problems
- > NetTutor
- > Self-Tests
- > e-Professors
- > Videos

‹ A › Multiplying Fractions In Problems 1–14, multiply (reduce answers to lowest terms).

1. $\frac{3}{4} \cdot \frac{7}{8}$ **2.** $\frac{2}{3} \cdot \frac{7}{3}$ **3.** $\frac{1}{6} \cdot \frac{6}{7}$ **4.** $\frac{5}{9} \cdot \frac{4}{5}$ **5.** $\frac{2}{5} \cdot \frac{5}{3}$

6. $\frac{6}{5} \cdot \frac{7}{6}$ **7.** $3 \cdot \frac{2}{5}$ **8.** $\frac{3}{4} \cdot 7$ **9.** $\frac{5}{6} \cdot \frac{3}{5}$ **10.** $\frac{7}{3} \cdot \frac{6}{7}$

11. $\frac{7}{5} \cdot \frac{15}{14}$ **12.** $\frac{2}{7} \cdot \frac{21}{8}$ **13.** $\frac{6}{7} \cdot \frac{14}{3}$ **14.** $\frac{21}{2} \cdot \frac{8}{7}$

‹ B › Multiplication with Mixed Numbers In Problems 15–24, multiply (reduce answers to lowest terms).

15. $1\frac{2}{3} \cdot \frac{6}{5}$ **16.** $2\frac{1}{4} \cdot \frac{4}{7}$ **17.** $\frac{9}{4} \cdot 3\frac{1}{9}$ **18.** $\frac{2}{15} \cdot 2\frac{1}{2}$ **19.** $2\frac{1}{3} \cdot 4\frac{1}{2}$

20. $2\frac{3}{5} \cdot 2\frac{1}{7}$ **21.** $3 \cdot 4\frac{1}{3}$ **22.** $5 \cdot 1\frac{2}{5}$ **23.** $5\frac{1}{6} \cdot 12$ **24.** $3\frac{1}{3} \cdot 6$

‹ C › Exponents and Fractions In Problems 25–36, multiply.

25. $\left(\frac{1}{3}\right)^2$ **26.** $\left(\frac{4}{5}\right)^2$ **27.** $\left(2\frac{1}{2}\right)^2$ **28.** $\left(1\frac{1}{4}\right)^2$

29. a. $\frac{3}{4} \times \frac{8}{9} \times \frac{1}{5}$ **b.** $\frac{5}{12} \times \frac{6}{7} \times \frac{7}{5}$

30. a. $\frac{4}{5} \times 2\frac{1}{2} \times 3$ **b.** $\frac{3}{8} \times 2\frac{1}{3} \times 4$

31. $\left(\frac{2}{3}\right)^2 \cdot \frac{3}{4}$ **32.** $\left(\frac{4}{5}\right)^2 \cdot \frac{7}{8}$

33. $\frac{14}{27} \cdot \left(\frac{3}{7}\right)^2$ **34.** $\frac{5}{12} \cdot \left(\frac{6}{5}\right)^2$ **35.** $\left(\frac{2}{3}\right)^3$ **36.** $\left(\frac{3}{5}\right)^3$

‹ D › Division of Fractions In Problems 37–48, divide (reduce answers to lowest terms).

37. $5 \div \frac{2}{3}$ **38.** $7 \div \frac{3}{5}$ **39.** $\frac{4}{5} \div 6$ **40.** $\frac{3}{4} \div 9$

41. $\frac{2}{3} \div \frac{6}{7}$ **42.** $\frac{3}{5} \div \frac{9}{10}$ **43.** $\frac{4}{5} \div \frac{8}{15}$ **44.** $\frac{3}{7} \div \frac{9}{14}$

45. $\frac{2}{3} \div \frac{5}{12}$ **46.** $\frac{1}{2} \div \frac{3}{4}$ **47.** $\frac{3}{4} \div \frac{3}{4}$ **48.** $\frac{9}{10} \div \frac{3}{5}$

‹ E › Division with Mixed Numbers In Problems 49–60, divide (reduce answers to lowest terms).

49. $\frac{3}{5} \div 1\frac{1}{2}$ **50.** $\frac{5}{8} \div 3\frac{1}{3}$ **51.** $3\frac{3}{4} \div \frac{3}{8}$ **52.** $1\frac{1}{5} \div \frac{3}{5}$

53. $6\frac{1}{2} \div 2\frac{1}{2}$ **54.** $1\frac{5}{8} \div 2\frac{7}{8}$ **55.** $3\frac{1}{8} \div 1\frac{1}{3}$ **56.** $2\frac{1}{2} \div 6\frac{1}{4}$

57. $3\frac{1}{8} \div 3\frac{1}{8}$ **58.** $10\frac{1}{2} \div 2\frac{1}{3}$ **59.** $1\frac{2}{3} \div 13\frac{3}{4}$ **60.** $4\frac{7}{10} \div 4\frac{7}{10}$

› Web IT go to mhhe.com/bello for more lessons

Web IT go to mhhe.com/bello for more lessons

⟨F⟩⟨G⟩ Applications Involving Multiplying and Dividing Fractions

61. *Area of pasture* A farm has an area of $\frac{2}{3}$ square mile and $\frac{3}{7}$ of the land is pasture. What area of the farm is pasture in square miles?

62. *Meatball recipe* A recipe for making meatballs to serve 100 people calls for 75 pounds of meat. Charlie Chef made $\frac{2}{3}$ of this recipe.

a. How many pounds of beef did he use?

b. About how many people will $\frac{2}{3}$ of the recipe serve?

63. *Number of attendees* Rosa invited 90 people to a party and $\frac{4}{5}$ of them came. How many people came to the party?

64. *Floating in space* Lt. Col. Aleksey Arkhipovich Leonov was the first person to leave an artificial satellite during orbit. He was in space for 20 minutes, and $\frac{3}{5}$ of this time he "floated" at the end of a line. How long did he float in space?

65. *Rain in Prince George* In Prince George it rains or snows on an average of $\frac{8}{15}$ of the days in November. How many days is that?

66. *Weight of the Lunar Rover on the moon* The weight of an object on the moon is $\frac{1}{6}$ of its weight on earth. How much did the Lunar Rover, weighing 450 pounds on earth, weigh on the moon?

67. *Screw penetration* On each turn a screw goes $\frac{3}{16}$ of an inch into the wood. How many turns are needed to make it go in $1\frac{1}{2}$ inches?

68. *Wallpaper* One sheet of wallpaper covers $4\frac{1}{2}$ feet of wall. How many sheets are needed to cover $24\frac{3}{4}$ feet of wall?

69. *Making vests* Pete has $10\frac{1}{2}$ yards of material. How many vests can he make if each one takes $\frac{5}{8}$ yard of material?

70. *Buying bonds* A bond is selling for $\$3\frac{1}{2}$. How many can be bought with \$98?

71. *Unit conversions* One rod equals $16\frac{1}{2}$ feet. How many feet is 40 rods?

72. *Unit conversions* Many horse races are 7 furlongs. If 40 rods is a furlong:

a. How many rods are in 7 furlongs?

b. If one rod is $16\frac{1}{2}$ feet, how many feet are in a furlong?

c. How many feet are in a 7-furlong race?

73. *Unit conversions* In the Old Testament, there is a unit of measurement called an *omer*. An omer is $2\frac{1}{5}$ liters. How many liters are in 5 omers?

74. *Unit conversions* In the Old Testament, there is a unit of weight called a *shekel*. A shekel is $11\frac{2}{5}$ grams. How many grams are in 10 shekels?

75. *Unit conversions* A gallon of gasoline weighs $6\frac{1}{5}$ pounds. If the content of your gas tank weighs $80\frac{3}{5}$ pounds, how many gallons of gasoline does the tank contain?

The following information will be used in Problems 76–80.

How many uses are in each Aquafresh® tube or pump?

Using 1 to 1.2 grams of toothpaste on the brush (numbers are approximate)

PUMPS	4.3 oz	90 brushings
	4.6 oz	110 brushings
	6.0 oz	140 brushings
	6.4 oz	160 brushings
TUBES	1.4 oz	37 brushings
	2.7 oz	73 brushings
	4.3 oz	100 brushings
	4.6 oz	120 brushings
	6.0 oz	150 brushings
	6.4 oz	170 brushings
	7.6 oz	200 brushings
	8.2 oz	220 brushings

Toothpaste The number of uses (brushings) in each Aquafresh tube or pump is shown. Find the grams of toothpaste per brushing.

76. 90 brushings in the 120-gram pump

77. 160 brushings in the 180-gram pump

78. 37 brushings in the 40-gram tube

79. 120 brushings in the 130-gram tube

80. 220 brushings in the 230-gram tube

Scales on maps The scale on a certain map is 1 inch = 36 miles. If the distance from Indianapolis to Dayton on the map is $2\frac{1}{4}$ inch, the approximate distance is $36 \times 2\frac{1}{4} = 36 \times \frac{9}{4} = 81$ miles.

81. Find the approximate distance from Indianapolis to Cincinnati, a distance of $2\frac{2}{3}$ inches on the map.

82. Find the approximate distance from Indianapolis to Terre Haute, a distance of $2\frac{1}{4}$ inches on the map.

83. Find the approximate distance from Indianapolis to Chicago, a distance of $4\frac{2}{3}$ inches on the map.

84. If the approximate distance between two cities is 108 miles, what is the distance on the map?

85. If the approximate distance between Indianapolis and Detroit is 240 miles, what is the distance on the map?

86. If the approximate distance between Indianapolis and Cleveland is 279 miles, what is the distance on the map?

Area In Problems 87–91, find the area of the envelope.

87. A-2 size measuring $4\frac{3}{8}$ inches by $5\frac{3}{4}$ inches.

88. A-7 size measuring $5\frac{1}{4}$ inches by $7\frac{1}{4}$ inches.

89. Business size 10 measuring $4\frac{1}{8}$ inches by $9\frac{1}{2}$ inches.

90. A U.S. postal envelope measuring 5 inches by $3\frac{1}{2}$ inches.

91. Maximum size international envelope is $9\frac{1}{4}$ inches by $4\frac{2}{3}$ inches.

92. *Area* The width of the screen is $13\frac{2}{5}$ inches and the height is $9\frac{1}{2}$ inches. What is its area?

93. *Area* The width of the screen is $10\frac{1}{2}$ inches and the height is $12\frac{4}{5}$ inches. What is its area?

94. *Area* The dimensions of the Palm™ organizer are $3\frac{1}{10}$ inches by $4\frac{1}{2}$ inches. What is its area?

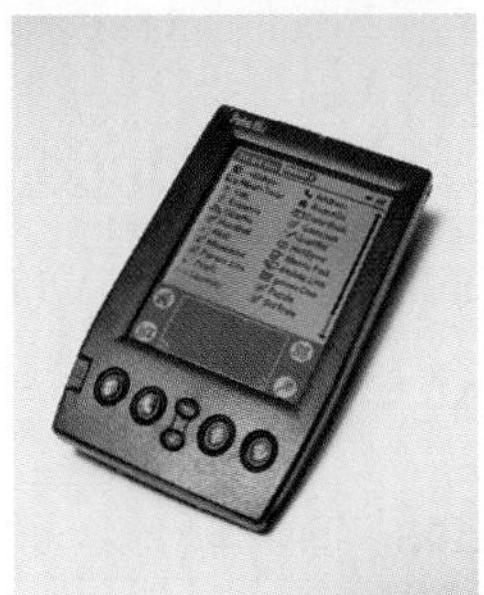

95. *Area* The area of a room is $15\frac{3}{4}$ square yards. If the room is $3\frac{1}{2}$ yards wide, how long is it?

96. *Area* A remodeling job calls for $65\frac{1}{2}$ square yards of carpet. How many remodeling jobs could you do with 655 square yards of carpet?

> > > Using Your Knowledge

By permission of John L. Hart FLP, and Creators Syndicate, Inc.

Web IT go to mhhe.com/bello for more lessons

Scale model A scale drawing or scale model is used to represent an object that is too large or small to be drawn actual size. Pictures in dictionaries show things to scale. For example, the picture of a bird may be $1\frac{1}{4}$ inches long. Actually the bird is $3\frac{1}{2}$ times as long. Thus, the bird is really

$$3\frac{1}{2} \times 1\frac{1}{4} = \frac{7}{2} \times \frac{5}{4} = \frac{35}{8} = 4\frac{3}{8} \text{ inches long}$$

97. *Blueprints* A landscape plan shows a flower bed $6\frac{1}{2}$ inches wide. If the scale on the plan is **1 in. = 4 ft,** what is the width of the actual flower bed?

98. *Blueprints* A set of drawings for an office building shows a conference room $7\frac{3}{4}$ inches long. If the scale is **1 in. = 6 ft,** what is the actual length of the conference room?

99. *Scale Model* An illustration of a honeybee is $4\frac{1}{2}$ centimeters long. If the scale is **1 cm = $\frac{1}{4}$ cm,** what is the actual size of the honeybee?

100. The picture of a bird in a dictionary is $1\frac{1}{2}$ inches long. The bird is actually $4\frac{1}{2}$ times as long. How long is the bird?

101. A bluebird is pictured with a length of $1\frac{1}{4}$ inches. The actual length is $2\frac{1}{2}$ times that of the picture. What is the actual length of the bird?

Write On

102. Write the procedure you use to multiply two fractions.

103. Write the procedure you use to multiply a fraction by 1.

104. Write the procedure you use to multiply a whole number by a fraction.

105. Write the procedure you use to divide one fraction by another fraction.

106. Write the procedure you use to divide a fraction by a nonzero whole number.

Concept Checker

Fill in the blank(s) with the correct word(s), phrase, or mathematical statement.

107. $\frac{a}{b} \cdot \frac{c}{d} =$ ____________

108. $\frac{a}{b} \div \frac{c}{d} =$ ____________

$\frac{a \cdot b}{c \cdot d}$

$\frac{a \cdot d}{b \cdot c}$

$\frac{a \cdot c}{b \cdot d}$

Mastery Test

Marinade:

1/4 cup balsamic vinegar
1/4 cup soy sauce
1/4 cup olive oil
3 cloves garlic, minced

109. The recipe for four servings of the marinade sauce used for Bello Burgers is above. Suppose you want to serve eight persons instead of four. How much balsamic vinegar do you need for the updated recipe?
Source: About.com

110. Multiply: $\frac{3}{5} \cdot \frac{2}{3}$

111. Multiply: $5 \cdot \frac{3}{4}$

112. Find: $\left(\frac{3}{4}\right)^3$

113. Find: $\left(\frac{3}{4}\right)^2 \cdot \left(1\frac{1}{2}\right)$

114. Divide: $\frac{3}{5} \div \frac{2}{3}$

115. Divide: $1\frac{3}{5} \div \frac{2}{5}$

116. Divide: $\frac{8}{5} \div 2\frac{2}{3}$

117. Find the area of a room that measures $3\frac{1}{3}$ yards by $4\frac{2}{3}$ yards.

Skill Checker

Write as a product of primes using exponents.

118. 84 **119.** 128 **120.** 72 **121.** 180 **122.** 105 **123.** 900

2.4 The Least Common Multiple (LCM)

Objectives

You should be able to:

A Find the Least Common Multiple (LCM) of two numbers.

B Find the Lowest Common Denominator (LCD) of two fractions and write them using the LCD as the denominator.

C Compare two fractions.

To Succeed, Review How To . . .

1. Write a number as a product of primes using exponents. (pp. 73–76)
2. Write a fraction as an equivalent one with a specified numerator or denominator. (pp. 120–122)

Getting Started

Planetary Alignments

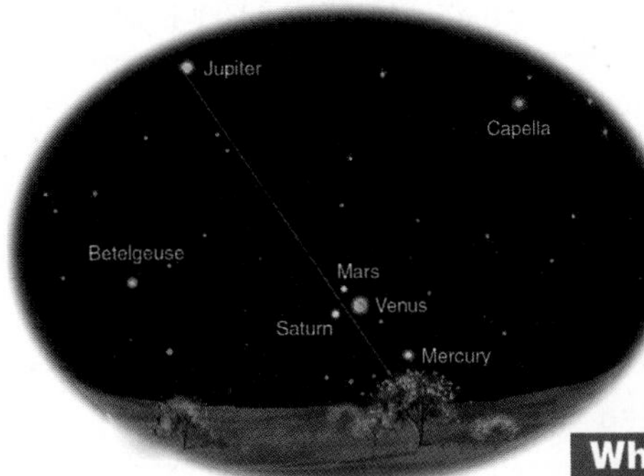

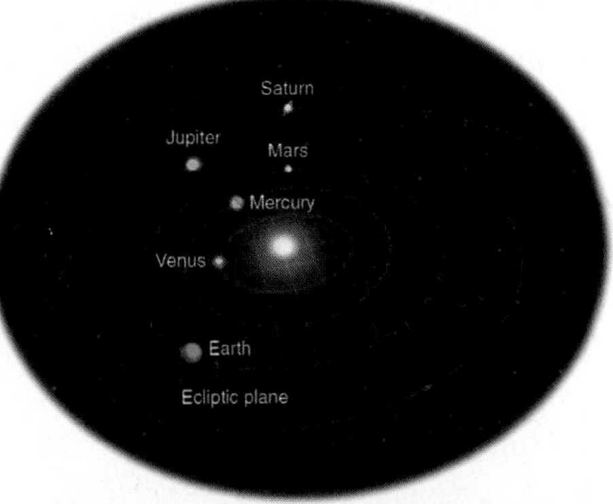

Why They Line Up

When: Monday, May 6

On May 6, Saturn, Mars, Venus, and Mercury fall into an almost straight line with Earth. Because the three dimensions of space are compressed into two when we view the night sky, these four planets appear to be at almost the same point in the sky. This "dimensional compression" also explains why all the planets stay close to the ecliptic line that cuts across the night sky. This line is really a plane in three dimensions, and the orbits of all planets lie pretty much along this plane.

Source: http://www.space.com

On May 5, 2000, and again on May 6, 2002, there was an unusual alignment of the planets Mercury, Venus, Mars, Jupiter, and Saturn as shown on the left, with their orbits shown on the right. When will this unusual alignment happen again? Astronomers say in 2040, 2060, and 2100. We will study the techniques for verifying this, but for now let us concentrate on Jupiter and Saturn. When will they line up again? It takes Jupiter 12 years to go around the sun once but Saturn takes 30 years! Let us look at the time (in years) it takes each of the planets to go around the sun 1, 2, 3, 4, and 5 orbits.

	1 orbit	2 orbits	3 orbits	4 orbits	5 orbits
Jupiter	12	24	36	48	60
Saturn	30	60	90	120	150

As you can see, after 60 years Jupiter would have made 5 orbits and Saturn 2 so it will take 60 years for Jupiter and Saturn to line up again! The number 60 is the **least common multiple (LCM)** of 12 and 30.

A > Finding the LCM of Two Numbers

The **least common multiple (LCM)** of two natural numbers is the *smallest* number that is a *multiple* of both numbers.

To find the LCM of two numbers, make a *list* of the multiples of each number, compare the lists, and find the **first** multiple that appears on both lists: that number is the **LCM** of the two numbers.

EXAMPLE 1 Finding the LCM of two numbers

Find the LCM of 8 and 12.

SOLUTION Write the multiples of 8 and of 12 and select the **first** multiple that appears on both lists.

Multiples of 8	8	16	24	32	40
Multiples of 12	12	24	36	48	60

Since the **first** multiple appearing on both lists is 24, the LCM of 8 and 12 is **24.**

PROBLEM 1

Find the LCM of 6 and 8.

EXAMPLE 2 Finding the LCM of two numbers

Find the LCM of 6 and 10.

SOLUTION

Multiples of 6	6	12	18	24	30	36
Multiples of 10	10	20	30	40	50	60

The **first** multiple appearing on both lists is 30, so the LCM of 6 and 10 is **30.**

PROBLEM 2

Find the LCM of 10 and 12.

You can save some time when finding the LCM of two numbers if you select the **larger** of the two numbers and list its multiples until you get a number that is a multiple of the smaller number. Thus, when finding the LCM of 6 and 10, select 10 (the larger of 6 and 10) and list the multiples of 10 until you get a multiple of 6 (the smaller). The multiples of 10 are 10, 20, **30** (Stop!). Since 30 is a multiple of 6, the LCM of 6 and 10 is 30 as before.

EXAMPLE 3 Finding the LCM of two numbers

Find the LCM of 9 and 12.

SOLUTION Using our short cut, list the multiples of 12 (the **larger** of the two given numbers) until you find a multiple of 9.

Multiples of 12 12 24 **36** (Stop!)

Since **36** is a multiple of 9, the LCM of 9 and 12 is 36.

PROBLEM 3

Find the LCM of 18 and 12.

The short cut we have used can save us even more time when the *larger* of the two given numbers is already a *multiple* of the smaller one. For example, if you are finding the LCM of 7 and 21 and follow the procedure, you list the multiples of 21, the larger number, and as soon as you do you realize that the very first multiple you list, the **21** is also a multiple of 7. Thus, the LCM of 7 and 21 is **21.** Similarly, the LCM of 10 and 30 is **30** and the LCM of 11 and 55 is **55.** This is because 30 is a multiple of 10 and 55 is a multiple of 11.

Answers to PROBLEMS

1. 24 **2.** 60 **3.** 36

EXAMPLE 4 **Finding the LCM of two numbers**
Find the LCM of 15 and 45.

SOLUTION Since 45 is a multiple of 15, the LCM of 15 and 45 is **45.**

PROBLEM 4
Find the LCM of 13 and 39.

B > Writing Fractions with the Lowest Common Denominator (LCD)

The graph shows the approximate expenses at a community college. The greatest expense is **food,** which is $\frac{3}{10}$ of the total. The next expense is **tuition,** 1/4 of the total. How much more is the food than the tuition? We can not exactly tell right now, because the whole is not divided into equal parts; that is, parts that are **common** to the whole. If we divide the whole into **20** equal parts, we can easily compare tuition and food. **Tuition** is $\frac{5}{20}$ and **food** is $\frac{6}{20}$, so the expenses for food are more than for tuition. But, where did the **20** come from? Remember **least common multiples?**

Expenses at a Community College

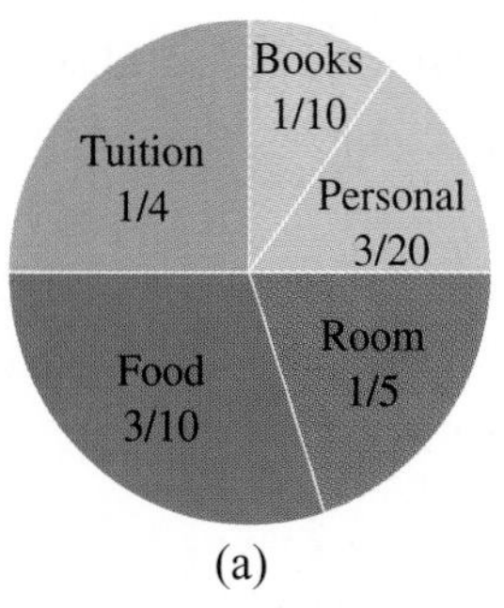

(a)

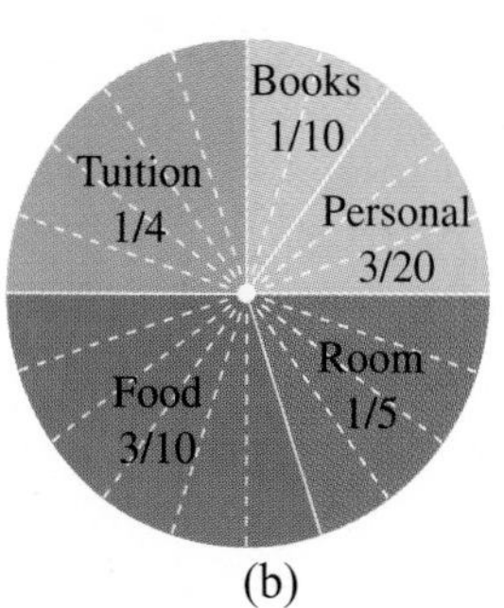

(b)

20 is the *least common multiple* of 10 and 4, the denominators for the tuition portion (1/4) and the food portion (3/10). In general, the **smallest** denominator that allows us to compare fractions directly—that is, having the same denominator—is the **lowest common denominator** (LCD) of the fractions. This number is also the **least common multiple** (LCM) of the denominators.

EXAMPLE 5 **Writing fractions with the LCD as denominator**
Personal and room expenses at a community college are $\frac{3}{20}$ and $\frac{1}{5}$ of the total expenses respectively. (See the diagram.)

a. Find the LCD of $\frac{3}{20}$ and $\frac{1}{5}$.

b. Write $\frac{3}{20}$ and $\frac{1}{5}$ using the LCD as the denominator.

SOLUTION

a. Since **20** is a multiple of 5, the LCD of $\frac{3}{20}$ and $\frac{1}{5}$ is **20.**

b. $\frac{3}{20}$ is already written using the LCD **20** as denominator.

To write $\frac{1}{5}$ with **20** as denominator, we have to multiply the denominator 5 (as well as the numerator 1), by 4 like this:

$$\frac{1}{5} = \frac{1 \cdot 4}{5 \cdot 4} = \frac{4}{20}$$

PROBLEM 5
Find the LCD of

a. $\frac{3}{10}$ and $\frac{1}{6}$.

b. Write $\frac{3}{10}$ and $\frac{1}{6}$ with the LCD as denominator.

Answers to PROBLEMS

4. 39 **5. a.** 30 **b.** $\frac{9}{30}, \frac{5}{30}$

EXAMPLE 6 Finding the LCD using multiples

a. Find the LCD of $\frac{3}{8}$ and $\frac{2}{5}$ using multiples.

b. Write $\frac{3}{8}$ and $\frac{2}{5}$ using the LCD as denominator.

SOLUTION

a. The multiples of 8 (the larger of the two denominators) are

8 16 24 32 40 Stop! 40 is a multiple of 5.

The LCD of $\frac{3}{8}$ and $\frac{2}{5}$ is **40**.

b. To write $\frac{3}{8}$ with a denominator of **40** we have to multiply the denominator 8 (as well as the numerator 3) by 5.

We have: $\frac{3}{8} = \frac{3 \cdot 5}{8 \cdot 5} = \frac{15}{40}$

Similarly, $\frac{2}{5} = \frac{2 \cdot 8}{5 \cdot 8} = \frac{16}{40}$

PROBLEM 6

a. Find the LCD of $\frac{3}{7}$ and $\frac{4}{5}$ using multiples.

b. Write $\frac{3}{7}$ and $\frac{4}{5}$ using the LCD as denominator.

Suppose you want to find the LCD of $\frac{1}{30}$ and $\frac{1}{54}$. To list the multiples of 54 is too time-consuming. Here is a second way to find the LCD of two fractions. (We are finding the LCD of $\frac{1}{30}$ and $\frac{1}{54}$.)

We first write 30 and 54 as products of primes.

Divide by 2. →	2⌊ 30	Divide by 2. →	2⌊ 54
Divide by 3. →	3⌊ 15	Divide by 3. →	3⌊ 27
Divide by 5. →	5⌊ 5	Divide by 3. →	3⌊ 9
Stop, we have 1.	1	Divide by 3. →	3⌊ 3
		Stop, we have 1.	1

Thus, $30 = 2 \cdot 3 \cdot 5$ Thus, $54 = 2 \cdot 3 \cdot 3 \cdot 3 = 2 \cdot 3^3$

It will help if you place the same primes vertically in a column.

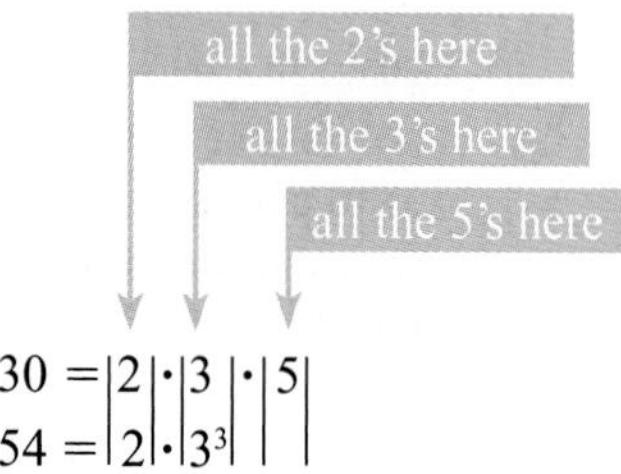

To find the LCD we must include 2 from the first column, 3^3 from the second column, and 5 from the third column. Thus the least common denominator is $2 \cdot 3^3 \cdot 5 = 270$. Here is the procedure.

PROCEDURE TO FIND THE LCD OF FRACTIONS USING PRIMES

1. Write each denominator as a product of primes using exponents.
2. Select the *highest* power of each prime which occurs in any factorization.
3. The product of the factors selected in Step 2 is the LCD.

Answers to PROBLEMS

6. a. 35 **b.** $\frac{15}{35}, \frac{28}{35}$

Some students prefer another method for finding the LCD. It works like this:

FINDING LCDs: ALTERNATE METHOD USING DIVISION

1. Write the denominators in a horizontal row and divide each number by a *prime* common to two or more of the numbers. (If any of the other numbers is *not* divisible by this prime, *circle* the number and carry it to the next line.)
2. Continue this process until no *prime* divides two of the quotients.
3. The LCD is the *product* of the primes and the numbers in the final line.

Step 1. Divide by 2. Divide by 3 (the next prime).

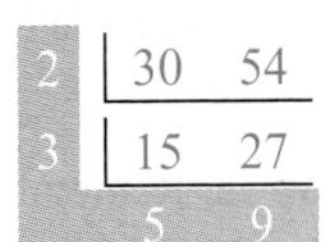

Step 2. No prime divides 5 and 9.

Step 3. The LCD is

$$2 \cdot 3 \cdot 5 \cdot 9 = 270$$

Note that in either case, the LCD is 270. We shall use both methods when finding LCDs.

EXAMPLE 7 Finding the LCD of fractions

Find the LCD of $\frac{1}{60}$ and $\frac{1}{18}$.

SOLUTION

METHOD 1

Step 1. Write 60 and 18 as products of primes.

$$60 = 2^2 \cdot 3 \cdot 5$$
$$18 = 2 \cdot 3^2$$

Step 2. Select each prime to the highest power to which it occurs (2^2, 3^2, and 5).

Step 3. The product of the factors from step 2 is the LCD:

$$2^2 \cdot 3^2 \cdot 5 = 180$$

METHOD 2

Step 1. Write the denominators in a row and divide by a prime common to two of the numbers.

```
2 | 60  18
3 | 30   9
    10   3
```

Step 2. Continue until no prime divides both quotients (no prime divides 10 and 3).

Step 3. The product of the primes and the numbers in the final line is the LCD:

$$2 \cdot 3 \cdot 10 \cdot 3 = 180$$

PROBLEM 7

Find the LCD of $\frac{1}{40}$ and $\frac{5}{12}$.

Answers to PROBLEMS

7. 120

The methods we have studied (multiples, primes, and division) can be used to find the *lowest common denominator* (LCD) of more than two fractions. For example, to find the LCD of $\frac{1}{10}$, $\frac{1}{12}$, and $\frac{1}{8}$ we use the short cut of Example 3 and list the multiples of the **largest** denominator, which is 12, until we get a number that is a multiple of 10 and 8.

The multiples of 12 are 12 24 36 48 60 72 84 96 108 120 (Stop!). Since 120 is a multiple of 10 and also a multiple of 8, the LCD of $\frac{1}{10}$, $\frac{1}{12}$, and $\frac{1}{8}$ is **120.**

Is there an easier way? Try the division method:

Step 1. Divide by 2.

Divide by 2.

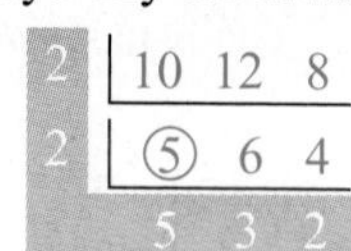

Step 2. Remember to circle the 5 because it is not divisible by 2 and carry it to the next line.

Step 3. The LCD is $2 \cdot 2 \cdot 5 \cdot 3 \cdot 2 =$ **120** as before.

EXAMPLE 8 Finding the LCD of three fractions

Find the LCD of $\frac{1}{6}$, $\frac{1}{10}$, and $\frac{1}{9}$.

SOLUTION In this case, it is easier to use the division method.

Step 1. Divide by 2. 2 | 6 10 ⑨

Divide by 3. 3 | 3 ⑤ 9

1 5 3

Step 2. Remember to circle the 9 and 5 and carry them to the next line.

Step 3. The LCD is $2 \cdot 3 \cdot 1 \cdot 5 \cdot 3 =$ **90.**

PROBLEM 8

Find the LCD of $\frac{1}{8}$, $\frac{1}{12}$, and $\frac{1}{14}$.

C > Comparing Fractions: Order

The LCD can be used to compare fractions. Here are the ingredients for the recipes of sweet rolls and buttermilk dough.

Sweet Rolls
3 packages of yeast
$\frac{3}{4}$ cup water
$\frac{3}{4}$ cup milk
$\frac{1}{2}$ cup sugar
$1\frac{1}{2}$ teaspoon salt
$\frac{1}{2}$ cup butter
2 eggs
$5\frac{1}{2}$ cups flour

Buttermilk Doughnuts
2 tablespoons shortening
$\frac{3}{4}$ cup sugar
2 eggs
4 cups flour
2 teaspoons baking powder
$\frac{1}{2}$ teaspoon cinnamon
1 cup buttermilk
$\frac{1}{2}$ teaspoon salt

Answers to PROBLEMS

8. 168

To find which recipe uses more salt is easy: the sweet rolls! You can convince yourself by writing $1\frac{1}{2}$ as $\frac{3}{2}$ and compare $\frac{3}{2}$ and $\frac{1}{2}$, two fractions with the same denominator. Here is the rule:

COMPARING FRACTIONS: SAME DENOMINATOR

To compare two fractions with the same denominator, compare the numerators. The one with the greater numerator is greater.

$\frac{1}{2}$ ▭▭

$\frac{3}{2}$ ▭▭ ▭▭

You can see that $\frac{3}{2}$ is greater than $\frac{1}{2}$.

Since $\frac{3}{2}$ has the numerator 3, which is greater than the numerator 1 in $\frac{1}{2}$ the $\frac{3}{2}$ is greater than $\frac{1}{2}$, and thus the sweet rolls use more salt.

Can you tell which recipe uses more sugar? Since $\frac{1}{2}$ and $\frac{3}{4}$ do **not** have the same denominator, we write both fractions as equivalent ones with the LCD as the denominator. Because 4 is a multiple of 2, the LCD of $\frac{1}{2}$ and $\frac{3}{4}$ is **4,** so we write $\frac{1}{2} = \frac{1 \cdot 2}{2 \cdot 2} = \frac{2}{4}$.

$$\frac{1}{2} = \frac{1 \cdot 2}{2 \cdot 2} = \frac{2}{4} \qquad \frac{1}{2} = \frac{2}{4}$$

$$\frac{3}{4}$$

Since the numerator in $\frac{3}{4}$, the 3, is **greater** than the numerator in $\frac{2}{4}$, the 2, we write $\frac{3}{4} > \frac{2}{4} = \frac{1}{2}$.

This means that the buttermilk dough has more sugar. Here is the rule to compare fractions with different denominators.

COMPARING FRACTIONS: DIFFERENT DENOMINATORS

To compare two fractions with different denominators, write both fractions with the LCD as their denominators. The fraction with the **greater** (larger) numerator is **greater** (larger).

EXAMPLE 9 Comparing fractions

Fill in the blank with $<$ or $>$ to make the resulting inequality true.

a. $\frac{5}{7}$ ________ $\frac{4}{7}$ **b.** $\frac{3}{5}$ ________ $\frac{4}{7}$

SOLUTION

a. The two fractions have the same denominator, but $\frac{5}{7}$ has a greater numerator than $\frac{4}{7}$. Thus $\frac{5}{7}$ is greater than $\frac{4}{7}$, that is, $\frac{5}{7} > \frac{4}{7}$.

b. We first have to write $\frac{3}{5}$ and $\frac{4}{7}$ as equivalent fractions with the LCD 35 as denominator.

$$\frac{3}{5} = \frac{3 \times ?}{5 \times 7} = \frac{3 \times 7}{5 \times 7} = \frac{21}{35}$$

$$\frac{4}{7} = \frac{4 \times ?}{7 \times 5} = \frac{4 \times 5}{7 \times 5} = \frac{20}{35}$$

Since $\frac{21}{35} > \frac{20}{35}$, $\frac{3}{5} > \frac{4}{7}$ ($\frac{3}{5}$ is greater than $\frac{4}{7}$)

PROBLEM 9

Fill in the blank with $<$ or $>$ to make the resulting inequality true.

a. $\frac{3}{17}$ ______ $\frac{2}{17}$

b. $\frac{1}{5}$ ______ $\frac{2}{9}$

Answers to PROBLEMS

9. a. $>$ **b.** $<$

Exercises 2.4

Web IT go to mhhe.com/bello for more lessons

⟨ A ⟩ Finding the LCM of Two Numbers In Problems 1–10, find the LCM of the numbers by listing the multiples.

1. 8 and 10 **2.** 6 and 10 **3.** 16 and 24

4. 21 and 28 **5.** 9 and 18 **6.** 30 and 60

7. 14 and 21 **8.** 80 and 120 **9.** 30, 15, and 60

10. 15, 20, and 30

⟨ B ⟩ Writing Fractions with the Lowest Common Denominator (LCD) In Problems 11–20, find the LCD of the fractions using multiples and write the fractions with the LCD as denominator.

11. $\frac{1}{3}$ and $\frac{1}{6}$ **12.** $\frac{2}{5}$ and $\frac{1}{15}$ **13.** $\frac{1}{21}$ and $\frac{1}{7}$ **14.** $\frac{2}{9}$ and $\frac{1}{3}$

15. $\frac{3}{4}$ and $\frac{1}{10}$ **16.** $\frac{7}{10}$ and $\frac{4}{15}$ **17.** $\frac{1}{6}$, $\frac{1}{12}$, and $\frac{1}{24}$ **18.** $\frac{7}{15}$, $\frac{3}{10}$, and $\frac{1}{6}$

19. $\frac{3}{5}$, $\frac{5}{8}$, and $\frac{7}{20}$ **20.** $\frac{2}{9}$, $\frac{7}{12}$, and $\frac{11}{24}$

In Problems 21–30, find the LCD using the decomposition of primes or the division method and write the fractions with the LCD as denominator.

21. $\frac{1}{18}$ and $\frac{1}{24}$ **22.** $\frac{3}{15}$ and $\frac{2}{45}$ **23.** $\frac{1}{32}$ and $\frac{1}{80}$ **24.** $\frac{2}{9}$ and $\frac{1}{12}$

25. $\frac{3}{4}$ and $\frac{3}{10}$ **26.** $\frac{7}{20}$ and $\frac{4}{15}$ **27.** $\frac{1}{6}$, $\frac{1}{12}$, and $\frac{1}{24}$ **28.** $\frac{7}{15}$, $\frac{3}{10}$, and $\frac{1}{6}$

29. $\frac{3}{5}$, $\frac{5}{8}$, and $\frac{7}{20}$ **30.** $\frac{2}{9}$, $\frac{7}{12}$, and $\frac{11}{24}$

⟨ C ⟩ Comparing Fractions: Order In Problems 31–34, find the greater of the two numbers.

31. $\frac{5}{8}$, $\frac{7}{8}$ **32.** $\frac{5}{9}$, $\frac{7}{9}$ **33.** $\frac{4}{11}$, $\frac{5}{11}$ **34.** $\frac{3}{7}$, $\frac{2}{7}$

In Problems 35–40, fill in the blank with $<$ or $>$ to make the resulting inequality true.

35. $\frac{2}{3}$ ___ $\frac{4}{5}$ **36.** $\frac{5}{8}$ ___ $\frac{1}{2}$ **37.** $1\frac{4}{7}$ ___ $1\frac{5}{7}$ **38.** $8\frac{3}{4}$ ___ $8\frac{7}{8}$ **39.** $11\frac{2}{7}$ ___ $11\frac{3}{8}$ **40.** $6\frac{1}{3}$ ___ $6\frac{2}{5}$

Applications

41. *Transportation* Do you use public transportation to go to school? If buses depart every 20 minutes and trains every 30 minutes and you just missed the bus and the train, how long do you have to wait before a bus and a train will be departing for your school at the same time?

42. *Transportation* Trains A, B, and C leave Grand Central Station every 10, 20, or 45 minutes, respectively. If A, B, and C just departed, what is the minimum time you have to wait before all three trains are available at the same time?

43. *Cicada* Do you know what a periodical *cicada* is? It is a type of insect sometimes called "locusts" even though they are unrelated to true locusts. The photo shows a 17-year *magicicada*. The 17 means that this species emerge every 17 years. There is also a 13-year *magicicada*. If both species emerge this year, how many years will it be before they emerge together again?

A 17-year cycle magicicada

44. *Cicada* Fortunately, cicadas have predators! Suppose predators emerge 3 years from now and you have a 15-year cicada cycle starting right now.

a. In how many years will the cicada face the predators?

b. If you have a 17-year *magicicada* cycle starting now, in how many years will they face the 3-year predators?

45. *Common cold* There is no cure for it, but here are some things you can do.

Grandma was right—chicken soup actually is good for a cold! One serving every 6 hours.

Zinc lozenges could help you get better sooner. One lozenge every 2 hours.

Aspirin, acetaminophen, and ibuprofen to relieve some symptoms: Two tablets every 4 hours.

You start your three medications (soup, lozenges, and aspirin) at 12 P.M. In how many hours will you have to take all three again?

46. *Common cold* If in addition to the soup (every 6 hours), the lozenges (every 2 hours), and the aspirin (every 4 hours) you take a 12-hour nasal spray and you start your medications at 12 P.M., in how many hours will you have to take all four medicines again?
Source: http://walmart.triaddigital.com/.

47. *Tamales and pastries* La Cubanita Restaurant prepares fresh tamales every 5 days. Pastries are freshly made every 4 days. Andreas had fresh tamales and pastries today. In how many days will the tamales and pastries be made fresh again?

48. *Tamales and pastries* The meat used for filling in the tamales and pastries is delivered every 2 days. If all items were fresh today, in how many days will they have fresh tamales (made fresh every 5 days), fresh pastries (made fresh every 4 days) filled with fresh meat (delivered fresh every 2 days)?

49. *Deliveries* The delivery for wraps is every 4 days but the ingredients for the sports supplement come every 3 days. If deliveries were made today, in how many days will fresh wraps and supplements be delivered?

50. *Deliveries* The delivery schedule for the products listed is as follows:

Smoothie:	Every 30 days
Wraps:	Every 4 days
Protein drinks:	Every 5 days
Sports supplements:	Every 3 days

If all products were delivered today, in how many days will all four products be delivered again?

JJ Smoothy

NOW OPEN

- **SMOOTHIES**
- **WRAPS**
- **PROTEIN DRINKS**
- **SPORTS SUPPLEMENTS**

Using Your Knowledge

Orbital Times The table showing the approximate orbital time for Mars, Jupiter, Saturn, and Uranus will be used in Problems 51–53.

	Mars	Jupiter	Saturn	Uranus
Orbital period (Earth years)	2	12	30	84

51. You remember the planet alignment from the Getting Started? As you can see from the table it takes about 12 years for Jupiter to orbit around the sun once but it takes Saturn 30 years to do so. If Mars takes about 2 years to orbit the sun and the last planetary alignment of Jupiter, Mars, and Saturn was in the year 2000, in what year will the alignment of the three planets happen again?

52. If Jupiter, Uranus, and Mars were aligned today, how many years would it take for them to align again?

53. If Saturn and Uranus were aligned today, how many years will it take for them to align again?

Write On

54. Write in your own words the procedure you use to find the LCM of three numbers by using the division method.

55. Which of the three methods shown in this section will be most efficient to find the LCM of:

a. 5, 10, and 20. Why?

b. 32 and 40. Why?

56. Write in your own words the criteria you use to determine which of the three methods to use when finding the LCM of three numbers.

57. Write in your own words what is the relationship between the LCD of several fractions and the LCM of the denominators of the fraction.

Concept Checker

Fill in the blank(s) with the correct word(s), phrase, or mathematical statement.

58. The ______________ of two numbers is the **smallest number** that is a **multiple** of both numbers.

59. The **LCD of two fractions** is also the ______________ of the **denominators** of the **fractions.**

60. To **compare** fractions with the **same** denominator, we have to **compare** the ______________.

61. To **compare** fractions with **different** denominators we have to write both fractions with the ______________ as their **denominator.**

denominator

numerators

LCD

LCM

Mastery Test

62. Fill in the blank with < or > to make the resulting inequality true: $\frac{4}{11}$ ____ $\frac{5}{11}$.

63. Fill in the blank with < or > to make the resulting inequality true: $\frac{3}{11}$ ____ $\frac{1}{4}$.

64. Find the LCM of 12 and 14.

65. Find the LCM of 15 and 45.

66. Find the LCM of 10, 3, and 14.

67. Write fractions equivalent to $\frac{3}{7}$ and $\frac{4}{5}$ using their LCD as denominator.

68. Find the LCD of $\frac{1}{40}$ and $\frac{1}{18}$.

69. Find the LCD of $\frac{1}{6}$, $\frac{1}{20}$, and $\frac{1}{9}$.

Skill Checker

70. Write $\frac{1}{8}$ and $\frac{1}{6}$ with a denominator of 24.

71. Write $\frac{5}{9}$ and $\frac{3}{8}$ with a denominator of 72.

72. Write $\frac{1}{8}$, $\frac{1}{12}$, and $\frac{1}{10}$ with a denominator of 120.

2.5 Addition and Subtraction of Fractions

Objectives

You should be able to:

A Add two fractions having the same denominator.

B Add two fractions with different denominators using the idea of a multiple to find the LCD.

C Use the LCD to add fractions.

D Use the LCD to subtract fractions.

E Find what fraction of a circle graph is represented by a given region.

To Succeed, Review How To . . .

1. Write a number as a product of primes using exponents. (pp. 73–76)
2. Write a mixed number as an improper fraction and vice versa. (p. 113)

Getting Started

The photo shows that 1 quarter plus 2 quarters equals 3 quarters. In symbols we have

$$\frac{1}{4} + \frac{2}{4} = \frac{1 + 2}{4} = \frac{3}{4}$$

Here is a diagram showing why.

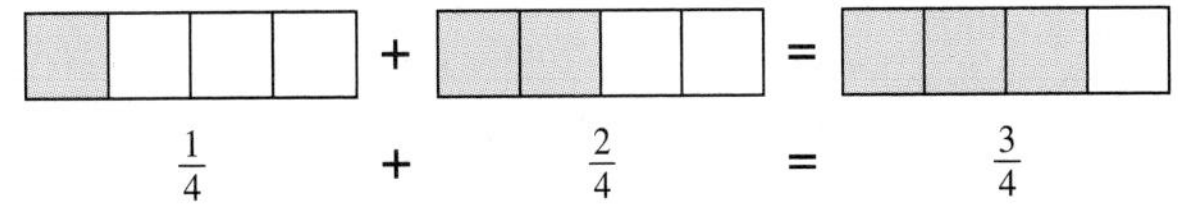

$$\frac{1}{4} \quad + \quad \frac{2}{4} \quad = \quad \frac{3}{4}$$

A › Adding Fractions with the Same Denominator

To add fractions with the *same* denominator, *add* the *numerators* and *keep* the *denominator*. Here is the rule:

ADDING FRACTIONS WITH THE SAME DENOMINATOR

For any numbers a, b, and c, where $b \neq 0$,

$$\frac{a}{b} + \frac{c}{b} = \frac{a + c}{b}$$

Add numerators a and c, keep the denominator b.

Thus,

$$\frac{1}{5} + \frac{3}{5} = \frac{1+3}{5} = \frac{4}{5}$$

$$\frac{3}{7} + \frac{2}{7} = \frac{3+2}{7} = \frac{5}{7}$$

$$\frac{1}{4} + \frac{5}{4} = \frac{1+5}{4} = \frac{6}{4} = \frac{3}{2}$$

Note that we reduced $\frac{6}{4}$ to $\frac{3}{2}$. When working with fractions (adding, subtracting, multiplying, or dividing them) you should *reduce* your answers if possible.

EXAMPLE 1 **Adding fractions with the same denominator**

Add.

a. $\frac{2}{5} + \frac{1}{5}$ **b.** $\frac{4}{9} + \frac{2}{9}$

SOLUTION

a. $\frac{2}{5} + \frac{1}{5} = \frac{2+1}{5} = \frac{3}{5}$ **b.** $\frac{4}{9} + \frac{2}{9} = \frac{4+2}{9} = \frac{6}{9} = \frac{2}{3}$

PROBLEM 1

Add.

a. $\frac{2}{11} + \frac{3}{11}$ **b.** $\frac{1}{8} + \frac{3}{8}$

B > Adding Fractions with Different Denominators

Now suppose we want to add $\frac{2}{5}$ and $\frac{1}{4}$. Since these two fractions do not have the same denominators, our rule does not work. We have to write $\frac{2}{5}$ and $\frac{1}{4}$ as equivalent fractions with the LCD as denominators so that we can use the rule.

To find the LCD of $\frac{2}{5}$ and $\frac{1}{4}$ look at the multiples of 5 until you find a multiple of 5 that is divisible by 4. The multiples of 5 are

5 10 15 20 (stop!)

20 is a multiple of 4

So the LCD of $\frac{2}{5}$ and $\frac{1}{4}$ is 20. Next, we write $\frac{2}{5}$ and $\frac{1}{4}$ using the LCD 20 as denominator

$$\frac{2 \cdot 4}{5 \cdot 4} = \frac{8}{20} \quad \text{and} \quad \frac{1 \cdot 5}{4 \cdot 5} = \frac{5}{20}$$

Thus, $\frac{8}{20}$ and $\frac{5}{20}$ are fractions equivalent to $\frac{2}{5}$ and $\frac{1}{4}$ and with the LCD as denominator, so they can be added. Thus,

$$\frac{2}{5} + \frac{1}{4} = \frac{2 \cdot 4}{5 \cdot 4} + \frac{1 \cdot 5}{4 \cdot 5} = \frac{8}{20} + \frac{5}{20} = \frac{8+5}{20} = \frac{13}{20}$$

You can also add fractions vertically, like this:

$$\begin{array}{rcrcr} \frac{2}{5} & = & \frac{2 \cdot 4}{5 \cdot 4} & = & \frac{8}{20} \\ +\frac{1}{4} & = & +\frac{1 \cdot 5}{4 \cdot 5} & = & +\frac{5}{20} \\ \hline & & & & \frac{13}{20} \end{array}$$

Answers to PROBLEMS

1. a. $\frac{5}{11}$ **b.** $\frac{4}{8} = \frac{1}{2}$

Adding Fractions with Different Denominators If we wish to add $\frac{3}{4} + \frac{1}{6}$, we first find the LCD of $\frac{3}{4}$ and $\frac{1}{6}$, which is 12, and write $\frac{3}{4}$ and $\frac{1}{6}$ with 12 as the denominator

$$\frac{3}{4} = \frac{3 \cdot 3}{4 \cdot 3} = \frac{9}{12}$$

and

$$\frac{1}{6} = \frac{1 \cdot 2}{6 \cdot 2} = \frac{2}{12}$$

Thus, $\frac{9}{12}$ and $\frac{2}{12}$ are equivalent to $\frac{3}{4}$ and $\frac{1}{6}$, respectively, and have the same denominator. Then

$$\frac{3}{4} + \frac{1}{6} = \frac{9}{12} + \frac{2}{12} = \frac{9+2}{12} = \frac{11}{12}$$

Note that we could have used $\frac{18}{24}$ and $\frac{4}{24}$ as our equivalent fractions with the same denominator. This was not done, however, because the fractions $\frac{9}{12}$ and $\frac{2}{12}$ had lesser denominators. When adding fractions, we *always* try to obtain the **L**owest **C**ommon **D**enominator (LCD). Suppose we insisted on using 24 as our denominator:

$$\frac{3}{4} = \frac{3 \cdot 6}{4 \cdot 6} = \frac{18}{24}$$

and

$$\frac{1}{6} = \frac{1 \cdot 4}{6 \cdot 4} = \frac{4}{24}$$

Thus,

$$\frac{3}{4} + \frac{1}{6} = \frac{18}{24} + \frac{4}{24}$$

$$= \frac{18+4}{24} = \frac{22}{24} = \frac{11}{12}$$

Of course, we got the same answer, but it was a lot more work!

Remember, to find the LCD of $\frac{1}{8}$ and $\frac{5}{6}$ list the multiple of 8 (the larger of the two denominators) until you find the first multiple of 6.

Multiples of 8: 8, 16, 24 ← A multiple of 6 (6 goes exactly into 24)

Not a multiple of 6 (6 does not go exactly into 8 or into 16)

Thus, the LCD of $\frac{1}{8}$ and $\frac{5}{6}$ is 24. In general, we have

FINDING THE LCD OF FRACTIONS

Check the multiples of the *greater* denominator until you get a multiple of the smaller denominator.

EXAMPLE 2 **Adding fractions with unlike denominators**

Add.

$$\frac{1}{8} + \frac{5}{6}$$

SOLUTION The LCD of $\frac{1}{8}$ and $\frac{5}{6}$ is 24 (try the multiples of 8: 8, 16, 24), and $24 = 8 \times 3 = 6 \times 4$. We write

$$\frac{1}{8} = \frac{1 \times 3}{8 \times 3} = \frac{3}{24}$$

$$\frac{5}{6} = \frac{5 \times 4}{6 \times 4} = \frac{20}{24}$$

Thus,

$$\frac{1}{8} + \frac{5}{6} = \frac{3}{24} + \frac{20}{24} = \frac{23}{24}$$

PROBLEM 2

Add.

$$\frac{3}{8} + \frac{1}{6}$$

Answers to PROBLEMS

2. $\frac{13}{24}$

EXAMPLE 3 **Adding fractions with unlike denominators**

Add.

$$\frac{7}{4} + \frac{1}{15}$$

SOLUTION We first find the LCD. The multiples of 15 (the larger of the two denominators) are

$$\underbrace{15, 30, 45,}_{\text{Not multiples of 4}}\ 60 \leftarrow \text{Multiple of 4 (4 goes into it)}$$

Thus the LCD of $\frac{7}{4}$ and $\frac{1}{15}$ is 60, and $\frac{7}{4} = \frac{105}{60}$ and $\frac{1}{15} = \frac{4}{60}$. So

$$\frac{7}{4} + \frac{1}{15} = \frac{105}{60} + \frac{4}{60} = \frac{109}{60} \quad \text{or} \quad 1\frac{49}{60}$$

PROBLEM 3

Add.

$$\frac{3}{4} + \frac{5}{9}$$

Here is the procedure we have used to add fractions with different denominators:

ADDING FRACTIONS WITH DIFFERENT DENOMINATORS

1. Find the LCD of the fractions (you can use multiples, product of primes, or division).
2. Write each fraction as an equivalent one with the LCD as denominator.
3. Add the fractions and reduce the answer, if possible.

We illustrate the procedure in Example 4, where we use multiples to find the LCD, and in Example 5, where we first find the LCD of $\frac{1}{8}$, $\frac{1}{12}$, and $\frac{1}{10}$ and then add the fractions.

C > Using the LCD to Add Fractions

EXAMPLE 4 **Finding the LCD using multiples and adding fractions**

Add.

$$\frac{1}{60} + \frac{5}{18}$$

SOLUTION The multiples of 60 are 60, 120, and 180, so the LCD of 60 and 18 is 180. We need to write $\frac{1}{60}$ and $\frac{5}{18}$ with a denominator of 180. To have a denominator of 180, we multiply the denominator of $\frac{1}{60}$ (and hence the numerator) by 3. Similarly, we multiply the numerator and denominator of $\frac{5}{18}$ by 10, obtaining

$$\frac{1}{60} = \frac{1 \cdot 3}{60 \cdot 3} = \frac{3}{180} \quad \text{and} \quad \frac{5}{18} = \frac{5 \cdot 10}{18 \cdot 10} = \frac{50}{180}$$

Thus,

$$\frac{1}{60} + \frac{5}{18} = \frac{3}{180} + \frac{50}{180} = \frac{53}{180}$$

(53 is prime, so this fraction cannot be reduced.)

PROBLEM 4

Add.

$$\frac{1}{40} + \frac{5}{12}$$

Can we add three fractions like $\frac{1}{8} + \frac{1}{12} + \frac{1}{10}$? Of course. We first illustrate how to find the LCD using all methods. (You choose the one you prefer.) Then we add the fractions.

Answers to PROBLEMS

3. $\frac{47}{36}$ or $1\frac{11}{36}$
4. $\frac{53}{120}$

Method 1

Step 1. Write 8, 12, and 10 as products of primes.

$$8 = 2^3$$
$$12 = 2^2 \cdot 3$$
$$10 = 2 \cdot 5$$

Step 2. Select each prime to the highest power to which it occurs (2^3, 3, 5).

Step 3. The product of the factors from Step 2 is the LCD—that is,

$$2^3 \cdot 3 \cdot 5 = 120$$

Method 2

Step 1. Divide by 2. Divide by 2.

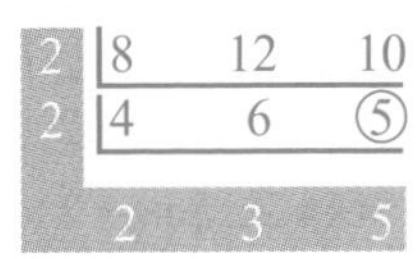

Step 2. Remember to circle the 5 and carry it to the next line.

Step 3. The LCD is

$$2 \cdot 2 \cdot 2 \cdot 3 \cdot 5 = 120$$

Method 3 Note that you can also find the LCD by listing the multiples of the *largest* denominator, which is 12, until you find a multiple of 12 that is divisible by 8 and 10.

Multiples of 12 are 12, 24, 36, 48, 60, 72, 84, 96, 108, and (120). Finally! 120 is divisible by 8 and by 10, so 120 is the LCD.

EXAMPLE 5 **Adding three fractions with unlike denominators**

Add.

$$\frac{1}{8} + \frac{1}{12} + \frac{1}{10}$$

SOLUTION The LCD is 120, so we write the fractions with 120 as denominator.

$$\frac{1}{8} = \frac{1 \cdot 15}{8 \cdot 15} = \frac{15}{120}, \quad \frac{1}{12} = \frac{1 \cdot 10}{12 \cdot 10} = \frac{10}{120}, \quad \frac{1}{10} = \frac{1 \cdot 12}{10 \cdot 12} = \frac{12}{120}$$

Thus,

$$\frac{1}{8} + \frac{1}{12} + \frac{1}{10}$$
$$= \frac{15}{120} + \frac{10}{120} + \frac{12}{120}$$
$$= \frac{15 + 10 + 12}{120} = \frac{37}{120}$$

PROBLEM 5

Add.

$$\frac{1}{8} + \frac{1}{12} + \frac{1}{9}$$

D > Subtraction of Fractions

Now that you know how to add, *subtraction* is no problem. All the rules we have mentioned still apply! Thus

$$\frac{5}{8} - \frac{2}{8} = \frac{5-2}{8} = \frac{3}{8}$$
$$\frac{7}{9} - \frac{1}{9} = \frac{7-1}{9} = \frac{6}{9} = \frac{2}{3}$$

The next example shows how to subtract fractions involving *different* denominators. As with addition, we find the LCD first.

EXAMPLE 6 **Subtracting fractions with unlike denominators**

Subtract.

a. $\frac{7}{12} - \frac{1}{18}$ **b.** $\frac{8}{15} - \frac{6}{25}$

SOLUTION

a. We first get the LCD of the fractions. Since the multiples of 18 are 18, 36, the LCD is 36.

(continued)

PROBLEM 6

Subtract.

a. $\frac{7}{12} - \frac{1}{10}$ **b.** $\frac{11}{15} - \frac{3}{20}$

Answers to PROBLEMS

5. $\frac{23}{72}$ **6. a.** $\frac{29}{60}$ **b.** $\frac{7}{12}$

METHOD 1

Step 1. $12 = |2^2| \cdot |3|$
$18 = |2| \cdot |3^2|$

Step 2. Select 2 to the highest power to which it occurs (2^2) and 3 to the highest power to which it occurs (3^2).

Step 3. The LCD is $2^2 \cdot 3^2 = 36$.

METHOD 2

Step 1. $2\ \underline{|12 \quad 18}$

Step 2. $3\ \underline{|\ 6 \quad 9}$

$\quad\quad 2 \quad 3$

Step 3. The LCD is
$2 \cdot 3 \cdot 2 \cdot 3 = 36$

We then write each fraction with 36 as the denominator.

$$\frac{7}{12} = \frac{7 \cdot 3}{12 \cdot 3} = \frac{21}{36} \quad \text{and} \quad \frac{1}{18} = \frac{1 \cdot 2}{18 \cdot 2} = \frac{2}{36}$$

Thus

$$\frac{7}{12} - \frac{1}{18} = \frac{21}{36} - \frac{2}{36} = \frac{21 - 2}{36} = \frac{19}{36}$$

b. To find the LCD of $\frac{8}{15}$ and $\frac{6}{25}$, write the multiples of 25, the larger of the two denominators, until you get a multiple that is divisible by 15. The multiples of 25 are

25 50 75

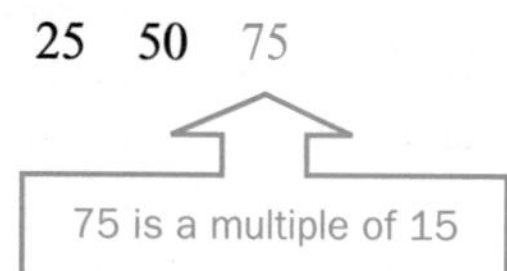

The LCD is 75.

Next, we write $\frac{8}{15}$ and $\frac{6}{25}$ using the LCD 75 as the denominator

$$\frac{8}{15} = \frac{8 \cdot 5}{15 \cdot 5} = \frac{40}{75} \quad \text{and} \quad \frac{6}{25} = \frac{6 \cdot 3}{25 \cdot 3} = \frac{18}{75}$$

Thus, $\frac{8}{15} - \frac{6}{25} = \frac{40}{75} - \frac{18}{75} = \frac{40 - 18}{75} = \frac{22}{75}$

EXAMPLE 7 Adding and subtracting fractions

Add and subtract.

a. $\frac{5}{9} + \frac{3}{8} - \frac{1}{12}$ **b.** $\frac{7}{8} - \frac{1}{3} + \frac{7}{12}$

SOLUTION

a. We first get the LCD of the fractions.

METHOD 1

Step 1.
$9 = \quad\quad\quad |3^2|$
$8 = |2^3|$
$12 = |2^2| \cdot |3|$

Step 2. Select 2 to the highest power to which it occurs (2^3) and 3 to the highest power to which it occurs (3^2).

Step 3. The LCD is $2^3 \cdot 3^2 = 72$.

METHOD 2

Step 1.

2	(9)	8	12
2	(9)	4	6
3	9	(2)	3
	3	2	1

Step 2. No prime divides 3, 2, and 1.

Step 3. The LCD is

$2 \cdot 2 \cdot 3 \cdot 3 \cdot 2 = 72$

PROBLEM 7

Add and subtract.

a. $\frac{3}{8} + \frac{1}{6} - \frac{2}{9}$ **b.** $\frac{7}{8} - \frac{1}{3} + \frac{11}{12}$

Answers to PROBLEMS

7. a. $\frac{23}{72}$ **b.** $\frac{35}{24} = 1\frac{11}{24}$

We now rewrite each fraction with a denominator of 72.

$$\frac{5}{9} = \frac{5 \cdot 8}{9 \cdot 8} = \frac{40}{72}, \quad \frac{3}{8} = \frac{3 \cdot 9}{8 \cdot 9} = \frac{27}{72}, \quad \frac{1}{12} = \frac{1 \cdot 6}{12 \cdot 6} = \frac{6}{72}$$

Thus

$$\frac{5}{9} + \frac{3}{8} - \frac{1}{12} = \frac{40}{72} + \frac{27}{72} - \frac{6}{72} = \frac{40 + 27 - 6}{72} = \frac{61}{72}$$

b. The LCD of $\frac{7}{8}$, $\frac{1}{3}$, and $\frac{7}{12}$ is 24

$$\frac{7}{8} = \frac{7 \cdot 3}{8 \cdot 3} = \frac{21}{24}$$

$$\frac{1}{3} = \frac{1 \cdot 8}{3 \cdot 8} = \frac{8}{24}$$

$$\frac{7}{12} = \frac{7 \cdot 2}{12 \cdot 2} = \frac{14}{24}$$

Thus,

$$\frac{7}{8} - \frac{1}{3} + \frac{7}{12} = \frac{21}{24} - \frac{8}{24} + \frac{14}{24} = \frac{27}{24} = \frac{9}{8} = 1\frac{1}{8}$$

E > Graphs and Fractions

A popular way of displaying information is by means of a pie chart (or circle graph), such as the one shown here, which illustrates the fraction of students with different eye colors.

Color of Students' Eyes

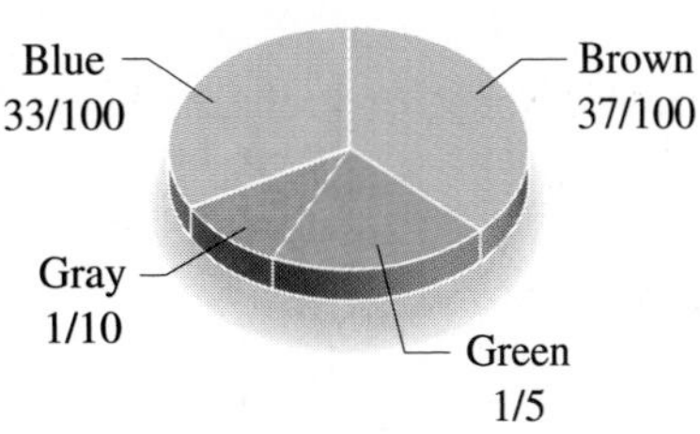

Source: Learn.co.uk.

EXAMPLE 8 Reading pie charts

a. What reduced fraction of the students have brown or blue eyes?

b. What reduced fraction of the students have gray or blue eyes?

SOLUTION

a. The fraction of the students with brown or blue eyes is

$$\frac{37}{100} + \frac{33}{100} = \frac{37 + 33}{100}$$

$$= \frac{70}{100} = \frac{7}{10}$$

b. The fraction of students with gray or blue eyes is

$$\frac{1}{10} + \frac{33}{100}$$

Write $\frac{1}{10}$ as $\frac{10}{100}$

$$= \frac{10}{100} + \frac{33}{100} = \frac{43}{100}$$

PROBLEM 8

What reduced fraction of the students have

a. brown or gray eyes?

b. green or blue eyes?

Answers to PROBLEMS

8. a. $\frac{47}{100}$ **b.** $\frac{53}{100}$

Exercises 2.5

⟨A⟩ Adding Fractions with the Same Denominator In Problems 1–10, add (reduce answers to lowest terms).

1. $\frac{1}{3}+\frac{1}{3}$ **2.** $\frac{1}{5}+\frac{2}{5}$ **3.** $\frac{1}{7}+\frac{4}{7}$

4. $\frac{1}{9}+\frac{7}{9}$ **5.** $\frac{2}{9}+\frac{4}{9}$ **6.** $\frac{3}{8}+\frac{5}{8}$

7. $\frac{1}{6}+\frac{5}{6}$ **8.** $\frac{2}{9}+\frac{10}{9}$ **9.** $\frac{3}{4}+\frac{5}{4}$

10. $\frac{6}{7}+\frac{8}{7}$

⟨B⟩ Adding Fractions with Different Denominators In Problems 11–26, add using multiples to find the LCD (reduce answers to lowest terms).

11. $\frac{1}{3}+\frac{1}{5}$ **12.** $\frac{1}{4}+\frac{1}{6}$ **13.** $\frac{1}{2}+\frac{1}{6}$ **14.** $\frac{7}{8}+\frac{3}{4}$

15. $\frac{1}{2}+\frac{4}{5}$ **16.** $\frac{5}{6}+\frac{3}{10}$ **17.** $\frac{4}{7}+\frac{3}{14}$ **18.** $\frac{1}{6}+\frac{11}{12}$

19. $\frac{1}{2}+\frac{3}{8}$ **20.** $\frac{5}{12}+\frac{1}{6}$ **21.** $\frac{1}{40}+\frac{1}{18}$ **22.** $\frac{5}{24}+\frac{7}{30}$

23. $\frac{2}{65}+\frac{3}{26}$ **24.** $\frac{7}{120}+\frac{11}{150}$ **25.** $\frac{7}{120}+\frac{1}{180}$ **26.** $\frac{1}{90}+\frac{7}{120}$

⟨C⟩ Using the LCD to Add Fractions In Problems 27–30, find the LCD and add the fractions.

27. $\frac{3}{10}+\frac{7}{20}+\frac{11}{60}$ **28.** $\frac{5}{9}+\frac{7}{12}+\frac{5}{18}$ **29.** $\frac{11}{14}+\frac{5}{6}+\frac{8}{9}$

30. $\frac{5}{36}+\frac{1}{80}+\frac{7}{90}$

⟨D⟩ Subtraction of Fractions In Problems 31–50, add and subtract as indicated (reduce answers to lowest terms).

31. $\frac{3}{7}-\frac{1}{7}$ **32.** $\frac{5}{8}-\frac{2}{8}$ **33.** $\frac{5}{6}-\frac{1}{6}$

34. $\frac{3}{8}-\frac{1}{8}$ **35.** $\frac{5}{12}-\frac{1}{4}$ **36.** $\frac{1}{3}-\frac{1}{6}$

37. $\frac{1}{2}-\frac{2}{5}$ **38.** $\frac{1}{4}-\frac{1}{6}$ **39.** $\frac{5}{20}-\frac{7}{40}$

40. $\frac{7}{10}-\frac{3}{20}$ **41.** $\frac{7}{8}-\frac{5}{12}$ **42.** $\frac{8}{15}-\frac{2}{25}$

43. $\frac{13}{60}-\frac{1}{48}$ **44.** $\frac{19}{24}-\frac{7}{60}$ **45.** $\frac{8}{9}-\frac{2}{9}-\frac{1}{9}$

46. $\frac{7}{11}-\frac{3}{11}-\frac{2}{11}$ **47.** $\frac{3}{4}+\frac{5}{12}-\frac{1}{6}$ **48.** $\frac{5}{6}+\frac{1}{9}-\frac{1}{3}$

49. $\frac{9}{2}-\frac{7}{3}$ **50.** $\frac{11}{5}-\frac{7}{4}$

51. A board $\frac{3}{4}$ inch thick is glued to another board $\frac{3}{8}$ inch thick. If the glue is $\frac{1}{32}$ inch thick, how thick is the result?

52. Candy Sweet bought $\frac{1}{4}$ pound of chocolate candy and $\frac{1}{2}$ pound of caramels. How many pounds of candy is this?

53. A father left $\frac{1}{4}$ of his estate to his daughter, $\frac{1}{2}$ to his wife, and $\frac{1}{8}$ to his son. How much of the estate remained?

54. A recent survey found that $\frac{3}{10}$ of the American people work long hours and smoke, while $\frac{1}{5}$ are overweight. Thus, the fraction of people who are neither of these is $1 - \frac{1}{5} - \frac{3}{10}$. What fraction is that?

55. Human bones are $\frac{1}{4}$ water, $\frac{3}{10}$ living tissue, and the rest minerals. The fraction of the bone that is minerals is $1 - \frac{1}{4} - \frac{3}{10}$. Find this fraction.

⟨ E ⟩ Graphs and Fractions The circle graph will be used in Problems 56–58.

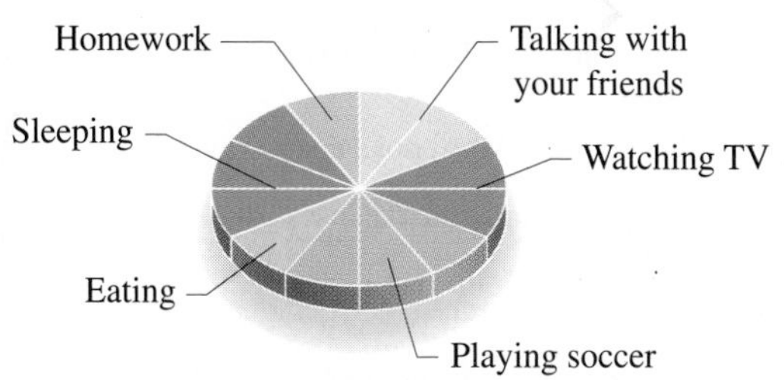

Source: Adapted from Learn.co.uk.

The circle graph is divided into 12 equal parts (slices).

56. What reduced fraction of the time was spent eating?

57. What reduced fraction of the time was spent watching TV?

58. What fraction of the time was spent doing homework?

The circle graph will be used in Problems 59–60.

The graph shows the fraction of the budget spent by a company on its employees.

Employee Expenses

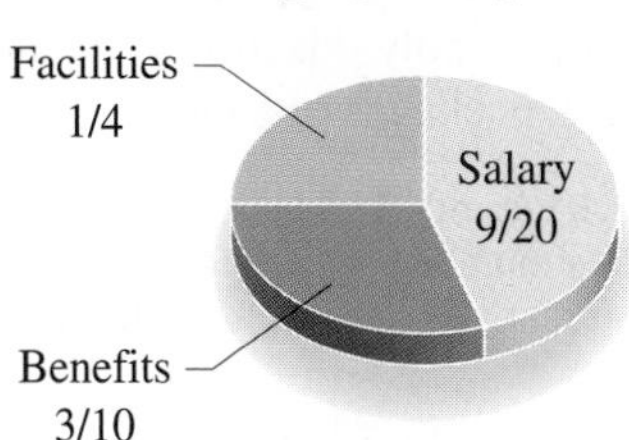

Source: Adapted from Visual Mining, Inc.

59. What fraction of the expenses was for benefits or salary?

60. What fraction of the expenses was for benefits or facilities?

The circle graph will be used in Problems 61–62.

The graph shows the fraction of the days in which it is sunny, snowy, rainy, or cloudy in a certain city.

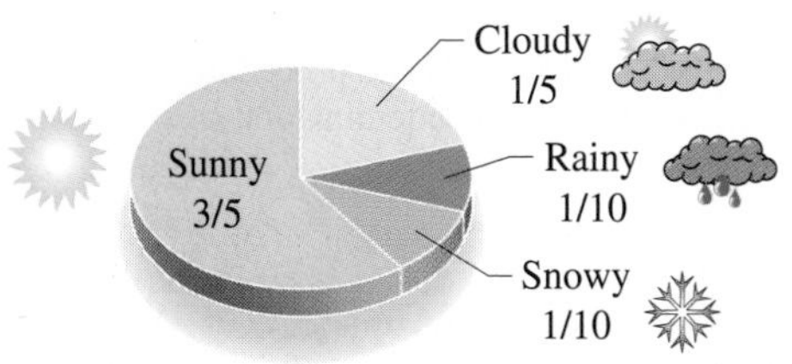

61. What reduced fraction of the days is rainy or snowy?

62. What reduced fraction of the days is rainy or cloudy?

The circle graph will be used in Problems 63–65.

The graph shows the mode of transportation used by people going to work in England.

Ways of Traveling to Work

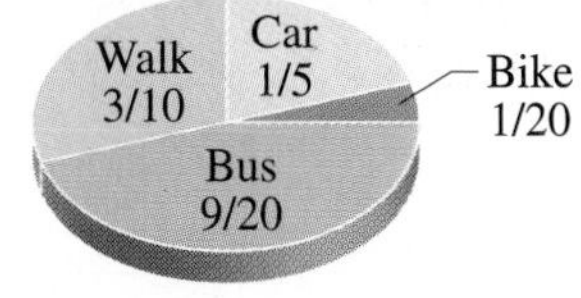

63. What fraction of the people walk or use a car?

64. What fraction of the people bike or use a car?

65. What fraction of the people do not walk?

Web IT go to mhhe.com/bello for more lessons

The circle graph shows the fraction of the money spent by a typical community college student in five different areas and will be used in Problems 66–70. The expenses total $3000.

Annual Expenses at a Community College

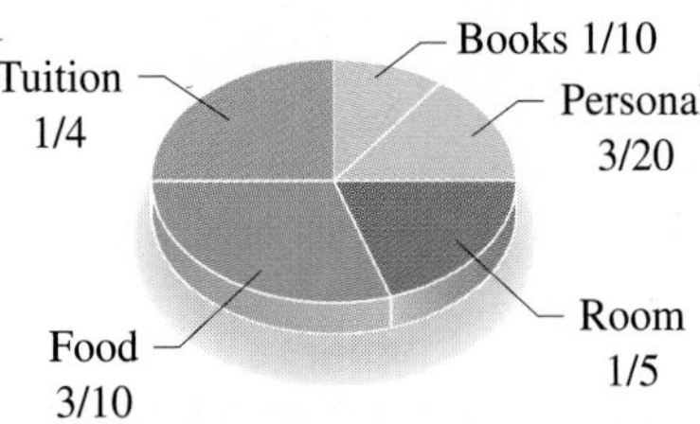

66. a. What fraction of the expenses go toward food and tuition?

b. What amount is for food and tuition?

67. a. What fraction of the expenses go toward food and room?

b. What amount is for food and room?

68. a. What fraction of the expenses go toward books and tuition?

b. What amount is for books and tuition?

69. a. What fraction of the expenses go toward books, personal, and room?

b. What amount is for books, personal, and room?

70. a. What fraction of the expenses go toward paying for everything except books?

b. What amount is for everything except books?

Using Your Knowledge

Hot Dogs, Buns, and LCDs In this section we learned how to find the LCD of several fractions by finding the multiples of the *larger* denominator until you get a multiple of the *smaller* denominators. We are going to apply this theory to purchasing hot dogs and buns! Have you noticed that hot dogs come 10 to a package but buns come in packages of 8 or 12?

71. What is the smallest number of packages of hot dogs (10 to a package) and buns (8 to a package) you must buy so that you have as many hot dogs as you have buns? (*Hint:* Think of *multiples.*)

72. If buns are sold in packages of 12 and hot dogs in packages of 10, what is the smallest number of packages of hot dogs and buns you must buy so that you have as many hot dogs as you have buns?

Write On

73. When adding $\frac{3}{4} + \frac{1}{6}$ we mentioned we could use the equivalent fractions $\frac{18}{24}$ and $\frac{4}{24}$, then do the addition. You can always use the product of the denominators as the denominator of the sum. Is this correct? Why or why not?

74. Write in your own words the process you prefer for finding the LCD of two fractions.

75. Write in your words the procedure you use to add fractions with different denominators.

76. Write in your own words the procedure you use to subtract fractions with different denominators.

Concept Checker

Fill in the blank(s) with the correct word(s), phrase, or mathematical statement.

77. $\frac{a}{c} + \frac{b}{c} =$ ____________

78. $\frac{a}{c} - \frac{b}{c} =$ ____________

$\frac{a+c}{b}$ $\frac{a-b}{c}$

$\frac{a+b}{c}$ $\frac{a-c}{b}$

Mastery Test

79. Find the LCD of $\frac{1}{30}$ and $\frac{1}{18}$.

80. Add: $\frac{1}{10} + \frac{7}{10}$

81. Add: $\frac{1}{8} + \frac{1}{6}$

82. Add: $\frac{1}{10} + \frac{7}{4}$

83. Perform the indicated operation: $\frac{1}{10} + \frac{1}{12} + \frac{3}{8}$

84. Subtract: $\frac{5}{12} - \frac{1}{18}$

85. Perform the indicated operation: $\frac{3}{10} + \frac{1}{12} - \frac{1}{8}$

86. The graph shows the concerns of a mythical chinchilla.

a. What fraction of the time is the chinchilla concerned about eating too much or too little?

b. What fraction of the time does the chinchilla either have no concerns or act weird?

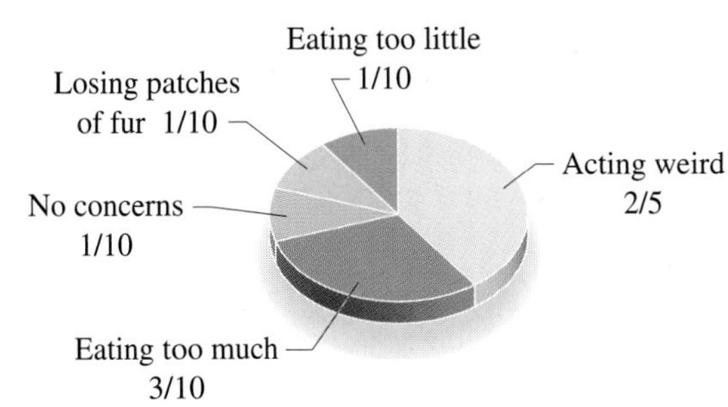

Skill Checker

Write as an improper fraction.

87. $3\frac{1}{5}$ **88.** $5\frac{3}{11}$ **89.** $6\frac{7}{8}$ **90.** $7\frac{10}{11}$

Write as a mixed number.

91. $\frac{10}{6}$ **92.** $\frac{45}{6}$

2.6 Addition and Subtraction of Mixed Numbers

Objectives

You should be able to:

A Add mixed numbers having the same denominator.

B Add mixed numbers with different denominators using the idea of a multiple to find the LCD.

C Subtract mixed numbers with different denominators using the idea of a multiple to find the LCD.

D Use the LCD of two or more fractions to add or subtract mixed numbers.

E Use addition of mixed numbers to find perimeters.

To Succeed, Review How To . . .

1. Write a number as a product of primes using exponents. (pp. 73–76)
2. Write a mixed number as an improper fraction and vice versa. (p. 113)

Getting Started

Prices & Yields

Bond	Price	Change	Yield	Change
2 yr	$99\frac{14}{32}$	$-\frac{2}{32}$	2.79	0.050
5 yr	$100\frac{8}{32}$	$-\frac{6}{32}$	3.94	0.040
10 yr	$100\frac{6}{32}$	$-\frac{10}{32}$	4.72	0.040
30 yr	$99\frac{19}{32}$	$-\frac{17}{32}$	5.4	0.040

Source: CNN/Money.

The price of a 2-year bond is $99\frac{14}{32}$ and the change in price is $-\frac{2}{32}$. The price *before* the change was $\frac{2}{32}$ higher or

$$99\frac{14}{32} + \frac{2}{32} = 99\frac{16}{32} = 99\frac{1}{2}$$

Mixed numbers can be added and subtracted. We first do addition with mixed numbers having the same denominator.

A > Adding Mixed Numbers with the Same Denominator

Can we add $3\frac{1}{5} + 1\frac{2}{5}$? Of course! You can change $3\frac{1}{5}$ and $1\frac{2}{5}$ to improper fractions first. Since

$$3\frac{1}{5} = \frac{5 \cdot 3 + 1}{5} = \frac{16}{5} \quad \text{and} \quad 1\frac{2}{5} = \frac{5 \cdot 1 + 2}{5} = \frac{7}{5}$$

we have

change

$$3\frac{1}{5} + 1\frac{2}{5} = \frac{16}{5} + \frac{7}{5} = \frac{16 + 7}{5} = \frac{23}{5} = 4\frac{3}{5}$$

change

You can also add $3\frac{1}{5}$ and $1\frac{2}{5}$ vertically, like this:

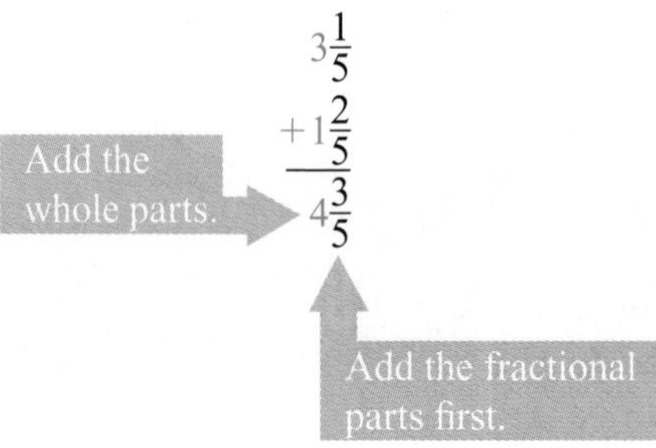

Note that when adding mixed numbers, the answer is given as a mixed number.

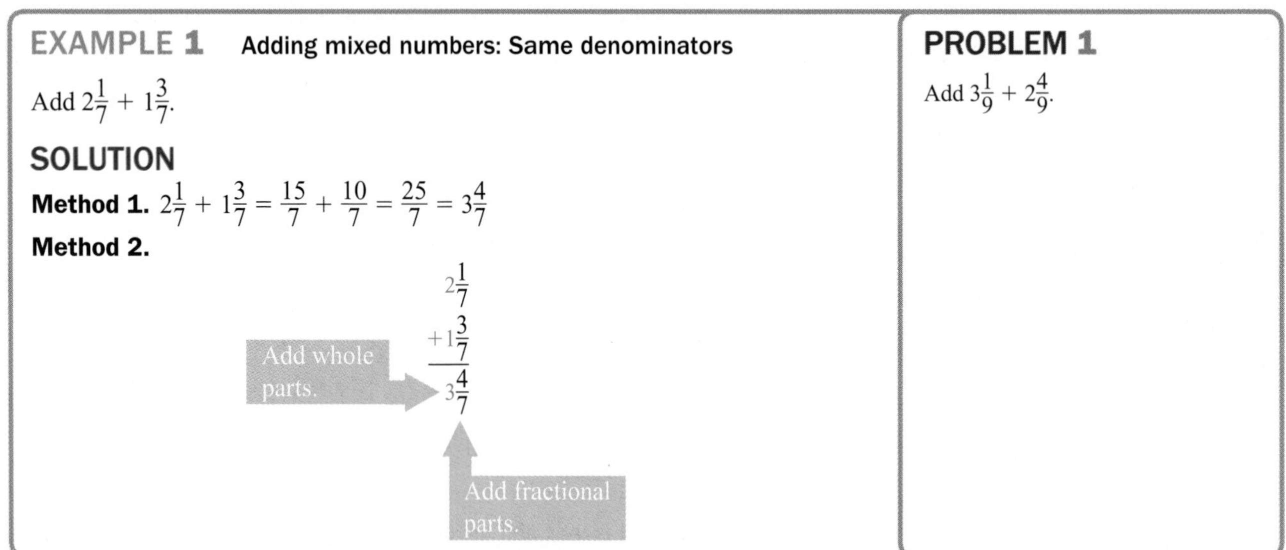

EXAMPLE 1 **Adding mixed numbers: Same denominators**

Add $2\frac{1}{7} + 1\frac{3}{7}$.

SOLUTION

Method 1. $2\frac{1}{7} + 1\frac{3}{7} = \frac{15}{7} + \frac{10}{7} = \frac{25}{7} = 3\frac{4}{7}$

Method 2.

$$\begin{array}{r} 2\frac{1}{7} \\ +1\frac{3}{7} \\ \hline 3\frac{4}{7} \end{array}$$

Add whole parts.

Add fractional parts.

PROBLEM 1

Add $3\frac{1}{9} + 2\frac{4}{9}$.

B > Adding Mixed Numbers with Different Denominators

If we have mixed numbers with different denominators, we must find the LCD first, as in Example 2.

Answers to PROBLEMS

1. $5\frac{5}{9}$

EXAMPLE 2 Adding mixed numbers with different denominators

Add $1\frac{3}{4} + \frac{2}{15}$.

SOLUTION We first find the LCD. The multiples of 15 (the larger of the two denominators) are

$$\underbrace{15, 30, 45,}_{\text{Not multiples of 4}}\ 60 \quad \longleftarrow \text{Multiple of 4 (4 goes into it)}$$

Thus, the LCD of $1\frac{3}{4}$ and $\frac{1}{15}$ is 60. Now $1\frac{3}{4} = \frac{7}{4}$. Thus, $\frac{7}{4} = \frac{105}{60}$ and $\frac{2}{15} = \frac{8}{60}$. So

$$1\frac{3}{4} + \frac{2}{15} = \frac{105}{60} + \frac{8}{60} = \frac{113}{60} = 1\frac{53}{60}$$

PROBLEM 2

Add $1\frac{3}{4} + \frac{1}{6}$.

EXAMPLE 3 Adding mixed numbers: Different denominators

Add $3\frac{3}{4} + 2\frac{5}{6}$.

SOLUTION

Method 1. Convert to improper fractions first.

$$3\frac{3}{4} = \frac{15}{4} \text{ and } 2\frac{5}{6} = \frac{17}{6}$$

The LCD of 4 and 6 is 12, so we rewrite the fractions with this denominator. (Note that 4 and 6 divide into 12.)

$$3\frac{3}{4} = \frac{15}{4} = \frac{15 \cdot 3}{4 \cdot 3} = \frac{45}{12} \text{ and } 2\frac{5}{6} = \frac{17}{6} = \frac{17 \cdot 2}{6 \cdot 2} = \frac{34}{12}$$

Thus,

$$3\frac{3}{4} + 2\frac{5}{6} = \frac{45}{12} + \frac{34}{12} = \frac{45 + 34}{12} = \frac{79}{12} = 6\frac{7}{12}$$

Method 2. Write the fractional parts using the LCD 12 as their denominator:

$$\begin{aligned} 3\frac{3}{4} &= 3\frac{9}{12} \quad \left(\frac{3}{4} = \frac{3 \cdot 3}{4 \cdot 3} = \frac{9}{12}\right) \\ +2\frac{5}{6} &= +2\frac{10}{12} \quad \left(\frac{5}{6} = \frac{5 \cdot 2}{6 \cdot 2} = \frac{10}{12}\right) \\ & \quad 5\frac{19}{12} = 5 + \frac{19}{12} \\ & \quad = 5 + 1\frac{7}{12} = 6\frac{7}{12} \end{aligned}$$

PROBLEM 3

Add $5\frac{1}{4} + 1\frac{5}{6}$.

C > Subtracting Mixed Numbers

The rules we mentioned for adding mixed numbers also apply to the subtraction of mixed numbers. We illustrate the procedure in Example 4.

EXAMPLE 4 Subtracting mixed numbers: Different denominators

Subtract $3\frac{1}{6} - 2\frac{5}{8}$.

PROBLEM 4

Subtract $4\frac{1}{6} - 3\frac{2}{9}$.

(continued)

Answers to PROBLEMS

2. $1\frac{11}{12}$ **3.** $7\frac{1}{12}$ **4.** $\frac{17}{18}$

SOLUTION The LCD of 6 and 8 is 24.

Method 1. First convert to improper fractions.

$$3\frac{1}{6} = \frac{19}{6} \quad \text{and} \quad 2\frac{5}{8} = \frac{21}{8}$$

Now, write $\frac{19}{6}$ and $\frac{21}{8}$ with 24 as a denominator by multiplying the numerator and denominator of $\frac{19}{6}$ by 4 and the numerator and denominator of $\frac{21}{8}$ by 3. We have

$$3\frac{1}{6} - 2\frac{5}{8} = \frac{19 \cdot 4}{6 \cdot 4} - \frac{21 \cdot 3}{8 \cdot 3}$$
$$= \frac{76}{24} - \frac{63}{24}$$
$$= \frac{13}{24}$$

Method 2. Write the fractional parts using the LCD 24 as the denominator.

$$3\frac{1}{6} = 3\frac{4}{24}\left(\frac{1}{6} = \frac{1 \cdot 4}{6 \cdot 4} = \frac{4}{24}\right)$$
$$-2\frac{5}{8} = -2\frac{15}{24}\left(\frac{5}{8} = \frac{5 \cdot 3}{8 \cdot 3} = \frac{15}{24}\right)$$

We cannot subtract $\frac{15}{24}$ from $\frac{4}{24}$. We have to borrow. Write $3\frac{4}{24}$ as $2 + \frac{24}{24} + \frac{4}{24} = 2\frac{28}{24}$.

We can then rewrite the problem as

$$3\frac{1}{6} = 3\frac{4}{24} = 2\frac{28}{24}$$
$$-2\frac{5}{8} = -2\frac{15}{24} = -2\frac{15}{24}$$
$$\frac{13}{24}$$

Note that the answer is the same with either method.

D › Addition and Subtraction of Mixed Numbers

Finally, we do a problem involving addition and subtraction with three mixed numbers.

EXAMPLE 5 **Adding and subtracting mixed numbers**

$1\frac{5}{9} + 2\frac{3}{10} - 1\frac{1}{12} =$ ________

SOLUTION We first find the LCD of $\frac{5}{9}$, $\frac{3}{10}$, and $\frac{1}{12}$.

METHOD 1

Step 1. Write the denominators as products of primes.

$$\begin{array}{r|c|c|c|c|} 9 = & & & 3^2 & \\ 10 = & 2 & \cdot & & 5 \\ 12 = & 2^2 & \cdot & 3 & \end{array}$$

Step 2. Select 2^2, 3^2, and 5.

Step 3. The LCD is

$$2^2 \cdot 3^2 \cdot 5 = 180.$$

METHOD 2

Step 1. Write the denominators in a horizontal row and divide by a prime divisor common to two or more numbers.

$$\begin{array}{r|rrr} 2 & ⑨ & 10 & 12 \\ \hline 3 & 9 & ⑤ & 6 \\ \hline & 3 & 5 & 2 \end{array}$$

Step 2. No prime divides 3, 5, and 2.

Step 3. The LCD is

$$2 \cdot 3 \cdot 3 \cdot 5 \cdot 2 = 180.$$

PROBLEM 5

$1\frac{3}{8} + 2\frac{3}{10} - 2\frac{1}{12} =$ ________

Answers to PROBLEMS

5. $1\frac{71}{120}$

In either case, the LCD is 180. We now rewrite each of the fractions as an improper fraction with a denominator of 180.

$$1\frac{5}{9} = \frac{9 \cdot 1 + 5}{9} = \frac{14}{9} = \frac{14 \cdot 20}{9 \cdot 20} = \frac{280}{180}$$

$$2\frac{3}{10} = \frac{10 \cdot 2 + 3}{10} = \frac{23}{10} = \frac{23 \cdot 18}{10 \cdot 18} = \frac{414}{180}$$

$$1\frac{1}{12} = \frac{12 \cdot 1 + 1}{12} = \frac{13}{12} = \frac{13 \cdot 15}{12 \cdot 15} = \frac{195}{180}$$

Thus,

$$1\frac{5}{9} + 2\frac{3}{10} - 1\frac{1}{12} = \frac{280}{180} + \frac{414}{180} - \frac{195}{180}$$

$$= \frac{280 + 414 - 195}{180}$$

$$= \frac{499}{180}$$ (Note that 499 ÷ 180 = 2 r 139.)

$$= 2\frac{139}{180}$$

E > Perimeter

As you recall from Section 1.3, the **perimeter** of an object (geometric figure) is the distance around the object. Here is the definition.

PERIMETER

The *distance* around an object is its *perimeter*.

EXAMPLE 6 Finding the perimeter

The dimensions of the family room are $21\frac{8}{12}$ feet by 15 feet. (Note that 21^8 means $21\frac{8}{12}$ feet.) How much baseboard molding do you need (red) for this room?

SOLUTION If we disregard the fact that the doors need no baseboard molding, we simply need to find the perimeter of the room, which is:

FAMILY ROOM
$21^8 \times 15^0$

$$21\frac{8}{12} \text{ feet} + 15 \text{ feet} + 21\frac{8}{12} \text{ feet} + 15 \text{ feet}$$

First note that $21\frac{8}{12}$ feet $= 21\frac{2}{3}$ feet. Thus, we need

$$21\frac{2}{3} \text{ feet} + 15 \text{ feet} + 21\frac{2}{3} \text{ feet} + 15 \text{ feet}$$

$$= 21\frac{2}{3} \text{ feet} + 21\frac{2}{3} \text{ feet} + 15 \text{ feet} + 15 \text{ feet}$$

$$= 42\frac{4}{3} \text{ feet} + 30 \text{ feet}$$

$$= \left(42 + 1\frac{1}{3}\right) \text{ feet} + 30 \text{ feet}$$

$$= 43\frac{1}{3} \text{ feet} + 30 \text{ feet}$$

$$= 73\frac{1}{3} \text{ feet of molding}$$

PROBLEM 6

How much molding is needed if the dimensions are 20^8 feet by 15 feet?

Answers to PROBLEMS

6. $71\frac{1}{3}$ ft

> Exercises 2.6

⟨ A ⟩ Adding Mixed Numbers with the Same Denominator In Problems 1–10, add. Reduce if possible.

1. $3\frac{1}{7} + 1\frac{3}{7}$

2. $3\frac{1}{9} + 4\frac{3}{9}$

3. $2\frac{1}{7} + \frac{3}{7}$

4. $5\frac{1}{9} + \frac{7}{9}$

5. $\frac{3}{8} + 5\frac{1}{8}$

6. $\frac{3}{8} + 2\frac{1}{8}$

7. $1\frac{3}{5} + 2\frac{4}{5}$

8. $2\frac{4}{7} + 5\frac{5}{7}$

9. $2 + 3\frac{1}{7}$

10. $3 + 4\frac{1}{8}$

⟨ B ⟩ Adding Mixed Numbers with Different Denominators In Problems 11–20, add. Reduce if possible.

11. $2\frac{3}{4} + \frac{2}{15}$

12. $2\frac{3}{5} + \frac{3}{8}$

13. $1\frac{3}{10} + 2\frac{11}{12}$

14. $1\frac{4}{5} + 3\frac{7}{9}$

15. $1\frac{3}{4} + 2\frac{5}{6}$

16. $2\frac{4}{5} + 3\frac{5}{6}$

17. $8\frac{1}{7} + 3\frac{1}{9}$

18. $6\frac{1}{8} + 5\frac{3}{7}$

19. $9\frac{1}{11} + 3\frac{1}{10}$

20. $7\frac{3}{8} + 1\frac{1}{9}$

⟨ C ⟩ Subtracting Mixed Numbers In Problems 21–34, subtract. Reduce if possible.

21. $3\frac{3}{7} - 1\frac{1}{7}$

22. $7\frac{5}{8} - 3\frac{3}{8}$

23. $4\frac{5}{6} - 3\frac{1}{6}$

24. $5\frac{3}{8} - 2\frac{1}{8}$

25. $3\frac{1}{12} - 1\frac{1}{4}$

26. $3\frac{1}{3} - 1\frac{5}{6}$

27. $3\frac{1}{2} - 2\frac{4}{5}$

28. $4\frac{1}{4} - 3\frac{5}{6}$

29. $4\frac{1}{20} - 3\frac{3}{40}$

30. $8\frac{3}{10} - 7\frac{9}{20}$

31. $3\frac{7}{8} - 1\frac{5}{12}$

32. $5\frac{8}{15} - 1\frac{2}{25}$

33. $3\frac{13}{60} - 3\frac{1}{48}$

34. $4\frac{19}{24} - 4\frac{7}{60}$

⟨ D ⟩ Addition and Subtraction of Mixed Numbers In Problems 35–44, add and subtract as indicated. Reduce if possible.

35. $3\frac{8}{9} + 1\frac{2}{9} - 1\frac{1}{9}$

36. $4\frac{7}{11} + 2\frac{3}{11} - 3\frac{2}{11}$

37. $3\frac{3}{4} + 1\frac{1}{12} - 1\frac{1}{6}$

38. $2\frac{5}{6} + 3\frac{1}{9} - 2\frac{1}{3}$

39. $4\frac{1}{2} - 2\frac{1}{3} + 3\frac{1}{4}$

40. $2\frac{1}{5} - 1\frac{3}{4} + 5\frac{1}{2}$

41. $3\frac{1}{65} + 10\frac{1}{26} - 1\frac{2}{65}$

42. $1\frac{7}{62} + 3\frac{1}{155} - 1\frac{3}{62}$

43.
$$\begin{array}{r} 14\frac{11}{45} \\ +7\frac{7}{60} \\ -3\frac{8}{45} \\ \hline \end{array}$$

44.
$$\begin{array}{r} 10\frac{3}{26} \\ +5\frac{1}{91} \\ -3\frac{1}{26} \\ \hline \end{array}$$

Applications

45. *Body temperature* The normal body temperature is $98\frac{6}{10}$ degrees Fahrenheit. Carlos had the flu, and his temperature was $101\frac{6}{10}$ degrees. How many degrees above normal is that?

46. *Human brain weight* The average human brain weighs approximately $3\frac{1}{8}$ pounds. The brain of the writer Anatole France weighed only $2\frac{1}{4}$ pounds. How much under the average was that?

47. *Human brain weight* As stated in Problem 46, the average human brain weighs approximately $3\frac{1}{8}$ pounds. The heaviest brain ever recorded was that of Ivan Sergeevich Turgenev, a Russian author. His brain weighed approximately $4\frac{7}{16}$ pounds. How much above the average is this weight?

48. *Ride length* Desi rode $\frac{3}{4}$ of a mile. She then rode $1\frac{2}{3}$ miles more. How far did she ride in all?

49. *Ingredients in recipe* A recipe uses $2\frac{1}{2}$ cups of flour and $\frac{3}{4}$ cup of sugar. What is the total number of cups of these ingredients?

50. *Thickness of board* A board $1\frac{3}{4}$ inch thick is glued to another board $\frac{5}{8}$ inch thick. If the glue is $\frac{1}{32}$ inch thick, how thick is the result?

51. *Packages weight* Sir Loin Stake, an English butcher, sold packages weighing $\frac{1}{4}$, $2\frac{1}{2}$, and 3 pounds. What was the total weight of the three packages?

52. *Working, smoking, and weight* A recent survey found that $\frac{3}{10}$ of the American people work long hours and smoke, while $\frac{1}{5}$ are overweight. Thus, the fraction of people who are neither of these is $1 - \frac{1}{5} - \frac{3}{10}$. What fraction is that?

53. *Human bones composition* Human bones are $\frac{1}{4}$ water, $\frac{9}{20}$ minerals, and the rest living tissue. The fraction of the bone that is living tissue is $1 - \frac{1}{4} - \frac{9}{20}$. Find this fraction.

54. *Newspaper expenditures* Americans spend $\$6\frac{1}{2}$ billion on daily newspapers and $\$3\frac{1}{10}$ billion on Sunday newspapers. How many billions of dollars are spent on newspapers?

55. *Average work hours: Canadians and Americans* Americans work an average of $46\frac{3}{5}$ hours per week, while Canadians work $38\frac{9}{10}$. How many more hours per week do Americans work?

56. *Household chore help from husbands* Do husbands help with the household chores? A recent survey estimated that husbands spend about $7\frac{1}{2}$ hours on weekdays helping around the house, $2\frac{3}{5}$ hours on Saturday, and 2 hours on Sunday. How many hours do husbands work around the house during the entire week?

57. *Hours worked* Pedro worked for $15\frac{1}{4}$ hours on Monday and $9\frac{2}{5}$ hours on Tuesday. How many hours did Pedro work altogether?

58. *Rain* Latasha is a weather observer for the U.S. Weather Service. She recorded $1\frac{1}{2}$ inches of rain on Saturday and $2\frac{1}{4}$ inches on Sunday. How many inches of rain did she record?

⟨E⟩ Perimeter In Problems 59–60, find the approximate amount of baseboard molding needed for the rooms shown (include door openings).

59.

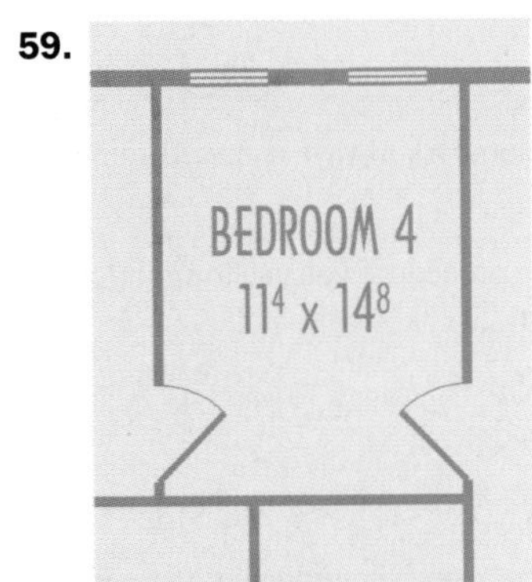

Note: 11^4 means $11\frac{4}{12}$ feet.

60.

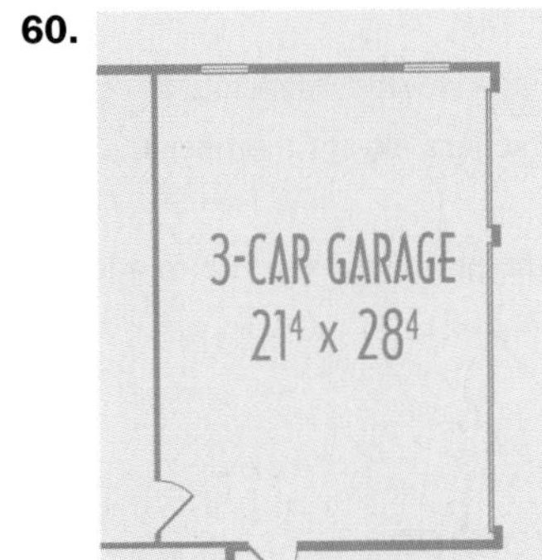

In Problems 61–64 write the addition or subtraction operation illustrated by the diagram and then add or subtract as indicated.

61.

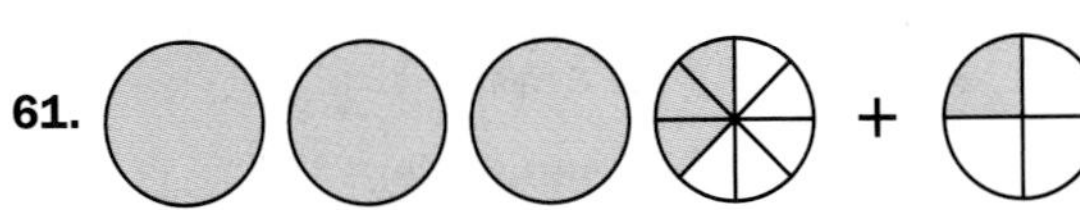

62.

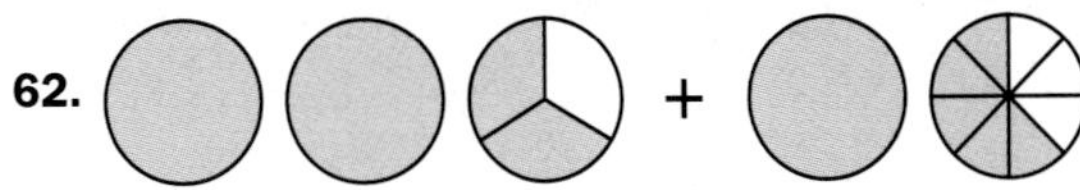

63.

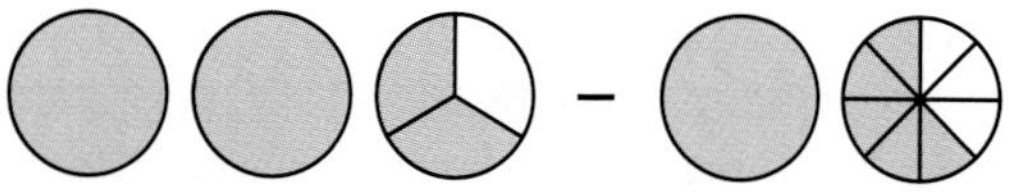

64.

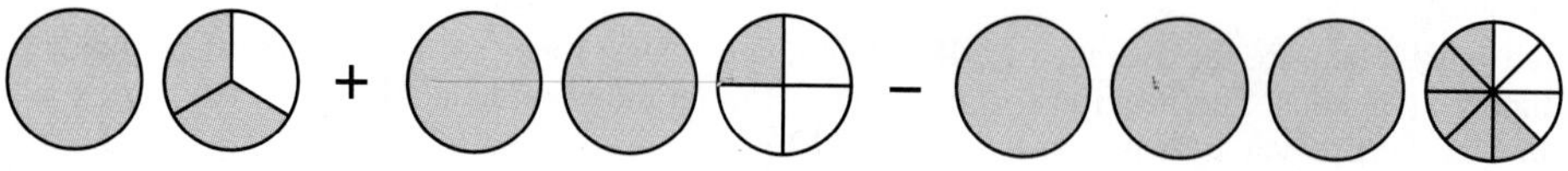

Using Your Knowledge

An Odd-Lot Problem Odd-lot brokers buy shares in round lots (100 shares) and break them up into odd lots to be sold to small investors. When selling less than 100 shares of stock (odd lots), odd-lot brokers charged $\frac{1}{8}$ point per share to their customers when stocks were sold in fractions. Thus, if your stock had a market value of $\$5\frac{1}{4}$, the broker would pay you

$$\$5\frac{1}{4} - \$\frac{1}{8} = \$5 + \$\frac{1}{4} - \$\frac{1}{8} = \$5 + \$\frac{2}{8} - \$\frac{1}{8}$$
$$= \$5\frac{1}{8} \text{ per share}$$

65. Find the odd-lot price of a stock whose market value is $\$3\frac{1}{4}$.

66. Find the odd-lot price of a stock whose market value is $\$2\frac{1}{4}$.

For stocks whose price is more than $55 per share, the price for this service is $\frac{1}{4}$ point per share. Thus, if the quoted price of a stock is $60, the price per share when buying less than 100 shares is

$$\$60 + \$\frac{1}{4} = \$60\frac{1}{4}$$

67. Find the price of one share of stock quoted at $\$62\frac{3}{8}$.

68. Find the price of one share of stock quoted at $\$57\frac{5}{8}$.

Write On

69. Is the sum of two proper fractions always a proper fraction? Explain and give examples.

70. Is the sum of two mixed numbers always a mixed number? Explain and give examples.

71. Write in your own words the procedure you use to add two mixed numbers.

72. Write in your own words the procedure you use to subtract one mixed number from another.

Concept Checker

Fill in the blank(s) with the correct word(s), phrase, or mathematical statement.

73. To **add** or **subtract** mixed numbers with **different denominators,** we must first find the ________________ of the **denominators.**

74. The **distance around** an object is the ________________ of the **object.**

GCD

LCD

area

perimeter

Mastery Test

75. Add: $2\frac{3}{4} + \frac{1}{15}$

76. Add: $3\frac{3}{7} + 2\frac{1}{7}$

77. Add: $2\frac{3}{4} + 3\frac{5}{6}$

78. Subtract: $3\frac{3}{4} - 1\frac{1}{15}$

79. Perform the indicated operations:

$2\frac{5}{9} + 3\frac{7}{10} - 4\frac{1}{12}$

80. Find the perimeter of the rectangle:

$15\frac{1}{2}$ ft

$30\frac{1}{4}$ ft

Skill Checker

Perform the indicated operations.

81. $\frac{5}{18} \cdot \frac{9}{10}$

82. $\frac{3}{10} \cdot \frac{6}{28}$

83. $\frac{4}{5} \div \frac{15}{32}$

84. $\frac{10}{33} \div \frac{25}{11}$

2.7 Order of Operations and Grouping Symbols

Objectives

A Simplify expressions containing fractions and mixed numbers using the order of operations.

B Remove grouping symbols within grouping symbols.

C Solve applications using the concepts studied.

To Succeed, Review How To . . .

1. Use the arithmetic facts (+, −, ×, ÷). (pp. 24, 37, 49, 62)
2. Evaluate an expression containing exponents. (pp. 77, 82–84)
3. Use the order of operations studied in Section 1.8 to simplify expressions. (pp. 82–84)

Getting Started

Do you exercise regularly? If you do, you probably take your pulse to ascertain what your heart rate is. To find your **ideal heart rate** (in beats per minute), subtract your age A from 205 and multiply the result by $\frac{7}{10}$. As you recall from Section 1.8, we use parentheses to indicate which operation we want to do first. In this case, we first want to subtract your age A from 205;

in symbols, $(205 - A)$.

Then, multiply the result by $\frac{7}{10}$, that is, $\frac{7}{10}(205 - A)$.

Thus, your **ideal heart rate** $= \frac{7}{10}(205 - A)$.

Now, suppose you are 25 years old. This means $A = 25$, and

$$\text{Ideal heart rate} = \frac{7}{10}(205 - 25)$$

To evaluate this last expression, we use the *order of operations* studied in Section 1.8. Thus, we do the operations inside the parentheses **first,** then multiply by $\frac{7}{10}$ like this:

$$\text{Ideal heart rate} = \frac{7}{10}(205 - 25)$$

$= \frac{7}{10}(180)$ Subtract 25 from 205.

$= \frac{7 \cdot 180}{10}$ Multiply 7 by 180.

$= \frac{1260}{10}$ $7 \cdot 180 = 1260$

$= 126$ Divide by 10.

This means that your **ideal heart rate** is 126 beats per minute. Is there an easier way? According to the *order of operations,* you could have divided 180 by 10, obtaining 18, and multiplied the 18 by 7. The result is the same, 126.

A › Order of Operations

The order of operations we used for whole numbers (Section 1.8) also applies to fractions and mixed numbers. These rules are restated here for your convenience.

ORDER OF OPERATIONS (PEMDAS)

1. Do all calculations inside *parentheses* and other grouping symbols ( ), [ ], { } first.
2. Evaluate all *exponential* expressions.
3. Do *multiplications* and *divisions* in order (as they occur) from left to right.
4. Do *additions* and *subtractions* in order (as they occur) from left to right.

EXAMPLE 1 Using the order of operations

Simplify.

a. $\frac{1}{2} \cdot \left(\frac{2}{3}\right)^2 - \frac{1}{18}$

b. $\left(\frac{1}{2}\right)^3 + \frac{3}{4} \cdot \frac{1}{2}$

SOLUTION

a. $\frac{1}{2} \cdot \left(\frac{2}{3}\right)^2 - \frac{1}{18}$

$= \frac{1}{2} \cdot \left(\frac{4}{9}\right) - \frac{1}{18}$ Do exponents first $\left(\frac{2}{3}\right)^2 = \frac{4}{9}$.

$= \frac{4}{18} - \frac{1}{18}$ Do $\times$, $\div$ in order from left to right $\frac{1}{2} \cdot \left(\frac{4}{9}\right) = \frac{4}{18}$.

$= \frac{3}{18}$ Do $+$, $-$ in order from left to right $\frac{4}{18} - \frac{1}{18} = \frac{3}{18}$.

$= \frac{1}{6}$ Reduce $\frac{3}{18}$ to $\frac{1}{6}$.

PROBLEM 1

Simplify.

a. $\frac{1}{3} \cdot \left(\frac{3}{2}\right)^2 - \frac{1}{12}$

b. $\left(\frac{1}{3}\right)^3 + \frac{2}{3} \cdot \frac{1}{9}$

Answers to PROBLEMS

1. a. $\frac{2}{3}$ b. $\frac{1}{9}$

b. $\left(\frac{1}{2}\right)^3 + \frac{3}{4} \cdot \frac{1}{2}$

$= \frac{1}{8} + \frac{3}{4} \cdot \frac{1}{2}$ Do exponents first $\left(\frac{1}{2}\right)^3 = \frac{1}{8}$.

$= \frac{1}{8} + \frac{3}{8}$ Do ×, ÷ in order from left to right $\frac{3}{4} \cdot \frac{1}{2} = \frac{3}{8}$.

$= \frac{4}{8}$ Do +, − in order from left to right $\frac{1}{8} + \frac{3}{8} = \frac{4}{8}$.

$= \frac{1}{2}$ Reduce $\frac{4}{8}$ to $\frac{1}{2}$.

EXAMPLE 2 Using the order of operations

Simplify.

a. $\frac{3}{4} \div \frac{1}{6} - \left(\frac{1}{2} + \frac{1}{5}\right)$ **b.** $8 \div \frac{1}{2} \cdot \frac{1}{2} \cdot \frac{1}{2} + \frac{1}{3} - 1$

SOLUTION

a. $\frac{3}{4} \div \frac{1}{6} - \left(\frac{1}{2} + \frac{1}{5}\right)$

$= \frac{3}{4} \div \frac{1}{6} - \left(\frac{7}{10}\right)$ Add inside parentheses: $\left(\frac{1}{2} + \frac{1}{5}\right) = \left(\frac{5}{10} + \frac{2}{10}\right) = \left(\frac{7}{10}\right)$.

$= \frac{9}{2} - \left(\frac{7}{10}\right)$ Do ×, ÷ in order from left to right $\frac{3}{4} \div \frac{1}{6} = \frac{3}{4} \cdot \frac{6}{1} = \frac{18}{4} = \frac{9}{2}$.

$= \frac{38}{10}$ Do +, − from left to right $\frac{9}{2} - \left(\frac{7}{10}\right) = \frac{45}{10} - \frac{7}{10} = \frac{38}{10}$.

$= \frac{19}{5}$ Reduce $\frac{38}{10}$ to $\frac{19}{5}$.

$= 3\frac{4}{5}$

b. $8 \div \frac{1}{2} \cdot \frac{1}{2} \cdot \frac{1}{2} + \frac{1}{3} - 1$

$= \frac{8}{1} \cdot \frac{2}{1} \cdot \frac{1}{2} \cdot \frac{1}{2} + \frac{1}{3} - 1$ Do ×, ÷ in order from left to right $8 \div \frac{1}{2} = \frac{8}{1} \cdot \frac{2}{1}$.

$= \frac{16}{1} \cdot \frac{1}{2} \cdot \frac{1}{2} + \frac{1}{3} - 1$ Do ×, ÷ in order from left to right $\frac{8}{1} \cdot \frac{2}{1} = \frac{16}{1}$.

$= 4 + \frac{1}{3} - 1$ Do ×, ÷ in order from left to right $\frac{16}{1} \cdot \frac{1}{2} \cdot \frac{1}{2} = \frac{16}{4} = 4$.

$= 4\frac{1}{3} - 1$ Do +, − in order from left to right $4 + \frac{1}{3} = 4\frac{1}{3}$.

$= 3\frac{1}{3}$ Do +, − in order from left to right $4\frac{1}{3} - 1 = 3\frac{1}{3}$.

PROBLEM 2

Simplify.

a. $\frac{3}{4} \div \frac{5}{6} - \left(\frac{1}{3} + \frac{1}{5}\right)$

b. $27 \div \frac{1}{3} \cdot \frac{1}{3} \cdot \frac{1}{3} + \frac{1}{2} - 1$

EXAMPLE 3 Using the order of operations

Simplify.

$$\left(\frac{1}{2}\right)^3 \div \frac{1}{4} \cdot \frac{1}{2} + \frac{1}{3}\left(\frac{5}{2} - \frac{1}{2}\right) - \frac{1}{3} \cdot \frac{1}{2}$$

SOLUTION

$\left(\frac{1}{2}\right)^3 \div \frac{1}{4} \cdot \frac{1}{2} + \frac{1}{3}\left(\frac{5}{2} - \frac{1}{2}\right) - \frac{1}{3} \cdot \frac{1}{2}$

$= \left(\frac{1}{2}\right)^3 \div \frac{1}{4} \cdot \frac{1}{2} + \frac{1}{3}(2) - \frac{1}{3} \cdot \frac{1}{2}$ Do operations inside parentheses: $\left(\frac{5}{2} - \frac{1}{2}\right) = \left(\frac{4}{2}\right) = (2)$.

$= \frac{1}{8} \div \frac{1}{4} \cdot \frac{1}{2} + \frac{1}{3}(2) - \frac{1}{3} \cdot \frac{1}{2}$ Do exponents: $\left(\frac{1}{2}\right)^3 = \frac{1}{8}$.

PROBLEM 3

Simplify.

$\left(\frac{1}{2}\right)^3 \div \frac{1}{8} \cdot \frac{1}{2} + \frac{1}{3}\left(\frac{3}{2} - \frac{1}{2}\right) - \frac{1}{3} \cdot \frac{1}{2}$

(continued)

Answers to PROBLEMS

2. a. $\frac{11}{30}$ **b.** $8\frac{1}{2}$ **3.** $\frac{2}{3}$

$= \frac{1}{2} \cdot \frac{1}{2} + \frac{1}{3}(2) - \frac{1}{3} \cdot \frac{1}{2}$ Do divisions: $\frac{1}{8} \div \frac{1}{4} = \frac{1}{8} \cdot \frac{4}{1} = \frac{1}{2}$.

$= \frac{1}{4} + \frac{2}{3} - \frac{1}{6}$ Do multiplications: $\frac{1}{2} \cdot \frac{1}{2} = \frac{1}{4}$; $\frac{1}{3}(2) = \frac{2}{3}$; $\frac{1}{3} \cdot \frac{1}{2} = \frac{1}{6}$.

$= \frac{11}{12} - \frac{1}{6}$ Do additions: $\frac{1}{4} + \frac{2}{3} = \frac{3}{12} + \frac{8}{12} = \frac{11}{12}$.

$= \frac{9}{12}$ Do subtractions: $\frac{11}{12} - \frac{1}{6} = \frac{11}{12} - \frac{2}{12} = \frac{9}{12}$.

$= \frac{3}{4}$ Reduce $\frac{9}{12}$ to $\frac{3}{4}$.

B › More Than One Set of Grouping Symbols

As you recall from Section 1.8, when grouping symbols occur within other grouping symbols (nested symbols), computations in the innermost grouping symbols are done first. We illustrate this in Example 4.

EXAMPLE 4 Using the order of operations

Simplify.

$$\frac{1}{5} \div 1\frac{1}{5} + \left\{12 \cdot \left(\frac{1}{2}\right)^2 - \left[\frac{1}{3} + \left(2\frac{1}{2} - \frac{1}{2}\right)\right]\right\}$$

SOLUTION

$\frac{1}{5} \div 1\frac{1}{5} + \left\{12 \cdot \left(\frac{1}{2}\right)^2 - \left[\frac{1}{3} + \left(2\frac{1}{2} - \frac{1}{2}\right)\right]\right\}$

$= \frac{1}{5} \div 1\frac{1}{5} + \left\{12 \cdot \left(\frac{1}{2}\right)^2 - \left[\frac{1}{3} + (2)\right]\right\}$ Subtract inside parentheses: $\left(2\frac{1}{2} - \frac{1}{2}\right) = (2)$.

$= \frac{1}{5} \div 1\frac{1}{5} + \left\{12 \cdot \left(\frac{1}{2}\right)^2 - 2\frac{1}{3}\right\}$ Add inside brackets: $\left[\frac{1}{3} + (2) = 2\frac{1}{3}\right]$.

$= \frac{1}{5} \div 1\frac{1}{5} + \left\{12 \cdot \left(\frac{1}{4}\right) - 2\frac{1}{3}\right\}$ Do exponents inside braces: $\left(\frac{1}{2}\right)^2 = \left(\frac{1}{4}\right)$.

$= \frac{1}{5} \div 1\frac{1}{5} + \left\{3 - 2\frac{1}{3}\right\}$ Multiply inside braces: $12 \cdot \left(\frac{1}{4}\right) = \left(\frac{12}{4}\right) = 3$.

$= \frac{1}{5} \div 1\frac{1}{5} + \left\{\frac{2}{3}\right\}$ Subtract inside braces: $3 - 2\frac{1}{3} = \frac{2}{3}$.

$= \frac{1}{6} + \frac{2}{3}$ Divide: $\frac{1}{5} \div 1\frac{1}{5} = \frac{1}{5} \div \frac{6}{5} = \frac{1}{5} \cdot \frac{5}{6} = \frac{1}{6}$.

$= \frac{5}{6}$ Add: $\frac{1}{6} + \frac{2}{3} = \frac{1}{6} + \frac{4}{6} = \frac{5}{6}$.

PROBLEM 4

Simplify.

$$\frac{1}{6} \div 1\frac{1}{6} + \left\{27 \cdot \left(\frac{1}{3}\right)^2 - \left[\frac{1}{3} + \left(2\frac{1}{3} - \frac{1}{3}\right)\right]\right\}$$

C › Applications: Averages

Suppose you score 8, 9, and 8 on three math quizzes. What is the average for the three quizzes? Here is the rule we need:

AVERAGES

To find the average of a set of numbers, add the numbers and divide by the number of addends.

The addends are the numbers to be added (8, 9, and 8).

Thus, to find the average of 8, 9, and 8, we add 8, 9, and 8 and divide by the number of addends, which is 3. The answer is

$$\frac{8 + 9 + 8}{3} = \frac{25}{3} = 8\frac{1}{3}$$

Answers to PROBLEMS

4. $\frac{17}{21}$

EXAMPLE 5 Calculating an average

Shroeder went fishing and caught four fish weighing $3\frac{1}{2}$, $5\frac{1}{4}$, $2\frac{1}{2}$, and $7\frac{1}{4}$ pounds, respectively. What is the average weight of the four fish?

SOLUTION To find the average, add the weights and divide by 4.

$$\frac{3\frac{1}{2} + 5\frac{1}{4} + 2\frac{1}{2} + 7\frac{1}{4}}{4}$$

To simplify the calculation, add the whole parts 3, 5, 2, and 7, obtaining 17, and then the fractional parts $\frac{1}{2} + \frac{1}{4} + \frac{1}{2} + \frac{1}{4} = 1\frac{1}{2}$.

Thus, we have: $\frac{3\frac{1}{2} + 5\frac{1}{4} + 2\frac{1}{2} + 7\frac{1}{4}}{4} = \frac{17 + 1\frac{1}{2}}{4} = \frac{18\frac{1}{2}}{4} = \frac{\frac{37}{2}}{4}$

$$= \frac{37}{2} \cdot \frac{1}{4} = \frac{37}{8} = 4\frac{5}{8}$$

This means that the average weight of the four fish is $4\frac{5}{8}$ pounds.

PROBLEM 5

Find the average weight of four fish weighing $5\frac{1}{4}$, $6\frac{1}{2}$, $4\frac{1}{4}$, and $3\frac{1}{2}$ pounds, respectively.

Boost your grade at mathzone.com!

- Practice Problems
- NetTutor
- Self-Tests
- e-Professors
- Videos

Exercises 2.7

‹A›‹B› Order of Operations

In Problems 1–25, simplify.

1. $\left(\frac{1}{2}\right)^2 \cdot \frac{1}{5} + \frac{1}{6}$
2. $\frac{1}{3} \cdot \left(\frac{1}{2}\right)^2 + \frac{1}{6}$
3. $\frac{1}{7} + \frac{1}{3} \cdot \left(\frac{1}{2}\right)^2$
4. $\frac{1}{6} + \left(\frac{1}{3}\right)^2 \cdot \frac{1}{2}$
5. $\frac{1}{7} \cdot \left(\frac{1}{2}\right)^3 - \frac{1}{56}$
6. $\frac{4}{9} \cdot \left(\frac{1}{2}\right)^2 - \left(\frac{1}{3}\right)^2$
7. $\frac{1}{2} - \frac{1}{3} \cdot \frac{1}{5}$
8. $\frac{1}{3} - \frac{1}{6} \cdot \frac{1}{5}$
9. $12 \div 6 - \left(\frac{1}{3} + \frac{1}{2}\right)$
10. $18 \div 9 - \left(\frac{1}{4} + \frac{1}{6}\right)$
11. $\frac{1}{3} \cdot \frac{1}{4} \div \frac{1}{2} + \left(\frac{5}{6} - \frac{1}{2}\right)$
12. $\frac{1}{3} \cdot \frac{1}{6} \div \frac{1}{2} + \left(\frac{4}{5} - \frac{1}{2}\right)$
13. $\frac{1}{6} \div \frac{1}{3} \cdot \frac{1}{3} \cdot \frac{1}{3} + \left(\frac{1}{4} - \frac{1}{9}\right)$
14. $\frac{1}{10} \div \frac{1}{2} \cdot \frac{1}{2} \cdot \frac{1}{2} + \left(\frac{2}{3} - \frac{1}{2}\right)$
15. $8 \div \frac{1}{2} \cdot \frac{1}{2} \cdot \frac{1}{2} - \left(\frac{1}{3} + \frac{1}{5}\right)$
16. $6 \div \frac{1}{3} \cdot \frac{1}{3} \cdot \frac{1}{3} - \left(\frac{1}{3} + \frac{1}{5}\right)$
17. $\frac{1}{10} \div \frac{1}{5} \cdot \frac{1}{2} + \frac{1}{8}\left(\frac{4}{5} - \frac{1}{2}\right) + \left(\frac{1}{8} \div \frac{1}{4}\right)$
18. $\frac{1}{15} \div \frac{1}{3} \cdot \frac{1}{3} + \frac{1}{2}\left(\frac{4}{5} - \frac{1}{2}\right) + \left(\frac{1}{8} \div \frac{1}{4}\right)$
19. $\frac{1}{5} \div \frac{1}{3} \cdot \frac{1}{3} + \frac{1}{2}\left(\frac{1}{2} - \frac{1}{5}\right) + \left(\frac{1}{8} \div \frac{1}{4}\right)$
20. $\frac{1}{5} \div \frac{1}{2} \cdot \frac{1}{2} + \frac{1}{2}\left(\frac{1}{2} - \frac{1}{5}\right) + \left(\frac{1}{8} \div \frac{1}{4}\right)$
21. $\frac{1}{20} \div \frac{1}{5} + \left\{\frac{1}{3} \div \frac{1}{4} - \left[\frac{1}{4} + \left(\frac{1}{3} - \frac{1}{5}\right)\right]\right\}$
22. $\left\{\frac{1}{4} \div \frac{1}{2} - \left[\frac{1}{3} + \left(\frac{3}{5} - \frac{1}{4}\right)\right]\right\} + \frac{1}{30} \div \frac{1}{6}$
23. $\frac{7}{30} \div \frac{1}{15} \cdot \left\{\frac{1}{10} \div \frac{1}{20} - \left[\frac{1}{2} \cdot \frac{1}{2} + \frac{1}{2}\right]\right\}$
24. $\frac{1}{30} \div \frac{1}{10} \cdot \left\{\frac{1}{2} \div \frac{1}{4} - \left[\frac{1}{3} \cdot \frac{1}{3} + \frac{1}{3}\right]\right\}$
25. $\frac{1}{4} \div \frac{1}{12} \cdot \frac{1}{6} + \left[\frac{1}{5}\left(\frac{1}{3} + \frac{1}{2}\right) - \frac{1}{6}\right] - \left(\frac{1}{3} + \frac{1}{2} \cdot \frac{1}{3}\right)$

‹C› Applications: Averages

26. *Rainfall in Tampa* The rainfall in Tampa, Florida, for the month of January was $\frac{1}{10}$ inch. A year later it was $2\frac{1}{2}$ inches. What was the average rainfall in January for the two years?

27. *Rainfall in Tampa* The rainfall in Tampa, Florida, for the month of February was $2\frac{9}{10}$ inches. A year later it was $2\frac{4}{5}$ inches. What was the average rainfall in February for the two years?

Answers to PROBLEMS

5. $4\frac{7}{8}$ pounds

Web IT go to **mhhe.com/bello** for more lessons

28. *Weight of ash* The approximate weight of one cubic foot of different varieties of ash (in pounds) is as follows:

Black ash: 40

Green ash: $41\frac{1}{2}$

White ash: 43

What is the average weight of one cubic foot of ash?

29. *Weight of hemlock* The approximate weight of one cubic foot of different varieties of hemlock (in pounds) is as follows:

Eastern: 29

Western: $32\frac{1}{2}$

Mountain: $32\frac{3}{5}$

What is the average weight of one cubic foot of hemlock?

30. *Top 3 box office moneymakers* The top three box office films of all time and their box office gross (in millions of dollars) are:

Titanic (1997): $600\frac{4}{5}$

Star Wars (1977): 460

E.T. (1982): $434\frac{9}{10}$

What is the average box office gross for these three films?

31. *Last 3 box office moneymakers* The last three films in the 100 top grossing films and their box office gross (in millions of dollars) are:

Rambo: First Blood (1985): $150\frac{2}{5}$

As Good As It Gets (1997): $148\frac{1}{2}$

Gremlins (1984): $148\frac{1}{5}$

a. What is the average box office gross for these three films?

b. What is the difference between the averages for the last three films in the top 100 and the first three films (Exercise 30)?

Source: EDI/Filmsources, *Variety*.

32. *Hours of sleep for shift workers* According to Shiftworker Online, the average number of hours of sleep per 24-hour period for shift workers is as follows:

Night shifts: $4\frac{3}{5}$

Evening shifts: $8\frac{1}{2}$

Day workers: $7\frac{1}{2}$

What is the average number of hours of sleep for these three types of workers?

33. *Nighttime pain* Here is the reported number of nights per month with nighttime pain and sleeplessness in different age groups.

18–34: $6\frac{4}{5}$ nights per month

35–49: $8\frac{1}{10}$ nights per month

50 and over: $10\frac{7}{10}$ nights per month

What is the average number of nights per month during which nighttime pain with sleeplessness occurs?

Source: Gallup Poll/National Sleep Foundation.

TV viewing The chart shows the average number of hours (per week) of TV viewing for different age groups in three Canadian cities. This will be used in Problems 34–40.

	Total Pop.	Children 2–11	Adolescents 12–17	Men 18 and over	Women 18 and over
Ontario	$20\frac{1}{10}$	$13\frac{4}{5}$	$12\frac{4}{5}$	$19\frac{4}{5}$	$23\frac{9}{10}$
Manitoba	$20\frac{9}{10}$	$14\frac{2}{5}$	$12\frac{4}{5}$	$21\frac{1}{10}$	$24\frac{9}{10}$
Saskatchewan	$20\frac{1}{2}$	$14\frac{3}{10}$	$12\frac{1}{2}$	$20\frac{1}{5}$	$25\frac{1}{10}$

Source: Statistics Canada.

Find the average number of hours per week of TV viewing for

34. The total population.

35. Children 2–11.

36. Adolescents 12–17.

37. Men 18 and over.

38. Women 18 and over.

39. On the average, which of the groups watches the most hours?

40. On the average, which of the groups watches the least hours?

Web IT go to mhhe.com/bello for more lessons

Fish weight Can you approximate the weight of a fish (in pounds) using a ruler? You can if you use the formulas below and remember that the length L and the girth G (distance around, or perimeter) of the fish must be measured in **inches.** If you don't have a flexible ruler to go around the fish, you can approximate the girth by doubling the height of the fish.

Formulas for the weight of the fish (in pounds)

$$\textbf{Bass} = \frac{L^2 \cdot G}{1200}$$

$$\textbf{Pike} = \frac{L^3}{3500}$$

$$\textbf{Trout} = \frac{L \cdot G^2}{800}$$

$$\textbf{Walleye} = \frac{L^3}{2700}$$

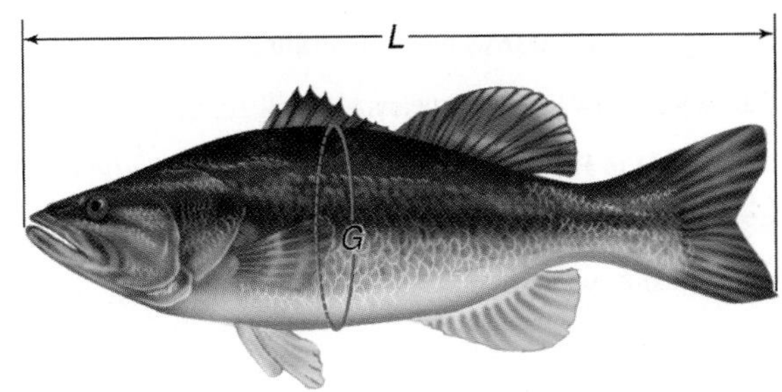

Source: http://dnr.wi.gov

41. Find the approximate weight of a bass 20 inches long and with a $15\frac{1}{2}$-inch girth.

42. Find the approximate weight of a pike 20 inches long.

43. Find the approximate weight of a trout 30 inches long and with a 25-inch girth.

44. Find the approximate weight of a walleye 24 inches long.

Using Your Knowledge

Divisors, Averages, Harmony, and Equations Consider the number 6.

a. 6 has 4 divisors: 6, 3, 2, and 1.

b. The average of the 4 divisors of 6 is $A_6 = \frac{6 + 3 + 2 + 1}{4} = \frac{12}{4} = 3$

c. The harmonic mean H_6 of the divisors is

$$H_6 = \frac{4}{\frac{1}{6} + \frac{1}{3} + \frac{1}{2} + \frac{1}{1}} = \frac{4}{\frac{1}{6} + \frac{2}{6} + \frac{3}{6} + \frac{6}{6}} = \frac{4}{\frac{12}{6}} = \frac{4}{1} \cdot \frac{6}{12} = 2$$

Now, drum roll here, $6 = A_6 \cdot H_6 = 3 \cdot 2$

Is this true for any whole number?

45. Consider the number 8.

a. Find the divisors of 8. (You should find 4 of them.)

b. Find the average of the 4 divisors of 8, that is, A_8.

c. Find the harmonic mean of the divisors, that is, H_8.

d. Is it true that $A_8 \cdot H_8 = 8$?

46. Repeat Problem 45 using the number 16. Remember, you must have $A_{16} \cdot H_{16} = 16$.

Write On

47. We have studied the order of operations twice: in this section, as it applies to fractions, and in Section 1.8 as it applied to whole numbers. Write in your own words the steps you use to simplify an expression using the order of operations.

48. When simplifying $16 \cdot \frac{3}{8}$, Maria multiplied 16 by 3 and then divided by 8. Her answer was 6. Ty divided 16 by 8 *first* then multiplied the result 2, by 3.

a. Write in your own words who is right, Maria or Ty?

b. Which procedure is easier, Maria's or Ty's?

Concept Checker

Fill in the blank(s) with the correct word(s), phrase, or mathematical statement.

49. The **P** in **PEMDAS** means to do all calculations inside __________ and other grouping symbols.

50. The **E** in **PEMDAS** means to **evaluate** all __________ expressions.

51. The **M** in **PEMDAS** means to do all __________ and **division** in order from left to right.

52. The **D** in **PEMDAS** means to do all **multiplication** and __________ in order from left to right.

53. The **A** in **PEMDAS** means to do all __________ and **subtraction** in order from left to right.

54. The **S** in **PEMDAS** means to do all addition and __________ in order from left to right.

subtraction
addition
division
multiplication
exponential
equivalent
parentheses

Mastery Test

55. Simplify: $\frac{3}{8} \div \frac{1}{12} - \left(\frac{1}{4} + \frac{1}{10}\right)$

56. Simplify: $9 \div \frac{1}{3} \cdot \frac{1}{3} \cdot \frac{1}{3} + \frac{1}{2} - 1$

57. Simplify: $\frac{1}{3} \cdot \left(\frac{3}{2}\right)^2 - \frac{1}{18}$

58. Simplify: $\left(\frac{1}{3}\right)^3 + \frac{2}{3} \cdot \frac{1}{9}$

59. Simplify: $\left(\frac{1}{3}\right)^3 \div \frac{1}{3} \cdot \frac{1}{9} + \frac{1}{2}\left(\frac{5}{3} - \frac{1}{3}\right) - \frac{1}{3} \cdot \frac{1}{2}$

60. Simplify: $\frac{1}{7} \div 1\frac{1}{7} + \left\{12 \cdot \left(\frac{1}{2}\right)^2 - \left[\frac{1}{4} + \left(2\frac{1}{2} - \frac{1}{2}\right)\right]\right\}$

61. Find the average of $2\frac{1}{2}$, $5\frac{1}{4}$, $3\frac{1}{2}$, and $4\frac{1}{4}$.

Skill Checker

Solve:

62. $x + 5 = 17$

63. $x + 7 = 13$

64. $10 - x = 3$

65. $15 = 5x$

66. $24 \div x = 6$

2.8 Equations and Problem Solving

Objectives

You should be able to:

A Translate word sentences and phrases into mathematical equations.

B Translate a given problem into a mathematical equation, then solve the equation.

C Solve applications involving the concepts studied.

To Succeed, Review How To . . .

1. Use the addition, subtraction, and division principles. (pp. 90–91)
2. Add, subtract, multiply, and divide fractions and mixed numbers. (pp. 130–133, 151–159)
3. Use the RSTUV procedure. (p. 92)

Getting Started

In the ad shown, it is claimed that 4 out of 5 people—that is, $\frac{4}{5}$ of the people—say Big John's beans taste better. If 400 people actually made that statement, can we find how many people were surveyed? To do this, we must learn how to translate the problem from words to mathematics.

4 out of 5 people say Big John's Beans taste better.

We bet 20¢ you'll agree.

A > Translating Mathematical Equations

PROCEDURE FOR SOLVING PROBLEMS

1. Read the problem carefully and decide what is asked for (the unknown).
2. Select □ or a letter to represent this unknown.
3. Translate the problem into an equation.

4. Use the rules studied to solve the equation.
5. Verify your answer.

How do we remember these steps? Look at the first letter in each sentence. We still call this the RSTUV method.

Read the problem.

Select the unknown.

Translate the problem.

Use the rules studied to solve.

Verify your answer.

As you can see from step 3 on the previous page, we have translated the word "of" as multiplication (·) and the words "this is" as =. Since these and other words will be used later, we give you a dictionary so that you can translate them properly.

Mathematics Dictionary

Word or phrase	Translation
Is, is equal to, equals, the same	$=$
Of, the product, times, multiple, multiplied by	$\times$ or $\cdot$
Add, more than, plus, sum, increased by, added to, more	$+$
Subtract, less than, minus, difference, decreased by, subtracted from, less	$-$
Divide, divided by, the quotient	$\div$
Double, twice, twice as much	$2\times$ or $2\cdot$
Half, half of, half as much	$\frac{1}{2}\times$ or $\frac{1}{2}\cdot$

Now let us use the dictionary to translate these phrases:

a. a number multiplied by 8

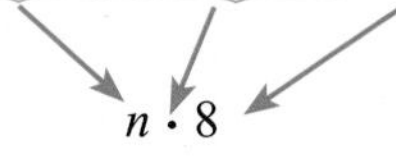

(You can use *any* letter in place of n.)

b. the sum of $\frac{1}{5}$ and a number

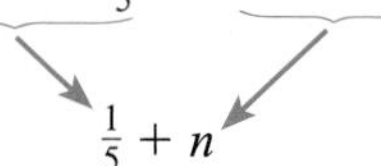

c. $\frac{1}{3}$ less than twice a number

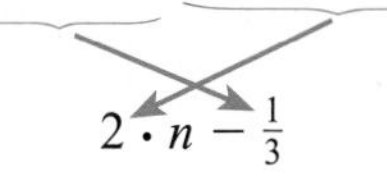

d. a number divided by 8

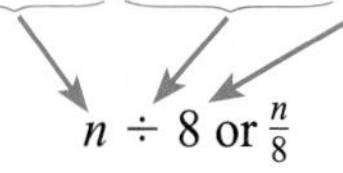

e. $\frac{3}{4}$ of a quantity is equal to 300

$$\frac{3}{4} \cdot q = 300$$

EXAMPLE 1 Translating words into mathematical equations

Translate into mathematical equations.

a. 7 more than a number is 8.

b. The difference between a number and 4 is 1.

c. Twice a number equals 10.

SOLUTION

a. $n + 7 = 8$

b. $n - 4 = 1$

c. $2 \cdot n = 10$

PROBLEM 1

Translate into mathematical equations.

a. 5 more than a number is 6.

b. The difference between a number and 8 is 2.

c. 3 times a number equals 12.

B > Solving Problems

How can we find the answers to (a), (b), and (c) in Example 1? We can solve by using the addition, subtraction, and multiplication principles we studied in Section 1.9. To make things easier, we shall state a more general principle here:

PRINCIPLES TO SOLVE EQUATIONS

If c is any number, the equation $a = b$ is equivalent to:

$a + c = b + c$ (Addition principle)

$a - c = b - c$ (Subtraction principle)

$a \cdot c = b \cdot c$ (Multiplication principle) ($c \neq 0$)

$a \div c = b \div c$ (Division principle) ($c \neq 0$)

or $\frac{a}{c} = \frac{b}{c}$

Thus, to find the answer to (a), $n + 7 = 8$, we *solve* this equation using the subtraction principle.

$$n + 7 = 8$$
$$n + \underbrace{7 - 7}_{0} = 8 - 7 \quad \text{Subtract 7. } 7 - 7 = 0$$
$$n = 1$$

To solve (b):

$$n - 4 = 1$$
$$n \underbrace{- 4 + 4}_{0} = 1 + 4 \quad \text{Add 4. } -4 + 4 = 4 - 4 = 0$$
$$n = 5$$

To solve (c):

$$2 \cdot n = 10$$
$$\frac{\cancel{2} \cdot n}{\cancel{2}} = \frac{10}{2} \quad \text{Divide by 2. } \frac{2}{2} = 1 \text{ and } 1 \cdot n = n$$
$$n = 5$$

Answers to PROBLEMS

1. a. $n + 5 = 6$ **b.** $n - 8 = 2$ **c.** $3 \cdot n = 12$

EXAMPLE 2 Translating and solving equations

Translate and solve.

a. A number x increased by $\frac{1}{7}$ gives $\frac{1}{2}$. Find x.
b. $\frac{1}{3}$ less than a number y is $\frac{2}{5}$. Find y.
c. The quotient of z and 5 is $\frac{3}{4}$. Find z.

SOLUTION

a. We translate the problem as:

A number x increased by $\frac{1}{7}$ gives $\frac{1}{2}$.

$$x + \frac{1}{7} = \frac{1}{2}$$

Since $\frac{1}{7}$ is being added to x, we subtract $\frac{1}{7}$ on both sides to obtain

$$x + \underbrace{\frac{1}{7} - \frac{1}{7}}_{0} = \frac{1}{2} - \frac{1}{7}$$

$$x = \frac{1}{2} - \frac{1}{7}$$

The LCD of 2 and 7 is 14, so we write

$$x = \frac{1 \cdot 7}{2 \cdot 7} - \frac{1 \cdot 2}{7 \cdot 2}$$

$$= \frac{7}{14} - \frac{2}{14} = \frac{5}{14}$$

b. Translating: $\frac{1}{3}$ less than a number y is $\frac{2}{5}$.

$$y - \frac{1}{3} = \frac{2}{5}$$

$$y \underbrace{- \frac{1}{3} + \frac{1}{3}}_{0} = \frac{2}{5} + \frac{1}{3} \quad \text{Add } \frac{1}{3}.$$

$$y = \frac{2}{5} + \frac{1}{3}$$

The LCD of 5 and 3 is 15, so we rewrite each of the fractions with a denominator of 15, obtaining

$$y = \frac{2 \cdot 3}{5 \cdot 3} + \frac{1 \cdot 5}{3 \cdot 5} = \frac{6}{15} + \frac{5}{15} = \frac{11}{15}$$

c. The quotient of z and 5 is $\frac{3}{4}$ is translated as

$$\frac{z}{5} = \frac{3}{4}$$

$$5 \cdot \frac{z}{5} = 5 \cdot \frac{3}{4} \quad \text{Multiply by 5.}$$

$$z = \frac{15}{4} = 3\frac{3}{4}$$

PROBLEM 2

Translate and solve.

a. A number n increased by $\frac{1}{4}$ is $\frac{3}{5}$. Find n.

b. $\frac{1}{3}$ less than a number m is $\frac{3}{5}$. Find m.

c. The quotient of q and 7 is $\frac{3}{5}$. Find q.

Here are some more problems involving translations. Don't forget to use the RSTUV method.

Answers to PROBLEMS

2. a. $n + \frac{1}{4} = \frac{3}{5}$; $n = \frac{7}{20}$ **b.** $m - \frac{1}{3} = \frac{3}{5}$; $m = \frac{14}{15}$ **c.** $\frac{q}{7} = \frac{3}{5}$; $q = \frac{21}{5} = 4\frac{1}{5}$

EXAMPLE 3 **Translating and solving equations**

Solve:

$\frac{3}{5}$ of what number is 12?

SOLUTION

$$\frac{3}{5} \cdot n = 12$$

$$\frac{\frac{3}{5} \cdot n}{\frac{3}{5}} = \frac{12}{\frac{3}{5}} \qquad \text{Divide by } \tfrac{3}{5}.$$

Thus,

$$n = \frac{12}{\frac{3}{5}} = 12 \div \frac{3}{5} = \overset{4}{\cancel{12}} \cdot \frac{5}{\underset{1}{\cancel{3}}} = 4 \cdot 5 = 20$$

CHECK Is $\frac{3}{5} \cdot 20 = 12$? Yes,

$$\frac{3}{\underset{1}{\cancel{5}}} \cdot \overset{4}{\cancel{20}} = 12$$

PROBLEM 3

$\frac{5}{6}$ of what number is 15?

EXAMPLE 4 **Translating and solving equations**

Solve:

What fraction of $1\frac{1}{2}$ is $2\frac{3}{4}$?

$$n \cdot 1\frac{1}{2} = 2\frac{3}{4}$$

SOLUTION

Since $1\frac{1}{2} = \frac{3}{2}$ and $2\frac{3}{4} = \frac{11}{4}$,

$$n \cdot \frac{3}{2} = \frac{11}{4}$$

Dividing by $\frac{3}{2}$,

$$\frac{n \cdot \frac{3}{2}}{\frac{3}{2}} = \frac{\frac{11}{4}}{\frac{3}{2}}$$

Thus,

$$n = \frac{11}{4} \div \frac{3}{2} = \frac{11}{4} \cdot \frac{2}{3} = \frac{11}{\underset{2}{\cancel{4}}} \cdot \frac{\overset{1}{\cancel{2}}}{3} = \frac{11}{6}$$

CHECK Is $\frac{11}{6} \times \frac{3}{2} = \frac{11}{4}$? Yes,

$$\frac{11}{\underset{2}{\cancel{6}}} \cdot \frac{\overset{1}{\cancel{3}}}{2} = \frac{11}{4}$$

PROBLEM 4

What fraction of $2\frac{1}{2}$ is $3\frac{1}{4}$?

EXAMPLE 5 **Translating and solving equations**

Find a number such that $\frac{3}{5}$ of it is $1\frac{3}{4}$.

SOLUTION

$$\frac{3}{5} \cdot n = 1\frac{3}{4}$$

$$\frac{3}{5} \cdot n = \frac{7}{4} \qquad \text{Write } 1\tfrac{3}{4} \text{ as } \tfrac{7}{4}$$

$$\frac{\frac{3}{5} \cdot n}{\frac{3}{5}} = \frac{\frac{7}{4}}{\frac{3}{5}} \qquad \text{Divide by } \tfrac{3}{5}.$$

PROBLEM 5

Find a number such that $\frac{2}{7}$ of it is $1\frac{1}{2}$.

Answers to PROBLEMS

3. 18 **4.** $\frac{13}{10}$ **5.** $\frac{21}{4}$

Thus,

$$n = \frac{7}{4} \div \frac{3}{5} = \frac{7}{4} \cdot \frac{5}{3} = \frac{35}{12}$$

CHECK Is $\frac{3}{5} \cdot \frac{35}{12} = \frac{7}{4}$? Yes,

$$\frac{\overset{1}{3}}{\underset{1}{\cancel{5}}} \cdot \frac{\overset{7}{\cancel{35}}}{\underset{4}{\cancel{12}}} = \frac{7}{4}$$

EXAMPLE 6 **Finding products**

The product of $3\frac{1}{2}$ and $1\frac{5}{7}$ is what number?

SOLUTION

$$3\frac{1}{2} \cdot 1\frac{5}{7} = n$$

or

$$\frac{7}{2} \cdot \frac{12}{7} = n$$

$$6 = n$$

PROBLEM 6

The product of $2\frac{1}{4}$ and $1\frac{1}{3}$ is what number?

C › Applications Involving Equations

The ideas presented in these examples can be used to solve problems of this type: If $7\frac{1}{2}$ ounces of tomato sauce cost 15¢, what will 10 ounces cost? To solve this problem, we make up a simpler problem that we can easily solve and then use it as a model. For example, if 8 apples cost 40¢, what will 10 apples cost?

Solution 8 apples cost 40¢

1 apple costs 5¢

10 apples cost 5 × 10 = 50¢

Now to solve the tomato sauce problem, we proceed similarly, using the problem on the left as a model.

8 apples cost 40¢ 1 apple costs 40¢ ÷ 8 = 5¢	$7\frac{1}{2} = \frac{15}{2}$ ounces cost 15¢ 1 ounce costs $15¢ \div \frac{15}{2} = \overset{1}{\cancel{15}} \cdot \frac{2}{\underset{1}{\cancel{15}}} = 2¢$
10 apples cost 5¢ × 10 = 50¢	10 ounces cost $2¢ \cdot 10 = 20¢$

What we have done here is to use one of the problem-solving techniques mentioned in Section 1.9: create a related problem or *model* that is easy to solve and follow that pattern to solve the given problem.

This problem can also be solved using ratios and proportions. These topics are covered in Chapter 4.

EXAMPLE 7 **Cost of detergent**

If 1 pint 6 ounces ($1\frac{3}{8}$ pints) of liquid detergent cost 66¢, what will 2 pints cost?

SOLUTION The solution follows the model already shown.

$1\frac{3}{8}$ pints cost 66¢

1 pint costs

$$66¢ \div \frac{11}{8} = \overset{6}{\cancel{66}}¢ \cdot \frac{8}{\underset{1}{\cancel{11}}} = 48¢$$

2 pints cost $48 \cdot 2¢ = 96¢$

PROBLEM 7

$1\frac{1}{4}$ pints of detergent cost 60¢. What will 2 pints cost?

Answers to PROBLEMS

6. 3 **7.** 96¢

EXAMPLE 8 Running and algebra

A runner runs $1\frac{1}{2}$ miles in $7\frac{1}{2}$ minutes. How far can she go in $10\frac{1}{2}$ minutes?

SOLUTION The solution follows the model already shown.

$$7\tfrac{1}{2} = \tfrac{15}{2} \text{ minutes to go}$$

$$1\tfrac{1}{2} = \tfrac{3}{2} \text{ miles}$$

In 1 minute, the runner goes

$$\frac{3}{2} \div \frac{15}{2} = \frac{\overset{1}{\cancel{3}}}{\underset{1}{\cancel{2}}} \cdot \frac{\overset{1}{\cancel{2}}}{\underset{5}{\cancel{15}}} = \frac{1}{5} \text{ mile}$$

In $10\frac{1}{2} = \frac{21}{2}$ minutes, the runner goes

$$\frac{1}{5} \cdot \frac{21}{2} = \frac{21}{10} = 2\frac{1}{10} \text{ miles}$$

PROBLEM 8

A runner runs $1\frac{1}{2}$ miles in $6\frac{1}{3}$ minutes. How far can he go in $12\frac{2}{3}$ minutes?

Now that we are familar with a model to help solve equations, we shall refer back to the *Getting Started* and solve the equation about the number of people who participated in a survey.

$$\frac{4}{5} \cdot p = 400$$

Divide both sides by $\frac{4}{5}$. This means

$$p = 400 \div \frac{4}{5} = \overset{100}{\cancel{400}} \cdot \frac{5}{\underset{1}{\cancel{4}}} = 500$$

Thus, 500 people were surveyed, which can be easily verified because $\frac{4}{5}$ of 500 is 400, that is, $\frac{4}{5} \times 500 = 400$.

Let us summarize what we have done:

a. We learned to translate words into equations.

b. We solved the equations using the four principles given.

c. We used simple models to solve applications.

We now use the RSTUV procedure to incorporate the techniques we've just mentioned.

EXAMPLE 9 Coins

Do you know what a Sacagawea dollar is? It is a dollar coin like the one shown. So that they could be used in coin machines, these dollars were designed with the same shape and weight as a Susan B. Anthony dollar. The Sacagawea dollar consists of an alloy that is 3/25 zinc, 7/100 manganese, 1/25 nickel, and the rest copper. What fraction of the Sacagawea dollar is copper?

SOLUTION Let us use the RSTUV procedure to solve this problem. The directions are in bold, the responses are below.

1. Read the problem. Remember, you may have to read the problem several times before you go on to step 2. The question is: What fraction of the Sacagawea dollar is copper?

PROBLEM 9

What fraction would be copper if the amount of manganese were increased to 9/100?

Answers to PROBLEMS

8. 3 miles **9.** $\frac{75}{100} = \frac{3}{4}$

2. Select the unknown. Since we want to find the fraction of the Sacagawea dollar that is copper, let the unknown (what we are looking for) be c.

3. Translate the problem into an equation or inequality. Write the essential information and translate into an *equation*. When we add the fractions (3/25, 7/100, 1/25, and c) we want to end with one (1).

3/25	7/100	1/25	the rest
zinc	manganese	nickel	copper

$$\frac{3}{25} + \frac{7}{100} + \frac{1}{25} + c = 1$$

4. Use the rules we have studied to solve the equation. To solve the equation, we have to simplify (add) first. The LCD of 25, 100, and 25 is 100, so we write all fractions with a denominator of 100.

$$\frac{3}{25} + \frac{7}{100} + \frac{1}{25} + c = 1$$

$$\frac{12}{100} + \frac{7}{100} + \frac{4}{100} + c = \frac{100}{100}$$

Add fractions.
$$\frac{23}{100} + c = \frac{100}{100}$$

Subtract $\frac{23}{100}$ from both sides.
$$c = \frac{100}{100} - \frac{23}{100}$$

$$= \frac{77}{100}$$

Thus, the fraction of the coin that is copper is $c = \frac{77}{100}$.

5. Verify the answer. To verify the answer, note that:

$$\frac{3}{25} + \frac{7}{100} + \frac{1}{25} + \frac{77}{100}$$

$$= \frac{12}{100} + \frac{7}{100} + \frac{4}{100} + \frac{77}{100} = \frac{100}{100} = 1$$

TRANSLATE THIS

The third step in the RSTUV procedure is to **TRANSLATE** the information into an equation. In Problems 1–10, **TRANSLATE** the sentence and match the correct translation with one of the equations A–O.

1. 8 more than a number is 17.
2. The difference between a number and 10 is 1.
3. The difference between 10 and a number is 1.
4. A number increased by $\frac{1}{8}$ gives $\frac{1}{3}$.
5. $\frac{1}{5}$ less than a number is $\frac{2}{5}$.
6. The quotient of a number and 6 is $\frac{2}{5}$.
7. The quotient of a 6 and a number is $\frac{2}{5}$.
8. $\frac{2}{3}$ of a number is 5.
9. What fraction of $1\frac{2}{3}$ is $3\frac{1}{5}$?
10. Find a number so that $\frac{4}{5}$ of it is $1\frac{2}{3}$.

A. $n + 17 = 8$
B. $10 - n = 1$
C. $n - \frac{1}{5} = \frac{2}{5}$
D. $\frac{6}{n} = \frac{2}{5}$
E. $n - \frac{2}{5} = \frac{1}{5}$
F. $n + \frac{1}{8} = \frac{1}{3}$
G. $3\frac{1}{5}n = 1\frac{2}{3}$
H. $\frac{2}{3} = 5n$
I. $n + 8 = 17$
J. $\frac{4}{5}n = 1\frac{2}{3}$
K. $\frac{n}{6} = \frac{2}{5}$
L. $\frac{1}{5} - n = \frac{2}{5}$
M. $1\frac{2}{3}n = 3\frac{1}{5}$
N. $n - 10 = 1$
O. $\frac{2}{3}n = 5$

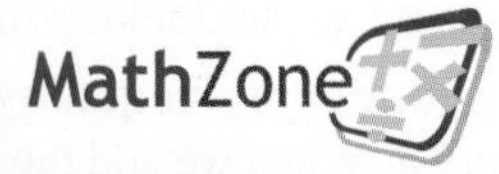

Boost your grade at mathzone.com!
- Practice Problems
- NetTutor
- Self-Tests
- e-Professors
- Videos

Exercises 2.8

Web IT go to mhhe.com/bello for more lessons

⟨A⟩ Translating Mathematical Equations

In Problems 1–20, translate into mathematical symbols, expressions, or equations.

1. Is
2. Plus
3. Of
4. Is equal to
5. Increased by
6. Multiplied by
7. Less
8. Half of a number
9. Twice a number
10. Subtracted
11. A number added to 5
12. 7 less than a number
13. A number less 7
14. The product of $1\frac{3}{4}$ and a number
15. The quotient of $\frac{3}{4}$ and a number is 5.
16. The quotient of a number and $\frac{3}{4}$ is 5.
17. Half of 3 times a number equals 2.
18. Twice a number increased by 2 is $\frac{8}{3}$.
19. Half a number decreased by 4 equals $\frac{3}{2}$.
20. A number subtracted from 8 is the same as the number.

⟨B⟩ Solving Problems

In Problems 21–40, translate and solve.

21. A number m increased by $\frac{1}{8}$ gives $\frac{3}{7}$. Find m.
22. A number n added to $\frac{1}{5}$ gives $\frac{3}{8}$. Find n.
23. $\frac{2}{5}$ more than a number p is $1\frac{3}{4}$. Find p.
24. $\frac{1}{5}$ less than a number x is $\frac{4}{7}$. Find x.
25. y decreased by $\frac{3}{4}$ yields $\frac{4}{5}$. Find y.
26. $\frac{3}{5}$ subtracted from z is $1\frac{7}{8}$. Find z.
27. u divided by 6 gives $3\frac{1}{2}$. Find u.
28. The quotient of r and 7 is $\frac{3}{5}$. Find r.
29. Three times a number t is $2\frac{1}{5}$. Find t.
30. Half of a number n is $1\frac{2}{3}$. Find n.
31. $1\frac{1}{2}$ of what number is $7\frac{1}{2}$?
32. $1\frac{5}{8}$ of what number is
33. What fraction of $1\frac{2}{3}$ is 4?
34. What fraction of $2\frac{1}{2}$ is 6?
35. What part of $2\frac{1}{2}$ is $6\frac{1}{4}$?
36. What part of $1\frac{1}{3}$ is $3\frac{1}{8}$?
37. $1\frac{1}{3}$ of what number is $4\frac{2}{3}$?
38. $3\frac{2}{5}$ of what number is $4\frac{1}{4}$?
39. $1\frac{1}{8}$ of $2\frac{1}{2}$ is what number?
40. $1\frac{1}{8}$ of $2\frac{2}{3}$ is what number?

⟨C⟩ Applications Involving Equations

(Use the RSTUV procedure!)

41. *Slicing bread* Laura baked a loaf of bread $22\frac{1}{2}$ inches long, then cut it into 12 big slices. How long would the loaf be if it had 16 slices of the same thickness as the 12 slices?

42. *Pancake recipe* A recipe for $2\frac{1}{2}$ dozen pancakes calls for $3\frac{1}{2}$ cups of milk. How many cups are needed to make $1\frac{1}{4}$ dozen pancakes?

43. *Cookie recipe* A recipe for $2\frac{1}{2}$ dozen cookies calls for $1\frac{7}{8}$ cups of flour. How many cups of flour are needed to make $1\frac{1}{3}$ dozen cookies?

44. *Making popcorn balls* Desi used $2\frac{2}{3}$ cups of popped popcorn to make two popcorn balls. How many cups would she need to make 5 popcorn balls?

45. *Banquet portions* At a certain banquet 20 people ate 15 pounds of ham. How much ham would have been needed to feed 32 people?

46. *Peat moss for rose bushes* Pete Moss uses 9 pounds of peat moss for his 27 rose bushes. If each bush gets an equal amount of peat, how many pounds would he need for 30 bushes?

47. *Running rate* A runner goes $1\frac{1}{2}$ kilometers in $4\frac{1}{2}$ minutes. At this pace how far can he go in $7\frac{1}{2}$ minutes?

48. *Product cost* If $6\frac{1}{2}$ ounces of a product cost 26¢, what will $3\frac{1}{2}$ ounces cost?

49. *Rice costs* If $3\frac{1}{4}$ pounds of rice cost 91¢, what will $2\frac{1}{2}$ pounds cost?

50. *Hourly salary* If you earn \$225 for $37\frac{1}{2}$ hours of work, how much should you earn for $46\frac{1}{2}$ hours of work at the same rate of pay?

U.S. and foreign patents The graph shows the fraction of utility (new and useful) patents granted in the United States and in foreign countries and will be used in Problems 51–52.

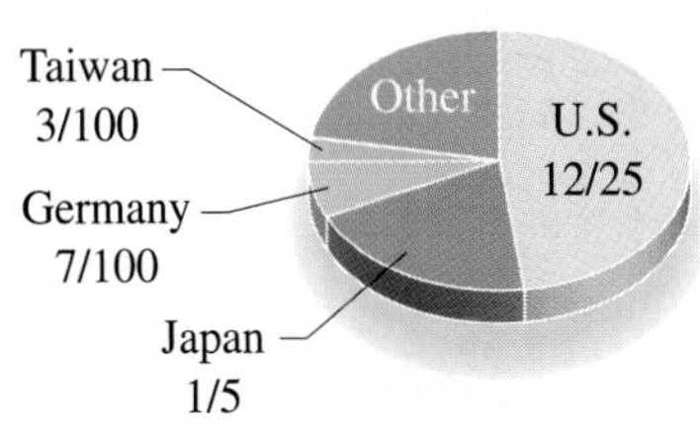

Source: U.S. Patents and Trademarks Office.

51. What fraction of the patents were foreign?

52. What fraction of the patents were granted to other count-ries?

53. *Celsius to Fahrenheit temperatures* If you know the temperature C in degrees Celsius, you can convert it to degrees Fahrenheit by using the formula

$$F = 1\frac{4}{5}C + 32$$

What is the temperature in degrees Fahrenheit when it is 25 degrees Celsius?

54. *Fahrenheit to Celsius temperature* On the other hand, if you want to convert from degrees Fahrenheit F, to degrees Celsius C, you can use the formula

$$C = \frac{5F - 160}{9}$$

What is the temperature in degrees Celsius when it is 68 degrees Fahrenheit?

55. *Fahrenheit temperature and cricket chirps* If you do not have a thermometer to find the temperature F in degrees Fahrenheit, you can always use the number of chirps c a cricket makes in one minute!

The formula is

$$F = \frac{c}{4} + 39$$

If a cricket makes 120 chirps in one minute, what is the temperature?

56. *Fahrenheit temperature and cricket chirps* There is another formula to find the temperature F in degrees Fahrenheit by counting the chirps c a cricket makes in one minute. The formula is

$$F = \frac{c - 40}{4} + 50$$

a. What is the temperature if a cricket makes 120 chirps in one minute?

b. Is the answer different from that in Exercise 55? By how much?

In Problems 57–64, follow the procedure of Example 9 to solve the problem.

57. *Mixing cement* General-purpose cement is good for just about everything except foundations and exposed paving. This cement is made by mixing $\frac{1}{6}$ part cement, $\frac{1}{3}$ part fine aggregate, and the rest coarse aggregate.

a. What fraction is cement?

b. What fraction is fine aggregate?

c. What fraction is coarse aggregate?

d. How many pounds of coarse aggregate would there be in a 50-pound bag of general-purpose cement?

58. *Best lemonade recipe* The so-called Best Lemonade Ever includes $1\frac{3}{4}$ cups of white sugar, 8 cups of water, and $1\frac{1}{2}$ cups of lemon juice.

a. How many cups is that?

b. If a cup is 8 ounces, how many ounces is that?

c. If you want to make 20 servings, how many ounces are in a serving?

59. *Watermelon punch recipe* A recipe for Watermelon Punch calls for

6 cups watermelon juice	$\frac{1}{4}$ cup raspberries
$\frac{1}{2}$ cup lemon juice	$\frac{1}{3}$ cup sugar

a. How many cups is that?

b. If a cup is 8 ounces, how many ounces is that?

c. If you want to make 10 servings, how many ounces are in a serving?

60. *Volume of water* A cubic foot (ft^3) of water weighs $62\frac{1}{2}$ pounds. If the water in a container weighs 250 pounds, how many cubic feet does the water occupy?

61. *Corvette gas tank* A cubic foot of gasoline weighs $42\frac{1}{2}$ pounds. The gasoline in the tank of a 57 Corvette weighs $138\frac{1}{8}$ pounds. About how many cubic feet does the tank of the Corvette hold?

62. *Corvette tank* A cubic foot of ethanol weighs $49\frac{1}{3}$ pounds. The ethanol in the tank of a specially designed Corvette weighs 222 pounds. About how many cubic feet does this specially designed tank hold?

63. *Weight of gasoline* A gallon of premium gasoline weighs $6\frac{3}{5}$ pounds. The capacity of the tank in a Corvette is 18 gallons. What is the weight of the gas in a Corvette with a full tank of gas?

64. *Weight of gasoline* A gallon of premium gasoline weighs about $6\frac{3}{5}$ pounds. The capacity of the tank in a Corvette is about 18 gallons. If the tank is $\frac{3}{4}$ full, what is the weight of the gas in the tank?

Using Your Knowledge

Unit Pricing You may have noticed that many supermarkets are using the idea of *unit pricing*. What this means is that each item carries a label stating its *unit price*. The *unit price* is the cost divided by the amount. In this case,

$$\frac{\text{cost}}{\text{ounces}}$$

The result is the cost per ounce. For example, if 3 ounces of tuna cost 75¢, the *unit price* is $75 \div 3 = 25$¢.

Now consider a $3\frac{1}{2}$ ounce can of tuna selling for 84¢. The *unit price* is

$$84 \div 3\frac{1}{2} = 84 \div \frac{7}{2} = \overset{12}{\cancel{84}} \cdot \frac{2}{\cancel{7}} = 24¢$$

That is, each ounce costs 24¢.

Now if another can of tuna containing $4\frac{1}{2}$ ounces costs 99¢, which can is the better buy? Since the first can of tuna costs 24¢ per ounce, $4\frac{1}{2}$ ounces would cost $4\frac{1}{2} \cdot 24 = \frac{9}{2} \cdot 24 = 108$¢, which is more than 99¢.

You can also solve this problem by finding the unit cost for the 99¢ can. This cost is

$$99 \div 4\frac{1}{2} = 99 \div \frac{9}{2} = 99 \cdot \frac{2}{9} = \frac{99}{1} \cdot \frac{2}{9} = 11 \cdot 2$$

or 22¢ per ounce. In either case, the second can is a better buy. Here is the summary of what we do to compare prices:

Step 1. Divide the price of the first item by the number of units it contains.

Step 2. Multiply the number obtained in step 1 by the number of units contained in the second item.

Step 3. Select the less expensive of the two items.

In the following problems, select the best buy.

65. Sardines selling at 45¢ for $4\frac{1}{2}$ ounces
Sardines selling at 66¢ for $5\frac{1}{2}$ ounces

66. Vienna sausage selling at 28¢ for $3\frac{1}{2}$ ounces
Vienna sausage selling at 33¢ for $5\frac{1}{2}$ ounces

67. Tuna selling at 70¢ for $3\frac{1}{2}$ ounces
Tuna selling at 98¢ for 4 ounces

68. Orange juice concentrate selling at 36¢ for $4\frac{1}{2}$ ounces
Orange juice concentrate selling at 39¢ for 5 ounces

69. Lemonade concentrate selling at 33¢ for $5\frac{1}{2}$ ounces
Lemonade concentrate selling at 29¢ for 5 ounces

Write On

70. Write in your own words the procedure you use to solve word problems.

71. Write a list of all the strategies you use to solve word problems.

72. When reading a word problem, what is the first thing you try to determine?

73. How do you verify an answer in a word problem?

Concept Checker

Fill in the blank(s) with the correct word(s), phrase, or mathematical statement.

74. According to the **addition principle** to solve equations, if c is any number, the equation $a = b$ is equivalent to ______________.

75. According to the **subtraction principle** to solve equations, if c is any number, the equation $a = b$ is equivalent to ______________.

76. According to the **multiplication principle** to solve equations, if c is any nonzero number, the equation $a = b$ is equivalent to ______________.

77. According to the **division principle** to solve equations, if c is any nonzero number, the equation $a = b$ is equivalent to ______________.

$a \div c = b \div c$

$b + a = c + a$

$a + c = b + c$

$c - a = b - a$

$a - c = b - c$

$a \cdot b = a \cdot c$

$a \cdot c = b \cdot c$

$b \div a = c \div a$

Mastery Test

78. Find a number such that $\frac{3}{5}$ of it is $3\frac{1}{4}$.

79. What fraction of $1\frac{1}{2}$ is $3\frac{3}{4}$?

80. Solve: $\frac{2}{3}$ of what number is 6?

81. Translate into symbols:

a. Three times a number equals 9.

b. The difference between a number and 5 is 2.

c. 8 more than a number is 7.

82. Translate and solve:

a. A number x increased by $\frac{1}{8}$ gives $\frac{1}{2}$. Find x.

b. $\frac{1}{4}$ less than a number y is $\frac{2}{5}$. Find y.

c. The quotient of z and 4 is $\frac{3}{5}$. Find z.

83. The product of $2\frac{1}{2}$ and $3\frac{5}{7}$ is what number?

84. One pint 8 ounces ($1\frac{1}{2}$ pints) of detergent cost 78 cents. What would 2 pints cost?

85. A runner runs $1\frac{1}{2}$ miles in $8\frac{1}{2}$ minutes. How far can she run in 17 minutes?

Skill Checker

86. Round 185 to the nearest hundred.

87. Round 185 to the nearest ten.

88. Round 3285 to the nearest thousand.

89. Simplify $8 \div 4 \cdot \frac{11}{2} - \left[3\left(\frac{5}{3} - \frac{1}{3}\right) + 1\right]$.

90. Simplify $4 \cdot \frac{3}{4} \div \frac{1}{3} + \left[\left(\frac{5}{8} - \frac{2}{7}\right) - \frac{1}{4}\right]$.

Collaborative Learning

Health professionals use many ratios to detect abnormalities. Form three different groups: Library, Internet, and Other. What is the leading cause of death in the United States? Heart disease! Let each of the groups find the answers to the following questions:

1. How many deaths a year are attributed to heart disease? What is the total number of annual deaths? What is the ratio of deaths attributed to heart disease to the total number of deaths? Do the answers of the groups agree? Why or why not?

There are several ratios associated with heart disease. Let us look at two of them, HDL/cholesterol and Waist/Hip.

2. Let each of the groups find out what the HDL/cholesterol ratio measures. What is the recommended value of this ratio?

<CONTINUED>

3. There is another noninvasive way to measure your cardiac health. This time, divide the groups into males and females. Complete the following table:

	Waist (w)	**Hips (h)**	**Ratio w/h**
Women			
Men			

For women the risk of heart disease increases when $w/h > \frac{8}{10} = \frac{4}{5} = 0.8$, for men, when $w/h > 1$.

As a matter of fact, when w/h is over 1.0 for men and 0.8 for women, the risk of heart attack or stroke is five to ten times greater than for persons with a lower ratio.

Research Questions

1. The earliest Egyptian and Greek fractions were usually unit fractions (having a numerator of 1). How were such fractions shown?
2. How were fractions indicated in ancient Rome?
3. It is probable that our method of writing common fractions is due essentially to the Hindus. Name the two Hindu mathematicians who wrote fractions as we do today, but without the bar.
4. The horizontal fraction bar was introduced by the Arabs and attributed to an Arab mathematician who lived around 1200. What was the name of the Arab mathematician?
5. Name the first European mathematician to use the fraction bar as it is used today.
6. Why was the diagonal fraction bar / (called a *solidus* or *virgules*) introduced and why?

Summary Chapter 2

Section	**Item**	**Meaning**	**Example**
2.1	Numerator	In the fraction $\frac{a}{b}$, a is the numerator.	The numerator of $\frac{2}{3}$ is 2.
	Denominator	In the fraction, $\frac{a}{b}$, b is the denominator.	The denominator of $\frac{2}{3}$ is 3.
2.1B	Proper fraction	A fraction whose numerator is less than the denominator.	$\frac{3}{4}$, $\frac{9}{11}$, and $\frac{16}{17}$ are proper fractions.
	Improper fraction	A fraction whose numerator is greater than or equal to its denominator.	$\frac{17}{16}$, $\frac{19}{19}$, and $\frac{1}{1}$ are improper fractions.

Section	Item	Meaning	Example
2.1C	Mixed number	The sum of a whole number and a proper fraction.	$5\frac{2}{3}$ and $10\frac{1}{4}$ are mixed numbers.
2.2	Equivalent fractions	Two fractions are equivalent if they are names for the same number.	$\frac{2}{4}$ and $\frac{1}{2}$ are equivalent.
2.3A	Product of two fractions	$\frac{a}{b} \cdot \frac{c}{d} = \frac{a \cdot c}{b \cdot d}$	$\frac{3}{5} \cdot \frac{2}{7} = \frac{3 \cdot 2}{5 \cdot 7} = \frac{6}{35}$
2.3D	Reciprocal of a fraction Division of fractions	The reciprocal of $\frac{a}{b}$ is $\frac{b}{a}$. $\frac{a}{b} \div \frac{c}{d} = \frac{a}{b} \cdot \frac{d}{c}$	$\frac{4}{5}$ and $\frac{5}{4}$ are reciprocals of each other. $\frac{2}{3} \div \frac{4}{5} = \frac{2}{3} \cdot \frac{5}{4} = \frac{10}{12} = \frac{5}{6}$
2.4A	LCM (least common multiple)	The LCM of two natural numbers is the *smallest* number that is a multiple of both numbers.	The LCM of 10 and 12 is 60 and the LCM of 82 and 41 is 82.
2.4B	LCD (lowest common denominator)	The LCD of two fractions is the LCM of the denominators of the fractions.	The LCD of $\frac{1}{4}$ and $\frac{1}{12}$ is 12 and the LCD of $\frac{1}{9}$ and $\frac{1}{12}$ is 36.
2.5A	Addition of fractions	$\frac{a}{b} + \frac{c}{b} = \frac{a + c}{b}$	$\frac{3}{7} + \frac{2}{7} = \frac{5}{7}$
2.5C	LCD of three fractions	The smallest number that is a multiple of all the denominators.	The LCD of $\frac{2}{5}$ and $\frac{3}{4}$ and $\frac{1}{10}$ is 20.
2.5D	Subtraction of fractions	$\frac{a}{b} - \frac{c}{b} = \frac{a - c}{b}$	$\frac{3}{7} - \frac{2}{7} = \frac{1}{7}$
2.6E	Perimeter	The distance around an object.	The perimeter of the rectangle is 9 inches. 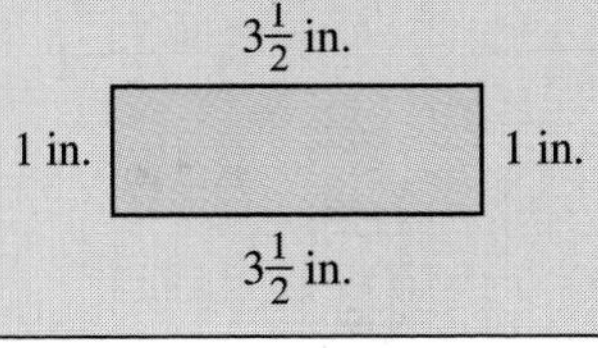
2.7	PEMDAS	The order of operations is **P** (calculations inside parentheses) **E** (exponential expressions) **M** (multiplications) **D** (divisions) **A** (additions) **S** (subtraction)	To evaluate $3 \cdot 2 \div 6 + (2 + 4) - 2^2$ **P:** $3 \cdot 2 \div 6 + (6) - 2^2$ **E:** $3 \cdot 2 \div 6 + (6) - 4$ **M:** $6 \div 6 + (6) - 4$ **D:** $1 + (6) - 4$ **A:** $7 - 4$ **S:** 3
2.8	Equivalent equations	Each of the following equations is equivalent to $a = b$ (c any number): $a + c = b + c$ $a - c = b - c$ $a \cdot c = b \cdot c$ (c not 0) $a \div c = b \div c$ (c not 0)	

Review Exercises Chapter 2

(If you need help with these exercises, look in the section indicated in brackets.)

1. ⟨ **2.1B** ⟩ *Classify as a proper or improper fraction.*

a. $\frac{9}{11}$ **b.** $\frac{0}{8}$ **c.** $\frac{3}{3}$

d. $\frac{5}{8}$ **e.** $\frac{11}{11}$

2. ⟨ **2.1C** ⟩ *Write as a mixed number.*

a. $\frac{22}{7}$ **b.** $\frac{18}{7}$ **c.** $\frac{29}{3}$

d. $\frac{14}{4}$ **e.** $\frac{19}{11}$

3. ⟨ **2.1D** ⟩ *Write as an improper fraction.*

a. $4\frac{1}{2}$ **b.** $3\frac{1}{9}$ **c.** $4\frac{2}{5}$

d. $8\frac{3}{14}$ **e.** $7\frac{7}{8}$

4. ⟨ **2.1E** ⟩ *The price of a stock is $80 per share. Find the P/E ratio of the stock if its earnings per share are:*

a. $10 **b.** $8 **c.** $20

d. $40 **e.** $16

5. ⟨ **2.2A** ⟩ *Find each missing number.*

a. $\frac{4}{3} = \frac{?}{6}$ **b.** $\frac{3}{5} = \frac{?}{25}$ **c.** $\frac{8}{9} = \frac{?}{27}$

d. $\frac{14}{21} = \frac{?}{42}$ **e.** $\frac{3}{9} = \frac{?}{54}$

6. ⟨ **2.2A** ⟩ *Find each missing number.*

a. $\frac{6}{21} = \frac{2}{?}$ **b.** $\frac{8}{10} = \frac{4}{?}$ **c.** $\frac{18}{24} = \frac{6}{?}$

d. $\frac{24}{48} = \frac{4}{?}$ **e.** $\frac{18}{30} = \frac{6}{?}$

7. ⟨ **2.2B** ⟩ *Reduce to lowest terms.*

a. $\frac{4}{8}$ **b.** $\frac{6}{9}$ **c.** $\frac{14}{35}$

d. $\frac{8}{28}$ **e.** $\frac{10}{95}$

8. ⟨ **2.2B** ⟩

a. Find the GCF of 12 and 36 and reduce $\frac{12}{36}$.

b. Find the GCF of 10 and 50 and reduce $\frac{10}{50}$.

c. Find the GCF of 18 and 45 and reduce $\frac{18}{45}$.

d. Find the GCF of 28 and 42 and reduce $\frac{28}{42}$.

e. Find the GCF of 51 and 34 and reduce $\frac{51}{34}$.

9. ⟨ **2.3A** ⟩ *Multiply (reduce answers to lowest terms).*

a. $\frac{1}{3} \cdot \frac{2}{7}$ **b.** $\frac{2}{5} \cdot \frac{5}{9}$

c. $\frac{3}{7} \cdot \frac{7}{9}$ **d.** $\frac{4}{5} \cdot \frac{15}{8}$

e. $\frac{7}{8} \cdot \frac{8}{7}$

10. ⟨ **2.3B** ⟩ *Multiply (reduce answers to lowest terms).*

a. $\frac{4}{7} \cdot 3\frac{1}{6}$ **b.** $\frac{3}{5} \cdot 3\frac{1}{3}$

c. $\frac{6}{7} \cdot 1\frac{3}{4}$ **d.** $\frac{9}{10} \cdot 2\frac{1}{4}$

e. $\frac{6}{7} \cdot 4\frac{2}{3}$

11. ⟨ **2.3C** ⟩ *Multiply.*

a. $\left(\frac{2}{5}\right)^2 \cdot \frac{5}{6}$ **b.** $\left(\frac{3}{2}\right)^2 \cdot \frac{4}{9}$

c. $\left(\frac{3}{2}\right)^2 \cdot \frac{8}{27}$ **d.** $\left(\frac{3}{2}\right)^2 \cdot \frac{14}{27}$

e. $\left(\frac{3}{2}\right)^2 \cdot \frac{8}{9}$

12. ⟨ **2.3D** ⟩ *Divide (reduce answers to lowest terms).*

a. $\frac{3}{4} \div \frac{6}{7}$ **b.** $\frac{3}{8} \div \frac{6}{7}$

c. $\frac{4}{5} \div \frac{5}{9}$ **d.** $\frac{5}{3} \div \frac{7}{9}$

e. $\frac{6}{7} \div \frac{12}{7}$

13. ⟨ **2.3E** ⟩ *Divide (reduce answers to lowest terms).*

a. $2\frac{1}{4} \div \frac{4}{5}$ **b.** $3\frac{1}{7} \div \frac{7}{8}$

c. $6\frac{1}{2} \div \frac{4}{13}$ **d.** $1\frac{1}{9} \div \frac{20}{27}$

e. $4\frac{1}{7} \div \frac{14}{15}$

14. ⟨ **2.3E** ⟩ *Divide (reduce answers to lowest terms).*

a. $\frac{3}{5} \div 1\frac{1}{5}$ **b.** $\frac{4}{7} \div 2\frac{3}{7}$

c. $\frac{3}{5} \div 3\frac{1}{5}$ **d.** $\frac{1}{7} \div 2\frac{1}{2}$

e. $\frac{2}{9} \div 3\frac{1}{8}$

15. ⟨ **2.3G** ⟩ *Find the area of a room with dimensions.*

a. $3\frac{1}{3}$ yards by $4\frac{2}{3}$ yards

b. $3\frac{1}{2}$ yards by $4\frac{1}{2}$ yards

c. $3\frac{1}{3}$ yards by $4\frac{1}{2}$ yards

d. $3\frac{1}{2}$ yards by $4\frac{1}{3}$ yards

e. $4\frac{1}{2}$ yards by $5\frac{1}{2}$ yards

16. ⟨ **2.4A** ⟩ *Find the LCM of each group of numbers.*

a. 8 and 12

b. 15 and 6

c. 18 and 12

d. 20 and 24

e. 54 and 180

17. ⟨ **2.4A** ⟩ *Find the LCM of*

a. 11 and 33

b. 17 and 34

c. 57 and 19

d. 40 and 10

e. 92 and 23

18. ⟨ **2.4B** ⟩ *Find the LCD and use it to write the fractions with the LCD as denominator.*

a. $\frac{7}{12}$ and $\frac{3}{16}$

b. $\frac{2}{15}$ and $\frac{5}{9}$

c. $\frac{5}{16}$ and $\frac{5}{18}$

d. $\frac{3}{7}$ and $\frac{4}{5}$

e. $\frac{5}{9}$ and $\frac{4}{15}$

19. ⟨ **2.4B** ⟩ *Find the LCD and use it to write the fractions with the LCD as denominator.*

a. $\frac{3}{4}, \frac{1}{2}$, and $\frac{5}{6}$

b. $\frac{5}{12}, \frac{1}{9}$, and $\frac{3}{8}$

c. $\frac{13}{16}, \frac{1}{18}$, and $\frac{11}{12}$

d. $\frac{1}{10}, \frac{3}{8}$, and $\frac{1}{12}$

e. $\frac{1}{5}, \frac{4}{9}$, and $\frac{1}{8}$

20. ⟨ **2.4C** ⟩ *Fill in the blank with $<$ or $>$ to make the resulting inequality true.*

a. $\frac{1}{3}$ ______ $\frac{3}{10}$

b. $\frac{2}{3}$ ______ $\frac{3}{7}$

c. $\frac{4}{5}$ ______ $\frac{5}{7}$

d. $\frac{2}{9}$ ______ $\frac{3}{7}$

e. $\frac{3}{8}$ ______ $\frac{5}{32}$

21. ⟨ **2.5A** ⟩ *Add (reduce answers to lowest terms).*

a. $\frac{1}{5} + \frac{2}{5}$

b. $\frac{2}{3} + \frac{1}{3}$

c. $\frac{3}{7} + \frac{1}{7}$

d. $\frac{2}{9} + \frac{1}{9}$

e. $\frac{7}{2} + \frac{9}{2}$

22. ⟨ **2.5B** ⟩ *Find the LCD and add (reduce answers to lowest terms).*

a. $\frac{1}{3} + \frac{5}{6}$

b. $\frac{1}{5} + \frac{1}{9}$

c. $\frac{3}{7} + \frac{5}{6}$

d. $\frac{1}{6} + \frac{9}{20}$

e. $\frac{2}{7} + \frac{3}{15}$

23. ‹ **2.5B** › *Find the LCD and add (reduce answers to lowest terms).*

a. $\frac{15}{4}+\frac{16}{3}$

b. $\frac{7}{2}+\frac{5}{3}$

c. $\frac{17}{4}+\frac{33}{16}$

d. $\frac{19}{9}+\frac{13}{3}$

e. $\frac{9}{8}+\frac{19}{9}$

24. ‹ **2.5D** › *Find the LCD of the fractions and then find each of the following.*

a. $\frac{5}{7}+\frac{1}{6}-\frac{1}{12}$

b. $\frac{3}{4}+\frac{1}{8}-\frac{1}{12}$

c. $\frac{5}{8}+\frac{3}{4}-\frac{1}{16}$

d. $\frac{4}{5}+\frac{1}{3}-\frac{2}{15}$

e. $\frac{2}{3}+\frac{3}{4}-\frac{1}{12}$

25. ‹ **2.5D** › *Find the LCD and subtract.*

a. $\frac{7}{8}-\frac{3}{4}$

b. $\frac{11}{12}-\frac{7}{18}$

c. $\frac{7}{12}-\frac{5}{16}$

d. $\frac{5}{7}-\frac{3}{5}$

e. $\frac{16}{27}-\frac{5}{24}$

26. ‹ **2.5E** › *Student music preferences*

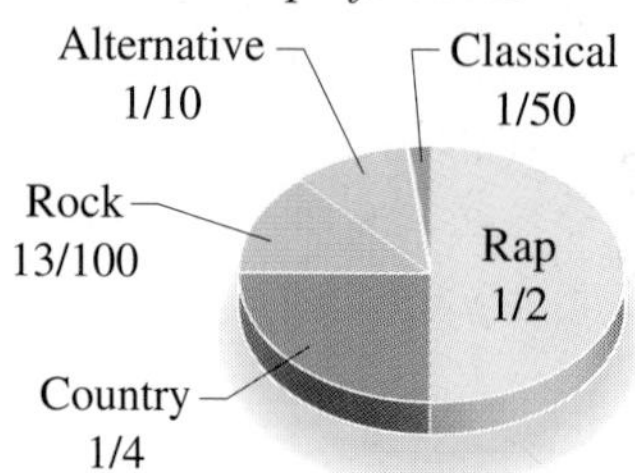

What fraction of the students prefer:

a. Rap or country?

b. Rock or country?

c. Alternative or rock?

d. Classical or rap?

e. Alternative or classical?

27. ‹ **2.6B** › *Add.*

a. $4\frac{1}{5}+3\frac{1}{6}$

b. $2\frac{1}{3}+3\frac{1}{12}$

c. $4\frac{4}{7}+3\frac{2}{8}$

d. $5\frac{1}{3}+2\frac{1}{9}$

e. $3\frac{5}{8}+5\frac{3}{12}$

28a. ‹ **2.6C** › *Subtract.*

a. $2\frac{7}{8}-2\frac{2}{3}$

b. $3\frac{1}{3}-1\frac{3}{5}$

c. $3\frac{1}{5}-2\frac{1}{3}$

d. $4\frac{3}{5}-3\frac{5}{8}$

e. $1\frac{7}{8}-1\frac{5}{9}$

28b. ‹ **2.6D** › *Add and subtract as indicated.*

a. $2\frac{5}{9}+3\frac{1}{8}-2\frac{1}{10}$

b. $3\frac{5}{9}+3\frac{1}{6}-2\frac{1}{10}$

c. $4\frac{5}{9}+3\frac{1}{12}-2\frac{1}{8}$

d. $5\frac{5}{9}+3\frac{1}{12}-2\frac{1}{6}$

e. $6\frac{5}{9}+3\frac{1}{8}-2\frac{1}{6}$

29. ‹ **2.6E** › *Find the perimeter of a room whose dimensions are:*

a. $4\frac{1}{4}$ yards by $5\frac{1}{2}$ yards

b. $3\frac{1}{2}$ yards by $4\frac{1}{3}$ yards

c. $4\frac{1}{3}$ yards by $5\frac{1}{2}$ yards

d. $3\frac{1}{2}$ yards by $5\frac{1}{3}$ yards

e. $3\frac{1}{6}$ yards by $2\frac{5}{6}$ yards

30. ⟨ **2.7A** ⟩ *Simplify:*

a. $\frac{1}{2} \cdot \left(\frac{2}{3}\right)^2 - \frac{1}{9}$

b. $\frac{1}{3} \cdot \left(\frac{3}{4}\right)^2 - \frac{1}{16}$

c. $\frac{1}{5} \cdot \left(\frac{5}{6}\right)^2 - \frac{1}{36}$

d. $\frac{1}{6} \cdot \left(\frac{6}{7}\right)^2 - \frac{1}{49}$

e. $\frac{1}{7} \cdot \left(\frac{7}{8}\right)^2 - \frac{1}{64}$

31. ⟨ **2.7A** ⟩ *Simplify:*

a. $4 \div \frac{1}{2} \cdot \frac{1}{2} \cdot \frac{1}{2} + \frac{1}{3} - 2$

b. $6 \div \frac{1}{2} \cdot \frac{1}{2} \cdot \frac{1}{2} + \frac{1}{3} - 3$

c. $8 \div \frac{1}{2} \cdot \frac{1}{2} \cdot \frac{1}{2} + \frac{1}{3} - 4$

d. $10 \div \frac{1}{2} \cdot \frac{1}{2} \cdot \frac{1}{2} + \frac{1}{3} - 5$

e. $12 \div \frac{1}{2} \cdot \frac{1}{2} \cdot \frac{1}{2} + \frac{1}{3} - 6$

32. ⟨ **2.7A** ⟩ *Simplify:*

a. $\left(\frac{1}{2}\right)^3 \div \frac{1}{3} \cdot \frac{1}{4} + \frac{1}{3}\left(\frac{7}{2} - \frac{1}{2}\right) - \frac{1}{3} \cdot \frac{1}{2}$

b. $\left(\frac{1}{2}\right)^3 \div \frac{1}{3} \cdot \frac{1}{4} + \frac{1}{3}\left(\frac{9}{2} - \frac{1}{2}\right) - \frac{1}{3} \cdot \frac{1}{2}$

c. $\left(\frac{1}{2}\right)^3 \div \frac{1}{3} \cdot \frac{1}{4} + \frac{1}{3}\left(\frac{5}{2} - \frac{1}{2}\right) - \frac{1}{3} \cdot \frac{1}{2}$

d. $\left(\frac{1}{2}\right)^3 \div \frac{1}{8} \cdot \frac{1}{2} + \frac{1}{3}\left(\frac{3}{2} - \frac{1}{2}\right) - \frac{1}{3} \cdot \frac{1}{2}$

e. $\left(\frac{1}{2}\right)^3 \div \frac{1}{4} \cdot \frac{1}{2} + \frac{1}{3}\left(\frac{11}{2} - \frac{1}{2}\right) - \frac{1}{3} \cdot \frac{1}{2}$

33. ⟨ **2.7B** ⟩ *Simplify*

a. $\frac{1}{6} \div 1\frac{1}{6} + \left\{16 \cdot \left(\frac{1}{2}\right)^2 - \left[\frac{1}{3} + \left(3\frac{1}{2} - \frac{1}{2}\right)\right]\right\}$

b. $\frac{1}{5} \div 1\frac{1}{5} + \left\{20 \cdot \left(\frac{1}{2}\right)^2 - \left[\frac{1}{3} + \left(4\frac{1}{2} - \frac{1}{2}\right)\right]\right\}$

c. $\frac{1}{4} \div 1\frac{1}{4} + \left\{24 \cdot \left(\frac{1}{2}\right)^2 - \left[\frac{1}{3} + \left(5\frac{1}{2} - \frac{1}{2}\right)\right]\right\}$

d. $\frac{1}{3} \div 1\frac{1}{3} + \left\{28 \cdot \left(\frac{1}{2}\right)^2 - \left[\frac{1}{3} + \left(6\frac{1}{2} - \frac{1}{2}\right)\right]\right\}$

e. $\frac{1}{2} \div 1\frac{1}{2} + \left\{32 \cdot \left(\frac{1}{2}\right)^2 - \left[\frac{1}{3} + \left(7\frac{1}{2} - \frac{1}{2}\right)\right]\right\}$

34. ⟨ **2.7C** ⟩ *Find the average weight of 4 fish weighing*

a. $3\frac{1}{2}$, $4\frac{1}{4}$, $2\frac{1}{2}$, and $7\frac{1}{4}$ pounds.

b. $4\frac{1}{2}$, $5\frac{1}{4}$, $3\frac{1}{2}$, and $8\frac{1}{4}$ pounds.

c. $5\frac{1}{2}$, $6\frac{1}{4}$, $4\frac{1}{2}$, and $9\frac{1}{4}$ pounds.

d. $6\frac{1}{2}$, $7\frac{1}{4}$, $5\frac{1}{2}$, and $10\frac{1}{4}$ pounds.

e. $7\frac{1}{2}$, $8\frac{1}{4}$, $6\frac{1}{2}$, and $11\frac{1}{4}$ pounds.

35. ⟨ **2.8A** ⟩ *Translate into mathematical equations.*

a. 8 more than a number is 10.

b. The difference between a number and 5 is 1.

c. Twice a number equals 12.

d. A number divided by 2 is 8.

e. The quotient of a number and 7 is 3.

36. ⟨ **2.8B** ⟩ *Translate and solve.*

a. A number p increased by $\frac{1}{6}$ gives $\frac{1}{3}$. Find p.

b. A number q increased by $\frac{1}{5}$ gives $\frac{1}{4}$. Find q.

c. A number r increased by $\frac{1}{4}$ gives $\frac{2}{5}$. Find r.

d. A number s increased by $\frac{1}{3}$ gives $\frac{5}{6}$. Find s.

e. A number t increased by $\frac{1}{2}$ gives $\frac{6}{7}$. Find t.

37. ⟨ **2.8B** ⟩ *Translate and solve.*

a. $\frac{1}{6}$ less than a number r is $\frac{2}{7}$. Find r.

b. $\frac{1}{5}$ less than a number s is $\frac{3}{7}$. Find s.

c. $\frac{1}{4}$ less than a number t is $\frac{4}{7}$. Find t.

d. $\frac{1}{3}$ less than a number u is $\frac{5}{7}$. Find u.

e. $\frac{1}{2}$ less than a number v is $\frac{6}{7}$. Find v.

38. ⟨ **2.8B** ⟩ *Translate and solve.*

a. The quotient of v and 3 is $\frac{2}{7}$. Find v.

b. The quotient of v and 4 is $\frac{3}{7}$. Find v.

c. The quotient of v and 5 is $\frac{4}{7}$. Find v.

d. The quotient of v and 6 is $\frac{5}{7}$. Find v.

e. The quotient of v and 7 is $\frac{6}{7}$. Find v.

39. ⟨ **2.8B** ⟩ *Find the number.*

a. $\frac{1}{2}$ of what number is 8?

b. $\frac{2}{3}$ of what number is 4?

c. $\frac{3}{5}$ of what number is 27?

d. $\frac{2}{7}$ of what number is 14?

e. $\frac{6}{5}$ of what number is 12?

40. ⟨ **2.8C** ⟩ *An alloy consists of 4 metals A, B, C, and D. What fraction of the alloy is D when A, B, and C are, respectively,*

a. $\frac{3}{25}, \frac{7}{100}, \frac{1}{25}$ **b.** $\frac{6}{25}, \frac{7}{100}, \frac{2}{25}$ **c.** $\frac{7}{25}, \frac{7}{100}, \frac{3}{25}$ **d.** $\frac{8}{25}, \frac{7}{100}, \frac{4}{25}$ **e.** $\frac{9}{25}, \frac{7}{100}, \frac{6}{25}$

> Practice Test Chapter 2

(Answers on page 196)

Visit www.mhhe.com/bello to view helpful videos that provide step-by-step solutions to several of the problems below.

1. Classify as a proper or improper fraction.

a. $\frac{3}{4}$ **b.** $\frac{5}{3}$ **c.** $\frac{8}{8}$

2. Write $\frac{23}{6}$ as a mixed number.

3. Write $2\frac{3}{7}$ as an improper fraction.

4. The price of a stock is \$60 per share and its earnings per share are \$5. What is the P/E ratio of the stock?

5. Fill in the blank: $\frac{4}{7} = \frac{\square}{21}$

6. Fill in the blank: $\frac{2}{5} = \frac{6}{\square}$

7. Reduce $\frac{14}{35}$ to lowest terms.

8. Find the GCF of 32 and 48, then reduce $\frac{32}{48}$.

9. Multiply: $\frac{1}{6} \cdot \frac{5}{7}$

10. Multiply: $\frac{3}{7} \cdot 2\frac{1}{6}$

11. Multiply: $\left(\frac{4}{3}\right)^2 \cdot \frac{1}{16}$

12. Divide: $\frac{2}{3} \div \frac{4}{7}$

13. Divide: $2\frac{1}{2} \div \frac{4}{3}$

14. Divide: $\frac{3}{2} \div 1\frac{1}{4}$

15. Find the area of a room whose dimensions are $4\frac{1}{3}$ yards by $5\frac{2}{3}$ yards.

16. Find the LCM of 15 and 20.

17. Find the LCM of 17 and 51.

18. Find the LCD of $\frac{7}{8}$ and $\frac{5}{12}$ and write $\frac{7}{8}$ and $\frac{5}{12}$ as equivalent fractions with the LCD as denominator.

19. Find the LCD of $\frac{7}{10}$, $\frac{3}{4}$, and $\frac{5}{8}$ and write $\frac{7}{10}$, $\frac{3}{4}$, and $\frac{5}{8}$ as equivalent fractions with the LCD as denominator.

20. Fill in the blank with $<$ or $>$ to make the resulting inequality true: $\frac{2}{3}$ ______ $\frac{4}{7}$

21. Add: $\frac{3}{7} + \frac{2}{7}$

22. Find the LCD and add: $\frac{2}{3} + \frac{1}{5}$

23. Find the LCD and add: $\frac{5}{3} + \frac{22}{7}$

24. Find the LCD of $\frac{6}{7}$, $\frac{5}{6}$, and $\frac{1}{12}$ and then find: $\frac{6}{7} + \frac{5}{6} - \frac{1}{12}$

25. Find the LCD and subtract: $\frac{7}{15} - \frac{3}{10}$

26. Here are the favorite sports for a group of students.

Favorite Sports

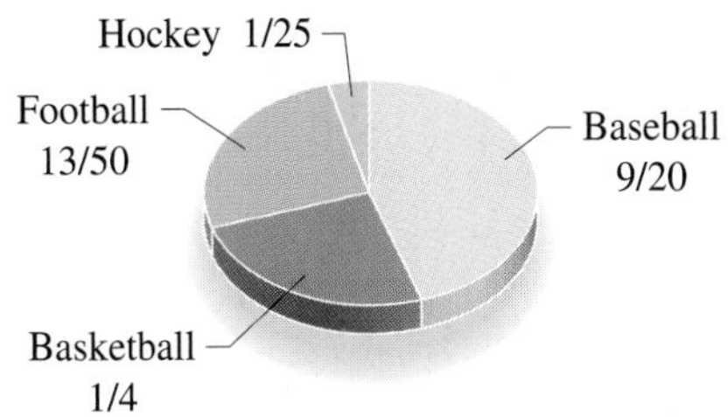

What fraction of the students chose either baseball or basketball?

27. Add: $3\frac{4}{5} + 2\frac{2}{3}$

28. Subtract:

a. $3\frac{5}{6} - 2\frac{1}{5}$ **b.** $2\frac{2}{7} - 1\frac{1}{4}$

29. Find the perimeter of a room whose dimensions are $3\frac{1}{3}$ yards by $5\frac{2}{3}$ yards.

30. Simplify: $\frac{1}{3} \cdot \left(\frac{3}{4}\right)^2 - \frac{1}{16}$

31. Simplify: $6 \div \frac{1}{2} \cdot \frac{1}{2} \cdot \frac{1}{2} + \frac{1}{3} - 2$

32. Simplify: $\left(\frac{1}{3}\right)^3 \div \frac{1}{3} \cdot \frac{1}{4} + \frac{1}{3}\left(\frac{5}{2} - \frac{1}{2}\right) - \frac{1}{3} \cdot \frac{1}{2}$

33. Simplify: $\frac{1}{3} \div 1\frac{1}{3} + \left\{12 \cdot \left(\frac{1}{2}\right)^2 - \left[\frac{1}{3} + \left(2\frac{1}{2} - \frac{1}{2}\right)\right]\right\}$

34. Pedro went fishing and caught 4 fish weighing $3\frac{1}{2}$, $5\frac{1}{4}$, $2\frac{1}{2}$, and $6\frac{1}{4}$ pounds, respectively. What is the average weight of the 4 fish?

35. Translate and solve: $\frac{1}{7}$ less than a number y is $\frac{2}{5}$. What is y?

36. $\frac{3}{5}$ of what number is 9?

37. What fraction of $1\frac{2}{3}$ is $2\frac{1}{9}$?

38. Find a number such that $\frac{4}{5}$ of it is $1\frac{1}{3}$.

39. The product of $1\frac{2}{3}$ and $2\frac{1}{5}$ is what number?

40. A proposed dollar coin is made an alloy consisting of $\frac{4}{25}$ zinc, $\frac{7}{100}$ manganese, $\frac{1}{25}$ nickel, and the rest copper. What fraction of the coin is copper?

Answers to Practice Test Chapter 2

Answer	If You Missed	Review		
	Question	Section	Examples	Page
1. **a.** Proper **b.** Improper **c.** Improper	1	2.1	2	112
2. $3\frac{5}{6}$	2	2.1	3	113
3. $\frac{17}{7}$	3	2.1	4	113
4. 12	4	2.1	5, 6, 7	114–115
5. 12	5	2.2	1	121–122
6. 15	6	2.2	2	122
7. $\frac{2}{5}$	7	2.2	3	123–124
8. GCF 16; $\frac{32}{48} = \frac{2}{3}$	8	2.2	1, 2, 3	121–124
9. $\frac{5}{42}$	9	2.3	1a	130
10. $\frac{13}{14}$	10	2.3	2a	131
11. $\frac{1}{9}$	11	2.3	4	133
12. $\frac{7}{6} = 1\frac{1}{6}$	12	2.3	5	135
13. $\frac{15}{8} = 1\frac{7}{8}$	13	2.3	6a	135
14. $\frac{6}{5} = 1\frac{1}{5}$	14	2.3	6b	135
15. $\frac{221}{9} = 24\frac{5}{9}$ square yards	15	2.3	9	136
16. 60	16	2.4	1–3	142
17. 51	17	2.4	4	143
18. LCD 24; $\frac{21}{24}, \frac{10}{24}$	18	2.4	5–7	143–145
19. LCD 40; $\frac{28}{40}, \frac{30}{40}, \frac{25}{40}$	19	2.4	8	146
20. $>$	20	2.4	9	147
21. $\frac{5}{7}$	21	2.5	1	152
22. LCD 15; $\frac{13}{15}$	22	2.5	2	153
23. LCD 21; $\frac{101}{21} = 4\frac{17}{21}$	23	2.5	3	154
24. LCD 84; $\frac{45}{28} = 1\frac{17}{28}$	24	2.5	5, 7	155–157
25. LCD 30; $\frac{1}{6}$	25	2.5	6	155–156
26. $\frac{14}{20} = \frac{7}{10}$	26	2.5	8	157
27. $\frac{97}{15} = 6\frac{7}{15}$	27	2.6	1, 2, 3	162–163
28. **a.** $\frac{49}{30} = 1\frac{19}{30}$ **b.** $1\frac{1}{28}$	28a, b	2.6	4	163–164
29. 18 yards	29	2.6	6	165
30. $\frac{1}{8}$	30	2.7	1	170–171
31. $\frac{4}{3} = 1\frac{1}{3}$	31	2.7	2	171
32. $\frac{19}{36}$	32	2.7	3	171–172
33. $\frac{11}{12}$	33	2.7	4	172
34. $4\frac{3}{8}$ pounds	34	2.7	5	173
35. $y - \frac{1}{7} = \frac{2}{5}; y = \frac{19}{35}$	35	2.8	2	179
36. 15	36	2.8	3	180
37. $\frac{19}{15} = 1\frac{4}{15}$	37	2.8	4	180
38. $\frac{5}{3} = 1\frac{2}{3}$	38	2.8	5	180–181
39. $\frac{11}{3} = 3\frac{2}{3}$	39	2.8	6	181
40. $\frac{73}{100}$	40	2.8	7, 8, 9	181–183

Cumulative Review Chapters 1–2

1. Write 438 in expanded form.

2. Write 900 + 80 + 4 in standard form.

3. Write the word name for 74,008.

4. Write six thousand seven hundred ten in standard form.

5. Round 8649 to the nearest hundred.

6. Add: 903 + 2776

7. Subtract: 652 − 498

8. Multiply: 137×319

9. Betty makes loan payments of \$310 each month for 12 months. What is the total amount of money paid?

10. Divide: $26\overline{)889}$

11. Write the prime factors of 24.

12. Write 180 as a product of primes.

13. Multiply: $2^3 \times 4 \times 7^0$

14. Simplify: $36 \div 6 \cdot 6 + 8 - 4$

15. Solve for m: $26 = m + 3$

16. Solve for x: $21 = 7x$

17. Classify $\frac{2}{3}$ as a proper or improper fraction.

18. Write $\frac{11}{2}$ as a mixed number.

19. Write $2\frac{1}{4}$ as an improper fraction.

20. Fill in the blank: $\frac{2}{3} = \frac{\square}{21}$

21. Fill in the blank: $\frac{2}{3} = \frac{18}{\square}$

22. Reduce $\frac{10}{12}$ to lowest terms.

23. Insert =, <, or > to make a true statement: $\frac{3}{4}$ ___ $\frac{5}{6}$

24. Multiply: $\frac{1}{2} \cdot 6\frac{1}{3}$

25. Multiply: $\left(\frac{7}{6}\right)^2 \cdot \frac{1}{49}$

26. Divide: $\frac{6}{7} \div 1\frac{1}{3}$

27. Find the LCD and add: $7\frac{1}{3} + 9\frac{3}{10}$

28. Subtract: $8\frac{1}{7} - 1\frac{8}{9}$

29. Translate and solve: $\frac{6}{7}$ less than a number z is $\frac{4}{9}$. What is z?

30. Find a number such that $\frac{9}{10}$ of it is $5\frac{1}{5}$.

31. $3\frac{1}{2}$ pounds of sugar cost 49 cents. How much will 8 pounds cost?

32. Find the perimeter of a room whose dimensions are $4\frac{1}{3}$ by $6\frac{2}{3}$ yards.

33. Find the area of a room whose dimensions are $4\frac{1}{3}$ by $6\frac{2}{3}$ yards.

34. Find the LCM of 16 and 20.

35. Find the LCM of 19 and 76.

36. Find the LCD of $\frac{7}{9}$ and $\frac{5}{12}$ and write $\frac{7}{9}$ and $\frac{5}{12}$ as equivalent fractions with the LCD as denominator.

37. Find the LCD of $\frac{7}{10}$, $\frac{5}{6}$, and $\frac{3}{5}$ and write $\frac{7}{10}$, $\frac{5}{6}$, and $\frac{3}{5}$ as equivalent fractions with the LCD as denominator.

Section

Chapter

3 *three*

Decimals

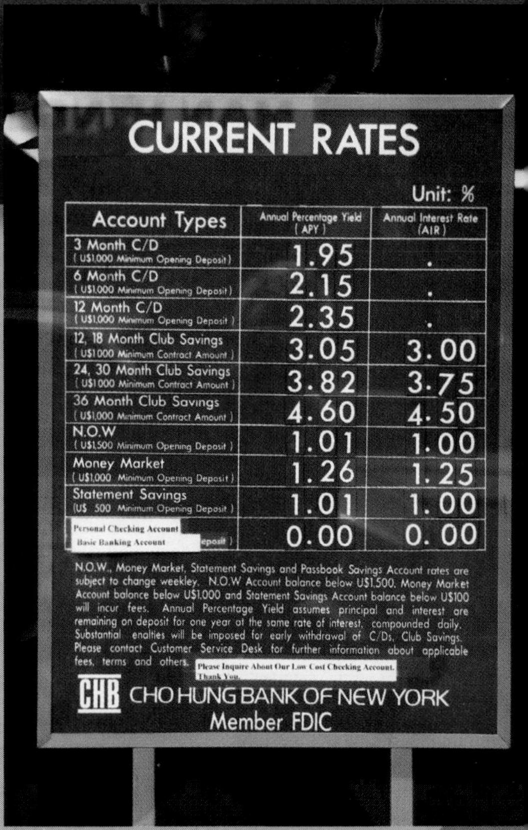

The Human Side of Mathematics

The introduction of a decimal system of money has made the use of **decimals** much more common. Even our stock market now uses decimals rather than fractions to quote the value of stocks.

Modern methods of writing decimals were invented less than 500 years ago, but their use can be traced back thousands of years. As far back as 1579 the Italian/French mathematician Francois Vieta called for the use of the decimal system in his book *Canon Mathematicus.*

The daily use of the decimal system was popularized by a book published in 1586 by Simon Stevin, a Dutch mathematician born in 1548. The book was aptly titled *The Thiende* (The Tenth). His aim: "To perform with an ease unheard of, all computations necessary between men by integers without fractions." This is one of our aims as well.

The notation used in our decimal system has evolved through time. The decimal point separating the whole part from the decimal part seems to have been the invention of Bartholomaeus Pitiscus, who used it in his trigonometrical tables in 1612. Our familiar decimals were used by John Napier, a Scottish mathematician, who developed the use of logarithms for carrying out complex calculations. The modern decimal point became standard in England in 1619, but many countries in Europe still use the decimal **comma** rather than the decimal **point.** (They write 3,1416 rather than 3.1416.) Whether you use commas or periods for the decimal, the objective is the same: to make clear where the ones column is!

3.1 Addition and Subtraction of Decimals

Objectives

You should be able to:

A › Write the word name for a decimal.

B › Write a decimal in expanded form.

C › Add two or more decimals.

D › Subtract one decimal from another.

To Succeed, Review How To . . .

1. Write the word name for a number. (pp. 5–6)
2. Write a number in expanded form. (pp. 3–4)
3. Work with the addition and subtraction facts. (pp. 24, 37–42)

Getting Started

Can you buy a dozen students with an ID for \$7.29? No, but if you are a student with an ID you can buy a dozen doughnuts for \$7.29! The \$7.29 contains the **decimal** part .29. In the **decimal** (the Latin word for *ten*) **system** we use the digits to count from 1 to 9; then we use the numbers 1 and 0 to express ten, the **base** of the system. The \$7.29 in the ad has a **decimal point** separating the **whole-number part** (7) from the **decimal part** (.29). You can think of \$7.29 as

\$7 **and** 29¢

or as

$$\$7 \textbf{ and } \$\frac{29}{100}$$

since 1 cent is one hundredths of a dollar ($\$\frac{1}{100}$).

A › Writing Word Names for Decimals

In Chapter 2 we represented a part of a whole by using **fraction notation.** We can also represent a part of a whole using **decimal notation.** A number written in **decimal notation** is simply called a **decimal.** The decimal number system was introduced in 1619 A.D. by the Scottish mathematician John Napier. A **decimal** consists of three parts:

1. The **whole**-number part.
2. A **period,** ., called the **decimal point.**
3. The **decimal** part.

Thus, in the decimal 247.83 the whole-number part is 247 and the decimal part is .83. When we use the decimal system we can write fractions without explicitly writing the numerator and denominator. For example, in the decimal system the decimal fraction $\frac{7}{10}$ is written as the **decimal** 0.7 and read as *seven tenths* or *zero point 7.* Note that the decimal 0.7 does not have a whole-number part, so we placed a zero to the left of the decimal point. In general, the **number of decimal places** in a number is the **number of digits to the right of the decimal point.** Thus, 0.19 has *two* decimal places and 57.568 has *three* decimal places. How many decimal places in a whole number such as 100? Technically, **0** but when dealing with money we usually write \$100 as \$100.00.

The place value chart used in Chapter 1 can be extended to include decimals so the place values are 100, 10, 1, $\frac{1}{10}$, $\frac{1}{100}$, and so on to help us write decimals in words as shown.

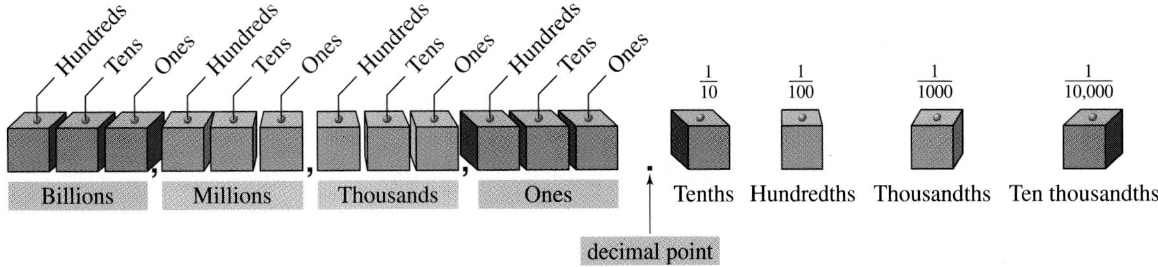

Notice that the names for the place values to the right of the decimal point end with **th.** A decimal fraction such as $\frac{9}{10}$ is read as ***nine tenths*** and $\frac{41}{100}$ is read as ***forty-one hundredths.*** How do we write 7.29 in words? Since 7 and $\frac{29}{100}$ can be written as $7\frac{29}{100}$, we can write 7.29 as seven and twenty-nine hundredths. Use the diagram to help in writing the word name for 284.356.

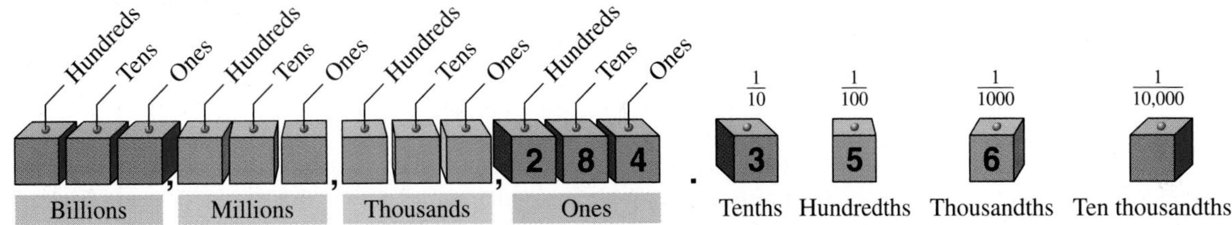

TO WRITE THE WORD NAME FOR A DECIMAL

1. Write the word name for the whole-number part (the number to the left of the decimal point).
2. Write the word *and* for the decimal point.
3. Write the word name for the number to the right of the decimal point, followed by the place value of the *last* digit.

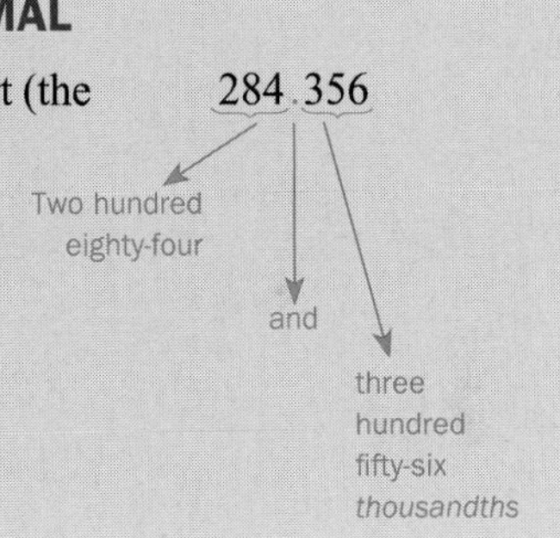

EXAMPLE 1 **Writing a number in words**

Give the word name for 187.93.

SOLUTION One hundred eighty-seven *and* ninety-three hundredths

PROBLEM 1

Write the word name for 147.17.

B > Writing Decimals in Expanded Form

Let us go back to \$7.29. You can think of \$7.29 as

$$\$7 + 2 \text{ dimes} + 9 \text{ cents}$$

or

$$\$7 + \frac{2}{10} + \frac{9}{100}$$

since a dime is one-tenth of a dollar (\$$\frac{1}{10}$) and a cent is one-hundredth of a dollar (\$$\frac{1}{100}$). When we write 7.29 as $7 + \frac{2}{10} + \frac{9}{100}$, we say that 7.29 is written in **expanded form.** Similarly, the decimal 284.356 can be written in expanded form like this:

$$2\,8\,4\,.\,3\,5\,6$$

$$200 + 80 + 4 + \frac{3}{10} + \frac{5}{100} + \frac{6}{1000}$$

Answers to PROBLEMS

1. One hundred forty-seven and seventeen hundredths

EXAMPLE 2 **Using expanded form**

Write 35.216 in expanded form.

SOLUTION

$$35.216 = 30 + 5 + \frac{2}{10} + \frac{1}{100} + \frac{6}{1000}$$

PROBLEM 2

Write 47.321 in expanded form.

C > Adding Decimals

Adding decimals is similar to adding whole numbers: line up the place values and add the numbers in each place value (hundredths, tenths, ones, tens, and so on), carrying if necessary. For example,

$$0.2 + 0.7 = 0.9$$

This is because $0.2 = \frac{2}{10}$ and $0.7 = \frac{7}{10}$. Thus,

$$0.2 + 0.7 = \frac{2}{10} + \frac{7}{10} = \frac{2+7}{10} = \frac{9}{10} = 0.9$$

If the addition is done using a vertical column as we did with whole numbers, we place the tenths digit in the tenths column and the ones (units) digits in the units column. We then proceed as shown in the diagram.

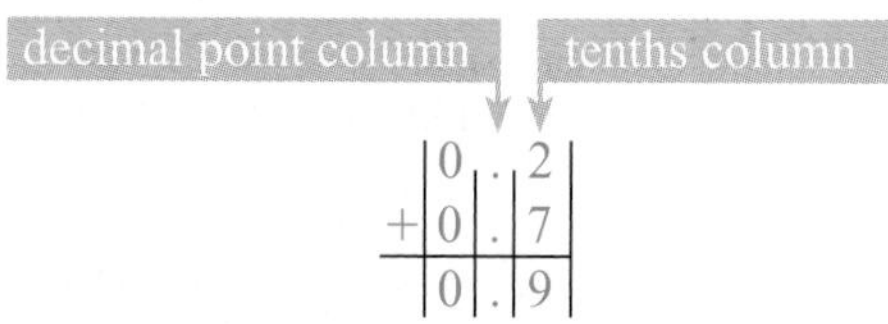

Note that 0.2 and 0.7 have *no* whole number parts (pure decimals) and are written with a 0 to the left of the decimal point.

Similarly, 3.2 + 4.6 is added like this:

Short Form	Long Form
3.2	$3 + \frac{2}{10}$
+ 4.6	$4 + \frac{6}{10}$
7.8	$7 + \frac{8}{10} = 7\frac{8}{10} = 7.8$

Of course, we use the short form to save time, making sure we line up (align) the decimal points by writing them in the *same* column and then making sure that digits of the same place value are in the same column. Here is the way to add:

$$4.13 + 5.24$$

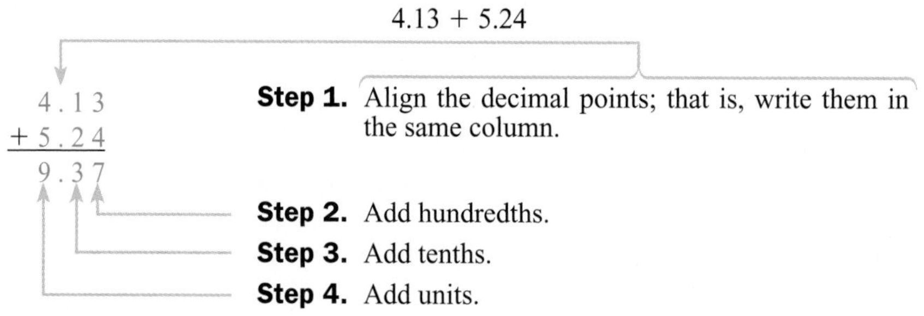

The result is 9.37.

Answers to PROBLEMS

2. $40 + 7 + \frac{3}{10} + \frac{2}{100} + \frac{1}{1000}$

EXAMPLE 3 Adding decimals

Add 5 + 12.15.

SOLUTION

$$\begin{array}{r} 5.00 \\ +12.15 \\ \hline 17.15 \end{array}$$

Step 1. Note that 5 = 5.

Step 2. Align the decimal points.

Step 3. Attach placeholder zeros to the 5, so that both addends have the same number of decimal digits.

Step 4. Add hundredths.

Step 5. Add tenths.

Step 6. Add units.

Step 7. Add tens.

PROBLEM 3

Add 4 + 12.4.

Note that 5 was written as 5.00. This is possible because

$$5.00 = 5 + \frac{0}{10} + \frac{0}{100}$$

Thus,

$$5 = 5.00$$

What happens if we have to carry when adding decimals, and what does it mean?

Short Form

$$\begin{array}{r} 4.8 \\ +3.7 \\ \hline 8.5 \end{array}$$

.8 + .7 = 1.5
write .5, carry 1

1 + 4 + 3 = 8

Expanded Form

$$\begin{array}{r} 4 + \frac{8}{10} \\ +3 + \frac{7}{10} \\ \hline 7 + \frac{8+7}{10} \end{array} = 7 + \frac{15}{10}$$

$$= 7 + \frac{10}{10} + \frac{5}{10}$$

$$= 7 + 1 + \frac{5}{10}$$

$$= 8 + \frac{5}{10}$$

$$= 8.5$$

Note that in the expanded form, $0.8 + 0.7 = \frac{8}{10} + \frac{7}{10} = \frac{15}{10}$. We carry **1** (regroup) because we write $\frac{15}{10} = \frac{10}{10} + \frac{5}{10} = \mathbf{1} + \frac{5}{10}$, leaving the $\frac{5}{10}$ in the first column and carrying the **1** to the ones' place.

Answers to PROBLEMS

3. 16.4

EXAMPLE 4 **Adding decimals**

Add 32.663 + 8.58.

SOLUTION

Short Form	Expanded Form
$\begin{array}{r} {\scriptstyle 11\ 1} \\ 32.663 \\ +\ 8.580 \\ \hline 41.243 \end{array}$	$\begin{aligned} &32 + \frac{6}{10} + \frac{6}{100} + \frac{3}{1000} \\ &+ 8 + \frac{5}{10} + \frac{8}{100} + \frac{0}{1000} \\ \hline &40 + \frac{11}{10} + \frac{14}{100} + \frac{3}{1000} \\ &= 40 + \boxed{\frac{10}{10} + \frac{1}{10}} + \frac{10}{100} + \frac{4}{100} + \frac{3}{1000} \\ &= 40 + \boxed{1 + \frac{1}{10}} + \frac{1}{10} + \frac{4}{100} + \frac{3}{1000} \\ &= 41 + \frac{2}{10} + \frac{4}{100} + \frac{3}{1000} \\ &= 41.243 \end{aligned}$

Of course, you should use the short form to save time!

PROBLEM 4

Add 49.28 + 7.921.

We can also add more than two numbers involving decimals as long as we continue to align the decimal points. For example, suppose you owe $3748.74 on a mortgage, $517.46 on a car, $229 on credit cards, and $550.51 on personal loans. How much do you owe? To find the answer we must *add* all these quantities. We do this in Example 5.

EXAMPLE 5 **Adding decimals**

Add $3748.74, $517.46, $229, and $550.51 to find how much you owe.

SOLUTION First we note that $229 = $229.00. (If you have $229, you have $229 and no cents; thus, $229 = $229.00.) We then proceed by steps:

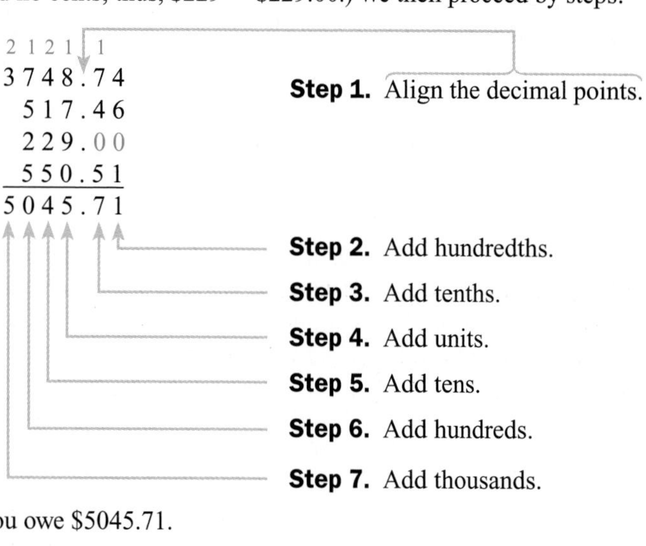

Thus, you owe $5045.71.

PROBLEM 5

A student owes $11,022.77 on his mortgage, $1521.54 on his car, $674 on credit cards, and $1618.73 on personal loans. How much does he owe?

Answers to PROBLEMS

4. 57.201 **5.** $14,837.04

D › Subtracting Decimals

Subtraction of decimals is like subtraction of whole numbers as long as you remember to align the decimal points and insert any placeholder zeros so that both numbers have the *same* number of decimal digits. For example, if you earned \$231.47 and you made a \$52 payment, you have \$231.47 − \$52. To find the answer we write.

$$\begin{array}{r} \$\,231.47 \\ -\ \ 52.00 \\ \hline 179.47 \end{array}$$

Align the decimal points.
Attach two zeros (\$52 is the same as \$52 and no cents or \$52.00).
The decimal point is in the same column.

Thus, you have \$179.47 left. You can check this answer by adding \$52 and \$179.47, obtaining \$231.47.

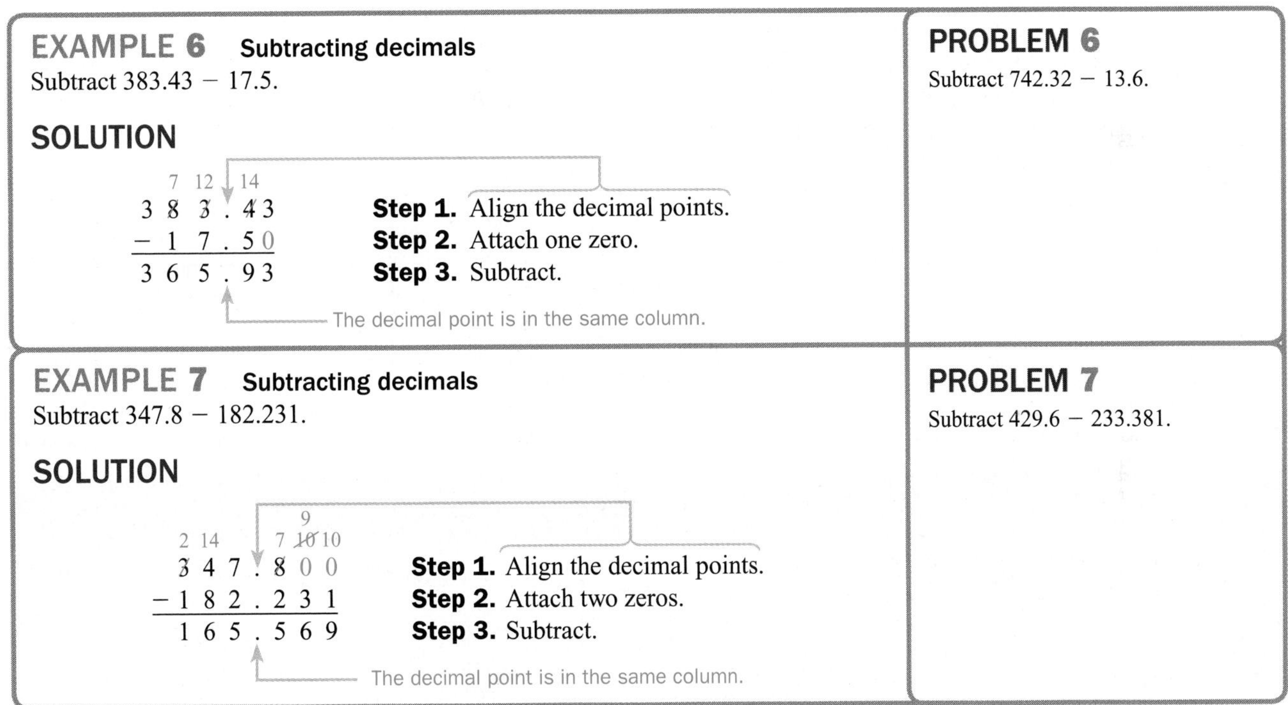

EXAMPLE 6 Subtracting decimals

Subtract 383.43 − 17.5.

SOLUTION

$$\begin{array}{r} 383.43 \\ -\ 17.50 \\ \hline 365.93 \end{array}$$

Step 1. Align the decimal points.
Step 2. Attach one zero.
Step 3. Subtract.

The decimal point is in the same column.

PROBLEM 6

Subtract 742.32 − 13.6.

EXAMPLE 7 Subtracting decimals

Subtract 347.8 − 182.231.

SOLUTION

$$\begin{array}{r} 347.800 \\ -\ 182.231 \\ \hline 165.569 \end{array}$$

Step 1. Align the decimal points.
Step 2. Attach two zeros.
Step 3. Subtract.

The decimal point is in the same column.

PROBLEM 7

Subtract 429.6 − 233.381.

Decimals are everywhere, and you have to know how to use them! Many interstate highways have signs (like the one shown) at the exits. For example, if you want to know how far you are from Shoney's Inn, you simply read the answer: 0.5. Of course, it is assumed that you understand that these distances are in miles. Thus, when you get off the interstate and take a left, you are 0.5 or $\frac{1}{2}$ mile from Shoney's Inn.

Answers to PROBLEMS

6. 728.72 **7.** 196.219

EXAMPLE 8 Subtracting decimals

Using the photo above, find the following:

a. How far is the Happy Traveler RV Park?

b. How far is it from the Holiday Inn to the Wingate Inn?

c. If you decide to walk from the Happy Traveler RV Park to the Holiday Inn, how far do you have to walk?

SOLUTION

a. The Happy Traveler RV Park is 0.8 miles to the right.

b. The Holiday Inn is 4.6 miles to the left and the Wingate Inn is 3.7 miles to the left, so the distance from the Holiday Inn to the Wingate Inn is 4.6 − 3.7 miles.

H(4.6) W(3.7)

Here is the subtraction:

$$\begin{array}{r} \overset{3}{\cancel{4}}.\overset{16}{\cancel{6}} \\ -\,3.7 \\ \hline 0.9 \end{array}$$

Align the decimal points.

Subtract.

Thus, the Holiday Inn is 0.9 mile from the Wingate Inn.

c. Note that to go to the Happy Traveler RV Park you have to go right, but the Holiday Inn is to the *left*. Thus, the distance from the Happy Traveler RV Park to the Holiday Inn is 0.8 + 4.6 miles.

Here is the addition:

H(4.6) RV(0.8)

$$\begin{array}{r} \overset{1}{0}.8 \\ +\,4.6 \\ \hline 5.4 \end{array}$$

Align the decimal points.

Add.

Thus, you have to walk 5.4 miles from the Happy Traveler RV Park to get to the Holiday Inn.

PROBLEM 8

a. How far is La Quinta Inn?

b. How far is it from the Holiday Inn to Shoney's Inn?

c. How far is it from the Happy Traveler RV Park to the Wingate Inn?

Answers to PROBLEMS

8. a. 3.5 mi to the left **b.** 4.1 mi **c.** 4.5 mi

Algebra Bridge

In this section we learned how to add or subtract decimals. In algebra, we add or subtract letters (variables). Look at the similarities.

Arithmetic	**Algebra**
0.2 + 0.7 = 0.9	$0.2y + 0.7y = 0.9y$
1.3 + 2.5 = 3.8	$1.3y^2 + 2.5y^2 = 3.8y^2$
12.81 + 3.05 = 15.86	$12.81y^3 + 3.05y^3 = 15.86y^3$

What will happen if we try $y + y^2$? Nothing! You cannot add y and y^2. Thus, to add $0.3y + 3.1y^2$ and $0.8y + 4.9y^2$, we first write the addition in a column with the exponents in **descending** order.

$$\begin{array}{r} 3.1y^2 + 0.3y \\ +\,4.9y^2 + 0.8y \\ \hline 8.0y^2 + 1.1y \end{array}$$

Add the y^2's. Add the y's.

Subtraction is done similarly. Thus, to subtract $0.4y + 4.5y^2$ from $0.6y + 7.8y^2$, we write both expressions in descending order in columns.

$$\begin{array}{r} 7.8y^2 + 0.6y \\ -(4.5y^2 + 0.4y) \\ \hline 3.3y^2 + 0.2y \end{array}$$

Subtract the y^2's. Subtract the y's.

Calculator Corner

If you use a calculator to add or subtract decimals, you do not have to worry about aligning the decimal point or entering the zero to the left of the decimal because the calculator will align the numbers for you.

Thus, to add 0.2 and 0.3, we press · 2 + · 3 ENTER. Moreover, to add 5 + 12.15 (Example 3), you simply press 5 + 1 2 · 1 5 ENTER, without having to write the 5 as 5.00.

The same idea works in subtraction. Thus, if you earned $231.47 and made a $52 payment, you have $231.47 − $52 left. To find how much you have left, simply press 2 3 1 · 4 7 − 5 2 ENTER, obtaining $179.47.

Exercises 3.1

⟨A⟩ Writing Word Names for Decimals In Problems 1–10, write the word name for the number.

1. 3.8 **2.** 9.6
3. 13.12 **4.** 46.78
5. 132.34 **6.** 394.05
7. 5.183 **8.** 9.238
9. 0.2172 **10.** 0.3495

⟨B⟩ Writing Decimals in Expanded Form In Problems 11–20, write in expanded form.

11. 3.21 **12.** 4.7 **13.** 41.38
14. 37.10 **15.** 89.123 **16.** 13.278
17. 238.392 **18.** 312.409 **19.** 301.5879
20. 791.354

⟨C⟩⟨D⟩ Adding Decimals and Subtracting Decimals

In Problems 21–60, add or subtract as required.

21. 0.4 + 0.1 **22.** 0.3 + 0.2 **23.** 0.6 + 0.9 **24.** 0.4 + 0.8
25. 0.3 − 0.1 **26.** 0.7 − 0.4 **27.** 8.3 − 5.2 **28.** 7.5 − 4.4
29. 5 − 3.2 **30.** 8 − 7.3 **31.** 9 − 4.1 **32.** 6 − 3.5
33. 3.8 − 1.9 **34.** 2.6 − 1.7 **35.** 1.1 − 0.8 **36.** 3.4 − 0.5
37. 12.23 + 9 **38.** 13.24 + 8 **39.** 4.6 + 18.73 **40.** 7.8 + 16.31
41. 17.35 − 8.4 **42.** 13 − 7.5 **43.** $648.01 + $341.06
44. $237.49 + $458.72 **45.** 72.03 + 847.124 **46.** 13.12 + 108.138
47. 104 + 78.103 **48.** 184 + 69.572 **49.** 0.35 + 3.6 + 0.127
50. 5.2 + 0.358 + 21.005 **51.** 27.2 − 0.35 **52.** 4.6 − 0.09
53. 19 − 16.62 **54.** 99 − 0.161 **55.** 9.43 − 6.406
56. 9.08 − 3.465 **57.** 8.2 − 1.356 **58.** 6.3 − 4.901
59. 6.09 + 3.0046 **60.** 2.01 +1.3045

Applications

61. *Trip length* The mileage indicator on a car read 18,327.2 at the beginning of a trip. At the end of the trip, it read 18,719.7. How long was the trip?

62. *Nicotine content* The highest nicotine content for a nonfilter cigarette is 1.7 milligrams. Another brand contains 0.8 milligram. What is the difference in the nicotine amounts?

63. *Total spending* A woman wrote checks for $18.47, $23.48, and $12.63. How much did she spend?

64. *Swimming records* In 1972 Mark Spitz established an Olympic swimming record in the 100-meter butterfly. He finished in 54.27 seconds. In 1988 Anthony Nesty finished in 53 seconds. How much faster was Nesty?

65. *Springboard records* In 1984 Greg Louganis won the springboard diving in the Olympics by scoring 754.41 points. He won again in 1988 by scoring 730.80. How many fewer points did he score in 1988?

66. *Change for a twenty* A man bought merchandise costing $6.84. He paid with a $20 bill. How much change did he receive?

Body composition The following information will be used in Problems 67–70.

By weight the average adult is composed of

43%	muscle	2.7%	liver	0.5%	kidneys
26%	skin	2.2%	brain	0.5%	heart
17.5%	bone	2.2%	intestines	0.2%	spleen
7%	blood	1.5%	lungs	0.1%	pancreas

However, an old-time song, named *16 Tons,* makes the following claim about the body of miners:

Some people say a man is made of mud,
Bone, skin, muscle, and blood.
Muscle and blood, skin and bone,
I owe my soul to the company store.

67. What percent of an average adult is composed of skin, muscle, blood, and bone?

68. Since the weight of all body parts must add up to 100%, if a person is made up of mud, skin, muscle, blood, and bone, what percent is mud?

69. There is another difficulty with these data. Add up all the percents and see what it is!

70. How much over 100% is the amount obtained in Problem 69?

Credit Do you know what your FICO (Fair Isaac Credit Organization) score is? "When you apply for credit lenders want to know what risk they'd take by loaning money to you. FICO scores are the credit scores most lenders use to determine your credit risk. You have three FICO scores, one for each of the three credit bureaus: Experian, TransUnion, and Equifax." *Source:* http://myfico.com/.

The higher your FICO® scores, the less you pay to buy on credit—no matter whether you're getting a home loan, a cell phone, a car loan, or signing up for credit cards. For example, on a $216,000 30-year, fixed-rate mortgage:

If Your FICO® Score Is	Your Interest Rate Is	. . . And Your Monthly Payment Is	And Your Total Payment Is
1. 760–850	6.2%	71. $1322.93	$476,258.10
2. 700–759	6.42%	72. $1353.92	$487,414.21
3. 680–699	6.6%	73. $1379.50	$496,623.15
4. 660–679	6.81%	74. $1409.60	$507,453.23
5. 640–659	7.24%	75. $1472.04	$529,929.30
6. 620–639	7.79%	76. $1553.43	$559,228.05

Source: http://www.myfico.com/.

In Problems 71 through 76, find the difference in *monthly payment* and the difference in *total payment* between persons in the specified categories:

71. Category 1 (760–850) and category 6 (620–639)

72. Category 1 (760–850) and category 5 (640–659)

73. Category 1 (760–850) and category 4 (660–679)

74. Category 1 (760–850) and category 3 (680–699)

75. Category 1 (760–850) and category 2 (700–759)

76. Category 2 (700–759) and category 3 (680–699)

Using Your Knowledge

Balancing Your Checkbook Here is a bank statement sent to a depositor. The balance on the statement is \$568.40. Unfortunately, this balance may be different from the one in the depositor's checkbook.

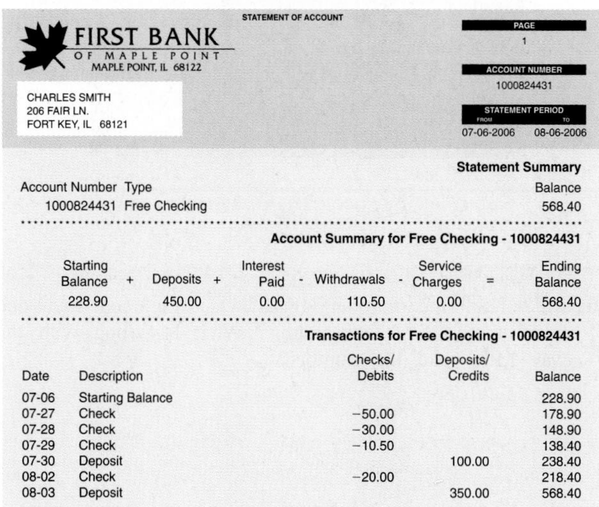

STATEMENT OF ACCOUNT

FIRST BANK OF MAPLE POINT
MAPLE POINT, IL 68122

PAGE 1

ACCOUNT NUMBER 1000824431

CHARLES SMITH
206 FAIR LN.
FORT KEY, IL 68121

STATEMENT PERIOD FROM 07-06-2006 TO 08-06-2006

Statement Summary

Account Number	Type	Balance
1000824431	Free Checking	568.40

Account Summary for Free Checking - 1000824431

Starting Balance	+ Deposits	+ Interest Paid	- Withdrawals	- Service Charges	= Ending Balance
228.90	450.00	0.00	110.50	0.00	568.40

Transactions for Free Checking - 1000824431

Date	Description	Checks/ Debits	Deposits/ Credits	Balance
07-06	Starting Balance			228.90
07-27	Check	−50.00		178.90
07-28	Check	−30.00		148.90
07-29	Check	−10.50		138.40
07-30	Deposit		100.00	238.40
08-02	Check	−20.00		218.40
08-03	Deposit		350.00	568.40

On the reverse side of the statement is a reconciliation form. It shows three outstanding checks for \$27.50, \$50.00, and \$10.00. (box below left)

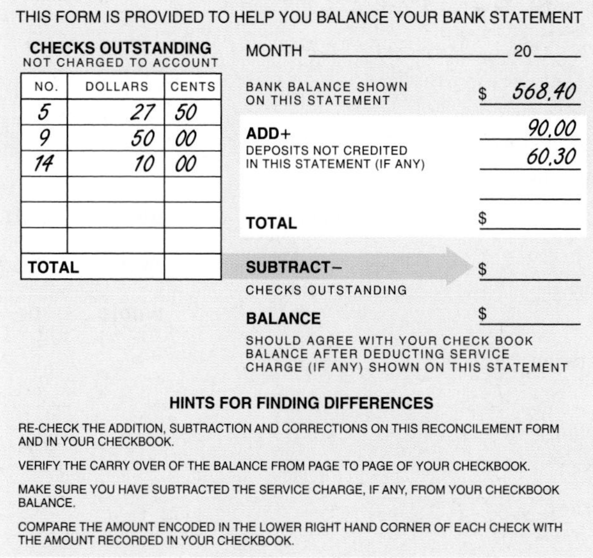

THIS FORM IS PROVIDED TO HELP YOU BALANCE YOUR BANK STATEMENT

CHECKS OUTSTANDING NOT CHARGED TO ACCOUNT

NO.	DOLLARS	CENTS
5	27	50
9	50	00
14	10	00
TOTAL		

MONTH ______ 20___

BANK BALANCE SHOWN ON THIS STATEMENT \$ 568.40

ADD+ DEPOSITS NOT CREDITED IN THIS STATEMENT (IF ANY) 90.00 60.30

TOTAL \$ ______

SUBTRACT– CHECKS OUTSTANDING \$ ______

BALANCE \$ ______

SHOULD AGREE WITH YOUR CHECK BOOK BALANCE AFTER DEDUCTING SERVICE CHARGE (IF ANY) SHOWN ON THIS STATEMENT

HINTS FOR FINDING DIFFERENCES

RE-CHECK THE ADDITION, SUBTRACTION AND CORRECTIONS ON THIS RECONCILEMENT FORM AND IN YOUR CHECKBOOK.

VERIFY THE CARRY OVER OF THE BALANCE FROM PAGE TO PAGE OF YOUR CHECKBOOK.

MAKE SURE YOU HAVE SUBTRACTED THE SERVICE CHARGE, IF ANY, FROM YOUR CHECKBOOK BALANCE.

COMPARE THE AMOUNT ENCODED IN THE LOWER RIGHT HAND CORNER OF EACH CHECK WITH THE AMOUNT RECORDED IN YOUR CHECKBOOK.

77. What is the sum of these checks?

The reconciliation form shows two deposits of \$90.00 and \$60.30 not credited.

78. What is the sum of \$90.00 and \$60.30?

The form directs the depositor to add the bank balance and the deposits not credited to get the total. (See preceding form.)(above right)

79. What is the total?

Then, to find the balance we subtract the outstanding checks from the total.

80. Find the balance.

Distance Answer questions 81–84.

81. How far is it to the Wingate Inn?

82. How far is it from Shoney's to La Quinta?

83. How far is it from La Quinta to the Holiday Inn?

84. How far is it from the Happy Traveler RV Park to La Quinta?

Distance Many interstate highways have signs like the one shown indicating how far food is!

85. How far is McDonald's?

86. How far is it from McDonald's to Perkins?

87. How far apart are Burger King and McDonald's?

88. If you decide to walk from the 76 station to Perkins, how far do you have to walk?

››› Write On

89. Pedro read the number 3805 as "three thousand eight hundred and five." What is wrong with the way Pedro read the number?

90. Milé read the number 18.105 as "Eighteen and one hundred and five thousandths." What is wrong with the way Milé read the number?

91. What is the difference between: Subtract "3 from 4.8" and "6.6 minus 4.8"? What answer do you get in each case?

››› Algebra Bridge

92. Add $0.4y + 9.2y^2$ and $0.7y + 8.2y^2$.

93. Add $3.4y^2 + 0.5y$ and $0.7y + 7.2y^2$.

94. Subtract $0.5y + 3.9y^2$ from $0.8y + 4.8y^2$.

95. Subtract $6.2y^2 + 0.4y$ from $7y^2 + 0.9y$.

››› Concept Checker

Fill in the blank(s) with the correct word(s), phrase, or mathematical statement.

decimal	**left**
right	**and**
whole	**or**

96. When writing the word name for a decimal, we use the **word** ______________ for the decimal point.

97. A decimal consists of three parts, the ______________ **number** part, the ______________ **point,** and the ______________ **part.**

98. The number of decimal places in a number is the number of **digits** to the ______________ of the decimal point.

99. **Fractions** such as $\frac{1}{10}$, $\frac{1}{100}$, and $\frac{1}{1000}$ are called ______________ fractions.

››› Mastery Test

100. Add: $7 + 13.18$

101. Add: $38.773 + 3.69$

102. Your bills for this month are: \$47.10 for the phone, \$59.49 for electricity, \$258.20 for the car payment, \$308 for the rent, and \$105.27 for miscellaneous expenses. What is the total for all your bills?

103. Write 41.208 in expanded form.

104. Write the word name for 283.98.

105. Subtract: $473.43 - 18.6$

106. Subtract: $458.9 - 293.342$

107. How far is it from  to ?

108. How far is it from  to ?

Skill Checker

Multiply.

109. 23×10

110. 235×100

111. 240×1000

112. Round 237 to the nearest ten.

113. Round 3487 to the nearest hundred.

114. Round 5480 to the nearest thousand.

3.2 Multiplication and Division of Decimals

Objectives

You should be able to:

A Find the product of (multiply) two decimals.

B Find the product of decimals involving a power of 10.

C Find the quotient of (divide) two decimals.

D Round a number to a specified number of decimal digits or places.

E Find the quotient of decimals involving a power of 10.

F Solve applications involving the concepts studied.

To Succeed, Review How To . . .

1. Write a decimal in expanded form. (pp. 201–202)
2. Multiply or divide a whole number by a power of 10. (p. 54)

Getting Started

The man in the picture sold three lollipops that cost 30¢ ($0.30) each. How much is the bill? To find the answer we must find

$$3 \times 0.30$$

which is 0.90, but do you know why? Since $0.30 = \frac{30}{100}$ and $3 = \frac{3}{1}$: we multiply

$$\frac{3}{1} \times \frac{30}{100} = \frac{90}{100} = 0.90$$

That is, the bill comes to 90¢. Note that we could simply have multiplied $3 \times 30 = 90$ and then placed the decimal point correctly in the result. To learn *where* to place the decimal point, look at these multiplications. The number in color indicates how many *decimal digits* (digits to the *right* of the decimal point) the number has.

$$\underbrace{0.3}_{1} \times \underbrace{7}_{0} = \frac{3}{10} \times 7 = \frac{21}{10} = 2\frac{1}{10} = 2 + \frac{1}{10} = \underbrace{2.1}_{1}$$

$$\underbrace{0.3}_{1} \times \underbrace{0.7}_{1} = \frac{3}{10} \times \frac{7}{10} = \frac{21}{100} = \underbrace{0.21}_{2}$$

$$\underbrace{0.3}_{1} \times \underbrace{0.07}_{2} = \frac{3}{10} \times \frac{7}{100} = \frac{21}{1000} = \underbrace{0.021}_{3}$$

$$\underbrace{0.3}_{1} \times \underbrace{0.007}_{3} = \frac{3}{10} \times \frac{7}{1000} = \frac{21}{10{,}000} = \underbrace{0.0021}_{4}$$

A › Multiplying Decimals

When you multiply decimals, the number of decimal digits in the product is the *sum* of the number of decimal digits in the factors. For example, 0.3 has *one* decimal digit and 0.0007 has *four;* the product,

$$\underbrace{0.3}_{\text{1 decimal digit}} \times \underbrace{0.0007}_{\text{4 decimals digits}} = \frac{3}{10} \times \frac{7}{10{,}000} = \frac{21}{100{,}000} = \underbrace{0.00021}_{1 + 4 = 5 \text{ decimal digits}}$$

has 1 + 4 = 5 decimal digits. Here is the rule used to multiply decimals:

TO MULTIPLY DECIMALS

1. *Multiply* the two decimal numbers *as if they were whole numbers.*
2. The *number of decimal digits* in the product is the *sum* of the number of decimal digits in the factors.

Now let us find the price of 3.5 and 5 pounds of roast selling at \$2.29 per pound. Following the rule just given, we have

Price for 3.5 Pounds

```
  2.29    2 decimal digits
 × 3.5    1 decimal digit
  1145
  687
 8.015    Count 2 + 1 = 3 decimal digits from the right in the answer.
```

The price is about \$8.02.

Price for 5 Pounds

```
  2.29    2 decimal digits
 ×   5    0 decimal digits
 11.45    Count 2 + 0 = 2 decimal digits from the right in the answer.
```

The price is \$11.45.

EXAMPLE 1 Multiplying decimals

Multiply: **a.** 2.31 × 4.2 **b.** 13.813 × 7.1

SOLUTION

a.

```
  2.31    2 decimal digits
 × 4.2    1 decimal digit
   462
  924
 9.702    Count 2 + 1 = 3 decimal digits.
```

b.

```
  13.813    3 decimal digits
 ×   7.1    1 decimal digit
   13813
  96691
 98.0723    Count 3 + 1 = 4 decimal digits.
```

PROBLEM 1

Multiply: **a.** 3.12 × 5.3 **b.** 12.172 × 5.1

Answers to PROBLEMS

1. a. 16.536 **b.** 62.0772

Sometimes we need to *prefix* zeros to (write zeros before) the product to obtain the required number of decimal digits. For example:

```
  0.005    3 decimal digits
×     3    0 decimal digits
   .015    3 + 0 = 3 decimal digits.
```

We need 3 decimal digits. Zero is inserted to obtain 3 decimal digits.

(Note that $0.005 \times 3 = \frac{5}{1000} \times 3 = \frac{15}{1000}$.)

The answer is written as 0.015

```
  0.016    3 decimal digits
×  0.23    2 decimal digits
     48
    32
 .00368    3 + 2 = 5 decimal digits.
```

2 zeros inserted here to obtain 5 decimal digits.

The answer is written as 0.00368

EXAMPLE 2 **Multiplying decimals**

Multiply:

a. 5.102×21.03 **b.** 5.213×0.0012

SOLUTION

a.

```
    5.102    3 decimal digits
  × 21.03    2 decimal digits
    15306
   5102
  10204
107.29506    3 + 2 = 5 decimal digits.
```

b.

```
   5.213    3 decimal digits
× 0.0012    4 decimal digits
   10426
   5213
.0062556    3 + 4 = 7 decimal digits.
```

The answer is written as 0.0062556

PROBLEM 2

Multiply:

a. 3.201×31.02 **b.** 4.132×0.0021

B › Multiplying by Powers of 10

In many cases we have to multiply decimals by **powers** of 10 (10, 100, 1000, etc.). The rules for doing this are very simple. See if you discover a pattern.

$$32.314 \times 10 = 323.14$$

$$32.314 \times 100 = 3231.4$$

$$32.314 \times 1000 = 32314.$$

Did you find the pattern?

Here is the general rule to multiply by powers of 10.

RULE FOR MULTIPLYING BY A POWER OF 10

To *multiply* a decimal number by 10, 100, 1000, or a higher power of 10, move the decimal point as many places to the *right* as there are *zeros* in the power of 10 being multiplied. (Sometimes you will need to attach additional zeros in order to move the decimal point.)

EXAMPLE 3 **Multiplying by powers of 10**

Multiply:

a. $41.356 \times 100 =$ ________ **b.** $32.3 \cdot 1000 =$ ________

c. $(0.417)(10) =$ ________

SOLUTION

a. $41.356 \times 100 = 4135.6$ Move the decimal point two places to the right. The answer is 4135.6.

PROBLEM 3

Multiply:

a. $58.12 \times 100 =$ ________

b. $43.1 \cdot 1000 =$ ________

c. $(10.296)(10) =$ ________

(continued)

Answers to PROBLEMS

2. a. 99.29502 **b.** 0.0086772 **3. a.** 5812 **b.** 43,100 **c.** 102.96

b. $32.3 \cdot 1000 = 32300$ Move the decimal point three places to the right and attach two additional zeros. The answer is 32,300.

c. $(0.417)(10) = 4.17$ Move the decimal point one place to the right. The answer should be written as 4.17.

EXAMPLE 4 Multiplying decimals at Burger King

The sign below says that items from the Burger King (BK) value menu cost .99 cents. What will be the cost of 100 items from the BK value menu?

SOLUTION To find the answer, we multiply .99 cents by 100.

$$.99 \text{ cents} \times 100 = (.99 \times 100) \text{ cents}$$
$$= 99. \text{ cents}$$ Move the decimal 2 places.
$$= 99 \text{ cents}$$

So, you can buy 100 BK value menu items for 99 cents!

PROBLEM 4

What will be the cost of 10 items from the BK value menu?

Do you see what the error in the sign is? What they really mean is that one BK value menu item is 99 cents (no decimal). This is a common error! Let us discuss the relationship between dollars and cents further.

EXAMPLE 5 Converting cents to dollars and vice versa

a. If you have 457 cents, how many dollars do you have?

b. If you have $5.48, how many cents do you have?

SOLUTION We solve these problems by substitution using two facts:

1. One cent is one hundredth of a dollar. 1 cent = $0.01

2. One dollar consists of 100 cents. $1 = 100 cents

Now, we use these two facts and substitution to solve the problems.

a. 457 cents = 457 ($0.01) = $4.57 (substituting)

Thus, if you have 457 cents, you have $4.57.

b. $5.48 = 5.48 dollars = 5.48 (100 cents) = 548 cents (substituting)

Thus, if you have $5.48 you have 548 cents.

PROBLEM 5

a. If you have 692 cents, how many dollars do you have?

b. If you have $7.92, how many cents do you have?

C › Dividing Decimals

We are now ready to do division of decimals. Actually, the division of decimal numbers is very similar to the division of whole numbers. For example, to find the cost per ounce of the tuna in the ad, we need to solve this division problem:

$$\frac{52}{6.5}$$ ← Price (in cents) / ← Number of ounces

Answers to PROBLEMS

4. 9.9¢ **5. a.** $6.92 **b.** 792 cents

If we multiply the numerator and denominator of this fraction by 10, we obtain

$$\frac{52}{6.5} = \frac{52 \times 10}{6.5 \times 10} = \frac{520}{65} = 8 \quad \text{(cents per ounce)}$$

Thus, $\frac{52}{6.5} = 8$, as can be easily checked, since $52 = 6.5 \times 8$. This problem can be shortened by using the following steps.

Step 1. Write the problem in the usual long division form. $6.5\overline{)52}$

Step 2. Move the decimal point in the divisor, 6.5, to the right until a whole number is obtained. (This is the same as multiplying 6.5 by 10.) $65.\overline{)52}$

Step 3. Move the decimal point in the dividend the *same* number of places as in Step 2. This is the same as multiplying the dividend 52 by 10. Attach zeros if necessary. $65.\overline{)52\,0}$

Step 4. Place the decimal point in the answer directly above the new decimal point in the dividend. $65\overline{)520.}$

Step 5. Divide exactly as you would divide whole numbers. The result is 8 cents per ounce, as before.

$$\begin{array}{r} 8. \\ 65\overline{)520.} \\ \underline{520} \\ 0 \end{array}$$

Here is another example: $\frac{1.28}{1.6}$.

Step 1. Write the problem in the usual long division form. $1.6\overline{)1.28}$

Step 2. Move the decimal point in the divisor to the right until a whole number is obtained. $16.\overline{)1.28}$

Step 3. Move the decimal point in the dividend the *same* number of places as in Step 2. $16.\overline{)1\,2.8}$

Step 4. Place the decimal point in the answer directly above the new decimal point in the dividend. $16\overline{)12.8}$

Step 5. Divide exactly as you would divide whole numbers.

$$\begin{array}{r} .8 \\ 16\overline{)12.8} \\ \underline{12.8} \\ 0 \end{array}$$

Thus

$$\frac{1.28}{1.6} = 0.8$$

EXAMPLE 6 Dividing decimals

Divide $\frac{2.1}{0.035}$.

SOLUTION

$0.035.\overline{)2\,100.}$ We moved the decimal point in the divisor (and also in the dividend) three places to the right. When doing this we had to attach two zeros to 2.1.

$$\begin{array}{r} 60. \\ 35\overline{)2100.} \\ \underline{210} \\ 00 \end{array}$$

We next place the decimal point in the answer directly above the one in the dividend and proceed in the usual manner. The answer is 60; that is,

$$\frac{2.1}{0.035} = 60$$

CHECK $0.035 \times 60 = 2.100$

PROBLEM 6

Divide $\frac{1.4}{0.035}$.

Answers to PROBLEMS

6. 40

Sometimes it is necessary to write one or more zeros in the quotient. We illustrate this procedure in Example 7.

EXAMPLE 7 Dividing decimals

Divide $\frac{0.0048}{12}$.

SOLUTION

$$\begin{array}{r} 0.0004 \\ 12\overline{)0.0048} \\ \underline{48} \\ 0 \end{array}$$

Zeros inserted

The divisor is already a whole number, so we place the decimal point in the answer directly above the one in the dividend and proceed as shown. Thus

$$\frac{0.0048}{12} = 0.0004$$

CHECK $12 \times 0.0004 = 0.0048$

PROBLEM 7

Divide $\frac{0.0065}{13}$.

D › Rounding Decimals

In Examples 6 and 7 the dividend (numerator) was exactly divisible by the divisor (denominator). If this is *not* the case, we must stop the division when a predetermined number of decimal digits is reached and *round* (approximate) the answer. For example, if three cans of soup cost 89¢, what is the cost per can approximated to the nearest cent? The cost will be

$$89 \div 3 \quad \text{or} \quad \begin{array}{r} 29 \\ 3\overline{)89} \\ \underline{6} \\ 29 \\ \underline{27} \\ 2 \end{array}$$

What do we do now? Since we have already obtained the whole part of the answer, we enter a decimal point after the 89 and continue the division until one decimal digit is obtained, as shown.

$$\begin{array}{r} 29.6 \\ 3\overline{)89.0} \\ \underline{6} \\ 29 \\ \underline{27} \\ 20 \\ \underline{18} \\ 2 \end{array}$$

We now approximate our answer, 29.6, to the nearest cent, that is, to 30¢. Thus, the cost per can will be 30¢. Here are the steps used in rounding numbers.

RULE FOR ROUNDING DECIMAL NUMBERS

Step 1. *Underline* the number of digits or places to which you are rounding.

Step 2. If the first number to the *right* of the underlined place is 5 or more, *add one* to the underlined number. Otherwise, *do not change* the underlined number.

Step 3. *Change* all the numbers to the *right* of the underlined number to zeros if they are to the *left* of the decimal point. Otherwise, simply delete them.

Answers to PROBLEMS

7. 0.0005

Here is the number 23.653 rounded to *two* and *one* decimal places, respectively.

23.6<u>5</u>3 becomes 23.65 The 3 is deleted because it is less than 5.

23.<u>6</u>53 becomes 23.7 The 6 is increased by 1, becoming 7, because the number to the right of the underlining is 5.

EXAMPLE 8 Rounding decimals

Round 234.851 to the specified place value.

a. the nearest ten. **b.** one decimal digit (the nearest tenth). **c.** two decimal digits (the nearest hundredth).

SOLUTION

a. Rounded to the nearest ten, 2<u>3</u>4.851 becomes 230.

b. Rounded to one decimal digit, 234.<u>8</u>51 becomes 234.9.

c. Rounded to two decimal digits, 234.8<u>5</u>1 becomes 234.85.

PROBLEM 8

Round 27.752 to the specified place value.

a. the nearest ten.

b. one decimal digit (the nearest tenth).

c. two decimal digits (the nearest hundredth).

The rule we have just developed can be used to round the answer in division problems. Here is how.

EXAMPLE 9 Rounding quotients

Divide 80 ÷ 0.14. (Round the answer to two decimal digits, the nearest hundredth.)

SOLUTION

0.14.)8000.

Move the decimal in the dividend and divisor two places to the right, attaching three zeros as shown, so we can round to the required two decimal digits.

```
      571.428
 14 )8000.000
     70
     100
      98
       20
       14
        60
        56
         40
         28
         120
         112
           8
```

Proceed as in the division of whole numbers, until the whole part of the answer, 571, is obtained. Since we are rounding to *two* decimal digits, attach three zeros to the 8000. and continue dividing until *three* decimal digits are obtained, as shown. The answer obtained, 571.4<u>2</u>8, when rounded to two decimal digits, becomes 571.43. (Since the digit to the *right* of the underlining, 8, was more than 5, we increased the last underlined digit, the 2, by 1.)

PROBLEM 9

Divide 56 ÷ 0.12.

(Round the answer to two decimal digits, the nearest hundredth.)

E › Dividing by Powers of 10

Division of decimals by powers of 10 is very easy. See if you discover the pattern:

$$346.31 \div 10 = 34.631$$
$$346.31 \div 100 = 3.4631$$
$$346.31 \div 1000 = 0.34631$$

Here is the general rule for dividing by a power of 10.

RULE FOR DIVIDING BY A POWER OF 10

To divide a decimal number by 10, 100, 1000, or a higher power of 10, move the decimal point as many places to the *left* as there are zeros in the divisor. (Sometimes it is necessary to prefix additional zeros in order to move the decimal point.)

Answers to PROBLEMS

8. a. 30 **b.** 27.8 **c.** 27.75

9. 466.67

EXAMPLE 10 **Dividing by powers of 10**

Divide:

a. 338.4 ÷ 100 **b.** 2.16 ÷ 1000

c. 3.16 ÷ 10

SOLUTION

a. 338.4 ÷ 100 = 3.384 — Move the decimal point *two* places to the left. The answer is 3.384.

b. 2.16 ÷ 1000 = 0.00216 — Move the decimal point *three* places to the left after prefixing two additional zeros. The answer is 0.00216.

c. 3.16 ÷ 10 = 0.316 — Move the decimal point one place to the left. The answer should be written as 0.316.

PROBLEM 10

Divide:

a. 352.9 ÷ 100

b. 3.27 ÷ 1000

c. 9.35 ÷ 10

F > Applications Involving Decimals

A booklet called *Conservation Payback,* published by Shell, uses division of decimals to find out the time (T) it takes to pay back the cost of an energy-saving measure. This is done by dividing the cost of undertaking the measure by the *amount saved the first year;* that is,

$$\text{T (time for payback)} = \frac{\text{cost}}{\text{amount saved}}$$

For example, an insulation blanket for your hot water heater costs $25. The amount saved in electricity the first year is $20. Thus, the time for payback is

$$\frac{\text{cost}}{\text{amount saved}} = \frac{25}{20} \quad \text{or} \quad 20\overline{)25.00} = 1.25$$

```
      1.25
 20 )25.00
     20
      5 0
      4 0
      1 00
      1 00
         0
```

That is, it takes 1.25 years to pay back the cost of the $25 blanket. We will use this formula in Example 11.

EXAMPLE 11 **Conservation by insulation**

It is estimated that insulating the walls of a house in Oregon (done by a contractor) costs $478 and will save $168 in heating costs the first year. To the nearest tenth of a year, how much time will it take to pay back the total cost?

SOLUTION According to the formula, we must divide the cost of insulating ($478) by the amount saved the first year ($168). We carry the division to two decimal places and then round to the nearest tenth.

```
        2.84
 168 )478.00
      336
      142 0
      134 4
        7 60
        6 72
          88
```

When the answer 2.84 is rounded to the nearest tenth, we obtain $2.\underline{8}4 \longrightarrow 2.8$. Thus it will take about 2.8 years to pay back the $478 cost.

PROBLEM 11

In Oregon a blanket to insulate a water heater costs $27. This measure can save $22 (in electricity) the first year. How long will it take (to the nearest tenth of a year) to pay back the cost of the insulation?

Answers to PROBLEMS

10. a. 3.529 **b.** 0.00327 **c.** 0.935 **11.** 1.2 years

Calculator Corner

Multiplying and dividing decimals using a calculator really simplifies things. You do not have to bother with the placement of the decimal point in the final answer at all! To do Example 2, part **a:** Multiply 5.102 by 21.03 by pressing 5 . 1 0 2 × 2 1 . 0 3 ENTER and the final answer, 107.29506, will appear in the display. Similarly, to complete the division 80 ÷ 0.14 (rounded to two decimal digits, as in Example 9), we enter 8 0 ÷ . 1 4 ENTER. The display shows 571.42857, giving 571.43 when rounded to two decimal digits.

› Exercises 3.2

Boost your grade at mathzone.com!
- Practice Problems
- NetTutor
- Self-Tests
- e-Professors
- Videos

‹ A › Multiplying Decimals In Problems 1–16, multiply.

1. 0.5 · 0.7 **2.** 0.9 · 0.2 **3.** 0.8 · 0.8 **4.** 0.7 · 0.9

5. 0.005 · 0.07 **6.** 0.012 · 0.3 **7.** 9.2 · 0.613 **8.** 0.514 · 7.4

9. 8.7 · 11 **10.** 78.1 · 108 **11.** 7.03 · 0.0035 **12.** 8.23 · 0.025

13. 3.0012 · 4.3 **14.** 6.1 · 2.013 **15.** 0.0031 · 0.82 **16.** 0.51 · 0.0045

‹ B › Multiplying by Powers of 10 In Problems 17–26, multiply.

17. 42.33 · 10 **18.** 36.37 · 10 **19.** 19.5 · 100 **20.** 18.3 · 100 **21.** 32.89 · 1000

22. 35.35 · 1000 **23.** 0.48 · 10 **24.** 0.37 · 10 **25.** 0.039 · 100 **26.** 0.048 · 100

‹ C › Dividing Decimals In Problems 27–36, divide.

27. $15\overline{)9}$ **28.** $48\overline{)6}$ **29.** $5\overline{)32}$ **30.** $8\overline{)36}$ **31.** 8.5 ÷ 0.005

32. 4.8 ÷ 0.003 **33.** 4 ÷ 0.05 **34.** 18 ÷ 0.006 **35.** 2.76 ÷ 60 **36.** 31.8 ÷ 30

‹ D › Rounding Decimals In Problems 37–46, round the numbers to the specified place value or indicated number of digits.

37. 34.8 to the nearest ten **38.** 505.6 to the nearest ten **39.** 96.87 to the nearest hundred

40. 241.2 to the nearest hundred **41.** 3.15 to one decimal digit **42.** 0.415 to two decimal digits

43. 7.81 to the nearest ten **44.** 7.81 to the nearest tenth **45.** 338.123 to the nearest hundredth

46. 338.123 to the nearest hundred

‹ E › Dividing by Powers of 10 In Problems 47–50, divide.

47. 7.8 ÷ 100 **48.** 3.5 ÷ 1000 **49.** 0.05 ÷ 100 **50.** 0.061 ÷ 1000

In Problems 51–60, divide and round the answer to two decimal digits, the nearest hundredth.

51. 1 ÷ 3 **52.** 20 ÷ 7 **53.** 0.06 ÷ 0.70 **54.** 0.05 ÷ 0.90 **55.** $12.243\overline{)2.8}$

56. $20\overline{)5.47}$ **57.** 8.156 ÷ 1000 **58.** 7.355 ÷ 100 **59.** $20\overline{)0.545}$ **60.** $60\overline{)0.386}$

‹ F › Applications Involving Decimals

61. *Cost of filling up* What is the cost of filling a 13.5-gallon gas tank, if gasoline costs $2.61 per gallon? Answer to the nearest cent.

62. *Cost of filling up* What is the cost of filling a 14.5-gallon gas tank, if gasoline costs $3.32 per gallon?

› Web IT *go to* **mhhe.com/bello** *for more lessons*

63. *Cost of operating a central air conditioner* The cost of operating a central air conditioner (used 24 hours a day) is about $0.67 per hour. How much does it cost to

a. operate the air for 24 hours?

b. operate the air for a month (30 days)?

64. *Cost of operating a fluorescent lightbulb* The cost of operating a 22-watt fluorescent lightbulb (used 12 hours a day) is about $0.0308 per hour. To the nearest cent, how much does it cost to

a. operate the bulb for 12 hours?

b. operate the bulb for 30 days (use the result from **a**)?

Gas meters In Section 1.1, we learned how to read electric meters. Gas meters are read in the same way, and the result is in therms.

65. If you use 30 therms of gas costing $1.8 per therm, what is the total cost of your gas?

66. If you use 50 therms of gas costing $1.27 per therm, what is the total cost of your gas?

67. Suppose you use 48 therms of gas. If the first 15 therms cost $1.09 per therm and the remainder cost $1.27 per therm, what is the total gas bill?

68. Suppose you use 50 therms of gas. If the first 15 therms cost $1.10 per therm and the remainder cost $1.30 per therm, what is the total gas bill?

69. *Van renting costs* The daily cost of renting a 15-foot van from Ryder® is $69.99 plus 49 cents per mile. If you rent a van for 3 days and travel 348 miles, what is the total rental cost?

70. *Van renting costs* A 20-foot truck rents for $79.99 per day plus 49 cents per mile. If you rent a truck for 3 days and travel 257 miles, what is the total rental cost?

Downloading times (text, pictures, video) The table gives the estimated download time for text, pictures, and videos. Since Web files on the Internet are compressed, to obtain the approximate real-world download time we divide these times by 2.

71. Find the estimated download time for text using a 9600 bps (bits per second) modem.

72. Find the estimated download time for a picture using a 14,400 bps modem.

73. Find the estimated download time for text using a 28,800 bps modem.

74. Find the estimated download time for video using a 28,800 bps modem.

Modem Speed	Text (2.2 KB)	Picture (300 KB)	Video (2.4 MB)
2400 bps	7.33 sec	16.6 min	2.42 hr
9600 bps	1.83 sec	4.17 min	33.3 min
14,400 bps	1.22 sec	2.78 min	22.2 min
28,800 bps	0.61 sec	1.39 min	11.1 min

75. What is the time difference when downloading a video with a 9600 bps modem as opposed to a 14,400 bps modem?

76. What is the time difference when downloading a picture with a 14,400 bps modem as opposed to a 28,800 bps modem?

77. *Price of cologne* Chantilly cologne concentrate sells for $2.83 an ounce. Find the price of 3 ounces. Round the answer to the nearest ten cents.

78. *Pints in 7 jeroboams* The largest bottle used in the wine and spirit trade is the jeroboam, containing 8.45 pints. How many pints are there in 7 of these bottles?

79. *Paper thickness* The thickness of one sheet of paper in a telephone directory is 0.0068 centimeter. How thick is the paper in a directory with 752 sheets?

80. TV Guide *circulation* The circulation of *TV Guide* at its height averaged 19,230,000 per week. If the price per copy was $0.75, how much money (on the average) did the sales amount to each week?

81. *Fuel economy* The world record for fuel economy on a closed-circuit course belongs to Ben Visser, driving a 1959 Opel Caravan station wagon. He traveled 376.59 miles on one gallon of gas. How far could he go on 3 gallons?

82. *Distance* Lionel Harrison drove the 1900 miles from Oxford, England, to Moscow in a Morris Minor fitted with a 62-gallon tank. If he used all the gas, how many miles per gallon did he get? (Round the answer to the nearest tenth.)

83. *Cola cost* Carlos paid $1.74 for 6 bottles of cola. What was the cost per bottle?

84. *Points per game* George Blanda played 326 professional football games. He scored 1919 points. How many points per game was that? (Round to the nearest tenth.)

85. *Gaining yards* The highest average yardage gain for a season belongs to Beattie Feathers. He gained 1004 yards in 101 carries. What was his average number of yards per carry? (Round the answer to the nearest tenth of a yard.)

86. *Fat loss* To lose a pound of fat, a person must burn about 3500 calories. Skiing fast uses up about 15 calories per minute. How long do you have to ski in order to lose a pound of fat, which is about 3500 calories? (Give the answer to the nearest minute.)

In Problems 87–91 use the formula preceding Example 11 to find the payback time. Give answers to the nearest tenth of a year.

	Conservation Measure	Cost (Do It Yourself)	Savings (1st Year)
87.	Add storm windows/door (Connecticut)	\$790	\$155
88.	Insulate basement walls (Connecticut)	\$621	\$360
89.	Caulk around windows (Texas)	\$41	\$18
90.	Increase attic insulation (Texas)	\$260	\$98
91.	Insulate floors (Oregon)	\$315	\$92

92. *Recycling* Recycling one aluminum can saves the equivalent of a half a gallon of gasoline. How many gallons of gasoline are saved if you recycle seven aluminum cans?

Source: http://members.aol.com

93. *Save on those diapers* A cloth diaper washed at home costs 3¢ per use. A disposable diaper costs 22¢ per use. How much do you save per use? If a typical baby uses 10,000 diapers, how much can you save (in dollars)?

Source: http://members.aol.com

94. *Worms* One pound of red worms can consume half a pound of food waste every day. How many pounds of food waste can 15 pounds of red worms consume? If you have 9 pounds of food waste, how many pounds of red worms do you need to consume the 9 pounds?

Source: http://www.ilacsd.org

Using Your Knowledge

Anthropology The relationship between the length of various bones and height is so precise that anthropological detectives with only a dried bone as a clue can determine about how tall a person was. The chart shows this relationship for persons whose height is between 60 and 85 inches.

Height (Inches)	
Male	**Female**
(2.89 × humerus) + 27.81	(2.75 × humerus) + 28.14
(3.27 × radius) + 33.83	(3.34 × radius) + 31.98
(1.88 × femur) + 32.01	(1.95 × femur) + 28.68
(2.38 × tibia) + 30.97	(2.35 × tibia) + 22.439

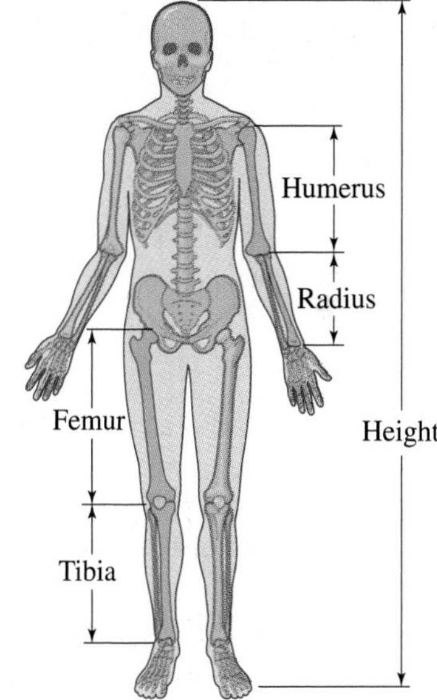

For example, suppose a 16.2-inch humerus bone from a human male is found. The man's former height is determined as follows:

$$(2.89 \times 16.2) + 27.81 = 46.82 + 27.81 = 74.63 \text{ inches}$$

(We rounded the product 2.89 × 16.2 to two decimal places.)

Anthropology Use the formulas in the chart to find the person's height. Round answers to the nearest tenth of an inch.

95. 16.2-inch tibia (male)

96. 12.5-inch tibia (female)

97. 8.25-inch radius (female)

98. 10.3-inch radius (male)

99. 18.8-inch femur (male)

100. 16-inch femur (female)

101. 15.9-inch humerus (male)

102. 12.75-inch humerus (female)

103. *Longest bone* The longest recorded human bone was the femur of the German giant Constantine. It measured 29.9 inches. Use the chart to find out his height. (He was actually 105 inches tall.) What's wrong? Read the last sentence above the chart.

104. *Height for Robert Wadlow* The femur of Robert Wadlow, the tallest man on record, measured 29.5 inches. How tall was he according to the table? (He was actually 107 inches tall.)

Write On

105. Write in your own words the procedure you use to round a decimal.

106. Write in your own words the procedure you use to change dollars to cents.

107. Write in your own words the procedure you use to change cents to dollars.

Concept Checker

Fill in the blank(s) with the correct word(s), phrase, or mathematical phrase.

108. When **multiplying** decimals, the number of decimal digits in the **product** is the _______ of the number of decimal digits in the factors.

109. To multiply a decimal number by **10, 100, 1000,** or a higher power of 10, move the decimal point as many places to the _______ as there are _______ in the power of 10 being multiplied.

110. When **rounding** a decimal number, if the first number to the _______ of the underlined place is **5 or more,** _______ one to the underlined number.

111. To **divide** a decimal number by 10, 100, 1000, or a higher power of 10, **move** the decimal point as many places to the _______ as there are zeros in the _______.

subtract	**sum**
divisor	**right**
left	**add**
zeros	

Mastery Test

112. Divide:

a. $449.6 \div 100$

b. $3.14 \div 1000$

c. $4.23 \div 10$

113. Divide: $90 \div 0.14$ (Round the answer to two decimal digits.)

114. Round 587.752 to

a. the nearest ten.

b. one decimal digit.

c. two decimal digits.

115. Divide: $\frac{0.0060}{12}$

116. Divide: $\frac{2.7}{0.045}$

117. Multiply:

a. 32.423×10

b. $48.4 \cdot 1000$

c. $(0.328)(100)$

118. Multiply:

a. 6.103×21.02 (Round the answer to two decimal digits.)

b. 3.214×0.0021 (Round the answer to two decimal digits.)

119. Multiply:

a. 4.41×3.2

b. 14.724×5.1

120. According to the Sacramento Municipal Utilities District, you can save $48 per year in electricity costs if you buy an Energy Star clothes dryer. If the dryer costs $360, how much time will it take to pay back the total cost?

Skill Checker

Find the missing number.

121. $\frac{2}{5} = \frac{?}{10}$

122. $\frac{3}{10} = \frac{?}{100}$

123. $\frac{2}{125} = \frac{?}{1000}$

3.3 Fractions and Decimals

Objectives

You should be able to:

A Write a fraction as an equivalent decimal.

B Write a terminating decimal as a fraction in reduced form.

C Write a repeating decimal as a fraction.

D Solve applications involving the concepts studied.

To Succeed, Review How To . . .

1. Write a fraction as an equivalent one with a specified denominator. (pp. 120–122)
2. Write a fraction in reduced form. (pp. 122–124)
3. Write the word name for a decimal. (p. 201)

Getting Started

The sign on the left shows the price of 1 gallon of gasoline using the fraction, $\frac{9}{10}$. However, the sign on the right shows this price as the decimal 0.9. If we are given a fraction, we can sometimes find its decimal equivalent by multiplying the numerator and denominator by a number that will cause the denominator to be a power of 10 (10, 100, 1000, etc.) and then writing the decimal equivalent. For example,

$$\frac{2}{5} = \frac{2 \cdot 2}{5 \cdot 2} = \frac{4}{10} = 0.4$$

$$\frac{3}{4} = \frac{3 \cdot 25}{4 \cdot 25} = \frac{75}{100} = 0.75$$

$$\frac{3}{125} = \frac{3 \cdot 8}{125 \cdot 8} = \frac{24}{1000} = 0.024$$

A › Writing Fractions as Decimals

These conversions can *always* be made by *dividing* the numerator by the denominator. Thus,

$$\frac{2}{5} = 2 \div 5 \quad \text{or} \quad \begin{array}{r} 0.4 \\ 5\overline{)2.0} \\ \underline{2\,0} \\ 0 \end{array}$$

That is, $\frac{2}{5} = 0.4$.

$$\frac{3}{4} = 3 \div 4 \quad \text{or} \quad \begin{array}{r} 0.75 \\ 4\overline{)3.00} \\ \underline{2\,8} \\ 20 \\ \underline{20} \\ 0 \end{array}$$

That is, $\frac{3}{4} = 0.75$.

$$\frac{3}{125} = 3 \div 125 \quad \text{or} \quad \begin{array}{r} 0.024 \\ 125\overline{)3.000} \\ \underline{2\,50} \\ 500 \\ \underline{500} \\ 0 \end{array}$$

That is, $\frac{3}{125} = 0.024$.

Here is the rule:

TO CONVERT A FRACTION TO A DECIMAL

Divide the numerator by the denominator.

EXAMPLE 1 Converting fractions to decimals

Write as a decimal.

a. $\frac{4}{5}$ **b.** $\frac{11}{40}$ **c.** $3\frac{11}{40}$

SOLUTION

a. $\frac{4}{5} = 4 \div 5$ or $5\overline{)4.0}$ quotient 0.8; $4\,0$; remainder 0. Hence $\frac{4}{5} = 0.8$.

b. $\frac{11}{40} = 11 \div 40$ or $40\overline{)11.000}$ quotient 0.275; $8\,0$, $3\,00$, $2\,80$, 200, 200, remainder 0. Hence $\frac{11}{40} = 0.275$.

c. Recall that $3\frac{11}{40} = 3 + \frac{11}{40}$

$= 3 + 0.275 = 3.275$

PROBLEM 1

Write as a decimal.

a. $\frac{3}{5}$ **b.** $\frac{3}{40}$ **c.** $5\frac{3}{40}$

In all the previous examples you obtained a **terminating decimal** for an answer; that is, when you divided the numerator by the denominator, you eventually got a remainder of 0. However, this is not always the case. For example, to write $\frac{1}{6}$ as a decimal, we proceed as before:

$\frac{1}{6} = 1 \div 6$ or $6\overline{)1.000}$ quotient 0.166; 6, 40, 36, 40, 36, remainder 4

If the division is carried further you will continue getting 6's in the quotient. The decimal equivalent for $\frac{1}{6}$ is a **repeating decimal.** The group of repeating digits is called the **repetend.** The answer can be written as

$$\frac{1}{6} = 0.1666\ldots$$

or by writing a bar called a vinculum over the repetend, like this:

$$\frac{1}{6} = 0.1\overline{6}$$

(The bar means the 6 repeats, so 6 is the repetend.)

EXAMPLE 2 Converting fractions to decimals

Write $\frac{1}{7}$ as a decimal.

PROBLEM 2

Write $\frac{2}{7}$ as a decimal.

Answers to PROBLEMS

1. a. 0.6 **b.** 0.075 **c.** 5.075

2. $0.\overline{285714}$

SOLUTION

$$\frac{1}{7} = 1 \div 7 \quad \text{or} \quad \begin{array}{r} 0.142857 \\ 7\overline{)1.0} \\ \underline{7} \\ 30 \\ \underline{28} \\ 20 \\ \underline{14} \\ 60 \\ \underline{56} \\ 40 \\ \underline{35} \\ 50 \\ \underline{49} \\ 1 \end{array}$$

Note that the remainder 1 is equal to the original dividend. This indicates that the quotient repeats itself. Thus, $\frac{1}{7} = 0.\overline{142857}$.

Note that in Example 2, we could have rounded the answer to two decimal digits. We then would obtain $\frac{1}{7} \approx 0.14$, where the sign $\approx$ means "is approximately equal to."

B › Writing Terminating Decimals as Fractions

We are now ready to convert decimals to fractions. Do you remember the word name for 0.2?

0.2 is two-tenths	Thus, $0.2 = \frac{2}{10} = \frac{1}{5}$
0.11 is eleven-hundredths	Thus, $0.11 = \frac{11}{100}$
0.150 is one hundred fifty thousandths	Thus, $0.150 = \frac{150}{1000} = \frac{3}{20}$

Note that 0.2 has 2 as numerator and 10 as denominator. Also, 0.11 has 11 as numerator and 100 as denominator, and 0.150 has 150 as numerator and 1000 as denominator. Here is the rule:

RULE FOR CONVERTING A TERMINATING DECIMAL TO A FRACTION

1. Write the digits to the *right* of the decimal point as the *numerator* of the fraction.
2. The *denominator* is a 1 *followed by as many zeros as there are decimal digits in the decimal.*
3. Reduce the fraction.

EXAMPLE 3 **Converting terminating decimals to fractions**

Write each decimal as a reduced fraction.

a. 0.025 **b.** 0.0175

SOLUTION

a. 0.025 is twenty-five thousandths. Thus

$$0.\underbrace{025}_{\text{3 digits}} = \frac{25}{1\underbrace{000}_{\text{3 zeros}}} = \frac{1}{40}$$

(continued)

PROBLEM 3

Write each decimal as a reduced fraction.

a. 0.050 **b.** 0.0350

Answers to PROBLEMS

3. a. $\frac{1}{20}$ **b.** $\frac{7}{200}$

b. 0.0175 is one hundred seventy-five ten-thousandths. Thus

$$\underbrace{0.0175}_{\text{4 digits}} = \frac{175}{\underbrace{10{,}000}_{\text{4 zeros}}} = \frac{7}{400}$$

In case the decimal has a whole number part, convert the decimal part to a fraction first and then add the whole number part. For example, to write 3.17 as a fraction, we write

$$3.17 = 3 + 0.17 = 3 + \frac{17}{100} = 3\frac{17}{100} = \frac{317}{100}$$

Note that 3.17 is three and seventeen hundredths; that is, $3.17 = 3\frac{17}{100} = \frac{317}{100}$. What do you think 8.91 is? $\frac{891}{100}$, of course. You can use this idea in Example 4.

EXAMPLE 4 **Converting terminating decimals to fractions**

Write each as a reduced fraction.

a. 2.19 **b.** 4.15

SOLUTION

a. 2.19 is two and nineteen hundredths. Thus,

$$2.19 = 2 + \frac{19}{100} = 2\frac{19}{100} = \frac{219}{100}$$

b. 4.15 is four and fifteen hundredths. Thus,

$$4.15 = 4 + \frac{\overset{3}{\cancel{15}}}{\underset{20}{\cancel{100}}} = 4 + \frac{3}{20} = 4\frac{3}{20} = \frac{83}{20}$$

PROBLEM 4

Write each as a reduced fraction.

a. 1.17 **b.** 4.35

C › Writing Repeating Decimals as Fractions

Can we write a repeating decimal as a fraction? Of course! Here are some examples; you can check them by division. See if you can find a pattern:

$$0.\overline{3} = \frac{3}{9} = \frac{1}{3}$$

$$0.\overline{61} = \frac{61}{99}$$

$$0.\overline{123} = \frac{123}{999} = \frac{41}{333}$$

The rule for making this conversion is given next. Note that this rule applies to decimals that repeat *immediately* after the decimal point—that is, *pure repeating decimals* such as $0.333\ldots = 0.\overline{3}$, $0.878787\ldots = 0.\overline{87}$—but not to decimals such as $0.1666\ldots = 0.1\overline{6}$.

RULE FOR CONVERTING A PURE REPEATING DECIMAL TO A FRACTION

1. Write the *repeating part* as the *numerator* of the fraction.
2. The *denominator* consists of *as many nines as there are digits in the repetend* (the part that repeats).

For example,

$$0.\overline{6} = \frac{6}{9} = \frac{2}{3}$$

One digit (the 6 in $0.\overline{6}$) — One nine (the 9 in the denominator)

Answers to PROBLEMS

4. a. $\frac{117}{100}$ **b.** $\frac{87}{20}$

$$0.\overline{16} = \frac{16}{99}$$

Two digits Two nines

EXAMPLE 5 Converting repeating decimals to fractions

Write each decimal as a reduced fraction.

a. $0.\overline{43}$ **b.** $0.\overline{102}$

SOLUTION

a. $0.\overline{43} = \frac{43}{99}$ **b.** $0.\overline{102} = \frac{102}{999} = \frac{34}{333}$

PROBLEM 5

Write each decimal as a reduced fraction.

a. $0.\overline{41}$ **b.** $0.\overline{105}$

D › Applications Involving Decimal Parts of Numbers

As we have shown, decimal numbers can be written as fractions and hence can represent a part of some quantity. For example, if we wish to find out what decimal part of 12 is 3, we proceed like this:

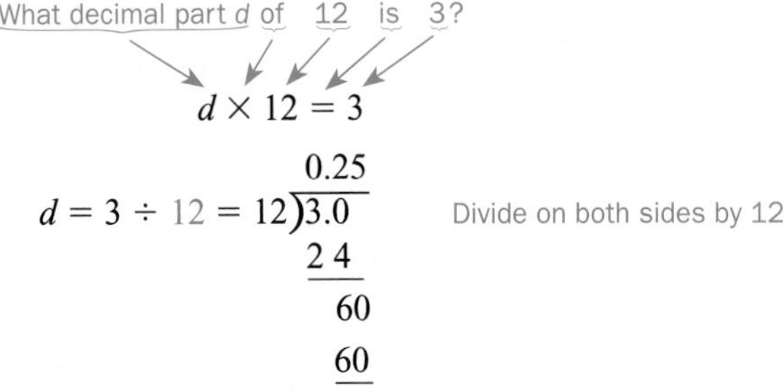

Thus, 3 is 0.25 of 12.

CHECK $0.25 \times 12 = 3$.

EXAMPLE 6 Finding the decimal part of a number

What decimal part of 16 is 5?

SOLUTION

What decimal part d of 16 is 5?

$$d \times 16 = 5$$

$$d = 5 \div 16 \quad \text{Divide on both sides by 16.}$$

$$= 16\overline{)5.0000} = 0.3125$$

$$\begin{array}{r} 4\,8 \\ \hline 20 \\ 16 \\ \hline 40 \\ 32 \\ \hline 80 \\ 80 \\ \hline 0 \end{array}$$

Thus, 5 is 0.3125 of 16, as can be easily checked, since $0.3125 \times 16 = 5$.

PROBLEM 6

What decimal part of 16 is 7?

Answers to PROBLEMS

5. a. $\frac{41}{99}$ **b.** $\frac{35}{333}$ **6.** 0.4375

Have you had breakfast at McDonald's lately? Example 7 will give you an idea of how much money is spent in doing so.

EXAMPLE 7 Finding the decimal part of a number

A restaurant brings in \$131,600 a year for breakfasts alone! If the total take is \$940,000 a year, what decimal part of the \$940,000 is for breakfasts?

SOLUTION In this problem, we want to know

What part of the \$940,000 is for breakfasts?

$$\square \times 940{,}000 = \$131{,}600$$

$$\square = \frac{131{,}600}{940{,}000} = \frac{1316}{9400} = \frac{329}{2350} = \frac{7}{50}$$

We now divide 7 by 50.

$$\begin{array}{r} 0.14 \\ 50\overline{)7.00} \\ \underline{5\,0} \\ 2\,00 \\ \underline{2\,00} \\ 0 \end{array}$$

Thus, 0.14 is the decimal part of the \$940,000 spent for breakfast. This means that \$0.14 of every dollar spent at this restaurant is for breakfast.

PROBLEM 7

A certain restaurant brings in \$128,000 a year for breakfasts. Its total sales amount to \$800,000 a year. What part of the \$800,000 is for breakfasts?

Calculator Corner

To convert a fraction to a decimal using a calculator is very simple. Thus, to write $\frac{4}{5}$ as a decimal (Example 1), we simply recall that $\frac{4}{5}$ means $4 \div 5$. Pressing  will give us the correct answer, 0.8. (Most calculators do not write the zero to the left of the decimal in the final answer.) You can also do Example 2 using division.

Exercises 3.3

A Writing Fractions as Decimals In Problems 1–10, write as a decimal.

1. $\frac{1}{2}$ **2.** $\frac{2}{5}$ **3.** $\frac{11}{16}$ **4.** $\frac{7}{40}$ **5.** $\frac{9}{20}$

6. $\frac{15}{16}$ **7.** $\frac{9}{10}$ **8.** $\frac{3}{100}$ **9.** $\frac{1}{4}$ **10.** $\frac{3}{8}$

Answers to PROBLEMS

7. 0.16

In Problems 11–20, write as a decimal (round to two decimal digits).

11. $\frac{5}{6}$ **12.** $\frac{7}{6}$ **13.** $\frac{3}{7}$ **14.** $\frac{7}{12}$ **15.** $\frac{8}{3}$

16. $\frac{11}{6}$ **17.** $\frac{1}{3}$ **18.** $\frac{2}{3}$ **19.** $\frac{2}{11}$ **20.** $\frac{12}{11}$

⟨B⟩ Writing Terminating Decimals as Fractions In Problems 21–28, write as a reduced fraction.

21. 0.8 **22.** 0.9 **23.** 0.19 **24.** 0.20 **25.** 0.030

26. 0.060 **27.** 3.10 **28.** 2.16

⟨C⟩ Writing Repeating Decimals as Fractions In Problems 29–34, write as a reduced fraction.

29. $0.\overline{5}$ **30.** $0.\overline{3}$ **31.** $0.\overline{21}$

32. $0.\overline{19}$ **33.** $0.\overline{11}$ **34.** $0.\overline{44}$

⟨D⟩ Applications Involving Decimal Parts of Numbers

Solve.

35. What decimal part of 8 is 3?

36. What decimal part of 16 is 9?

37. What decimal part of 1.5 is 37.5?

38. What decimal part of 2.3 is 36.8?

39. Find a number such that 0.25 of it is 1.2.

40. Find a number such that 0.5 of it is 1.6.

41. Find 2.5 of 14.

42. Find 0.33 of 60.

Applications

43. *Batting averages* The batting average of a baseball player is obtained by dividing the number of hits by the number of times at bat. Find the batting average of a player who has gotten 1 hit in 3 at-bats (round the answer to the nearest thousandth).

44. *Batting averages* Find the batting average of a player with 5 hits in 14 at-bats (round the answer to the nearest thousandth). (*Hint:* See Problem 43.)

45. *Football completions* On September 21, 1980, Richard Todd attempted 59 passes and completed 42. His completion average was 42 ÷ 59. Write this number as a decimal rounded to the nearest tenth.

46. *Football completions* Find the completion average for Ken Anderson, who completed 20 out of 22 passes in a game between Cincinnati and Pittsburgh on November 10, 1974. Round the answer to the nearest tenth.

Source: Football.com.

47. *Wheat cost* If 11 pounds of wheat cost $24, what is the cost per pound? (Round the answer to the nearest cent.)

48. *Rainy days in Hawaii* In Mt. Waialeale, Hawaii, $\frac{9}{10}$ of the days of the year are rainy. Write $\frac{9}{10}$ as a decimal.

49. *Sunny days in Little Rock* In Little Rock, Arkansas, $\frac{5}{8}$ of the days of the year are sunny. Write $\frac{5}{8}$ as a decimal.

50. *A headache problem* Here are some interesting statistics about headaches.

a. Migraine headaches strike $\frac{1}{8}$ of all Americans. Write $\frac{1}{8}$ as a decimal.

b. Two-thirds of the victims are women. Write $\frac{2}{3}$ as a decimal. (Round the answer to the nearest hundredth.)

c. If both parents suffer migraine headaches, three-quarters of their children will. Write $\frac{3}{4}$ as a decimal.

51. *Sleeping pill consumption* Sleeping pills are taken by $\frac{1}{18}$ of all Americans at least once a week. Write this fraction as a decimal. (Round the answer to the nearest thousandth.)

52. *False teeth* One American in six wears a full set of false teeth. Write $\frac{1}{6}$ as a decimal, rounded to the nearest hundredth.

53. *Toilet flushing* The average American home uses 60 gallons of water a day, of which 29 are spent in flushing the toilet. Write $\frac{29}{60}$ as a decimal. (Round the answer to the nearest hundredth.)

54. *Dirty words in college* The average college student throws in a dirty word for every 11 clean ones. Write $\frac{1}{11}$ as a decimal. (Round the answer to the nearest thousandth.)

Web IT go to mhhe.com/bello for more lessons

55. *Home runs by the Babe* On the average, Babe Ruth hit a home run $\frac{1}{12}$ of the times he came to bat. Write this fraction as a decimal.

56. *House buying* A family spent 2.5 times its annual income to buy a house. If the family bought a $75,000 house, what was its annual income?

57. *House buying* A family spent 2.3 times its annual income to buy a house. If the house cost $92,000, what was the family's annual income?

58. *Annual income* A family spent $125,000 to buy a house. This amount represented 2.5 times its annual income. What was the family's annual income?

59. *Annual income* A family spent $69,000 to buy a house. This amount represented 2.3 times its annual income. What was the family's annual income?

60. *Oil consumption* In a recent year, 0.25 of all the petroleum consumed worldwide on a particular day was used by the United States. The amount consumed by the United States that day amounted to 19.7 million barrels. How many barrels were consumed worldwide that day?

FICO scores Do you know what your FICO (Fair Isaac Credit Organization) score is? "When you apply for credit—whether for a credit card, a car loan, or a mortgage—lenders want to know what risk they'd take by loaning money to you. FICO scores are the credit scores most lenders use to determine your credit risk. You have three FICO scores, one for each of the three credit bureaus: Experian, TransUnion, and Equifax."

Source: http://www.myfico.com

The diagram shows the fraction that makes up each of the categories in your FICO score. In Problems 71–74, write the fraction for the specified category as a decimal.

61. Payment history

62. Amounts owed

63. Length of credit history

64. New credit

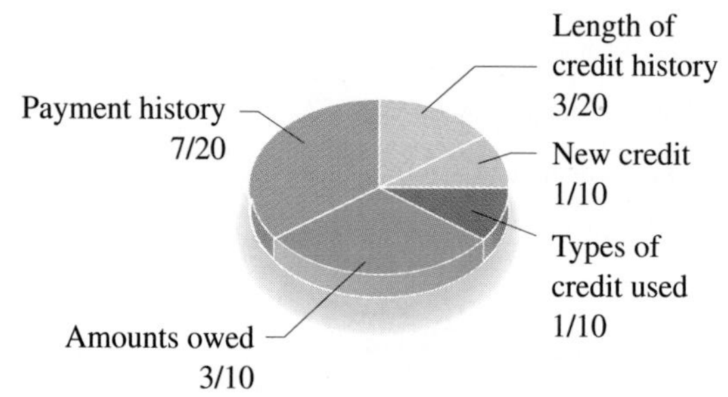

››› Using Your Knowledge

Many products (such as cereal, milk, etc.) list their nutrition information per serving. For example, Product 19 contains 3 grams of protein. If the recommended daily allowance (RDA) of protein is 70 grams, Product 19 provides $\frac{3}{70}$ of your daily protein needs. In the following problems find the fraction of the protein RDA (70 grams) provided by the given product. Then write your answer as a decimal, to two digits.

65. Special K, 4 grams per serving

66. Cornflakes, 2 grams per serving

67. Spinach (1 cup), 5 grams

68. Froot Loops, 1 gram per serving

69. 1 egg, 7 grams

Bolting In The chart shows the drill bit size (column 2) to be used to tap a screw or bolt of a certain size (column 1). If you use the closest fractional size to a $\frac{3}{64}$-inch bit (third row), the chart shows the decimal inches equivalency (column 4) is **.0469** (as you will discover, this is only an approximation).

(1)	(2)	(3)	(4)
To Tap This Size Screw or Bolt:	**Use This Drill Bit:**	**(Closest Fractional:)**	**Decimal Inches**
0-80 NF*	3/64"	3/64"	**.0469**
1-72 NF	#53	1/16"	.0595
3-48 NC*	#47	5/64"	.0785
4-48 NF	#42	3/32"	.0935

Source: http://www.korit.com/tapndrill.htm.

* NF means Natural fine thread; NC means National coarse thread.

70. Find the *exact* measurement for a $\frac{3}{64}$-inch bit written as a decimal, then round your answer to four decimal places.

71. Find the *exact* measurement for a $\frac{1}{16}$-inch bit written as a decimal, then round your answer to four decimal places.

72. Find the *exact* measurement for a $\frac{5}{64}$-inch bit written as a decimal, then round your answer to four decimal places.

73. Find the *exact* measurement for a $\frac{3}{32}$-inch bit written as a decimal, then round your answer to four decimal places.

Terminating	Repeating
$\frac{7}{8} = \frac{7}{2 \cdot 2 \cdot 2} = 0.875$	$\frac{1}{3} = \frac{1}{3} = 0.\overline{3}$
$\frac{3}{4} = \frac{3}{2 \cdot 2} = 0.75$	$\frac{2}{3} = \frac{2}{3} = 0.\overline{6}$
$\frac{5}{8} = \frac{5}{2 \cdot 2 \cdot 2} = 0.625$	$\frac{1}{6} = \frac{1}{2 \cdot 3} = 0.1\overline{6}$
$\frac{1}{2} = \frac{1}{2} = 0.5$	$\frac{1}{9} = \frac{1}{3 \cdot 3} = 0.\overline{1}$
$\frac{3}{8} = \frac{3}{2 \cdot 2 \cdot 2} = 0.375$	$\frac{1}{12} = \frac{1}{2 \cdot 2 \cdot 3} = 0.08\overline{3}$
$\frac{1}{5} = \frac{1}{5} = 0.2$	$\frac{1}{7} = \frac{1}{7} = 0.\overline{142857}$
$\frac{1}{10} = \frac{1}{2 \cdot 5} = 0.1$	$\frac{1}{11} = \frac{1}{11} = 0.\overline{09}$

74. Here are some fractions and their decimal equivalents.

Look at the denominators of the terminating fractions. Now look at the denominators of the repeating fractions. Can you make a conjecture (guess) about the denominators of the fractions that terminate?

Write On

75. Write in your own words the procedure you use to convert a fraction to a decimal.

76. Write in your own words the procedure you use to convert a decimal to a fraction. Does the same procedure apply to terminating as well as repeating fractions? What is the difference?

Concept Checker

Fill in the blank(s) with the correct word(s), phrase, or mathematical statement.

77. To convert a **fraction** to a **decimal,** divide the ______________ by the ______________.

78. The first step in converting a terminating decimal to a fraction is to write the digits to the ______________ of the decimal point as the ______________ of the fraction.

79. The first step in converting a **pure repeating decimal** to a **fraction** is to write the repeating part as the ______________ of the fraction.

80. When changing a **pure repeating decimal** to a **fraction** the ______________ consists of as many ______________ as there are digits in the ______________.

nines

left

repetend

numerator

denominator

right

Mastery Test

81. A bookstore sells \$4800 worth of mathematics books. If total sales are \$8000, what decimal part of sales are mathematics books?

82. What decimal part of 16 is 7?

83. Write as a reduced fraction:

a. 0.41 **b.** 0.303

84. Write as a reduced fraction:

a. 3.19 **b.** 2.15

85. Write as a reduced fraction:

a. 0.035 **b.** 0.0375

86. Write $\frac{3}{7}$ as a decimal

87. Write as a decimal:

a. $\frac{3}{5}$ **b.** $\frac{9}{40}$

Skill Checker

Fill in the blank with < or > so that the resulting inequality is true.

88. $\frac{3}{10}$ ______ $\frac{4}{13}$ **89.** $\frac{3}{11}$ ______ $\frac{2}{7}$ **90.** $\frac{4}{9}$ ______ $\frac{3}{7}$ **91.** $\frac{5}{7}$ ______ $\frac{5}{6}$

3.4 Decimals, Fractions, and Order of Operations

Objectives

You should be able to:

A Compare two or more decimals and determine which is larger.

B Compare fractions and decimals and determine which is larger.

C Solve applications involving fractions, decimals, and the order of operations.

To Succeed, Review How To . . .

1. Write a fraction as a decimal. (pp. 223–225)
2. Divide a decimal by a whole number. (p. 216)
3. Understand the meaning of and notation for pure repeating decimals. (p. 226)
4. Work with the order of operations. (pp. 82–84)

Getting Started

Which of the two Pepsi offers is better? To answer this question, we can find the price of each can by dividing \$2.00 by 8 and \$2.99 by 12 and then comparing the individual price of each can.

$$\begin{array}{r} 0.25 \\ 8\overline{)2.00} \\ \underline{1\,6} \\ 40 \\ \underline{40} \\ 0 \end{array} \qquad \begin{array}{r} 0.24 \\ 12\overline{)2.99} \\ \underline{2\,4} \\ 59 \\ \underline{48} \\ 11 \end{array}$$

Since 0.25 > 0.24, 12 cans for \$2.99 is a better deal.

A > Comparing Decimals

What about comparing 0.251 and 0.25? We can do this by writing 0.25 as 0.250 and noting that 0.251 > 0.250. We inserted a placeholder 0 at the end of 0.25 to make it 0.250 so that both decimals had three decimal digits. In general, to compare two decimals we use the following procedure:

TO COMPARE DECIMALS

Make sure all decimals have the same number of decimal digits (include extra placeholder 0's to the right of the decimal point if necessary). Compare corresponding digits starting at the left until two of the digits are different. The number with the larger digit is the larger of the decimals.

Thus, to compare 3.14, 3.15, and 3.1, we first write 3.1 as 3.10. We then write the three decimals as

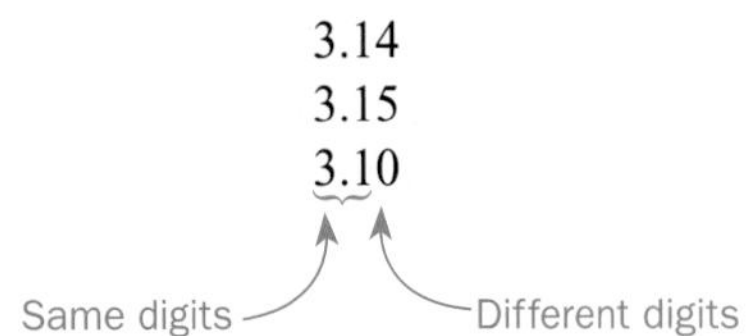

Since the 5 is the greatest digit in the hundredths place, 3.15 is the greatest decimal. Thus,

$$3.15 > 3.14 > 3.10$$

EXAMPLE 1 Comparing decimals

Arrange in order of decreasing magnitude (from largest to smallest): 5.013, 5.01, 5.004.

SOLUTION First, write a 0 at the end of 5.01 so that all three numbers have three decimal digits. Then write:

5.013
5.010
5.004

Thus

$$5.013 > 5.010 > 5.004$$

(Note that 5.013 and 5.010 both have a 1 in the hundredths place, so they are both larger than 5.004. Moreover, since 5.013 has a 3 in the thousandths place and 5.010 has a 0, 5.013 is greater.)

PROBLEM 1

Arrange in order of decreasing magnitude: 6.024, 6.02, and 6.002.

How can we compare $3.\overline{14}$, $3.1\overline{4}$, and 3.14? Since the bar indicates the part that repeats,

$$3.\overline{14} = 3.141414\ldots$$

$$3.1\overline{4} = 3.144444\ldots$$

$$3.14 = 3.140000 \quad \text{(Note added 0's)}$$

The first three columns (starting at the left) have the same numbers, 3, 1, and 4, respectively. In the fourth (thousandths) column, the largest number is the 4 in $3.1\overline{4}$. Thus, $3.1\overline{4}$ is the greatest decimal, followed by $3.\overline{14}$ and 3.14; that is,

$$3.1\overline{4} > 3.\overline{14} > 3.14$$

EXAMPLE 2 Comparing decimals

Write in order of decreasing magnitude: $3.14\overline{5}$, 3.145, and $3.1\overline{45}$.

SOLUTION We rewrite the numbers as:

$$3.14\overline{5} = 3.14555\ldots$$

$$3.145 = 3.14500$$

$$3.1\overline{45} = 3.14545\ldots$$

Starting from the left, the first four columns have the same numbers (3, 1, 4, and 5). In the fifth column, the greatest number is 5, followed by 4 and then by 0, corresponding to $3.14\overline{5}$, $3.1\overline{45}$, and 3.145, respectively. Thus,

$$3.14\overline{5} > 3.1\overline{45} > 3.145$$

PROBLEM 2

Write in order of decreasing magnitude: $2.\overline{145}$, $2.1\overline{45}$, and 2.145.

Answers to PROBLEMS

1. $6.024 > 6.02 > 6.002$

2. $2.1\overline{45} > 2.\overline{145} > 2.145$

B > Comparing Fractions and Decimals

Are you watching your intake of fats? If you eat a McDonald's hamburger weighing 100 grams, 11 of these grams will be fat; that is, $\frac{11}{100}$ of the hamburger is fat. On the other hand, 0.11009 of a Burger King hamburger is fat. Which hamburger has more fat? To compare these numbers, we write $\frac{11}{100}$ as a decimal, adding three zeros so that we can compare the result with 0.11009. We write:

$$\frac{11}{100} = 0.11000$$

and

$$0.11009 = 0.11009$$

Thus, $0.11009 > 0.11000$; that is, the Burger King hamburger has proportionately more fat.

Here is the rule:

TO COMPARE A DECIMAL AND A FRACTION

Write the fraction as a decimal and compare it with the given number.

EXAMPLE 3 Finding the water in the Coke

In a court case called *The U.S. v. Forty Barrels and Twenty Kegs of Coca-Cola,* the first analysis indicated that $\frac{3}{7}$ of the Coca-Cola was water. Another analysis showed that 0.41 of the Coca-Cola was water. Which of the two analyses indicated more water in the Coke?

SOLUTION We must compare $\frac{3}{7}$ and 0.41. If we divide 3 by 7, we obtain

$$\begin{array}{r} 0.42 \\ 7\overline{)3.00} \\ \underline{2\,8} \\ 20 \\ \underline{14} \\ 6 \end{array}$$

(We stop here because the answer has two decimal places, so we can compare 0.42 with 0.41.)

Since $\frac{3}{7} \approx 0.42 > 0.41$, the first analysis showed more water in the Coke.

PROBLEM 3

A chemical analysis indicated that $\frac{4}{7}$ of a substance was oil. A second analysis showed that 0.572 of the substance was oil. Which of the two analyses showed more oil?

EXAMPLE 4 Comparing land areas

The chart shows the fraction of the land area covered by each continent.

a. Africa covers approximately $\frac{1}{5}$ of the land area of Earth. Which is larger, $\frac{1}{5}$ or 0.203?

b. South America covers 0.089 of the land area of Earth. Which is smaller, 0.089 or $\frac{9}{100}$?

Continent	Total Land Area
The World	1.00
Asia (plus the Middle East)	0.30
Africa	0.203
North America	0.163
South America	0.089
Antarctica	$\frac{9}{100}$
Europe	$\frac{7}{100}$
Australia (plus Oceania)	$\frac{5}{100}$

Source: Enchanted Learning.

PROBLEM 4

a. North America covers approximately $\frac{4}{25}$ of the land area of Earth. Which is larger, $\frac{4}{25}$ or 0.163?

b. Europe covers 0.067 of the land area of Earth. Which is smaller, 0.067 or $\frac{7}{100}$?

Answers to PROBLEMS

3. The second analysis

4. a. 0.163 **b.** 0.067

SOLUTION

a. To compare $\frac{1}{5}$ and 0.203, we rewrite $\frac{1}{5}$ as a decimal. Dividing 1 by 5 we have:

$$\frac{1}{5} = 0.20$$

We then write $\quad \frac{1}{5} = 0.200$

and $\quad 0.203$

The first two columns after the decimal (the 2 and the 0) are the same, but in the thousandths column, we have a 3 for the 0.203.

Thus, $\quad 0.203 > \frac{1}{5}$

This means that 0.203 is larger than $\frac{1}{5}$.

b. Writing $\frac{9}{100}$ as a decimal, we obtain:

$$\frac{9}{100} = 0.090$$

and $\quad 0.089$

Thus, $\quad 0.089 < \frac{9}{100}$

which means that 0.089 is smaller than $\frac{9}{100}$.

Obesity, Cancer Linked

Obesity may be linked to one in six cancer deaths, a new study says. The study used body mass index, which measures weight against height. A BMI of 18.5 to 24.9 was considered normal, and a BMI of 30 or above was considered obese.

Percent increase in death rates from all cancers based on body mass index score

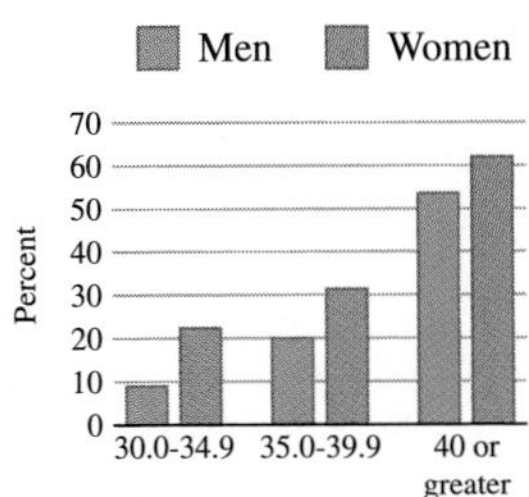

Body mass index (BMI) is equal to:

$$\frac{\text{Weight (pounds)}}{\text{Height (inches)}^2} \times 703$$

Source: Tampa Tribune, 4/24/03

C > Applications and Order of Operations

As you can see from the chart to the left, there is a relationship between obesity (being overweight) and cancer. How do we measure obesity? By using the body mass index (BMI) as defined in the margin. A BMI of 18.5 to 24.9 is normal, but if the BMI is over 30 the person is considered obese. Now, suppose a person is 74 inches tall and weighs 148 pounds. Is the person obese?

According to the formula,

$$\text{BMI} = \frac{\text{weight}}{\text{height}^2} \times 703 = \frac{148}{74^2} \times 703 = \frac{\cancel{2} \cdot \cancel{74}}{\cancel{74} \cdot \cancel{74}} \times 19 \cdot \cancel{37} = 19$$

Since 19 is between 18.5 and 24.9 the person is normal.

Note that in this case 74 was a factor of 148 in the numerator and $2 \times 37 = 74$, so the computation was simple. For more complicated calculations of the form $\frac{a}{b} \times n$ you can use one of these two methods:

Method 1. Convert $\frac{a}{b}$ to a decimal and multiply by n.

Method 2. Write n as $\frac{n}{1}$ and multiply by $\frac{a}{b}$.

Of course, you should do the calculations using the order of operations for decimals, which, fortunately, are the same rules as those used for fractions. Here they are for your convenience:

ORDER OF OPERATIONS (PEMDAS)

1. Do all calculations inside *parentheses* and other grouping symbols ( ), [ ], { } first.
2. Evaluate all *exponential* expressions.
3. Do *multiplications* and *divisions* in order from left to right.
4. Do *additions* and *subtractions* in order from left to right.

Let us use all these concepts in Example 5.

EXAMPLE 5 Finding BMI (Body Mass Index)

Find the BMI (to the nearest whole number) for a man 70 inches tall and weighing 250 pounds.

SOLUTION

$$\text{BMI} = \frac{\text{weight}}{\text{height}^2} \times 703 = \frac{250}{70^2} \times 703 = \frac{\overset{25}{\cancel{250}}}{70 \cdot \underset{7}{\cancel{70}}} \times 703$$

First, divide numerator and denominator of $\frac{250}{70 \cdot 70}$ by 10 to obtain $\frac{25}{70 \cdot 7}$. Then it is probably easier to use Method 2, multiply 25 by 703 and divide the product, 17,575, by 490. We obtain

$$\begin{array}{r} 35.8 \\ 490\overline{)17575.0} \\ \underline{1470} \\ 2875 \\ \underline{2450} \\ 4250 \\ \underline{3920} \\ 330 \end{array}$$

To the nearest whole number 35.8 is 36. Thus, the BMI is about 36. Note that according to the graph, a BMI of 35–39.9 for a man indicates a 20 percent increase in death rates for all cancers.

PROBLEM 5

Find the BMI (to the nearest whole number) for a woman 60 inches tall and weighing 180 pounds.

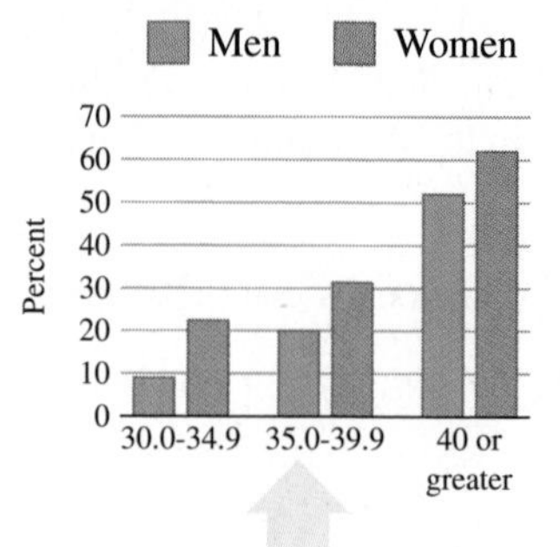

Now, a reminder: if there are several operations involved, follow the order of operations given on page 235. Let us see how.

EXAMPLE 6 Order of operations involving fractions

Simplify $\left(\frac{1}{2}\right)^3 \div \frac{1}{4} \cdot \frac{1}{2} + \frac{1}{3}\left(\frac{5}{2} - \frac{1}{2}\right)$.

SOLUTION

$\left(\frac{1}{2}\right)^3 \div \frac{1}{4} \cdot \frac{1}{2} + \frac{1}{3}\left(\frac{5}{2} - \frac{1}{2}\right)$

$= \left(\frac{1}{2}\right)^3 \div \frac{1}{4} \cdot \frac{1}{2} + \frac{1}{3}(2)$ Do operations inside parentheses: $\left(\frac{5}{2} - \frac{1}{2}\right) = \left(\frac{4}{2}\right) = (2)$

$= \frac{1}{8} \div \frac{1}{4} \cdot \frac{1}{2} + \frac{1}{3}(2)$ Do exponents: $\left(\frac{1}{2}\right)^3 = \frac{1}{8}$

$= \frac{1}{2} \cdot \frac{1}{2} + \frac{1}{3}(2)$ Do division: $\frac{1}{8} \div \frac{1}{4} = \frac{1}{8} \cdot \frac{4}{1} = \frac{1}{2}$

$= \frac{1}{4} + \frac{2}{3}$ Do multiplication: $\frac{1}{2} \cdot \frac{1}{2} = \frac{1}{4}; \frac{1}{3}(2) = \frac{2}{3}$

$= \frac{11}{12}$ Do addition: $\frac{1}{4} + \frac{2}{3} = \frac{3}{12} + \frac{8}{12} = \frac{11}{12}$

PROBLEM 6

Simplify $\left(\frac{1}{2}\right)^3 \div \frac{1}{4} \cdot \frac{1}{2} + \frac{1}{3}\left(\frac{5}{2} - \frac{3}{2}\right)$.

We will practice more with the order of operations in Problems 31–46.

Calculator Corner

As we mentioned in the Calculator Corner for Section 3.3, you can convert a fraction to a decimal by dividing the numerator by the denominator. Then you can compare the resulting decimal to a given fraction. Thus, in Example 3, the fraction $\frac{3}{7}$ can be converted to a decimal by pressing 3 ÷ 7 ENTER. The calculator will give the answer 0.428571428, which you can immediately compare with 0.41.

Answers to PROBLEMS

5. 35 **6.** $\frac{7}{12}$

> Exercises 3.4

⟨A⟩ Comparing Decimals

In Problems 1–16, arrange in order of decreasing magnitude using the > sign.

1. 66.066, 66.06, and 66.606

2. 0.301, 0.311, and 0.31

3. 0.501, 0.51, and 0.5101

4. 0.60, 0.6007, and 0.607

5. 9.099, 9.909, and 9.999

6. 2.8031, 2.8301, and 2.8013

7. 7.043, 7.403, and 7.430

8. 3.351, 3.35, and 3.6

9. $3.1\overline{4}$, $3.\overline{14}$, and 3.14

10. 2.87, $2.\overline{87}$, and $2.8\overline{7}$

11. 5.1, $5.\overline{1}$, and 5.12

12. 8.9, $8.\overline{9}$, and 8.99

13. 0.333, $0.\overline{3}$, and 0.33

14. $0.9\overline{9}$, 0.999, and 0.99

15. 0.88, $0.\overline{8}$, and $0.\overline{81}$

16. $3.\overline{7}$, 3.77, and $3.\overline{78}$

⟨B⟩ Comparing Fractions and Decimals

In Problems 17–26, determine which is greater.

17. $\frac{1}{9}$ or 0.111

18. 0.23 or $\frac{2}{9}$

19. 0.1666 or $\frac{1}{6}$

20. 0.83 or $\frac{5}{6}$

21. 0.285 or $\frac{2}{7}$

22. 0.43 or $\frac{3}{7}$

23. $0.1\overline{4}$ or $\frac{1}{7}$

24. $\frac{3}{7}$ or $0.4\overline{2}$

25. $0.\overline{9}$ or $\frac{1}{11}$

26. $\frac{1}{13}$ or $0.0\overline{7}$

27. Here are the heights of three of the tallest people in the world:

Name	Feet	Inches
Sulaiman Ali Nashnush	8	$\frac{1}{25}$
Gabriel Estavao Monjane	8	$\frac{3}{4}$
Constantine	8	0.8

Arrange them from tallest to shortest.

⟨C⟩ Applications and Order of Operations

28. *Batting champions* The batting champions for the American League from 2002 to 2005 and their averages are as follows:

Year	American League	AVG	Team
2002	Manny Ramirez	0.349	Boston
2003	Bill Mueller	0.326	Boston
2004	Ichiro Suzuki	0.372	Seattle
2005	Michael Young	0.331	Texas

In what year was the average highest? Lowest?

29. *Finding BMI* Find the BMI (to the nearest whole number) of a person 59 inches tall and weighing 150 pounds. *Hint:* $\text{BMI} = \frac{\text{weight}}{\text{height}^2} \times 703$

30. *Finding BMI* Find the BMI (to the nearest whole number) of a person 61 inches tall and weighing 100 pounds.

In Problems 31–52, use the order of operations.

31. $\frac{3}{4} \cdot 16.64$

32. $\frac{3}{5} \cdot 218.14$

33. $2\frac{3}{5} \cdot 4.65$

34. $1\frac{3}{4} \cdot 120.8$

35. $\frac{10}{9} \cdot 8.25$

36. $\frac{11}{6} \cdot 8.12$

37. $4\frac{3}{4} - 1.75$

38. $40\frac{3}{5} - 18.42$

39. $\frac{2}{7} \cdot 0.2135 + \frac{7}{4} \cdot 0.248$

40. $\frac{4}{5} \cdot 685.1 - \frac{5}{8} \cdot 98.08$

41. $\frac{3}{4} \cdot 0.92 - 0.96 \cdot \frac{3}{8}$

42. $\frac{2}{3} \cdot 0.81 - 0.48 \cdot \frac{5}{8}$

Web IT go to mhhe.com/bello for more lessons

43. $\frac{4}{3} \cdot 128.1 - 31.5 \div \frac{7}{5}$

44. $\frac{5}{6} \cdot 0.2136 - 0.105 \div \frac{3}{5}$

45. $14.05 \div \frac{5}{8} + \frac{2}{5} \cdot 15.5$

46. $28.07 \div \frac{7}{3} + \frac{5}{6} \cdot 24.06$

47. $12 \div 6 - \left(\frac{1}{3} + \frac{1}{2}\right)$

48. $18 \div 9 - \left(\frac{1}{4} + \frac{1}{6}\right)$

49. $\frac{1}{3} \cdot \frac{1}{4} \div \frac{1}{2} + \left(\frac{5}{6} - \frac{1}{2}\right)$

50. $\frac{1}{3} \cdot \frac{1}{6} \div \frac{1}{2} + \left(\frac{4}{5} - \frac{1}{2}\right)$

51. $\frac{1}{6} \div \frac{1}{3} \cdot \frac{1}{3} \cdot \frac{1}{3} + \left(\frac{1}{4} - \frac{1}{9}\right)$

52. $\frac{1}{10} \div \frac{1}{2} \cdot \frac{1}{2} \cdot \frac{1}{2} + \left(\frac{2}{3} - \frac{1}{2}\right)$

Using Your Knowledge

Decimals are used to measure several physical properties, such as weight and heat.

53. *Weight of Metals* Here are the weights of several metals in pounds per cubic inch. Arrange them from *heaviest* to *lightest.*

Brass	0.3103
Cadmium	0.3121
Copper	0.3195
Nickel	0.3175

54. *Specific Heat* The **specific heat** of a substance is the amount of heat required to raise the temperature of 1 gram of the substance 1°C. Here are the specific heats of several substances. Arrange the substances from *least* to *greatest* in terms of specific heat. (The lower the specific heat, the less heat is needed to raise the temperature of the substance a given amount.)

Boron	0.309
Sugar	0.30
Wood	0.33

Write On

55. Two common approximations for the number π are $\frac{22}{7}$ and 3.141593. How would you determine which of these two approximations is greater?

56. Fill in the blank with > or < to make the resulting statement true:

a. 3.141592 ______ π

b. 3.141593 ______ π

Write in your own words the reasons for your answers.

Concept Checker

Fill in the blank(s) with the correct word(s), phrase, or mathematical statement.

57. To compare a decimal and a fraction, write the fraction as a ______________ and compare it with the given number.

58. When using the order of operations with decimals, we first do all calculations inside __________.

59. The E in **PEMDAS** means that we do all ______________ expressions.

60. When using the order of operations, we must do them as they occur from ______________ to ______________.

right **decimal**
left **parentheses**
exponential
powers

Mastery Test

61. Find the BMI (to the nearest whole number) for a person 70 inches tall and weighing 230 pounds.

Hint: $\text{BMI} = \frac{\text{weight}}{\text{height}^2} \times 703$

62. An encyclopedia claims that $\frac{7}{10}$ of the surface of the earth is covered by oceans. Another encyclopedia says that 0.71 of the earth is covered by oceans. Which is greater, $\frac{7}{10}$ or 0.71?

63. Three-sevenths of the students in class A are women. The fraction of women in class B is 0.4. Which class has more women?

64. Write in order of *increasing* magnitude (smallest to largest):

$7.14\overline{6}$ 7.146 $7.1\overline{46}$

65. Write in order of *decreasing* magnitude (largest to smallest):

8.015 8.01 8.005

Skill Checker

Find:

66. $3.4 \cdot 8.12$

67. $4.5 \cdot 6.31$

68. $0.6 \div 0.03$

69. $3.5 \div 0.07$

70. $3.8 + 3.2 \cdot 4 \div 2$

71. $3.1(4.5 - 0.5) + 3.2 \div 2 \cdot 4$

3.5 Equations and Problem Solving

Objectives

You should be able to:

A Solve equations involving decimals.

B Solve word problems involving decimals.

To Succeed, Review How To . . .

1. Use the principles and rules used to solve equations. (pp. 90–91)
2. Use the RSTUV method to solve word problems. (p. 92)
3. Add, subtract, multiply, and divide decimals. (pp. 202–206, 212–216)

Getting Started

Gas range has 170° valve rotation for more precise flame setting

CUT $100

Low as **$654.99** white $19 MONTHLY

The ad says that $100 is cut from the price if you buy the gas range for $654.99. How much did it cost before? $100 *more*—that is, $754.99. If we let c be the old cost, we know that:

The old cost was cut by $100 to $654.99:

$$c - 100 = 654.99$$

$$c - 100 + 100 = 654.99 + 100 \quad \text{Add 100 to both sides.}$$

$$c = 754.99$$

We have illustrated the fact that you can add a number like 100 to both sides of an equation and obtain an equivalent equation.

Equations such as $c - 100 = 654.99$ can be solved using the idea of an equivalent equation. We restate the principles involved here for your convenience.

A › Solving Equations Involving Decimals

PRINCIPLES FOR SOLVING EQUATIONS

The equation $a = b$ is *equivalent* to:

$a + c = b + c$ Addition principle

$a - c = b - c$ Subtraction principle

$a \cdot c = b \cdot c$ Multiplication principle ($c \neq 0$)

$a \div c = b \div c$ Division principle ($c \neq 0$)

or $\frac{a}{c} = \frac{b}{c}$ ($c \neq 0$)

In effect, these principles tell us that to solve an equation, we may add or subtract the same number c on both sides and multiply or divide both sides by the same nonzero number c.

EXAMPLE 1 **Solve by subtracting a number**

Solve: $x + 7.2 = 9.5$.

SOLUTION Since we wish to have x by itself on the left side (it now has 7.2 added to it), we *subtract* 7.2 from both sides. Here is the procedure:

$x + 7.2 = 9.5$ Given.

$x + 7.2 - 7.2 = 9.5 - 7.2$ Subtract 7.2 from both sides.

$x = 9.5 - 7.2$ Subtract 7.2 from 9.5.

$= 2.3$

PROBLEM 1

Solve $y + 6.4 = 9.8$.

EXAMPLE 2 **Solve by adding a number**

Solve: $y - 8.2 = 3.9$.

SOLUTION This time we *add* 8.2 to both sides like this:

$y - 8.2 = 3.9$ Given.

$y - 8.2 + 8.2 = 3.9 + 8.2$ Add 8.2 to both sides.

$y = 3.9 + 8.2$ Add 3.9 and 8.2.

$= 12.1$

PROBLEM 2

Solve $x - 5.4 = 6.7$.

EXAMPLE 3 **Solve by dividing by a number**

Solve: $3.6 = 0.6z$.

SOLUTION Since z is being *multiplied* by 0.6, we must *divide* both sides by 0.6. We have:

$3.6 = 0.6z$ Given.

$3.6 \div 0.6 = 0.6z \div 0.6$ Divide both sides by 0.6.

$3.6 \div 0.6 = z$ Divide 3.6 by 0.6.

$6 = z$

$$\begin{array}{r} 6 \\ 0.6\overline{)3.6} \\ \underline{3.6} \\ 0 \end{array}$$

Thus, the solution is $6 = z$, or $z = 6$.

PROBLEM 3

Solve $6.3 = 0.9z$.

EXAMPLE 4 **Solve by multiplying by a number**

Solve: $6 = \frac{m}{3.2}$.

SOLUTION To have the m by itself on the right, we must multiply both sides by 3.2. Here is the procedure:

PROBLEM 4

Solve $5 = \frac{n}{2.4}$.

Answers to PROBLEMS

1. $y = 3.4$ **2.** $x = 12.1$

3. $z = 7$ **4.** $n = 12$

$$6 = \frac{m}{3.2} \quad \text{Given.}$$
$$3.2 \cdot 6 = \frac{m}{3.2} \cdot 3.2 \quad \text{Multiply both sides by 3.2.}$$
$$3.2 \cdot 6 = m \quad \text{Multiply 3.2 by 6.}$$
$$19.2 = m$$

B › Solving Word Problems Involving Decimals

Do you own stock? Has it gone up or down in value? The Dow Jones Industrial Average tracks the prices of 30 leading stocks and is supposed to indicate the trends for individual stocks.

EXAMPLE 5 Stock market losses

On October 19, 1987, the Dow Jones Industrial Average (DJIA) lost a record 508.32 points, to close at 1738.42. What was the DJIA the day before?

SOLUTION To solve this problem, we shall use the RSTUV method we studied in Sections 1.9 and 2.8.

1. Read the problem. Read the problem (several times if necessary).

2. Select the unknown. Select a letter (say d) to represent the unknown, the DJIA on October 18, 1987.

3. Translate the problem into an equation. Translate the problem into an equation.

The DJIA lost 508.32 points to close at 1738.42.

$$d - 508.32 = 1738.42$$

4. Use the rules we have studied to solve the equation. Use the rules studied to solve the equation.

$$d - 508.32 = 1738.42 \quad \text{Given.}$$
$$d - 508.32 + 508.32 = 1738.42 + 508.32 \quad \text{Add 508.32 to both sides.}$$
$$d = 1738.42 + 508.32 \quad \text{Add 1738.42 + 508.32.}$$
$$d = 2246.74$$

That is, the DJIA was 2246.74 the day before.

5. Verify the answer. Verify the answer. Is it true that $2246.74 - 508.32 = 1738.42$? Yes! Thus, the answer is correct. The DJIA is now over 10,000!

PROBLEM 5

On October 20, 1987, the DJIA gained 102.27 points, to close at 1840.69. What was the DJIA the day before?

EXAMPLE 6 Travel distances

You enter I-75 at exit 4 (Miami Gardens Dr.) traveling north. What is the mile marker number after you travel the specified number of miles.

a. 30 miles

b. 120 miles

c. d miles

d. If you have a flat tire at marker 256.8, how many miles m have you traveled?

PROBLEM 6

Find the marker number if you enter I-75 at exit 111 (Immokalee R.) and travel north for the specified number of miles.

a. 50 miles

b. 75 miles

c. m miles

d. If you have a flat tire at mile marker 263.4, how many miles m have you traveled?

(continued)

Answers to PROBLEMS

5. 1738.42 6. a. 161 b. 186 c. $111 + m$ d. 152.4

SOLUTION

1. Read the problem. Read the problem very carefully. There are several things you should know: Miles are marked consecutively and in ascending order as you travel north, that is, if you start at exit 4, the next marker (which may also be an exit) is 5, then 6, and so on.

2. Select the unknown. The unknown is the mile marker number.

3. Translate the problem into an equation or inequality.

a. If you start at 4 and go 30 miles, you are at marker $4 + 30 = 34$.

b. If you start at 4 and go 120 miles, you are at marker $4 + 120 = 124$.

c. If you start at 4 and go d miles, you are at marker $4 + d$. (See the pattern?)

d. If you start at 4 and travel m miles, you are at marker $4 + m$, which is given as marker 256.8.

4. Use the rules we have studied to solve the equation.

Thus,

$$\begin{aligned} 4 + m &= 256.8 \\ 4 + m - 4 &= 256.8 - 4 \quad \text{Subtract 4.} \\ m &= 252.8 \end{aligned}$$

So you have traveled 252.8 miles.

5. Verify the answer. The verification is left for you.

Sometimes you must use several of the principles to solve equations, as illustrated next.

EXAMPLE 7 Calculating bills

A customer received a cell phone bill for \$85.40 (before taxes and other charges). How many extra minutes did the customer use?

Plan Choices	Monthly Access	Monthly Airtime Allowance (in minutes)	Per Minute Rate After Allowance
America's Choice[sm] 300	\$35	300	\$0.45

SOLUTION

1. Read the problem. Read the problem carefully. Note that the bill is for \$85.40 and consists of two parts: the \$35 monthly access *plus* extra minutes after the 300-minute allowance. The minutes cost \$0.45 each.

2. Select the unknown. Let m be the number of extra minutes. Since they cost \$0.45 each, the cost for the m extra minutes is $\$0.45 \cdot m$.

3. Translate the problem into an equation or inequality. As we mentioned in step 1, the bill consists of

The monthly access *plus* the cost of the extra minutes

$$\$35 \quad + \quad 0.45 \cdot m$$

Since the bill is for \$85.40, we have the equation

$$\$35 + 0.45 \cdot m = \$85.40$$

PROBLEM 7

If the bill is for \$62.90, how many extra minutes did the customer use?

Answers to PROBLEMS

7. 62

4. Use the rules we have studied to solve the equation.

$$35 + 0.45 \cdot m - 35 = 85.40 - 35 \quad \text{Subtract 35.}$$
$$0.45m = 50.40$$
$$m = \frac{50.40}{0.45} \quad \text{Divide by 0.45.}$$
$$m = \frac{5040}{45} \quad \text{Multiply numerator and denominator by 100.}$$
$$m = 112$$

Thus, the number of extra minutes used is 112.

5. Verify the answer. The total bill is \$35 plus $\$0.45 \cdot 112$.

= \$35 + \$50.40 or \$85.40

TRANSLATE THIS

The third step in the RSTUV procedure is to **TRANSLATE** the information into an equation. In Problems 1–10 **TRANSLATE** the sentence and match the correct translation with one of the equations A–O.

1. The average cost for tuition and fees at a 2-year public college is \$2191. This represents a 5.4% increase from last year's cost L.

Source: www.CollegeBoard.com.

2. The average 2-year public college student receives grant aid (G) that reduces the average tuition and fees of \$2191 to a net price of \$400.

Source: www.CollegeBoard.com.

3. According to the Census Bureau a person with a bachelor's degree earns 62% more on average than those with only a high school diploma. If a person with a high school diploma makes h dollars a year, what is the pay b for a person with a bachelor's degree?

Source: www.CollegeBoard.com.

4. At 2-year public institutions, the average tuition and fees is \$2191, which is \$112 more than last year's cost L. What was the cost of tuition and fees last year?

Source: www.CollegeBoard.com.

5. At American University, the cost per credit for undergraduate courses is \$918. What is the cost C for an undergraduate taking h credits?

Source: American University.

- **A.** $T + 3.8 = 103.8$
- **B.** $2191 - G = 400$
- **C.** $P = 39 + 0.45m$
- **D.** $D = 10.9391P$
- **E.** $2191 = L + 5.4\%L$
- **F.** $b = h + 62\%h$
- **G.** $C = 13.75h$
- **H.** $C = 918h$
- **I.** $G - 2191 = 400$
- **J.** $h = b + 62\%h$
- **K.** $C = 51W$
- **L.** $P = 10.9391D$
- **M.** $W = 51C$
- **N.** $2191 = L + 112$
- **O.** $D = 10.9391P$

6. The monthly access fee for a cell phone plan is \$39 plus \$0.45 for each minute after a 300-minute allowance. If Joey used m minutes over 300, write an equation for the amount P Joey paid.

7. If the body temperature T of a penguin were 3.8°F warmer, it would be as warm as a goat, 103.8°F.

8. The Urban Mobility Institute reports that commuters say they would be willing to pay \$13.75 an hour to avoid traffic congestion. What would be the cost C for h hours of avoiding traffic congestion?

9. Since the price of gas is higher, commuters are willing to pay more than \$13.75 an hour to avoid traffic congestion, say W dollars per hour. South Florida travelers lose 51 hours per year due to traffic congestion. Write an equation that will give the total cost C for Florida travelers to avoid losing any time to traffic congestion.

10. As of this writing, one U.S. dollar is worth 10.9391 Mexican pesos. Write the formula to convert D dollars to P Mexican pesos.

Exercises 3.5

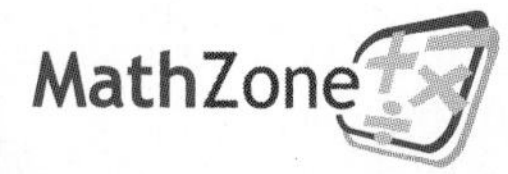

⟨A⟩ Solving Equations Involving Decimals

In Problems 1–16, solve the equation.

1. $x + 8.2 = 9.7$

2. $x + 7.3 = 9.8$

3. $y + 3.6 = 10.1$

4. $y + 4.5 = 10.2$

5. $z - 3.5 = 2.1$

6. $y - 4.9 = 11.3$

7. $z - 6.4 = 10.1$

8. $z - 3.7 = 12.3$

9. $4.2 = 0.7m$

10. $5.6 = 0.8m$

11. $0.63 = 0.9m$

12. $0.72 = 0.9m$

13. $7 = \frac{n}{3.4}$

14. $4 = \frac{n}{3.6}$

15. $3.1 = \frac{n}{4}$

16. $4.5 = \frac{n}{3}$

⟨B⟩ Solving Word Problems Involving Decimals

In Problems 17–44, use the RSTUV method to solve.

17. *Grade point averages (GPAs)* The grade point average (GPA) of a student went up by 0.32. If the student's GPA is 3.21 now, what was it before?

18. *Stock prices* The price of a stock went up by \$0.32. If the stock now costs \$7.31, what was its price before?

19. *Temperature of polar bears* If the body temperature of a polar bear were 4.7°F warmer, it would be as warm as a goat, w103.8°F. What is the body temperature of a polar bear?

20. *Temperature of arctic gulls* If the body temperature of an arctic gull were increased by 5.9°C, it would be as warm as a goat, 39.9°C. What is the body temperature of the arctic gull?

21. *Garbage* The average American produces 2.7 pounds more garbage per day now than in 1960. If 2.5 pounds of garbage were produced by each person per day in 1960, how many pounds of garbage are produced per day by each person now?

22. *Eating vegetables* The average American eats 59.6 more pounds of vegetables than fruits a year. If the average American consumes 192.1 pounds of vegetables, what is the corresponding consumption of fruit?

23. *Losing billions* In January 1980 the Hunt brothers had \$4.5 billion in silver bullion. By March the price had fallen by \$4 billion. How much was their silver worth in March? (Bunker Hunt's only comment to their staggering loss was "A billion isn't what it used to be.")

24. *Business failures in the UK* The number of business failures this year in the UK is predicted to increase by 5100 to 25,000 next year. How many business failures were there in the UK this year?

25. *Decreasing Latin America exports* In a period of seven years, U.S. exports to Latin America decreased by \$8.061 billion to \$27.969 billion. How many billions were exported to Latin America before?

26. *Composite ACT scores* Between 1998 and 2001 the composite score on the ACT decreased by 0.2 points to 20.8. What was the composite score on the ACT in 1998?

27. *Social Security deductions* To find the Social Security deduction from your wages, multiply your wages by 0.0751. If the Social Security deduction on a check amounted to \$22.53, what was the amount of the wages?

28. *FUTA tax* To find the Federal Unemployment Tax (FUTA) deduction for an employer, multiply the total of the salaries paid by the employer by 0.062. If the FUTA tax totaled \$145.08, how much was the total of the salaries?

29. *Mileage deductions* The Internal Revenue Service allows a \$0.36 deduction for each mile driven for business purposes. If a person claimed a \$368.28 deduction, how many miles had the person driven?

30. *Health insurance* The first step in figuring the health insurance deduction for self-employed individuals in 2002 is to multiply the annual amount of health insurance premiums by 0.70. If this product is \$3348.20, what was the total cost of health insurance premiums to the nearest dollar?

31. *Converting dollars to Euros* As of this writing, one Euro (the official currency of 12 European countries) was worth \$1.19 (dollars). To convert dollars to Euros, use the formula $E = \frac{D}{1.19}$. If you are in England and have \$50, how many Euros do you have?

32. *Paying for dinner in Greece* Next, you travel to Greece and your dinner costs \$142.80 (dollars before tax). How many Euros is that?

33. *Wine by the liter in France* You buy a 2-liter bottle of wine in France. If one liter is 0.95 quarts, use the formula $Q = 0.95L$ to find how many quarts of wine you have. Answer to one decimal place.

34. *Buying oil by the liter* You buy 5 liters of oil for your car. How many quarts is that? Answer to one decimal place.

35. *Operating costs and mileage traveled* It is estimated that the operating costs (gas, oil, maintenance, and tires) for a standard six-cylinder automobile are 10.8 cents per mile, that is, \$0.108. If the cost amounts to \$885.60, how many miles were traveled?

36. *Total cost of operating a car and miles traveled* It is estimated that the total cost of operating an automobile is about \$0.45 per mile. If the cost of operating an automobile amounts to \$6840, how many miles were traveled?

37. *Renting a truck in Canada* Discount Auto of Canada rents a pickup truck for \$39.99 per day plus \$0.12 per kilometer. If your bill for a one-day rental comes to \$60.63 (in dollars, before taxes), how many kilometers did you drive?

38. *Renting a car from the university* The University of Minnesota Fleet Services will rent a full-size sedan for \$40 a day plus \$0.15 per mile. If the charges for one day amount to \$74.20 (before taxes and other charges), how many miles did you drive?

39. *Partial payment for total cost at a public university* Annual tuition and fees at a public university are 0.30 of the total annual cost C. Your \$9000 partial payment consists of a \$6000 loan plus the $0.30C$. How much is the total annual cost C?

40. *Partial payment for total cost at a private university* Board charges at a private university are 0.10 of the total annual cost C. Your \$17,100 partial payment consists of a \$15,200 scholarship plus the $0.10C$. How much is the total annual cost C?

41. *Movie trivia* Quick, which movie do you think has brought in the most money (to the nearest million) at the box office?

a. If you add the domestic gross income of this top movie (T) and the domestic gross of *Star Wars* (S), you will get \$1062 million. Translate this statement into an equation.

b. If the top-grossing movie T made \$140 million more than *Star Wars* (S), translate this statement into an equation.

c. *Star Wars* actually grossed \$461 million. Find the gross for the top movie T. OK, the T is for *Titanic*.

42. *Movie Trivia* The top-grossing movie mentioned in Problem 41 is not the best-ranking movie (defined as the movie that ranked number 1 the most weekends). The number of weekend rankings at number 1 for the best-ranking movie (B) is one more than the number 1 weekend rankings T for *Titanic*.

a. Translate this statement into an equation.

b. If *Titanic* came in number 1 for 15 weekends, find B? By the way, the movie is *E.T.*

43. *Music* The worldwide sales of the two best-selling albums of all time reach \$92 million. If one album sells T millions and the other B millions,

a. Translate this statement into an equation.

b. T sold \$10 million more than B. Translate this into an equation.

c. Substitute the equation you obtained in part **b** into the equation in part **a**, and solve for B. The albums are *Thriller* and *Back in Black*.

Source: Wikepedia Disclaimer: The *world's best-selling album* cannot be listed officially, since there is no international body to count global record sales.

44. *Music* The Recording Industry Association of America says that the total number of copies of the two best-selling albums of all time reached 54 million. If one album sold E million copies and the other T million copies:

a. Translate this statement into an equation.

b. E sold 2 million more copies than T. Translate this into an equation.

c. Substitute the equation you obtained in part **b** into the equation in part **a** and solve for T. The albums are *Thriller* and the *Eagles Greatest Hits*.

Source: http://www.infoplease.com/ipea/A0151020.html.

Web IT go to mhhe.com/bello for more lessons

> > > Using Your Knowledge

The distance D (in miles) traveled in time T (in hours) by an object moving at a rate R (in miles per hour) is given by $D = R \cdot T$.

45. A car travels 137.5 miles in 2.5 hours. How fast was it traveling?

46. A car travels 227.5 miles in 3.5 hours. How fast was it traveling?

47. In 2003, Gil de Ferran won the Indianapolis 500 by posting a speed of about 156 miles per hour. How many hours did it take to cover the 500 miles? (Give the answer to the nearest tenth of an hour.)

48. In 2003 Mark Martin won the Coca-Cola 600 by posting a speed of about 138 miles per hour. How many hours did it take to cover the 600 miles? (Give the answer to three decimal digits.)

> > > Write On

49. Write in your own words the procedure you use to solve word problems.

50. Write in your own words how you decide what the unknown is in a word problem.

Concept Checker

Fill in the blank(s) with the correct word(s), phrase, or mathematical statement.

51. The **addition** principle states that the equation $a = b$ is equivalent to the equation ______. $a \div c = b \div c$

52. The **subtraction** principle states that the equation $a = b$ is equivalent to the equation ______. $a + c = b + c$

53. The **multiplication** principle states that the equation $a = b$ is equivalent to the equation ______. $a - c = b - c$

54. The **division** principle states that the equation $a = b$ is equivalent to the equation ______. $a \cdot c = b \cdot c$

Mastery Test

55. Solve: $8 = \frac{n}{2.3}$

56. Solve: $4.5 = 0.5r$

57. Solve: $x - 7.2 = 2.9$

58. Solve: $z + 5.2 = 7.4$

59. A student withdrew $304.57 from her account. If the balance (the money left) in her account is $202.59, how much money did she have in the account originally?

60. If you enter Interstate 75 at exit 49 traveling north, find the mile marker number after you travel

a. 40 miles.

b. 120 miles.

c. *m* miles.

d. If your car breaks down and you use the phone at marker 256.8, how many miles have you traveled since you entered the interstate?

Skill Checker

Perform the indicated operations.

61. $\frac{1}{7} + \frac{2}{5}$

62. $\frac{3}{8} + \frac{5}{12}$

63. $5\frac{3}{7} - 3\frac{4}{5}$

64. $6\frac{3}{8} - 5\frac{4}{7}$

Collaborative Learning

Planets	Time of Revolution Around the Sun
Mercury	88 days
Venus	224.7 days
Earth	365.25 days
Mars	687 days
Jupiter	11.86 years
Saturn	29.46 years
Uranus	84.01 years
Neptune	164.8 years
Pluto*	248.5 years

* Pluto is not classified as a planet at this time.

Weight	Multiplier	Factor
Earth weight	×	0.38
Earth weight	×	0.91
Earth weight	×	1.00
Earth weight	×	0.38
Earth weight	×	2.60
Earth weight	×	1.10
Earth weight	×	0.90
Earth weight	×	1.20
Earth weight	×	0.08

Suppose you want to become younger (in age) and trimmer (in weight). We can help you with that, but it will require a little travel. Let us start with the age issue. On Earth, a year is 365 days long but the year on other planets is dependent on the distance the planet is away from the sun. The first table shows the time it takes each of the planets to go around the sun. Now, suppose you are 18 years old. What is your Mercurian age? Since Mercury goes around the sun every 88 days, you would be much older! (You have been around more!) As a matter of fact your Mercurian age M would be $M = 18 \times \frac{365}{88} \approx 75$ years old.

In which planet will you become youngest? We will let you find out for yourself. *Hint:* The denominator has to be in days, for example, 88 days.

Now, for the weight issue. Suppose you weigh 130 pounds on Earth. How much would you weigh on Mercury? Since your weight is dependent on the laws of gravity of the planet you are visiting, your weight will change by the factor shown in the second table.

On Mercury, your weight would be: $130 \times 0.38 \approx 49$ pounds.

On which planet would your weight be the highest? You find out!

Now, divide into three groups. The first group will handle the first three planets (Mercury, Venus, Earth), the second group will take Mars, Jupiter, and Saturn and the third group will have Uranus, Neptune, and Pluto.

Answer the following questions regarding your group and planet.

1. Find your age and weight on the planets assigned to your group.
2. In which of your assigned planets are you older? Younger?
3. What is the oldest age in all groups? The youngest?
4. What is the heaviest weight in all groups? The lightest?
5. If A_p is the age on your planet, A_e your age on Earth and T_p the time of one revolution around the sun, based on your calculations, find a formula for A_p.
6. If W_p is the weight on your planet, W_e your age on Earth and F_p the factor assigned to your planet, find a formula for W_p.

See, you can be younger and trimmer if you do a little travel.

You can check some of your work at http://www.exploratorium.edu/ronh/weight/.

Research Questions

1. Where did the decimal point first appear, in what year, and who used the decimal point?
2. Write a brief description of the decimal system we use, how it works, and how decimals are used.
Reference: http://www.infoplease.com
3. Write a brief description of the Dewey decimal system, how it works, and for what it is used.
Reference: http://www.mtsu.edu
4. Write a paragraph about Bartholomeus Pitiscus and the ways in which he used the decimal point.
5. Write a paragraph explaining where our decimal system comes from and its evolution throughout the years.
Reference: http://www.scit.wlv.ac.uk

Summary Chapter 3

Section	Item	Meaning	Example
3.1A	Word names	The word name for a number is the number written in words.	The word name for 4.7 is four and seven tenths.
3.1B	Expanded form	Numeral written as a sum indicating the value of each digit.	$78.2 = 70 + 8 + \frac{2}{10}$
3.2B	Multiplying by powers of 10	A product involving 10, 100, 1000, and so on as a factor.	$93.78 \times 100 = 9378$
3.2D	Rounding decimals	Writing a decimal to a specified number of digits or places.	6.37 rounded to the nearest tenth is 6.4.
3.2E	Dividing by powers of 10.	A quotient whose divisor is 10, 100, 1000, and so on.	$386.9 \div 1000 = 0.3869$

(continued)

Section	Item	Meaning	Example
3.3B	Terminating decimal	A number whose decimal part ends.	3.5, 4.7123, and 3.1415 are terminating decimals.
3.3C	Repeating decimal	A number whose decimal part repeats indefinitely.	4.333 . . . and $92.\overline{63}$ are repeating decimals.
3.4A	Arranging in order of decreasing magnitude	Arranging numbers from largest to smallest by using the > (greater than) sign.	3.7 > 3.6 > 3.5
3.5A	Equivalent equations	Two equations are equivalent if they have the same solution.	$x + 9 = 10$ and $x + 9 - 9 = 10 - 9$ are equivalent.

Review Exercises Chapter 3

(If you need help with these exercises, look in the section indicated in brackets.)

1. ‹ **3.1A** › *Give the word name.*

a. 23.389

b. 22.34

c. 24.564

d. 27.8

e. 29.67

2. ‹ **3.1B** › *Write in expanded form.*

a. 37.4

b. 59.09

c. 145.035

d. 150.309

e. 234.003

3. ‹ **3.1C** › *Add.*

a. $8.51 + 13.43$

b. $9.6457 + 15.78$

c. $5.773 + 18.0026$

d. $6.204 + 23.248$

e. $9.24 + 14.28$

4. ‹ **3.1C** › *Add.*

a. $35.6 + 3.76$

b. $43.234 + 4.8$

c. $22.232 + 5.43$

d. $33.23 + 7.89$

e. $39.4217 + 8.34$

5. ‹ **3.1D** › *Subtract.*

a. $332.45 - 17.649$

b. $342.34 - 18.9$

c. $323.32 - 45.045$

d. $365.35 - 17.8$

e. $43.56 - 19.9$

6. ‹ **3.2A** › *Multiply.*

a. $3.14 \cdot 0.012$

b. $2.34 \cdot 0.14$

c. $3.45 \cdot 0.9615$

d. $2.345 \cdot 0.016$

e. $3.42 \cdot 0.3$

7. ‹ **3.2B** › *Multiply.*

a. $0.37 \cdot 1000$

b. $0.049 \cdot 100$

c. $0.25 \cdot 10$

d. $4.285 \cdot 1000$

e. $0.945 \cdot 1000$

8. ‹ **3.2C** › *Divide.*

a. $\frac{21.35}{0.35}$ **b.** $\frac{57.33}{0.91}$

c. $\frac{3.864}{0.042}$ **d.** $\frac{2.9052}{0.36}$

e. $\frac{3.7228}{0.041}$

9. ‹ **3.2D** › *Round to the nearest tenth.*

a. 329.67

b. 238.34

c. 887.362

d. 459.43

e. 348.344

10. ⟨**3.2D**⟩ *Divide and round answers to two decimal digits.*

a. $80 \div 15$

b. $90 \div 16$

c. $48 \div 7$

d. $84 \div 13$

e. $97 \div 14$

11. ⟨**3.2E**⟩ *Divide.*

a. $3.12 \div 100$

b. $4.18 \div 1000$

c. $32.1 \div 100$

d. $82.15 \div 10$

e. $472.3 \div 100$

12. ⟨**3.2F**⟩ *Find the price per can.*

a. 6 cans for $1.92

b. 6 cans for $2.04

c. 6 cans for $1.80

d. 6 cans for $1.68

e. 6 cans for $1.62

13. ⟨**3.3A**⟩ *Write as a decimal.*

a. $\frac{3}{5}$ **b.** $\frac{9}{10}$

c. $\frac{5}{2}$ **d.** $\frac{3}{16}$

e. $\frac{7}{8}$

14. ⟨**3.3A**⟩ *Write as a decimal.*

a. $\frac{1}{3}$ **b.** $\frac{5}{6}$

c. $\frac{2}{3}$ **d.** $\frac{2}{7}$

e. $\frac{1}{9}$

15. ⟨**3.3B**⟩ *Write as a reduced fraction.*

a. 0.38 **b.** 0.41

c. 0.6 **d.** 0.03

e. 0.333

16. ⟨**3.3B**⟩ *Write as a reduced fraction.*

a. 2.33 **b.** 3.47

c. 6.55 **d.** 1.37

e. 2.134

17. ⟨**3.3C**⟩ *Write as a reduced fraction.*

a. $0.\overline{45}$ **b.** $0.\overline{08}$

c. $0.\overline{080}$ **d.** $0.\overline{004}$

e. $0.\overline{011}$

18. ⟨**3.3D**⟩

a. What decimal part of 16 is 4?

b. What decimal part of 5 is 3?

c. What decimal part of 12 is 6?

d. What decimal part of 8 is 6?

e. What decimal part of 16 is 3?

19. ⟨**3.3D**⟩ *A person's rent was $300 per month. What decimal part went for rent if the monthly expenses were*

a. $1500

b. $1200

c. $2100

d. $1000

e. $1800

20. ⟨**3.4A**⟩ *Arrange in order of decreasing magnitude and write using the > sign.*

a. 1.032, 1.03, 1.003

b. 2.032, 2.03, 2.003

c. 3.033, 3.03, 3.032

d. 4.05, 4.052, 4.055

e. 5.003, 5.03, 5.033

21. ⟨**3.4A**⟩ *Arrange in order of decreasing magnitude and write using the > sign.*

a. $1.21\overline{6}$, 1.216, $1.2\overline{16}$

b. 2.336, $2.33\overline{6}$, $2.3\overline{36}$

c. $3.21\overline{6}$, $3.2\overline{16}$, 3.216

d. $4.5\overline{42}$, $4.54\overline{2}$, 4.542

e. $5.1\overline{23}$, 5.123, $5.12\overline{3}$

22. **⟨3.4B⟩** *Fill in the blank with $<$ or $>$ so that the resulting inequality is true.*

a. $\frac{1}{11}$ ________ 0.09

b. $\frac{2}{11}$ ________ 0.18

c. $\frac{3}{11}$ ________ 0.28

d. $\frac{4}{11}$ ________ 0.37

e. $\frac{5}{11}$ ________ 0.45

23. **⟨3.4C⟩** *Find the $BMI = \frac{weight}{height^2} \times 703$ (to the nearest whole number) for a person who is*

a. 60 inches tall and weighs 150 pounds.

b. 65 inches tall and weighs 150 pounds.

c. 65 inches tall and weighs 200 pounds.

d. 60 inches tall and weighs 200 pounds.

e. 70 inches tall and weighs 160 pounds.

24. **⟨3.5A⟩** *Solve.*

a. $x + 3.6 = 7.9$

b. $x + 4.6 = 6.9$

c. $x + 5.4 = 5.9$

d. $x + 6.3 = 9.9$

e. $x + 7.2 = 9.9$

25. **⟨3.5A⟩** *Solve.*

a. $y - 1.4 = 5.9$

b. $y - 1.5 = 6.2$

c. $y - 7.42 = 5.9$

d. $y - 4.2 = 5.8$

e. $y - 7.8 = 3.7$

26. **⟨3.5A⟩** *Solve.*

a. $4.5 = 0.9y$

b. $5.6 = 0.8y$

c. $3.6 = 0.9y$

d. $7.2 = 0.9y$

e. $4.8 = 0.6y$

27. **⟨3.5A⟩** *Solve.*

a. $6 = \frac{z}{4.1}$

b. $7 = \frac{z}{5.1}$

c. $2 = \frac{z}{5.4}$

d. $6.2 = \frac{z}{7.1}$

e. $7 = \frac{z}{8.1}$

28. **⟨3.5B⟩**

a. The price of an item decreased by \$4.28 to \$39.95. What was the price of the item before?

b. The price of an item decreased by \$3.01 to \$39.95. What was the price of the item before?

c. The price of an item decreased by \$7.73 to \$39.95. What was the price of the item before?

d. The price of an item decreased by \$6.83 to \$39.95. What was the price of the item before?

e. The price of an item decreased by \$9.55 to \$39.95. What was the price of the item before?

Practice Test Chapter 3

(Answers on page 252)

Visit www.mhhe.com/bello to view helpful videos that provide step-by-step solutions to several of the problems below.

1. Give the word name for 342.85.

2. Write 24.278 in expanded form.

3. $9 + 12.18 =$ _____

4. $46.654 + 8.69 =$ _____

5. $447.42 - 18.5 =$ _____

6. $5.34 \cdot 0.013 =$ _____

7. $0.43 \cdot 1000 =$ _____

8. $\frac{252}{0.42} =$ _____

9. Round 349.851 to the nearest tenth.

10. $70 \div 0.15 =$ _____ (Round answer to two decimal digits.)

11. $4.18 \div 1000 =$ _____

12. Carlos paid \$1.92 for 6 bottles of cola. What was the price per bottle?

13. Write $\frac{5}{8}$ as a decimal.

14. Write $\frac{1}{6}$ as a decimal.

15. Write 0.035 as a reduced fraction.

16. Write 3.41 as a reduced fraction.

17. Write $0.\overline{36}$ as a reduced fraction.

18. What decimal part of 12 is 9?

19. A person's rent is \$600 per month. If the person's total monthly expenses are \$1800, what decimal part goes for rent?

20. Arrange in order of decreasing magnitude and write using the > sign.

$$8.21\overline{6} \quad 8.216 \quad 8.\overline{216}$$

21. Fill in the blank with < or > so that the resulting inequality is true.

$$\frac{6}{7} \text{ \_\_\_\_\_ } 0.86$$

22. Find the BMI $= \frac{\text{weight}}{\text{height}^2} \times 703$ (to the nearest whole number) for a person who is 70 inches tall and weighs 175 pounds.

23. Solve $x + 3.6 = 8.9$.

24. Solve $2.7 = 0.3y$.

25. Solve $8 = \frac{z}{3.1}$.

Answers to Practice Test Chapter 3

Answer	If You Missed	Review		
	Question	Section	Examples	Page
1. Three hundred forty-two and eighty-five hundredths	1	3.1	1	201
2. $20 + 4 + \frac{2}{10} + \frac{7}{100} + \frac{8}{1000}$	2	3.1	2	202
3. 21.18	3	3.1	3	203
4. 55.344	4	3.1	4, 5	203–204
5. 428.92	5	3.1	6, 7	205
6. 0.06942	6	3.2	1, 2	212–213
7. 430	7	3.2	3	213–214
8. 600	8	3.2	6, 7	215–216
9. 349.9	9	3.2	8	217
10. 466.67	10	3.2	9	217
11. 0.00418	11	3.2	10	218
12. \$0.32 or 32¢	12	3.2	11	218
13. 0.625	13	3.3	1	224
14. $0.1\overline{6}$	14	3.3	2	224–225
15. $\frac{7}{200}$	15	3.3	3	225–226
16. $\frac{341}{100}$	16	3.3	4	226
17. $\frac{4}{11}$	17	3.3	5	227
18. 0.75	18	3.3	6	227
19. $0.\overline{3}$	19	3.3	7	228
20. $8.21\overline{6} > 8.2\overline{16} > 8.216$	20	3.4	1, 2	233
21. $<$	21	3.4	3, 4	234–235
22. 25	22	3.4	5	236
23. $x = 5.3$	23	3.5	1	240
24. $y = 9$	24	3.5	3	240
25. $z = 24.8$	25	3.5	4	240–241

Cumulative Review Chapters 1–3

1. Write 300 + 90 + 4 in standard form.

2. Write three thousand, two hundred ten in standard form.

3. Write the prime factors of 20.

4. Write 60 as a product of primes.

5. Multiply: $2^2 \times 5 \times 5^0$

6. Simplify: $49 \div 7 \cdot 7 + 8 - 5$

7. Classify $\frac{5}{4}$ as proper or improper.

8. Write $\frac{11}{2}$ as a mixed number.

9. Write $5\frac{3}{8}$ as an improper fraction.

10. $\frac{2}{3} = \frac{?}{27}$

11. $\frac{5}{7} = \frac{25}{?}$

12. Multiply: $\frac{1}{2} \cdot 5\frac{1}{6}$

13. Multiply: $\left(\frac{7}{2}\right)^2 \cdot \frac{1}{49}$

14. Divide: $\frac{45}{4} \div 2\frac{7}{9}$

15. Add: $6\frac{2}{3} + 8\frac{3}{8}$

16. Subtract: $13\frac{1}{3} - 1\frac{3}{4}$

17. Translate and solve: $\frac{6}{7}$ less than a number z is $\frac{3}{5}$. What is z?

18. Find a number such that $\frac{7}{8}$ of it is $3\frac{1}{2}$.

19. $3\frac{1}{2}$ pounds of sugar cost 49 cents. How much will 7 pounds cost?

20. Give the word name for 135.64.

21. Write 94.478 in expanded form.

22. Add: 46.654 + 9.69

23. Subtract: 241.42 − 12.5

24. Multiply: $5.98 \cdot 1.9$

25. Divide: $\frac{663}{0.39}$

26. Round 249.851 to the nearest ten.

27. Divide: $10 \div 0.13$ (Round answer to two decimal digits.)

28. Carlos paid $3.04 for 8 bottles of cola. What was the price per bottle?

29. Write $\frac{5}{6}$ as a decimal.

30. Write 0.35 as a reduced fraction

31. Write $0.\overline{78}$ as a reduced fraction.

32. What decimal part of 15 is 3?

33. Teri's rent is $200 per month. If Teri's total monthly expenses are $2000, what fraction goes for rent?

34. Arrange in order of decreasing magnitude and write using the > sign:

5.314 $5.31\overline{4}$ $5.\overline{314}$

35. Insert = , <, or > to make a true statement:

0.26 ________ $\frac{13}{20}$

36. Solve for x: $x + 2.1 = 9.4$

37. Solve for y: $3.2 = 0.4y$

38. Solve for z: $7 = \frac{z}{4.8}$

Section

Chapter

4 four

Ratio, Rate, and Proportion

The Human Side of Mathematics

Leonardo who? Not Da Vinci, but Fibonacci, a man of few faces (no portrait of his face exists) but many names. Leonardo of Pisa, Leonardo Pisano, Pisano, Leonardo Fibonacci, or simply Fibonacci were a few. Fibonacci means "of the family of Bonacci" or "son of Bonacci," but he sometimes called himself Leonardo Bigollo, "good for nothing or one of lesser importance" in the Venetian dialect or "traveler" in the Tuscan version.

His book, the *Liber Abaci* (*Book of Calculations*) published in 1202 began with the statement: "The nine Indian figures are 9 8 7 6 5 4 3 2 1. With these nine figures and with the sign of 0 any number may be written." In Chapter 12 Fibonacci stated "the rabbit problem," whose solution included a sequence of numbers now called the Fibonacci sequence. The sequence is:

$$1, 1, 2, 3, 5, 8, 13, 21, 34, 55 \ldots.$$

How do you construct such a sequence? It is easy!

*To get the **next** number, **add** the **last two.***

Thus,

$1 + 1 = 2$ (Add the first two numbers to get the third.)

$1 + 2 = 3$ (Add the 2nd and 3rd numbers to get the 4th.)

$2 + 3 = 5$ (Add the 3rd and 4th numbers to get the 5th.)

$3 + 5 = 8$ (Add the 4th and 5th numbers to get the 6th.)

and so on. Want to learn more about it? Try the research questions at the end of the chapter!

4.1 Ratio and Proportion

Objectives

You should be able to:

A Write a ratio as a reduced fraction.

B Write a proportion using the proper equation.

C Determine if two pairs of numbers are proportional.

D Solve a proportion.

E Solve applications involving the concepts studied.

To Succeed, Review How To . . .

1. Reduce a fraction to lowest terms. (pp. 122–124)
2. Solve equations. (pp. 89–91)

Getting Started

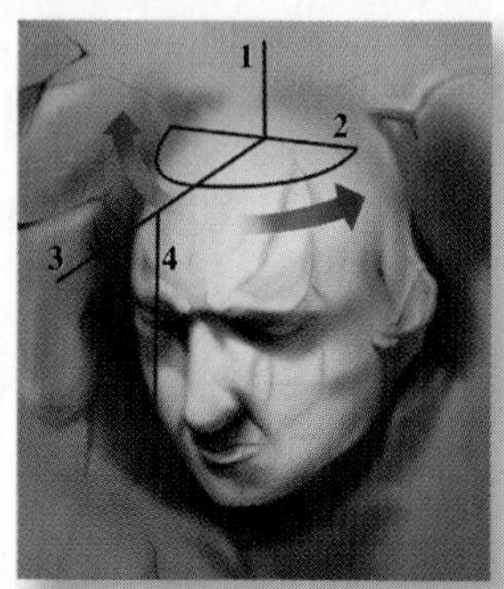

The photo shows sculptor Gutzon Burglum measuring a scale model of the presidents' busts carved on Mount Rushmore. The models were created on a 1 to 12 (1:12) inch **ratio,** that is, 1 inch on the model would equal 1 foot (12 inches) on the mountain. The transfer of the measurements was made using the "pointing machine" shown.

A Writing Ratios as Reduced Fractions

We use ratios every day. For example, the student–teacher ratio in your school may be 25 to 1 (25:1) and the ratio of men to women in your class may be 12 to 15.

RATIO

A ratio is a quotient of two numbers.

The ratio 12 to 15 may also be written as the **fraction** $\frac{12}{15}$ or $\frac{4}{5}$. When ratios are written as fractions, they are written in **reduced** form. Here are three ways to write ratios.

RATIO NOTATION

Ratios are denoted by each of the following notations:

$$a \text{ to } b \qquad a{:}b \qquad \frac{a}{b}$$

EXAMPLE 1 Expressing ratios as fractions

Express each of the following ratios as a fraction in reduced form.

a. 1 to 6 **b.** 2 to 10 **c.** 4 to 22

SOLUTION

a. The ratio 1 to 6 is written as $\frac{1}{6}$.

b. The ratio 2 to 10 is written as $\frac{2}{10} = \frac{1}{5}$.

c. The ratio 4 to 22 is written as $\frac{4}{22} = \frac{2}{11}$.

PROBLEM 1

Express each of the following ratios as a fraction in reduced form.

a. 1 to 7 **b.** 3 to 15 **c.** 4 to 18

Note that in Example 1(b), $\frac{2}{10} = \frac{1}{5}$; that is, the pair of numbers 2 and 10 has the same ratio as the pair of numbers 1 and 5.

EXAMPLE 2 Ratio of cars to people in the United States

Is traffic heavy in your city? The ratio of cars to people in the United States is 780 per 1000. Write this ratio as a fraction in reduced form.

SOLUTION

$$\frac{780}{1000} = \frac{78 \cdot \cancel{10}}{100 \cdot \cancel{10}} = \frac{78}{100} = \frac{39}{50}$$

PROBLEM 2

The ratio of cars to people in the Netherlands is 418 to 1000. Write this ratio in reduced form.

EXAMPLE 3 Ratio of circumference to diameter

The ratio of circumference (distance around) to diameter (distance across the top) of a soda can is 7.26 inches to 2.31 inches. Write this ratio as a fraction in reduced form.

SOLUTION

The ratio is

$$\frac{7.26}{2.31}$$

Since it is easier to reduce a fraction without decimals, multiply the numerator and denominator of the fraction by 100 and then reduce:

$$\frac{7.26}{2.31} = \frac{7.26 \cdot 100}{2.31 \cdot 100} = \frac{726}{231} = \frac{242}{77} = \frac{22}{7}$$

The ratio of circumference to diameter of any cylindrical can is the number π (pi), which is about 3.1416. It is also approximated by $\frac{22}{7}$.

PROBLEM 3

The ratio of circumference to diameter of a paint can is 10.78 to 3.43. Write this ratio as a fraction in reduced form.

EXAMPLE 4 Ratio of the wage gap

The wage gap is a statistical indicator often used as an index of the status of women's earnings relative to men's.

a. Women earn 73 cents for every dollar (100 cents) men make. Write the ratio of 73 to 100 as a fraction in reduced form.

b. African-American women earn 64 cents for every dollar earned by Caucasian men. Write the ratio of 64 to 100 as a fraction in reduced form.

SOLUTION

a. $\frac{73}{100}$ **b.** $\frac{64}{100} = \frac{\cancel{4} \cdot 16}{\cancel{4} \cdot 25} = \frac{16}{25}$

PROBLEM 4

a. In 1963, women earned 58 cents for every dollar earned by men. Write the ratio of 58 to 100 as a reduced fraction.

b. Hispanic women earn 52 cents for every dollar earned by white men. Write the ratio of 52 to 100 as a fraction in reduced form.

Answers to PROBLEMS

1. a. $\frac{1}{7}$ **b.** $\frac{1}{5}$ **c.** $\frac{2}{9}$ **2.** $\frac{209}{500}$ **3.** $\frac{22}{7}$ **4. a.** $\frac{29}{50}$ **b.** $\frac{13}{25}$

EXAMPLE 5 **Aspect ratio and HDTV**

The aspect ratio of the picture in a high-definition television (HDTV) is 16:9 and is defined as the ratio of the width to the height of the TV picture. A Panasonic television is $32\frac{1}{2}$ wide and $22\frac{2}{3}$ high.

a. What is the ratio of the width to the height of the set?

b. Do the dimensions follow the ratio 16:9?

SOLUTION

a. The width ($32\frac{1}{2}$) to height ($22\frac{2}{3}$) ratio is

$$\frac{32\frac{1}{2}}{22\frac{2}{3}} = \frac{\frac{65}{2}}{\frac{68}{3}} = \frac{65}{2} \cdot \frac{3}{68} = \frac{195}{136}$$

b. When written as a fraction, 16:9 is $\frac{16}{9} \approx 1.8$ but $\frac{195}{136} \approx 1.4$, so the dimensions of the television do not follow the ratio 16:9.

PROBLEM 5

A television set is $32\frac{1}{2}$ inches wide and $20\frac{2}{3}$ inches high.

a. What is the ratio of the width to the height of the set?

b. Do the dimensions follow the ratio 16:9?

B > Writing Proportions

An equation that states that two ratios are equal is called a **proportion.** Thus, a proportion can be written as

WRITING PROPORTIONS

$$\frac{a}{b} = \frac{c}{d}$$

Read as "*a* is to *b* as *c* is to *d*."

Thus, the proportion 3 is to 4 as 6 is to 8 is written as

$$\frac{3}{4} = \frac{6}{8}$$

Similarly, the proportion 2 is to 9 as x is to 3 is written as

$$\frac{2}{9} = \frac{x}{3}$$

EXAMPLE 6 **Writing proportions**

Write each of the following proportions as an equation.

a. 5 is to 10 as 1 is to 2

b. 5 is to 6 as 15 is to x

SOLUTION

a. 5 is to 10 as 1 is to 2 is written as

$$\frac{5}{10} = \frac{1}{2}$$

b. 5 is to 6 as 15 is to x is written as

$$\frac{5}{6} = \frac{15}{x}$$

PROBLEM 6

Write each of the following proportions as an equation.

a. 4 is to 12 as 1 is to 3

b. 5 is to 8 as 10 is to y

We have already mentioned that the proportion a is to b as c is to d can be written as

$$\frac{a}{b} = \frac{c}{d}$$

Answers to PROBLEMS

5. a. $\frac{195}{124}$ **b.** No

6. a. $\frac{4}{12} = \frac{1}{3}$ **b.** $\frac{5}{8} = \frac{10}{y}$

which is an example of an **equation.** If we multiply both sides of this equation by bd, we obtain

$$(bd)\frac{a}{b} = (bd)\frac{c}{d}$$

or

$$(\not{b}d)\frac{a}{\not{b}} = (b\not{d})\frac{c}{\not{d}}$$

That is,

$$ad = bc$$

We can remember this more easily by using the following rule:

THE CROSS-PRODUCT RULE

$$\text{If } \frac{a}{b} = \frac{c}{d} \quad \text{then} \quad ad = bc$$

That is, the **cross products** formed from the proportion are equal.

$$\frac{a}{b} \times \frac{c}{d} \rightarrow ad = bc$$

For example,

$$\frac{3}{4} = \frac{18}{24}$$

means

$$3 \cdot 24 = 4 \cdot 18$$

and

$$\frac{6}{7} = \frac{x}{14}$$

means

$$6 \cdot 14 = 7 \cdot x$$

This process is known as cross multiplying.

C › Determining if Numbers Are Proportional

We can use the cross product to find out if two pairs of numbers are proportional. Thus, to determine if the pair of numbers 10, 8 and 5, 4 are proportional, we write

$$\frac{10}{8} \stackrel{?}{=} \frac{5}{4} \qquad 10 \cdot 4 \stackrel{?}{=} 8 \cdot 5$$

Since the cross products are equal ($10 \cdot 4 = 40$ and $8 \cdot 5 = 40$), the two pairs are proportional, that is,

$$\frac{10}{8} = \frac{5}{4}$$

EXAMPLE 7 **Official width to length for the flag**

There is a law stating that the ratio of width to length for the American flag should be 10 to 19. One of the largest flags measures 210 by 411 feet. Are the pairs of numbers 10, 19 and 210, 411 proportional?

SOLUTION We write

$$\frac{10}{19} \stackrel{?}{=} \frac{210}{411} \qquad 10 \cdot 411 \stackrel{?}{=} 19 \cdot 210$$

But $10 \cdot 411 = 4110$ and $19 \cdot 210 = 3990$; thus, the cross products are not equal, so the two pairs are *not* proportional. The flag does not satisfy the 10 to 19 official flag ratio.

PROBLEM 7

The World's Largest Flag (Superflag) at the Hoover Dam (see photo) measures 255 feet by 505 feet. Do these measurements satisfy the 10 to 19 official flag ratio?

Answers to PROBLEMS

7. No, $10 \cdot 505 \neq 19 \cdot 255$

D › Solving Proportions

To solve proportions, we can use cross products. For example, the proportion $\frac{3}{2} = \frac{9}{x}$ can be solved as follows:

$3x = 2 \cdot 9$ Find the cross products.

$x = \frac{2 \cdot 9}{3}$ Divide both sides by 3.

$x = \frac{18}{3}$ Multiply 2 by 9.

$x = 6$ Divide 18 by 3.

You can check that 6 is the solution by replacing x with 6 in the original equation and then cross multiplying. We write

$$\frac{3}{2} = \frac{9}{6}$$

Since $3 \cdot 6 = 2 \cdot 9$ (both products are 18), the solution $x = 6$ is correct.

EXAMPLE 8 Solving proportions

Solve the proportions:

a. $\frac{6}{x} = \frac{2}{5}$ **b.** $\frac{x}{4} = \frac{9}{8}$ **c.** $\frac{5}{7} = \frac{x}{14}$

SOLUTION

a. $\frac{6}{x} = \frac{2}{5}$

$6 \cdot 5 = 2 \cdot x$ Cross multiply.

$6 \cdot 5 = 2x$

$\frac{6 \cdot 5}{2} = x$ Divide *both* sides by 2 $\left(\frac{2x}{2} = x\right)$.

$\frac{30}{2} = x$ Multiply 6 by 5.

$15 = x$ Divide 30 by 2 $\left(\frac{30}{2} = 15\right)$.

The solution is $x = 15$.

b. $\frac{x}{4} = \frac{9}{8}$

$8x = 4 \cdot 9$ Cross multiply.

$x = \frac{4 \cdot 9}{8}$ Divide *both* sides by 8 $\left(\frac{8x}{8} = x\right)$.

$x = \frac{36}{8}$ Multiply 4 by 9.

$x = \frac{9}{2}$ Simplify by dividing the numerator and denominator by 4.

The solution is $x = \frac{9}{2}$.

c. $\frac{5}{7} = \frac{x}{14}$

$5 \cdot 14 = 7x$ Cross multiply.

$\frac{5 \cdot 14}{7} = x$ Divide both sides by 7 $\left(\frac{7x}{7} = x\right)$.

$\frac{70}{7} = x$ Multiply 5 by 14.

$10 = x$ Divide 70 by 7 $\left(\frac{70}{7} = 10\right)$.

The solution is $x = 10$.

PROBLEM 8

Solve the proportions:

a. $\frac{8}{x} = \frac{4}{5}$

b. $\frac{x}{6} = \frac{7}{8}$

c. $\frac{6}{7} = \frac{x}{14}$

Answers to PROBLEMS

8. a. $x = 10$ **b.** $x = \frac{21}{4}$ **c.** $x = 12$

E > Applications Involving Proportions

EXAMPLE 9 **Solving proportions**

A worker in an assembly line takes 3 hours to produce 28 parts. At that rate how many parts can she produce in 9 hours?

SOLUTION The rate of production is 28 parts per 3 hours, that is, $\frac{28}{3}$. If she produced x parts in 9 hours, the rate would be $\frac{x}{9}$. Since we wish to maintain the rate $\frac{28}{3}$,

$$\frac{x}{9} = \frac{28}{3} \begin{matrix}\leftarrow \text{Parts} \\ \leftarrow \text{Hours}\end{matrix}$$

$$3x = 9 \cdot 28$$

$$3x = 252$$

$$x = \frac{252}{3} = 84$$

Thus, the worker should produce 84 parts in 9 hours.

PROBLEM 9

A worker in a factory produces 32 parts in 5 hours. At that rate how many parts can he produce in 20 hours?

An important application of ratio and proportion is the creation of a scale drawing or model, a representation of an object that is to be drawn or built to actual size. The scale gives the relationship between the measurements of the drawing or model and the measurements of the real object. For example, Mini Me is Dr. Evil's clone and is said to be $\frac{1}{8}$ of Dr. Evil's size, so the scale is

$$1 \text{ inch} = 8 \text{ inches}$$

How tall is Dr. Evil? We shall find out next.

EXAMPLE 10 **Solving proportions**

If Mini Me is 32 inches tall and $\frac{1}{8}$ of Dr. Evil's size, how tall is Dr. Evil?

SOLUTION Since the scale is 1 inch (Mini Me) = 8 inches (Dr. Evil), and Mini Me is 32 inches tall, we write the proportion

$$\frac{\text{Mini Me}}{\text{Dr. Evil}} = \frac{1 \text{ inch}}{32 \text{ inches}} = \frac{8 \text{ inches}}{E}$$

$$1 \cdot E = 32 \cdot 8 \quad \text{Cross multiply.}$$

$$E = 256 \quad \text{Multiply 32 by 8.}$$

Thus, Dr. Evil has to be 256 inches tall (21 ft $\frac{1}{3}$ ft!), which is clearly impossible.

PROBLEM 10

If Mini Me is 32 inches tall and $\frac{3}{7}$ of Dr. Evil's size, how tall is Dr. Evil?

> Exercises 4.1

< A > Writing Ratios as Reduced Fractions In Problems 1–10, write the given ratio as a fraction in reduced form.

1. 3 to 8 **2.** 4 to 17 **3.** 5 to 35 **4.** 8 to 64 **5.** 32 to 4

6. 40 to 8 **7.** 11 to 3 **8.** 13 to 9 **9.** 0.5 to 0.15 **10.** 0.6 to 0.24

Answers to PROBLEMS

9. 128 **10.** $74\frac{2}{3}$ in.

Web IT go to mhhe.com/bello for more lessons

⟨ B ⟩ Writing Proportions

In Problems 11–18, write each proportion as an equation.

11. 1 is to 4 as 5 is to 20

12. 5 is to 16 as 10 is to 32

13. a is to 3 as b is to 7

14. 7 is to a as 8 is to b

15. a is to 6 as b is to 18

16. a is to b as 6 is to 8

17. 3 is to a as 12 is to b

18. 9 is to 18 as a is to b

⟨ C ⟩ Determining if Numbers are Proportional

In Problems 19–28, determine if the pairs of numbers are proportional.

19. 3, 4 and 5, 6

20. 6, 12 and 4, 8

21. 6, 9 and 8, 12

22. 7, 3 and 9, 4

23. 0.3, 5 and 3, 50

24. 0.9, 7 and 4, 30

25. 6, 1.5 and 8, 2

26. 3, 1.5 and 4, 0.2

27. 3, 1.2 and 5, 0.2

28. 5, 2.5 and 6, 3

⟨ D ⟩ Solving Proportions

In Problems 29–50, solve for x in the given proportion.

29. 3 is to 4 as 6 is to x

30. 5 is to 8 as 10 is to x

31. 9 is to 10 as 18 is to x

32. 5 is to 7 as x is to 21

33. 8 is to 5 as x is to 30

34. 4 is to 3 as x is to 12

35. 12 is to x as 4 is to 5

36. 18 is to x as 6 is to 5

37. 20 is to x as 4 is to 5

38. x is to 20 as 9 is to 10

39. x is to 21 as 2 is to 3

40. x is to 22 as 4 is to 2

41. $\frac{x}{16} = \frac{3}{4}$

42. $\frac{x}{21} = \frac{5}{3}$

43. $\frac{15}{x} = \frac{5}{3}$

44. $\frac{14}{x} = \frac{7}{6}$

45. $\frac{8}{9} = \frac{x}{16}$

46. $\frac{7}{2} = \frac{x}{4}$

47. $\frac{14}{22} = \frac{7}{x}$

48. $\frac{18}{30} = \frac{6}{x}$

49. $\frac{3.5}{4} = \frac{x}{7}$

50. $\frac{4.5}{6} = \frac{x}{8}$

In Problems 51–60, write a proportion that could be used to solve for the variable, then solve.

51. 2 pairs of Bill Blass shorts for \$26
8 pairs of Bill Blass shorts for \$$d$

52. 2 pairs of Body Code shorts for \$25
5 pairs of Body Code shorts for \$$d$

53. 5 Power Bar Energy Bites for \$4
8 Power Bar Energy Bites for \$$x$

54. 12-pack soda product for \$1.98
8-pack soda product for \$$x$

55. \$4.38 for a 4-pound roast
\$$y$ for a 6-pound roast

56. \$18 for 2 pounds of grouper fillet
\$$y$ for 3 pounds of grouper fillet

57. \$9 for 3 pounds of top round steak
\$$z$ for 5 pounds of top round steak

58. \$3.60 for 5 dozen Bic Round Stick pens
\$2.16 for x dozen Bic Round Stick pens

59. \$6.95 for a 5-pack of Sanford Uniball pens
\$5.56 for an x-pack of Sanford Uniball pens

60. \$2.88 for a 6-pack of Staples Outflow pens
\$1.92 for an x-pack of Staples Outflow pens

⟨ E ⟩ Applications Involving Proportions

61. *Ratio of cars to people in Bermuda* The ratio of cars to people in Bermuda is 340 per 1000. Write this ratio as a fraction in reduced form.

62. *Ratio of cars to people in Haiti* The ratio of cars to people in Haiti is 4 per 1000. Write this ratio as a fraction in reduced form.

63. *Business failures* In a recent year, the failure rate for businesses in the United States was 114 per 10,000. Write this ratio as a fraction in reduced form.

64. *Business failures during the Depression* The highest failure rate for businesses in the United States occurred in 1932 during the Depression. The ratio was 154 per 10,000. Write this ratio as a fraction in reduced form.

65. *Dollar bill dimensions* The dimensions of a dollar bill are 6.15 inches long by 2.61 inches wide. Write the ratio 6.15 to 2.61 as a fraction in reduced form.

66. *German emergency notes* The smallest German emergency notes were issued in 1920 and measured 0.70 by 0.72 inch. Write the ratio 0.70 to 0.72 as a fraction in reduced form.

67. *Pupil–teacher ratio in Sweden* The lowest pupil–teacher ratio in the world is that of Sweden, 18 to 1. If a school in Sweden has 900 students, how many teachers does it have?

68. *Pupil–teacher ratio in Upper Volta* The highest pupil–teacher ratio in the world is that of the Upper Volta, about 600 to 1. If a school in the Upper Volta has 1800 students, how many teachers does it have?

69. *Property tax rate* The property tax in a certain state is $9 for every $1000 of assessed valuation (9 mills). What is the tax on a property assessed at $40,000?

70. *Property taxes* If the property tax rate is 12 mills ($12 per thousand), what is the tax on a property assessed at $45,000?

71. *Scale airplane models* A plastic model of a 747 airplane is constructed to a scale of 1 to 144. If the wingspan of the model is 40 centimeters, how long is the actual wingspan on a 747 airplane? Note that 100 centimeters equal 1 meter.

72. *Airplane models* If the length of the model airplane mentioned in Problem 71 is 48 centimeters, what is the actual length of a 747 airplane?

73. *Ratio of declining to advancing stocks* On a certain day the ratio of stocks declining in price to those advancing in price was 5 to 2. If 300,000 stocks declined in price, how many advanced in price?

74. *Stocks* On a certain day the ratio of stocks advancing in price to those declining in price was 4 to 3. If 150,000 stocks advanced in price, how many declined in price?

This graph will be used in Problems 75–76.

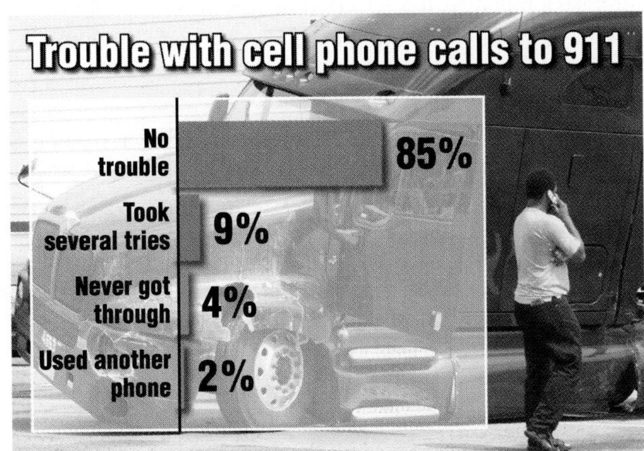

Source: Data from Anne R. Carey and Keith Carter, USA TODAY.

75. *911 calls* As you can see from the graph, 85 out of 100 calls to 911 had no trouble.

a. Write the ratio 85 out of 100 as a reduced fraction.

b. On a certain day, 500 calls were made to 911. How many calls to 911 would you expect to have no trouble? (See the graph.)

c. On a certain day, 102 callers reported some trouble calling 911. How many total calls were made to 911?

76. *911 calls* As you can see from the graph, 9 out of 100 calls to 911 took several tries.

a. Write the ratio 9 out of 100 as a fraction.

b. On a certain day, 300 calls were made to 911. How many calls to 911 would you expect to have taken several tries?

c. On a certain day, 72 callers reported they had to make several tries to reach 911 (see the graph). How many total calls were made to 911?

This graph will be used in Problem 77.

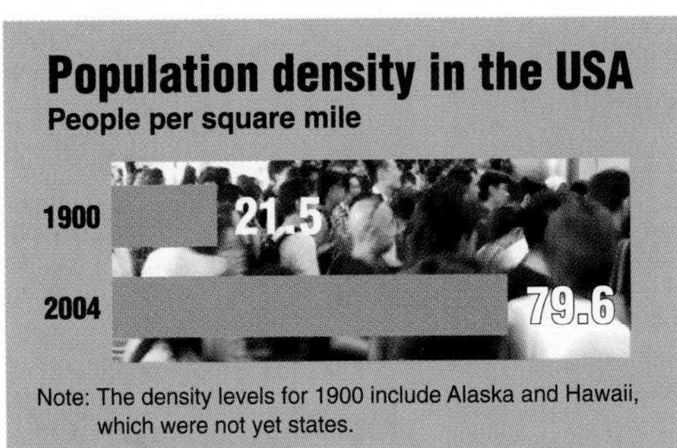

Source: Data from Anne R. Carey and Keith Carter, USA TODAY.

77. *Population density of Lake Wobegon* In 2004, the population density of the United States reached 80 people per square mile (see graph). How many people would you expect around Lake Wobegon, Minnesota, an area of 50 square miles?

78. *Atlantis population density* According to Plato's description, the population of the lost city of Atlantis was between 6 and 10 million people. If the area of Atlantis was 86,400 square kilometers and the population is assumed to have been 8,640,000 inhabitants, what was the population density for Atlantis?

Sources: atlantishistory.com; Eden—The Andrew Collins website.

79. *Population density in Macau* One of the most densely populated cities is Macau (a special administrative region of the People's Republic of China) with about 446,250 inhabitants living on 21 square kilometers of land. What is the population density per square kilometer in Macau?

80. *Population density in Monaco* Another densely populated city is Monaco with 31,980 inhabitants living on 1.95 square kilometers of land. What is the population density per square kilometer in Monaco?

81. *Aspect ratio (width to height)* The aspect ratio of a computer screen is 4:3.

a. Write 4:3 as a fraction.

b. If the screen is 32 inches wide, what is the screen's height?

82. *Aspect ratio (width to height) of HDTV* The aspect ratio of a television set is 16:9.

a. Write 16:9 as a fraction.

b. If the screen is 27 inches high, what is the screen's width?

Web IT go to mhhe.com/bello for more lessons

83. *Challenge Index* The Challenge Index is the total number of Advanced Placement (AP) or International Baccalaureate (IB) tests given at a school in May, divided by the number of seniors graduating in June.

a. Find the Challenge Index written as a reduced fraction for New Trier High in Winnetka, Illinois, which gave 1918 AP tests and graduated 970 seniors.

b. Write the answer to part **a** as a decimal rounded to three places.

84. *Challenge Index* The Challenge Index is the total number of AP or IB tests given at a school in May, divided by the number of seniors graduating in June.

a. Find the Challenge Index written as a reduced fraction for Ottawa Hills High in Toledo, Ohio, which gave only 154 AP tests but also graduated only 78 seniors.

b. Write the answer to part **a** as a decimal rounded to three places.

Source: For Problems 83 and 84: http://www.msnbc.msn.com.

››› Using Your Knowledge

Ratios are used in business to measure **liquidity** (the ability to pay short-term debts), **profit,** and **solvency** (ability to pay debts). Here is the definition of some of these ratios:

$$\text{current ratio} = \frac{\text{current assets}}{\text{current liabilities}}$$

$$\text{net profit margin} = \frac{\text{net profits}}{\text{net sales}}$$

$$\text{debt-to-assets ratio} = \frac{\text{total liabilities}}{\text{total assets}}$$

Suppose a business has $15,000 in current assets, $8000 in current liabilities, $60,000 in total assets, and $60,000 in total liabilities.

85. Write the current ratio for the business as a fraction in reduced form.

86. Write the debt-to-assets ratio for the business as a fraction in reduced form.

87. The net sales of a business amounted to $50,000. If net profits were $8000, what was the net profit margin?

››› Write On

88. Explain in your own words the difference between a ratio and a proportion.

89. Explain in your own words the procedure you use to solve a proportion.

90. Explain in your own words how proportions could be used in using a recipe to cook a favorite dish.

››› Concept Checker

Fill in the blank(s) with the correct word(s), phrase, or mathematical statement.

91. A **ratio** is a __________ of two numbers.

92. The **ratio** of a to b can be written **three** ways: __________, __________ and __________.

93. An **equation** stating that two ratios are **equal** is called a __________.

94. If $\frac{a}{b} = \frac{c}{d}$ then $ad = bc$ is called the __________ rule.

$\frac{a}{b}$	$a:b$
$b:a$	cross product
ratio	$\frac{b}{a}$
a to b	b to a
proportion	quotient

››› Mastery Test

95. The ratio of registered vehicles to people in Madison, Wisconsin, is 170,000 to 210,000. Write this ratio as a fraction in reduced form.

96. The ratio of bicycles to registered vehicles in Madison, Wisconsin, is 3 to 2. If there are 170,000 registered vehicles in Madison, how many bicycles do they have?

97. Write each proportion as an equation:

a. 2 is to 5 as 4 is to 10

b. 5 is to 8 as 15 is to x

98. Express as a fraction in reduced form:

a. 3 to 6

b. 8 to 44

c. 1 to 7

99. The ratio of width to length for an American flag is 10 to 19. A flag measuring 50 by 95 is made. Does this flag satisfy the 10 to 19 ratio?

100. Solve the proportion:

a. $\frac{6}{x} = \frac{3}{2}$

b. $\frac{3}{x} = \frac{5}{6}$

c. $\frac{5}{7} = \frac{x}{14}$

101. A worker can produce 10 parts in 3 hours. At that rate, how long would it take this worker to produce 30 parts?

102. A statue of Professor Math is 30 inches tall. If the statue is $\frac{3}{7}$ as tall as Professor Math, how tall is Professor Math?

Skill Checker

Reduce to lowest terms.

103. $\frac{350}{1000}$

104. $\frac{3.5}{100}$

105. $\frac{4.8}{3.2}$

106. $\frac{6.4}{3.2}$

4.2 Rates

Objectives

You should be able to:

A Write a rate as a reduced fraction with the correct units.

B Write unit rates and use them to compare prices.

To Succeed, Review How To . . .

1. Round numbers. (p. 216)
2. Reduce a fraction to lowest terms. (pp. 122–124)
3. Write a fraction as a decimal. (pp. 223–225)

Getting Started

The table gives the *rates* for printing a classified ad. In the preceding section, we used *ratios* to compare *like quantities* such as circumference to diameter (both in inches) or width to length (both in feet). When a ratio is used to compare *unlike quantities,* such as cost per day, or miles per gallon, we call it a **rate.**

Local Noncontract Classified Rates (per-day rates)

	Noncommissionable	
Days	**3-Line Min. Line Rate**	**4-InchMin. Inch Rate**
Sunday only	$5.09	$71.22
1 day	$4.49	$62.86
2–5 days	$3.69	$51.74
6–10 days	$3.48	$48.78
11–15 days	$3.30	$46.14
16+ days	$3.09	$43.26

Source: Newspaper Agency Corporation.

A › Writing Rates

RATES

A ***rate*** is a ratio used to compare two ***different*** kinds of measures.

According to the chart in the *Getting Started,* suppose you wish to run an ad one time on Sunday (first line in the table). The cost will be $5.09 per line. This *rate* is written as $\frac{\$5.09}{\text{line}}$. (Note that *per* indicates division.) Similarly, if your car travels 200 miles on 8 gallons of gas, the miles-per-gallon rate can be written: $\frac{200 \text{ miles}}{8 \text{ gallons}}$, or 25 $\frac{\text{miles}}{\text{gallon}}$, or 25 miles per gallon, or 25 mi/gal, or 25 mpg.

EXAMPLE 1 Writing a rate

One of the highest rates ever offered a writer was $30,000 to Ernest Hemingway for a 2000-word article on bullfighting. Find the rate per word. (Per-word rates for magazine writers are around $1.60/word!) (The article was published by *Sports Illustrated*.)

SOLUTION We want to find how many dollars per word, so we divide:

$$\frac{30{,}000 \text{ dollars}}{2000 \text{ words}} = \$15 \text{ per word}$$

PROBLEM 1

John Creasey published 564 books over a 41-year period (1932–1973). To the nearest whole number, how many books per year is that?

EXAMPLE 2 Writing a rate

What is the miles-per-gallon rate for your car? Two automotive engineers named Craig Henderson and Bill Green traveled 1759 miles on 17 gallons of gas (from Dodger Stadium to Vancouver, B.C.). What was their miles-per-gallon rate? (Round the answer to the nearest whole number.)

SOLUTION To find miles per gallon, we divide 1759 by 17.

```
     103.4
17)1759.0
   17
    059
     51
      80
      68
      12
```

Thus, their rate was about 103 miles per gallon (rounded from 103.4).

PROBLEM 2

Lionel Harrison and E. A. Ferguson drove 1900 miles from London to Moscow on 62 gallons of gas. What was their mile-per-gallon rate? (Round the answer to the nearest whole number.)

EXAMPLE 3 Writing a rate

An 18-pound bag of lawn food covers 5000 square feet of lawn. To the nearest whole number, what is the rate of coverage in square feet per pound?

SOLUTION Here we want square feet per pound, so we divide 5000 by 18 (remember *per* means divide). We have

```
     277.7
18)5000.0
   36
   140
   126
    140
    126
     140
     126
      14
```

Thus, the rate of coverage is about 278 square feet per pound $\left(\frac{278 \text{ square feet}}{1 \text{ pound}}\right)$.

PROBLEM 3

An 18-pound bag of lawn food costs $6. To the nearest cent, what is the cost per pound?

EXAMPLE 4 Writing a rate

Natasha Bello is a Pharm. D intern at a drugstore. She earned $861 the first two weeks in the summer.

a. What was her rate of pay per week, that is, how much did she make each week?

b. If she worked a total of 82 hours, what was her hourly rate of pay?

PROBLEM 4

Randy works in the mathematics department and earned $612 the first two weeks in the summer.

a. How much did he make each week?

b. If he worked a total of 72 hours, what was his hourly rate of pay?

Answers to PROBLEMS

1. 14 **2.** 31 **3.** 33¢ **4. a.** $306 per week **b.** $8.50 per hour

SOLUTION

a. The rate of pay *per week* is the ratio of the money earned ($861) divided by the length of time she worked (2 weeks).

$$\frac{\$861}{2 \text{ weeks}} = 430.50 \frac{\text{dollars}}{\text{week}} \quad \text{or} \quad \$430.50 \text{ per week}$$

b. The rate of pay *per hour* is the ratio of the money earned ($861) divided by the number of hours she worked (82 hours).

$$\frac{\$861}{82 \text{ hours}} = 10.50 \frac{\text{dollars}}{\text{hour}} \quad \text{or} \quad \$10.50 \text{ per hour}$$

$$\begin{array}{r} 10.5 \\ 82\overline{)861.0} \\ \underline{82} \\ 41\ 0 \\ \underline{41\ 0} \\ 0 \end{array}$$

B › Unit Rates

Most supermarkets help you compare prices by posting the **unit price** for the items.

UNIT RATE

A **unit rate** is a rate in which the denominator is 1.

Thus, if a 5-pound bag of potatoes costs 90¢, the cost per pound is

$$\frac{90¢}{5 \text{ pounds}} = \frac{18¢}{1 \text{ pound}}$$

and the unit price is 18¢ per pound. Note that per pound indicates per *one* pound.

EXAMPLE 5 Calculating the unit price

A 30-ounce jar of gourmet popping corn costs $3.99. What is the unit price in cents per ounce?

SOLUTION

$$\frac{\$3.99}{30 \text{ ounces}} = \frac{399¢}{30 \text{ ounces}}$$

$$\begin{array}{r} 13.3 \\ 30\overline{)399.0} \\ \underline{30} \\ 99 \\ \underline{90} \\ 9\ 0 \\ \underline{9\ 0} \\ 0 \end{array}$$

Thus, the unit price is $\frac{13.3¢}{1 \text{ ounce}}$ = 13.3¢ per ounce.

PROBLEM 5

A 10-ounce jar of microwave popping corn costs $1.89. What is the unit price in cents per ounce?

Answers to PROBLEMS

5. 18.9¢

Unit prices will help you to find the best buy when comparing prices. Thus, if the $1\frac{1}{2}$-pound container of brand X gourmet popcorn is selling for \$1.80, is it a better buy than the popcorn of Example 5? To be able to compare prices, we must find the total number of ounces in brand X. Since 1 pound = 16 ounces, $1\frac{1}{2}$ pounds = 16 ounces + 8 ounces = 24 ounces. The price per ounce of brand X is

$$\frac{180¢}{24 \text{ ounces}} = \frac{7.5¢}{1 \text{ ounce}} = 7.5¢ \text{ per ounce}$$

Thus, brand X is a better buy.

EXAMPLE 6 Calculating the unit price

A 12-ounce bottle of gourmet popcorn oil sells for \$1.99. The 24-ounce brand X bottle costs \$3.99. Which is the better buy?

You can reason like this:

Gourmet : 12 ounces sell for \$1.99.

Brand X : 24 ounces (*twice* as much) should cost *twice* as much, or \$2 · 1.99 = \$3.98. But brand X sells for \$3.99 (too much); the gourmet popcorn is a better buy. But can you prove it? Let's see.

SOLUTION We compare unit prices to the nearest hundredth of a cent.

$$\text{Gourmet: } \frac{199¢}{12 \text{ ounces}}$$

```
    16.58
12)199.00
   12
    79
    72
     7 0
     6 0
       1 00
         96
          4
```

Thus, Gourmet $\approx \frac{16.58¢}{1 \text{ ounce}}$ = 16.58¢ per ounce.

$$\text{Brand X: } \frac{399¢}{24 \text{ ounces}}$$

```
    16.625
24)399.000
   24
   159
   144
    15 0
    14 4
       60
       48
       120
       120
         0
```

Thus, brand X $\approx \frac{16.63¢}{1 \text{ ounce}}$ = 16.63¢ per ounce.

Thus, the gourmet popcorn oil at 16.58¢ per ounce is a better buy.

PROBLEM 6

A 29-ounce jar of spaghetti sauce costs \$1.09. The 24-ounce bottle costs 89¢. Which is the better buy?

Answers to PROBLEMS

6. 24-ounce bottle

› Exercises 4.2

‹ A › Writing Rates In Problems 1–10, write the indicated rates.

1. *Rate of pay per hour* A student is paid at the rate of $38 for 8 hours of work. What is the rate per hour?

2. *Construction worker's pay* According to the Bureau of Labor Statistics, the average construction worker makes $760 a week and works 40 hours. What is the rate per hour?

3. *Calculating miles per gallon* A car uses 12 gallons of gas on a 258-mile trip. How many miles per gallon is that?

4. *Miles per gallon* A student traveled 231 miles in going from Houston to Dallas. If she used 12 gallons of gas, how many miles per gallon did her car get?

5. *Plane's rate of travel* A plane flew from Boston to Chicago in 1.5 hours. If the air distance between these two cities is 840 miles, what was the plane's rate of travel?

6. *Plane's rate of travel* The air distance between New York and Dallas is 1375 miles. If a plane takes $2\frac{1}{2}$ hours to make the flight, what is the plane's rate of travel?

7. *More flying* One of the fastest flights in the continental United States was made by Captain Robert G. Sowers. He flew 2459 miles from Los Angeles to New York in 2 hours. What was his rate of travel?

8. *Marathon runner* One of the fastest marathon runners is Balaine Desimo, who completed 26 miles in 2.1 hours. To the nearest mile, what was his rate of travel?

9. *Tortilla rate of consumption* Tom Nall ate 74 tortillas in 30 minutes at Mariano's Mexican Restaurant in Dallas, Texas. To the nearest tenth, what was his per-minute rate of consumption?

10. *Refinery rate of production* The Amerada Hess refinery in St. Croix produced 345,000 barrels of oil a day. What is the rate of production in barrels per hour?

‹ B › Unit Rates In Problems 11–20, write each rate as a unit rate.

11. *Calories per gram of hamburgers* A McDonald's hamburger has 263 calories and weighs 100 grams. Find the number of calories per gram.

12. *Calories per gram of hamburgers* A Burger King hamburger has 275 calories and weighs 110 grams. Find the number of calories per gram.

13. *Rate of cars serviced in one hour* A car dealership services 90 cars in an average day. If the service department operates 7.5 hours each day, at what rate per hour are cars serviced?

14. *Monthly sales per person* A car dealership has 16 salespersons and sold 104 cars during the last month. What was the monthly number of sales per salesperson?

15. *Bamboo growth rate* Which is the fastest-growing plant you know? A bamboo plant grew 36 inches in a 24-hour period. What was the hourly growth rate for this plant?

16. *Daily growth rate for tomato plants* At the Tsukuba Science Expo in Japan it was announced that a *single* tomato plant produced 12,312 tomatoes in 347 days. To the nearest tenth, what was the daily rate at which tomatoes were growing on this plant?

17. *Caloric content of avocados* The fruit with the *highest* caloric value is the avocado: $1\frac{1}{2}$ pounds of edible avocado contain 1110 calories. What is the caloric rate per edible pound?

18. *Caloric content of cucumbers* The fruit with the *lowest* caloric value is the cucumber. A $1\frac{1}{2}$-pound cucumber has 109.5 calories. What is the caloric rate per pound of cucumbers?

19. *Scoring rate for Wilt Chamberlain* Wilt Chamberlain scored 31,419 points in 1045 basketball games. To the nearest tenth, at what rate was he scoring per game?

20. *Points per season* Kareem Abdul-Jabbar scored 38,987 points in 20 seasons. To the nearest whole number, how many points per season did he score?

Enrollment costs Have you taken a course to enhance your personal or professional skills? Here are some courses, their length, and their cost. In Problems 21–26, find the cost per hour to the nearest cent.

	Course	Cost	Hours
21.	Business Survival Skills	$872	84
22.	Introduction to Microsoft Word	$87	8
23.	Using Websites to Improve Your Bottom Line	$87	4
24.	Who Moved my Cheese?©	$97	4
25.	PC Repair and Troubleshooting	$895	35
26.	Microsoft Windows Professional	$698	21

27. *Animal facts* Elephants eat more than 1400 pounds of food (hay, grain, fruit, and vegetables) a week. How many pounds of food per day can an elephant consume?

28. *Animal facts* An elephant drinks 770 gallons of water each week. How many gallons a day can an elephant drink?

29. *Animal facts* An American alligator lives about 40 years and, during that time, grows about 3000 teeth. How many teeth per year is that?

30. *Animal facts* A llama can travel 60 miles in three days. How many miles per day can a llama travel?

Unit prices In Problems 31–36, find each unit price (to the nearest cent) and determine which size has the lowest unit price.

31.

Product	Size	Price
Close-Up Classic Red Gel Regular	6 oz	$2.29
Close-Up Classic Red Gel Regular	4 oz	$1.59

32.

Product	Size	Price
Smooth 'N Shine Instant Repair Spray-On Polisher	4 oz	$4.99
Frizz-Ease Relax Wave-Maker Styling Spray Ripple Effect	6 oz	$6.99

33.

Product	Size	Price
Walgreens Fruit Chews Candy Assorted	26 oz	$1.99
Walgreens Gum Drops	11 oz	$0.99

34.

Product	Size	Price
Nabisco SnackWells Cookies Creme Sandwich	7.75 oz	$2.79
Nabisco Sociables Baked Savory Crackers	8 oz	$2.99

35.

Product	Size	Price
Dry Idea Anti-Perspirant & Deodorant Roll-On Powder Fresh	2.5 oz	$4.49
Lady Mitchum Anti-Perspirant & Deodorant Roll-On Powder Fresh Scent	1.5 oz	$3.79

36.

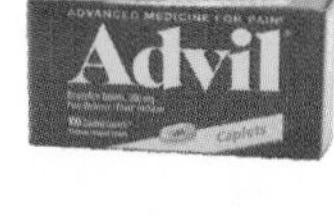

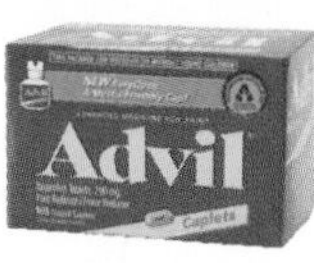

Product	Size	Price
2386720 Advil Caplet	100 CT	$7.99
2754968 Advil Caplet	165 CT	$11.99

37. *Is cheaper always better?* As a consumer, you probably believe that cheaper is *always* better. If you do, read on.

a. Dermassage dishwashing liquid costs $1.31 for 22 ounces. To the nearest cent, what is the cost per ounce?

b. White Magic dishwashing liquid costs $1.75 for 32 ounces. To the nearest cent, what is the cost per ounce?

c. Based on price alone, which is the better buy, Dermassage or White Magic?

But how much do you use per wash? *Consumer Reports* estimated that it costs 10¢ for 10 washes with Dermassage and 18¢ for the same number of washes with White Magic. Dermassage is a better buy!

38. *Is cheaper always better?* A&P wool-washing liquid costs 79¢ for 16 ounces. Ivory Liquid is $1.25 for 22 ounces.

a. To the nearest cent, what is the price per ounce of the A&P liquid?

b. To the nearest cent, what is the price per ounce of the Ivory Liquid?

c. Based on price alone, which is the better buy?

But wait. How much do you have to use? According to *Consumer Reports,* it costs 17¢ for 10 washes with the A&P liquid, but only 12¢ for 10 washes with Ivory Liquid!

For an article on these and other comparisons, see the *Consumer Reports* issue on washing liquid costs.

39. *Completion rate* The University of Minnesota defines the completion rate as the quotient of hours completed divided by hours attempted. For example, if you have completed **30** credit hours out of **40** attempted your completion ratio is

$$\frac{30}{40} = \frac{3}{4} = \frac{3 \cdot 25}{4 \cdot 25} = \frac{75}{100}$$

This is a 75% (per hundred) completion rate.

a. Find the completion rate written as a reduced fraction for a student who completed 40 credit hours out of 50 attempted.

b. Write the answer in **a** with a denominator of 100.

c. What is the student's completion rate as a percent?

Source: http://www.umn.edu.

	Fall	Spring	Total
Attempted credits	22	18	40
Failures, incompletes, no passes and withdrawals	2	8	10
Completed credits	20	10	

30 completed/40 attempted = 75% completion rate.

40. *Completion rates*

a. Find the completion rate written as a reduced fraction for a student who completed 45 credit hours out of 60 attempted.

b. Write the answer in **a** with a denominator of 100.

c. What is the student's completion rate as a percent?

41. *Decimal completion rates*

a. Find the completion rate to two decimal places of a student who completed 50 credit hours out of 60 attempted.

b. Write the answer in **a** with a denominator of 100.

c. What is the student's completion rate as a percent?

42. *Decimal completion rates*

a. Find the completion rate to three decimal places of a student who completed 70 credit hours out of 80 attempted.

b. Write the answer in **a** with a denominator of 100.

c. What is the student's completion rate as a percent?

43. *Completion rate* Illinois State University requires a 67% completion rate in order to be eligible for financial aid.

a. Find the completion rate to three decimal places of a student who completed 40 credit hours out of 60 attempted.

b. Write the answer in **a** with a denominator of 100.

c. What is the student's completion rate as a percent?

d. Does the student qualify for financial aid?

Source: http://www.policy.ilstu.edu.

44. *Completion rate*

a. Find the completion rate to three decimal places of a student who completed 60 credit hours out of 90 attempted

b. What is the student's completion rate as a percent?

c. What is the student's completion rate if part **a** is first rounded to two decimal places?

d. Remember, to qualify for financial aid you need a minimum completion rate of 67%. Does the student in part **b** qualify?

e. Does the student in part **c** qualify?

See how important rounding is?

Web IT go to **mhhe.com/bello** for more lessons

>>> Using Your Knowledge

Classified Rates Look at the rate chart at the beginning of this section. It tells you how much you have to pay per line for a specified period of time. For example, if you run an ad for 15 days (3-line minimum), the rate is \$3.30 per line. This means that if your ad has 5 lines and it runs for 15 days, the cost will be 5 · \$3.30 · 15 = \$247.50. Now, let us do the reverse process.

45. Suppose you paid \$348 for a 10-line ad running for 10 days.

a. How much did you pay each day?

b. What was your line-per-day rate?

46. Suppose you paid \$168 for a 7-line ad running 8 days.

a. How much did you pay per day?

b. What was your line-per-day rate?

>>> Write On

47. Is a ratio always a rate? Explain and give examples.

48. Is a rate always a ratio? Explain and give examples.

49. Write in your own words the procedure you use to find a unit rate.

50. When you say that your car gets 25 miles per gallon, is 25 miles per gallon

a. a ratio? **b.** a rate? **c.** a unit rate?

Concept Checker

Fill in the blank(s) with the correct word(s), phrase, or mathematical statement.

51. A ______________ is a **ratio** used to compare two **different** kinds of measures.

52. A **rate** in which the denominator is 1 is called a ______________ rate.

fraction	**simplified**
rate	**unit**
proportion	

Mastery Test

53. A 6-ounce tube of toothpaste costs \$2.30. What is the unit price in cents per ounce? (Answer to one decimal place.)

54. A can of spray costs \$3.99 for 4 ounces. Another brand costs \$5.99 for 6 ounces. Which can is the better buy?

55. A student earned \$508.40 in a 2-week period.

a. What was the rate of pay per week?

b. If the student worked 82 hours, what was the rate of pay per hour?

56. A 50-pound bag of Turf Supreme fertilizer covers 8000 square feet. To the nearest whole number, what is the rate of coverage in square feet per pound?

57. Mikisha traveled from Tampa to Miami, a distance of 280 miles, on one tank of gas. If her tank holds about 13 gallons of gas, what was her mile-per-gallon rate to the nearest whole number?

58. The school magazine offers you \$2600 to write a 1500-word article dealing with spring-break activities. To the nearest cent, what is the rate per word?

Skill Checker

Solve.

59. $18 = 3x$

60. $15 = 8x$

61. $6 \cdot 5 = 14x$

62. $7 \cdot 8 = 12x$

63. $6x = 9 \cdot 10$

64. $9x = 4 \cdot 6$

4.3 Word Problems Involving Proportions

Objective

You should be able to:

A Solve word problems involving proportions.

To Succeed, Review How To . . .

1. Reduce a fraction to lowest terms. (pp. 122–124)
2. Use the RSTUV method to solve word problems. (p. 92)
3. Solve a proportion. (pp. 260–261)

Getting Started

The length of each line is *proportional* to the one above it. More precisely, each line is 1.618034 . . . times as long as the previous one.

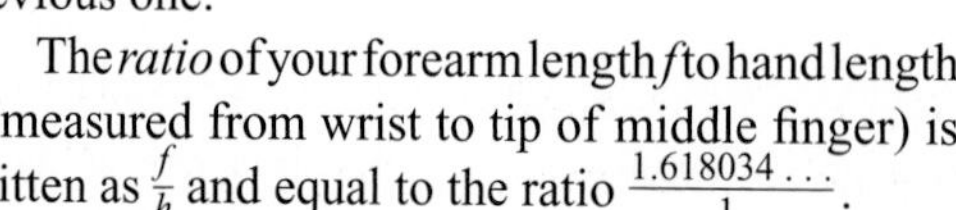

The *ratio* of your forearm length f to hand length h (measured from wrist to tip of middle finger) is written as $\frac{f}{h}$ and equal to the ratio $\frac{1.618034\ldots}{1}$.

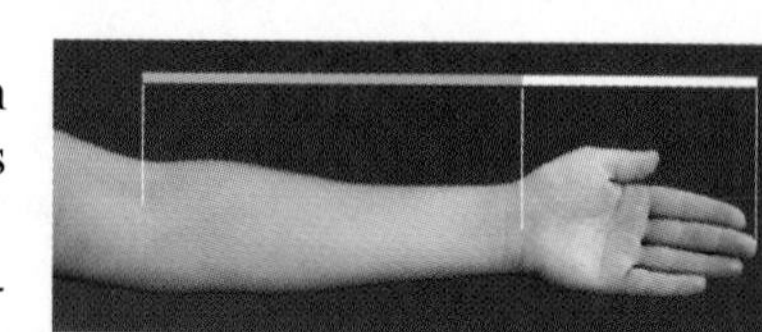

We say that your forearm and hand are *proportional*. As you recall, an equality of ratios is called a *proportion*.

Here $\dfrac{f}{h} = \dfrac{1.618034\ldots}{1}$

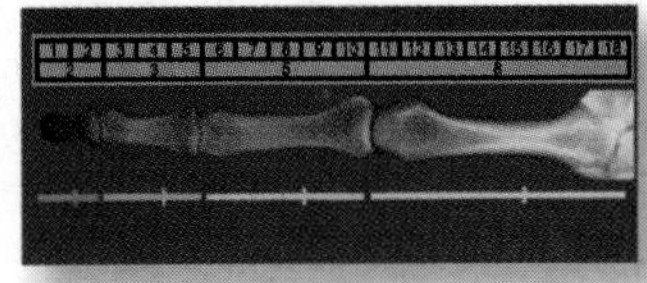

Now, suppose the length of your hand is 7 inches, what is the approximate length f of your forearm? You can find out by using the RSTUV procedure we have studied, but your answer should be about 11.3 inches. Does this work all the time? Find out for yourself. Measure the length h of your hand and predict the length f of your forearm. Is the prediction close to the actual length? Is $\frac{f}{h}$ close to 1.618034…?

We shall practice using the RSTUV procedure in solving proportions next.

A > Word Problems Involving Proportions

EXAMPLE 1 Solving proportions: Rye grass coverage

A pound of rye grass seed covers 120 square feet of lawn. How many pounds are needed to seed a lawn measuring 60 feet by 50 feet (3000 square feet)?

SOLUTION We use the RSTUV method discussed in Section 2.8.

1. Read the problem. Read the problem and decide what it asks for. (We want to know how many pounds of seed are needed.)

2. Select the unknown. Select a letter to represent the unknown. (Let p be the number of pounds needed.)

3. Translate into an equation. Translate the problem into an equation. 1 pound seeds 120 square feet $\left(\frac{1 \text{ pound}}{120 \text{ square feet}}\right)$; p pounds seed 3000 square feet $\left(\frac{p \text{ pounds}}{3000 \text{ square feet}}\right)$. Thus,

$$\frac{1}{120} = \frac{p}{3000}$$

4. Use the rules to solve. Use the rules studied to solve the equation.

$$1 \cdot 3000 = 120p \quad \text{Cross multiply.}$$

$$\frac{1 \cdot 3000}{120} = p \quad \text{Divide both sides by 120.}$$

$$25 = p \quad \text{Simplify.}$$

Thus 25 pounds are needed.

5. Verify the answer. Verify your answer. If we substitute 25 for p in step 3, we have:

$$\frac{1}{120} = \frac{25}{3000}$$

Cross multiplying, $1 \cdot 3000 = 120 \cdot 25$, which is a true statement. Thus, our answer is correct.

PROBLEM 1

A pound of Bahia grass seed covers 100 square feet of lawn. How many pounds are needed to seed a lawn measuring 90 feet by 50 feet (4500 square feet)?

To make the work easier, we will shorten some of the steps in the next examples.

EXAMPLE 2 Solving proportions: Computer trouble

Have you had trouble with a computer lately? A recent study indicated that 2 out of 5 families run into trouble with a computer in the course of a year. If 3000 families were surveyed, how many families would have run into trouble with the computer?

PROBLEM 2

If a study revealed that 2 out of 5 families run into trouble with a computer and it is known that 300 families had trouble with the computer, how many families were surveyed?

(continued)

Answers to PROBLEMS

1. 45 pounds **2.** 750

SOLUTION

1. Read. Read the problem.

2. Select unknown. Select f to represent the number of families that ran into trouble.

3. Translate. Translate: Note that 3000 is the *total* number of families surveyed. Thus

$$\frac{2}{5} = \frac{f}{3000}$$

4. Use cross products. Use cross products:

$$2 \cdot 3000 = 5f$$

$$\frac{2 \cdot 3000}{5} = f \quad \text{Divide by 5.}$$

$$\frac{6000}{5} = f \quad \text{Multiply.}$$

$$1200 = f \quad \text{Divide.}$$

That is, 1200 families ran into trouble.

5. Verify. Verification is left for you.

EXAMPLE 3 Solving proportions: Protein in diet

Do you have enough protein in your diet? Females are supposed to have 44 grams every day. If 2 tablespoons of peanut butter provide 8 grams of protein, how many tablespoons does a female need to have 44 grams of protein?

SOLUTION

1. Read. Read the problem.

2. Select unknown. Select t to be the number of tablespoons needed.

3. Translate. Translate: 2 tablespoons provide 8 grams $\left(\frac{2 \text{ tablespoons}}{8 \text{ grams}}\right)$; t tablespoons provide 44 grams $\left(\frac{t \text{ tablespoons}}{44 \text{ grams}}\right)$.

$$\frac{2}{8} = \frac{t}{44}$$

4. Use cross products. Use cross products:

$$2 \cdot 44 = 8t$$

$$\frac{2 \cdot 44}{8} = t \quad \text{Divide both sides by 8.}$$

$$\frac{88}{8} = t \quad \text{Multiply.}$$

$$11 = t \quad \text{Divide.}$$

Thus, 11 tablespoons of peanut butter are needed to obtain 44 grams of protein.

5. Verify. Verification is left for you.

PROBLEM 3

Males need 56 grams of protein every day. If 2 tablespoons of peanut butter provide 8 grams of protein, how many tablespoons does a male need to have 56 grams of protein?

EXAMPLE 4 Solving proportions: Buying stocks

The cost of 50 shares of 3M (MMM) stock is \$106.25. At this rate, how many shares can you buy with \$425?

SOLUTION

1. Read. Read the problem.

2. Select unknown. Select n to be the number of shares you can buy with \$425.

3. Translate. Translate: 50 shares cost \$106.25 $\left(\frac{50 \text{ shares}}{\$106.25}\right)$, n shares cost \$425 $\left(\frac{n}{\$425}\right)$. Thus,

$$\frac{50}{106.25} = \frac{n}{425}$$

PROBLEM 4

The cost of 50 shares of Comp-U-Check is \$137.50. How many shares can you buy with \$1100?

Answers to PROBLEMS

3. 14 tablespoons **4.** 400

4. Use cross products. Use cross products:

$$50 \cdot 425 = 106.25n$$

$$\frac{50 \cdot 425}{106.25} = n \quad \text{Divide both sides by 106.25.}$$

$$\frac{21{,}250}{106.25} = n \quad \text{Multiply.}$$

$$200 = n \quad \text{Divide.}$$

Thus, 200 shares can be bought with $425.

5. Verify. Verification is left for you.

Proportions are used in medicine for calculating dosages. Suppose you know the adult dose for a certain medicine. How can you find the dose suitable for children? "The only rule based on scientific principles and the one which should be used" is Clark's Rule.

The rule states that the ratio of the weight of the child W_c to the average weight of an adult W_a is proportional to the children's dose c divided by the adult dose a, that is,

$$\text{Clark's Rule} \quad \frac{W_c}{W_a} = \frac{c}{a}$$

Source: Texas Health Science Technology Education.

We use this rule in Example 5.

EXAMPLE 5 Solving proportions: Children's dose of antibiotics

The recommended dose of antibiotics for a 150-pound adult is 3 pills. What is the equivalent dose suitable for a child weighing 50 pounds?

SOLUTION You can reason like this:

150 pound adult: 3 pills

50 pound ($\frac{1}{3}$ weight): $\frac{1}{3}$(3 pills) = 1 pill

But can you prove it? Let's see.

1. Read the problem. Read the problem. We have to use Clark's Rule.

2. Select the unknown. The unknown is the dose of pills for a child; call it c.

3. Translate the problem into an equation or inequality. According to Clark's Rule,

$$\frac{W_c}{W_a} = \frac{c}{a}$$

W_c (the weight of the child) = 50

W_a (the weight of the adult) = 150

a (the adult dose) = 3 pills

4. Use the rules we have studied to solve the equation. Substitute these values into the equation:

$$\frac{50}{150} = \frac{c}{3}$$

$$3 \cdot 50 = 150c \quad \text{Cross multiply.}$$

$$150 = 150c \quad \text{Multiply 3 by 50.}$$

$$1 = c \quad \text{Divide by 150.}$$

Thus, the equivalent dose of antibiotics for a 50-pound child is 1 pill.

5. Verify the answer. To verify the answer, note that the ratio of the weight of the child to the weight of the adult is $\frac{50}{150}$ or $\frac{1}{3}$. In the same manner, the ratio of pills taken by the child to pills taken by the adult is 1 to 3 or $\frac{1}{3}$.

PROBLEM 5

What is the dose for a child weighing 75 pounds?

Answers to PROBLEMS

5. $1\frac{1}{2} = 1.5$ pills

Finally, we shall consider strength, not your algebra strength, but the strength of different animals! Elephants can carry up to 25% of their own weight on their backs, camels about 20%, and leaf-cutting ants about 3 times their own weight, but rhinoceros beetles can carry about 850 times their body weight. How strong is that? We shall see in Example 6.

Source: Data from edHelper.com.

EXAMPLE 6 Solving proportions: Rhinoceros beetle

A rhinoceros beetle weighs 30 grams and can carry 850 times its body weight, that is, 25,500 grams. If a person could carry proportionally as much as the rhinoceros beetle, how much could a 60-kilogram (kg) student carry?

SOLUTION

1. Read the problem. We want to find how much a 60-kilogram student can carry.

2. Select the unknown. Let W be the weight of the student. The ratio of body weight to carrying weight for the beetle is $\frac{30}{25{,}500}$. For the student the ratio is $\frac{60}{W}$.

3. Translate the problem into an equation or inequality. We want the weights to be proportional, so

$$\frac{30}{25{,}500} = \frac{60}{W}$$

4. Use the rules we have studied to solve the equation.

Cross multiply: $30W = 60 \cdot 25{,}500$

Divide by 30: $W = 51{,}000$ kg

Since 1 kilogram is about 2.2 pounds, the 51,000 kilograms represents more than 100,000 pounds! Strong indeed.

5. Verify the answer. If students could carry 850 times their weight (like the beetle), a 60-kilogram student could carry $850 \cdot 60 = 51{,}000$ kg as stated.

PROBLEM 6

Proportionally, how much could a 90 kilogram football player carry?

› Exercises 4.3

Boost your grade at mathzone.com!
- Practice Problems
- NetTutor
- Self-Tests
- e-Professors
- Videos

‹ A › Word Problems Involving Proportions In Problems 1–30, use the RSTUV procedure to solve.

1. *Lawn coverage* A pound of weed and feed covers 200 square feet. How many pounds are needed to cover a lawn measuring 60 feet by 50 feet (3000 square feet)?

2. *Fertilizing the lawn* A pound of fertilizer covers 250 square feet. How many pounds are needed to fertilize a lawn measuring 70 feet by 50 feet (3500 square feet)?

3. *Advancing stocks* On a recent day, 2 out of 5 stocks in the New York Stock Exchange (NYSE) advanced in price. If 1900 issues were traded, how many advanced in price?

4. *Declining stocks* On a recent day, 3 out of 5 stocks in the NYSE declined in price. If 1500 issues were traded, how many declined in price?

5. *Male RDA of protein* The recommended daily allowance (RDA) of protein for males in the 15–18 age range is 56 grams per day. Two ounces of cheddar cheese contain 14 grams of protein. How many ounces of cheddar cheese are needed to provide 56 grams of protein?

6. *Protein RDA for children* The RDA of protein for children 7–10 is 30 grams per day. One cup of whole milk contains 8 grams of protein. How many cups of milk are needed to provide the 30 grams RDA?

Answers to PROBLEMS

6. 76,500 kg

7. *Buying stocks* Fifty shares of Titan Corp. stock cost $112.50. How many shares can be bought with $562.50?

8. *Buying Mega stocks* Three shares of Mega stock cost $144.75. How many shares can be bought with $1544?

9. *Showering for sleep* It has been estimated that a 10-minute shower is worth $1\frac{1}{2}$ hours of sleep. Using this scale, how long do you have to shower to get 6 hours of sleep?

10. *Mail* It has been estimated that 3 out of 4 pieces of mailed advertising are opened and glanced at. Using this scale, how many pieces of advertising should you send if you want 900 persons to open and glance at your ad?

11. *Getting the votes* It has been estimated that for every 10 volunteers working in a political campaign, the candidate gets 100 votes on election day. A candidate has 770 volunteers. How many votes can this candidate expect to get on election day?

12. *Waiter, Waiter!* To make sure your dinner party runs smoothly, you will need 3 waiters for every 20 guests. How many waiters should you hire for a party of 80?

13. *Drinking water* You can maintain your body fluid level by drinking four gulps of water every 20 minutes during prolonged exercise. How many gulps do you need for a 2-hour workout?

14. *Pages in a 2000-word document* A full, double-spaced typewritten page will have about 250 words on it if typed using a pica typefont. How many pages would there be in a double-spaced 2000-word paper typed using a pica typefont?

15. *Pages in a 6600-word paper* A full, double-spaced typewritten page will have about 330 words on it if typed using an elite typefont. How many pages would there be in a double-spaced 6600-word paper typed using an elite typefont?

16. *Inflation and spending* A writer for the *New York Times* estimates that a 1-point increase in the inflation rate will add $1.3 billion to federal spending in the first year. If the inflation rate increases 3.5 points, how much money will be added to federal spending in the first year?

17. *Interest rates* A 1-point increase in interest rates adds $2.3 billion a year to federal spending. If interest rates went up 2.5 points, how much would that add to federal spending?

18. *Windows needed for the superinsulated house* A superinsulated home must have 12 square feet of windows for every 100 square feet of floor space. How many square feet of windows are needed for a superinsulated 1700-square-foot home?

19. *Growing mesquite grass* It takes 1725 pounds of water to grow 1 pound of mesquite grass (used for cattle feeding). How many pounds of water are needed to grow a 50-pound bale of mesquite grass?

20. *New plant and equipment spending* U.S. manufacturers spent $150 billion on new plants and equipment but $200 billion on mergers and acquisitions. Kohlberg, Kravis, and Roberts acquired Nabisco for about $20 billion. How many billions should they expect to spend on new plants and equipment?

The serving size for each of the pizzas shown is $\frac{1}{4}$ of the pizza. Which one do you think has the least fat?

Source: California Project Lean.

21. *Pizza* Charley's pizza has 13 grams of fat per serving. How many grams of fat are in $\frac{3}{4}$ of the pizza?

22. *Pizza* The Peppy pizza has 21 grams of fat per serving. How many grams of fat are in $\frac{1}{2}$ of the pizza?

23. *Pizza* Garden Delight Pizza has 7 grams of fat per serving. How many grams of fat are in the whole pizza?

24. *Overweight Americans* 54 out of 100 Americans consider themselves overweight. If there are 290 million Americans, how many consider themselves overweight?

25. *Global Internet population* In 2004, 175 million people in the United States were part of the global Internet population. If the 175 million people represent $\frac{1}{4}$ of the global Internet population, how many people are there in the global Internet population?

26. *Leaf-cutting ant loads* A leaf-cutting ant weighing 1.5 grams can carry a leaf weighing 4.5 grams. If a person could proportionally carry as much weight, how much could a 60-kilogram student carry?

27. *Making nail polish remover* The formula for an acetone (nail polish remover) molecule is C_3H_6O. This means that for every 3 carbon (C) atoms, there are 6 hydrogen (H) atoms and one oxygen (O) atom in the molecule. How many carbon atoms must combine with 720 hydrogen atoms to form acetone molecules?

28. *Polish remover* Referring to Problem 27, how many oxygen atoms are needed to combine with 660 hydrogen atoms to form acetone molecules?

29. *Map distances to scale* The scale in a map is 1 inch = 20 miles. What is the actual distance between two towns that are 3.5 inches apart on the map?

30. *Map spacing* Two towns are 300 miles apart. If the scale on the map is 1 inch = 20 miles, how far apart are the towns on the map?

Web IT go to mhhe.com/bello for more lessons

Credit scores Do you remember a discussion about FICO (Fair Isaac Credit Organization) scores? Your FICO scores are the credit scores most lenders use to determine your credit risk.

FICO ® Scores Affect Your Monthly Payments

If Your FICO ® Score Is	Your Interest Rate Is
760–850	6.28%

In Problems 31–34, assume that you have an excellent FICO score between 760 and 850. Your interest **rate** is 6.28% and your monthly payment is dependent on how much money you borrow. The table shows the *rate* you pay each month per *thousand* dollars borrowed on a 15-year loan (**$8.59**) and on a 30-year loan (**$6.18**). Remember, your *rate* of payment is based on how many *thousands* you borrow. If you are borrowing $210,000, you are borrowing $21 (thousand), and if you are borrowing $107,000, you are borrowing $107 (thousand).
Source: http://www.myfico.com/.

15-Year Loan	30-Year Loan
$8.59 a month	$6.18 a month

31. What will your monthly payment be on a $100,000, 15-year loan?

32. What will your monthly payment be on a $175,000, 30-year loan?

33. What will your monthly payment be on a $210,000, 30-year loan?

34. What will your monthly payment be on a $110,000, 15-year loan?

Using Your Knowledge

Wildlife Populations Have you seen scientists or bird lovers banding or tagging animals? This procedure is used to estimate the size of wildlife populations. Here is how it works. First, a number of animals are captured, tagged, and released. Later, a different group is captured, and the ratio of tagged animals to the number of animals captured is determined. From this ratio the size of the population can be estimated. For example, a research team tagged 60 birds for identification. Later, they captured 240 birds and found 15 tagged. Can they estimate the number n of birds in the population? They know that the ratio of tagged birds to the total number of birds is 15 to 240. They also know that this ratio was 60 to n. Thus,

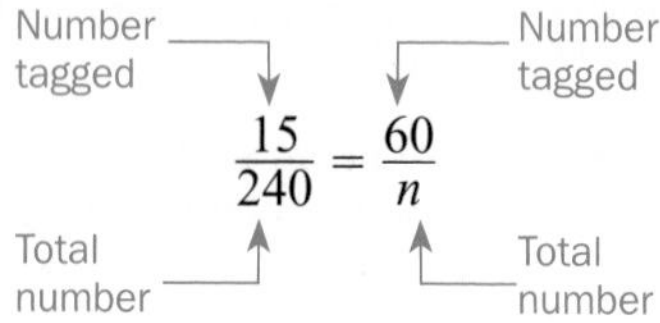

$$\frac{15}{240} = \frac{60}{n}$$

Solving the proportion they obtain $n = 960$.

Follow this procedure to solve Problems 35 and 36.

35. Researchers in the Vienna Woods tagged 300 birds. In a later sample they found that 12 out of 980 birds were tagged. About how many birds were there in the Vienna Woods?

36. In Muddy Lake 600 fish were tagged. Later, 12 out of 480 fish were found to be tagged. About how many fish were in Muddy Lake?

Write On

37. We solved proportions such as $\frac{x}{2} = \frac{1}{2}$ by the method of cross products. Can you solve $\frac{x}{2} = \frac{1}{2} + 1$ by using cross products? Explain.

38. Clark's Rule helps you to find the suitable child's dose for a medicine based on the adult dose. Does the rule tell you how to administer the medicine? Explain.

Concept Checker

Fill in the blank(s) with the correct word(s), phrase, or mathematical statement.

39. When using the **RSTUV** procedure, step **T** means to ______________ the problem.

40. To **solve a proportion** we use the ______________ **rule.**

guessing **cross product**
transpose **algebra**
translate

Mastery Test

41. According to Clark's Rule $\frac{W_c}{W_a} = \frac{c}{a}$, Where W_c is the weight of the child, W_a is the weight of the adult, a is the adult dose, and c is the child's dose. If the adult dose (150-pound person) of cough syrup is 6 teaspoons daily, what is a 50-pound child's dose?

42. A leaf-cutting ant weighing 1.5 grams can carry a leaf weighing 3 grams. If a person could proportionally carry as much weight, how much could a 70-kilogram student carry?

43. *Stocks* Fifty shares of Mega stock cost $212.50. How many shares can you buy with $850?

44. *Protein* If a 120-pound woman needs 50 grams of protein a day and a 4-ounce serving of chicken provides 30 grams of protein, how many ounces of chicken are needed to provide the required 50 grams of protein?

45. *College Algebra students* Two out of three students at a college passed College Algebra with a C or better. If 600 students were taking College Algebra, how many passed with a C or better?

46. *Rye seed* Perennial rye seed covers 250 square feet per pound. How many pounds are needed to seed a lawn measuring 50 feet by 100 feet (5000 square feet)?

Skill Checker

Round 245.92

47. to the nearest tenth.

48. to the nearest unit.

49. to the nearest ten.

50. to the nearest hundred.

Collaborative Learning

There is a relationship between the numbers in the Fibonacci sequence, the so-called Golden Section numbers, and $\frac{1}{89}$. Here is some information you will need to do this activity:

The Fibonacci numbers are: 1, 1, 2, 3, 5, 8, 13, . . .

The Golden Section numbers are: 0.6180339887 . . . 1.6180339887 . . . and $\frac{1}{89} = 0.01123595\ldots$

Form three groups of students.

Group 1. Find the ratio of numbers *preceding* each other in the Fibonacci sequence and write them as decimals to six places. Here are some:

$$\frac{1}{1} = 1.000000 \qquad \frac{3}{2} = 1.500000$$

$$\frac{2}{1} = 2.000000 \qquad \frac{5}{3} = 1.666666$$

What number is the pattern approaching?

Group 2. Find the ratio of numbers *following* each other in the Fibonacci sequence and write them as decimals to six places. Here are some:

$$\frac{1}{1} = 1.000000 \qquad \frac{2}{3} = 0.666666$$

$$\frac{1}{2} = 0.500000 \qquad \frac{3}{5} = 0.600000$$

What number is the pattern approaching?

Group 3. Arrange the numbers in the Fibonacci sequence 1, 1, 2, 3, 5, . . . as decimals like this:

0.01 0.001 0.0002 0.00003 0.000005

Find the next 5 terms and the sum of the 10 terms obtained. What number is the pattern approaching? Pick any *four* consecutive Fibonacci numbers, say, **1 2 3 5**

Group 1. Find the product of the *first* and *last* numbers, 1 × 5 = **5** this time.

Group 2. Find *twice* the product of the two middle numbers, **2** × (**2** × **3**) = **12** this time.

<CONTINUED>

Group 3. Get the answers from Groups 1 and 2 and square them.

$$5^2 = 25 \quad \text{and} \quad 12^2 = 144$$

ALL GROUPS: Make a conjecture about writing the sum of the two answers as a number squared. Repeat the process with four different consecutive Fibonacci numbers. Does the conjecture still work?

Research Questions

1. The *Human Side of Mathematics* at the beginning of this chapter mentions several names for Leonardo Fibonacci. Find at least two more names used for Fibonacci.
2. Find at least two different interpretations of the name Fibonacci.
3. Find the meaning of the word *bigollo* in Italian.
4. The *Human Side of Mathematics* mentions the *Liber Abaci,* a book written by Fibonacci. Find the titles of three other books he wrote and describe their contents.
5. The *Human Side of Mathematics* mentions "the rabbit problem." State the problem and indicate how the Fibonacci sequence relates to it.
6. There are at least five more famous problems by Fibonacci in the *Liber Abaci.* State three of them.
7. In what year did Fibonacci die?
8. There is a set of numbers called the Golden Section numbers associated with the Fibonacci sequence. What are these numbers, and how are they associated with the Fibonacci sequence?

Summary Chapter 4

Section	Item	Meaning	Example
4.1A	Ratio	A quotient of two numbers	The ratio 2 to 3 (also written as 2:3 or $\frac{2}{3}$)
4.1B	Proportion	An equation stating that two ratios are equal	$\frac{a}{b} = \frac{c}{d}$ is a proportion.
4.2	Rates	A ratio used to compare unlike quantities	$\frac{21 \text{ miles}}{3 \text{ gallons}}$
4.2B	Unit rate	A rate in which the denominator is 1	$\frac{28 \text{ miles}}{1 \text{ gallon}}$
4.3A	RSTUV	Method used to solve word problems (**R**ead, **S**elect a variable, **T**ranslate, **U**se the rules studied to solve, **V**erify the answer)	

Review Exercises Chapter 4

(If you need help with these exercises, look in the section indicated in brackets.)

1. **‹4.1A›** *Express each ratio as a fraction in lowest terms:*

a. 1 to 10

b. 2 to 10

c. 3 to 10

d. 4 to 10

e. 5 to 10

2. **‹4.1A›** *The ratio of cars to people for different countries is given. Write each ratio as a fraction in reduced form.*

a. Canada, 425 cars per 1000 persons

b. Belgium, 325 cars per 1000 persons

c. Austria, 300 cars per 1000 persons

d. Cyprus, 150 cars per 1000 persons

e. Taiwan, 40 cars per 1000 persons

3. **‹4.1A›** *The approximate ratio of circumference (distance around) to diameter (distance across the top) of a soda can is given. Write this ratio as a fraction in reduced form.*

a. 14.52 inches to 4.62 inches

b. 14.58 inches to 4.64 inches

c. 14.60 inches to 4.65 inches

d. 14.64 inches to 4.66 inches

e. 14.68 inches to 4.67 inches

4. **‹4.1A›** *The male-to-female ratio for different countries is given. Write this ratio as a fraction in reduced form.*

a. Kuwait, 58 to 42

b. Guam, 55 to 45

c. Pakistan, 52 to 48

d. Malaysia, 48 to 52

e. Grenada, 46 to 54

5. **‹4.1B›** *Write the following proportions as equations.*

a. 3 is to 7 as 6 is to x

b. 4 is to 7 as 12 is to x

c. 5 is to 7 as x is to 21

d. 6 is to 7 as 33 is to x

e. 7 is to 35 as 5 is to x

6. **‹4.1C›** *Determine if the following pairs of numbers are proportional.*

a. 2, 3 and 4, 5

b. 8, 10 and 4, 5

c. 5, 6 and 12, 15

d. 12, 18 and 2, 3

e. 9, 12 and 3, 4

7. **‹4.1D›** *Solve the proportion:*

a. $\frac{x}{4} = \frac{1}{2}$ **b.** $\frac{x}{6} = \frac{1}{2}$

c. $\frac{x}{8} = \frac{1}{2}$ **d.** $\frac{x}{10} = \frac{1}{2}$

e. $\frac{x}{12} = \frac{1}{2}$

8. **‹4.1D›** *Solve the proportion:*

a. $\frac{2}{x} = \frac{2}{5}$ **b.** $\frac{4}{x} = \frac{2}{5}$

c. $\frac{6}{x} = \frac{2}{5}$ **d.** $\frac{10}{x} = \frac{2}{5}$

e. $\frac{12}{x} = \frac{2}{5}$

9. **‹4.1D›** *Solve the proportion:*

a. $\frac{x}{4} = \frac{9}{2}$ **b.** $\frac{x}{4} = \frac{9}{12}$

c. $\frac{x}{4} = \frac{9}{18}$ **d.** $\frac{x}{4} = \frac{9}{36}$

e. $\frac{x}{4} = \frac{9}{6}$

10. **‹4.1D›** *Solve the proportion:*

a. $\frac{5}{7} = \frac{x}{7}$ **b.** $\frac{5}{7} = \frac{x}{14}$

c. $\frac{5}{7} = \frac{x}{28}$ **d.** $\frac{5}{7} = \frac{x}{35}$

e. $\frac{5}{7} = \frac{x}{42}$

11. **〈4.1E〉** *A worker in an assembly line takes 3 hours to produce 27 parts. At that rate how many parts can she produce in:*

a. 1 hour? **b.** 2 hours?

c. 4 hours? **d.** 5 hours?

e. 6 hours?

12. **〈4.2A〉** *A famous writer was paid $12,000 for an essay. What was her rate per word, if the essay had:*

a. 2000 words? **b.** 3000 words?

c. 4000 words? **d.** 6000 words?

e. 8000 words?

13. **〈4.2A〉** *A student took a 400-mile car trip. To the nearest whole number, what was the miles-per-gallon rate if the amount of gas used was:*

a. 18 gallons? **b.** 19 gallons?

c. 20 gallons? **d.** 21 gallons?

e. 22 gallons?

14. **〈4.2A〉** *A bag of fertilizer covers 5000 square feet of lawn. To the nearest whole number, what is the rate of coverage (in square feet per pound) if the bag contains:*

a. 20 pounds of fertilizer?

b. 22 pounds of fertilizer?

c. 24 pounds of fertilizer?

d. 25 pounds of fertilizer?

e. 50 pounds of fertilizer?

15. **〈4.2B〉** *A jar of popcorn costs $2.39. To the nearest cent, what is the unit price in cents per ounce if the jar contains:*

a. 24 ounces? **b.** 16 ounces?

c. 32 ounces? **d.** 40 ounces?

e. 48 ounces?

16. **〈4.2B〉** *A 12-ounce bottle of generic popcorn oil sells for $1.39. Which is the better buy if the brand X popcorn oil contains 24 ounces and costs:*

a. $2.75? **b.** $2.76?

c. $2.77? **d.** $2.74?

e. $2.79?

17. **〈4.3A〉** *A pound of grass seed covers 120 square feet of lawn. Find how many pounds are needed to seed a lawn with the following measurements:*

a. 60 feet by 30 feet (1800 square feet).

b. 70 feet by 30 feet (2100 square feet).

c. 90 feet by 24 feet (2160 square feet).

d. 60 feet by 45 feet (2700 square feet).

e. 50 feet by 60 feet (3000 square feet).

18. **〈4.3A〉** *A survey indicated that three out of five doctors used brand X aspirin. Find how many used brand X if:*

a. 3000 doctors were surveyed.

b. 4000 doctors were surveyed.

c. 5000 doctors were surveyed.

d. 6000 doctors were surveyed.

e. 8000 doctors were surveyed.

19. **〈4.3A〉** *The RDA of protein for males is 56 grams per day. Find how many ounces of a certain product are needed to provide the 56 grams of protein if it is known that*

a. 3 ounces of the product provide 8 grams of protein.

b. 4 ounces of the product provide 8 grams of protein.

c. 5 ounces of the product provide 8 grams of protein.

d. 6 ounces of the product provide 8 grams of protein.

e. 7 ounces of the product provide 8 grams of protein.

20. **〈4.3A〉** *The cost of 50 shares of Fly-by-Night Airline is $87.50. Find how many shares you can buy with:*

a. $350.

b. $700.

c. $612.50.

d. $787.50.

e. $1050.

Practice Test Chapter 4

(Answers on page 284)

Visit www.mhhe.com/bello to view helpful videos that provide step-by-step solutions to several of the problems below.

1. Express each ratio as a fraction in lowest terms:

a. 2 to 7 **b.** 3 to 18 **c.** 10 to 58

2. The ratio of cars to people in Australia is 485 per 1000. Write this ratio as a fraction in reduced form.

3. The approximate ratio of circumference (distance around) to diameter (distance across the top) of a soda can is 15.55 inches to 4.95 inches. Write this ratio as a fraction in reduced form.

4. The male-to-female ratio in India is 54 to 46. Write this ratio as a fraction in reduced form.

5. Write the following proportions.

a. 2 is to 7 as 6 is to 21

b. 5 is to 7 as 15 is to x

6. There is a law stating that the ratio of width to length for the American flag should be 10 to 19. A flag measured 40 by 78 feet. Are the pairs of numbers 10, 19 and 40, 78 proportional?

7. Solve the proportion $\frac{x}{2} = \frac{5}{20}$.

8. Solve the proportion $\frac{6}{x} = \frac{2}{5}$.

9. Solve the proportion $\frac{x}{8} = \frac{9}{12}$.

10. Solve the proportion $\frac{5}{7} = \frac{x}{28}$.

11. A worker in an assembly line takes 3 hours to produce 26 parts. At that rate how many parts can she produce in 9 hours?

12. A famous writer was paid $12,000 for a 2000-word article. Find the rate per word.

13. A student traveled 300 miles on 17 gallons of gas. What was the miles-per-gallon rate? (Round the answer to the nearest whole number.)

14. An 18-pound bag of lawn food covers 3000 square feet of lawn. To the nearest whole number, what is the rate of coverage in square feet per pound?

15. A 30-ounce jar of popcorn costs $2.49. What is the unit price in cents per ounce? (Answer to the nearest cent.)

16. A 12-ounce bottle of gourmet popcorn oil sells for $1.39. The 24-ounce brand X bottle costs $2.77. Which is the better buy?

17. A pound of grass seed covers 120 square feet of lawn. How many pounds are needed to seed a lawn measuring 60 feet by 40 feet (2400 square feet)?

18. A survey indicated that 3 out of 7 doctors used brand X aspirin. If 2100 doctors were surveyed, how many used brand X?

19. The RDA of protein for males is 56 grams per day. Two ounces of a certain product provide 4 grams of protein. How many ounces of the product are needed to provide 56 grams of protein?

20. The cost of 50 shares of Fly-by-Night Airline is $87.50. How many shares can you buy with $875?

Answers to Practice Test Chapter 4

Answer	If You Missed	Review		
	Question	Section	Examples	Page
1. **a.** $\frac{2}{7}$ **b.** $\frac{1}{6}$ **c.** $\frac{5}{29}$	1	4.1	1	257
2. $\frac{97}{200}$	2	4.1	2	257
3. $\frac{311}{99}$	3	4.1	3	257
4. $\frac{27}{23}$	4	4.1	4	257
5. **a.** $\frac{2}{7} = \frac{6}{21}$ **b.** $\frac{5}{7} = \frac{15}{x}$	5	4.1	6	258
6. No	6	4.1	7	259
7. $x = \frac{1}{2}$	7	4.1	8b	260
8. $x = 15$	8	4.1	8a	260
9. $x = 6$	9	4.1	8b	260
10. $x = 20$	10	4.1	8c	260
11. 78	11	4.1	9, 10	261
12. $6	12	4.2	1	266
13. 18	13	4.2	2	266
14. 167	14	4.2	3	266
15. 8¢	15	4.2	5	267
16. Brand X, the 24-oz bottle	16	4.2	6	268
17. 20 lb	17	4.3	1	273
18. 900	18	4.3	2	273–274
19. 28	19	4.3	3	274
20. 500	20	4.3	4	274–275

Cumulative Review Chapters 1–4

1. Write nine thousand, eight hundred ten in standard form.

2. Write the prime factors of 56.

3. Multiply: $2^3 \times 7 \times 4^0$

4. Simplify: $25 \div 5 \cdot 5 + 7 - 3$

5. Classify $\frac{9}{7}$ as proper or improper.

6. Write $\frac{39}{4}$ as a mixed number.

7. Write $7\frac{2}{3}$ as an improper fraction.

8. Multiply: $\frac{1}{2} \cdot 3\frac{1}{7}$

9. Multiply: $\left(\frac{3}{7}\right)^2 \cdot \frac{1}{9}$

10. Divide: $\frac{8}{5} \div 2\frac{2}{7}$

11. Add: $7\frac{2}{3} + 1\frac{3}{5}$

12. Subtract: $5\frac{1}{4} - 1\frac{7}{8}$

13. Translate and solve: $\frac{7}{9}$ less than a number c is $\frac{1}{2}$. What is c?

14. Find a number such that $\frac{11}{12}$ of it is $7\frac{1}{10}$.

15. Give the word name for 241.35.

16. Write 44.874 in expanded form.

17. Add: $36.454 + 9.69$

18. Subtract: $342.42 - 13.5$

19. Multiply: $0.554 \cdot 0.15$

20. Divide: $\frac{135}{0.27}$

21. Round 449.851 to the nearest ten.

22. Divide: $10 \div 0.13$ (Round answer to two decimal digits.)

23. Write $\frac{7}{12}$ as a decimal.

24. Write 0.15 as a reduced fraction.

25. Write $0.\overline{84}$ as a reduced fraction.

26. What decimal part of 12 is 9?

27. Arrange in order of decreasing magnitude and write using the $>$ sign: $6.435 \quad 6.43\overline{5} \quad 6.4\overline{35}$

28. Insert $=$, $<$, or $>$ to make a true statement: 0.89 ______ $\frac{7}{20}$

29. Solve for x: $x + 2.5 = 6.5$

30. Solve for y: $2.1 = 0.3y$

31. Solve for z: $9 = \frac{z}{6.9}$

32. The ratio of cars to people in Austria is 495 to 1000. Write this ratio as a fraction in reduced form.

33. Write the following proportion: 6 is to 2 as 54 is to x.

34. There is a law stating that the ratio of width to length for the American flag should be 10 to 19. Is a flag measuring 50 by 97 feet of the correct ratio?

35. Solve the proportion: $\frac{j}{5} = \frac{6}{150}$

36. Solve the proportion: $\frac{20}{c} = \frac{4}{3}$

37. A worker in an assembly line takes 9 hours to produce 25 parts. At that rate how many parts can she produce in 36 hours?

38. A salesperson traveled 600 miles on 17 gallons of gas. How many miles per gallon did the salesperson get? (Round to the nearest whole number.)

39. A 24-ounce jar of peanut butter costs \$2.89. What is the unit price in cents per ounce? (Answer to the nearest cent.)

40. A pound of lawn food covers 120 square feet of lawn. How many pounds are needed to cover a lawn measuring 80 by 60 feet (4800 square feet)?

41. The protein RDA for males is 56 grams per day. Two ounces of a certain product provide 4 grams of protein. How many ounces of the product are needed to provide 48 grams of protein?

42. The cost of 80 shares of Fly-by-Night Airline is \$87.50. How many shares can you buy with \$875.00?

Section

Chapter 5 *five* Percent

The Human Side of Mathematics

What is a percent? It is a way of expressing ratios in terms of whole numbers. In this chapter you will learn that converting a ratio or fraction of the form $\frac{a}{b}$ to a percent is as easy as 1, 2, 3. Here are the steps:

1. Divide a by b.
2. Multiply by 100.
3. Append the % sign. For example, $\frac{3}{4} = 0.75 \times 100 = 75\%$

But where did the % sign come from? It evolved from a symbol introduced in an anonymous Italian manuscript of 1425, where the author used the symbol P c° instead of "per cent" or "by the hundred." By about 1650 the symbol evolved to $\frac{o}{o}$ and our modern %.

The % symbol has been used since the end of the fifteenth century in interest computations, profit-and-loss documents, and taxes. As a matter of fact, when the Roman emperor Augustus levied a tax on all goods sold at auction, the official rate was 1%. As arithmetic books appeared near the end of that century, the use of percent became well established. If you remember your Roman numerals, recall that X stands for 10 and C for 100, so Giorgio Chiarino (1481) used "XX. per .C." for 20 percent and "VIII in X perceto" to mean 8 to 10 percent.

5.1 Percent Notation

Objectives

You should be able to:

A › Convert a percent to a decimal.

B › Convert a decimal to a percent.

C › Convert a percent to a fraction.

D › Convert a fraction to a percent.

E › Solve applications involving the concepts studied.

To Succeed, Review How To . . .

1. Write a ratio as a reduced fraction. (pp. 122–124, 256–258)
2. Divide a number by 100. (pp. 213, 217)

Getting Started

The ad says you can get a condominium for "just 5% down." The symbol % (read "percent") means *per hundred*. Thus, the ad states that for every \$100 the condominium costs, the buyer will have to pay \$5 at the time of purchase.

The word **percent** comes from the Latin phrase "*per centum,*" which means "*per hundred*" The "*per*" indicates a division and the "*centum*" indicates that the division will be by 100. We use the symbol % to indicate a percent. Thus, instead of saying "5 per hundred" we simply write 5%. As you recall, the 5% can be written in at least three ways:

1. As the **ratio** 5 to 100
2. As the **fraction** $\frac{5}{100}$
3. As the **decimal** 0.05

In this section, we will learn how to convert from one of these forms to another.

A › Converting from Percent to Decimal Notation

Since % means per hundred, numbers written as percents can be converted to decimals by dividing by 100, that is, by moving the decimal point two places to the left (see Objective E in Section 3.2). Look for the patterns in the conversions.

$$37\% = \frac{37}{100} = 0.37$$

$$4\% = \frac{4}{100} = 0.04$$

$$129\% = \frac{129}{100} = 1.29$$

Thus, to change a percent to a decimal, we use this rule:

CONVERTING A PERCENT TO A DECIMAL

Move the decimal point in the number *two* places to the *left* and omit the % symbol.

For example,

$$93\% = .93 = 0.93$$
$$41\% = .41 = 0.41$$
$$147\% = 1.47 = 1.47$$

EXAMPLE 1 Converting percents to decimals

Write as a decimal.

a. 49% **b.** 23.7%

SOLUTION

a. $49\% = .49 = 0.49$ **b.** $23.7\% = .237 = 0.237$

PROBLEM 1

Write as a decimal.

a. 47% **b.** 493%

If the given percent involves a fraction, we first write the fractional part as a decimal and then move the decimal two places to the left. Thus, to write $12\frac{1}{2}\%$ as a decimal, we write

$$12\tfrac{1}{2}\% = 12.5\% = .125 = 0.125$$

Write $\frac{1}{2}$ as .5. Move the decimal point two places to the left.

EXAMPLE 2 Converting percents to decimals

Write as a decimal.

a. $3\frac{1}{4}\%$ **b.** $18\frac{2}{3}\%$

SOLUTION

a. $3\frac{1}{4}\% = 3.25\% = .0325 = 0.0325$

b. Since $\frac{2}{3} = 0.666\ldots = 0.\overline{6}$,

$$18\tfrac{2}{3}\% = 18.\overline{6}\% = .18\overline{6} = 0.18\overline{6}$$

PROBLEM 2

Write as a decimal.

a. $2\frac{1}{8}\%$ **b.** $5\frac{1}{3}\%$

B › Converting from Decimal to Percent Notation

Since % means *per hundred,* to convert a decimal to a percent, we first convert the decimal to hundredths and then write the hundredths as a percent. For example.

$$0.17 = \frac{17}{100} = 17\%$$
$$0.02 = \frac{2}{100} = 2\%$$
$$4.11 = \frac{411}{100} = 411\%$$

Answers to PROBLEMS

1. a. 0.47 **b.** 4.93

2. a. 0.02125 **b.** $0.05\overline{3}$

Here is the rule we use.

CONVERTING A DECIMAL TO A PERCENT

Move the decimal point *two* places to the *right* and attach the % symbol.

Thus

$$0.43 = 043.\% = 43\%$$
$$0.09 = 009.\% = 9\%$$

EXAMPLE 3 Converting decimals to percents

Write as a percent.

a. 0.05 **b.** 4.19 **c.** 81.2

SOLUTION

a. $0.05 = 005.\% = 5\%$

b. $4.19 = 419.\% = 419\%$

c. $81.2 = 8120.\% = 8120\%$

PROBLEM 3

Write as a percent.

a. 0.07 **b.** 3.14 **c.** 71.8

C › Converting from Percent to Fraction Notation

Since % means *per hundred,*

$$5\% = \frac{5}{100} = \frac{1}{20}$$
$$7\% = \frac{7}{100}$$
$$23\% = \frac{23}{100}$$
$$4.7\% = \frac{4.7}{100}$$
$$134\% = \frac{134}{100} = \frac{67}{50}$$

Thus, we can convert a number written as a percent to a fraction by using the following rule:

CONVERTING A PERCENT TO A FRACTION

Write the *number* over 100, *reduce* the fraction, and omit the % sign.

EXAMPLE 4 Converting percents to fractions

Write as a fraction.

a. 49% **b.** 75%

SOLUTION

a. $49\% = \frac{49}{100}$

b. $75\% = \frac{75}{100} = \frac{3}{4}$

PROBLEM 4

Write as a fraction.

a. 41% **b.** 25%

Answers to PROBLEMS

3. a. 7% **b.** 314% **c.** 7180% **4. a.** $\frac{41}{100}$ **b.** $\frac{1}{4}$

In case the number involves a fraction or a decimal, we follow a similar procedure. Thus,

$$12\tfrac{1}{2}\% = \frac{12\tfrac{1}{2}}{100} = \frac{\tfrac{25}{2}}{100} = \frac{25}{2} \div 100 = \frac{25}{2} \times \frac{1}{100} = \frac{25}{200} = \frac{1}{8}$$

$$16.5\% = \frac{16.5}{100} = \frac{16.5 \times 10}{100 \times 10} = \frac{165}{1000} = \frac{33}{200}$$

Note that 16.5 has *one* decimal digit, so we multiplied by 10.

EXAMPLE 5 Converting percents to fractions

Write as a fraction.

a. $5\tfrac{1}{2}\%$ **b.** 15.55%

SOLUTION

a. $5\tfrac{1}{2}\% = \frac{5\tfrac{1}{2}}{100} = \frac{\tfrac{11}{2}}{100} = \frac{11}{200}$

b. $15.55\% = \frac{15.55}{100} = \frac{15.55 \times 100}{100 \times 100} = \frac{1555}{10{,}000} = \frac{311}{2000}$

PROBLEM 5

Write as a fraction.

a. $3\tfrac{1}{3}\%$ **b.** 4.5%

D › Converting from Fraction to Percent Notation

How do we write a fraction as a percent? If the fraction has a denominator that is a factor of 100, it is easy. To write $\frac{1}{5}$ as a percent, we first multiply numerator and denominator by a number that will make the denominator 100. Thus,

$$\frac{1}{5} = \frac{1 \times 20}{5 \times 20} = \frac{20}{100} = 20\%$$

Similarly,

$$\frac{3}{4} = \frac{3 \times 25}{4 \times 25} = \frac{75}{100} = 75\%$$

Note that in both cases the denominator of the fraction was a factor of 100.

EXAMPLE 6 Converting a fraction to a percent

Write $\frac{4}{5}$ as a percent.

SOLUTION $\frac{4}{5} = \frac{4 \times 20}{5 \times 20} = \frac{80}{100} = 80\%$

PROBLEM 6

Write $\frac{2}{5}$ as a percent.

In Example 6 the denominator of $\frac{4}{5}$, the 5, was a *factor* of 100. Suppose we wish to change $\frac{1}{6}$ to a percent. The problem here is that 6 is *not* a factor of 100. Don't panic! We can write $\frac{1}{6}$ as a percent by dividing the numerator 1 by the denominator 6. Divide 1 by 6, continuing the division until we have two decimal digits:

```
   0.16
6 )1.00
    6
    40
    36
     4   remainder
```

Answers to PROBLEMS

5. a. $\frac{1}{30}$ **b.** $\frac{9}{200}$

6. 40%

The answer is 0.16 with remainder 4; that is,

$$\frac{1}{6} = 0.16\frac{4}{6} = 0.16\frac{2}{3} = \frac{16\frac{2}{3}}{100} = 16\frac{2}{3}\%$$

Similarly, $\frac{2}{3}$ can be written as a percent by dividing 2 by 3, obtaining

$$\begin{array}{r} 0.66 \\ 3\overline{)2.00} \\ \underline{1\,8} \\ 20 \\ \underline{18} \\ 2 \end{array} \quad \text{remainder}$$

Thus,

$$\frac{2}{3} = 0.66\frac{2}{3} = \frac{66\frac{2}{3}}{100} = 66\frac{2}{3}\%$$

Here is the rule we need.

CONVERTING A FRACTION TO A PERCENT

Divide the *numerator* by the *denominator* (carry the division to two decimal places), convert the resulting decimal to a percent by *moving the decimal point two places to the right,* and attach the % symbol.

EXAMPLE 7 Converting a fraction to a percent

Write $\frac{5}{8}$ as a percent.

SOLUTION Dividing 5 by 8, we have

$$\begin{array}{r} 0.62 \\ 8\overline{)5.00} \\ \underline{4\,8} \\ 20 \\ \underline{16} \\ 4 \end{array}$$

Thus,

$$\frac{5}{8} = 0.62\frac{4}{8} = 0.62\frac{1}{2} = 0.62\frac{1}{2}\% = 62\frac{1}{2}\%$$

PROBLEM 7

Write $\frac{3}{16}$ as a percent.

E > Applications Involving Percents

EXAMPLE 8 Converting percents to fractions or decimals.

A report from the National Research Council recommends that we get more than 55% of calories from carbohydrates. Write 55% in each form.

a. As a reduced fraction **b.** As a decimal

SOLUTION

a. $55\% = \frac{55}{100} = \frac{11}{20}$

b. $55\% = .55 = 0.55$

PROBLEM 8

The same report recommends that we reduce total fats to less than 30% of all calories. Write 30% in each form.

a. As a reduced fraction

b. As a decimal

Answers to PROBLEMS

7. $18\frac{3}{4}\%$ **8. a.** $\frac{3}{10}$ **b.** 0.30

As a final helpful item, here are some of the most-used decimal-fraction-percent equivalents.

Decimal-Fraction-Percent Equivalents

Fraction	$\frac{1}{8}$	$\frac{1}{6}$	$\frac{1}{5}$	$\frac{1}{4}$	$\frac{1}{3}$	$\frac{3}{8}$	$\frac{2}{5}$	$\frac{1}{2}$
Decimal	0.125	$0.1\overline{6}$	0.2	0.25	$0.\overline{3}$	0.375	0.4	0.5
Percent	$12\frac{1}{2}\%$	$16\frac{2}{3}\%$	20%	25%	$33\frac{1}{3}\%$	$37\frac{1}{2}\%$	40%	50%

Fraction	$\frac{3}{5}$	$\frac{5}{8}$	$\frac{2}{3}$	$\frac{3}{4}$	$\frac{4}{5}$	$\frac{5}{6}$	$\frac{7}{8}$	1
Decimal	0.6	0.625	$0.\overline{6}$	0.75	0.8	$0.8\overline{3}$	0.875	1.0
Percent	60%	$62\frac{1}{2}\%$	$66\frac{2}{3}\%$	75%	80%	$83\frac{1}{3}\%$	$87\frac{1}{2}\%$	100%

You should try to become familiar with the table before attempting the exercises.

Where can you get all the carbohydrates from Example 8? You can get them at Mimi's Cafe!

EXAMPLE 9 Converting fractions, decimals, and percents

Mimi's Cafe has a 300-person seating capacity. If 60 persons can sit in the Garden Room,

a. What fraction of the persons can sit in the Garden Room?

b. What percent of the persons can sit in the Garden Room?

c. Write the fraction of the persons that can sit in the Garden Room as a decimal.

SOLUTION

a. 60 out of 300 can sit in the Garden Room, that is,

$$\frac{60}{300} = \frac{1}{5} \text{ of the persons.}$$

b. To convert $\frac{1}{5}$ to a decimal, divide 1 by 5, or look at the chart following Example 8.

In either case, $\frac{1}{5} = 20\%$.

c. To write $\frac{1}{5}$ as a decimal, divide 1 by 5 or look at the chart.

In either case, $\frac{1}{5} = 0.20$.

PROBLEM 9

Beto's Mexican Restaurant has a 200-person seating capacity. If 45 persons can sit in the Guadalajara Room:

a. What fraction of the persons can sit in the Guadalajara Room?

b. What percent of the persons can sit in the Guadalajara Room?

c. Write the fraction of the persons that can sit in the Guadalajara Room as a decimal.

Calculator Corner

Some calculators have a special percent key, [%], that will automatically change percents to decimals. Thus, to do Example 4(a): Change 49% to a decimal by pressing [4] [9] [%]. The display shows the decimal representation 0.49. Similarly, to change 23.7% to a decimal, key in [2] [3] [.] [7] [%] and the answer 0.237 will be displayed.

If you *do not* have a [%] key, you can still use the rule given in the text; that is, divide the number by 100. Thus, to convert 23.7% to a decimal, enter [2] [3] [.] [7] [÷] [1] [0] [0] [ENTER]. The answer, 0.237, will be displayed.

Answers to PROBLEMS

9. a. $\frac{9}{40}$ **b.** $22\frac{1}{2}\%$ **c.** 0.225

Exercises 5.1

Web IT go to mhhe.com/bello for more lessons

⟨A⟩ Converting from Percent to Decimal Notation In Problems 1–12, write as a decimal.

1. 3% **2.** 2% **3.** 10% **4.** 15% **5.** 300%

6. 100% **7.** $12\frac{1}{4}\%$ **8.** $10\frac{1}{4}\%$ **9.** 11.5% **10.** 0.09%

11. 0.3% **12.** 0.1%

⟨B⟩ Converting from Decimal to Percent Notation In Problems 13–22, write as a percent.

13. 0.04 **14.** 0.06 **15.** 0.813 **16.** 0.312 **17.** 3.14

18. 9.31 **19.** 1.00 **20.** 2.1 **21.** 0.002 **22.** 0.314

⟨C⟩ Converting from Percent to Fraction Notation In Problems 23–36, write as a (reduced) fraction.

23. 30% **24.** 40% **25.** 6% **26.** 2% **27.** 7%

28. 19% **29.** $4\frac{1}{2}\%$ **30.** $2\frac{1}{4}\%$ **31.** $1\frac{1}{3}\%$ **32.** $5\frac{2}{3}\%$

33. 3.4% **34.** 6.2% **35.** 10.5% **36.** 20.5%

⟨D⟩ Converting from Fraction to Percent Notation In Problems 37–50, write as a percent.

37. $\frac{3}{5}$ **38.** $\frac{4}{25}$ **39.** $\frac{1}{2}$ **40.** $\frac{3}{50}$ **41.** $\frac{5}{6}$

42. $\frac{1}{3}$ **43.** $\frac{3}{8}$ **44.** $\frac{7}{8}$ **45.** $\frac{4}{3}$ **46.** $\frac{7}{6}$

47. $\frac{81}{100}$ **48.** $\frac{10}{100}$ **49.** $\frac{3}{20}$ **50.** $\frac{7}{20}$

⟨E⟩ Applications Involving Percents

51. *Success in your first year* Which college courses are most important to a student's first-year success? The results are in the chart.

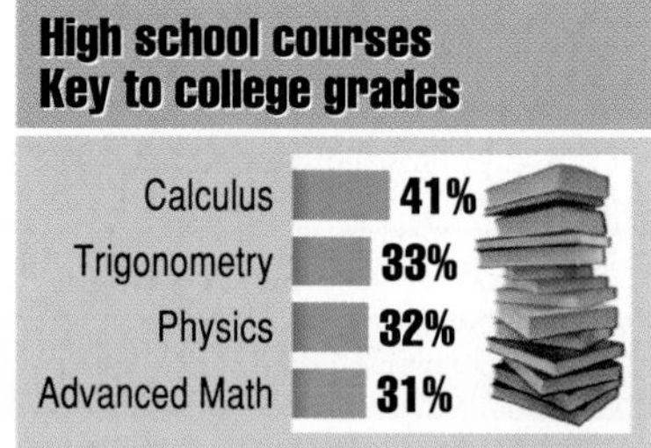

Source: ACT Inc., Information for Life Transitions.

a. Write 41% as a fraction and as a decimal.

b. Write 33% as a fraction and as a decimal.

c. Write 32% as a reduced fraction and as a decimal.

d. Write 31% as a fraction and as a decimal.

52. *Popular cell phone features* Which are the most popular cell phone features? The results are in the chart.

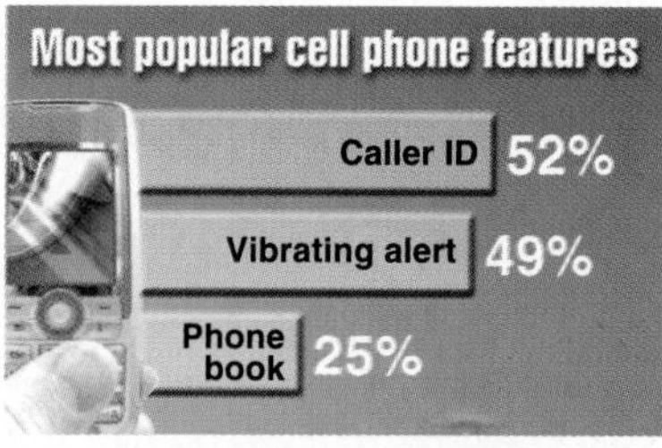

Source: Sony Ericson Mobile Communications.

a. Write 52% as a reduced fraction and as a decimal.

b. Write 49% as a fraction and as a decimal

c. Write 25% as a reduced fraction and as a decimal.

53. *Cell phones* Now that you know the desirable features in a cell phone, where are cell phones not welcome? The chart tells you!

Source: Opinion Research Corporation for Cingular Wireless.

a. Write 44% as a reduced fraction and as a decimal.

b. Write 20% as a reduced fraction and as a decimal

c. Write 15% as a reduced fraction and as a decimal.

54. *Bathroom time* How long do you spend in the bathroom? Adults ages 30 to 70 spend an hour a day, almost two weeks a year, in the bathroom. Look at the chart to see what they are doing.

Source: Yankelovich Partners.

a. Write 53% as a fraction and as a decimal.

b. Write 47% as a fraction and as a decimal.

c. Write 33% as a fraction and as a decimal.

55. *Internet use* Do you use the Internet? Assume there are currently 300 million people in the United States, and 210 million of them use the Internet.

a. What reduced fraction of the people use the Internet?

b. What percent of the people use the Internet?

c. Write the percent of people using the Internet as a decimal rounded to the nearest hundredth.

56. *Internet use* Of the 210 million people that use the Internet, 145 million use it at home.

a. What fraction of the people use the Internet at home?

b. What percent of the people use the Internet at home? (Answer to the nearest percent.)

c. Write the percent of people using the Internet at home as a decimal rounded to the nearest hundredth.

57. *Internet use* Of the 210 million people that use the Internet, 165 million have made a purchase online:

a. What fraction of the people made a purchase online?

b. What percent of the people made a purchase online? (Answer to the nearest percent.)

c. Write the fraction of people that made a purchase online as a decimal rounded to the nearest hundredth.

58. *Radio station ownership* What radio station do you listen to and who owns it? According to a National Telecommunications and Information Administration survey, there are about 10,000 AM and FM stations in the United States. If 170 of these stations are owned by African-Americans,

a. What reduced fraction of the stations is that?

b. What percent of the stations is that? (Answer to the nearest percent.)

c. Write the fraction of stations owned by African-Americans as a decimal

59. *Radio station ownership* African-Americans currently own 100 out of 4800 commercial AM stations in the United States.

a. What reduced fraction of the stations is that?

b. What percent of the stations is that? (Answer to the nearest percent.)

c. Write the fraction of stations owned by African-Americans as a decimal rounded to the nearest hundredth.

60. *Radio station ownership* There are 10,000 AM and FM radio stations in the United States. Hispanics currently own 130 of them.

a. What reduced fraction of the stations is that?

b. What percent of the stations is that? (Answer to the nearest percent.)

c. Write the fraction of AM stations owned by Hispanics in part **a** as a decimal.

61. *Radio station ownership* Hispanics currently own 85 out of 4800 commercial AM stations in the United States.

a. What reduced fraction of the stations is that?

b. What percent of the stations is that? (Answer to the nearest percent.)

c. Write the fraction of stations owned by Hispanics in part **a** as a decimal rounded to the nearest thousandth.

Source: National Telecommunications and Information Administration.

62. *HAP (hazardous air pollution)* Do you have a Hazardous Air Pollutant (HAP) Inventory in your county or state? In Hillsborough County, about 14,000 tons of HAPs were emitted. If about 7000 came from point sources such as power plants:

a. What reduced fraction of the HAPs came from point sources?

b. What percent of the HAPs came from point sources?

c. Write the percent of the HAPs coming from point sources as a decimal.

Web IT go to mhhe.com/bello for more lessons

Web IT go to mhhe.com/bello for more lessons

63. *HAP* If about 4200 tons of the 14,000 tons (see Problem 62) came from mobile sources such as on-road vehicles:

a. What reduced fraction of the HAPs came from mobile sources?

b. What percent of the HAPs came from mobile sources?

c. Write the percent of the HAPs coming from mobile sources as a decimal.

Source: Environmental Protection Agency.

64. *Surveys* In a mythical survey of 300 chinchillas, the number expressing the concerns listed are as follows:

a. Eating too little, 30. What reduced fraction is that? What percent is that?

b. Losing fur patches, 33. What reduced fraction is that? What percent is that?

c. Had no problems, 34. What reduced fraction is that? What percent is that (to the nearest percent)?

d. Eating too much, 87. What reduced fraction is that? What percent is that?

e. Acting weird, 126. What reduced fraction is that? What percent is that?

If you actually want to see a pie chart for these outcomes, go to http://www.one38.org/1000/piecharts/issue03/chinchilla.html.

Online poll The following information will be used in Problems 65–68.

An Excite Today Online Poll asked the question: Have you ever taken medication to help you fall asleep? The results are shown below.

Yes 52% (4463 votes)

No 46% (3993 votes)

I'm not sure 0% (44 votes)

Current number of voters: 8500

Source: Excite.com http://poll.excite.com/poll/results.jsp?cat_id=1&poll_id=7

65. Of the respondents, 52% said they have taken medication to help them fall asleep:

a. Write 52% as a reduced fraction.

b. Write 52% as a decimal.

66. Of the respondents, 46% said they have *not* taken medication to help them fall asleep:

a. Write 46% as a decimal.

b. Write 46% as a reduced fraction.

67. In the survey, 52% of the respondents answered *Yes,* 46% answered *No,* and 0% were *not sure.* Do the totals add to 100%? Explain.

68. In the survey, 4463 respondents out of 8500 answered Yes and 3993 answered No.

a. Write the number of respondents answering Yes as a fraction (do not reduce).

b. Convert the fraction in part **a** to a percent rounded to the nearest tenth.

c. Write the fraction of respondents answering No as a fraction (do not reduce).

d. Convert the fraction in part **c** to a percent rounded to the nearest tenth.

e. What about the 44 respondents out of 8500 who were not sure? Write the number of respondents not sure as a percent rounded to the nearest tenth.

f. Do the totals add to 100% now?

>>> Using Your Knowledge

The Pursuit of Happiness Recently the magazine *Psychology Today* conducted a survey about happiness. Here are some conclusions from that report:

69. 7 out of 10 people said they had been happy over the last six months. What percent of the people is that?

70. 70% expected to be happier in the future than now. What fraction of the people is that?

71. 0.40 of the people felt lonely.

a. What percent of the people is that?

b. What fraction of the people is that?

72. Only 4% of the men were ready to cry. Write 4% as a decimal.

73. Of the people surveyed, 49% were single. Write 49% as

a. a fraction. **b.** a decimal.

Do you wonder how they came up with some of these percents? They used their knowledge. You do the same and fill in the spaces in the following table, which refers to the marital status of the 52,000 people surveyed. For example, in the first line 25,480 persons out of 52,000 were single. This is

$$\frac{25{,}480}{52{,}000} = 49\%$$

	Marital Status	Number	Percent
	Single	25,480	49%
74.	Married (first time)	15,600	___
75.	Remarried	2,600	___
76.	Divorced, separated	5,720	___
77.	Widowed	520	___
78.	Cohabiting	2,080	___

Write On

79. The chart following Example 8 shows the most used decimal-fraction-percent equivalents. In real life, percents are the most used. Explain in your own words why you think that is.

80. List at least three activities in your everyday life in which you use percents.

Concept Checker

Fill in the blank(s) with the correct word(s), phrase, or mathematical statement.

divide	**attach**
left	**reduce**
omit	**multiply**
two	**percent**
right	**fraction**
100	**%**

81. To convert a **percent** to a **decimal,** move the decimal point ______ places to the ______ and omit the % symbol.

82. To convert a **decimal** to a **percent,** move the decimal point ______ places to the ______ and attach the % symbol.

83. To convert a **percent** to a **fraction,** follow these steps:

a. Write the number over ______

b. ______ the resulting fraction

c. ______ the % symbol

84. To convert a **fraction** to a **percent,** follow these steps:

a. ______ the numerator by the denominator

b. Convert the resulting decimal to a ______

c. Attach the ______ symbol

Mastery Test

85. Among taxpayers who owed the IRS money, 82% said they would use money from their savings or checking accounts to pay their taxes. Write 82% as

a. a reduced fraction. **b.** a decimal.

86. Write as a percent:

a. $\frac{3}{8}$ **b.** $\frac{3}{5}$

87. Write as a reduced fraction:

a. $6\frac{1}{2}\%$ **b.** 6.55%

88. Write as a reduced fraction:

a. 19% **b.** 80%

89. Write as a percent:

a. 0.06 **b.** 6.19 **c.** 42.2

90. Write as a decimal:

a. $2\frac{1}{2}\%$ **b.** $10\frac{2}{3}\%$

91. Write as a decimal:

a. 38% **b.** 29.3%

Skill Checker

92. Multiply 0.60×40

93. Multiply $\frac{1}{3} \times 87$

94. Solve $R \times 60 = 30$

95. Solve $0.30x = 12$

96. Solve $50x = 20$

97. Solve $15 = 45x$

98. Solve $81{,}500R = 8150$

5.2 Percent Problems

Objectives

You should be able to:

A Find the percentage when the base and rate are given.

B Find the percent when the base and percentage are given.

C Find the base when the percentage and percent are given.

D Solve applications involving the concepts studied.

To Succeed, Review How To . . .

1. Write a percent as a fraction or as a decimal. (pp. 288, 290–291)
2. Use the **RSTUV** procedure to solve word problems. (p. 92)
3. Solve equations involving decimals. (pp. 240, 241)

Getting Started

In the cartoon, the chairman of the board of Mogul Oil wishes to have a 300% increase in profit. If he had a profit of $10 million last year, by how much does he want to increase his profit? Since $300\% = \frac{300}{100} = 3.00 = 3$, he wishes to increase his profit by

$$300\% \text{ of } 10 \quad \text{(million)}$$

$$3 \times 10 = 30 \quad \text{(million)}$$

Thus, he wishes to increase his profit by $30 million.

FUNKY WINKERBEAN **Tom Batiuk**

A › Finding the Percentage

As you can see from the example in the *Getting Started,* there are three quantities involved:

1. The **base,** or **total** (the standard used for comparison purposes)
2. The **percentage** (the part being compared with the base or total)
3. The **percent,** or **rate** (the part indicating the ratio of the percentage to the base)

Note that the *percent,* or *rate,* is usually written using the symbol %, but the *percentage* is a number. In the preceding example the *base,* or *total,* is $10 million, the *percentage* is 30, and the *percent* is 300%. In this section, you will be asked to find the percentage P, the percent, or rate, R, or the base B when the other two are given. To do this, we will use the **RSTUV** procedure studied in Sections 1.9, 2.8, and 3.5 along with the following dictionary.

Word	Translation	Symbol
of	multiply	$\cdot$ or $\times$
is	equal	$=$
what	variable or letter	$x, y, \square$
percent	decimal or fraction	$\times \frac{1}{100}$ or 0.01

Let us solve a problem: 40% of 60 is what number? Here are the steps:

Step 1. Read the problem.

Step 2. Select a variable to represent the unknown. The *base* is 60 and the *percent* is 40%; we need to know the percentage P.

Step 3. Translate the problem into an equation.

40% of 60 is what number? is translated as

$0.40 \times 60 = \square$

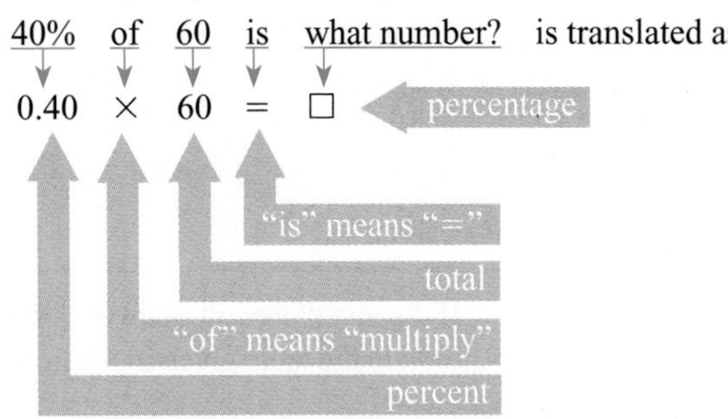

Step 4. Use the rules studied to solve the equation. (Remember to write all percents as decimals or fractions *before* solving the equation.)

$$0.40 \times 60 = P \quad \text{Given.}$$
$$24 = P \quad \text{Multiply.}$$

Thus, 40% of 60 is 24.

Step 5. Verify that the answer is correct.

Since $\frac{24}{60} = 0.40 = 40\%$, our answer is correct.

EXAMPLE 1 Finding the percentage of a number

60% of 30 is what number?

SOLUTION

Step 1. Read the problem.

Step 2. Select a variable to represent the unknown. Let P be the unknown.

Step 3. Translate into an equation.

60% of 30 is what number?

$0.60 \times 30 = P$

Step 4. Use the rules studied to solve the equation.

$$0.60 \times 30 = P \quad \text{Given.}$$
$$18 = P \quad \text{Multiply.}$$

Thus, 60% of 30 is 18.

Step 5. Verify that the answer is correct.

PROBLEM 1

70% of 40 is what number?

If the given percent involves a fraction, try to write the fractional part as a decimal by dividing numerator by denominator as shown in Example 2.

Answers to PROBLEMS

1. 28

EXAMPLE 2 Finding the percentage of a number

Find $12\frac{1}{2}\%$ of 80.

SOLUTION

Step 1. Read the problem.

Step 2. Select a variable to represent the unknown. Let P be the unknown.

Step 3. Translate into an equation.

$$12\frac{1}{2}\% \text{ of } 80 \text{ is what number?}$$

$$0.12\frac{1}{2} \times 80 = P$$

$$0.125 \times 80 = P$$

Step 4. Use the rules studied to solve the equation.

$$0.125 \times 80 = P \quad \text{Given.}$$

$$10.000 = P \quad \text{Multiply.}$$

Thus, $12\frac{1}{2}\%$ of 80 is 10.

Step 5. Verify that the answer is correct.

PROBLEM 2

Find $8\frac{1}{2}\%$ of 60.

If the given percent involves a fraction that cannot be converted to a terminating decimal, look in the table given in Section 5.1 for its fractional equivalent, as shown in Example 3.

EXAMPLE 3 Finding the percentage of a number

What is $33\frac{1}{3}\%$ of 84?

SOLUTION

Step 1. Read the problem.

Step 2. Select a variable to represent the unknown. Let P be the unknown.

Step 3. Translate into an equation.

$$33\frac{1}{3}\% \text{ of } 84 \text{ is what number?}$$

$$\frac{1}{3} \times 84 = P$$

From the table in Section 5.1, $33\frac{1}{3}\% = \frac{1}{3}$ because

$$33\frac{1}{3}\% = \frac{\frac{100}{3}}{100} = \frac{100}{3} \cdot \frac{1}{100} = \frac{1}{3}$$

Step 4. Use the rules studied to solve the equation.

$$\frac{1}{3} \times 84 = P \quad \text{Given.}$$

$$28 = P \quad \text{Multiply.}$$

Thus, $33\frac{1}{3}\%$ of 84 is 28.

Step 5. Verify that the answer is correct.

PROBLEM 3

What is $66\frac{2}{3}\%$ of 90?

B > Finding the Percent

Here is another type of problem. The prices in the ad start at \$81,500 and you can pay \$8150 down. What *percent* is that? This problem can be stated in different ways. Here are some of them:

8150 is what percent of 81,500?

Find what percent of 81,500 is 8150.

What percent of 81,500 is 8150?

Answers to PROBLEMS

2. 5.1 **3.** 60

As before, we use the **RSTUV** procedure to solve the problem.

Step 1. **R**ead the problem.

Step 2. **S**elect a variable to represent the unknown. Let R be the unknown.

Step 3. **T**ranslate into an equation.

What percent of 81,500 is 8150?

$$R \times 81{,}500 = 8150$$

Step 4. **U**se the rules studied to solve the equation.

$$R \times 81{,}500 = 8150 \quad \text{Given.}$$

$$R = \frac{8150}{81{,}500} = \frac{1}{10} \quad \text{Divide by 81,500.}$$

Since we are looking for a percent, we must convert $\frac{1}{10}$ to a percent. Thus

$$R = \frac{1}{10} = \frac{1 \cdot 10}{10 \cdot 10} = \frac{10}{100} = 10\%$$

Thus, \$8150 is 10% of \$81,500.

Step 5. **V**erify the answer.
Is it true that 10% of 81,500 is 8150? Yes
($0.10 \times 81{,}500 = 8150$). Thus, our answer is correct.

EXAMPLE 4 Finding the percent of a number

Find what percent of 40 is 20.

SOLUTION

Step 1. **R**ead the problem.

Step 2. **S**elect a variable to represent the unknown. Let R be the unknown.

Step 3. **T**ranslate into an equation.

What percent of 40 is 20?

$$R \times 40 = 20$$

Step 4. **U**se the rules studied to solve the equation.

$$R \times 40 = 20 \quad \text{Given.}$$

$$R = \frac{20}{40} = \frac{1}{2} \quad \text{Divide by 40.}$$

Since we are looking for a percent, we must convert $\frac{1}{2}$ to a percent. Thus

$$R = \frac{1}{2} = \frac{1 \cdot 50}{2 \cdot 50} = \frac{50}{100} = 50\%$$

Thus, 50% of 40 is 20.

Step 5. **V**erify the answer.
Is it true that 50% of 40 is 20? Yes
($0.50 \times 40 = 20$). Thus, our answer is correct.

PROBLEM 4

Find what percent of 30 is 3.

SAVE 20% thru wednesday
Free-moving
Body Suit and Jeans
both machine washable
REGULAR PRICE 30.00 each

C › Finding the Base

Suppose you save \$6 when buying the article in the ad. Verify the regular (base) price of the article. To find the answer, we use the **RSTUV** procedure.

Answers to PROBLEMS

4. 10%

Step 1. Read the problem.

Step 2. Select a variable to represent the unknown. Let B be the unknown.

Step 3. Translate the problem. Since you save \$6, which is equivalent to the 20%, the equation would be

$$20\% \text{ of } B = 6$$

$$0.20 \times B = 6$$

Step 4. Use the rules studied to solve the equation.

$$0.20 \times B = 6 \quad \text{Given.}$$

$$B = \frac{6}{0.20} \quad \text{Divide by 0.20.}$$

$$= \frac{6 \cdot 100}{0.20 \cdot 100} \quad \text{Multiply the numerator and denominator by 100.}$$

$$= \frac{600}{20} = 30 \quad \text{Simplify.}$$

(You get the same answer if you use long division to divide 6 by 0.20.)

Thus, the original (regular) price of the article was \$30.

Step 5. Verify the answer.

Is it true that 20% of 30 is 6? Yes

($0.20 \times 30 = 6$). Thus, our answer is correct.

There are different ways of stating problems in which the base is the unknown. For example, the statement "10 is 40% of what number?" can be translated as

10 is 40% of a number. Find the number.

Find a number so that 40% of it is 10.

40% of what number is equal to 10?

We solve for the base in Example 5.

EXAMPLE 5 Finding the base

10 is 40% of what number?

SOLUTION As before, we use the **RSTUV** method.

Step 1. Read the problem.

Step 2. Select a variable to represent the unknown. Let B be the unknown.

Step 3. Translate into an equation.

10 is 40% of what number?

$$10 = 0.40 \times B$$

Step 4. Use the rules studied to solve the equation.

$$10 = 0.40 \times B \quad \text{Given.}$$

$$B = \frac{10}{0.40} \quad \text{Divide by 0.40.}$$

$$= \frac{10 \cdot 100}{0.40 \cdot 100} \quad \text{Multiply by 100.}$$

$$= \frac{1000}{40} = 25 \quad \text{Simplify.}$$

Thus, 10 is 40% of 25.

Step 5. Verify the answer.

It is true that $10 = 0.40 \times 25$.

Thus, our answer is correct.

PROBLEM 5

50 is 20% of what number?

Answers to PROBLEMS

5. 250

D > Applications Involving Percents

EXAMPLE 6 Calculating the price of an item

The sales tax on an item was \$3.64. If the tax rate is 4%, what was the price of the item?

SOLUTION

Step 1. Read the problem.

Step 2. Select a variable to represent the unknown. Let B dollars be the unknown.

Step 3. Translate into an equation: We know that \$3.64 is 4% of the price, so

$$3.64 = 0.04 \times B$$

Step 4. Use the rules studied to solve the equation.

$3.64 = 0.04 \times B$ Given.

$B = \frac{3.64}{0.04}$ Divide by 0.04.

$= \frac{3.64 \cdot 100}{0.04 \cdot 100}$ Multiply the numerator and denominator by 100.

$= \frac{364}{4} = 91$ Simplify.

So the price was \$91.

Step 5. Verify the answer.
It is true that $\$3.64 = 0.04 \times 91$.
Thus, our answer is correct.

PROBLEM 6

The sales tax on an item was \$3.28. If the tax rate is 8%, what was the price of the item?

EXAMPLE 7 Percent problems and college costs

Natasha is a student at the University of Wisconsin. Tuition and fees for out-of-state students are about \$10,500. She has a scholarship that pays 58% of the \$10,500.

a. How much does the scholarship pay and how much does Natasha have to contribute?

b. Natasha has \$4000 saved to pay her part of the tuition and fees from part **a.** What percent of her part of the tuition and fees will she be paying?

c. The state of Wisconsin claims that in-state students only pay 38% of the real cost of tuition and fees. If in-state tuition and fees cost about \$4500, what is the cost to the state per student?

SOLUTION

a. The scholarship pays 58% of the \$10,500

Translation: $0.58 \times 10{,}500 = \$6090$

Verify this by multiplying 0.58 by 10,500. Be careful with the decimal.

Since the scholarship pays \$6090, she pays the rest, that is, Natasha pays $\$10{,}500 - \$6090 = \$4410$.

b. She will be paying \$4000 out of the \$4410, that is, $\frac{4000}{4410} = 91\%$ (to the nearest percent).

Verify this by dividing 4000 by 4410. Write the answer to the nearest percent.

PROBLEM 7

Doug is a student at the University of South Florida. Out-of-state fees at USF are about \$7000. He has a scholarship that pays 80% of the \$7000.

a. How much does the scholarship pay and how much does Doug have to contribute?

b. Doug has \$1000 to pay his part of the tuition from part **a.** What percent of his part of the tuition and fees will he be paying?

c. If tuition and fees cost \$1400 and this represents 25% of the state's total cost of tuition and fees, what is the state's cost per student?

(continued)

Answers to PROBLEMS

6. \$41

7. a. \$5600; \$1400

b. 71% (to the nearest percent)

c. \$5600

c. According to the problem:

Students pay 38% of tuition and fees, which is \$4500

Translation: $0.38 \times T = 4500$ Let T stand for tuition and fees.

Divide by 0.38. $T = \dfrac{4500}{0.38}$

Multiply numerator, denominator by 100. $= \dfrac{4500 \cdot 100}{0.38 \cdot 100} = \dfrac{450{,}000}{38}$

Divide 450,000 by 38. $= \$11{,}842.11$ Divide 450,000 by 38 (to the nearest cent).

Thus, it costs the state \$11,842 per student.

TRANSLATE THIS

The third step in the RSTUV procedure is to **TRANSLATE** the information into an equation. In Problems 1–10 **TRANSLATE** the sentence and match the correct translation with one of the equations A–O.

1. 70% of 80 is a number P.

2. $12\frac{1}{2}$% of 200 is a number P.

3. $33\frac{1}{3}$% of 60 is what number P?

4. What percent R of 250 is 25?

5. What percent R of 80 is 40?

6. 30% of B is 40.

7. 20 is 50% of what number B?

8. 0.48 is 6% of the price P of an item.

9. The tax T on an item represents 10% of the cost C of the item.

10. The tax T on an item is the product of its price P and its tax rate R.

A. $\frac{1}{3}\% \times 60 = P$	**B.** $0.80 \times 70 = P$	**C.** $R \times 80 = 40$	**D.** $0.70 \times 80 = P$	**E.** $R \times 40 = 80$
F. $0.125 \times 200 = P$	**G.** $20 = 50 \times B$	**H.** $0.30 \times B = 40$	**I.** $R \times 250 = 25$	**J.** $0.48 = 0.06 \times P$
K. $20 = 0.50 \times B$	**L.** $T = 0.10 \times C$	**M.** $\frac{1}{3} \times 60 = P$	**N.** $T = P \times R$	**O.** $P = T \times R$

Calculator Corner

If your calculator has a special % key, you can do some of the problems in this section a little bit faster. For example, to do Example 2: Find $12\frac{1}{2}$% of 80 by pressing 1 2 . 5 % × 8 0 ENTER and the display will show the correct answer, 10. Note that the calculator *does not* do the thinking for you. You still have to know that $\frac{1}{2} = 0.5$ and that "of" means to "multiply." As a matter of fact, the other two types of problems in the section *have to be* "set up," using the procedure given in the text, *before* the calculator is used. To do Example 2, without a % key, you press 0 . 1 2 5 × 8 0 ENTER, obtaining the same answer, 10.

Exercises 5.2

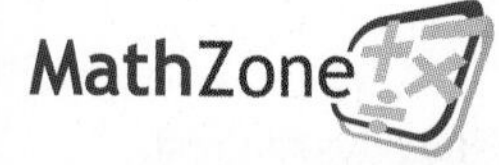

⟨A⟩ Finding the Percentage In Problems 1–14, find the percentage.

1. Find 40% of 80.

2. Find 15% of 60.

3. 150% of 8 is what number?

4. 35% of 60 is what number?

5. What is 20% of \$15?

6. What is 95% of 40?

7. Find 60% of 48.

8. Find 120% of \$30.

9. $12\frac{1}{2}$% of 40 is ________.

10. $8\frac{1}{4}$% of 50 is ________.

11. Find 3.5% of 60.

12. Find 110.5% of 30.

13. Find $16\frac{2}{3}$% of 120.

14. Find $83\frac{1}{3}$% of 90.

⟨B⟩ Finding the Percent In Problems 15–32, find the percent.

15. 315 is what percent of 3150?

16. 15 is what percent of 60?

17. 8 is what percent of 4?

18. 7 is what percent of 21?

19. Find what percent of 50 is 5.

20. Find what percent of 40 is 8.

21. What percent of 40 is 5?

22. What percent of 15 is 10?

23. $5\frac{1}{2}$ is what percent of 22?

24. $2\frac{1}{4}$ is what percent of 27?

25. 3 is ________ % of 5.

26. 20 is ________ % of 30.

27. 50 is ________ % of 25.

28. 30 is ________ % of $7\frac{1}{2}$.

29. \$50 is ________ % of \$60.

30. \$90 is ________ % of \$270.

31. \$0.75 is ________ % of \$4.50.

32. \$0.25 is ________ % of \$1.50.

⟨C⟩ Finding the Base In Problems 33–50, find the base.

33. 30% of what number is 60?

34. 15% of what number is 45?

35. $2\frac{1}{4}$% of what number is 9?

36. $5\frac{1}{2}$% of what number is 11?

37. 20 is 40% of what number?

38. 60 is 150% of what number?

39. 15 is $33\frac{1}{3}$% of a number. Find the number.

40. 20 is $66\frac{2}{3}$% of a number. Find the number.

41. Find a number so that 120% of it is 20.

42. Find a number so that $3\frac{1}{4}$% of it is $6\frac{1}{2}$.

43. 40% of what number is $7\frac{1}{2}$?

44. 140% of what number is 70?

45. 4.75% of what number is 38?

46. 3.25% of what number is 39?

47. 100% of what number is 3?

48. 8% of what number is 3.2?

49. $12\frac{1}{2}$% of what number is 37.5?

50. $33\frac{1}{3}$% of what number is $66\frac{2}{3}$?

⟨D⟩ Applications Involving Percents The graph will be used in Problems 51–52.

51. The U.S. population in 2000 was about 282 million. To the nearest million, how many kids were there in 2000?

52. By the year 2020, the U.S. population will swell to 386 million. To the nearest million, how many are estimated to be kids?

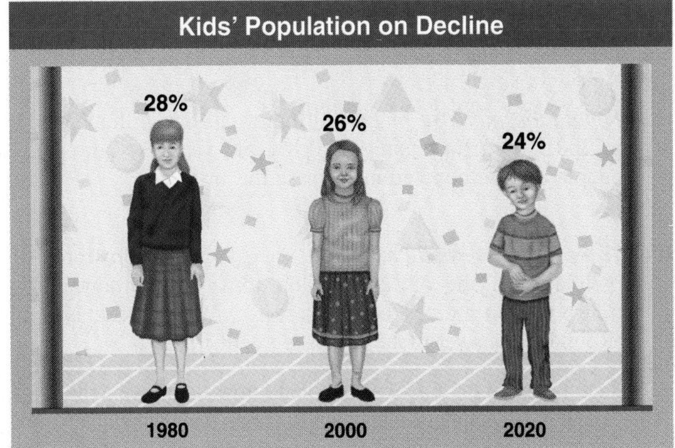

Source: Data from childstats.gov

Saving money The graph will be used in Problems 53–56.

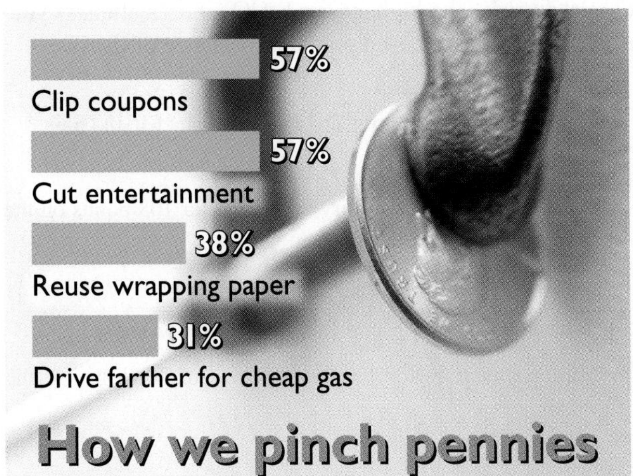

Source: Data from Progressive Insurance.

53. If there were 40 members in a family, how many would you expect to clip coupons? (Answer to the nearest whole number.)

54. If there were 4000 students in a small college, how many would you expect to cut entertainment?

55. According to the graph, 38% of the people surveyed reuse wrapping paper. If 95 persons actually reuse wrapping paper, how many persons were surveyed?

56. According to the graph, 31% of the people surveyed drive farther for cheap gas. If 93 people actually drive farther to get cheap gas, how many persons were surveyed?

Web IT go to mhhe.com/bello for more lessons

Saving for college The graph will be used in Problems 57–60.

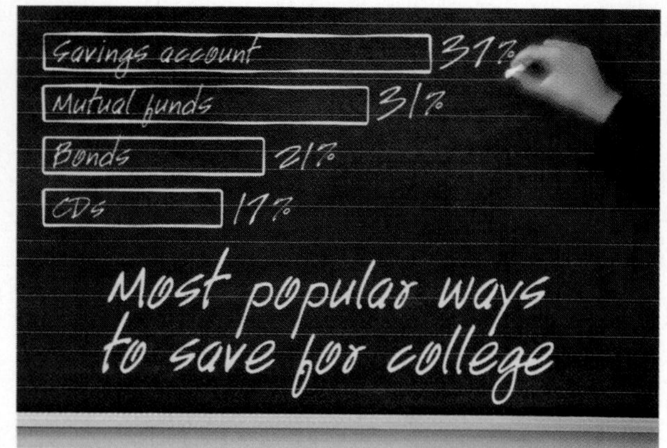

Source: Data from USA TODAY research.

57. 280 parents were surveyed as to the manner in which they saved for their children's college education.

a. If 112 parents used a savings account to save for college, what percent of the 280 is that?

b. Is the percent obtained in part **a** more or less than the 37% predicted by the graph?

58. 550 parents were surveyed as to the manner in which they saved for their children's college education.

a. If 176 parents used mutual funds to save for college, what percent of the 550 is that?

b. Is the percent obtained in part **a** more or less than the 31% predicted by the graph?

59. If 153 of the parents saved for college by using CDs (certificates of deposit) and 153 represents 17% of the total number of parents surveyed, how many parents were surveyed?

60. If 105 of the parents saved for college using bonds and 105 represents 21% of the total number of parents surveyed, how many parents were surveyed?

61. *Oil reserves* North America has approximately 5% of the world's oil reserves. This amount represents 50 billion barrels. What are the world's oil reserves?

62. *Gas reserves* Western Europe has approximately 3% of the world's gas reserves. This amount represents 60 trillion cubic meters. What are the world's gas reserves?

63. *Coal reserves* Eastern Europe and the former USSR have approximately 30% of the world's coal reserves. This amount represents 300,000 short tons. What are the world's coal reserves?

64. *Population increase in Bangladesh* In 2007, Bangladesh had a population of 150 million. Its growth rate was estimated at 2% per year. By how many people would the population increase the first year?

65. *Population increase in India* India had one billion people in 2001. Its growth rate was 1.8% per year. By how many people did the population increase the first year?

66. *Carbon monoxide released* In a recent year about 190 million tons of pollutants were released in the United States. Of this amount 47% was carbon monoxide. How many tons is that?

67. *Transportation pollutants* Of the 190 million tons of pollutants in the air, 50 million tons were produced by transportation sources (cars, buses, etc.). What percent is that? (Round the answer to the nearest percent.)

68. *Percent correct in a test* On a test consisting of 60 items a student got 40 correct. What percent did she get correct? (Round the answer to the nearest percent.)

69. *Sugar and water in Coke* Do you know who invented Coca-Cola? John Pemberton did, in 1885. In 1909 an analysis of Coca-Cola revealed that 48% of the drink was sugar and 41% was water. How much sugar and how much water was there in a 16-ounce bottle of Coca-Cola?

70. *Regular price of a sale article* The price of an article on sale was 90% of the regular price. If the sale price was $18, what was the regular price?

FICO scores are the credit scores most lenders use to determine your credit risk. The higher your FICO® scores, the less you pay to buy on credit. The average (mean) FICO score is 723. The percents shown are based on the importance of the five categories for the general population.

What's in your FICO® Scores?

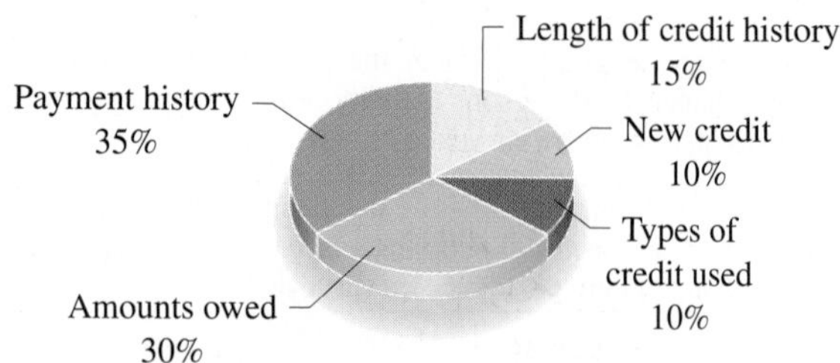

Source: http://www.myfico.com/

FICO scores In problems 71–74, find your FICO (Fair Isaac Company) score if

71. Your length of credit history accounted for 108 points on the score.

72. Your new credits accounted for 70 points on the score.

73. The amounts owed accounted for 213 points on the score.

74. Your payment history accounted for 262.5 points on the score.

Using Your Knowledge

Credit In many cases it is necessary to borrow money to expand your purchasing power. To use this money you pay a fee called a **finance charge.** Let us assume you wish to borrow \$100, to be repaid at \$25 per month plus interest of 1.5% per month on the *unpaid* balance. (The unpaid balance the first month is \$100, the second month, \$100 − \$25 = \$75, and so on.) Here is what you pay:

	Interest	Total Payment
Month 1	$0.015 \times 100 = \$1.50$	$25 + 1.50 = \$26.50$
Month 2	$0.015 \times 75 \approx 1.13$	$25 + 1.13 = 26.13$
Month 3	$0.015 \times 50 = 0.75$	$25 + 0.75 = 25.75$
Month 4	$0.015 \times 25 \approx 0.38$	$25 + 0.38 = 25.38$
		Total Payment \$103.76

Thus, the loan company might set up four equal payments of \$25.94 per month. Note that you have paid \$3.76 (\$103.76 − \$100) on the \$100 for 4 months.

75. Suppose a bank lends you \$200 to be repaid at \$50 per month plus interest at 3% of the unpaid balance. What is the total interest you pay? (*Hint:* It is *not* \$6.)

76. If the loan is to be paid in four equal payments, how much would your payments be per month?

77. What percent of the \$200 is the total interest?

78. Which loan is better, this one or another loan charging 15% annually?

The following credit cost table may help you to make decisions about loans and even save you same money in the future. Use the knowledge you gain from it.

Credit Cost Table Here is a summary of the current costs of various types of credit generally available, together with recommendations concerning each.

Type of Credit	Usual Cost Rate per Annum	Recommendation
College education loans qualifying for federal subsidy	3%	By all means!
College education loans with state or private guarantee, or with security, but without subsidy	7%	OK for good students.
First-mortgage home loans, open-end monthly payment plan	$4\frac{1}{2}$ to 12%	Good for most families with permanent, stable employment. Will clear the home from debt in 20 years or less, for rent money.
Unsecured home improvement loans from S. & L. assn's and banks	10 to 12%	Better than first-mortgage refinancing for major improvements to homes with old, low-rate mortgage that will not secure advances. Otherwise not recommended except in cases of extreme need.
Credit union loans	6 to 14%	Cheapest and most satisfactory form of small credit and auto financing.
Revolving credit plans of retail stores, including bank charge card plans	18 to 22%	OK, if you *must;* check for specials and offers. But use them sparingly.
30-day retail credit, any plan, including charge cards	None if paid in 30 days	OK for convenience, if the price of the goods is right and you *know* you can pay the bill when it comes. But check prices against cash stores.
Automobile loans	5% and up	Beware the high cost of required insurance. Better drive the old car until you save enough money to trade up with cash.

(continued)

Type of Credit	Usual Cost Rate per Annum	Recommendation
Appliance loans	18% and up	Cash will usually buy it for 20% to 30% less. Save and shop!
Personal loans—confidential—no questions asked	30% and up	Not recommended.
Loans from friends, relatives, or employers	?	Not recommended.
Payday loans, loan sharks, unlicensed	The sky's the limit	No!!!

Write On

79. How many basic types of percent problems are there? Explain the procedure you use to solve each type.

80. Write in your own words the difference between *percent* and *percentage*.

Concept Checker

Fill in the blank(s) with the correct word(s), phrase, or mathematical statement.

81. When working percent problems, the __________ is the **standard** used for **comparison** purposes.

82. The **part** being compared to the **base** or **total** is called the __________.

83. The **part** indicating the **ratio** of the **percentage** to the **base** is the __________.

84. The **percent** is usually written using the symbol __________ but the **percentage** is a __________.

number **percentage**

base **%**

percent

Mastery Test

85. The tution and fees at a college are $1500. If financial aid pays for 53% of the tuition and fees, how much is paid by financial aid? How much do you have to pay?

86. The sales tax on an item was $1.48. If the tax rate is 8%, what was the price of the item?

87. 20 is 40% of what number?

88. A bicycle costs $126. If the $126 price includes a 30% discount, what was the original price of the bicycle?

89. Find $33\frac{1}{3}\%$ of 180.

90. What percent of 40 is 10?

91. Find $12\frac{1}{2}\%$ of 160.

92. 60% of 90 is what number?

Skill Checker

Solve the following proportions.

93. $\frac{20}{100} = \frac{n}{80}$

94. $\frac{P}{100} = \frac{30}{90}$

95. $\frac{40}{100} = \frac{20}{W}$

96. $\frac{30}{100} = \frac{60}{n}$

5.3 Solving Percent Problems Using Proportions

Objectives

You should be able to:

A > Find the percentage when the base and rate are given.

B > Find the percent when the base and percentage are given.

C > Find the base when the percentage and percent are given.

To Succeed, Review How To . . .

1. Write a percent as a fraction. (pp. 290–291)
2. Solve proportions. (pp. 260–261)

Getting Started

25% Off the Slides (Sandals)

The slides are on sale for 25% off. 25% can be written as the ratio of 25 to 100, that is, $\frac{25}{100}$.

The ratios 1 to 4, 50 to 200, and 75 to 300 are all equivalent to the ratio 25 to 100, that is,

$$\frac{\text{Percent}}{100} \rightarrow \frac{25}{100} = \frac{1}{4} = \frac{50}{200} = \frac{75}{300} \leftarrow \frac{\text{Part}}{\text{Whole}}$$

We will study how to solve percent problems using proportions in this section.

Entire Stock of
ADIDAS SLIDES

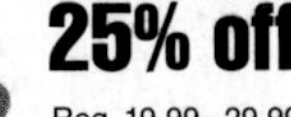

A > Finding the Percentage Using Proportions

To solve percent problems using proportions, we proceed by steps:

Step 1. Write the percent ratio as the fraction $\frac{\text{Percent}}{100}$

Step 2. Write a second ratio as $\frac{\text{Part}}{\text{Whole}}$

Step 3. Set up the proportion $\boxed{\frac{\text{Percent}}{100} = \frac{\text{Part}}{\text{Whole}}}$

PERCENT:	**The number with the percent sign (%)**
PART:	**The number with the word *is* (numerator)**
WHOLE:	**The number with the word *of* (denominator)**

As in Section 5.2, there are three types of problems:

1. Finding the percentage (Part)
2. Finding the percent
3. Finding the base or total (Whole)

EXAMPLE 1 **Finding the percentage using a proportion**

What is 20% of 80?

Note: The question can be written equivalently as
Find 20% of 80, or
20% of 80 is what number?

SOLUTION We solve using the three steps suggested:

Step 1. Write the percent ratio as the fraction $\frac{20}{100}$

Step 2. Write the second ratio as $\frac{\text{Part}}{\text{Whole}}$

Note that the number with the word *of* is 80, so 80 goes in the denominator. If we let n be the part, the second ratio is $\frac{n}{80}$.

Step 3. Set up the proportion $\boxed{\frac{20}{100} = \frac{n}{80}}$

To solve the proportion:

1. Cross multiply. $20 \times 80 = 100 \times n$
2. Divide by 100. $\frac{20 \times 80}{100} = n$
3. Simplify. $\frac{1600}{100} = 16 = n$

Thus, 20% of 80 is 16.

Verification: 20% of 80 = 0.20 × 80, which is indeed 16.

PROBLEM 1

30% of 90 is what number?

B > Finding the Percent Using Proportions

EXAMPLE 2 **Finding the percent of a number using proportions**

30 is what percent of 90?

Note: The question can be written equivalently as:
Find what percent of 90 is 30.
What percent of 90 is 30?

SOLUTION We solve using the three steps suggested:

Step 1. Write the percent ratio as the fraction $\frac{\text{Percent}}{100}$

Step 2. Write the second ratio as $\frac{\text{Part}}{\text{Whole}}$

Note that the number with the word *of* is 90. It is the whole, so 90 goes in the denominator.

30 appears with the word *is*. It is the numerator. So the second ratio is $\frac{30}{90}$.

Step 3. Set up the proportion $\boxed{\frac{\text{Percent}}{100} = \frac{30}{90}}$

To solve the proportion:

1. Cross multiply. $\text{Percent} \times 90 = 100 \times 30$
2. Divide by 90. $\text{Percent} = \frac{100 \times 30}{90}$
3. Simplify. $\text{Percent} = \frac{100}{3} = 33\frac{1}{3}$

Thus, 30 is $33\frac{1}{3}$% of 90.

Verification: $33\frac{1}{3}$% of 90 = $\frac{1}{3} \times 90$, which is indeed 30.

PROBLEM 2

Find what percent of 45 is 15.

Answers to PROBLEMS

1. 27 **2.** $33\frac{1}{3}$%

C > Finding the Base Using Proportions

EXAMPLE 3 **Finding the base or total using proportions**

Find a number so that 40% of it is 20.

Note: The question can be written equivalently as:
20 is 40% of what number?
40% of what number is 20?

SOLUTION We solve using the three steps suggested:

Step 1. Write the percent ratio as the fraction $\frac{40}{100}$

Step 2. Write the second ratio as $\frac{\text{Part}}{\text{Whole}}$

Since 20 appears next to the word *is,* 20 is the numerator.

The second ratio is $\frac{20}{\text{Whole}}$.

Step 3. Set up the proportion $\boxed{\frac{40}{100} = \frac{20}{\text{Whole}}}$

To solve the proportion:

1. Cross multiply. $40 \times \text{Whole} = 100 \times 20$
2. Divide by 40. $\text{Whole} = \frac{100 \times 20}{40}$
3. Simplify. $\text{Whole} = \frac{2000}{40} = 50$

Thus, we found a number (50) so that 40% of it is 20.

Verification: 40% of 50 = 0.40 × 50, which is indeed 20.

PROBLEM 3

50 is 40% of what number?

> Exercises 5.3

< A > Finding the Percentage Using Proportions In Problems 1–14, find the percentage using proportions.

1. Find 30% of 80.
2. Find 15% of 80.
3. 150% of 20 is what number?
4. 25% of 60 is what number?
5. What is 20% of $30?
6. What is 95% of 80?
7. Find 60% of 45.
8. Find 120% of $40.
9. $12\frac{1}{2}$% of 60 is ______.
10. $8\frac{1}{4}$% of 60 is ______.
11. 3.5% of 60 is ______.
12. 110.5% of 80 is ______.
13. Find $16\frac{2}{3}$% of 132.
14. Find $83\frac{1}{3}$% of 102.

< B > Finding the Percent Using Proportions In Problems 15–32, find the percent using proportions.

15. 325 is what percent of 3250?
16. 15 is what percent of 90?
17. 16 is what percent of 4?
18. 7 is what percent of 28?
19. Find what percent of 50 is 10.
20. Find what percent of 40 is 5.
21. What percent of 40 is 8?
22. What percent of 25 is 10?
23. $6\frac{1}{2}$ is what percent of 26?
24. $2\frac{1}{4}$ is what percent of 27?
25. 4 is ______ % of 5.
26. 20 is ______ % of 40.
27. 50 is ______ % of 25.
28. 45 is ______ % of $7\frac{1}{2}$.
29. $50 is ______ % of $60.
30. $90 is ______ % of $3600.
31. $0.75 is ______ % of $4.50.
32. $0.25 is ______ % of $1.50.

Answers to PROBLEMS

3. 125

> Web IT go to mhhe.com/bello for more lessons

Web IT go to mhhe.com/bello for more lessons

⟨C⟩ Finding the Base Using Proportions In Problems 33–54, find the base using proportions.

33. 30% of what number is 90?

34. 15% of what number is 60?

35. $2\frac{1}{4}$% of what number is 18?

36. $5\frac{1}{2}$% of what number is 22?

37. 60 is 40% of what number?

38. 30 is 150% of what number?

39. 30 is $33\frac{1}{3}$% of a number. Find the number.

40. 20 is $66\frac{2}{3}$% of a number. Find the number.

41. Find a number so that 120% of it is 42.

42. Find a number so that $3\frac{1}{4}$% of it is $6\frac{1}{2}$.

43. 40% of what number is $7\frac{1}{2}$?

44. 140% of what number is 70?

45. 4.75% of what number is 76?

46. 3.25% of what number is 78?

47. 100% of what number is 49?

48. 8% of what number is 3.2?

49. $12\frac{1}{2}$% of what number is 37.5?

50. $33\frac{1}{3}$% of what number is $66\frac{2}{3}$?

51. $16\frac{2}{3}$% of what number is 8?

52. $66\frac{2}{3}$% of what number is $33\frac{1}{3}$?

53. 50 is $62\frac{1}{2}$% of a number; what is the number?

54. 60 is $37\frac{1}{2}$% of a number; what is the number?

⟩⟩⟩ Using Your Knowledge

Cardiac Health How can you measure your cardiac health? By looking at the ratio $\frac{w}{h}$, where w is your waist size and h is your hip size.

For women: If $\frac{w}{h} > \frac{8}{10}$, the risk of heart disease increases. For men: If $\frac{w}{h} > 1$, the risk of heart disease increases.

55. A woman has a 24-inch waist. What h (hip size) will make her risk of heart disease increase? (*Hint:* Solve $\frac{24}{h} > \frac{8}{10}$ as a proportion.)

56. A woman has a 30-inch waist. What h (hip size) will make her risk of heart disease increase?

57. A man has a 32-inch waist. What h (hip size) will make his risk of heart disease increase? (*Hint:* Solve $\frac{32}{h} > 1$ as a proportion.)

58. A man has a 40-inch waist. What h (hip size) will make his risk of heart disease increase?

⟩⟩⟩ Write On

59. Which method would you use to solve percent problems: the method used in Section 5.2 or proportions? Explain why.

60. In Section 5.2, we have the formula $P = R \times B$. If the rate is written as a fraction with a denominator of 100, explain how you can derive the proportion $\frac{\text{Percent}}{100} = \frac{\text{Part}}{\text{Whole}}$ from $P = R \times B$.

⟩⟩⟩ Concept Checker

Fill in the blank(s) with the correct word(s), phrase, or mathematical statement.

61. To solve percent problems using **proportions,** we first write the **percent ratio** as the **fraction** ________.

62. To solve percent problems using **proportions,** we set up the **proportion** $\frac{\text{Percent}}{100} = \frac{\_\_\_\_\_\_}{\text{Whole}}$.

Percent/100

100/Percent

Part

Percent

⟩⟩⟩ Mastery Test

63. Set up a proportion and solve: What is 40% of 80?

64. Set up a proportion and solve: 30 is what percent of 40?

65. Set up a proportion and solve: $3\frac{1}{2}$% of what number is 21?

Skill Checker

Write as a decimal.

66. $66\frac{2}{3}\%$ **67.** $5\frac{1}{2}\%$ **68.** $8\frac{1}{4}\%$

69. $16\frac{2}{3}\%$ **70.** $33\frac{1}{3}\%$

5.4 Taxes, Interest, Commissions, and Discounts

Objectives

You should be able to:

A Find the sales tax and the total cost of an item.

B Find the simple interest in a transaction.

C Find the compound interest in a transaction.

D Find the commission on a sale.

E Find the discount and the sale price of an item.

F Solve applications involving percents.

To Succeed, Review How To . . .

1. Find a percent of a number. (pp. 300–301, 310)
2. Round to the nearest cent. (pp. 216–217)

Getting Started

The man in the cartoon is ruined by taxes. A tax is an amount of money collected by the government and calculated as a percent of some total amount.

A › Sales Tax

The **sales tax** in a city in Alabama is 4%, which means that if you buy an item costing $60, the tax would be

$$4\% \text{ of } \$60$$

$$0.04 \times \$60 = \$2.40$$

In general, the sales tax is equal to the product of the sales tax rate and the list price.

SALES TAX

Sales Tax = Sales Tax Rate × List Price

That is, the tax is \$2.40.

Since the sales tax is always added to the buyer's cost, the total cost of the item is given by the sum of the list price and the sales tax.

TOTAL COST

Total cost = List price + Sales tax

Thus, total cost = \$60 + \$2.40
= \$62.40

EXAMPLE 1 Finding the sales tax

The sales tax in Florida is 6%. How much would you pay for a pair of slacks costing \$18.50?

SOLUTION Since

$$\begin{array}{c} 6\% \text{ of } 18.50 \\ \downarrow \quad \downarrow \quad \downarrow \\ 0.06 \times 18.50 = 1.1100 \end{array}$$

The total cost is \$18.50 + \$1.11 = \$19.61.

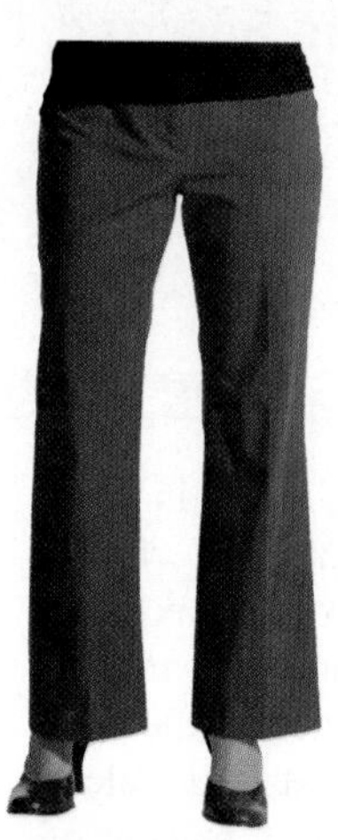

PROBLEM 1

If the sales tax is 5%, how much would you pay for a pair of shoes costing \$33?

B › Simple Interest

The idea of *percent* is also used in banking. For example, if you deposit \$300 in a bank paying $5\frac{1}{4}\%$ annually on the amount deposited, then your **simple interest** (the amount the bank pays you for using your money) at the end of the year is computed by using the formula

$$I = P \times R \times T$$

simple interest (I), principal (P), rate (R), time (T)

In our example,

$$I = 300 \times 5\tfrac{1}{4}\% \times 1$$

300: principal (amount deposited); $5\frac{1}{4}\%$: interest rate per year; 1: time (years)

Thus,

$$I = \$300 \times 0.0525 \times 1 = \$15.7500 = \$15.75$$

Of course, you can use the same idea to *borrow* money from the bank. For example, if you borrow \$600 at 12% annual interest for 4 months, the interest you have to pay is

$$I = P \times R \times T$$

$$I = \$600 \times 12\% \times \frac{4}{12}$$

Note that the time must be in *years* and 4 months is $\frac{4}{12}$ or $\frac{1}{3}$ of a year.

$$= \$600 \times 0.12 \times \frac{1}{3}$$

$$= \$24$$

Answers to PROBLEMS

1. \$34.65

EXAMPLE 2 Finding the simple interest I

Find the simple interest earned on \$1600 invested at 9.5% annual rate for 3 months.

SOLUTION

$$I = P \times R \times T = \$1600 \times 9.5\% \times \frac{1}{4}$$ 3 months is $\frac{3}{12} = \frac{1}{4}$ year.

$$= \$1600 \times 0.095 \times \frac{1}{4} = \$38$$

PROBLEM 2

Find the simple interest paid on \$1200 borrowed at 8.5% for 4 months.

C › Compound Interest

The process of computing interest on the earned interest is called compounding the interest. To illustrate the difference between simple and compound interest, suppose you invest \$100 for one year at 7% simple interest. At the end of the year, your interest will be $I = PRT$, where $P = \$100$, $R = 7\% = 0.07$, and $T = 1$, or $I = 100 \times 0.07 \times 1 = \$7$.

Now, suppose you invest the same \$100 at 7% compounded quarterly (four times a year). Although the annual rate is 7% = 0.07, the time is $\frac{1}{4}$ of a year, so that the rate each quarter is $\frac{7\%}{4} = 0.0175$. The compound interest calculation for each quarter is as follows.

\$100	Principal
× 0.0175	Rate per quarter
\$1.75	Interest (1st quarter)
\$101.75	Principal (\$100 + \$1.75)
× 0.0175	Rate per quarter
\$1.780625	Interest (2nd quarter)
\$103.53	Principal (\$101.75 + 1.78)
× 0.0175	Rate per quarter
\$1.811775	Interest (3rd quarter)
\$105.34	Principal (\$103.53 + \$1.81)
× 0.0175	Rate per quarter
\$1.84345	Interest (4th quarter)

Thus, the total compound interest for the four quarters is
$\$1.75 + \$1.78 + \$1.81 + \$1.84 \approx \$7.18$.

EXAMPLE 3 Finding the compound interest

Find the total compound interest and the final compound amount for \$100 invested at 8% compounded semiannually (twice a year) for 2 years.

SOLUTION Here $P = \$100$, $r = 8\%$, and the semiannual rate is $\frac{8\%}{2} = 4\% = 0.04$.

\$100	Principal
× 0.04	Semiannual rate
\$4.00	Interest (1st 6 months)
\$104	Principal (\$100 + \$4)
× 0.04	Semiannual rate
\$4.16	Interest (2nd 6 months)
\$108.16	Principal (\$104 + \$4.16)
× 0.04	Semiannual rate
\$4.3264	Interest (3rd 6 months)
\$112.49	Principal (\$108.16 + \$4.33)
× 0.04	Semiannual rate
\$4.4996	Interest (4th 6 months)

The total compound interest is \$4 + \$4.16 + \$4.33 + \$4.50 = \$16.99, and the final compound amount is \$100 + \$16.99 = \$116.99.

PROBLEM 3

Find the total compound interest and the final compound amount for \$100 invested at 6% compounded semiannually for 2 years.

Answers to PROBLEMS

2. \$34 **3.** \$12.55; \$112.55

In Example 3, we multiplied the principal P by the rate per period. The rate per period i is defined as follows:

Rate per period i

$$i = \frac{\text{Nominal rate}}{\text{Number of periods per year}}$$

Thus, 8% compounded semiannually yields $i = \frac{8\%}{2} = 4\%$ per period.

The computations in Example 3 can be shortened if we use the following formula to find the compound amount A.

Compound amount A

$$A = P(1 + i)^n$$

In this formula, A is the compound amount, P is the principal, i is the rate per period, and n is the number of periods.

Here is the derivation. Let I be the compound interest, P be the original principal, i be the rate per period, and A_1 be the compound amount at the end of the first period.

$$\begin{aligned} I &= Pi && \text{Interest for the first period} \\ A_1 &= P + I \\ &= P + Pi && \text{Substitute } Pi \text{ for } I. \\ &= P(1 + i) && \text{Use the distributive property.} \end{aligned}$$

After the end of the second period, the compound amount A_2 is

$$\begin{aligned} A_2 &= A_1 + A_1 i \\ &= A_1(1 + i) && \text{Use the distributive property.} \\ &= P(1 + i)(1 + i) && \text{Substitute } P(1 + i) \text{ for } A_1. \\ &= P(1 + i)^2 && \text{Substitute } (1 + i)^2 \text{ for } (1 + i)(1 + i) \end{aligned}$$

If we continue this procedure, after n periods A_n will be $A_n = P(1 + i)^n$.

FORMULA FOR THE COMPOUND AMOUNT *A*

The compound amount *A* for a rate *i* per period applied to a principal *P* for *n* periods is given by

$$A = P(1 + i)^n$$

Now let's solve Example 3 using this formula. Thus, $P = \$100$, $i = \frac{8\%}{2} = 4\% = 0.04$, and $n = 2 \times 2 = 4$ (two years, semiannually).

$$\begin{aligned} A &= \$100(1 + 0.4)^4 \\ &= \$100(1.04)^4 \\ &= \$100 \times 1.16985856 \\ &\approx \$116.99 \end{aligned}$$

EXAMPLE 4 Finding the compound amount using the formula

Find the compound amount and the compound interest when \$5000 is invested at 6% compounded semiannually for 3 years.

SOLUTION Here $P = \$5000$, $i = \frac{6\%}{2} = 3\% = 0.03$, and $n = 6$ periods.

$$\begin{aligned} A &= \$5000(1 + 0.03)^6 \\ &= \$5000(1.03)^6 \\ &= \$5000 \times (1.03 \times 1.03 \times 1.03 \times 1.03 \times 1.03 \times 1.03) \\ &\approx \$5000(1.19405230) \\ &\approx \$5970.26 \end{aligned}$$

The compound interest $I = A - P = \$5970.26 - \$5000 = \$970.26$.

PROBLEM 4

Find the compound amount and the interest when \$10,000 is invested at 8% compounded semiannually for 2 years.

D > Commissions

Percents are also used as incentives for obtaining more sales from salespersons. This is done by giving salespeople a share of the sales, called a **commission.** This commission is stated as a percent of the sales income. In Florida, for example, real estate people receive a 7% commission on each sale. Thus if they sell a house valued at \$95,000, the commission is

$$\underbrace{7\%}\ \text{of}\ \$95{,}000$$
$$\downarrow \quad \downarrow \quad \downarrow$$
$$0.07 \times 95{,}000 = \$6650$$

In general, the commission is equal to the product of the commission rate and the sale price.

COMMISSION

Commission = Commission Rate × Sale Price

EXAMPLE 5 Finding commissions

The Move-and-Go Car Lot pays salespeople an 8% commission on their sales. What would the commission be on the sale of a used car costing \$2499.95?

SOLUTION The commission is 8% of \$2499.95:

$$0.08 \times \$2499.95 = \$199.9960$$

If we round \$199.9960, we find the commission to be \$200.

PROBLEM 5

Find the commission on a \$1599.95 sale if the commission rate is 6%.

E > Discounts

As we have mentioned before, percents are used to attract buyers to stores. A sale at "20% off," that is, a **discount** of 20%, means that regular prices have been *reduced* by $20\% = \frac{20}{100} = \frac{1}{5}$. Thus, if the **regular,** or **list,** price of an article is \$50 and the article is offered on sale at a 20% discount, the *discount* is

$$\underbrace{20\%}\ \text{of}\ \$50$$
$$\downarrow \quad \downarrow \quad \downarrow$$
$$0.20 \times \$50 = \$10.00$$

$$\begin{aligned} \text{Sale price} &= \text{Regular price} - \text{Discount} \\ &= \$50 - \$10 = \$40 \end{aligned}$$

Thus, the sale price of the article is \$40.

Answers to PROBLEMS

4. \$11,698.59; \$1698.59
5. \$96

EXAMPLE 6 **Finding discount and sale price**

Find the discount and the sale price of an article regularly selling for \$12.50, which is being advertised at 10% off.

SOLUTION The discount is 10% of \$12.50:

$$0.10 \times \$12.50 = \$1.25$$

Thus, the sale price is \$12.50 − \$1.25 = \$11.25.

PROBLEM 6

Find the discount and sale price of an \$8.50 article selling at 20% off.

F > Applications Involving Percents

Do you know what the letters in the word *tips* stand for? They stand for "**t**o **i**nsure **p**rompt **s**ervice." At restaurants, diners tip anywhere from 10% to 25% of the total bill. We illustrate how to figure the tip next.

EXAMPLE 7 **Calculating the tip**

The total bill for a meal comes to \$38.50. If you want to leave a 15% tip, how much tip should you leave?

SOLUTION We have to find 15% of 38.50.

$$0.15 \times \$38.50 = \$5.775$$
$$= \$5.78 \quad \text{(To the nearest cent)}$$

(For practical reasons, you would probably leave a \$5.75 or a \$6 tip.)

PROBLEM 7

The total bill for a meal is \$60.70. If you want to tip 20%, how much should the tip be? (Answer to the nearest cent.)

Calculator Corner

A calculator is very convenient when doing taxes, interest, commissions, and discounts. Let us take each of the problems separately. For example, to solve Example 1: Find how much you have to pay for a pair of pants costing \$18.50 if the tax is 6% by simply pressing 1 8 . 5 0 + 6 % ENTER. The final answer, \$19.61, will be displayed.

If your calculator does not have a % key, you may have to use the **store** M+ and **recall** MR keys. As their names imply, these keys *store* numbers in their memory and then *recall* them. Suppose you wish to solve Example 1 with a calculator. You press 1 8 . 5 0 M+ × 0 . 0 6 ENTER. Here the calculator has **stored** the 18.50 and taken 6% of 18.50. The display shows 1.11. Now, press + MR ENTER and the calculator will add the number stored (18.50) to the 1.11, giving you \$19.61, as before.

If your calculator has a y^x key (a **power key**), then the quantity $(1.03)^6$ appearing in Example 4 can be obtained as follows: 1 . 0 3 y^x 6 ENTER. You will get 1.19405230.

> Exercises 5.4

< A > Sales Tax In Problems 1–8, solve the problems involving sales tax.

1. The retail sales tax in Alabama is 4%. Find the total cost of a \$3500 used car.

2. The retail sales tax in Florida is 6%. Find the total cost of a lamp selling for \$15.50.

3. The retail sales tax in Georgia is 4%. What is the total cost of an item selling for \$12.50?

4. One of the most expensive cars ever built, the Presidential Lincoln Continental Executive, cost \$500,000. If a 7% excise tax had to be paid when buying this car, what was the tax on the car?

Answers to PROBLEMS

6. Discount: \$1.70; sale price: \$6.80 **7.** \$12.14

5. One of the highest prices paid for a bottle of wine of any size was $13,200. If the sales tax was 5%, how much was the tax and the bottle of wine?

6. The sales tax on a car was $150, and the sales tax rate was 5%. What was the purchase price of the car before taxes?

7. The purchase price of a ring was $1500, before taxes. The sales tax amounted to $60. What was the sales tax rate?

8. A man paid $3520 in income taxes. His annual salary is $16,000. What percent of his income went for taxes?

‹B› Simple Interest In Problems 9–16, find the simple interest.

	Principal	Annual Rate of Interest	Time
9.	$200	12%	1 year
10.	$250	16%	3 years
11.	$400	15%	$\frac{1}{2}$ year
12.	$1200	9%	3 months
13.	$1500	15%	2 months
14.	$900	13%	4 months
15.	$300	9%	60 days*
16.	$900	16%	30 days*

* Assume a year has 360 days.

17. In one year the highest bank interest rate was that of Brazil at 106%.

a. How much simple interest would you earn if you deposited $500 for one year at this rate?

b. How much simple interest would you pay on a two-year, $2500 loan using the 106% rate?

18. The funds rate (interest that banks charge each other) was at a 46-year low of 1% in June 2004.

a. How much simple interest would you earn if you deposited $600 for one year at this rate?

b. How much simple interest would you pay on a two-year, $4000 loan at the 1% rate?

19. Ms. Garcia had a $1000 bond paying 8% annual interest. How much simple interest did she receive in two years?

20. Mr. Liang borrowed $150 at 10% interest for 2 years. How much simple interest did he pay?

‹C› Compound Interest In Problems 21–25, use the compound interest formula, $A = P(1 + i)^n$.

21. A small computer software company invests $5000 at 10% compounded semiannually for 2 years. What will the compound amount be at the end of this period?

22. A 3-year, $1000 certificate of deposit yields 12% interest compounded annually. How much money will it be worth at the end of the 3 years?

23. Douglas Thornton invested $50,000 at 10% compounded semiannually for a period of 2 years. What amount should he receive at the end of the 2 years?

24. Maria Rodriguez deposited $1000 in her account April 1. The bank paid 6% interest compounded monthly.

a. What was her balance June 1?

b. On June 1, Maria made a deposit and her new balance was $1500. How much was her deposit?

25. Jorge Riopedre bought a $500 certificate of deposit on May 1. If the certificate paid 12% interest compounded monthly, what was his balance July 1?

‹D› Commissions In Problems 26–30, solve the commission problems.

26. Howard Hues, a paint salesperson, earns a 10% commission rate. What is his commission if he sells $4800 worth of paint?

27. Dan Driver, a car salesperson, earns an 8% commission rate. How much commission money did he get when he sold a $3500 car?

28. Connie Dominion, a real estate agent, sells a house for $37,000. Her commission rate is 3.5%. How much is her commission?

29. Toy Auto Sales pays their salespersons $500 a month plus a 2% commission on each sale. If a person sold only one car for $6300, what was that person's total salary for the month?

30. One of the highest-priced diamonds ever sold at an auction was bought by Cartier of New York for $1,050,000. If the salesperson got a 2% commission, how much was her commission?

‹E› Discounts In Problems 31–35, solve the problems involving discounts.

31. An article regularly sells for $14.50. It is on sale at 10% off. How much do you have to pay for it now?

32. A lamp is offered on sale at 20% off its regular $15 price. What is the sale price of the lamp?

33. A scratched refrigerator regularly priced at $460 was sold at a 10% discount. What was its new price?

34. Mary Smith owed $500 on her account. She paid off the whole amount, receiving a 3% discount for early payment. How much was the discount?

35. An invoice for $250 offers a 2% discount if paid within 60 days. How much is the discount?

Web IT go to mhhe.com/bello for more lessons

⟨F⟩ Applications Involving Percents

36. *Tipping* The total bill for a meal comes to \$30.40. If you decide to leave a 10% tip, how much should you leave? (Answer to the nearest cent.)

37. *Tipping* A restaurant bill amounted to \$80.30. If service was exceptionally good and you decided to tip 25%, how much was the tip? (Answer to the nearest cent.)

38. *Tipping* A meal for two at a fancy restaurant comes to \$73.40. You decide to tip 15% and split the bill.

a. How much is the tip?

b. To the nearest cent, how much should each person pay?

39. *Tipping* A birthday meal for eight comes to \$82.50. A 15% service charge is added for parties of eight or more. To the nearest cent, how much is the service charge?

40. *Tipping tips* Suppose your bill in a restaurant amounted to \$54. If you want to leave a 10% tip, you simply move the decimal point one place left in \$54, obtaining \$5.40: the tip.

a. What is the tip if you want to leave 20%?

b. Since 20% = **2** × 10%, a quicker way will be to **double** the \$5.40. What is the tip then? Do you get the same answer as in part **a?**

41. *Tipping tips* You received good service and you are going to leave a 25% tip on your \$54 bill.

a. What is the tip?

b. Is there a shortcut to get the 25% tip? Of course! Double the 10% (the \$5.40) and then take half of it ($\frac{1}{2}$ of 10% = 5%). Your tip is now 20% + 5% or 25%. Write an expression that would result in the 25% tip.

c. Simplify the expression that you created in part **b.** Do you get the same answer as in **a?**

42. *Tipping tips* Suppose your dinner bill came to \$74. Fill in the blanks in the table with an expression that will calculate the tip, and then simplify the expression. Check by taking 10%, 15%, 20%, and 25% of \$74.

10% tip	
15% tip	
20% tip	
25% tip	

43. *Tipping tips* There is another tipping shortcut based on the tax added to the bill. For example, suppose the dinner bill was \$62 and a 6% tax was automatically

a. What is the tax?

b. If you want to leave a 12% tip, how much money would you leave as a tip?

c. If you want to leave an 18% tip, how much money would you leave as a tip?

44. *Tipping tips* Suppose your dinner bill was \$80 and a 7% tax was automatically

a. What is the tax?

b. If you want to leave a 14% tip, how much money would you leave as a tip?

c. If you want to leave a 21% tip, how much money would you leave as a tip?

45. *Tipping tips* Suppose your dinner bill was \$58 plus 8%

a. How much is the tax?

b. If you want to leave a 16% tip, how much money would you leave?

c. If you want to leave a 12% tip, how much money would you leave?

⟩⟩⟩ Using Your Knowledge

Taxable Income Sooner or later (probably sooner) you will have to pay federal income taxes to the government. These taxes are paid on your **taxable income,** which is arrived at by following the steps in federal income tax Form 1040. After finding your taxable income, you may be required to use schedule Y-1.

Schedule Y-1—Married Filing Jointly or Qualifying Widow(er)	If Taxable Income		Then: The Tax Is		
	Is Over	**But Not Over**	**This Amount**	**Plus This %**	**Of the Excess Over**
	\$0	\$14,000	\$0.00	10%	\$0.00
	\$14,000	\$56,800	\$1,400.00	15%	\$14,000
	\$56,800	\$114,650	\$7,820.00	25%	\$56,800
	\$114,650	\$174,700	\$22,282.50	28%	\$114,650
	\$174,700	\$311,950	\$39,096.50	33%	\$174,700
	\$311,950	—	\$84,389.00	35%	\$311,950

To get the most current tax rates, go to http://www.irs.gov.

Suppose your taxable income is \$12,000. Your tax is 10% of this amount (first line, 0 to \$14,000):

$$0.10 \times \$12{,}000 = \$1200$$

If your taxable income is \$40,000, you fall between \$14,000 and \$56,800 (second line). Your tax is

$$\begin{aligned} &\$1400 + 15\% \text{ of the amount over } \$14{,}000 \\ &= \$1400 + 0.15 \times (\$40{,}000 - \$14{,}000) \\ &= \$1400 + 0.15 \times \$26{,}000 \\ &= \$1400 + \$3900 \\ &= \$5300 \end{aligned}$$

Taxable Income In Problems 46–51, find the tax for each taxable income.

46. \$20,000

47. \$25,000

48. \$50,000

49. \$60,000

50. \$80,000

51. \$100,000

Write On

52. What are the three variables (factors) used when calculating simple interest?

53. Write in your own words which would be better for you: to take a 20% discount on an item, or to take 10% off and then 10% off the reduced price.

54. Which investment is better for you: \$10,000 invested at 5% compounded semiannually or \$10,000 invested at 4% compounded monthly. Explain why.

55. Most people give 10%, 15%, or 20% of the total bill as a tip. (See Problem 42.)

a. In your own words, give a rule to find 10% of any amount. (*Hint:* It is a matter of moving the decimal point.)

b. Based on the rule in part **a,** state a rule that can be used to find 15% of any amount.

c. Based on the rule in part **a,** state a rule that can be used to find 20% of any amount.

Concept Checker

Fill in the blank(s) with the correct word(s), phrase, or mathematical statement. The formula $A = P(1 + i)^n$ will be used in Problems 56–59.

56. In the formula, A represents the ________.

57. In the formula, P is the ________.

58. In the formula, i is the rate per ________.

59. The rate per period i is the quotient of the **nominal** rate and the **number of periods** per ________.

month	**interest**
period	**principal**
amount	**year**

Mastery Test

60. The total bill for a meal is \$78.50. If you want to leave a 15% tip, how much money (to the nearest cent) should you leave as a tip?

61. Find the discount and the sale price of a book selling for \$80 and being advertised at 12% off.

62. Beverly is a real estate agent in Florida, and sold a house valued at \$105,000. If her commission rate is 7%, how much commission did she get?

63. Find the compound amount and the compound interest when \$10,000 is invested at 4% compounded semiannually for 2 years.

64. Find the simple interest paid on a 6%, 3-month, \$1200 loan.

65. The state sales tax in California is 7.25%. Find the total cost (price plus tax) of a pair of running shoes costing \$120.

Skill Checker

In Problems 66–71, give the answer to the nearest percent.

66. What percent of 52 is 30?

67. What percent of 48 is 80?

68. Find what percent of 62 is 30.

69. Find what percent of 72 is 40.

70. 15 is what percent of 22?

71. 25 is what percent of 30?

5.5 Applications: Percent of Increase or Decrease

Objectives

You should be able to:

A Solve percent problems involving percent increase or decrease.

B Solve applications involving percent increase or decrease.

To Succeed, Review How To . . .

Find what percent of one number is another. (pp. 300–301, 310)

Getting Started

Is your college tuition going up? The chart shows the projected annual growth in undergraduate tuition and fees at 4-year public and private colleges. What was the *increase* for public tuition and fees over the 10-year period? We shall find out next.

If costs keep growing at this pace, tuition and fees will more than double over the next 10 years.

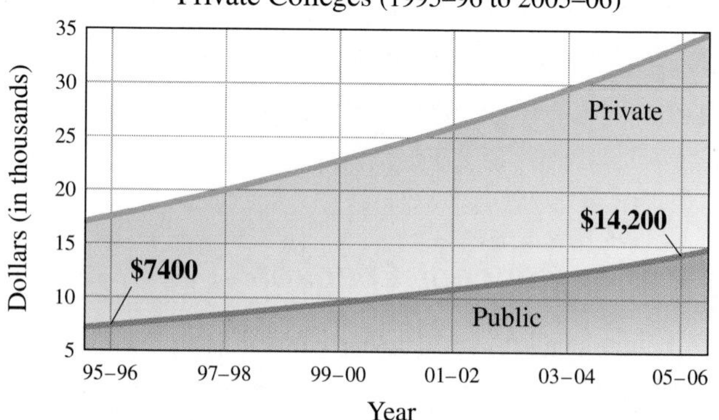

Source: Data from Prudential Financial.

A › Percent of Increase or Decrease

According to the graph in the *Getting Started,* to find the increase over the 10-year period, look at the cost in 1995–1996 **($7400)** and the cost in 2005–2006 **($14,200).**

The **increase** is **$14,200 − $7400 = $6800.**

The percent of increase over 1995–1996 is

$$\frac{\$6800}{\$7400} = \frac{34}{37} \approx 92\%$$

(Increase → $6800; Over 95–96 → $7400)

$$\begin{array}{r} 0.918 \\ 37\overline{)34.000} \\ \underline{33\ 3} \\ 70 \\ \underline{37} \\ 330 \\ \underline{296} \\ 34 \end{array}$$

Here is the rule we have used:

TO FIND A PERCENT OF INCREASE OR DECREASE

1. Find the *amount* of increase (or decrease).
2. Divide by the original amount.
3. Write the answer to step 2 as a percent.

EXAMPLE 1 Calculating percent increase

In a 4-year period, sneaker sales rose from \$1.8 billion to \$3 billion. Find the percent of increase to the nearest percent.

SOLUTION We follow the three steps given:

Step 1. The amount of increase is

$$3.0 - 1.8 = 1.2$$

Step 2. The original amount is 1.8. We divide the amount of increase (1.2) by 1.8:

```
        0.666
1.8)1.2 000
    1 0 8
      1 20
      1 08
        120
        108
         12
```

Step 3. Write 0.666 as 67%.

PROBLEM 1

Last year sneaker sales went from \$3 billion to \$4.3 billion. Find the percent of increase to the nearest percent.

Percents of increase and decrease are used in business and other fields. For example, the Dow Jones Industrial Average (DJIA) tracks the prices of 30 leading stocks and is supposed to indicate trends for individual stocks. On October 19, 1987, the market had its largest percent drop ever (up to this time). We calculate the percent of decrease in stock prices next.

EXAMPLE 2 Calculating percent decrease

On October 19, 1987, the DJIA went down from 2246 to 1738. To the nearest percent, what was the percent of decrease?

SOLUTION As before, we use three steps:

Step 1. The *amount* of decrease is:

$$2246 - 1738 = 508$$

Step 2. We divide the decrease (508) by the original amount (2246):

```
         0.226
2246)508.000
     449 2
      58 80
      44 92
      13 880
      13 476
         404
```

Step 3. Write 0.226 as 23%.

Thus, the market declined by 23%.

PROBLEM 2

The DJIA's biggest one-day slide occurred on September 17, 2001. On that day, the DJIA went down from 9606 to 8921.

To the nearest whole number, what was the percent of decrease?

Note: The decline that day was larger than the 508 points of October 19, 1987. Percentage-wise, however, the October 1987 decline was much larger.

Answers to PROBLEMS

1. 43% **2.** 7%

B > Applications Involving Percent Increase or Decrease

Do you spend a lot on entertainment? In 2002, baby-boom households spent an average of \$2100 annually on entertainment. (This includes fees and admissions to parks, movies, plays, etc., and the cost of TVs, stereos, and other electronic equipment.) Are these costs going to increase? Read on.

EXAMPLE 3 Calculating percent increase

It is estimated that the amount spent on entertainment will *increase* by $33\frac{1}{3}\%$ in the next decade. If we are spending \$2100 annually now, how much will we be spending in a decade?

SOLUTION This time, we have to find the *amount* of increase. Since we spend \$2100 now, in a decade we will be spending

$$
\begin{aligned}
&2100 + 33\tfrac{1}{3}\% \text{ of } 2100 \\
&= 2100 + \frac{1}{3} \times 2100 \\
&= 2100 + \frac{2100}{3} \\
&= 2100 + 700 \\
&= 2800
\end{aligned}
$$

We will be spending \$2800 in a decade.

PROBLEM 3

The total amount spent on corporate travel and entertainment will increase by 21% in a 2-year period.

If \$95 billion was spent this year, how many billions of dollars will be spent in 2 years? (Answer to the nearest billion.)

EXAMPLE 4 Calculating percent increase

Does it really pay to get a college education? Look at the graph! The brown (bottom) bars represent the annual earnings for all workers. If you were not a high school graduate, your annual earnings would be \$18,900. If you have a bachelor's degree, your annual earnings would be \$45,400. What is the percent increase in annual salary of having a bachelor's degree as compared to not being a high school graduate?

SOLUTION To find the percent increase, look at the earnings for a worker who is not a high school graduate (\$18,900) and the earnings for a worker with a bachelor's degree (\$45,400).

The difference is \$45,400 − \$18,900 = \$26,500.

The percent increase is

Increase → \$26,500
Not high school → \$18,900

$$\frac{\$26{,}500}{\$18{,}900} \approx 140\%$$

```
             1.402
18900 )26 500.000
       18 900
        7 600 0
        7 560 0
           40 000
           37 800
            2 200
```

Thus, the percent increase in pay for a bachelor's degree over not having a high school diploma is about 140%. That is, a worker with a bachelor's degree makes 140% more (\$26,500) than a worker who is not a high school graduate.

PROBLEM 4

Find the percent increase (to the nearest percent) in annual salary between a worker who is not a high school graduate and a worker with a master's degree.

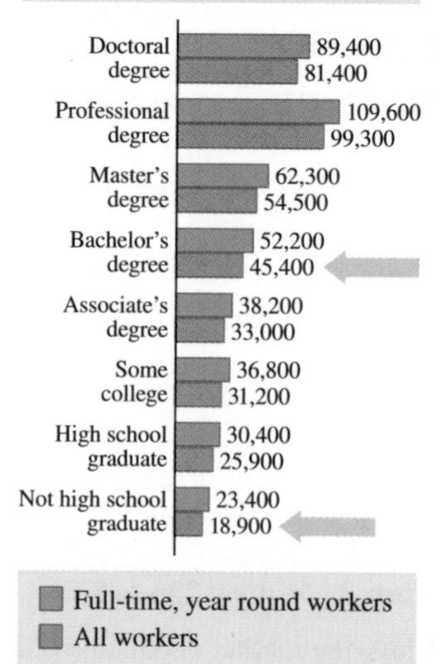

Answers to PROBLEMS

3. \$115 billion **4.** 188%

Now, suppose you want to buy a new car but you do not want to pay full price; instead, your goal is to pay the wholesale price of the car. How can you do it? First, you can go to one of the popular websites and find the base (invoice) price and the MSRP (manufacturer's suggested retail price) of the car you want. The information in the table is for a Honda Accord Coupe. To the nearest \$100, the invoice price is \$20,300 and the MSRP is \$22,600. What is the percent increase between these two prices? We will find out in Example 5.

Source: Data from Carprices.com.

2006	Honda	Accord
EX 2dr	Coupe	(2.4L 4cyl 5M)
MSRP		**Invoice**
\$22,550		\$20,299

EXAMPLE 5 Calculating percent increase

Find the percent increase between the invoice price of \$20,300 and the MSRP of \$22,600.

SOLUTION

This time the difference is: \$22,600 − \$20,300 = \$2300

(Wholesale ↑ \$22,600; MSRP ↑ \$20,300)

The percent increase over the base is: $\dfrac{\$2300}{\$20{,}300} \approx 11\%$ (check this)

Thus, the dealer is making about 11% (the percent increase) over the base price.

PROBLEM 5

Find the percent increase between the \$25,200 invoice price of a car and its MSRP of \$27,700. (Round the answer to the nearest percent.)

If you buy the car in Example 5, how much can you expect to get for it at the end of the year? It is estimated that a new car loses anywhere from 15% to 20% of its value the first year. This decline in price is called **depreciation.** Thus, if you bought the car in Example 5 for the invoice price of \$20,300, at the end of one year the car would depreciate between 15% of \$20,300 (\$3045) and 20% of \$20,300 (\$4060). Because of the depreciation, you should expect to get between \$17,255 (\$20,300 − \$3045) and \$16,240 (\$20,300 − \$4060) for the car one year later.

Let us look at an example with real data next.

EXAMPLE 6 Calculating percent decrease: Depreciation

The base price of a brand-new Ford Taurus is about \$19,000, according to Kelley Blue Book. The wholesale price of a Ford Taurus with just 100 miles on it is \$15,400, a drop of \$3600 from its original price. What is the percent of depreciation for the car?

SOLUTION

The difference is: \$19,000 − \$15,400 = \$3600

(New ↑ \$19,000; Used ↑ \$15,400)

The percent decrease over the base is: $\dfrac{\$3600}{\$19{,}000} \approx 19\%.$

Thus, the depreciation after just putting 100 miles on the car is a whopping 19%.

Source: Bankrate.com.

PROBLEM 6

The price of a car decreases from \$20,000 to \$15,000 after just 500 miles of use. What is the percent depreciation for the car?

Answers to PROBLEMS

5. 10% **6.** 25%

› Exercises 5.5

‹ A › Percent of Increase or Decrease

In Problems 1–13, give the answer to the nearest percent.

1. A computer programmer received a $4000 raise. If her previous salary was $20,000, what was the percent of increase in salary?

2. Felix Perez received a $1750 annual raise from the state. If his salary was $25,000 before the raise, what percent raise did he receive?

3. In one 2-year period, the number of households reached by ESPN went from 60 to 70 million. Find the percent of increase in households reached.

4. In one 2-year period, the number of households reached by ESPN went from 50 to 60 million. What percent of increase in households reached is that?

5. It is estimated that the number of airline passengers at Chicago's O'Hare Airport will increase from 72 million to 73 million in 1 year. What is the predicted percent of increase in passengers at O'Hare?

6. The estimated number of airline passengers at Atlanta Airport in the year 2005 will be 85 million, up from the present 80 million. What is the predicted percent of increase in passengers for Atlanta?

7. During a 2-year period, construction workers' average hourly earnings rose from $18.27 to $19.02. What is the percent increase?

8. During a 2-year period, government workers' salaries rose from $20,687 to $21,187. What is the percent increase?

9. In the first 3 months of one year, the DJIA rose from 10,000 to 10,400. What was the percent of increase?

10. The 1929 stock market crash sent prices from a weekly peak of 380.3 in August 28, 1929, to a weekly bottom of 41.6 on July 4, 1932. To the nearest percent, what was the decline in prices?

11. During the year, the number of mishandled baggage complaints per 1000 passengers at the 10 largest airlines declined from 4.5 in February to 3.85 in March. Find the percent of decrease in complaints.

12. According to a United Airlines executive vice president, a flight from New York La Guardia Airport to Chicago O'Hare Airport took 130 minutes 5 years ago. Today, with aircraft just as fast, and traveling the same distance, it takes 140 minutes. What is the percent of increase in the time it takes to travel from New York to Chicago?

13. During the first 6 months of the year, the number of mishandled baggage complaints per 1000 passengers at United Airlines decreased from 5.07 to 3.66. Find the percent of decrease in complaints.

‹ B › Applications Involving Percent Increase or Decrease

Give all answers to the nearest percent.

14. *Increase in Supercenters* Have you been to one of those all-purpose stores called a Supercenter? In an 8-year period the number of Supercenters grew from 150 to 1200. What was the percent of increase of Supercenters in the 8-year period?

Source: ACNielsen.

15. *Growth of dollar stores* From 1993 to 2006, the number of dollar stores grew from 3650 to 16,000. What was the percent increase of dollar stores between 1993 and 2006?

Source: ACNielsen.

16. *Population increase* The total U.S. population will increase from 288 million in 2005 to 404 million in 2050. What is the percent of increase?

17. *Increase in Hispanic population* The Hispanic population in the United States will increase from 38 million in 2005 to 98 million in 2050. What is the percent increase?

18. *Dating increases* The largest paid content category on the Internet is Personals/Dating. The revenue in that area rose from $72 million to $302 million.

a. What is the percent increase?

b. If you have a website dealing with Personals/Dating and your revenue was $10,000, what would you expect your revenue to be next year based on the answer to part **a?**

19. *Increase in online content* The number of U.S. consumers paying for online content rose from 10 million to 13 million.

a. What is the percent increase?

b. If you have a website with 300 paying customers, how many customers would you expect next year based on the answer to part **a?**

Web IT go to mhhe.com/bello for more lessons

20. *Increase in Turbo Tax* During the month of April, taxpayers rush to Turbo Tax to learn more about their taxes. The number of visitors to Turbo Tax grew from 5.7 million to 6.8 million in one year.

a. What is the percent increase?

b. If you have an online tax site that has 200 visitors, what would you expect the number of visitors to be next year based on the answer to part **a?**

21. *Hospital website uses* According to a Manhattan Research study, 10.3 million online consumers have used a hospital website in the past 3 months, up from 3 million last year.

a. What is the percent increase?

b. If you are the webmaster of a hospital website that had 100,000 visitors in the past 3 months, how many visitors would you expect next year based on the answer to part **a?**

22. *Online travel plans* Do you make your travel plans on the Web? The Travel Industry Association indicates that around 64 million Americans now research their travel options online, an increase from the 12 million who did so in 1997.

a. What is the percent increase?

b. If you have an online travel booking site that did 7000 bookings in 1997, how many would you expect this year based on the answer to part **a?**

23. *Shopping mall visits* According to the research firm comScore Media Metrix, college students' visits to online shopping sites have grown dramatically since September. During the week ending November 10, 4.8 million online college students visited shopping malls, compared to 3.7 million for the week ending September 1.

a. What was the percent increase in visitors?

b. If you have an online shopping site with 6000 visitors for the week ending September 1, how many visitors would you expect during the week ending November 10 based on the answer to part **a?**

24. *Online use from work* In a survey by the Pew Internet and American Life Project, it was determined that 55 million Americans now go online from work, up from 43 million who went online two years earlier.

a. What was the percent increase for the 2-year-period?

b. If you expect the same increase as in part **a** for your company and there are 500 employees going online from work this year, how many would you expect to go online from work 2 years from now?

25. *CD rates* An ad at Bankrate.com states that you can invest \$10,000 in a CD at a 1.5% annual rate.

a. How much will your \$10,000 investment be worth at the end of one year?

b. If your investment at the end of one year is reduced by a 28% tax, how much will your \$10,000 investment be worth after taxes after one year?

26. *Increase in Super Bowl prices* Official Super Bowl ticket prices increased from \$375 to \$400 in a recent year.

a. What is the percent increase?

b. If the price increases by the same percent next year, what would you expect the price of a ticket to be next year?

27. *Scalper rates* Scalpers usually sell tickets at a much higher price than the official \$400 rate (see Problem 26). The table shows ticket prices for a recent Super Bowl on different days. (Game day is Sunday.)

Friday	\$1750
Saturday	\$1700
Game day	\$1500

a. What is the percent increase between Friday's price and the official \$400 rate?

b. What is the percent decrease in price between Friday and Saturday?

c. What is the percent decrease in price between Saturday and game day (Sunday)?

28. *Consolidation loans* According to the Financial Aid.com calculator, if you have \$20,000 in student loans with a \$217 monthly payment, you can decrease your monthly payment to \$123 with a consolidation loan at 3%.

a. What is the percent of decrease in the payment?

b. The information about the consolidation states that the rate is 3%. Is this necessarily a better deal? Explain.

29. *Consolidation loans* For this problem, look at the following information:

	Amount	Rate	Pmt	Time	Interest
Original	\$20,000	5.5%	\$217	10 yr	\$6040
Consolidated	\$20,000	3%	\$123	17.3 yr	\$5534.80

a. What is the percent decrease in the interest paid between the original and the consolidated loan?

b. What is the percent increase in the number of years between the original and the consolidated loan?

With this information, reexamine Problem 28(b).

30. *Cooling bills* During hot weather, if you raise the thermostat from 72° to 76°, your cooling bill will decrease by 40%. If your cooling bill was \$140 and you did raise the thermostat from 72° to 76°, what would you expect the cooling bill to be?

31. *Verbal GRE scores* One of the criteria for accepting students to graduate school is a GRE (Graduate Record Exam) score. A student takes the verbal portion of the GRE and obtains a score of 120. The student decides to take the test again and this time scores 140. To the nearest percent, what is the increase in the student's score?

32. *Quantitative GRE scores* A student scores 115 on the quantitative portion of the GRE (Graduate Record Exam). The student retakes the exam and this time scores 125. To the nearest percent, what is the increase in the student's score?

Refer to the table for questions 33–34.

Year	Amount
2000	271
2001	289
2002	313
2003	233
2004	277

33. *The cost of entertainment* The table shows the annual amount of money spent on entertainment (fees and admissions) for persons under age 25.

a. The increase from 2000 to 2001 was $18. What is the percent increase to the nearest percent?

b. In what years did the biggest increase in spending occur?

c. To the nearest percent, what was the increase?

34. *Decrease in the cost of entertainment* Refer to the table and answer the following questions.

a. In what years was there a decrease in the amount spent on entertainment?

b. To the nearest percent, what was the decrease in spending?

Source: Bureau of Labor

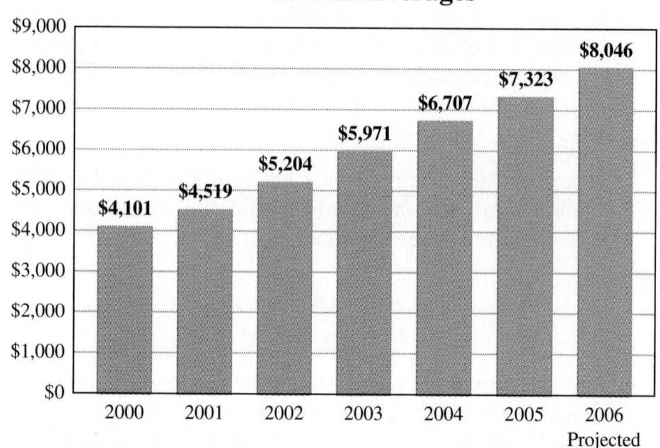

Source: Hewitt Health Value Initiative, http://was4.hewitt.com/hewitt/resource/newsroom/pressrel/2005/10-10-05_charts. pdf.

35. *Total health care costs for employees* Refer to the graph to answer the following questions.

a. Find the increase in total health care costs from 2005 to 2006.

b. To the nearest percent, what was the increase in health care costs from 2005 to 2006?

Using Your Knowledge

College Tuition Fees Problems 36–40 refer to the graph in the *Getting Started* at the beginning of this section. Give the answers to the nearest percent.

36. The cost of tuition and fees for private schools in 1995–1996 was about $16,900. In 2005–2006, it is about $34,900. What is the percent increase?

37. What was the percent difference between college costs at a public college ($7400) and a private college ($16,900) in 1995–1996?

38. What is the percent difference between college costs at a public college ($14,200) and a private college ($34,900) in 2005–2006?

39. Did the cost of public colleges more than double in the period 1995–1996 to 2005–2006? Find the difference in cost between 1995–1996 and 2005–2006 and answer the question.

40. Did the cost of private colleges more than double in the period 1995–1996 to 2005–2006? (See Problems 36 and 38.) Find the difference in cost between 1995–1996 and 2005–2006 and answer the question.

Write On

41. Suppose you have a \$10,000 loan at 5% for 10 years with a \$106 monthly payment. You are offered a 3% loan with a \$100 monthly payment for 10 years. Is this a better loan? Explain. (*Hint:* The interest on the 5% loan is \$2720; the interest on the 3% loan is \$2000.)

42. The 3% loan with a \$100 monthly payment in Problem 41 can be paid in 115 monthly payments. Is this loan better than the original loan at 5% for 10 years? Explain.

43. Write in your own words the factors you would consider when comparing two loans.

Concept Checker

Fill in the blank(s) with the correct word(s), phrase, or mathematical statement.

increase	**original**
decrease	**money**

44. The **first step** in finding a percent of increase or decrease is to **find the amount** of ________ or ________.

45. The **second step** to find a percent of increase or decrease is to **divide** the amount of increase or decrease by the ________ amount.

Mastery Test

Give the answer to the nearest percent.

46. The base price of a car is \$23,500. The dealer is selling it for \$25,000. What is the percent increase over the base price?

47. The base price for a car is \$25,000. The wholesale price for the same car after just using it for 100 miles is \$22,000. What is the percent of depreciation for the car?

48. The wholesale price of a car is \$20,000. The MSRP is \$22,000. What is the percent increase between the wholesale price and the MSRP?

49. The average annual earnings for a high school graduate are about \$26,000. For a person with an associate's degree, they are \$33,000. What is the percent increase in annual salary between a high school graduate and a person with an associate's degree?

50. A student estimated that the amount to be spent on entertainment had to be increased by 20% each month in order to satisfy current needs. If the student is spending \$200 a month on entertainment now, how much will the student be spending on entertainment next month?

51. On a certain month, the DJIA went down from 9000 to 8000. To the nearest percent, what was the percent of decrease?

52. From 1999 to 2000, the total amount for Pell Grants awarded to students went from \$43.5 billion to \$46.9 billion. Find the percent of increase to the nearest percent.

Skill Checker

Give the answer to the nearest percent.

53. Find the interest on \$3550 at 18% for 1 month.

54. Find the interest on \$150,000 at 7.5% for 1 month.

55. Find 1.5% of \$260.

56. Find the compound amount A at 4% if \$1000 is compounded monthly for 2 months.

57. Find the compound amount A at 8% if \$1000 is compounded monthly for 2 months.

5.6 Consumer Credit

Objectives

You should be able to solve application problems involving:

A › Credit cards.

B › Student loans.

C › Mortgages.

To Succeed, Review How To . . .

1. Solve percent problems. (pp. 298–302, 309–311)
2. Perform basic operations on fractions and decimals. (pp. 130–133, 151–157, 202–206, 212–216)

Getting Started

Which Card Is Better for You?

It depends on many factors such as the interest rate (the interest you pay the card company), the annual fee (how much you pay the card company for using the card), and the grace period (the interest-free time a lender allows between the transaction date and the billing date, usually 30 days). To learn the basics, look at the websites!

A › Credit Cards

How can you save money on your credit card? Look at the three terms we introduced in Getting Started and use them to your advantage. Here are three suggestions for savings:

1. Get the *lowest* possible interest rate (even 0% is sometimes available).
2. Get a card with *no* annual fee.
3. Pay off your entire balance within the *grace period* every month.

Let us see how all this works in practice.

EXAMPLE 1 Credit card comparisons

Suppose you have a card with a \$25 annual fee and an 18% annual percentage rate (APR). If the average monthly balance on your card is \$500 and you can get a different card with a 14% APR and no annual fee, how much will your savings be if you change to the second card?

SOLUTION You will be saving

4% (18% − 14%) of \$500, or \$20, in interest and the \$25 annual fee.

The total savings amount to \$20 + \$25 = \$45.

PROBLEM 1

Suppose you have a card with a \$50 annual fee and a 10% APR. If your average monthly balance is \$1000 and you can get a card with a 14% APR and no annual fee, how much will your total savings be?

Answers to PROBLEMS

1. \$10

EXAMPLE 2 Credit card comparisons

If the average monthly balance on your card is \$500, your annual card fee is \$20, and your APR is 14%, does it make sense to change to another card with no annual fee and a 19% APR?

SOLUTION

With the first card you pay: 14% of \$500 or \$70, plus the \$20 fee, \$90 in all.

If you change to the second card, you pay 19% of \$500, or \$95.

It is better to stay with the first card.

PROBLEM 2

If the average monthly balance on your card is \$1000, your annual fee is \$50, and your APR is 12%, does it make sense to change to another card with no annual fee and an 18% APR?

EXAMPLE 3 Balance transfers

Desiree charged her vacation expenses of \$3550 to her credit card, which had an APR of 18%. She then decided not to make any additional purchases with this card until the balance was paid off.

a. If the minimum payment on this card is 2% of the balance, find the minimum payment.

b. Find the amount of interest and the amount applied to reduce the principal when the 2% minimum payment is made.

c. Desiree received a special offer card with a 10% APR and no annual fee. How much of the 2% minimum payment would be interest and how much would be applied to reduce the principal with the new card?

d. Compare the amounts by which the principal is reduced with the 18% and the 10% cards.

SOLUTION

a. The minimum payment is 2% of \$3550 $= 0.02 \times 3550 =$ \$71.

b. The amount of interest for \$3550 at 18% for one month* is

$$0.18 \times 3550 \times \frac{1}{12} = \$53.25$$

Since she made a \$71 payment, the amount going to reduce the principal is \$71 − \$53.25 = \$17.75.

c. With the 10% card, the interest for the month is

$$0.10 \times 3550 \times \frac{1}{12} = \$29.58$$

The amount going to reduce the principal is \$71 − \$29.58 = \$41.42.

d. With the 18% card, the principal is reduced by \$17.75 (she owes \$3550 − \$17.75 = \$3532.25). With the 10% card, the principal is reduced by \$41.42 (she owes \$3550 − \$41.42 or \$3508.58).

*Technically, the interest on credit cards is calculated daily using the DPR (Daily Periodic Rate), which would be $\frac{0.18}{365}$ here, but the difference for the total interest for the month is very small.

PROBLEM 3

Repeat Example 3 if the balance was \$3000 and the original APR was 15%.

Now, suppose you receive this notice in the mail:

> Limited-time 0% APR†
> Your credit line has been increased to \$26,200!

with this fine print added at the end:

> †Effective on or after the first day following your statement Closing Date in October, the Daily Periodic Rate ("DPR") for new Cash Advances and for new Purchases posting to your account through your **April** statement Closing Date is 0% (corresponding **ANNUAL PERCENTAGE RATE ("APR")** of 0%). Thereafter, the DPR for these promotional Cash Advance balances will be 0.035589% (corresponding **APR** of 12.99%), and the DPR for these promotional Purchase balances will be 0.035589% (corresponding **APR** of 12.99%).

Answers to PROBLEMS

2. No

3. a. \$60

b. \$37.50; \$22.50

c. \$25; \$35

d. \$22.50 with the 15% card, \$35 with the 10% card

> **Important Reminder:** The transaction fee for credit card access checks, including the enclosed checks, is 3% of each transaction (Min. \$5, Max. \$50). See your Credit Card Agreement for any other applicable transaction fees.

You have two credit cards with \$1000 and \$500 balances, respectively, each charging 9% APR, and you want to be debt free in 2 years! Should you take this deal to fulfill your debt-free goal? Before you do, read the fine print. The limited time at 0% is 6 months (from end of October to April). After that, your interest rate on the new card will be 12.99% (which we round to 13%). Let us compare the total amounts you would pay. Warning: You need a calculator!

PROBLEM SOLVING

1. Read the problem. You have two credit cards now, and you want to compare their total costs for a 2-year period to the cost of the new card at 0% interest for 6 months and 13% thereafter for the remaining 18 months (24 months total).

2. Select the unknown. We have two unknowns: (1) the amount to pay on the two old (9%) credit cards and (2) the amount to pay on the new 13% credit card. Both actions should take 2 years.

3. Translate the problem into an equation or inequality. We will use the compound amount formula. To figure the amount we have to pay for the two 9% credit cards and the new 13% card and compare the results, we use the formula for the compound amount $A = P(1 + i)^n$.

4. Use the rules we have studied and a calculator to solve the problem. For the two old cards, $P = \$1500(\$1000 + \$500)$, $n = 24$, $i = \frac{9\%}{12}$

Thus, $A = 1500\left(1 + \frac{0.09}{12}\right)^{24} = \$1794.62$

For the new card, you do not pay interest for 6 months, and then you pay 13% for 18 months. $A = 1500\left(1 + \frac{0.13}{12}\right)^{18} = \$1821.06$

Thus, the new card with the 0% introductory rate is more expensive! (\$1821.06 against \$1794.62.) But there is more. To pay off the two 9% cards you have to write two checks, one for \$1000 and one for \$500. The Important Reminder tells you that the fee is 3% of each transaction, that is, \$30 and \$15, an additional \$45 cost. Definitely, stay with the old cards unless you are able to pay off the new card in 6 months at 0% interest.

5. Verify the answer. The verification is left to you.

Now that you know how to recognize the best credit card deals, let us concentrate on how to handle the card you have. As we mentioned, there are two things that can save you money: the interest rate (APR) you pay and the size of your monthly payment. They are both initially controlled by the credit card company, not you. Here are some typical rates and monthly payments:

Unpaid Balance	**Monthly Rate**
0–\$500	$\frac{18\%}{12} = 1.5\%$ monthly
Over \$500	$\frac{12\%}{12} = 1\%$ monthly
New Balance	**Minimum Payment**
Less than \$200	\$10
More than \$200	5% of the new balance

Let us see how this works.

EXAMPLE 4 Finding balances, finance charges, and payments

Khan received a statement from his credit card company. His previous balance was \$280. He made a \$20 payment and charged an additional \$60. If he pays 1.5% of the unpaid balance as a finance charge, find

a. the unpaid balance.
b. the finance charge.
c. the new balance.
d. the minimum payment (see table).

SOLUTION

a. The unpaid balance is \$280 − \$20 = \$260.
b. The finance charge is 1.5% of \$260 = 0.015 × 260 = \$3.90.
c. The new balance is

Unpaid balance	\$260.00
Finance charge	\$3.90
Purchases	\$60.00
New balance	\$323.90

d. Since the new balance is over \$200, the card company determines that the minimum payment should be

5% of \$323.90 = 0.05 × 323.90 = \$16.1950 or \$16.20

PROBLEM 4

Cecilia had a \$350 previous balance on her credit card. She made a \$50 payment and charged an additional \$200. If she pays 1.5% of the unpaid balance as a finance charge, find

a. the unpaid balance.
b. the finance charge.
c. the new balance.
d. the minimum payment.

B › Student Loans

The terminology and ideas of credit cards are very similar to those used when dealing with student loans. The federal loan program for students attending school at least half-time is called the Stafford loan program, and interest rates are variable (they are reset each July 1) but are much lower than on credit cards (as low as 2.82%, capped at 8.25%). There are two types of Stafford loans: subsidized (based on need, with the interest paid by the federal government while the student is in school) and unsubsidized. The term of payment can be as high as 10 years (120 months) with payments starting 6 months after graduation.

EXAMPLE 5 Full Stafford loan

Panfilo successfully graduated from college after getting the full amounts allowed for Stafford loans: \$2625 as a freshman, \$3500 as a sophomore, \$5500 as a junior, and \$5500 as a senior, a total of \$17,125 at 3% for 10 years. His first payment was \$165.36.

a. How much interest and how much principal are in the first payment?
b. If his interest rate had been the maximum 8.25%, how much more of the first payment would have gone to interest as compared with his 3% loan and the \$165.36 payment?
c. How much more total interest would be paid on an 8.25%, 10-year loan with a \$204.10 monthly payment as compared to the 3% loan and the \$165.36 payment over the same 10-year period?

SOLUTION

a. We use the formula $I = P \times R \times T$, where $P = \$17{,}125$, $R = 3\%$, and $T = \frac{1}{12}$ ($\frac{1}{12}$ of a year is 1 month).

PROBLEM 5

Latasha graduated from college also, but her Stafford loans amounted to \$15,000 at 4% for 10 years. Her payment was \$151.87.

a. How much interest and how much principal are in the first payment?
b. If her interest rate had been the maximum 8.25%, how much of the first payment would have gone to interest as compared with her 4% loan and the \$151.87 payment?
c. How much more total interest would be paid on an 8.25%, 10-year loan with a \$183.98 monthly payment as compared to the 4% loan and the \$151.87 payment over the same 10-year period?

(continued)

Answers to PROBLEMS

4. a. \$300 **b.** \$4.50 **c.** \$504.50 **d.** \$25.23 **5. a.** \$50 interest, \$101.87 principal **b.** \$53.13 **c.** \$3853.20

The interest for 1 month is

$$I = 17{,}125 \times 0.03 \times \frac{1}{12} = \$42.81$$

The difference between the payment of \$165.36 and the interest of \$42.81, that is, \$165.36 − \$42.81 = \$122.55, goes to reduce the principal.

b. The interest at 8.25% for 1 month would be

$$I = 17{,}125 \times 0.0825 \times \frac{1}{12} = \$117.73$$

The difference in interest between the 3% loan and the 8.25% loan is \$117.73 − \$42.81 = \$74.92.

c. The amount paid at the 3% rate is \$165.36 × 120 = \$19,843.20. Since the loan is for \$17,125, the interest is \$19,843.20 − \$17,125 or \$2718.20.

The amount paid at the 8.25% rate is \$204.10 × 120 = \$24,492. On a \$17,125 loan, the interest is \$24,492 − \$17,125 = \$7367. Thus, at the 8.25% rate, the interest is \$7367 − \$2718.20 = \$4648.80 more than with the 3% rate.

The amount owed after the first month is \$17,125 − \$122.55 = \$17,002.45.

The interest is the difference between the total paid (\$19,843.20) and the amount of the loan (\$17,125).

C › Mortgages (Home Loans)

A **mortgage** is a long-term loan (usually to buy a house) that you can get from a bank, savings union, mortgage broker, or online lender. Because mortgages involve such large amounts, they are paid over longer periods, usually 15 to 30 years. The breakdown of each payment (the amount that goes toward principal, interest, etc.) changes over time, and is detailed in an **amortization table** like the one shown for a 30-year, \$150,000 mortgage at a fixed 7.5% rate.

Amortization Table for a \$150,000, 30-Year Mortgage at 7.5%

Here's how principal and interest change over the life of a loan

Payment Number	Principal Balance	Payment Amount	Interest Paid	Principal Applied	New Balance
1	\$150,000	\$1048.82	\$937.50	\$111.32	\$149,888.68
60	\$142,086.93	\$1048.82	\$888.04	\$160.78	\$141,926.15
120	\$130,426.14	\$1048.82	\$815.16	\$233.66	\$130,192.48
240	\$88,851.22	\$1048.82	\$555.32	\$493.50	\$88,357.72
359 (next to last)	\$2078.14	\$1048.82	\$12.99	\$1035.83	\$1042.31

Source: Data from Bankrate.com.

EXAMPLE 6 Interest on a 30-year mortgage

Suppose you have a \$150,000, 30-year mortgage at 7.5% with a monthly payment of \$1048.82.

a. What would the interest be in the first month?

b. How much of the \$1048.82 payment would go toward paying the \$150,000 principal?

c. Look at the amortization table and determine what the principal balance and the new balance are on the next-to-last payment.

PROBLEM 6

Suppose you have a \$100,000, 30-year mortgage at 6% with a monthly payment of \$600.

a. What would the interest be in the first month?

b. How much of the \$600 payment would go toward paying the \$100,000 principal?

Answers to PROBLEMS

6. a. \$500 **b.** \$100

SOLUTION

a. The interest is $I = P \times R \times T$, where $P = \$150{,}000$, $R = 7.5\%$, and T is $\frac{1}{12}$ ($\frac{1}{12}$ of a year is 1 month).

Thus, $I = 150{,}000 \times 0.075 \times \frac{1}{12} = \$937.50$. (Exciting—it is just like in the amortization table!)

b. The payment is \$1048.82 and the interest is \$937.50, so the amount that would go toward paying the principal is \$1048.82 − \$937.50 = \$111.32. (Wow, just like the table again!)

c. The principal balance for payment 359 is \$2078.14. The new balance is \$1042.31. When you pay that, you are paid up!

Exercises 5.6

⟨A⟩ Credit Cards In Problems 1–7, follow the procedure in Example 4 to complete the table.

	Previous Balance	Payment	Unpaid Balance	Finance Charge (1.5%)	Purchases	New Balance
1.	\$100	\$10	___	___	\$50	___
2.	\$300	\$190	___	___	\$25	___
3.	\$134.39	\$25	___	___	\$73.98	___
4.	\$145.96	\$55	___	___	\$44.97	___
5.	\$378.93	\$75	___	___	\$248.99	___
6.	\$420.50	\$100	___	___	\$300	___
7.	\$500	\$300	___	___	\$250	___

In Problems 8–16, use the table below to determine:

a. the finance charge for the month.
b. the new balance.
c. the minimum monthly payment.

Unpaid Balance	Monthly Rate
0–\$500	$\frac{18\%}{12} = 1.5\%$ monthly
Over \$500	$\frac{12\%}{12} = 1\%$ monthly
New Balance	**Minimum Payment**
Less than \$200	\$10
More than \$200	5% of the new balance

	Previous Balance	New Purchases
8.	\$50.40	\$173
9.	\$85	\$150
10.	\$154	\$75
11.	\$344	\$60
12.	\$666.80	\$53.49
13.	\$80.45	\$98.73
14.	\$34.97	\$50
15.	\$55.90	\$35.99
16.	\$98.56	\$45.01

Web IT go to mhhe.com/bello for more lessons

Web IT go to mhhe.com/bello for more lessons

⟨B⟩ Student Loans In Problems 17–18, solve the Stafford loan problems.

17. After graduation, the balance on Marcus's Stafford loan was $20,000, to be paid at 3% for 10 years (120 payments). His first payment was $193.12.

a. How much interest and how much principal are in the first payment?

b. If his interest rate had been the maximum 8.25%, how much more of the first payment would have gone to interest as compared with his 3% loan and the $193.12 payment?

c. How much more total interest would be paid on an 8.25%, 10-year loan with a $245.31 monthly payment as compared to the 3% loan and the $193.12 payment over the same 10-year period?

18. Frances had a $25,000 Stafford loan at 2.85% to be paid after graduation with 120 payments (10 years) of $239.67.

a. How much interest and how much principal are in the first payment?

b. If her interest rate had been the maximum 8.25%, how much more of the first payment would have gone to interest as compared with her 2.85% loan and the $239.67 payment?

c. How much more total interest would be paid on an 8.25%, 10-year loan with a $306.63 monthly payment as compared to the 2.85% loan and the $239.67 payment over the same 10-year period?

The chart shows the estimated standard repayment chart for several loan amounts at different interest rates. In Problems 19–22,

a. Find the interest for the first month for the 6% loans of $10,000, $18,500, $20,000, and $23,000.

b. Find the amount of the first payment that will go to pay the principal for each of the four loans.

c. Find the total amount of interest that will be paid for each of the four loans.

Estimated Standard Repayment Chart

	Interest Rate:[1]		5.00%	6.00%	7.00%	8.25%
	Loan Amount	Number of Payments[2]	Approximate Monthly Payments[3]			
19.	$10,000	120	$106	$111	$116	$123
20.	$18,500	120	$196	$205	$215	$227
21.	$20,000	120	$212	$222	$232	$245
22.	$23,000	120	$244	$255	$267	$282

1. Rate is as shown. 2. (120) but terms may vary. 3. At least $50.

Source: Educaid.com

⟨C⟩ Mortgages (Home Loans) In Problems 23–24, answer the questions about mortgages.

23. Athanassio and Gregoria Pappas want a 30-year loan at 7% to buy a small $50,000 house in Tarpon Springs.

a. If they can get a loan for 95% of the value of the house, what will be the amount of the loan?

b. The difference between the loan amount and the $50,000 is the down payment. How much would that be?

c. If the monthly payment is $316, how much of the first payment is principal and how much is interest?

24. Tim and Frances Johnston wish to obtain a 30-year, $40,000 loan at 9% to buy a house in Georgia.

a. If they can get a loan for 95% of the value of the house, what will be the amount of the loan?

b. The difference between the loan amount and the $40,000 is the down payment. How much would that be?

c. If the monthly payment is $305, how much of the first payment is principal and how much is interest?

The charts show loan amortization schedules for the principal shown (30-year loans). In Problems 25–28,

a. Find the interest paid, principal applied, and new balance on line 1.

b. Find the interest paid, principal applied, and new balance on line 2.

Loan Amortization

	Number	Principal Balance	Payment Amount	Interest Paid: 6%	Principal Applied	New Balance
25.	1	$100,000.00	$599.55	___	___	___
	2	$99,900.45	$599.55	___	___	___

26.

Loan Amortization

Number	Principal Balance	Payment Amount	Interest Paid: 5%	Principal Applied	New Balance
1	$100,000.00	$536.82	____	____	____
2	$99,879.85	$536.82	____	____	____

27.

Loan Amortization

Number	Principal Balance	Payment Amount	Interest Paid: 7%	Principal Applied	New Balance
1	$150,000.00	$997.95	____	____	____
2	$149,877.05	$997.95	____	____	____

28.

Loan Amortization

Number	Principal Balance	Payment Amount	Interest Paid: 8%	Principal Applied	New Balance
1	$200,000.00	$1467.53	____	____	____
2	$199,865.80	$1467.53	____	____	____

Some people (not you, of course) may think that if you get a 30-year mortgage for $125,000 the interest you pay is found using the simple interest formula $I = P \times R \times T$, where I is the interest, P the principal, R the rate, and T the time in years (see Example 5).

29. Find the simple interest when P = $125,000, $R = 0.06$, and $T = 30$.

(*Note:* This answer is *not* correct, since a mortgage uses *compound* instead of *simple* interest.) To see how much interest you *really* have to pay, you must consult the graph.

30. To find the actual interest paid on a 30-year, $125,000 loan say at 10%, look at the graph. The top of the bar labeled 10% shows the interest: $269,907. However, that is not all you pay, you must pay the $125,000 principal as well, for a total of $394,907.

a. Refer to the graph and find the interest paid on a 30-year, $125,000 loan at 9%.

b. Find the total amount (interest plus principal) on a 30-year, $125,000 loan at 9%.

Total Interest Paid on a 30-Year $125,000 Loan at Various Rates

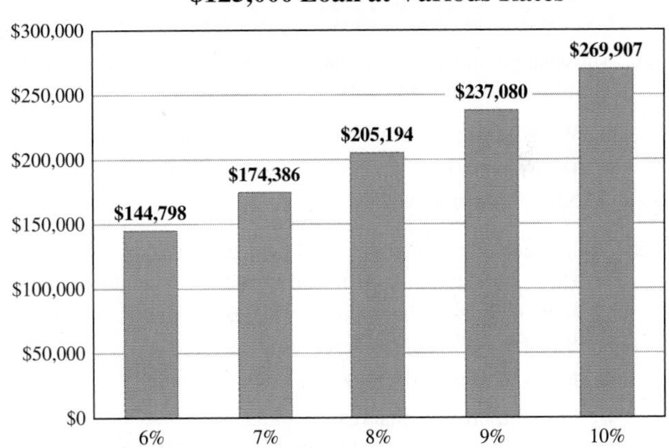

Source: http://michaelbluejay.com/house/loan.html

31. a. Refer to the graph and find the interest paid on a 30-year, $125,000 loan at 6%.

b. Find the total amount (interest plus principal) on a 30-year, $125,000 loan at 6%.

32. a. Refer to the graph and find the interest paid on a 30-year, $125,000 loan at 8%.

b. Find the total amount (interest plus principal) on a 30-year, $125,000 loan at 8%.

Web IT go to mhhe.com/bello for more lessons

Using Your Knowledge

Interest We mentioned in Example 3 that the difference in interest when calculating monthly charges using $\frac{1}{12}$ of a year as opposed to daily charges of $\frac{1}{365}$ of a year was very small. How small? Use your knowledge to find out!

33. Suppose you want to find the interest $I = P \times R \times T$ on a \$1000 loan at 6% and you assume that T (one month) is $\frac{1}{12}$ of a year. What is the interest to the nearest cent?

34. Repeat Problem 33 if $T = \frac{30}{365}$. What is the difference in the interest?

Write On

35. When using the formula $I = P \times R \times T$ for simple interest and the time is one month, what is T? Why?

36. When using the formula $I = P \times R \times T$ for simple interest and the time is one day, what is T? Why?

37. If we assume that a year has 360 days, the time is 90 days, and we are using the formula $I = P \times R \times T$, what is T in reduced form? Why?

38. There is a procedure to find the interest on any amount for 60 days at 6% if we assume that the year has 360 days. For example, the interest on \$1000 for 60 days at 6% is

$$I = P \times R \times T = 1000 \times \frac{6}{100} \times \frac{60}{360} = \$10.00$$

Write in your own words what the rule should be.

Concept Checker

Fill in the blank(s) with the correct word(s), phrase, or mathematical statement.

39. A **mortgage** is a long-term ________.

40. The money derived from a **mortgage** is usually used to buy a ________.

car **house**
debt **loan**

Mastery Test

41. Suppose you have a \$200,000, 30-year mortgage at 6% with a monthly payment of \$1200.

a. What will the interest be in the first month?

b. How much of the \$1200 payment will go toward paying the \$200,000 principal?

42. Raja had a \$25,000 Stafford loan at 3% for 10 years. His first payment was \$240.

a. How much interest and how much principal are in the first payment?

b. If his interest rate had been the maximum 8.25%, how much more of the first payment would have gone to interest as compared with his 3% loan and the \$240 payment?

c. How much more total interest would be paid on an 8.25%, 10-year loan with a \$306 monthly payment as compared to the 3% loan and the \$240 payment over the same 10-year period?

43. Mida received a statement from her credit card company. Her previous balance was \$200. She made a \$50 payment and charged an additional \$60. If she pays 1.5% of the unpaid balance as a finance charge,

a. Find the unpaid balance.

b. Find the finance charge.

c. Find the new balance.

d. Find the minimum payment (see table).

New Balance	Minimum Payment
Less than \$200	\$10
More than \$200	5% of the new balance

Skill Checker

In Problems 44–47, find:

44. $\frac{1}{2}$ of 50 = __

45. $\frac{1}{2}$ of 100 = __

46. $3 \cdot 100$ = ___

47. $7 \cdot 100$ = ___

Collaborative Learning

Do you know what the ten hottest careers for college graduates are? The 10 fastest-growing occupations between 2002 and 2012 are shown in the table and are either technology related or medical/health related. If you are trying to decide on a career, exploring the available opportunities could help with your choice. To find a fast-growing career, divide into groups and do the following:

1. Separate these 10 careers into two categories—Technical or Medical. Fill in the last column (% I) with the percent increase for that career. The percent increase is found by dividing the difference of the two numbers by the original number.
2. Answer the following questions:
   a. Of the two categories, which field seems to have higher percents of increase?
   b. In the technology field, which career has the highest percent of increase? The lowest?
   c. In the medical field, which career has the highest percent of increase? The lowest?
   d. Besides the availability of jobs, what other information would you want to know about a career?

10 Fastest-Growing Occupations for College Grads

Occupation	2002	2012	% I
Network systems and data communications analysts	186	292	
Physician assistants	63	94	
Medical records and health information technicians	147	216	
Computer software engineers, applications	394	573	
Computer software engineers, systems software	281	409	
Physical therapist assistants	50	73	
Fitness trainers and aerobics instructors	183	264	
Database administrators	110	159	
Veterinary technologists and technicians	53	76	
Dental hygienists	148	212	

Source: United States Bureau of Labor Statistics.

Research Questions

The chart shows the total U.S. population and the African-American population according to the U.S. census (conducted every 10 years).

1. To the nearest percent, what was the African-American population in each of the years listed?
2. In what year was the percent of African-Americans highest?
3. What was the highest percent decrease in the African-American population? In what time period?
4. What was the highest percent increase in the African-American population? In what time period?

(continued)

Year	Total U.S. Population	African-American Population
1860	31,400,000	4,400,000
1880	50,100,000	6,500,000
1900	76,000,000	9,100,000
1920	105,700,000	10,500,000
1940	131,700,000	13,200,000
1960	179,300,000	17,900,000
1980	226,500,000	27,200,000
2000	281,400,000	36,400,000

Source: U.S. Census data. All numbers are rounded.

5. Now, some real research! Make a table like this one and answer questions 1–4 about the Hispanic population.

Hint: You can find some of the information by doing a Web search on "census statistics."

Summary Chapter 5

Section	Item	Meaning	Example
5.1	Percent	Per hundred	$9\% = \frac{9}{100}$
5.2	Base (total)	The standard used for comparison purposes	$B = \frac{P}{R}$; B the base, P the percentage, R the rate.
	Percentage	The part being compared with the base or total	$P = B \times R$
	Percent (rate)	The part indicating the ratio of the percentage to the base	$R = \frac{P}{B}$
5.3	Solving percent problems using proportions	To solve percent problems using proportions, set up the proportion $\frac{\text{Percent}}{100} = \frac{\text{Part}}{\text{Whole}}$	**1.** To find 40% of 50, set up the proportion $\frac{40}{100} = \frac{n}{50}$ and solve the proportion. **2.** To find what percent of 50 is 10, set up the proportion $\frac{\text{Percent}}{100} = \frac{10}{50}$ then solve it. **3.** To find a number so that 30% of it is 60, set up the proportion $\frac{30}{100} = \frac{60}{\text{Whole}}$ then solve it.
5.4A	Total cost	List price + sales tax	The total cost of an item listed at \$200 when the tax rate is 5% is: $200 + 200 \times 0.05 = \$210$.

Section	Item	Meaning	Example
5.4B	Simple interest	The amount paid for using money $I = P \times R \times T$	The amount paid for borrowing \$100 at 6% simple interest for one year is $I = 100 \times \frac{6}{100} \times 1$ $= \$6$
5.4C	Compound interest Total compound amount A for a rate i **per period** applied to a principal P for n periods is given by	Computing interest on the earned interest $A = P(1 + i)^n$	\$100 invested at 6% compounded semiannually for one year earns \$6.09 in interest. The compound amount A for a \$100, three-year deposit compounded at 4% semiannually is $A = 100\left(1 + \frac{4\%}{2}\right)^6$
5.4D	Commission	A share of the sales price earned by a salesperson	A 5% commission on a \$300 sale is $\$300 \times 0.05 = \$15$.
5.4E	Discount	A reduction in the regular price of an item	A 2% discount on a \$500 item amounts to $\$500 \times 0.02 = \$10$.
5.5A	Percent increase/ decrease	The percent a given quantity increases or decreases	If an item increases in price from \$10 to \$12, the percent increase is $\frac{12-10}{10} = \frac{2}{10} = 20\%$.
5.5B	Depreciation	The loss in value of an item	A \$20,000 car is valued at \$18,000 after two years. The depreciation of the car is $\frac{20{,}000 - 18{,}000}{20{,}000} = \frac{2000}{20{,}000} = 10\%$.
5.6A	APR	Annual Percentage Rate	The APR on credit cards varies from 6% to 18%.
5.6B	Stafford loan	A type of federal student loan	Rates vary and are adjusted every July but cannot exceed 8.25%.
5.6C	Mortgage	A long-term loan (usually to buy a house) that you can get from a bank, credit union, mortgage broker, or online lender	You may have a 30-year, 6% mortgage on your home (terms and interest rates vary).

> Review Exercises Chapter 5

(If you need help with these exercises, look in the section indicated in brackets.)

1. ‹**5.1A**› *Write as a decimal.*

a. 39% **b.** 1% **c.** 13%

d. 101% **e.** 207%

2. ‹**5.1A**› *Write as a decimal.*

a. 3.2% **b.** 11.2% **c.** 71.4%

d. 17.51% **e.** 142.5%

3. ‹**5.1A**› *Write as a decimal.*

a. $6\frac{1}{4}\%$ **b.** $71\frac{1}{2}\%$ **c.** $5\frac{3}{8}\%$

d. $17\frac{1}{4}\%$ **e.** $52\frac{1}{8}\%$

4. ‹**5.1A**› *Write as a repeating decimal.*

a. $6\frac{1}{3}\%$ **b.** $8\frac{2}{3}\%$ **c.** $1\frac{1}{6}\%$

d. $18\frac{5}{6}\%$ **e.** $20\frac{1}{9}\%$

5. **⟨5.1B⟩** *Write as a percent.*

a. 0.01 **b.** 0.07 **c.** 0.17

d. 0.91 **e.** 0.83

6. **⟨5.1B⟩** *Write as a percent.*

a. 3.2 **b.** 1.1 **c.** 7.9

d. 9.1 **e.** 4.32

7. **⟨5.1C⟩** *Write as a reduced fraction.*

a. 17% **b.** 23% **c.** 51%

d. 111% **e.** 201%

8. **⟨5.1C⟩** *Write as a reduced fraction.*

a. 10% **b.** 40% **c.** 15%

d. 35% **e.** 42%

9. **⟨5.1C⟩** *Write as a reduced fraction.*

a. $16\frac{2}{3}\%$ **b.** $33\frac{1}{3}\%$ **c.** $62\frac{1}{2}\%$

d. $83\frac{1}{3}\%$ **e.** $87\frac{1}{2}\%$

10. **⟨5.1D⟩** *Write as a percent.*

a. $\frac{3}{8}$ **b.** $\frac{5}{8}$ **c.** $\frac{1}{16}$

d. $\frac{3}{16}$ **e.** $\frac{5}{16}$

11. **⟨5.2A⟩** *Find the percentage.*

a. 60% of 30 is what number?

b. 70% of 140 is what number?

c. 40% of 80 is what number?

d. 30% of 90 is what number?

e. 35% of 105 is what number?

12. **⟨5.2A⟩** *Find the percentage.*

a. $12\frac{1}{2}\%$ of 80

b. $40\frac{1}{2}\%$ of 60

c. $15\frac{1}{2}\%$ of 250

d. $10\frac{2}{3}\%$ of 300

e. $24\frac{3}{4}\%$ of 7000

13. **⟨5.2B⟩** *Find the percent.*

a. What percent of 20 is 5?

b. What percent of 50 is 10?

c. What percent of 60 is 20?

d. What percent of 80 is 160?

e. What percent of 5 is 1?

14. **⟨5.2B⟩** *Find the percent.*

a. Find what percent of 40 is 30.

b. Find what percent of 50 is 20.

c. Find what percent of 20 is 40.

d. Find what percent of 30 is 20.

e. Find what percent of 60 is 10.

15. **⟨5.2B⟩** *Find the percent.*

a. 20 is what percent of 40?

b. 30 is what percent of 90?

c. 60 is what percent of 80?

d. 90 is what percent of 60?

e. 30 is what percent of 80?

16. **⟨5.2C⟩** *Find the base.*

a. 30 is 50% of what number?

b. 20 is 40% of what number?

c. 15 is 75% of what number?

d. 20 is 60% of what number?

e. 60 is 90% of what number?

17. **⟨5.2C⟩** *Find the base.*

a. 60 is 40% of a number. Find the number.

b. 90 is 30% of a number. Find the number.

c. 40 is 80% of a number. Find the number.

d. 75 is 5% of a number. Find the number.

e. 42 is 50% of a number. Find the number.

18. **⟨5.2C⟩** *Find a number so that:*

a. 40% of the number is 10.

b. 50% of the number is 5.

c. 70% of the number is 140.

d. 65% of the number is 195.

e. 16% of the number is 40.

19. **⟨5.3A⟩** *Set up a proportion and find the number.*

a. What is 30% of 40?

b. What is 40% of 72?

c. What is 50% of 94?

d. What is 60% of 50?

e. What is 70% of 70?

20. **⟨5.3B⟩** *Set up a proportion and find the percent.*

a. 80 is what percent of 800?

b. 22 is what percent of 110?

c. 28 is what percent of 70?

d. 90 is what percent of 180?

e. 80 is what percent of 40?

21. ⟨**5.3C**⟩ *Set up a proportion and find the number.*

a. 20 is 10% of what number?

b. 30 is 12% of what number?

c. 45 is 90% of what number?

d. 50 is 25% of what number?

e. 63 is 18% of what number?

22. ⟨**5.4A**⟩ *In each problem find the total price for the given tax rate and price.*

	Tax Rate (%)	Price
a.	6	$20
b.	4	50
c.	5	18
d.	6.5	80
e.	4.5	300

23. ⟨**5.4B**⟩ *Find the simple interest.*

	Amount	Annual Rate (%)	Time (Years)
a.	$100	10	2
b.	250	12	3
c.	300	15	2
d.	600	9	2
e.	3000	8.5	3

24. ⟨**5.4B**⟩ *Find the simple interest.*

	Amount	Annual Rate (%)	Time (Months)
a.	$600	10	2
b.	600	12	3
c.	250	8	6
d.	450	9	8
e.	300	13	10

25. ⟨**5.4C**⟩ *If interest is compounded semiannually, find the compound amount and the interest when $10,000 is invested at*

a. 4% for two years.

b. 6% for two years.

c. 8% for two years.

d. 10% for two years.

e. 12% for two years.

26. ⟨**5.4D**⟩ *Find the commission.*

	Sale	Commission Rate (%)
a.	$100	8
b.	250	6
c.	17,500	7
d.	300	6.5
e.	700	5.5

27. ⟨**5.4E**⟩ *Find the sale price for the given article.*

	Regular Price	Discount Rate (%)
a.	$35	40
b.	40	50
c.	60	$33\frac{1}{3}$
d.	90	10
e.	540	15

28. ⟨**5.5A**⟩ *Find the percent of increase of sales (to the nearest percent) for a company whose sales increased from $2 billion to*

a. $3 billion.

b. $3.1 billion.

c. $3.2 billion.

d. $3.3 billion.

e. $3.4 billion.

29. ⟨**5.5A**⟩ *To the nearest percent, what is the percent of decrease in the price of gold currently selling at $400 an ounce if the new price is*

a. $395 an ounce.

b. $390 an ounce.

c. $385 an ounce.

d. $380 an ounce.

e. $375 an ounce.

30. ⟨**5.5A**⟩ *ABC Manufacturing's current sales are $48 million annually. Find the projected sales for next year if sales are to increase by*

a. $16\frac{2}{3}\%$

b. $33\frac{1}{3}\%$

c. $37\frac{1}{2}\%$

d. $66\frac{2}{3}\%$

e. $83\frac{1}{3}\%$

31. **‹5.6A›** *If the finance charge is 1.5% of the unpaid balance, find the unpaid balance, the finance charge, and the new balance.*

	Previous Balance	Payment	Additional Charges
a.	$100	$20	$50
b.	$150	$30	$60
c.	$200	$40	$70
d.	$300	$100	$100
e.	$500	$200	$300

32. **‹5.6B›** *Based on the information in the following table, find*

How much interest and how much principal are in the first payment on the 3% loan.

If the interest rate were the maximum 8.25%, how much more of the first payment would go to interest as compared with the 3% loan and the corresponding payment?

How much more total interest would be paid on an 8.25%, 10-year loan with the payment shown as compared to the 3% loan and the payment shown?

	Stafford Loan	Rate	Term	1st Pmt (at 3%)	1st Pmt (at 8.25%)
a.	$5000	3%	10 yr	$48.28	$61.33
b.	$10,000	3%	10 yr	$96.56	$122.65
c.	$15,000	3%	10 yr	$144.84	$183.98
d.	$20,000	3%	10 yr	$193.12	$245.31
e.	$25,000	3%	10 yr	$241.40	$306.63

Practice Test Chapter 5

Visit www.mhhe.com/bello to view helpful videos that provide step-by-step solutions to several of the problems below.

(Answers on page 346)

1. a. Write 37% as a decimal.
b. Write 17.8% as a decimal.

2. a. Write $5\frac{1}{4}\%$ as a decimal.
b. Write $15\frac{2}{3}\%$ as a repeating decimal.

3. a. Write 0.09 as a percent.
b. Write 5.1 as a percent.

4. a. Write 11% as a reduced fraction.
b. Write 60% as a reduced fraction.

5. Write $12\frac{1}{2}\%$ as a reduced fraction.

6. Write $\frac{7}{8}$ as a percent.

7. 70% of 50 is what number?

8. Find $10\frac{1}{2}\%$ of 80.

9. What is $33\frac{1}{3}\%$ of 96?

10. Find what percent of 50 is 30.

11. 40 is 50% of what number?

12. Set up a proportion and find 10% of 40.

13. Set up a proportion and find what percent of 80 is 20.

14. Set up a proportion and find 40% of what number is 10.

15. If the sales tax rate in one city is 5%, find the total price paid for a pair of shoes that cost $16.60.

16. Find the simple interest earned on $500 invested at 6.5% for 2 years.

17. Find the simple interest earned on $1200 invested at 5.5% for 3 months.

18. Find the compound amount and the interest when $10,000 is invested at 6% compounded semiannually for two years.

19. What is the commission of a salesperson on a $400 sale if her commission rate is 15%?

20. Find the sale price of an article regularly selling for $19.50 that is being advertised at 30% off.

21. The sales of children's sneakers rose from $2 billion to $2.4 billion. Find the percent of increase to the nearest percent.

22. In a certain week, the London stock market went down from 1740 to 1653. What was the percent of decrease to the nearest percent?

23. A company estimates that expenses will increase by $33\frac{1}{3}\%$ during the next year. If expenses are currently $81 million, how much will the company be spending next year?

24. A student received a statement from a credit card company. The previous balance was $480, a $50 payment was made, and an additional $100 was charged to the account. If the finance charge is 1.5% of the unpaid balance, find
a. the unpaid balance.
b. the finance charge.
c. the new balance.

25. A student had a $20,000 loan at 4% for 10 years. The first payment was $202.49.
a. How much interest and how much principal are in the first payment?
b. If the interest rate were the maximum 8.25%, how much more of the first payment would have gone to interest as compared with the 4% loan and $202.49 payment?
c. How much more total interest would be paid on an 8.25%, 10-year loan with a $245.31 monthly payment as compared to the 4% loan and the $202.49 payment over the same 10-year period?

Answers to Practice Test Chapter 5

Answer	If You Missed	Review		
	Question	Section	Examples	Page
1. **a.** 0.37 **b.** 0.178	1	5.1	1	289
2. **a.** 0.0525 **b.** $0.15\overline{6}$	2	5.1	2	289
3. **a.** 9% **b.** 510%	3	5.1	3	290
4. **a.** $\frac{11}{100}$ **b.** $\frac{3}{5}$	4	5.1	4	290
5. $\frac{1}{8}$	5	5.1	5	291
6. $87\frac{1}{2}\%$	6	5.1	6, 7	291–292
7. 35	7	5.2	1	299
8. 8.4	8	5.2	2	300
9. 32	9	5.2	3	300
10. 60%	10	5.2	4	301
11. 80	11	5.2	5	302
12. $\frac{10}{100} = \frac{\text{Part}}{40}$; Part = 4	12	5.3	1	310
13. $\frac{\text{Percent}}{100} = \frac{20}{80}$; Percent = 25%	13	5.3	2	310
14. $\frac{40}{100} = \frac{10}{\text{Whole}}$; Whole = 25	14	5.3	3	311
15. \$17.43	15	5.4	1	314
16. \$65	16	5.4	2	315
17. \$16.50	17	5.4	2	315
18. \$11,255.09; \$1255.09	18	5.4	3, 4	315, 317
19. \$60	19	5.4	5	317
20. \$13.65	20	5.4	6	318
21. 20%	21	5.5	1	323
22. 5%	22	5.5	2	323
23. \$108 million	23	5.5	3–6	324–325
24. **a.** \$430 **b.** \$6.45 **c.** \$536.45	24	5.6	1–4	330–331, 333
25. **a.** \$66.67 int.; \$135.82 principal **b.** \$70.83 **c.** \$5138.40	25	5.6	5–6	333–335

Cumulative Review Chapters 1–5

1. Write six thousand, five hundred ten in standard form.

2. Simplify: $36 \div 6 \cdot 6 + 7 - 3$

3. Classify $\frac{7}{6}$ as proper or improper.

4. Write $\frac{31}{4}$ as a mixed number.

5. Write $6\frac{4}{9}$ as an improper fraction.

6. Multiply: $\left(\frac{5}{6}\right)^2 \cdot \frac{1}{25}$

7. Divide: $\frac{10}{7} \div 8\frac{1}{3}$

8. Translate and solve: $\frac{3}{4}$ less than a number x is $\frac{1}{3}$. What is x?

9. Find a number such that $\frac{10}{11}$ of it is $7\frac{1}{9}$.

10. Give the word name for 342.41.

11. Write 34.773 in expanded form.

12. Subtract: $641.42 - 14.5$

13. Multiply: $5.94 \cdot 1.5$

14. Divide: $\frac{189}{0.27}$

15. Round 749.851 to the nearest tenth.

16. Divide: $10 \div 0.13$ (Round answer to two decimal digits.)

17. Write $0.\overline{12}$ as a reduced fraction.

18. What decimal part of 12 is 3?

19. Arrange in order of decreasing magnitude and write using the > sign:
$4.293 \quad 4.29\overline{3} \quad 4.\overline{293}$

20. Insert =, <, or > to make a true statement:
0.25 ____ $\frac{13}{20}$

21. Solve for x: $x + 2.3 = 6.2$

22. Solve for y: $1.4 = 0.2y$

23. Solve for z: $5 = \frac{z}{4.2}$

24. The ratio of cars to people in New Zealand is 480 to 1000. Write this ratio as a fraction in reduced form.

25. Write the following proportion: 5 is to 9 as 40 is to x.

26. There is a law stating that "the ratio of width to length for the American flag should be 10 to 19." Is a flag measuring 50 by 97 feet in the correct ratio?

27. Solve the proportion: $\frac{f}{4} = \frac{5}{80}$

28. Solve the proportion: $\frac{12}{f} = \frac{2}{3}$

29. A student traveled 300 miles on 16 gallons of gas. How many miles per gallon did the student get? (Round to the nearest whole number.)

30. A 24-ounce jar of peanut butter costs \$2.79. What is the unit price in cents per ounce? (Answer to the nearest cent.)

31. A pound of grass seed covers 3500 square feet of lawn. How many pounds are needed to cover a lawn measuring 200 by 70 feet (14,000 square feet)?

32. The protein RDA for males is 56 grams per day. Four ounces of a certain product provide 2 grams of protein. How many ounces of the product are needed to provide 56 grams of protein?

33. Write 12% as a decimal.

34. Write $7\frac{1}{4}\%$ as a decimal.

35. Write 0.03 as a percent.

36. 10% of 80 is what number?

37. What is $66\frac{2}{3}\%$ of 54?

38. What percent of 32 is 16?

39. 12 is 60% of what number?

40. The sales tax rate in a certain state is 5%. Find the total price paid for a pair of shoes that costs \$16.

41. Find the simple interest earned on \$400 invested at 8.5% for 2 years.

Section

Chapter

6 six

Statistics and Graphs

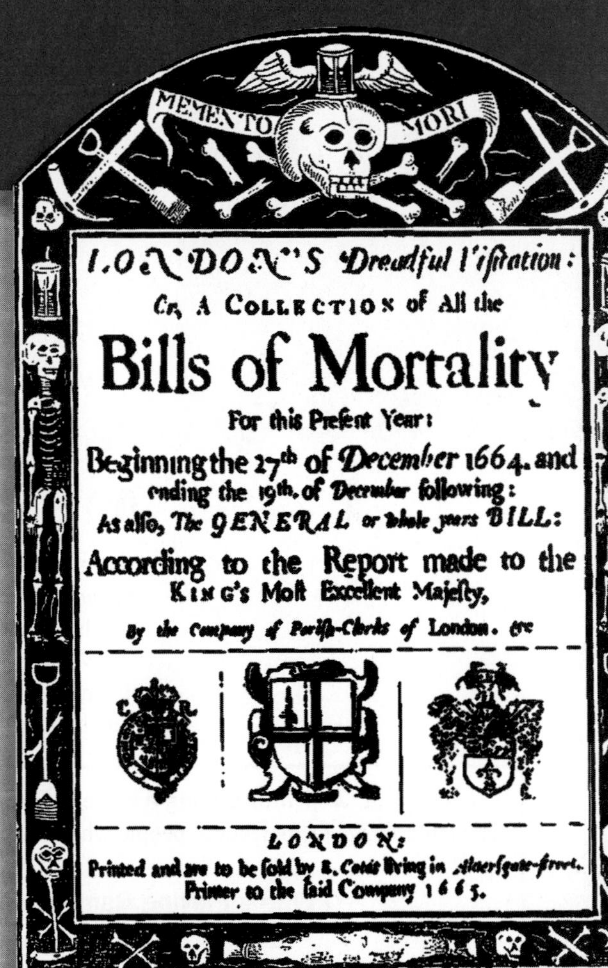

MEMENTO MORI

LONDON'S Dreadful Visitation:
Or, A COLLECTION of All the
Bills of Mortality
For this Prefent Year:
Beginning the 27th of December 1664. and
ending the 19th. of December following:
As alfo, The GENERAL or whole years BILL:
According to the Report made to the
KING's Moft Excellent Majefty,
By the Company of Parifh-Clerks of London. &c.

LONDON:
Printed and are to be fold by E. Cotes living in Aldersgate-ftreet.
Printer to the faid Company 1665.

The Human Side of Mathematics

Statistical analysis was born in London, where in 1662 John Graunt published a remarkable book, *Natural and Political Observations upon the Bills of Mortality*. The causes of death (weird diseases such as jawfaln, King's-Evil, planet, and tissick) were reported in the Bills of Mortality, which were published regularly starting in 1629, and were presented as London's "dreadful visitations."

After this perhaps morbid beginning for statistical analysis, many mathematicians, among them such famous ones as Pierre-Simon Laplace (1749–1827) and Carl Friedrich Gauss (1777–1855), made important contributions to the basic ideas of statistics. Furthermore, the analysis of numerical data is fundamental in so many different fields that one could make a long list of scientists in such areas as biology, geology, genetics, and evolution who contributed greatly to this study. The well-known names of Charles Darwin (1809–1882) and Gregor Mendel (1822–1884) would surely be included in this list.

6.1 Tables and Pictographs

Objectives

You should be able to:

A Read and interpret the information in a table.

B Read and interpret the information in a pictograph.

To Succeed, Review How To . . .

1. Find the percent of a number. (pp. 300–301, 310)
2. Determine what percent of a whole is a given number. (pp. 300–301, 310)

Getting Started

Which is your favorite fast-food restaurant? A national telephone survey of 1000 adults was conducted by Rasmussen Reports. The results can be shown several different ways. One way is to write how many persons regarded each of the four restaurants as favorable or unfavorable. A better and more efficient way would be to write the percent of adults who viewed each restaurant as favorable or unfavorable as shown in the table.

	Favorable	Unfavorable
Wendy's	73%	19%
McDonald's	66%	27%
Burger King	63%	29%
Kentucky Fried Chicken	62%	30%

Source: Data from Rasmussen Reports.

Note: Percents do not add up to 100% because of rounding or no response.

A > Reading and Interpreting Tables

Can we make any conclusions from the above data? The restaurant that was viewed as the *most* favorable among all adults surveyed was Wendy's, with a 73% rating. Which was the least *unfavorable?* Wendy's again!

Many practical situations require a compact and accurate summary of the information obtained. A **table** will present information in an efficient way by using rows and columns. Here is an example.

EXAMPLE 1 Interpreting a table: Computer games

The table shows the percent of people who play either computer or video games.

Who's Playing Games?

	Computer Games	Video Games
Under 18	29.7%	37.9%
18–35	28.7%	39.5%
35+	41.6%	22.7%
Male	58.1%	71.5%
Female	41.9%	28.5%

Source: Data from ESA.

PROBLEM 1

Refer to the table and answer the following questions.

a. What percent of the people under 18 play video games?

b. What percent of the people 35 and over play computer games?

c. Who plays more video games, males or females?

Answers to PROBLEMS

1. a. 37.9% **b.** 41.6%
c. Males, 71.5% to 28.5%

a. What percent of the people under 18 play computer games?
b. What percent of the people 35 and over play video games?
c. Who plays more computer games, males or females?

SOLUTION

a. The *first* row under "Computer Games" (column 2) indicates that 29.7% of the people under 18 play computer games.
b. The *third* row under "Video Games" (column 3) indicates that 22.7% of the people 35 and over play video games.
c. The percent of males who play "Computer Games" (column 2, row 4) is 58.1%, while the percent of females is 41.9%. Thus, more males play computer games.

EXAMPLE 2 Interpreting a table

The table shows the number of crashes classified by severity (fatal, injury, or property damage) and by the month.

Crash Severity

	Fatal	Injury	Property Damage	Total Crashes
Month	**Number**	**Number**	**Number**	**Number**
January	2,935	147,000	413,000	562,000
February	2,591	144,000	332,000	478,000
March	2,869	154,000	336,000	493,000
April	3,015	150,000	320,000	473,000
May	3,426	160,000	335,000	499,000
June	3,320	157,000	342,000	503,000
July	3,490	155,000	333,000	491,000
August	3,584	157,000	325,000	485,000
September	3,233	156,000	337,000	495,000
October	3,417	162,000	375,000	541,000
November	3,102	158,000	399,000	560,000
December	3,271	164,000	434,000	601,000
Total	38,253	1,862,000	4,281,000	6,181,000

Source: Vehicle miles traveled, Federal Highway Administration, *Traffic Volume Trends* (June 2005).

a. In what month did the most **fatal** crashes occur?
b. In what month did the fewest **injuries** occur?
c. In what month did the most **crashes** occur?
d. In what month did the fewest **crashes** occur?

SOLUTION

a. The second column shows the number of fatal crashes. The highest number in that column is **3584** and it occurred in **August.**
b. The third column shows the number of injuries. The smallest number in the column is **144,000** occurring in **February.**
c. The last column shows the total number of crashes. In that column the highest number is **601,000** occurring in **December.**
d. The fewest crashes showing in the last column occurred during **April, 473,000** crashes.

PROBLEM 2

Referring to the table:

a. In what month did the most accidents result in property damage?
b. In what month did the fewest accidents result in property damage?
c. In what month did the number of total crashes exceed 600,000?
d. How many injury accidents occurred in March?

Answers to PROBLEMS

2. a. December **b.** April **c.** December **d.** 154,000

B > Reading and Interpreting Pictographs

The information in the preceding tables can also be summarized by using a **pictograph,** a type of graph that uses symbols to represent numerical data in a statistical or financial graph, with each value represented by a *proportional* number or size of pictures. Thus, if we want to show the fact that 62 people like Kentucky Fried Chicken but 30 do not, we can make a pictograph that looks like this:

 = 10 people who like KFC

 = 10 people who do not like KFC

EXAMPLE 3 Interpreting a pictograph: SARS infections

The pictograph shows the actual number of people infected with SARS (Sudden Acute Respiratory Syndrome) or known dead at Mount Sinai Hospital, a prayer group, and York Central Hospital.

a. How many people were infected at the prayer group?

b. How many people in the prayer group died?

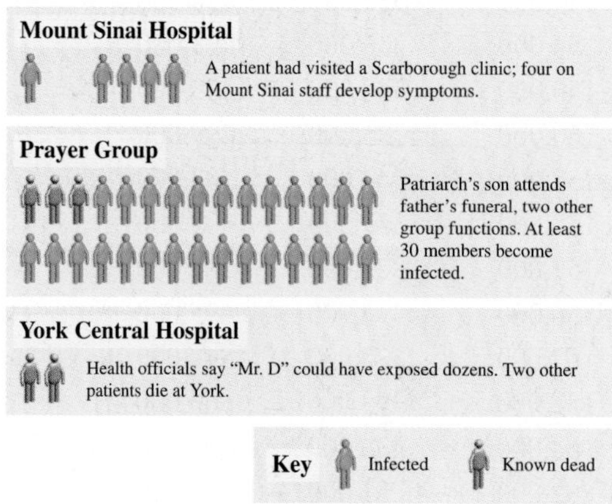

Source: Data from *Time,* May 5, 2003.

SOLUTION

a. The prayer group has 27 (red) infected people.

b. The prayer group has 3 (gray) known dead.

PROBLEM 3

Referring to the pictograph:

a. How many people in the staff were infected at Mount Sinai Hospital?

b. How many died at York Central Hospital?

EXAMPLE 4 Interpreting a pictograph: Cola preferences

The pictograph shows the number of students attending the specified event whose number-one soda choice was ACE cola.

If [can] represents 100 students, how many students chose ACE cola as their number-one choice at the specified event?

a. at the dance

b. at the party

PROBLEM 4

Referring to the pictograph, how many students chose ACE cola as their number-one choice at the game?

Answers to PROBLEMS

3. a. 4 **b.** 2 **4.** 250

SOLUTION

a. students chose ACE cola as their number-one choice at the dance. Since = 100 students, = $3 \cdot 100 = 300$ students.

b. Since = 100 students, = $\frac{1}{2}$ of 100 = 50 students. Thus, = $300 + 50 = 350$ students.

EXAMPLE 5 Interpreting a pictograph: Online courses

Have you taken a course online? The information in the pictograph is based on a survey of 144 students done by the University of Phoenix.

Executives or business owners

 =10 

Middle managers

 =10 

Technical or licensed professionals

 =10    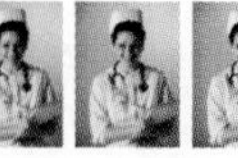 

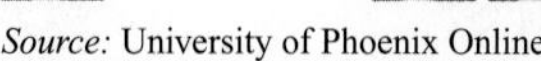

Source: University of Phoenix Online.

a. How many students are executives or business owners?

b. Which category has the most students?

c. How many students are technical or licensed professionals?

SOLUTION

a. Since =10 executives or business owners, means that there are 20 executives or business owners.

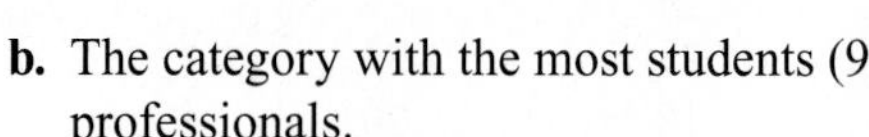

b. The category with the most students (9  and ) is technical or licensed professionals.

c. There are 9 = 90 and = 4. Thus, there are 94 technical or licensed professionals. Note that the survey consisted of 144 students, so = 4.

PROBLEM 5

Referring to the pictograph:

a. How many students are middle managers?

b. Which category has the fewest students?

Answers to PROBLEMS

5. a. 30

b. Executives or business owners

> Exercises 6.1

‹ A › Reading and Interpreting Tables

Restaurants Refer to the table for Exercises 1–4.

1. Which restaurant was viewed as the *most* favorable among the investors?

2. Which restaurant was viewed as the most unfavorable among the investors?

3. What percent of the investors viewed McDonald's as favorable?

4. What percent of the investors viewed Burger King as unfavorable?

Table for Problems 1–4

Among Investors	Favorable	Unfavorable
Wendy's	75%	18%
McDonald's	69%	26%
Kentucky Fried Chicken	66%	28%
Burger King	62%	31%

Source: Data from Rasmussen Reports.

Software The table shows the number of students and professors who pay for the software they download.

5. How many students pay every time?

6. How many students never pay?

7. How many professors pay every time?

8. How many more professors than students pay every time?

9. In what category is the number of students and professors the same?

10. Which category has the largest difference between students and professors?

Table for Problems 5–10

	Students	Professors
Every time	22	94
Most times	42	42
Seldom	52	32
Never	84	32

Source: Data from Ipsos and ClickZ Network.

E-mail The table shows the number and type of spam (unsolicited "junk" e-mail sent to large numbers of persons to promote products or services) received by the same 200 persons in July and August.

11. Which type of spam has the largest increase from July to August?

12. Which type of spam has the smallest (positive) increase from July to August?

13. Which types of spam stayed the same in July and August?

14. Which types of spam declined from July to August?

15. Which type of spam has the largest decline from July to August?

16. Which type of spam has the smallest decline from July to August?

Table for Problems 11–16

Type of Spam	July	August
Internet	14	22
Other	28	32
Scams	18	20
Products	40	40
Spiritual	2	2
Financial	30	28
Leisure	16	14
Adult	28	24
Health	24	18

Source: Data from Erightmail's Probe Network and ClickZ Network.

‹ B › Reading and Interpreting Pictographs

Hours worked Refer to the clock pictograph on page 355 for Problems 17–22.

17. How many hours did the classified employees work per week?

18. How many hours did the professional-managerial employees work per week?

The pictograph represents the number of hours worked per week by classified, professional-managerial, and administrative employees, respectively. Assume each symbol represents 10 hours.

19. How many hours did the administrative employees work per week?

20. Which employees worked the most hours per week?

21. Which employees worked the fewest hours per week?

22. If employees get overtime pay for hours worked over 40, which employee category gets overtime pay?

Internet The following information will be used in Problems 23–30.

"Digital Media Universe" (DMU) measurements from Nielsen//NetRatings that allow for more accurate tracking of Internet users revealed high U.S. traffic figures for Microsoft, AOL Time Warner, Yahoo!, and Google for a recent month. Each symbol represents 10 million users.

Parent

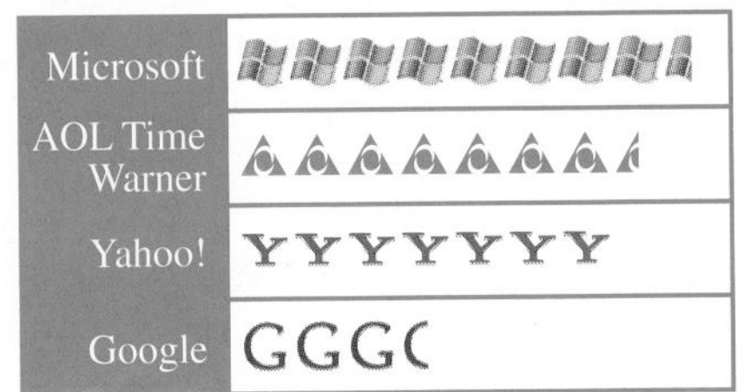

23. Which company had the most users?

24. Which company had the fewest users?

25. How many users did Yahoo have?

26. How many users did AOL have?

27. How many users did Google have?

28. What was the difference in the number of users between Microsoft and AOL?

29. What was the difference in the number of users between AOL and Yahoo?

30. What was the difference in the number of users between Yahoo and Google?

Food The graph shows the percent of people that prefer the indicated Peeps shapes and will be used in Problems 31–36. Each of the marshmallow Peeps represents 10%.

Source: Data from marshmallowpeeps.com.

31. What percent of the people prefer chicks?

32. Which is the most favorite shape?

33. Which is the least preferred shape?

34. What percent of the people prefer Snowmen?

35. What is the percent difference between the most favorite and the least favorite shape?

36. If 500 people were surveyed, how many preferred Snowmen?

Web IT go to mhhe.com/bello for more lessons

Using Your Knowledge

Nutrition The knowledge we have gained in this section can be used to follow sound nutritional habits. Once in a while, however, we may deviate from the plan and eat cheeseburgers! Here is the nutritional information for five different restaurant cheeseburgers:

Nutritional Info	Burger King	Del Taco	Jack in the Box	McDonald's	Wendy's
Calories	360	330	360	330	310
Fat	17 g	13 g	18 g	14 g	12 g
Sodium	805 mg	870 mg	740 mg	800 mg	820 mg
Cholesterol	50 mg	35 mg	60 mg	45 mg	45 mg

Source: Data from Fast Food Source.com.

37. Which restaurant's cheeseburger has the fewest calories?

38. Which restaurant's cheeseburger has the most calories?

39. If you are on a low-sodium diet, which cheeseburger should you select?

40. If you are on a low-fat diet, which cheeseburger should you select?

Write On

41. Write in your own words the advantages and disadvantages of representing data using pictographs.

42. Write in your own words the advantages of using a table instead of a pictograph to present data.

Concept Checker

Fill in the blank(s) with the correct word(s), phrase, or mathematical statement.

43. A ______ presents information **by using rows and columns.**

44. A _________ is a type of graph that **uses symbols to represent the numerical data.**

pictograph **bar graph**
table **line graph**

Mastery Test

Buying power The table illustrates the buying power by race and will be used in Problems 45–47.

45. Which race had the most buying power in 2003?

46. Which race will have the least buying power in 2007?

47. What would be the difference in buying power between 2007 and 2003 for Asians?

U.S. Buying Power by Race (billions of dollars)

	2003	2007
White	6756.9	8504.8
Black	687.7	921.3
American Indian	45.2	63.1
Asian	344.2	526.0
Other	254.9	406.5
Multiracial	125.8	164.6

Source: Data from Selig Center.

Transportation The pictograph shows the transportation method used by students to get to school and will be used in Problems 48–52. Each symbol represents one student in a class.

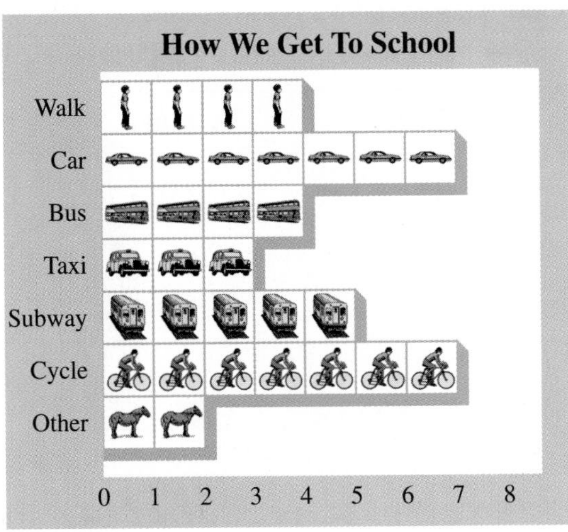

Source: Data from SPA.

48. Which is the least-used form of transportation?

49. How many students come by car?

50. How many students walk?

51. Find the difference between the number of students who cycle and those who walk.

52. If each symbol represents 100, how many students come by taxi?

Skill Checker

In Problems 53–58, find:

53. 2% of 725.

54. 4% of 725.

55. 70% of 725.

56. \$498 − \$401.

57. 34% − 9%.

58. 28% − 8%.

6.2 Bar and Line Graphs

Objectives

You should be able to:

A Read and interpret the information in bar graphs.

B Draw bar graphs.

C Read and interpret the information in line graphs.

D Draw line graphs.

To Succeed, Review How To . . .

Add, subtract, multiply, and divide whole numbers. (pp. 24, 37, 49, 62)

Getting Started

The **bar graphs** here compare at a glance the outcome of giving a medicine to patients. A recent milestone study regarding hormone replacement therapy for relieving the effects of menopause showed that there were more *risks* than *benefits* in the treatment. The people who conducted the survey persuaded doctors to stop the experiments by displaying the statistics using bar and line graphs. Let us look at the first graph labeled Risks. In this graph, the red bars (women who took the medicine) are always *longer* than the blue bars (women who took a placebo: fake medicine), indicating that more women taking the medicine had heart attacks, strokes, breast cancer, and blood clots. For example, the red bar in the heart attack category is about 37 units long (see the scale on the left: 0, 10, 20, 30, 40, 50, 60), while the blue bar is about 30 units long. This means that 37 women (out of 10,000) suffered a heart attack while taking the medicine but only 30 (out of 10,000) suffered heart attacks when taking the placebo. We shall look at the numbers for strokes, breast cancer, and blood clots in the examples.

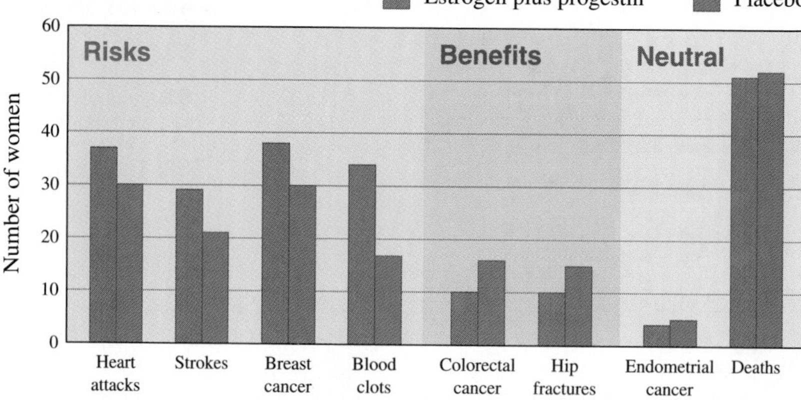

A › Reading and Interpreting Bar Graphs

A **bar graph** is a convenient way of comparing different categories by using bars whose lengths are *proportional* to the number of items in the category. In the graph labeled **Risks,** the first category is Heart Attacks and the second category is Strokes. If you look at the vertical scale 0, 10, 20, 30, 40, 50, 60, you can see that the red bar over the Stroke category is about 29 units long, while the blue bar is about 21 units long. This means that 29 women (out of 10,000) suffered a stroke while taking the medicine but only 21 (out of 10,000) had a stroke while taking the placebo.

EXAMPLE 1 Interpreting bar graphs

Refer to the graph labeled **Risks** in order to answer the following questions.

a. How many women (out of 10,000) reported breast cancer while taking the medicine?

b. How many women (out of 10,000) reported breast cancer while taking the placebo?

c. How many more women reported breast cancer while taking the medicine?

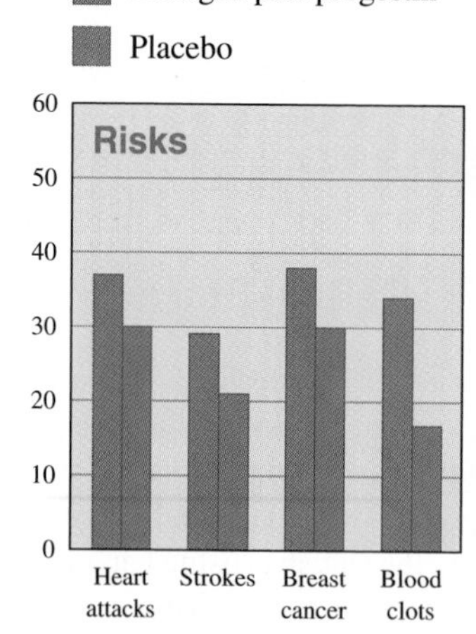

Source: Data from Journal of the American Medical Association.

SOLUTION

a. The red bar over the Breast cancer category, representing the women who took the medicine, is about 38 units long. This means that 38 women reported breast cancer while taking the medicine.

b. The blue bar is 30 units long. This means that 30 women reported breast cancer while taking the placebo.

c. $38 - 30 = 8$ more women (out of 10,000) reported breast cancer while taking the medicine.

PROBLEM 1

Refer to the graph labeled **Risks** to answer the following questions.

a. How many women (out of 10,000) reported blood clots while taking the medicine?

b. How many women (out of 10,000) reported blood clots while taking the placebo?

c. How many more women reported blood clots while taking the medicine?

We have used vertical bars in our graphs. Sometimes, horizontal bars are also used. In such cases, the *categories* (what kind of item) may be in the vertical or y-axis while the *frequencies* (how many items?) may be in the horizontal or x-axis. Learning all this may earn you a vacation. So, where will you go? Let us look at Example 2.

Answers to PROBLEMS

1. a. About 34 **b.** About 17 **c.** 17

EXAMPLE 2 Interpreting bar graphs

The horizontal bar graph tells you where vacationers are headed. Both the categories (Scenic drive, Beach or lake, Big city, and Small town) and the frequencies are shown horizontally.

a. Which is the most frequent destination and what percent of the people select it?

b. Which is the least frequent destination and what percent of the people select it?

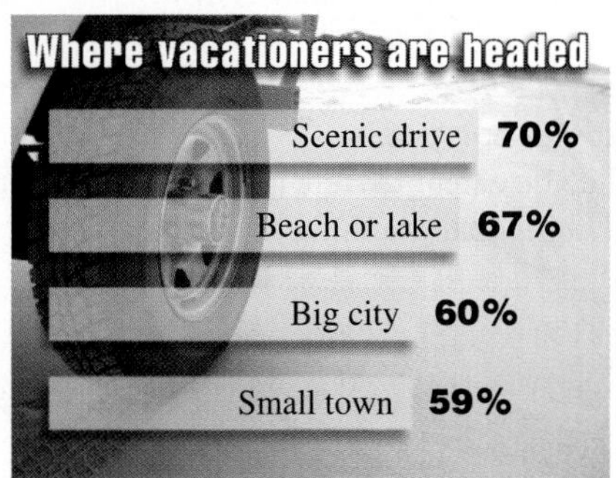

Source: Data from Travel Industry Association of America (TIA).

SOLUTION

a. The most frequent destination is the scenic drive, chosen by 70% of the people.

b. The least frequent destination is a small town, and 59% of the people select it.

PROBLEM 2

Refer to the graph and answer the following questions.

a. Which is the second most frequent destination, and what percent of the people select it?

b. Which is the second least frequent destination, and what percent of the people select it?

B › Drawing Bar Graphs

y
Frequencies
Categories
x

Now that we know how to read and interpret bar graphs, we should be able to draw our own. (After you learn how to do it, you can use commercial software, such as Excel, to draw it for you!) Let us start with vertical bar graphs, the ones in which the categories are on the horizontal or x-axis and the frequencies appear on the vertical or y-axis. They should look like the one shown here.

Now, you may be embarrassed if you cannot do vertical bar graphs, but what would embarrass you the most on your first date? Read on and find out.

EXAMPLE 3 Drawing a vertical bar graph

What would embarrass you the most on your first date? Draw a vertical bar graph using the following data.

Bad breath (BB)	40%
Acne breakout (AB)	23%
Fly open (FO)	22%
Greasy hair (GH)	14%

Source: Data from Wirthlin Worldwide for Listerine.

SOLUTION

To identify the graph, we label it "First Date Blunders." We have four categories represented in the *horizontal* axis. The *frequencies* go from 14 to 40, so we make the vertical axis go from 0 to 50 at 10-unit intervals as shown.

The bars are 40, 23, 22, and 14 units long, corresponding to the percents given in the table.

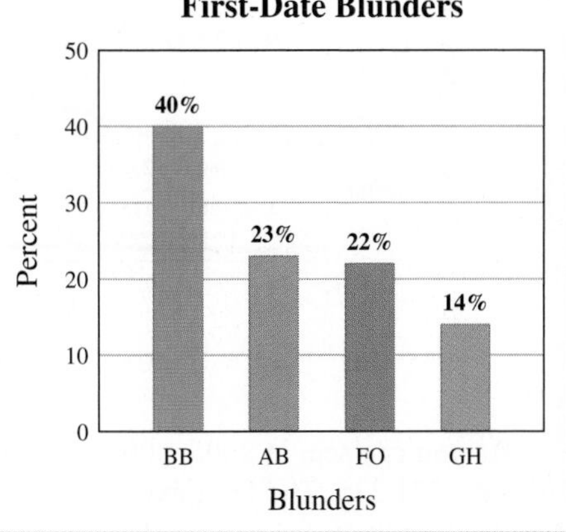

PROBLEM 3

Are you stressed out yet? What caused it? According to a survey by CyberPulse for Wrigley Healthcare's Surpass, the most common causes for stress are as shown:

Money (M)	49%
Family responsibilities (FR)	22%
Work deadlines (WD)	22%
Commuting (C)	7%

Draw a vertical bar graph using this data.

Answers to PROBLEMS

2. a. Beach or lake; 67%

b. Big city; 60%

3.

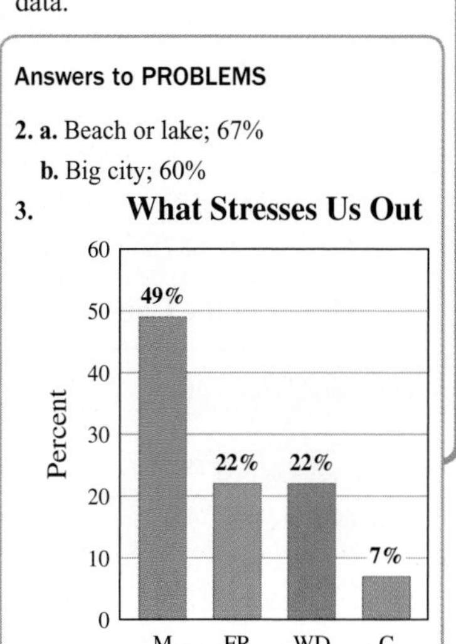

What about horizontal bar graphs? The procedure for drawing such graphs is very similar, and we illustrate it next.

EXAMPLE 4 Drawing a horizontal bar graph

Have you had the flu lately? Where did you catch it? Here are the places where germs thrive in the office. Draw a *horizontal* bar graph for the data. ,

Telephones (T)	29%
Doorknobs (D)	28%
Restrooms (R)	24%

Source: Data from Opinion Research for Kimberly Clark.

SOLUTION We title the graph "Where Office Germs Thrive." This time we place the categories (T, D, and R) on the vertical axis and the frequencies on the horizontal axis. Because the frequencies range from 24 to 29, we make the horizontal scale go from 0 to 40 at 10-unit intervals. To make the work easier, we insert vertical dashed lines at 10-unit intervals. The graph is shown.

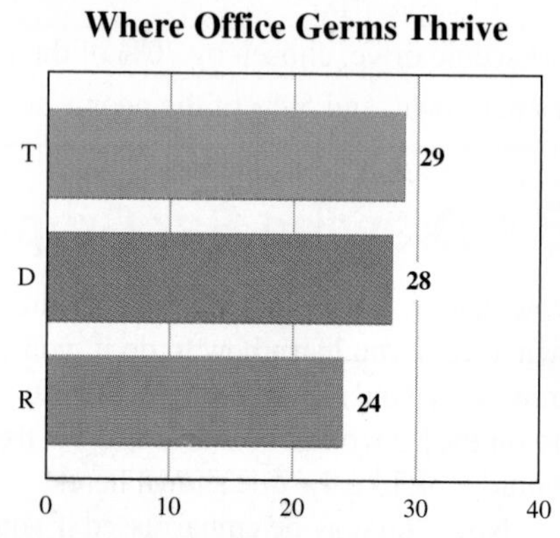

PROBLEM 4

What things do drivers keep in their cars? Here is the answer, according to a survey by Roper ASW done for Allstate.

Tapes/CDs (TCD)	72%
Umbrellas (U)	59%
Money (M)	36%
Clothes (C)	26%

Draw a horizontal bar graph for this data.

C › Reading and Interpreting Line Graphs

As we mentioned before, bar graphs are most useful when we wish to *compare* the frequency of different categories. If we want to show a trend or a change over time, then we use **line graphs.** Thus, to compare tuition and fee charges in 4-year private, 4-year public, and 2-year public colleges from 1975 to 2006, we use the line graphs shown.

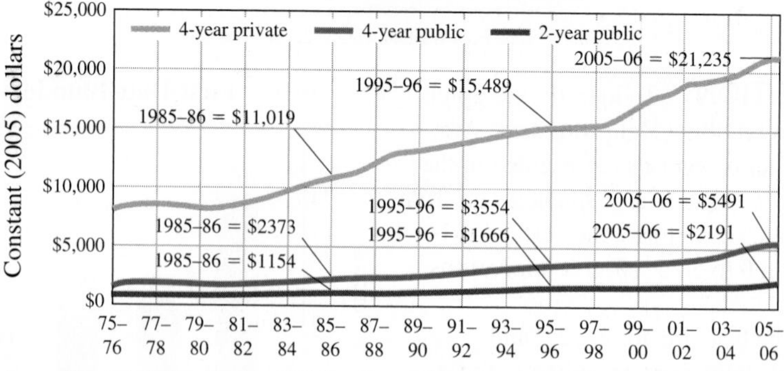

Source: Trends in College Pricing, 2005, http://www.ed.gov/.

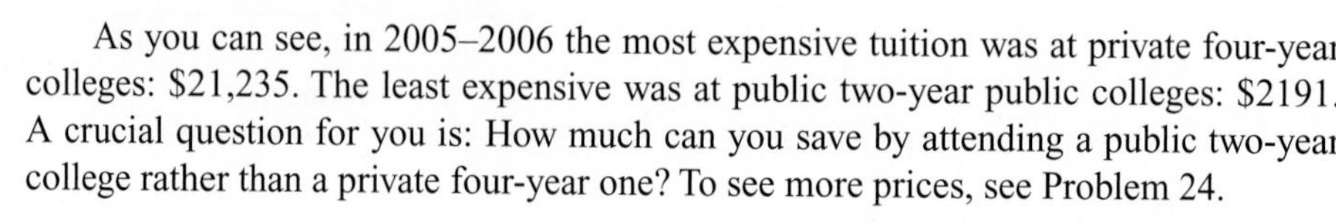

As you can see, in 2005–2006 the most expensive tuition was at private four-year colleges: $21,235. The least expensive was at public two-year public colleges: $2191. A crucial question for you is: How much can you save by attending a public two-year college rather than a private four-year one? To see more prices, see Problem 24.

Answers to PROBLEMS

4.

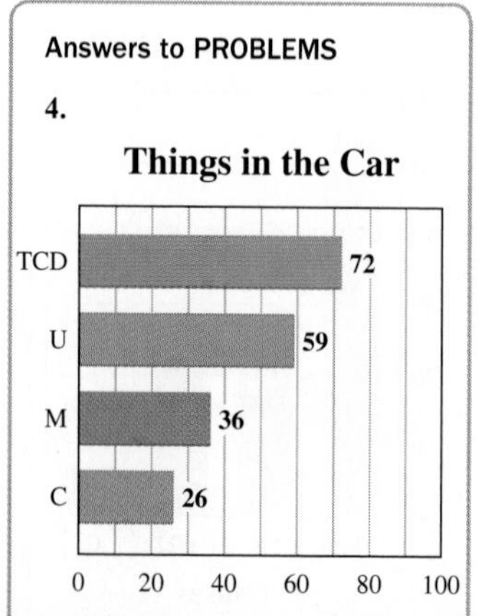

EXAMPLE 5 Interpreting line graphs

Remember the study mentioned in the *Getting Started* section? The four line graphs **below** relate the *risk* of some illnesses (*vertical* scale, from 0 to 0.03) and the *length* of time the women have taken the medicine (*horizontal* scale, from 0 to 7 years).

a. Study the graphs and determine in which years (to the nearest year) the placebo group (blue) had fewer **heart attacks** than the medicine group (red).

b. Study the graphs and determine in which years (to the nearest year) the placebo group had fewer **blood clots** than the medicine group.

c. Which was the only condition in which the patients taking the medicine fared better than the ones taking the placebo?

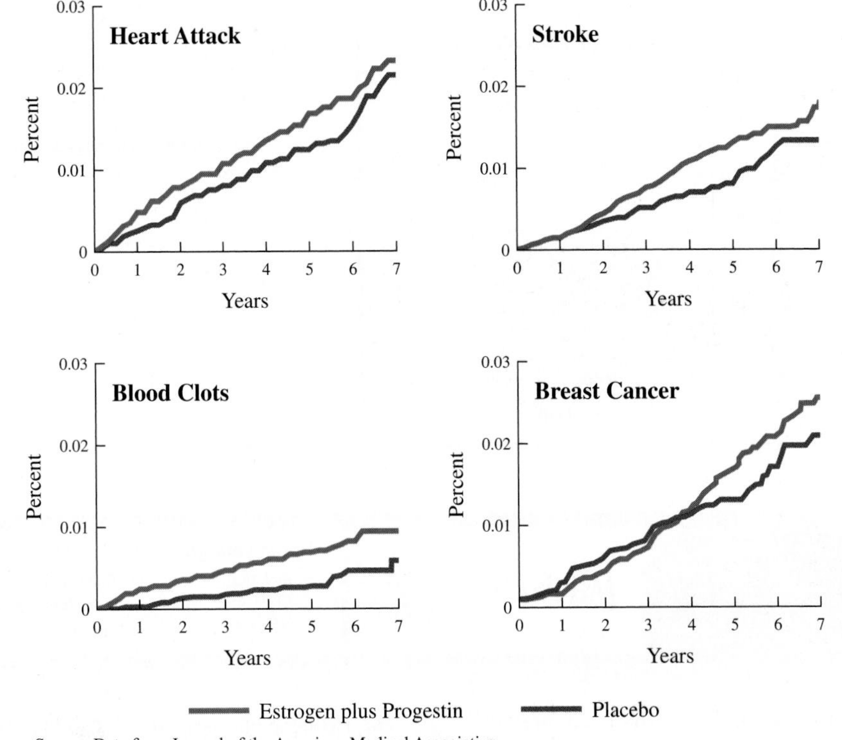

Source: Data from Journal of the American Medical Association.

SOLUTION

a. From 0 to 7

b. From 0 to 7

c. Breast cancer (but only years 1–4)

PROBLEM 5

Refer to the graphs and answer the following questions.

a. In which year (to the nearest year) did the placebo group have fewer strokes than the medicine group?

b. In which year (to the nearest year) did the placebo group have fewer breast cancers than the medicine group?

c. In what years and in which condition were the patients taking the medicine better off than the ones taking the placebo?

D > Drawing Line Graphs

Now that we know how to interpret line graphs, we should be able to draw our own. We shall do so next.

Answers to PROBLEMS

5. a. 1 to 7 **b.** 4 to 7

c. Breast cancer patients are better off when taking the medicine in years 1 to 4.

EXAMPLE 6 Drawing line graphs

Do you buy CDs in the store or download them from the Internet? A survey by PriceWaterhouseCoopers revealed that Americans spent \$10 million downloading CDs from the Internet in 2003. Here are the figures (in millions) for the next four years. Make a line graph of this data.

Year	Millions
2004	\$ 30
2005	\$125
2006	\$300
2007	\$600

SOLUTION The categories (years 2004, 2005, 2006, and 2007) are on the horizontal axis and the amounts (\$30, \$125, \$300, and \$600 million) are on the vertical axis. For convenience, we use a \$100 million scale on the vertical axis. To graph the point corresponding to 2004, we start at 2004, go up 30 units, and graph the point. For 2005, we go to 2005, go up 125 units, and graph the point. We do the same for 2006 and 2007. Finally, we join the points with line segments as shown.

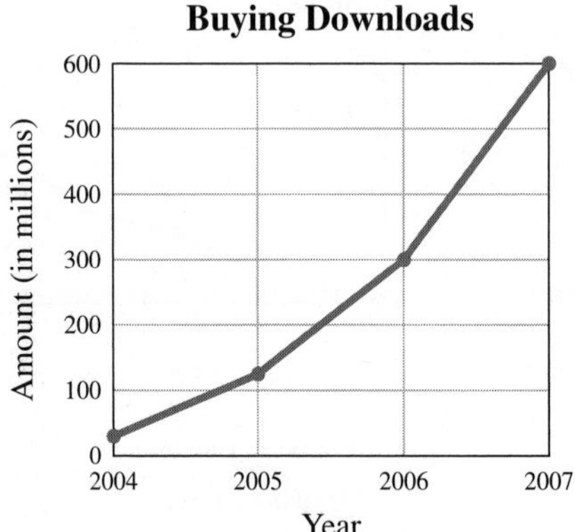

PROBLEM 6

Is there more or less crime in the United States today? A Gallup poll conducted in four successive years indicated that the percent of people who thought there was more crime in the United States today was as shown in the following table. Make a line graph of this data using a vertical scale of 40, 50, 60, and 70.

Year	Percent
2000	47
2001	43
2002	62
2003	60

Source: Data from The Gallup Organization.

Exercises 6.2

Web IT go to mhhe.com/bello for more lessons

⟨ A ⟩ Reading and Interpreting Bar Graphs In Problems 1–13, answer the questions by interpreting the bar graphs.

1. *Medicine* Refer to the graph labeled **Benefits.** How many women (out of 10,000) reported colorectal cancer

a. while taking the medicine (red)?

b. while taking the placebo (blue)?

c. Which is better, to take the medicine or to take the placebo?

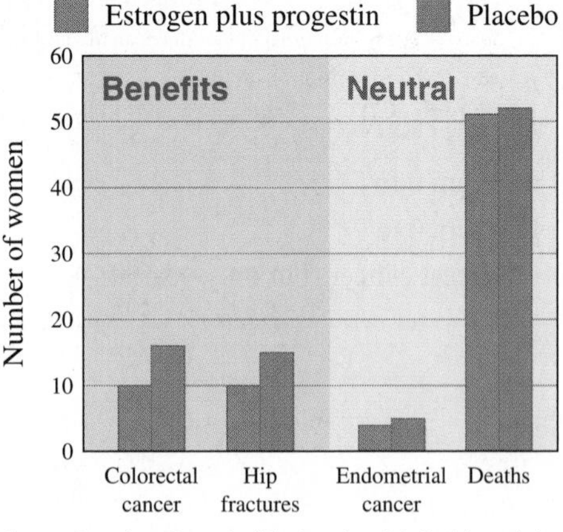

Source: Data from Journal of the American Medical Association.

Answers to PROBLEMS

6.

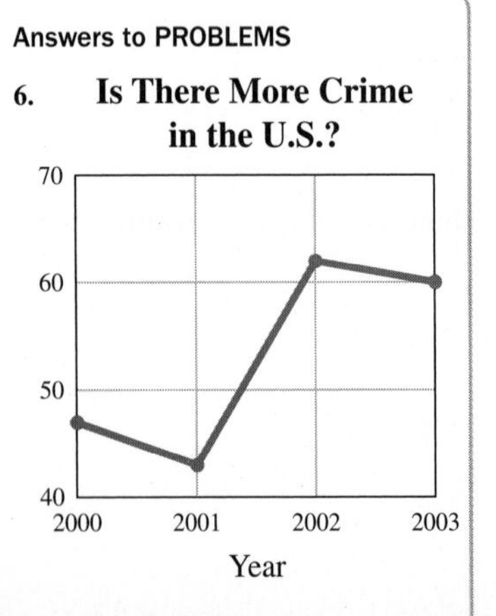

2. *Medicine* Refer to the graph labeled **Benefits.** How many women (out of 10,000) reported hip fractures

a. while taking the medicine (red)?

b. while taking the placebo (blue)?

c. Which is better, to take the medicine or to take the placebo?

3. *Medicine* Refer to the graph labeled **Neutral.** How many women (out of 10,000) reported endometrial cancer

a. while taking the medicine (red)?

b. while taking the placebo (blue)?

c. Which is better, to take the medicine or to take the placebo?

4. *Medicine* Refer to the graph labeled **Neutral.** How many women (out of 10,000) died

a. while taking the medicine (red)?

b. while taking the placebo (blue)?

c. Which is better, to take the medicine or to take the placebo?

Problems 5–7 refer to the graph below:

Back to School Spending

Source: Data from ICR for Capital One.

5. *Spending* Which of the three categories spends the most going back to school? How much do they spend?

6. *Spending* Which of the three categories spends the least going back to school? How much do they spend?

7. *Spending* What is the difference in spending between college and middle school students?

Problems 8–10 refer to a survey by *Business Week* of 725 employees.

8. *Stress* How great is the amount of stress you feel at work? The 725 employees surveyed by *Business Week* had these answers:

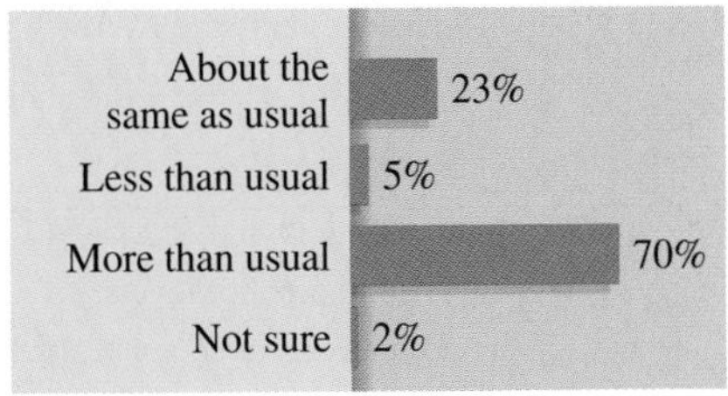

a. Which was the most common response?

b. What percent felt "About the same as usual"?

c. What percent was "Not sure"?

d. How many out of the 725 were not sure if stress affected their health? (Round the answer up.)

9. *Stress* The same 725 employees were asked if the stress was affecting their health. Here are the results:

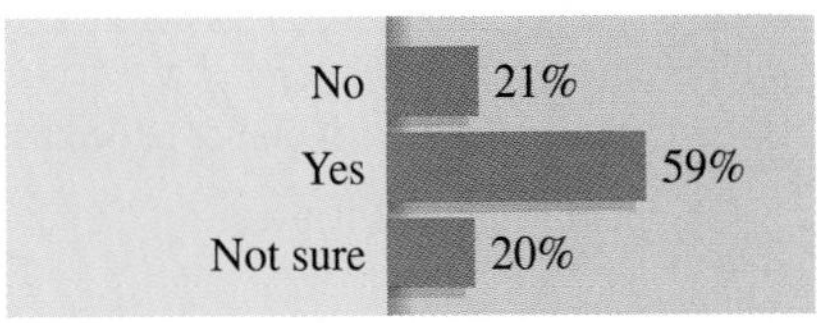

a. What percent thought the stress was affecting their health?

b. What percent was not sure?

c. How many out of the 725 were not sure if stress affected their health?

Web IT go to mhhe.com/bello for more lessons

Web IT go to mhhe.com/bello for more lessons

10. *Education* Here is the educational attainment of the 725 employees surveyed:

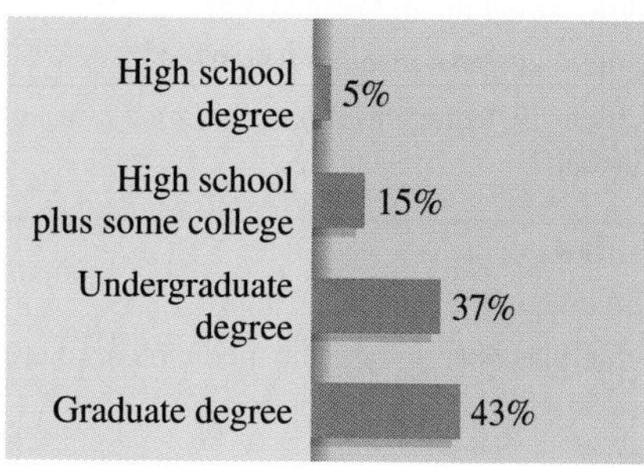

Source: Business Week Online.

a. Which category had the most employees?

b. What percent of the employees had only a high school degree?

c. What percent of the employees had graduate degrees?

d. How many of the 725 employees had graduate degrees?

11. *Traffic fatalities* The graph shows the number of traffic fatalities and the age groups of the decedents.

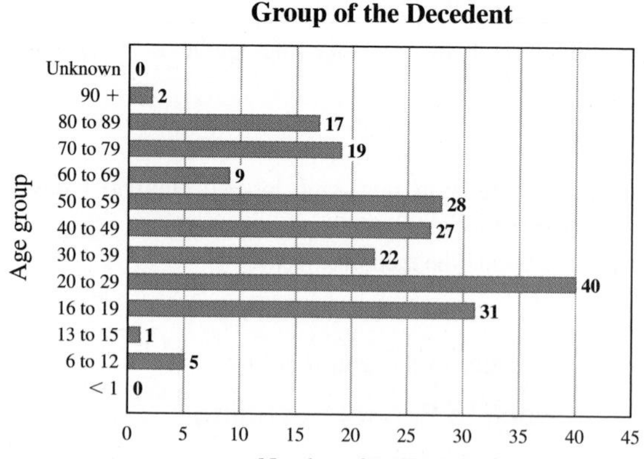

a. Which age group has the most fatalities?

b. Which age group has the fewest fatalities? Why do you think that is?

c. Are there more fatalities involving people that are less than 50 years old or more than 50 years old?

d. Which age group had only two fatalities? Why do you think that is?

12. *Traffic fatalities* At what time do fatal accidents occur? Look at the graph!

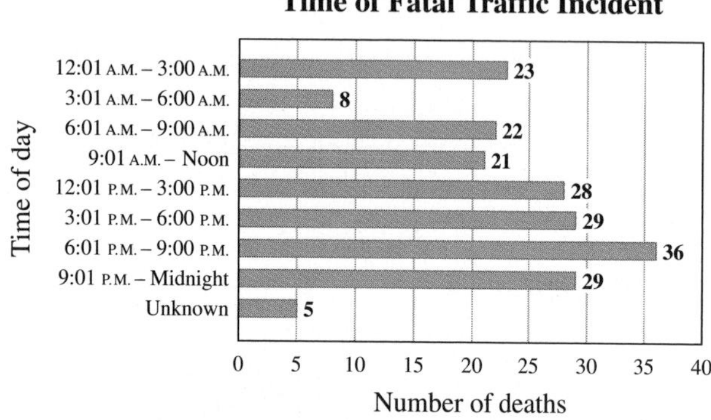

Find the number of fatalities between:

a. 12:01 A.M. and 3:00 A.M.

b. 3:01 A.M. and 6:00 A.M.

c. What is the most likely time period for a fatal traffic incident?

d. Aside from "unknown," what is the least likely time period for a fatal traffic incident?

13. *Blood alcohol levels* The bar graph shows the number of traffic fatalities and the blood alcohol level (BAL) of the driver.

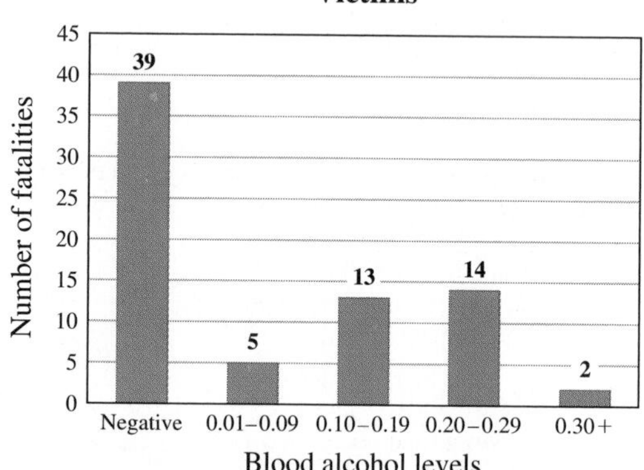

Source: Data from King County Government, Seattle, WA.

a. What was the number of fatalities with a negative (0%) BAL?

b. In many states, a person is legally drunk if their BAL is 0.10 or more. How many people were legally drunk? (In some other states 0.08 or more is legally drunk.)

c. What was the most prevalent BAL for the people who were legally drunk? How many persons had that BAL?

For the latest reports, go to http://www.metrokc.gov.

⟨B⟩ Drawing Bar Graphs In Problems 14–18, answer the questions by drawing bar graphs.

14. *Movies* Do you go to the movies often? The survey shows the percent of people who go at least once a month.

Categories Age Bracket	Frequencies Percent
18–24	83%
25–34	54%
35–44	43%
45–54	37%
55–64	27%
65–up	20%

Source: Data from TELENATION/MarketFacts, Inc.

a. Draw a vertical bar graph for the data.

Movie Attendance by Age Group

b. What age bracket goes to the movies the most frequently?

c. What age bracket goes to the movies the least?

15. *Phone calls* How many unwanted calls do you get daily? The number of unwanted calls received by the given percent of the people is shown.

Categories	Frequencies
0	15%
1–2	41%
3–5	28%
6–up	12%

Source: Data from Bruskin/Goldring Research for Sony Electronics.

a. Draw a vertical bar graph for the data.

Unwanted Calls

b. Which is the most common number of calls received?

c. What percent of the people received no unwanted calls?

16. *Military* Which branch of the military has the most women? A Defense Department survey shows the following numbers:

Categories	Frequencies
Air Force	19.4%
Army	15.4%
Navy	14.4%
Marines	6%

a. Draw a horizontal bar graph of the data.

Women in the Armed Forces

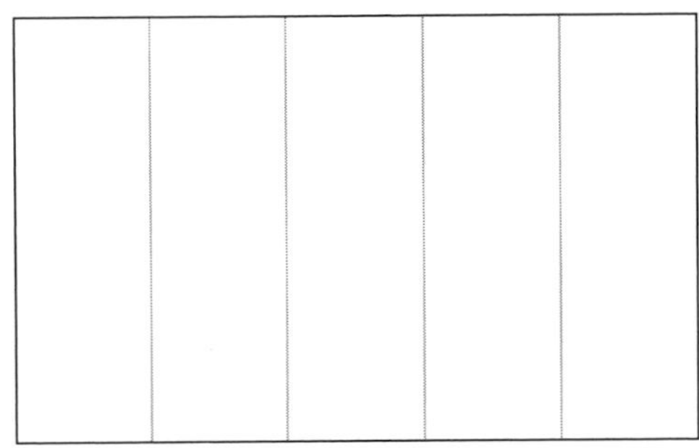

b. Which branch has the highest percent of women?

c. Which branch has the lowest percent of women?

d. Can you use the information to find out if there are more women in the Air Force than in the Army? Explain.

Web IT go to mhhe.com/bello for more lessons

17. *Travel* Which is the most expensive city for business travel? A survey by Runzheimer International showed that the price for three meals and overnight lodging in business-class hotels and restaurants is as follows:

Categories	Cost
London	\$498
Geneva	\$410
Moscow	\$407
Manhattan	\$401

a. Draw a horizontal bar graph for the data.

Daily Travel Expenses

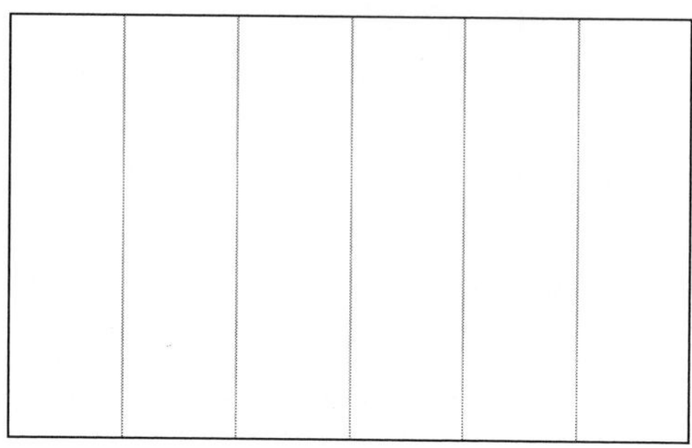

b. Which is the most expensive city?

c. Which is the least expensive city?

d. What is the price difference between the most expensive and least expensive cities?

18. *Internet* Who has the most Internet knowledge? A survey of *USA Today* adult respondents answering the question "Who has the most Internet knowledge?" revealed the following data:

Categories	Percent
Kids	72%
Adults	21%
Both the same	2%

a. Draw a horizontal bar graph for the data.

Who Has the Most Internet Knowledge?

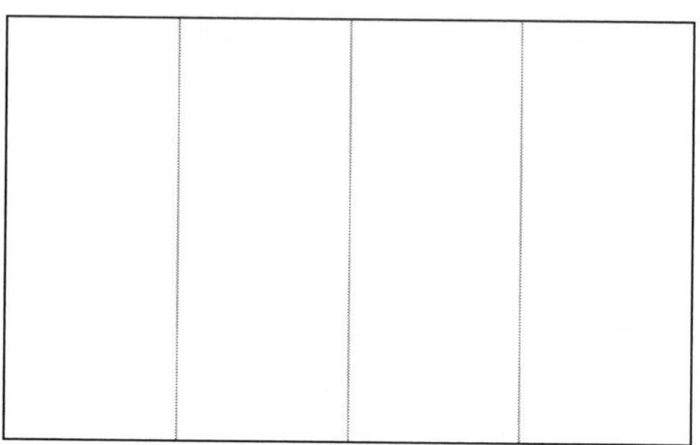

b. According to the survey, who has the most knowledge?

c. What percent of the people think that kids and adults have the same knowledge?

‹ C › Reading and Interpreting Line Graphs

In Problems 19–24, answer the questions by interpreting the line graphs.

19. *Unit price* The graph shows the average price per pound of fresh, whole chicken in five recent years.

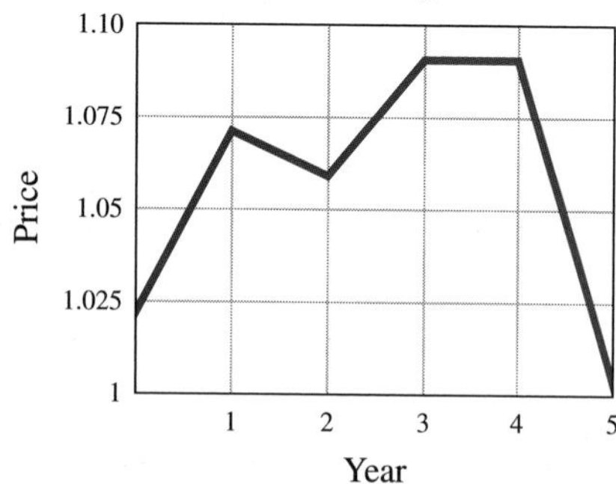

a. In which years was the price highest?

b. In what year was the price lowest?

c. In what year was the price about \$1 a pound?

20. *Unit price* The graph shows the average price per pound of whole wheat bread in five recent years.

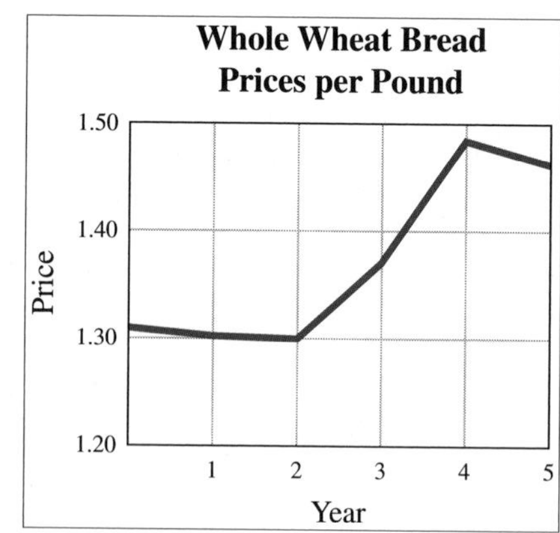

Source: Data from Bureau of Labor Statistics.

a. In what year was the price highest?

b. In what year was the price lowest?

c. In what year was the price about \$1.30 a pound?

21. *Sports* The rate of boys (blue) and girls (red) participating in high school sports is shown.

a. What was the boys' rate of participation in 2001–2003?

b. What was the girls' rate of participation in 2001–2003?

c. What was the difference in the rate of participation between boys and girls in 2001–2003?

d. What was the difference in the rate of participation between boys and girls in 1980–1982?

e. Was the difference between the boys' and girls' rates of participation greater in 1980–1982 or in 2001–2003?

Although more boys continue to play high school sports than girls, the gender gap is getting smaller. A USA TODAY analysis has found that the rate of high school girls playing varsity sports continues on a slow but steady rise. The overall boys' participation rate has been flat during the same period.

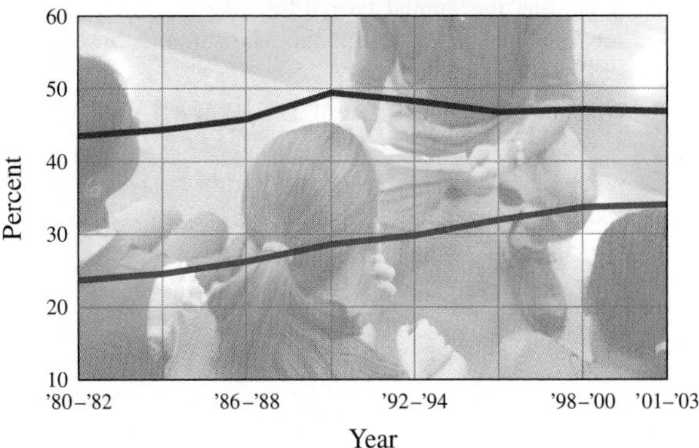

Average participation rates over three-year periods, calculated by comparing the average number of participants during the three years, divided by the average population of that gender, ages 14–17. If athletes participate in two or more sports during a given year, they are double or triple counted.

22. *Crime* Refer to the graph and answer the following questions.

Is There More Crime or Less Crime in the United States Today?

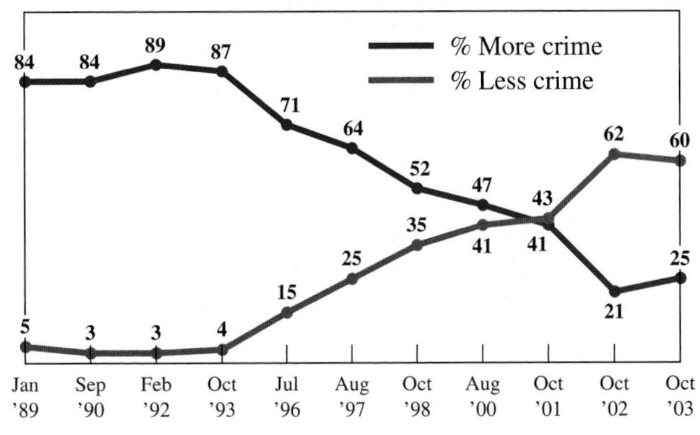

a. In October 2003, what percent of the people thought there was more crime in the United States? Less crime?

b. In what time interval was the percent of people thinking that there was more crime in the United States decreasing?

c. In what interval was the percent of people thinking that there was less crime in the United States increasing?

23. *Crime* What about your neighborhood? Are you afraid to walk alone at night? Here are the results of a recent Gallup poll:

Is There an Area Near Where You Live Where You Are Afraid to Walk Alone at Night?

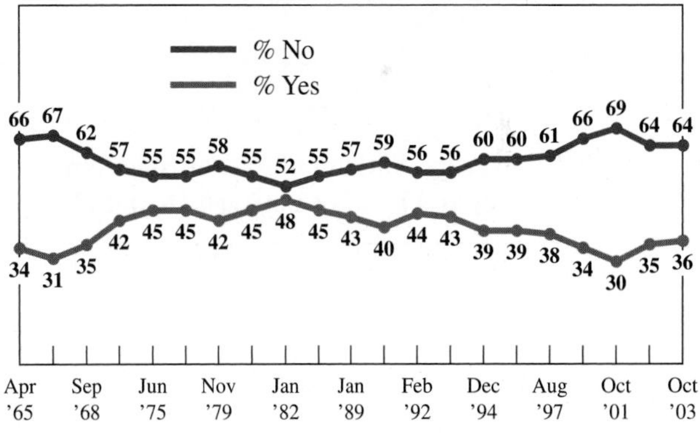

Source: Data from The Gallup Organization.

a. In what year was the percent of people afraid to walk alone at night highest? Lowest?

b. What is the percent difference between the people who were afraid to walk alone and those who were not in October 2003?

c. In what year did people feel safest?

24. *Tuition expenses* The graph shows the annual cost of tuition, fees, and room and board (TFRB) for 4-year private institutions (blue line) and 4-year public institutions (red line).

a. What was the annual cost of TFRB for a 4-year private institution in 1985–86?

b. What was the annual cost of TFRB for a 4-year public institution in 1985–86?

c. What was the annual cost difference in TFRB between private and public institutions in 1985–86?

d. What was the annual cost of TFRB for a 4-year private institution in 2005–06?

e. What was the annual cost of TFRB for a 4-year public institution in 2005–06?

f. What was the annual cost difference in TFRB between private and public institutions in 2005–06?

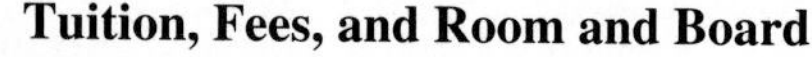

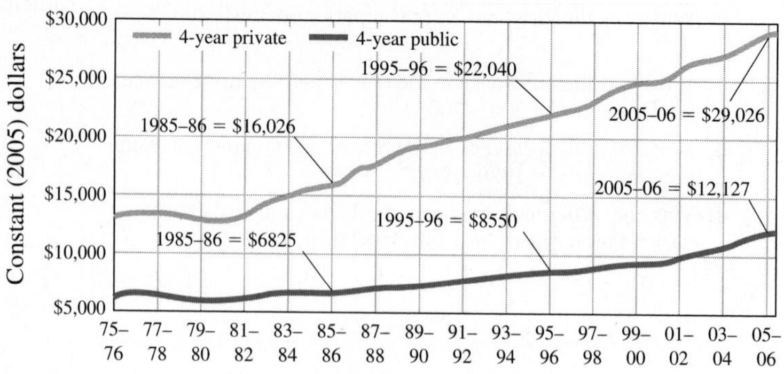

Source: Table 4 and data online (collegeboard.com/trends).

‹ D › Drawing Line Graphs In Problems 25–40, make a line graph for the given data representing a recent 5-year period. (Data: U.S. Department of Labor, Bureau of Labor Statistics at http://www.bls.gov.)

How do you compare with these averages?

25. *Clothing* The table shows the average amount of money spent on apparel by men between 16 and 25. Graph the data using 1–5 as the years and 200–300 as the amounts at $10 intervals.

1	247
2	221
3	209
4	294
5	262

Average Amount of Money Spent on Apparel by Men Between 16 and 25

26. *Clothing* The table shows the average amount of money spent on apparel by women between 16 and 25. Graph the data using 1–5 as the years and 350–450 as the amounts at $10 intervals.

1	434
2	382
3	377
4	405
5	359

Average Amount of Money Spent on Apparel by Women Between 16 and 26

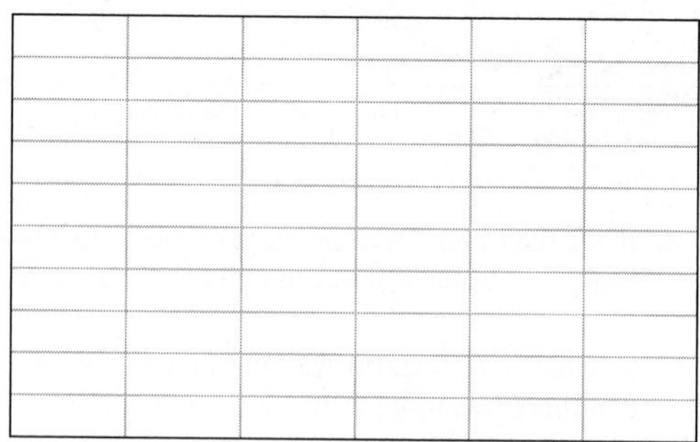

27. *Food in general* The table shows the average amount of money spent on food by persons under 25. Graph the data using 1–5 as the years and 2600 to 4000 as the amounts at $200 intervals.

1	2838
2	3075
3	3354
4	3213
5	3724

Average Amount of Money Spent on Food by Persons Under 25

28. *Food at home* The table shows the average amount of money spent on food at home by persons under 25. Graph the data using 1–5 as the years and 2500 to 3000 as the amounts at $100 intervals.

1	2758
2	2547
3	2890
4	2951
5	2936

Average Amount of Money Spent on Food at Home by Persons Under 25

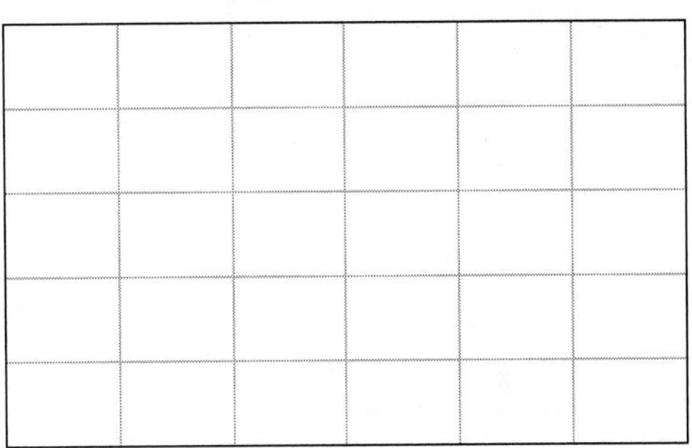

29. *Fresh fruit* The table shows the average amount of money spent on fresh fruit by persons under 25. Graph the data using 1–5 as the years and 120 to 150 as the amounts at $5 intervals.

1	137
2	124
3	136
4	146
5	142

Average Amount of Money Spent on Fresh Fruit by Persons Under 25

30. *Vegetables* The table shows the average amount of money spent on vegetables by persons under 25. Graph the data using 1–5 as the years and 120 to 150 as the amounts at $5 intervals.

1	124
2	130
3	146
4	148
5	148

Average Amount of Money Spent on Vegetables by Persons Under 25

Web IT go to mhhe.com/bello for more lessons

Web IT go to mhhe.com/bello for more lessons

31. *Housing* The table shows the average amount of money spent on housing by persons under 25. Graph the data using 1–5 as the years and 5000 to 8000 as the amounts at $500 intervals.

1	5860
2	6151
3	6585
4	7109
5	7585

Average Amount of Money Spent on Housing by Persons Under 25

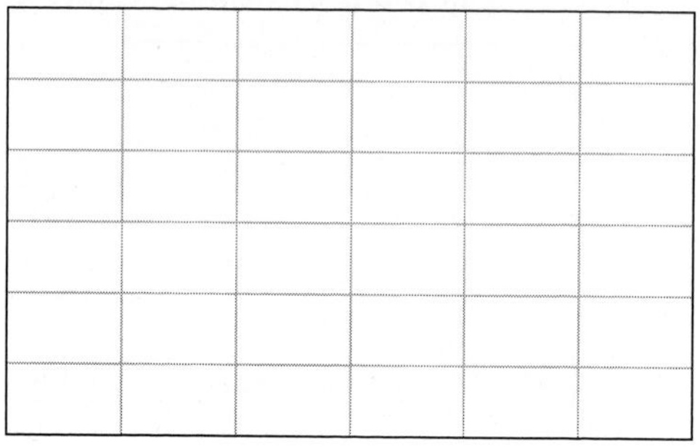

32. *Housing* The table shows the average amount of money spent on housing by persons between 25 and 34. Graph the data using 1–5 as the years and 11,000 to 14,000 as the amounts at $500 intervals.

1	11,774
2	12,015
3	12,519
4	13,050
5	13,828

Average Amount of Money Spent on Housing by Persons Between 25 and 34

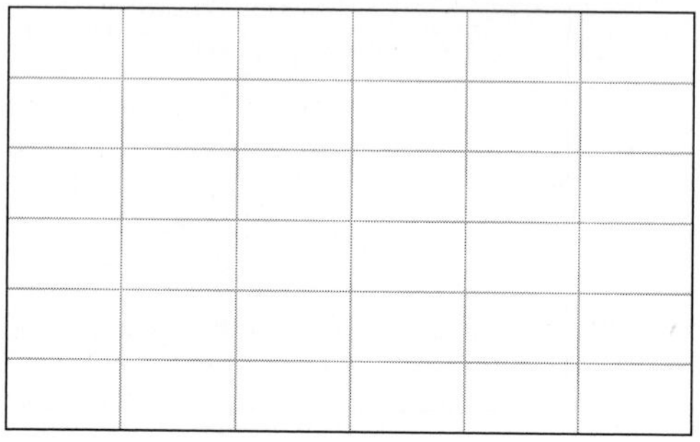

33. *Entertainment* The table shows the average amount of money spent on entertainment by persons under 25. Graph the data using 1–5 as the years and 900 to 1200 as the amounts at $50 intervals.

1	1051
2	974
3	1149
4	1091
5	1152

Average Amount of Money Spent on Entertainment by Persons Under 25

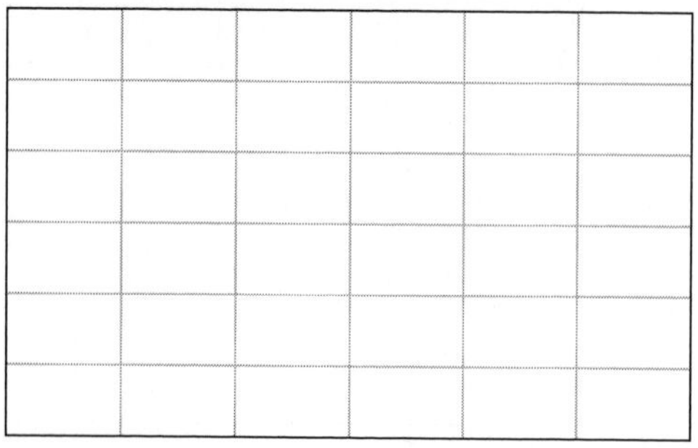

34. *Entertainment* The table shows the average amount of money spent on entertainment by persons between 25 and 34. Graph the data using 1–5 as the years and 1700 to 2100 as the amounts at $50 intervals.

1	1865
2	1757
3	1776
4	1876
5	2001

Average Amount of Money Spent on Entertainment by Persons Between 25 and 34

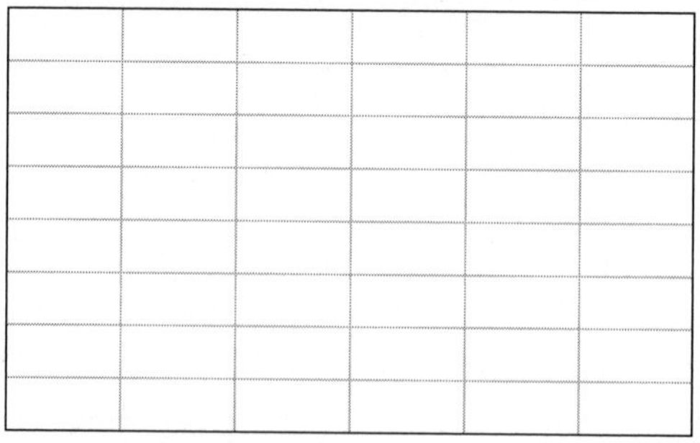

35. *Health care* The table shows the average amount of money spent on health care by persons under 25. Graph the data using 1–5 as the years and 400 to 600 as the amounts at $50 intervals.

1	425
2	445
3	551
4	504
5	530

Average Amount of Money Spent on Health Care by Persons Under 25

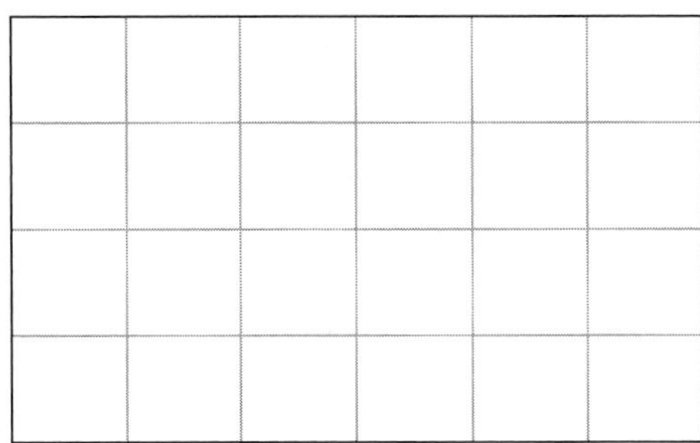

36. *Health care* The table shows the average amount of money spent on health care by persons between 25 and 34. Graph the data using 1–5 as the years and 1000 to 1300 as the amounts at $50 intervals.

1	1236
2	1185
3	1170
4	1256
5	1286

Average Amount of Money Spent on Health Care by Persons Between 25 and 34

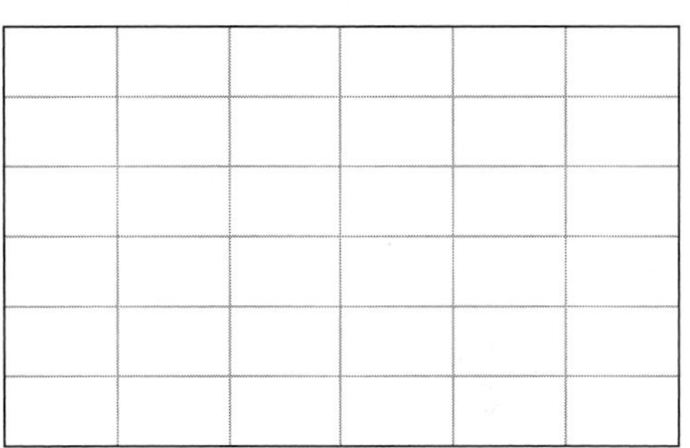

37. *Wages-salaries* The table shows the average amount of annual wages-salaries earned by persons between 25 and 34. Graph the data using 1–5 as the years and 37,000 to 47,000 as the amounts at $1000 intervals.

1	37,455
2	38,548
3	39,372
4	42,770
5	46,301

Average Amount of Annual Wages-Salaries Earned by Persons Between 25 and 34

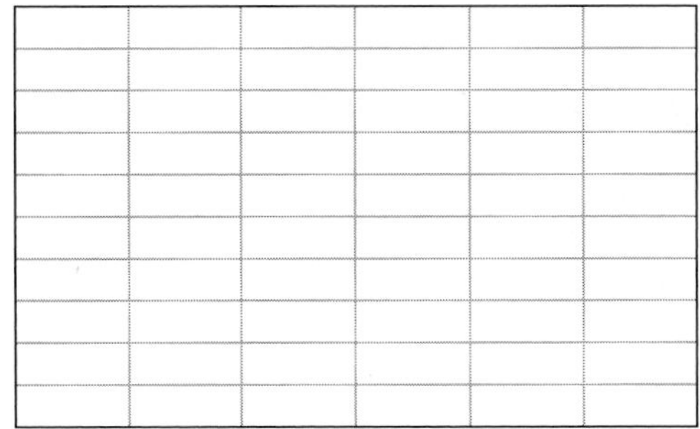

38. *Wages-salaries* The table shows the average amount of annual wages-salaries earned by persons under 25. Graph the data using 1–5 as the years and 12,000 to 18,000 as the amounts at $1000 intervals.

1	13,098
2	14,553
3	16,210
4	16,908
5	17,650

Average Amount of Annual Wages-Salaries Earned by Persons Under 25

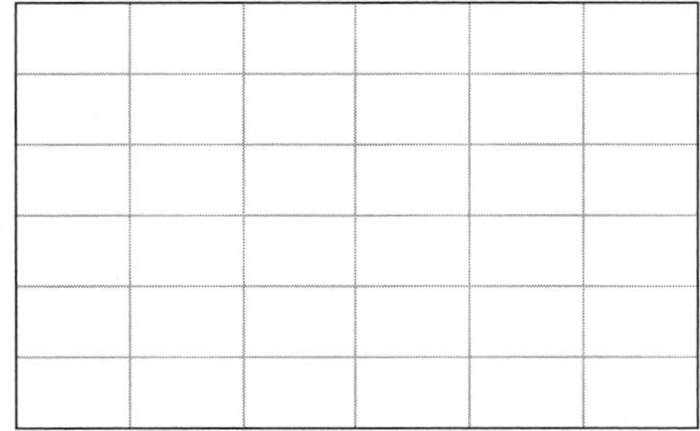

39. *Income tax* The table shows the average amount of annual federal income taxes paid by persons between 25 and 34. Graph the data using 1–5 as the years and 2100 to 2700 as the amounts at $100 intervals.

1	2567
2	2588
3	2316
4	2205
5	2266

Average Amount of Annual Federal Income Taxes Paid by Persons Between 25 and 34

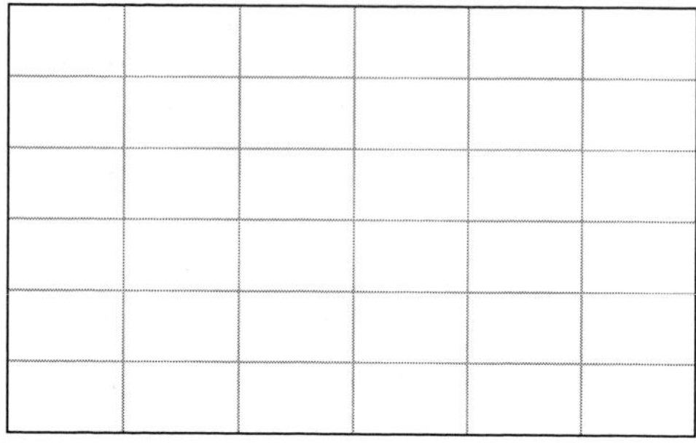

40. *Income tax* The table shows the average amount of annual federal income taxes paid by persons under 25. Graph the data using 1–5 as the years and 200 to 800 as the amounts at $100 intervals.

1	516
2	673
3	630
4	696
5	319

Average Amount of Annual Federal Income Taxes Paid by Persons Under 25

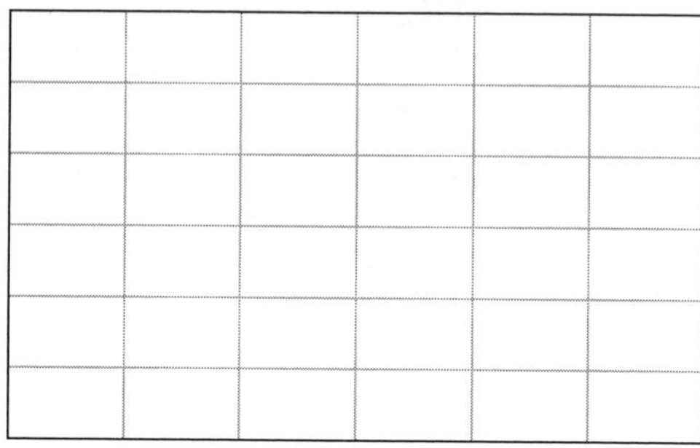

Using Your Knowledge

Misuses of Statistics In this section, we have shown an honest way of depicting statistical data by means of a bar graph. But you can lie with statistics! Here is how. In a newspaper ad for a certain magazine, the circulation of the magazine was as shown in the graph. The heights of the bars in the diagram seem to indicate that sales in the first nine months were tripled by the first quarter of the next year (a whopping 200% rise in sales!).

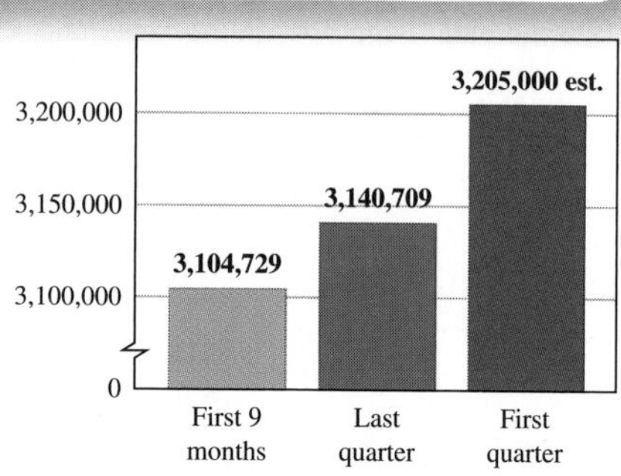

41. Find the approximate percent increase in sales from the first nine months to the first quarter of the next year. Was it 200%?

42. What was the approximate increase in the number of magazines sold?

Write On

43. Explain in your own words what is wrong with the graph on the misuses of statistics in *Using Your Knowledge.*

44. Explain in your own words how statistics can be misused or misleading. Concentrate on examples involving bar and line graphs!

Concept Checker

Fill in the blank(s) with the correct word(s), phrase, or mathematical statement.

45. **A bar graph** compares different categories by using bars whose lengths are ________________ to the number of items in the category.

46. If we want to show a **trend** or **change over time,** we use ________________ graphs.

bar

line

proportional

equal

Mastery Test

47. *Insurance* The table shows the average annual amount of vehicle insurance paid by a driver less than 25 years old in five successive years. Graph the data using 1–5 as the years and 350 to 500 as the amounts at $25 intervals.

1	383
2	408
3	449
4	449
5	479

Average Amount of Vehicle Insurance Paid by a Driver Less Than 25 Years Old

48. *Toothpaste* The table gives the percent of males squeezing the toothpaste tube from the bottom, depending on their age. Draw a vertical bar graph for the data using the age group as the categories and the percents as the frequency at 10-unit intervals.

Age	Bottom
21–34	37%
35–44	33%
45–54	10%
55+	10%

Percent of Males Squeezing the Toothpaste Tube from the Bottom

49. *Age ranges* The table shows the projected percent of age ranges in the United States for the year 2050. Draw a horizontal bar graph of the data using the age brackets as the categories and the percents as the frequencies at 10-unit intervals.

Age	Percent
0–14	19%
15–64	62%
65+	19%

Projected Percent of Age Ranges in the United States for the Year 2050

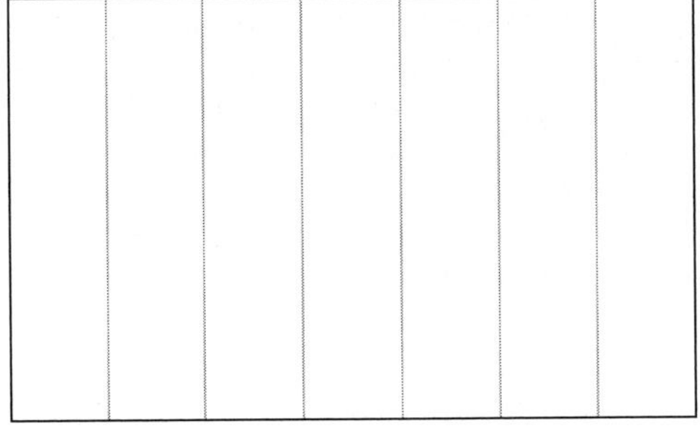

50. *Customer preferences* The bar graph shows the customers' preferences for other services desired from ATMs.

a. Which other service is the most desired?

b. Which is the least-desired service?

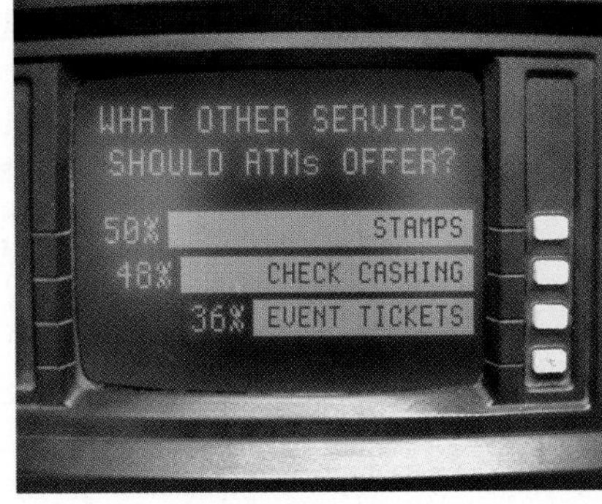

51. *Economics* Are economic conditions getting better or worse? Use the line graph to answer the following questions.

a. What percent of the people answered better on October 6–8?

b. What percent of the people answered worse on January 10–14?

c. When did the most people think the economy was getting worse?

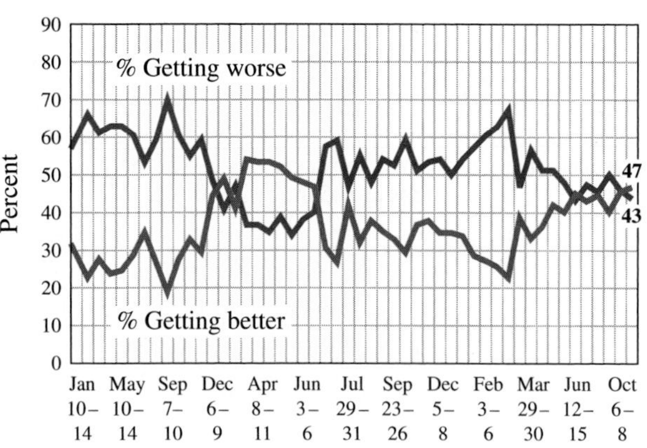

Skill Checker

In Problems 52–56, find:

52. $\frac{1}{2}$ of 33

53. 22% of 500.

54. 4% of 500.

55. 5% of 500.

56. 8% of 500.

6.3 Circle Graphs (Pie Charts)

Objectives

You should be able to:

A Read and interpret the information in circle graphs.

B Draw circle graphs involving numbers or percents.

To Succeed, Review How To . . .

1. Find the percent of a number. (pp. 300–301, 310)
2. Determine what percent of a whole is a given number. (pp. 300–301, 310)

Getting Started

Can you predict the probability of rainfall using a **circle graph** or **pie chart?** The Bureau of Meteorology does it in Australia. Rainfall predictions can be given using numerical data, in writing or as a circle graph. For example, the chart shown here gives the probability of dry, wet, or normal weather based on the type of year (El Niño, normal, or La Niña)

El Niño Year	Normal Year	La Niña Year
50% dry	33.3% dry	17% dry
17% wet	33.3% wet	50% wet
33% normal	33.3% normal	33% normal

If you know you are having an El Niño year (first column), what can you say about rainfall? The probability that it will be dry is about 50%, wet 17%, and normal 33%. Can you predict what will happen if you are having a La Niña year?

The information can also be summarized using a circle graph. For example, the third circle graph shows the information for a La Niña year. The easiest category to graph is wet because wet represents 50% or half of the circle. The top half of the circle shows dry (17%) and normal (33%). Because 17 is about one-half of 33, the tan region representing the dry weather is about half the size of the normal region. We will learn how to interpret and construct circle graphs in this section.

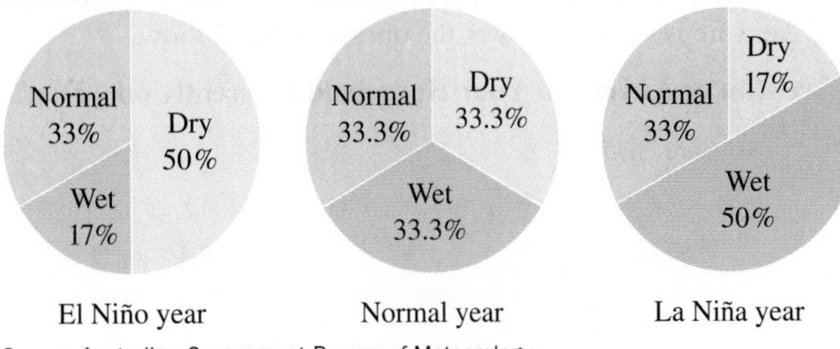

Source: Australian Government Bureau of Meteorology.

A › Reading and Interpreting Circle Graphs

As we have seen in the *Getting Started,* a **circle graph** or **pie chart** is a type of graph that shows what percent of a quantity is used in different categories or the ratio of one category to another. Circle graphs are used when exact quantities are less important than the relative size of the categories. What can these graphs describe? Let us follow the hypothetical routine of a student. You get up and have to decide if you are going to have breakfast.

EXAMPLE 1 Interpreting a circle graph

The circle graph shows how often people eat breakfast.

Source: Data from The Quaker Oats Co.

a. What percent of the people eat breakfast rarely?

b. Which category occupies the largest sector of the graph? What percent of the graph does it occupy?

c. If 500 people were surveyed, how many would eat breakfast most days?

SOLUTION

a. 21% of the people eat breakfast rarely.

b. The largest sector is "Every day." It occupies 38% of the graph.

c. 22% of the people eat breakfast most days. If 500 people were surveyed, we would need to find 22% of 500 = $0.22 \cdot 500 = 110$. Thus, 110 of the 500 people surveyed would eat breakfast most days.

PROBLEM 1

Refer to the circle graph and answer the following questions.

a. What percent of the people eat breakfast every day?

b. Which category occupies the smallest sector of the graph? What percent of the graph does it occupy?

c. If 500 people were surveyed, how many would eat breakfast some days?

It is possible that you did not eat any breakfast because you are on a diet. How many people are on a diet? Let us see.

Answers to PROBLEMS

1. a. 38%
b. Some days; 19%
c. 95

EXAMPLE 2 **Interpreting a circle graph**

The graph is divided into three categories identified by color. Yes is blue, No is purple, and Not sure is yellow. Answer the questions that follow.

Are You or Anyone in Your Household Currently on a Diet?

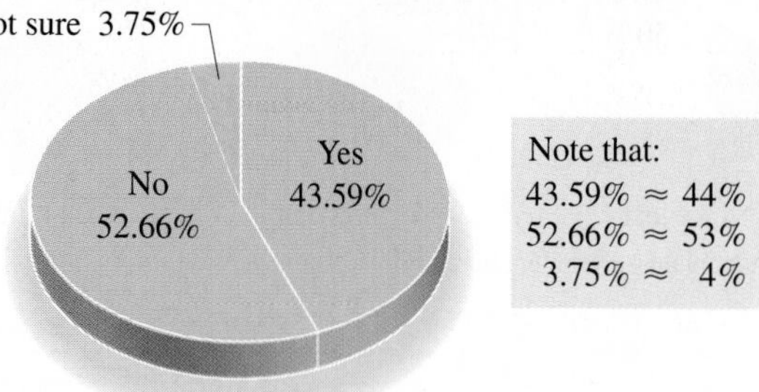

Source: Data from Insight Express.

a. What percent of the households contain people who are currently on diets?
b. Are there more households with people on diets or not on diets?
c. If 500 people were surveyed, how many would not be sure?

SOLUTION

a. The blue category, representing households with people who are on diets, occupies 43.59% or about 44% of the total circle. Thus, 44% of the households contain people who are on diets.
b. About 44% of the households contain people on diets and about 53% contain people who are not. Thus, there are more households containing people *not* on diets.
c. About 4% of the people are not sure. If 500 people were surveyed, this would represent 4% of $500 = 0.04 \cdot 500 = 20$ households.

PROBLEM 2

Refer to the pie chart and answer the following questions to the nearest percent

a. What percent of the households contain people who are currently not on diets?
b. What is the difference between the percent of households with people who are not on diets and households with those who are?
c. If 500 people were surveyed, how many households would contain people on diets?

Now that you have eaten your breakfast, you are ready to go to work. How will you get there? Example 3 will discuss the possibilities.

EXAMPLE 3 **Interpreting a circle graph**

The circle graph shows the percent of people using different forms of transportation to commute to work.

a. Which is the most common form of commuting? What percent of the people do it?
b. Which is the second most common form of commuting?
c. In a group of 500 commuters, how many would you expect to use public transit?

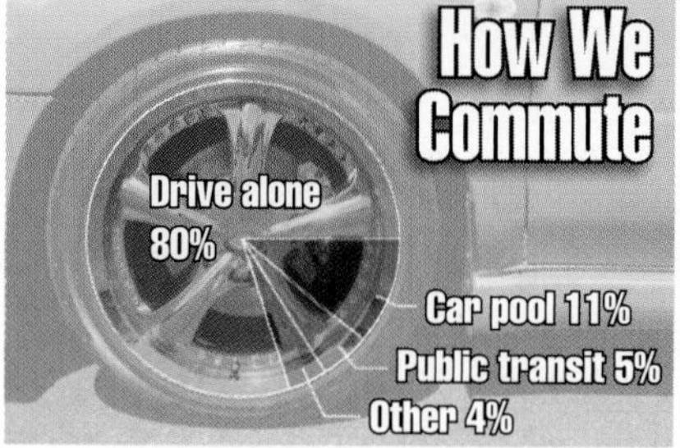

Source: Data from National Public Transportation Survey.

SOLUTION

a. The most common form of commuting is to drive alone. This is done by 80% of the commuters.
b. The second most common form of commuting is carpooling.
c. 5% of the commuters use public transit. If there were 500 commuters, we would expect 5% of $500 = 0.05 \cdot 500 = 25$ commuters to use public transit.

PROBLEM 3

Refer to the circle graph and answer the following questions.

a. What percent of the commuters carpool?
b. What percent of the commuters use other ways of commuting?
c. In a group of 500 commuters, how many would you expect to drive alone?

Answers to PROBLEMS

2. a. 53% **b.** 9% **c.** 220 **3. a.** 11% **b.** 4% **c.** 400

B > Drawing Circle Graphs

Now that you know how to read and interpret circle graphs dealing with percents, you have to know how to draw them yourself or use a software program to do it for you! The idea behind making circle graphs is similar to the one we used to represent fractions. As you recall from Chapter 2, the fraction $\frac{3}{4}$ can be represented by using a rectangle divided into four equal parts and shading three of these parts as shown:

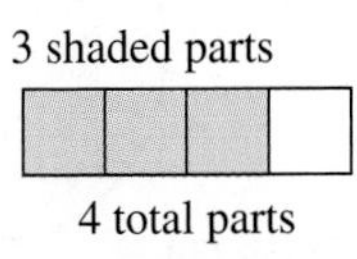

If instead of a rectangle we were shading a circle to represent the fraction $\frac{3}{4}$, we would divide a circle into four parts and shade three of them as shown:

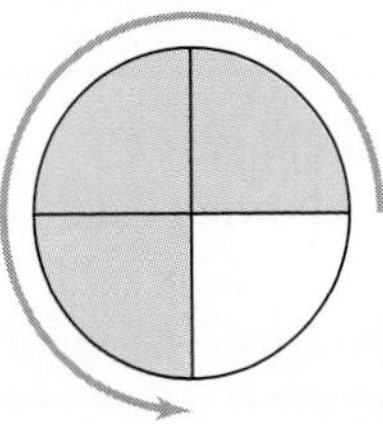

Remember those commuters who drive alone in Example 3? Perhaps you have your own car and you are one of them! We can make a circle graph of the expenses associated with owning a car. To do so, follow these steps.

Step 1. Make a circle.

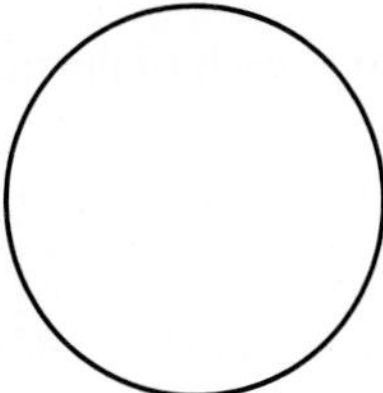

Step 2. Divide it into two equal parts.

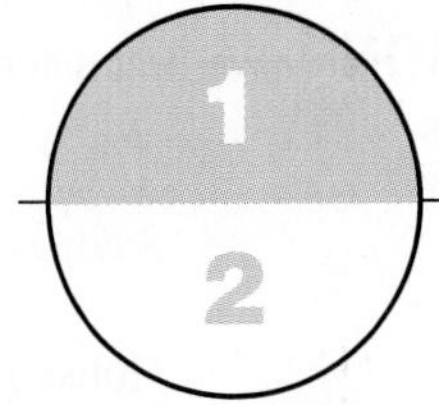

Step 3. Subdivide each of the two parts into five equal parts.

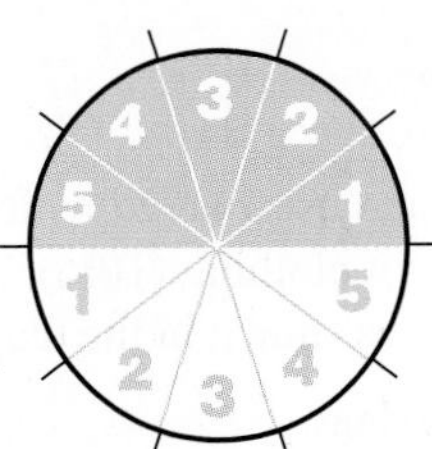

Step 4. Each of the subdivisions represents $\frac{1}{10}$, or 10%.

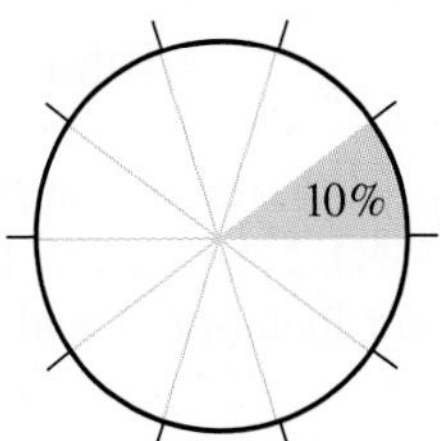

Now, suppose your automobile expense categories last year were as follows:

Payments	$2500
Gas, oil	1500
Repairs and maintenance	1000
Total	$5000

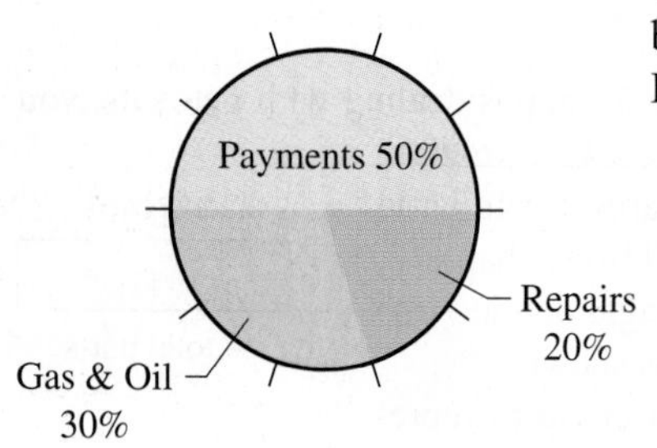

To make a circle graph corresponding to this information, we first note that the difference between this problem and the ones we solved before is that the *percents* are not given. However, they can be obtained by comparing each expense to the total as follows:

Payments	$\frac{2500}{5000} = \frac{1}{2} = 50\%$
Gas, oil	$\frac{1500}{5000} = \frac{3}{10} = 30\%$
Repairs and maintenance	$\frac{1000}{5000} = \frac{1}{5} = 20\%$

We then make a circle and divide it into 10 equal regions, using 5 for payments, 3 for gas and oil, and 2 for repairs and maintenance as shown in the margin.

EXAMPLE 4 Drawing a circle graph

Make a circle graph to show the following data:

Family Budget (Monthly)	
Savings	$ 300
Housing	500
Clothing	200
Food	800
Other	200
Total	$2000

SOLUTION We first determine what *percent* of the total each of the items represents.

Savings	$\frac{300}{2000} = \frac{3}{20} = 15\%$
Housing	$\frac{500}{2000} = \frac{1}{4} = 25\%$
Clothing	$\frac{200}{2000} = \frac{1}{10} = 10\%$
Food	$\frac{800}{2000} = \frac{2}{5} = 40\%$
Other	$\frac{200}{2000} = \frac{1}{10} = 10\%$

We then make a circle, divide it into 10 equal regions, and use $1\frac{1}{2}$ regions for savings (15%), $2\frac{1}{2}$ regions for housing (25%), 1 region for clothing (10%), 4 regions for food (40%), and 1 region for other (10%), as shown in the diagram:

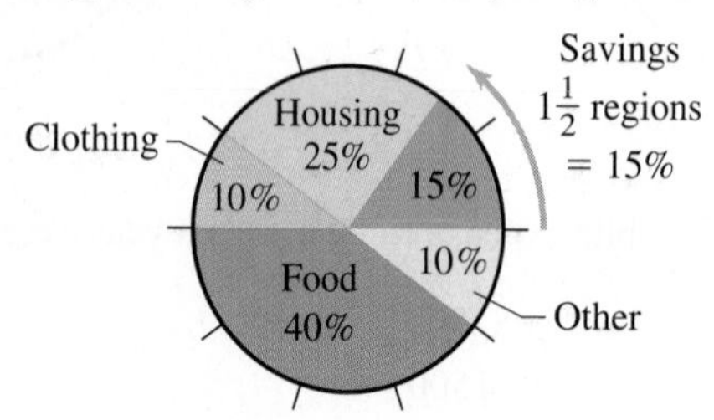

PROBLEM 4

Make a circle graph to show the following data:

Family Budget (Monthly)	
Savings (S)	$ 150
Housing (H)	375
Clothing (C)	225
Food (F)	450
Other (O)	300
Total	$1500

Answers to PROBLEMS

4.

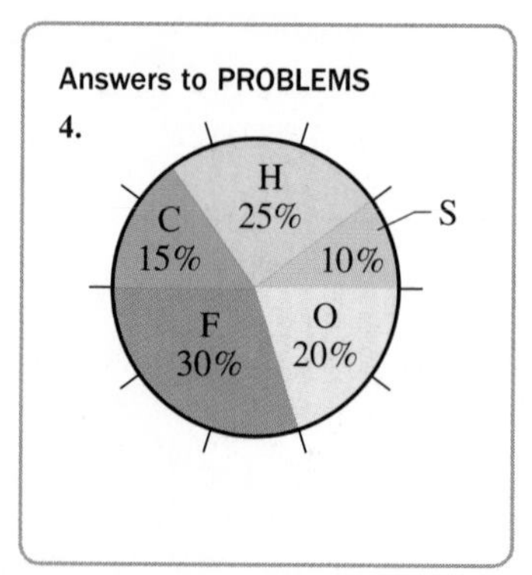

› Exercises 6.3

Web IT go to mhhe.com/bello for more lessons

‹ A › Reading and Interpreting Circle Graphs In Problems 1–13, answer the questions about the circle graphs.

1. *Transportation* Do you have a job? Which method of transportation do you use to get there? The circle graph shows the different modes of transportation used by people going to work in England.

Ways of Traveling to Work

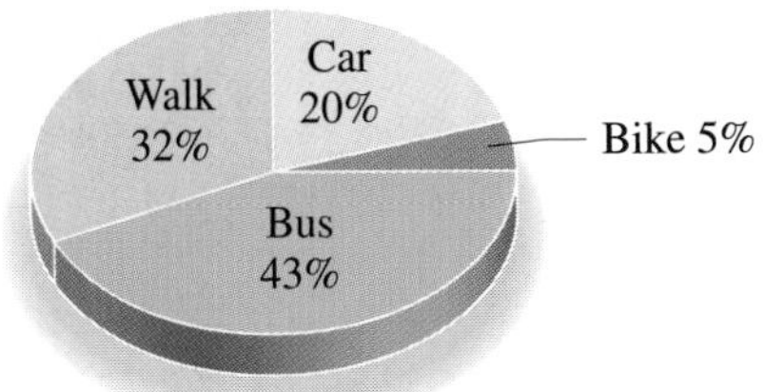

a. What is the preferred mode of transportation?

b. What is the least-preferred mode of transportation?

c. In Dallas-Fort Worth, about 91% of the people drive to work. What is the percent difference of people driving to work between Dallas-Fort Worth and England?

2. *Daily Routine* The circle graph is divided into 12 equal parts.

One Student's Daily Routine

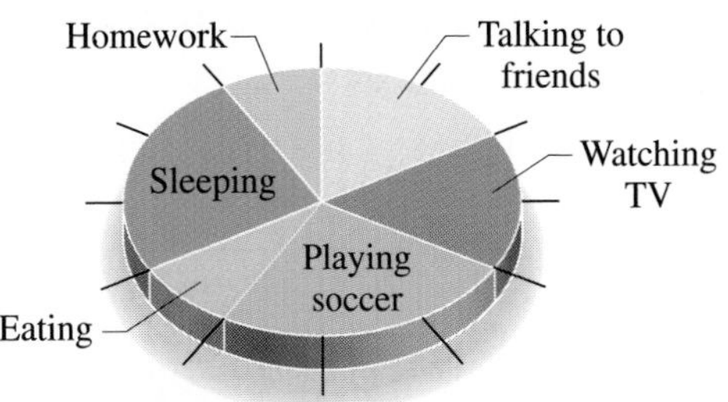

Source: Data from learn.com.uk.

a. How many parts (slices) were spent sleeping?

b. How many parts (slices) were spent watching TV?

c. Which activities took the most time?

d. Which activities took the least time?

e. What fraction of the time was spent eating? Remember that the pie has 12 equal parts (slices).

f. What fraction of the time was spent doing homework?

3. *Cheese* The circle graph shows the percents of different types of cheese produced.

Types of Cheese Produced

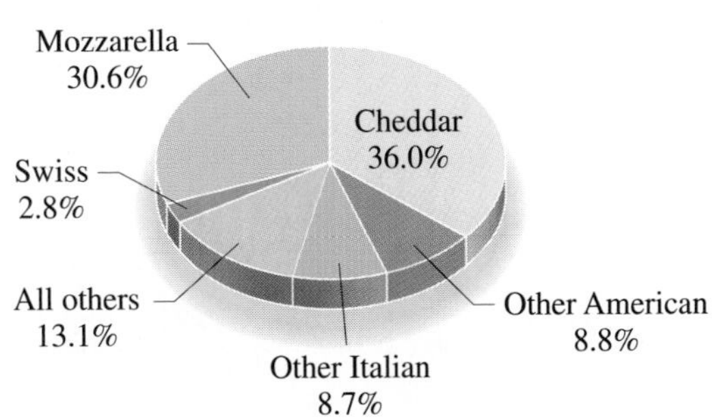

Source: Data from U.S. Department of Agriculture.

a. Which type of cheese was produced the most?

b. Which type of cheese was produced the least?

c. If you assume that the cheese that is produced the most is also the most popular, which is the second most popular cheese?

4. *Sea lion* What does a Stellar sea lion eat? The circle graph tells you.

What Does a Stellar Sea Lion Eat?

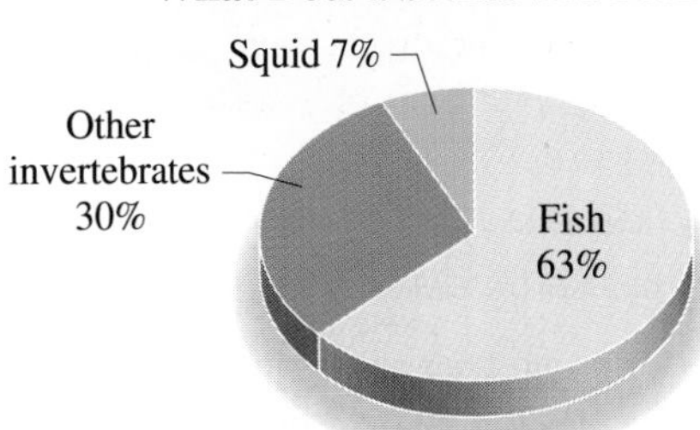

If you are in charge of feeding the Stellar sea lions in the zoo:

a. Which food would you stock the most?

b. If you buy 100 pounds of sea lion feed, how many pounds should be squid?

c. If you buy 200 pounds of Stellar sea lion feed, how many pounds should be squid?

Web IT go to mhhe.com/bello for more lessons

d. "Other invertebrates" means octopuses, shrimp, and crabs, preferably in the same amounts. If you buy 300 pounds of Stellar sea lion feed, how many pounds of crab should it contain?

e. A male Stellar sea lion weighs about 2200 pounds and eats about 200 pounds of food each day. How many pounds of fish would he eat?

f. A female Stellar sea lion, on the other hand, weighs about 600 pounds and eats 50 pounds of food each day. How many pounds of squid does she eat each day?

g. How many pounds of shrimp does a female Stellar sea lion eat every day?

To read more about Stellar sea lions go to http://tinyurl.com/r1r3.

5. *Water use* The circle graph shows the average indoor water use in Portland, Oregon.

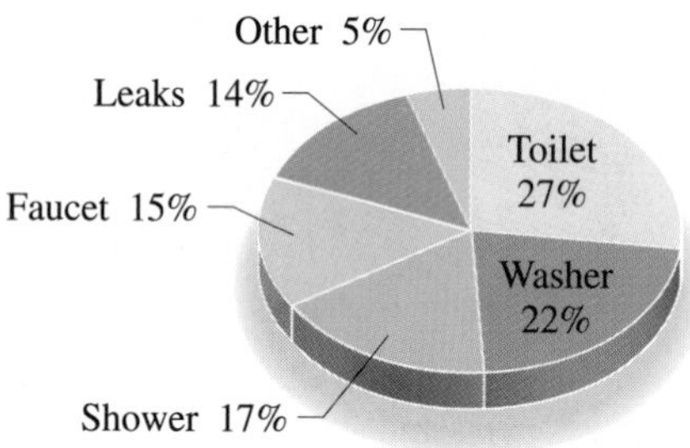

Source: Data from Portland Water Bureau.

a. Where is the most water used?

b. If you use 500 gallons of water, how much would be used for showering?

c. What uses more water, the faucet or leaks?

d. If "other" uses take 10 gallons of water, how much water would be used by the faucets?

6. *Pizza* The pie chart shows the percent of ingredients (by weight) used in making a sausage pizza.

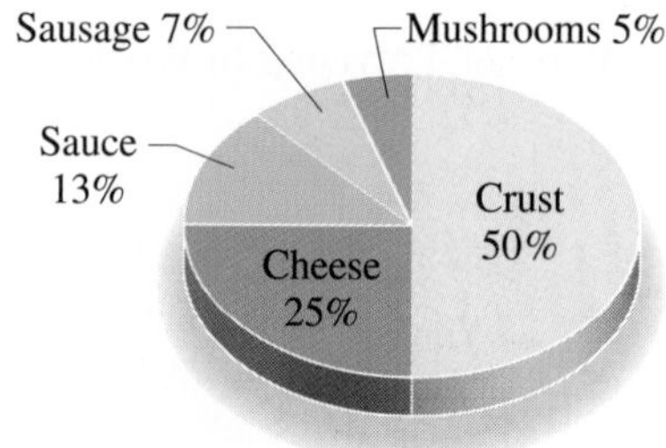

a. What percent of the pizza is crust?

b. What percent of the pizza is cheese?

c. Which ingredient makes the smallest part of the pizza by weight?

d. If you estimated that a pizza weighs four pounds, how many pounds would be crust and how many pounds would be cheese?

e. If you were to make 100 of these four-pound pizzas, how many pounds of cheese would you need?

7. *Trash* Have you looked in your trash lately? You have an average trash can if your percents are like those shown.

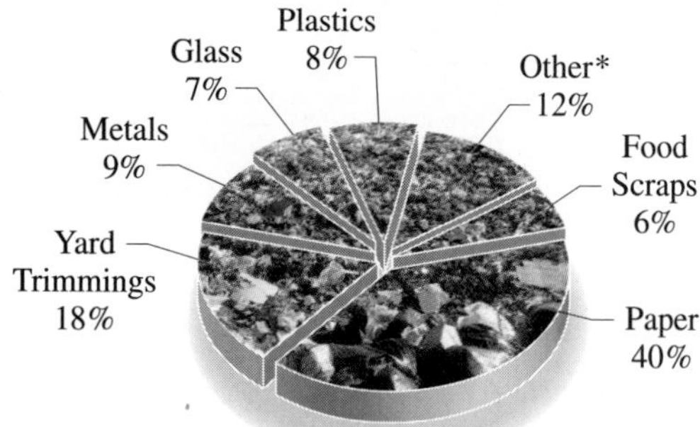

*(e.g., rubber, leather, textiles, wood, miscellaneous inorganic wastes)

Source: Data from U.S. Environmental Protection Agency.

a. What is the most prevalent item in your trash?

b. Which is the second most prevalent item in your trash?

c. If you have 50 pounds of trash, how many pounds of paper would you expect?

d. How many pounds of trimmings?

Actually, you probably recycle and do not have as much paper!

8. *Job satisfaction* The circle graph shows the job satisfaction of 500 workers, 20 to 29 years-old.

Source: Data from Insight Express.

a. What percent of the people are happy with their careers? How many is that?

b. What percent of the people thought they took the wrong career path? How many is that?

c. What percent of the people dislike the job but like their career path? How many is that?

9. *Energy* The circle graph shows the breakdown of how the world produces its energy.

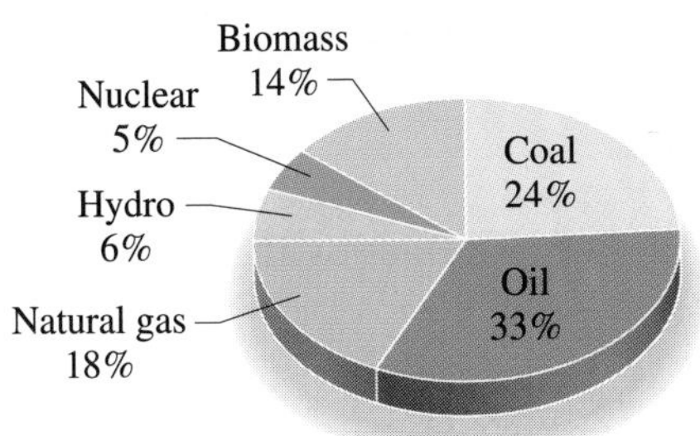

Source: Data from University of Michigan.

a. Which energy source produces the most energy?

b. Which energy source produces the least energy?

c. Fossil fuels (coal, oil, and natural gas) emit greenhouse gases when burned. Which of these three fossil fuels produces the least energy?

10. *Government* Refer to the circle graph showing where the federal government dollar comes from.

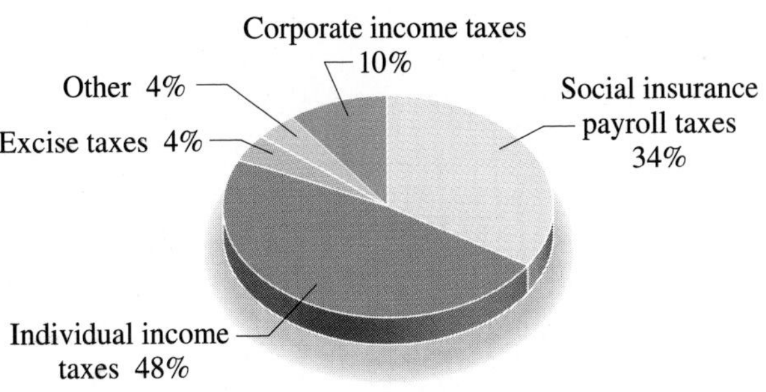

Source: Data from Economic Report of the President.

a. From what source does the federal government get the most money? What percent?

b. What is the second-largest source of money for the federal government? What percent?

c. If the total amount generated (coming in) amounts to $1700 billion, how much is generated by corporate income taxes?

Expenses The following circle graph will be used in Problems 11–13.

11. Which is the greatest expense?

12. Which is the second-largest expense?

13. Which is the smallest expense?

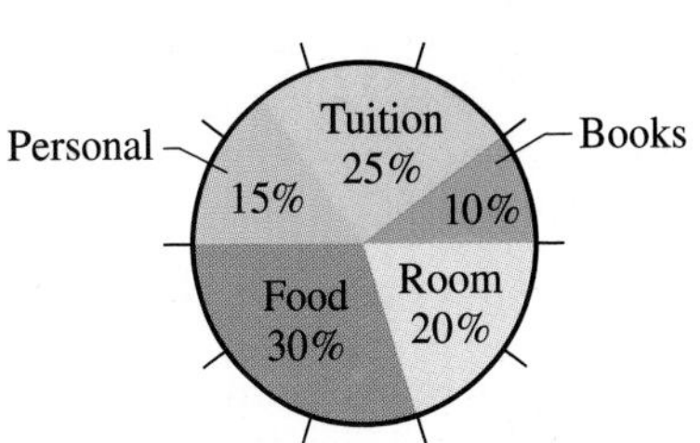

⟨B⟩ Drawing Circle Graphs In Problems 14–24 make a circle graph for the given data.

14.

Automobile Expenses

Payments	$3000
Gas, oil	1000
Repairs	500
Tires	200
Insurance	300
Total	$5000

Automobile Expenses

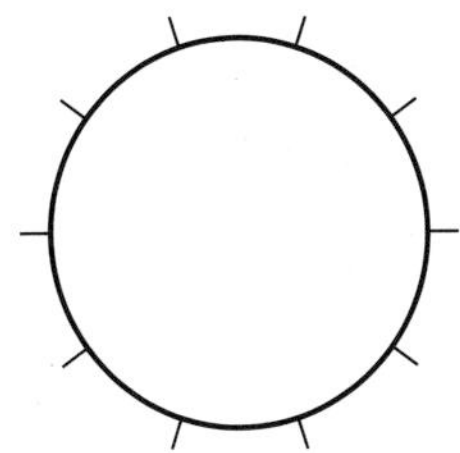

15.

Family Budget (Monthly)

Savings	$ 150
Housing	250
Clothing	100
Food	400
Other	100
Total	$1000

Family Budget (Monthly)

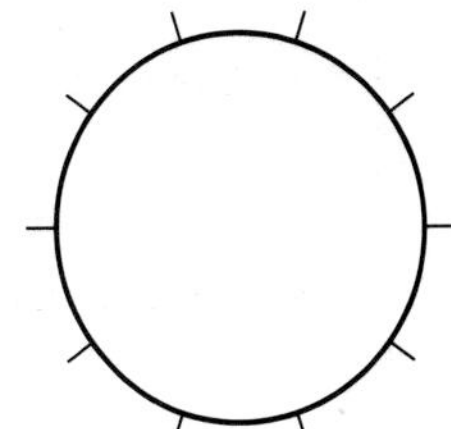

Web IT go to mhhe.com/bello for more lessons

Web IT go to mhhe.com/bello for more lessons

16.

School Expenses (per Semester)	
Room	$ 600
Food	1050
Tuition	690
Travel	450
Books	210

School Expenses

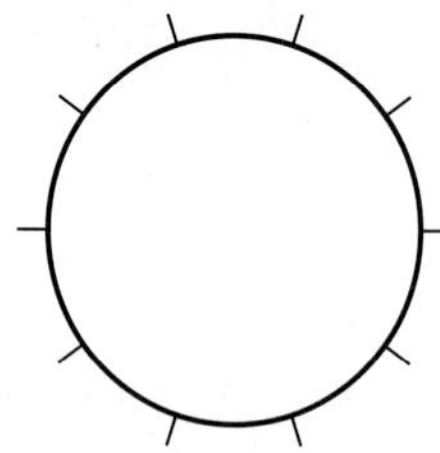

17.

Chores Done by Husbands (Weekly)	
Housework	36 minutes
Kitchen work	12 minutes
Family care	24 minutes
Shopping	24 minutes

Chores Done by Husbands

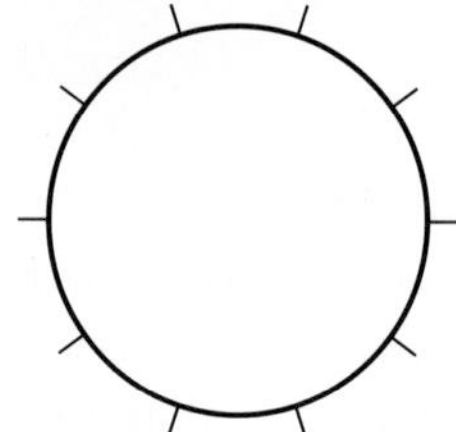

18. *Vacations* According to a survey of 800 persons made by the U.S. Travel Data Center, the most likely places to spend our summer vacations are:

City	248
Ocean	208
Small town	176
Mountains	80
Lakes	40
National parks	48

Places to Spend Summer Vacation

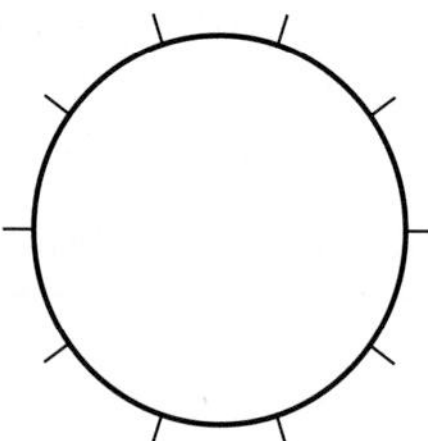

19. *Cars* According to a Roper Report, 9 out of 10 people in the United States belong to a family in which one member owns a car. Here is what they own:

Midsize	630
Full-size	522
Compact	378
Pickup	144
Other	126
Total	1800

Type of Car Owned by Family with a Car

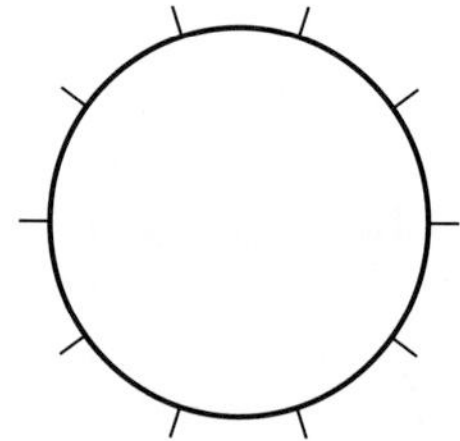

20. *Food* In a recent survey of 2000 people made by the Snackfood Association, the favorite flavors of potato chips were as follows:

Regular salted	1320
Barbecue	248
Sour cream and onion	206
Cheddar cheese	78
Regular unsalted	36
Other	112

Favorite Flavors of Potato Chips

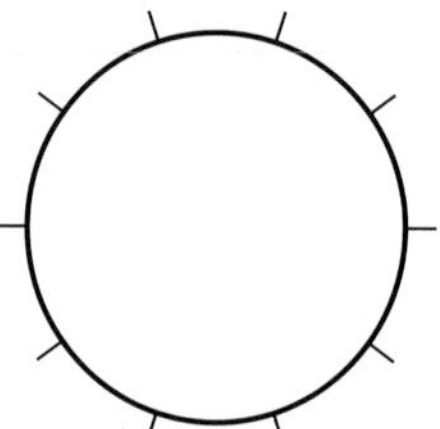

21. *Internet* How do you use the Internet? A survey of 1062 students conducted by BSA-Ipsos indicated that the number of students using the Internet for personal, school, and work was as follows:

Personal use:	670
School use:	360
Work use:	32

Round to the nearest percent.

Student Internet Use

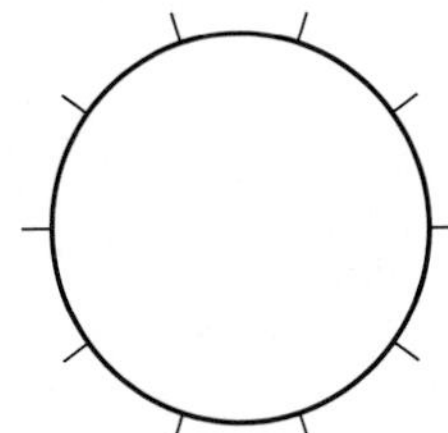

Source: http://www.definetheline.com.

22. *Internet* How do your professors use the Internet? The number of academics using the Internet for personal, school, and other uses according to a survey of 200 academics conducted by BSA-Ipsos was as follows:

Personal use:	22
Other use:	0
Work use:	178

Round to the nearest percent.

Academics Internet Use

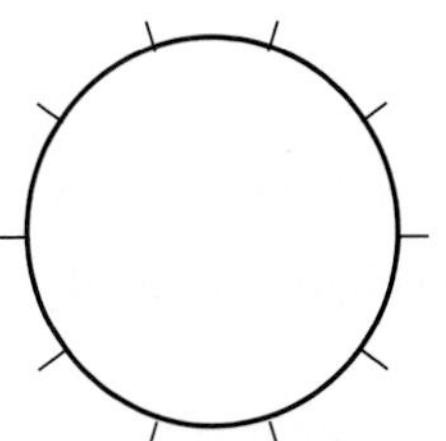

Source: http://www.definetheline.com.

23. *Software* Should you be punished for using unlicensed or pirated software? The number of academics calling for the indicated punishment in a BSA-Ipsos survey of 200 academics is as shown:

No computing resources:	100
Academic probation:	30
No penalty:	28
Fined:	24
Suspended:	10
Not sure:	8

Round to the nearest percent.

Punishment for Using Unlicensed or Pirated Software

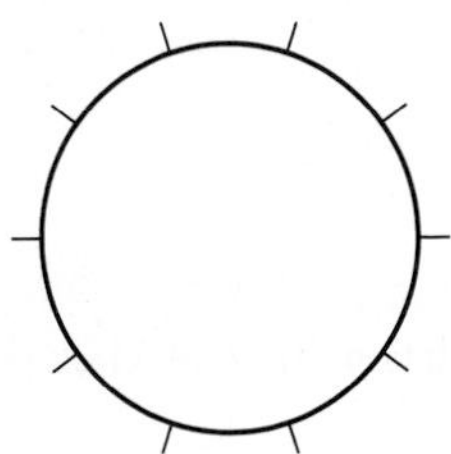

Source: http://www.definetheline.com.

24. *Software* How do students feel about using unlicensed or pirated software? The number of students calling for the indicated punishment in the BSA-Ipsos survey of 1062 students is as shown:

No penalty:	350
No computer resources:	319
Academic probation:	127
Suspended:	32
Not sure:	234

Round to the nearest percent.

Punishment for Using Unlicensed or Pirated Software

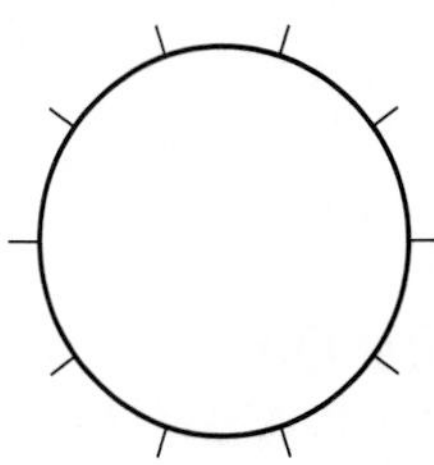

Source: http://www.definetheline.com.

Web IT go to mhhe.com/bello for more lessons

Using Your Knowledge

Water Pollution Some of the circle graphs we have studied have been divided into 10 equal regions, each representing 10%. There is another way of drawing these graphs. As you may know, a circle has 360 degrees (written 360°). Now, suppose you wish to make a circle graph for the following data:

Sources of Water Pollution	
Industrial	60%
Urban sewage	25%
Agriculture	15%

How many degrees will correspond to each category?

For industrial pollution we need	60% of 360°
or	$0.60 \times 360° = 216°$
For urban sewage we need	25% of 360°
or	$0.25 \times 360° = 90°$
For agriculture we have	15% of 360°
or	$0.15 \times 360° = 54°$

You can then use an instrument called a **protractor** to measure these degrees, mark the corresponding regions in the circle, and finish the graph.

Now, the sources of air pollution are as shown in the table.

Transportation	40%
Fuel combustion	20%
Industry	15%
Other	25%

Use your knowledge to find the number of degrees corresponding to each of the following categories.

25. Transportation

26. Fuel combustion

27. Industry

28. Other

Write On

29. Write a description of a circle graph (pie chart) in your own words.

30. What is the relation between circle graphs and percents?

Concept Checker

Fill in the blank(s) with the correct word(s), phrase, or mathematical statement.

31. A ________ **graph** is a type of graph that shows what percent of a quantity is used in different categories.

32. A **circle graph** can also show the ________ of one **category** to another.

bar **circle**
line **relationship**
ratio

Mastery Test

The graph will be used in Exercises 33–35.

The average American worker stays late at work three to five days a week. The graph shows the percent of workers who say they stay late at work the number of hours in the graph.

33. What percent of the workers stay late for one hour?

34. What percent of the workers stay late for two hours?

35. If your company has 500 employees, how many would you expect to say they worked late for three hours?

Source: Data from Management Recruiters International.

36. When 1700 adults age 20 and older were asked how often they barbecued, the results were as follows:

170 said never.

510 said two to three times a month.

425 said less than once a month.

340 said once a month.

255 said once a week or more.

Make a circle graph for the data.

How Often Do You Barbecue?

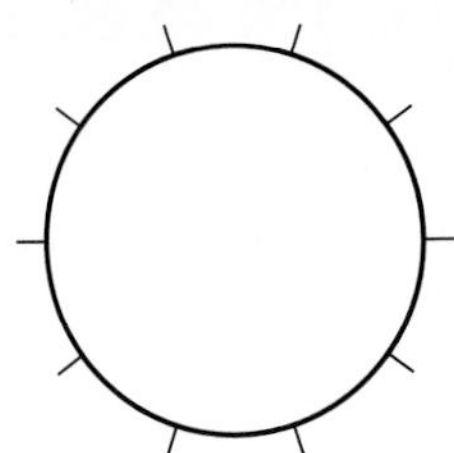

Skill Checker

In Problems 37–40, find:

37. $\frac{2.41 + 2.51 + 2.61}{3}$

38. $\frac{208 + 179 + 150}{3}$

39. $\frac{1.41 + 1.45 + 1.51 + 1.63}{4}$

40. $\frac{157 + 116 + 104 + 99 + 89}{5}$

6.4 Mean, Median, and Mode

Objectives

You should be able to:

A › Find the mean of a set of numbers.

B › Find the median of a set of numbers.

C › Find the mode of a set of numbers.

D › Solve applications involving the mean, median, and mode.

To Succeed, Review How To . . .

1. Add, subtract, multiply, and divide decimals. (pp. 202–206, 212–216)
2. Round numbers. (pp. 216–217)

Getting Started

What is the *average* price of gas in your area? At the Amoco station, the **average** price is the sum of the three prices divided by 3, that is,

$$\frac{\$2.51 + \$2.64 + \$2.74}{3} = \frac{\$7.89}{3} = \$2.63$$

Now, what is the number in the middle **(median)** for the Amoco gas? It is $2.64. What about the median for the four prices in the Hess gas? Is there one? Next, is there a price that occurs more than the others do? If you look at all the gas prices shown, the number 2.51 occurs twice; it is the **mode** for the prices. We make these ideas precise next.

A › Finding the Mean

The most common average for a set of *n* numbers is the *mean* or *average*. The **mean** is a statistic (a number that describes a set of data) that measures *central tendency,* a sort of *center* for a set of numbers. Here is the definition.

MEAN

The mean (average) of a set of *n* numbers is the sum of the numbers divided by the number *n* of elements in the set.

Thus, the average price of Hess non-diesel gas (regular, plus, premium) is

$$\frac{\$2.41 + \$2.51 + \$2.61}{3} = \frac{\$7.53}{3} = \$2.51$$

EXAMPLE 1 Calculating the mean

The chart illustrates the five top-selling record albums of all time. What is the average (mean) number of albums sold?

Top-Selling Record Albums of All Time

28 Million	• *Eagles Their Greatest Hits 1971–1975,* Eagles (Elektra)
26 Million	• *Thriller,* Michael Jackson (Epic)
23 Million	• *The Wall,* Pink Floyd (Columbia)
22 Million	• *Led Zeppelin IV,* Led Zeppelin (Swan Song)
21 Million	• *Greatest Hits Volumes I & II,* Billy Joel (Columbia)

Source: Data from Fact Monster.

SOLUTION To find the mean of the five numbers, we add the numbers and divide by 5. Thus, the mean is

$$\frac{28 + 26 + 23 + 22 + 21}{5} = \frac{120}{5} = 24 \text{ million copies.}$$

PROBLEM 1

The chart shows the next five best-selling record albums of all time. What was the mean number of albums sold?

19 Million	• *Back in Black,* AC/DC (Elektra)
19 Million	• *The Beatles,* The Beatles (Capitol)
19 Million	• *Come On Over,* Shania Twain (Mercury Nashville)
18 Million	• *Rumours,* Fleetwood Mac (Warner Bros.)
17 Million	• *The Bodyguard* (Soundtrack), Whitney Houston (Arista)

EXAMPLE 2 Calculating the mean

Have you been drinking your milk? How many calories did you consume? If we allow you to count chocolate milk, the chart shows the number of calories per cup in four types of milk. Find the average (mean) number of calories per cup.

How Milk Measures up in Calories

Source: Data from USA TODAY.

SOLUTION There are four types of milk and they contain 208, 179, 158, and 150 calories per cup. Thus, the mean average number of calories is:

$$\frac{208 + 179 + 158 + 150}{4} = \frac{695}{4} = 173.75 \approx 174 \text{ calories}$$

Remember that the symbol $\approx$ means "approximately equal." We rounded the answer to the nearest calorie, obtaining 174.

PROBLEM 2

Here is the caloric content of four types of hamburgers:

Generic large (no mayo)	511
McDonald's Big Mac	590
Burger King Supreme	550
Whataburger	607

Find the average (mean) number of calories in these hamburgers.

Answers to PROBLEMS

1. 18.4 million

2. $564.5 \approx 565$ calories

One of the most important averages at this time is your grade point average (GPA). Calculating your grade point average is an example of a **weighted mean.** First, each class you take carries a number of credit hours, and each grade you make is assigned a weight as follows: A: 4 points; B: 3 points; C: 2 points; D: 1 point; F: 0 points. You earn points by multiplying the number of credit hours in a class by the weight of your grade. For example, if you make an A (4 points) in a 3-hour class, you get $4 \times 3 = 12$ points. For a C (2 points) in a 4-hour class, you get $2 \times 4 = 8$ points. To find your GPA you divide the number of points earned by the number of credit hours you are taking. Here is an example from Oklahoma State University.

EXAMPLE 3 Calculating a weighted mean

Suppose you are taking five courses with the credit hours, grades earned, and points shown. What is your GPA?

Course	Credit Hours	Grade Earned	Points Earned
A&S 1111	1	A (4)	$1 \times 4 = 4$
ENGL 1113	3	A (4)	$3 \times 4 = 12$
PSYC 1113	3	B (3)	$3 \times 3 = 9$
HIST 1103	3	B (3)	$3 \times 3 = 9$
CHEM 1314	4	C (2)	$4 \times 2 = 8$
	14 credit hours		42 points earned

SOLUTION As you can see from the table, the number of credit hours in each course is in the second column (14 total), the grade earned and the corresponding points in the third column [A(4), A(4), B(3), B(3), and C(2)], and the points earned, the product of the numbers in columns 2 and 3, in the last column (42 total).

$$\text{Your GPA is the: } \frac{\text{Points earned}}{\text{Credit hours}} = \frac{42}{14} = 3.$$

Note that in general, the GPA may not be a whole number.

PROBLEM 3

Find the GPA for a student at the University of Texas with the courses, grades, and credit hours shown. Round the answer to two decimal places.

Course	Grade	Hours	Grade Points
English	A	3 sem. hours $\times$ 4 pts.	= 12
Mathematics	D	3 sem. hours $\times$ 1 pt.	= 3
History	A	3 sem. hours $\times$ 4 pts.	= 12
Chemistry	B	4 sem. hours $\times$ 3 pts.	= 12
Kinesiology	B	1 sem. hour $\times$ 3 pts.	= 3

Source: Data from University of Texas at Brownsville.

B > Finding the Median

As we mentioned in the *Getting Started* section, there is another type of "average" or measure of central tendency: the median.

Here is the official definition.

MEDIAN

The median of a set of numbers is the middle number when the numbers are arranged in ascending (or descending) order and there is an odd number of items. If there is an even number of items, the median is the average of the two middle numbers.

For the Amoco gas, the middle price is easy to find: 2.64. If we use the first three prices in the Hess sign, the middle price is 2.51. But suppose you want to find the middle price for the four prices: 2.41, 2.51, 2.61, and 2.45. Two things to notice:

1. The prices are not in order.
2. The number of prices is even, so there is no middle price.

Answers to PROBLEMS

3. 3.00

We can fix both problems!

1. Write the prices in order (ascending or descending).
2. To find the middle, find the average of the two middle prices.

Like this:

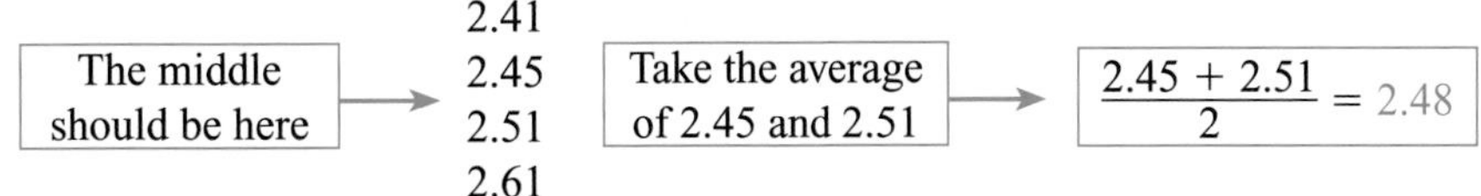

Thus, the "middle number" of 2.41, 2.45, 2.51 and 2.61, the *median,* is 2.48.

EXAMPLE 4 **Calculating the median**

How much do you spend on health care? The amounts spent the last four years for persons under 25 are as shown. What is the median amount of money spent on health care for persons under 25?

1	445
2	551
3	504
4	530

Source: Data from Bureau of Labor Statistics.

SOLUTION Note that the numbers are not in order, so we rewrite them in ascending order:

445 504 530 551

$$\frac{504 + 530}{2} = 517$$

Since we have an *even* number of items (4), take the average of the two middle numbers. The result is the median.

The median is $517.

PROBLEM 4

The amount spent the last four years for persons between 25 and 34 are as shown. What is the median amount of money spent on health care by persons between 25 and 34?

1	1185
2	1170
3	1256
4	1286

C > Finding the Mode

We mentioned one more measure of central tendency, the number that occurs the most frequently, called the **mode.** If we look at the prices at the Amoco station, no price is more prevalent than any of the other prices. The same is true of the gas prices at Hess. Now, let us look at the prices at *both* gas stations:

2.51, 2.64, 2.74, 2.41, 2.51, 2.61, 2.45

There is a number that occurs twice: 2.51. This number is called the mode. Here is the definition:

MODE

The mode of a set of numbers is the number in the set that occurs most often.

1. **If there is no number that occurs most often, the set has no mode.**
2. **If several numbers occur an equal number of times, the data set will have several modes.**

Answers to PROBLEMS

4. $1220.50

EXAMPLE 5 Finding the mode

Find the mode of 3, 8, 7, 5, 8.

SOLUTION The number that occurs most often (twice) is 8, so 8 is the mode.

PROBLEM 5

Find the mode of:

7 4 5 4 3

Note that a set of numbers can only have one mean and one median but could have *more than one* mode. For example, the set

9 3 4 9 3 7

has two modes, 9 and 3. On the other hand, the set

9 3 4 5 2 7

has no mode.

D › Applications Involving Mean, Median, and Mode

We have introduced three measures of central tendency. The following shows how they compare:

1. **The Mean** The most commonly used of the three measures.

   Good: A set of data always has a unique mean, which takes account of each item of the data.

   Bad: Finding the mean takes the most calculation of the three measures.

   Bad: Sensitivity to extreme values. For instance, the mean of the data 2, 4, 6, and 8 is $\frac{20}{4} = 5$, but the mean of 2, 4, 6, and 48 (instead of 8) is $\frac{60}{4} = 15$, a shift of 10 units toward the extreme value of 48.

2. **The Median** *Good:* The median always exists and is unique.

   Good: Requires very little computation and is not sensitive to extreme values.

   Bad: To find the median, the data must be arranged in order of magnitude, and this may not be practical for large sets of data.

   Bad: Failure to take account of each item of data. Hence, in many statistical problems, the median is not a reliable measure.

3. **The Mode** *Good:* Requires no calculation.

   Good: The mode may be most useful. For example, suppose a shoe manufacturer surveys 100 women to see which of three styles, A, B, or C, each one prefers. Style A is selected by 30 women, style B by 50, and style C by 20. The mode is 50, and there is not much doubt about which style the manufacturer will feature.

   Bad: The mode may not exist, as in the case of the data 2, 4, 6, and 8 (there is no mode).

Answers to PROBLEMS

5. 4

EXAMPLE 6 Finding the mean, median, and mode

The payrolls of the 10 best-paid teams in Major League Baseball, approximated to the nearest million, are in the last column of the table.

a. Find the mean of the salaries.
b. Find the median of the salaries.
c. Find the mode of the salaries.
d. Why is the mean higher than the median?

Team	Payroll
1. Yankees	\$198,662,180 ≈ 199 million
2. Red Sox	\$120,100,524 ≈ 120 million
3. Angels	\$103,625,333 ≈ 104 million
4. White Sox	\$102,875,667 ≈ 103 million
5. Mets	\$100,901,085 ≈ 101 million
6. Dodgers	\$99,176,950 ≈ 99 million
7. Cubs	\$94,841,167 ≈ 95 million
8. Astros	\$92,551,503 ≈ 93 million
9. Braves	\$92,461,852 ≈ 92 million
10. Giants	\$90,862,063 ≈ 91 million

Source: http://msn.foxsports.com/mlb/story/5476130.

PROBLEM 6

Here are the 2006/2007 salaries for six members of the Los Angeles Lakers basketball team, approximated to the nearest million.

Player	Salary
Kobe Bryant	\$19,490,625 ≈ 19 million
Lamar Odom	\$13,524,000 ≈ 14 million
Vladimir Radmanovic	\$5,632,200 ≈ 6 million
Andrew Bynum	\$2,172,000 ≈ 2 million
Brian Cook	\$3,500,000 ≈ 4 million

Source: Data from Hoopshype.

a. Find the mean of the approximated salaries.
b. Find the median of the approximated salaries.
c. Find the mode of the salaries.
d. Why is the mean higher than the median?

SOLUTION

a. The mean of the 10 salaries is

$$\frac{199 + 120 + 104 + 103 + 101 + 99 + 95 + 93 + 92 + 91}{10} = \frac{1097}{10} \approx 110 \text{ million}$$

b. The numbers are

199 120 104 103 101 (5 numbers) | 99 95 93 92 91 (5 numbers)

$$\text{Median: } \frac{101 + 99}{2} = \frac{200}{2} = 100$$

c. There is no mode.

d. The mean (110) is higher than the median (100) because of the high salaries for the New York Yankees. If the New York Yankees' \$199 million is excluded from the calculations, the mean is about \$100 million, an amount equal to the median, and this man would be very happy to save the \$99 million!

Calculator Corner

If you have a calculator with [Σ+] (read "sigma plus") and [$\bar{x}$] (x bar) keys, you are in luck. The calculation for the mean is automatically done for you. First, place the calculator in the statistics mode (press MODE STAT or ADD STAT).

To find the mean of the numbers in Example 2, press 2 0 8 [Σ+] 1 7 9 [Σ+] 1 5 8 [Σ+] 1 5 0 [Σ+] and finally [$\bar{x}$]. The answer 173.75 will automatically appear. If you do not have the [Σ+] and [$\bar{x}$] keys, use parentheses and enter (2 0 8 + 1 7 9 + 1 5 8 + 1 5 0) ÷ 4 ENTER and then you will get the 173.75.

Answers to PROBLEMS

6. a. 9 million **b.** 6 million **c.** No mode **d.** Because Kobe Bryant pulls up the average salary.

› Exercises 6.4

‹ A › Finding the Mean In Problems 1–4, find the mean (average) of the numbers.

1. 1, 5, 9, 13, 17
2. 1, 3, 9, 27, 81, 243
3. 1, 4, 9, 16, 25, 36
4. 1, 8, 27, 64, 125

‹ B › Finding the Median In Problems 5–8, find the median of the numbers.

5. 1, 5, 9, 13, 17
6. 1, 3, 9, 27, 81, 243
7. 1, 9, 36, 4, 25, 16
8. 1, 27, 8, 125, 64

‹ C › Finding the Mode In Problems 9–12, find the mode(s) if there is one.

9. 5, 89, 52, 5, 52, 98
10. 29, 25, 22, 25, 52, 8
11. 2, 7, 11, 8, 11, 8, 11
12. 2, 8, 8, 8, 1, 2, 2, 2, 2

‹ D › Applications Involving Mean, Median, and Mode

Gas prices The following table will be used in Problems 13–14. Round answers to the nearest cent.

International Gasoline Prices

Most Expensive		Least Expensive	
Location	**Price per Gallon**	**Location**	**Price per Gallon**
Hong Kong	$5.34	Caracas, Venezuela	$0.28
London, England	$4.55	Jakarta, Indonesia	$0.74
Paris, France	$4.41	Cairo, Egypt	$0.75
Amsterdam, Neth	$4.38	Kuwait City, Kuwait	$0.77
Seoul, S. Korea	$4.35	Manama, Bahrain	$0.82

Source: Runzheimer International.

13. **a.** Find the mean, median, and mode of the five most expensive prices.
b. Are the mean, median, and mode close?
c. Which of the three do you think is most representative of the prices?
d. Delete Hong Kong and answer part **c** again.

14. **a.** Find the mean, median, and mode of the five least expensive prices.
b. Are the mean, median, and mode close?
c. Which of the three do you think is most representative of the prices?
d. Delete Caracas and answer part **c** again.

15. *GPA* Find the GPA for a student taking 13 hours at Mary Washington College. Weight of grade: A = 4 points, B = 3 points, C = 2 points, D = 1 point, F = 0 points, ZC = 0 points. Round answers to two decimal places.

Course	Grade	Credits Attempted	Credits Earned
MATH0111	A	3.0	3.0
ENGL0101	B	3.0	3.0
PSYC0101	D	3.0	3.0
BIOLO121	F	4.0	0.0
BIOLO121LB	ZC	0.0	0.0
Totals:		**13.0**	**9.0**

Source: Data from Mary Washington College.

16. *GPA* Find the GPA (to the nearest hundredth) for a student taking 13 hours at Montgomery College. Weight of grade: A = 4 points, B = 3 points, C = 2 points, D = 1 point, F = 0 points. Round answers to two decimal places.

Course	Credit Hrs (CH)	Grade
PY102	3	A
SO101	3	C
BI101	4	B
PE129	1	A
HE107	2	B
	13	

Source: Data from Montgomery College.

17. *Average annual salary* The table shows the average annual salary for five consecutive years for a person with an associate degree.

a. Find the mean, median, and mode of the person's salary for the five years.

b. Which is the most representative of the person's salary: the mean, the median, or the mode?

1	39,468
2	39,276
3	40,827
4	46,778
5	49,733

18. *Average annual salary* The table shows the average annual salary for five consecutive years for a person with less than a high school diploma.

a. Find the mean, median, and mode of the person's salary for the five years. (Compare with the person in Problem 17!)

b. Which is the most representative of the person's salary: the mean, the median, or the mode?

1	20,484
2	19,935
3	21,611
4	22,679
5	23,845

Source: Bureau of Labor Statistics.

19. *Major League Baseball (MLB) salaries* The table shows the 2006 salaries for the 10 best-paid MLB baseball players.

a. Round the salaries to the nearest million.

b. Find the mean, median, and mode of the rounded salaries.

c. Which is the most representative of the salaries: the mean, the median, or the mode?

Top 10 MLB Player Salaries

Player	Position	Team	Salary
1. Alex Rodriguez	3B	Yankees	$25,680,727
2. Derek Jeter	SS	Yankees	$20,600,000
3. Jason Giambi	1B	Yankees	$20,428,571
4. Barry Bonds	OF	Giants	$20,000,000
5. Jeff Bagwell	1B	Astros	$19,369,019
6. Mike Mussina	P	Yankees	$19,000,000
7. Manny Ramirez	OF	Red Sox	$18,279,238
8. Todd Helton	1B	Rockies	$16,600,000
9. Andy Pettitte	P	Astros	$16,428,416
10. Magglio Ordonez	OF	Tigers	$16,200,000

Source: http://msn.foxsports.com.

20. *Annual salaries* Who makes more money, the baseball players or university presidents? The table shows the 10 best-paid presidents at public universities in the United States.

a. Round the salaries to the nearest thousand dollars.

b. Find the mean, median, and mode of the rounded salaries.

c. Which is the most representative of the salaries: the mean, the median, or the mode?

d. What is the difference between the mean and median of baseball players (Problem 19) and college presidents?

Shirley Ann Jackson, Rensselaer Polytechnic Institute	$891,400
Gordon Gee, Vanderbilt University	$852,023
Judith Rodin, University of Pennsylvania	$845,474
Arnold J. Levine, Rockefeller University	$844,600
William R. Brody, Johns Hopkins University	$772,276
Michael R. Ferrari, Texas Christian University	$667,901
Steven B. Sample, University of Southern California	$656,420
Jon Westling, Boston University	$656,098
Richard C. Levin, Yale University	$654,452
Constantine N. Papadakis, Drexel University	$650,886

Source: Data from The Chronicle of Higher Education.

Web IT go to mhhe.com/bello for more lessons

Web IT go to mhhe.com/bello for more lessons

21. *Major League salaries* The table shows the salaries for the 5 most overpaid players in Major League Baseball according to Foxsports.

Source: http://msn.foxsports.com.

Player	Position	Team	Salary
Chan Ho Park	Pitcher	Padres	\$15.3 million
Eric Milton	Pitcher	Reds	\$9.8 million
Jason Kendall	Catcher	A's	\$11.57 million
Kaz Matsui	2nd base	Mets	\$8.058 million
Sean Casey	1st base	Pirates	\$8.5 million

a. Round the salaries to the nearest million.

b. Find the mean, median, and mode of the rounded salaries. Round answers to one decimal place.

c. Which is the most representative of the salaries: the mean, the median, or the mode?

22. *Major League salaries* What is the salary difference between the overpaid and the underpaid players in Major League Baseball? The table shows the salaries for the 5 most underpaid players in Major League Baseball according to Foxsports.

Source: http://msn.foxsports.com.

Player	Position	Team	Salary
David Ortiz	Designated Hitter	Red Sox	\$6.5 million
Chris Carpenter	Pitcher	Cardinals	\$5 million
Trevor Hoffman	Pitcher	Padres	\$4.5 million
Melvin Mora	3rd base	Orioles	\$4 million
Tony Clark	1st base	Diamondbacks	\$1.034 million

a. Round the salaries to the nearest million.

b. Find the mean, median, and mode of the rounded salaries. Round answers to one decimal place.

c. Which is the most representative of the salaries: the mean, the median, or the mode?

d. What is the difference between the mean and median of overpaid baseball players (Problem 21) and underpaid baseball players?

23. *Annual salaries* What college degree will earn you the most money? At the moment, engineering majors are seeing the most cash. The average salary in four engineering fields is as shown. Find the mean and median for the four salaries.

Source: http://www.shreveporttimes.com.

Chemical engineering	\$55,900
Electrical engineering	\$52,899
Mechanical engineering	\$50,672
Civil engineering	\$44,999

24. *Annual salaries* Another field bringing increased salaries is accounting. Of course, it depends where the job is. K-Force, a staffing firm based in Tampa, Florida, reports the following entry-level salaries for general-accounting staff members in four cities.

Boston	\$41,600
New York	\$51,800
Chicago	\$48,700
San Francisco	\$53,800

Source: http://www.collegejournal.com.

a. Find the mean and median for the salaries in the four cities.

b. According to the *College Journal* "Recent accounting graduates received an average annual starting salary of \$46,188." Is this closer to the mean or to the median in part **a?**

25. *Internet* The chart shows the number of visitors (in thousands) to five different Internet sites.

a. Find, to one decimal place, the mean and median of the number of visitors (in thousands) to these five sites from **home.**

b. Find, to one decimal place, the mean and median of the number of visitors (in thousands) to these five sites from **work.**

Unique U.S. Visitors to Flower, Gift, and Greetings Sites by Location, January 2006

	Total Unique Visitors (000)			
	All Locations	Home	Work	University
1. AmericanGreetings.com	8,688	5,560	2,784	528
2. Hallmark	6,669	3,601	2,745	518
3. FTD.com	1,935	1,175	678	173
4. 1-800-flowers.com	1,753	961	803	76
5. Martha Stewart sites	1,546	904	608	108

Source: http://www.clickz.com.

Using Your Knowledge

Misuses of Statistics We have just studied three measures of central tendency: the mean, the median, and the mode. All these measures are frequently called **averages.** Suppose that the chart shows the salaries at Scrooge Manufacturing Company.

Boss
$100,000

Boss's son
$50,000

Boss's assistant
$25,000

Boss's secretaries
$10,000

Workers
$6000 each

26. Scrooge claims that the workers should not unionize; after all, he says, the *average* salary is $22,500. Can you discover what average this is?

27. Many Chevitz, the union leader, claims that Scrooge Manufacturing really needs a union. Just look at their salaries! A meager $6000 on the *average*. Can you discover what average he means?

28. B. Crooked, the politician, wants both union and management support. He says that the workers are about *average* as far as salary is concerned. The company's *average* salary is $8000. Can you discover what average B. Crooked has in mind?

Write On

29. Write in your own words what is meant by the median of a set of scores. Is the median a good measure of a set of scores? Give examples.

30. Write in your own words what is meant by the mode of a set of scores. Is the mode a good measure of a set of scores? Is it possible to have more than one mode in a set of scores? Is it possible that there is no mode for a set of scores? Give examples.

Concept Checker

Fill in the blank(s) with the correct word(s), phrase, or mathematical statement.

31. The __________ of a set of ***n*** numbers is the **sum** of the numbers **divided by** the number ***n*** of elements in the set.

32. For an __________ number of items, the __________ of a set of numbers is the **middle number** when the numbers are arranged in order.

33. For an __________ number of items, the __________ of a set of numbers is the __________ of the **two middle numbers.**

34. The __________ of a set of numbers is the **number that occurs the most often** in the set.

35. When no number occurs most often in a set of numbers, the set has no __________.

36. A set of numbers can only have __________ **mean** and __________ **median,** but it could have **more than one** __________.

smallest
one
mean
even
odd
mode
median
average
largest

Mastery Test

The table shows the number and type of spam (unsolicited "junk" e-mail sent to large numbers of people to promote products or services) received by the same 200 persons in July and August.

37. Find the mean number of spams received in July.

38. Find the median number of spams received in July.

39. Find the mode (if it exists) of the number of spams received in July.

40. Name the categories in which the number of spams received did not change from July to August.

41. Which category increased the most from July to August?

Type of Spam	July	August
Internet	14	22
Other	28	32
Scams	18	20
Products	40	40
Spiritual	2	2
Financial	30	28
Leisure	16	14
Adult	28	24
Health	24	18

Source: Data from Brightmail Probe Network.

42. Find the GPA of a student taking 13 credit hours at Raritan Valley Community College and earning the grades and points shown. Weight of grade: A = 4 points, B = 3 points, C = 2 points, D = 1 point, F = 0 points. Answer to two decimal places.

Course	Credits	Grade
History	3	W (–)
English	3	A (4)
Psychology	3	C (2)
French	4	D (1)

Note: W means the student withdrew. No credits or points counted.

Skill Checker

In Problems 43–47, Multiply:

43. $60 \cdot \frac{1}{12}$

44. $11 \cdot \frac{1}{12}$

45. $10{,}560 \cdot \frac{1}{5280}$

46. $7.75 \cdot 12$

47. $8.75 \cdot 12$

Collaborative Learning

Form two groups. One group will investigate **race** discrimination complaints, and the other will investigate **sex** discrimination complaints, as shown in the following tables.

Race-Based Charges

	FY 1998	FY 1999	FY 2000	FY 2001	FY 2002
Complaints	28,820	28,819	28,945	28,912	29,910
Resolutions	35,716	35,094	33,188	32,077	33,199
Monetary benefits (in millions)	\$32.2	\$53.2	\$61.7	\$86.5	\$81.1

Sex-Based Charges

	FY 1998	FY 1999	FY 2000	FY 2001	FY 2002
Complaints	24,454	23,907	25,194	25,140	25,536
Resolutions	31,818	30,643	29,631	28,602	29,088
Monetary benefits (in millions)	\$58.7	\$81.7	\$109.0	\$94.4	\$94.7

Source: Data from U.S. Equal Opportunity Commission.

Answer the questions.

1. What was the average number of complaints for racial discrimination and for sex discrimination from 1998 to 2002?
2. What was the average number of resolutions for racial discrimination and for sex discrimination from 1998 to 2002?
3. What were the average monetary benefits for 1998 to 2002?

Discussion Which average monetary benefit per resolved case was greater, race or sex discrimination? Why?

Research Questions

1. Go to the library, look through newspapers and magazines, and find some examples of bar and circle graphs. Are there any distortions in the drawings? Discuss your findings.
2. Write a brief report about the contents of John Graunt's *Bills of Mortality*.
3. Write a paragraph about how statistics are used in sports.
4. Write a report on how surveys are used to determine the ratings and rankings of television programs by organizations such as A.C. Nielsen.
5. Discuss the Harris and Gallup polls and the techniques and types of statistics used in their surveys.
6. Prepare a report or an exhibit on how statistics are used in medicine, psychology, and/or business.
7. Research and write a report on how Gregor Mendel, Sir Francis Galton, and Florence Nightingale used statistics in their work.

Summary Chapter 6

Section	Item	Meaning	Example
6.1B	Pictograph	A type of graph that uses symbols to represent numerical data.	If each symbol represents 100 hamburgers sold, [three hamburger symbols] means that 300 hamburgers have been sold.
6.2A	Bar graph	A graph used to compare different categories by using bars whose lengths are proportional to the number of items in the category. The graph shows that most adults want to live to be 100.	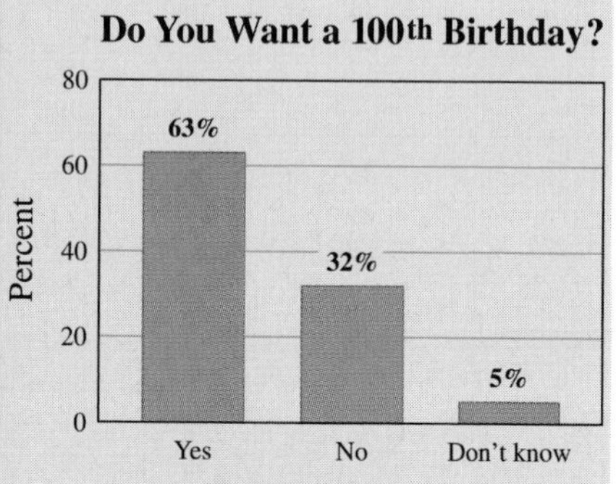 **Do You Want a 100th Birthday?** Percent; Yes 63%, No 32%, Don't know 5%. *Source:* Data from USATODAY.com.
6.2C	Line graph	A type of graph that usually shows a trend or change over time. The line graph shows the satisfaction or dissatisfaction of Americans with health care costs.	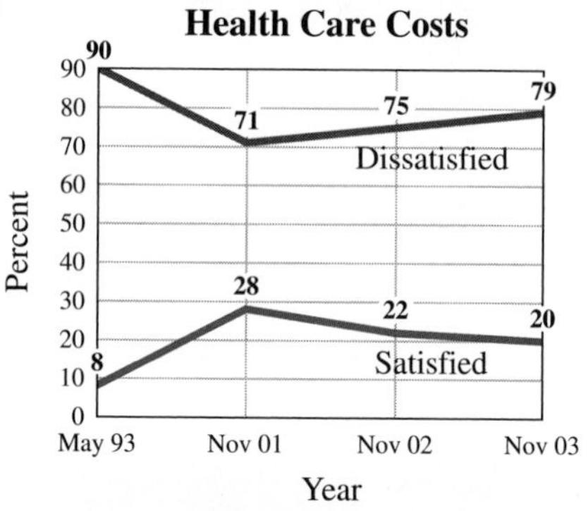 **Health Care Costs** Percent by Year; Dissatisfied: May 93 90, Nov 01 71, Nov 02 75, Nov 03 79; Satisfied: May 93 8, Nov 01 28, Nov 02 22, Nov 03 20. *Source:* Data from The Gallup Organization.
6.3A	Circle graph (pie chart)	A type of graph that shows what percent of a quantity is used in different categories or the ratio of one category to another.	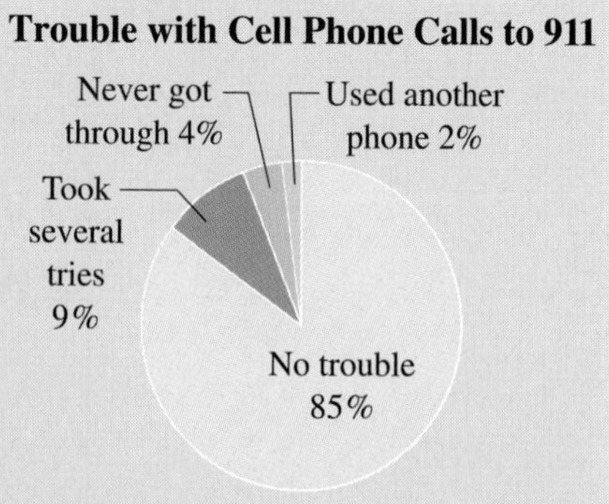 **Trouble with Cell Phone Calls to 911** Never got through 4%; Used another phone 2%; Took several tries 9%; No trouble 85%. The circle graph shows the percent of people who had trouble when calling 911.

Section	Item	Meaning	Example
6.4A	Mean	The mean of a set of n numbers is the sum of the numbers divided by n.	The mean of 7, 6, and 8 is $\frac{7+6+8}{3} = 7$
6.4A	Weighted mean	The weighted mean of a set of n numbers is the sum of the products formed by multiplying each number by its assigned weight divided by the sum of all the weights.	Finding a grade point average (GPA).
6.4B	Median	The median of a set of numbers is the middle number when the numbers are arranged in ascending order.	The median of 5, 8, and 15 is 8. The median of 5, 8, 10, and 15 is $\frac{8+10}{2} = 9$.
6.4C	Mode	The number in a set that occurs most often. If no such number exists, there is no mode.	The mode of 3, 4, 5, 7, 4 is 4. The set 3, 4, 5, 6 has no mode.

› Review Exercises Chapter 6

If you need help with these exercises, look in the section indicated in brackets.

1. ‹ **6.1A** › *The table shows how much consumers are willing to pay for digital music and the potential revenue that can be derived from it.*

Price Expectations of Digital Music

Price of digital song	$0.01	$0.50	$0.99	$1.49	$1.99
Percent of market reached	89%	82%	74%	42%	39%
Potential revenue	N/A	$992.20	$1772.89	$1514.44	$1878.16

Source: Data from ClickZ Network.

a. What percent of the market will be reached if the price per song is $1.49?

b. What percent of the market will be reached if the price per song is $1.99?

c. What is the potential revenue when the price is $1.49 per song?

d. What is the potential revenue when the price is $1.99 per song?

e. Which price per song would reach the highest percentage of the market?

2. ‹ **6.1B** › *The pictograph shows the number of hamburgers sold at Burger King, Del Taco, Jack in the Box, McDonald's, and Wendy's. Each symbol represents 100 hamburgers.*

a. Which brand sold the most?

b. Which brand sold the second most?

c. How many Jack in the Box hamburgers were sold?

d. How many McDonald's hamburgers were sold?

e. How many Del Taco hamburgers were sold?

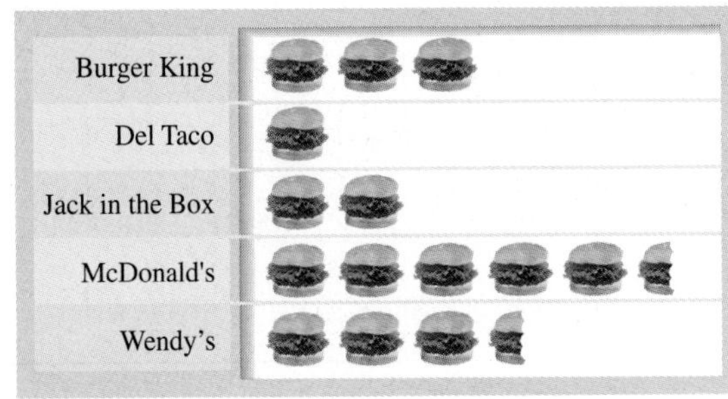

3. ⟨ **6.2A** ⟩ *The bar graph shows the percent of people who have downloaded music files from the Internet for playback at another time.*

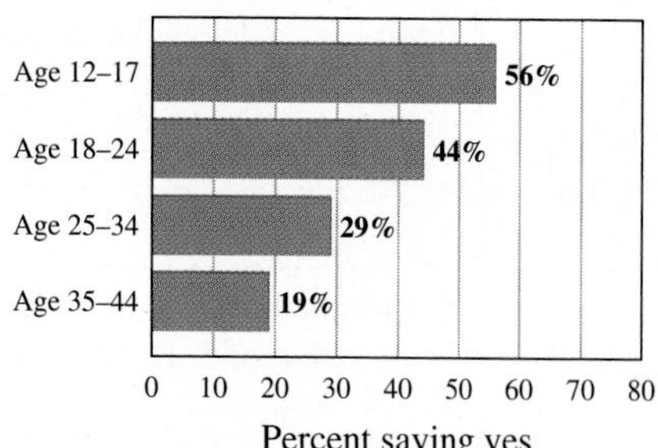

Source: Data from Edison Media Research.

a. Which of the four age brackets has the second-highest percent of people who have downloaded music files?

b. Which of the four age brackets has the next-to-the-lowest percent of people who have downloaded music files?

c. What percent of the 18–24 age bracket downloaded music files?

d. What is the percent difference of downloaded music between the 25–34 and the 35–44 age brackets?

e. What is the percent difference of downloaded music between the age categories with the highest and lowest percent of downloads?

4. ⟨ **6.2B** ⟩ *In the same survey, the percents of consumers who have downloaded music and said they are purchasing more music are as follows:*

Age	Percent
12–17	25%
18–24	25%
25–34	22%
35–44	28%

Use the age brackets as the categories and the percents from 0 to 50 as the frequencies and make a vertical bar graph for the data.

Consumers Who Have Downloaded Music and Said They Purchased More Music

5. ⟨ **6.2B** ⟩ *The table shows the number of cell phone users from 2001 to 2005.*

Year	Millions
2001	120
2002	140
2003	160
2004	170
2005	180

Use the years as the categories and the numbers (in millions) from 100 to 200 at 10-unit intervals as the frequencies and make a horizontal bar graph for the data.

Number of Cell Phone Users by Year

6. ⟨ **6.2C** ⟩ *The graph shows the total and senior populations (in millions) in the United States from 1950 to 2050.*

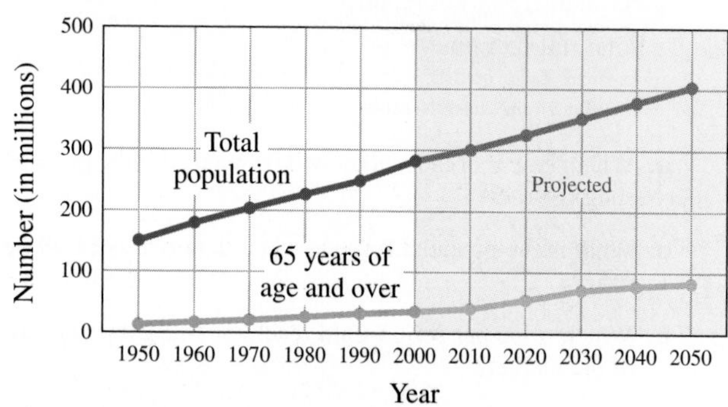

Source: Data from Centers for Disease Control and Prevention.

What was or would be the approximate total population:

a. in 1960?

b. in 1970?

c. in 1990?

d. in 2010?

e. in 2020?

7. **〈6.2C〉** *Refer to the graph in Problem 6. What was or would be the approximate 65 years of age and older population in:*

a. 1960

b. 1980

c. 2020

d. 2040

e. 2050

8. **〈6.2D〉** *The chart shows the projected online spending (in billions) for college students, teens, and kids. Make a line graph of the spending for kids using a 0 to 1 billion vertical scale using 0.2-unit intervals.*

Online Spending Forecast (in billions)

	College Students	Teens	Kids
2003	$4.5	$1.7	$0.2
2004	$5.5	$2.6	$0.4
2005	$6.4	$3.6	$0.7
2006	$7.4	$4.8	$1.0

Source: Data from Jupiter Research, October 2001.

Online Spending Forecast for Kids

9. **〈6.3A〉** *The circle graph shows the percent of people who believe that the most popular place for proposing is as shown in the graph.*

Favorite Places for Proposing

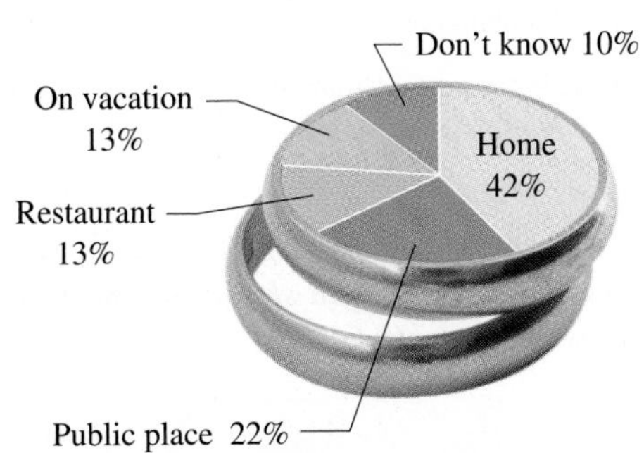

Source: Data from StrategyOne for Korbel Champagne.

a. What percent of the people chose "restaurant"?

b. What percent of the people chose "home"?

c. What percent of the people chose "public place"?

d. Which was believed to be the most popular place for proposing?

e. What was the percent difference between the people who chose home as the most popular place and those who chose a public place?

10. **〈6.3A〉** *Refer to the circle graph in Problem 9. Suppose 1000 people were surveyed. How many people would choose:*

a. Don't know?

b. On vacation?

c. Restaurant?

d. Public place?

e. Home?

11. **〈6.3B〉** *Make a circle graph to show the ways we commute to work.*

Category	Percent
Drive alone (DA)	70%
Car pool (CP)	10%
Public transit (PT)	5%
Other (O)	15%

Ways We Commute to Work

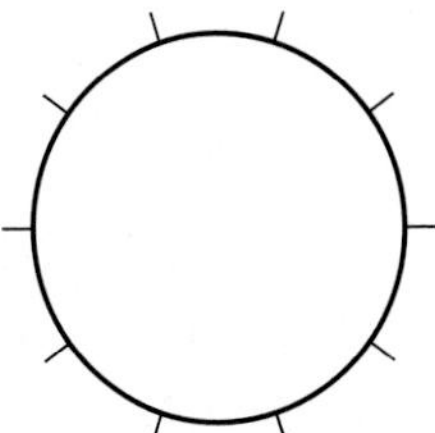

12. ⟨ **6.4A, B, C** ⟩ *The table shows the price (in cents) of five different fast-food chain hamburgers.*

Category	Price (cents)
Burger King	79
Del Taco	99
McDonald's	99
Jack in the Box	99
Wendy's	89

Source: Data from Fast Food Source.com.

a. Find the mean price for the five hamburgers.

b. Find the median price for the five hamburgers.

c. Find the mode for the price of the five hamburgers.

d. Find the mean price of the first four hamburgers.

e. Find the median price of the first four hamburgers.

13. ⟨ **6.4A, B, C** ⟩ *The table shows the fat content (in grams) of five different fast-food chain hamburgers.*

Category	Fat (grams)
Burger King	17
Del Taco	13
McDonald's	18
Jack in the Box	14
Wendy's	12

Source: Data from Fast Food Source.com.

a. Find, to the nearest gram, the mean fat content for the five hamburgers.

b. Find the median fat content for the five hamburgers.

c. Find the mode, if it exists, for the fat content of the five hamburgers.

d. Find, to the nearest gram, the mean fat content of the first four hamburgers.

e. Find, to the nearest gram, the median fat content of the first four hamburgers.

14. ⟨ **6.4A** ⟩ *A student is taking five courses with the credit hours and points earned shown. To one decimal place, what is the student's GPA?*

Course	Credit Hours	Grade (Points Earned)
V	3	C(2)
W	3	B(3)
X	4	C(2)
Y	1	A(4)
Z	3	A(4)

15. ⟨ **6.4A** ⟩ *A student is taking five courses with the credit hours and points earned shown. To one decimal place, what is the student's GPA?*

Course	Credit Hours	Grade (Points Earned)
V	3	B(3)
W	2	D(1)
X	3	C(2)
Y	4	A(4)
Z	3	A(4)

> Practice Test Chapter 6

Visit www.mhhe.com/bello to view helpful videos that provide step-by-step solutions to several of the problems below. (Answers on page 405)

1. The table shows how much consumers are willing to pay for digital music and the potential revenue that can be derived from it.

Price Expectations of Digital Music

Price of digital song	$0.01	$0.50	$0.99	$1.49	$1.99
Percentage of market reached	89%	82%	74%	42%	39%
Potential revenue	N/A	$992.20	$1772.89	$1514.44	$1878.16

Source: Data from ClickZ Network.

a. What percent of the market will be reached if the price per song is $0.99?

b. What is the potential revenue when the price is $0.99 per song?

2. The pictograph shows the number of hamburgers sold at Burger King, Del Taco, Jack in the Box, McDonald's, and Wendy's. Each symbol represents 100 hamburgers.

a. Which brand sold the most?

b. How many Wendy's hamburgers were sold?

3. The bar graph shows the percent of people who have downloaded music files from the Internet for playback at another time.

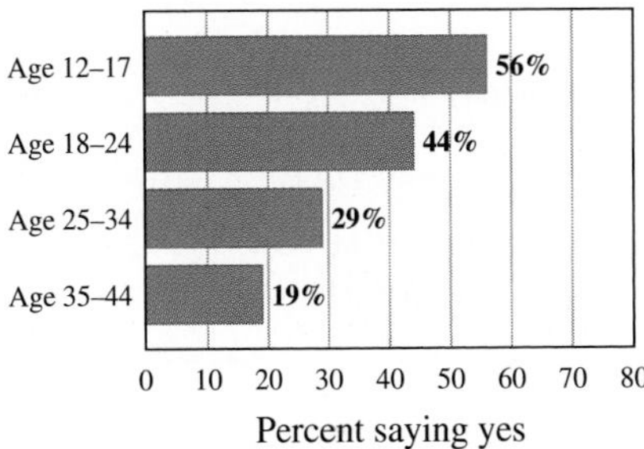

Source: Data from Edison Media Research.

a. Which of the four age brackets has the lowest percent of people who have downloaded music files?

b. What percent of the 25–34 age bracket downloaded music files?

4. In the same survey, the percents of consumers who have downloaded music and said they are purchasing more music are as follows:

Age	Percent
12–17	20%
18–24	20%
25–34	22%
35–44	25%

Use the age brackets as the categories and the percents from 0 to 40 as the frequencies and make a vertical bar graph for the data.

5. The graph shows the total and senior populations (in millions) from 1950 to 2050.

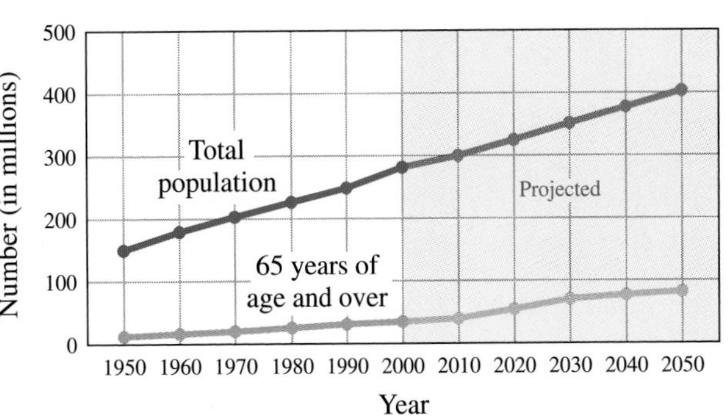

Source: Data from Centers for Disease Control and Prevention.

a. What is the approximate projected total population for the year 2030?

b. What is the approximate projected 65 years of age and over population for the year 2020?

6. The chart shows the projected online spending (in billions) for college students, teens, and kids.

Online Spending Forecast (in billions)

	College Students	Teens	Kids
2003	$4.5	$1.7	$0.2
2004	$5.5	$2.6	$0.4
2005	$6.4	$3.6	$0.7
2006	$7.4	$4.8	$1.0

Source: Data from Jupiter Research, October 2001.

Make a line graph of the spending for college students using a 1–10 billion vertical scale.

7. The circle graph shows the percent of people who believe that the most popular place for proposing is as shown in the graph.

Favorite Places for Proposing

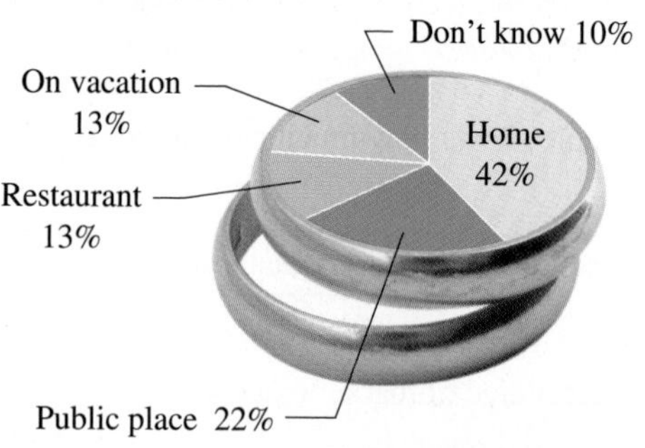

Source: Data from StrategyOne for Korbel Champagne.

a. What percent of the people chose "on vacation"?

b. If 500 people were surveyed, how many would have chosen "on vacation"?

8. Based on the information given in the following table, make a circle graph to show the ways we commute to work.

Category	Percent
Drive alone (DA)	80%
Car pool (CP)	10%
Public transit (PT)	5%
Other (O)	5%

Source: Data from National Public Transportation Survey.

9. The table shows the milligrams (mg) of cholesterol in five different fast-food chain hamburgers.

a. Find the mean number of grams of cholesterol for the hamburgers.

b. Find the median number of grams of cholesterol for the hamburgers.

c. Find the mode(s) of the number of grams of cholesterol for the hamburgers.

Category	Cholesterol (mg)
Burger King	50
Del Taco	35
McDonald's	60
Jack in the Box	45
Wendy's	45

Source: Data from Fast Food Source.com.

10. A student is taking five courses with the credit hours and points earned shown. To one decimal place, what is the student's GPA?

Course	Credit Hours	Grade (Points Earned)
V	3	B(3)
W	3	B(3)
X	4	C(2)
Y	1	A(4)
Z	3	A(4)

Answers to Practice Test Chapter 6

Answer	If You Missed Question	Review Section	Review Examples	Review Page
1. a. 74% **b.** $1772.89	1	6.1	1	350–351
2. a. McDonald's **b.** 350	2	6.1	3–5	352–353
3. a. 35–44 age bracket **b.** 29	3	6.2	1, 2	358–359
4. Purchasing More Music (bar graph: Percent vs. age 12–17: 20%, 18–24: 20%, 25–34: 22%, 35–44: 25%)	4	6.2	3, 4	359–360
5. a. About 350 million **b.** About 50 million	5	6.2	5	361
6. Online Spending Forecast (line graph: Spending (in billions) vs. Year, 2003–2006)	6	6.2	6	362
7. a. 13% **b.** 65	7	6.3	1, 2, 3	375–376
8. Ways to Commute to Work (circle graph: Drive alone 80%, Car pool 10%, Public transit 5%, Other 5%)	8	6.3	4	378
9. a. 47 **b.** 45 **c.** 45	9	6.4	1, 2, 4, 5	387, 389–390
10. 3.0	10	6.4	3	388

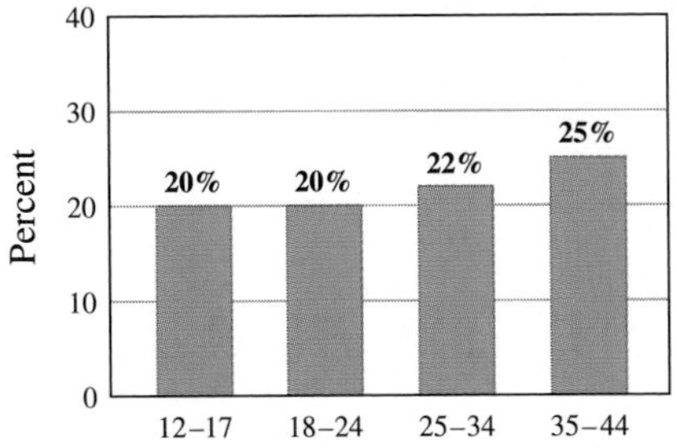

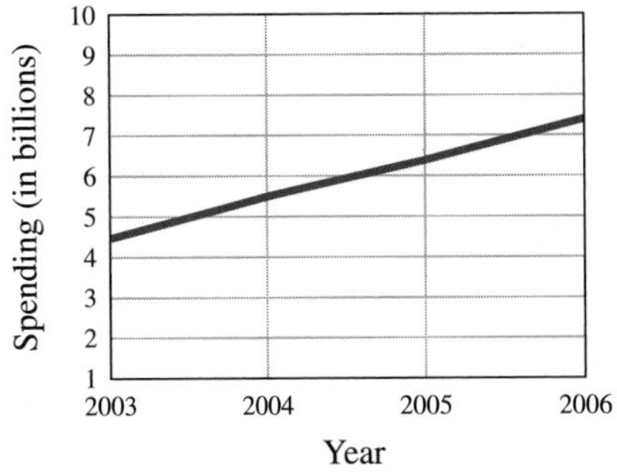

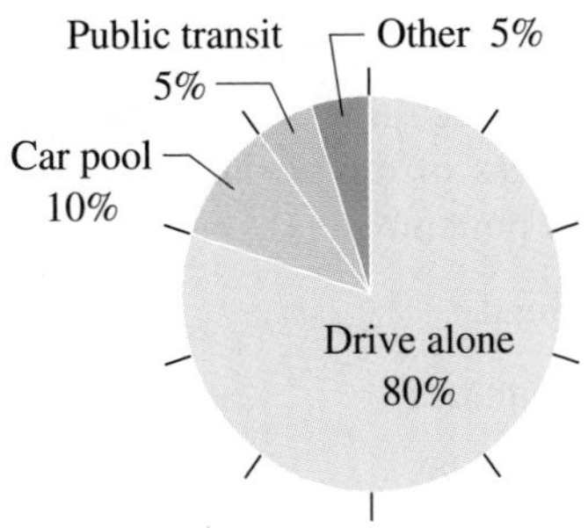

Cumulative Review Chapters 1–6

1. Write six thousand, five hundred ten in standard form.
2. Simplify: $4 \div 2 \cdot 2 + 9 - 6$
3. Classify $\frac{2}{9}$ as proper or improper.
4. Write $\frac{25}{8}$ as a mixed number.
5. Write $2\frac{1}{6}$ as an improper fraction.
6. Multiply: $\left(\frac{6}{5}\right)^2 \cdot \frac{1}{36}$
7. Divide: $\frac{15}{2} \div 4\frac{1}{6}$
8. Translate and solve: $\frac{3}{4}$ less than a number z is $\frac{3}{8}$. What is z?
9. Find a number such that $\frac{8}{9}$ of it is $4\frac{1}{4}$.
10. Give the word name for 352.51.
11. Write 64.175 in expanded form.
12. Subtract: $541.42 - 12.5$
13. Multiply: $59.9 \cdot 0.013$
14. Divide: $\frac{126}{0.21}$
15. Round 749.851 to the nearest tenth.
16. Divide: $80 \div 0.13$ (Round the answer to two decimal digits.)
17. Write $0.\overline{78}$ as a reduced fraction.
18. What decimal part of 30 is 9?
19. Arrange in order of decreasing magnitude and write the inequality using the $>$ sign: $9.568 \quad 9.56\overline{8} \quad 9.5\overline{68}$
20. Insert $=$, $<$, or $>$ to make a true statement. 0.53 ____ $\frac{19}{20}$
21. Solve for x: $x + 3.7 = 7.9$
22. Solve for y: $4.5 = 0.5y$
23. Solve for z: $4 = \frac{z}{3.3}$
24. The ratio of cars to people in New Zealand is 280 to 1000. Write this ratio as a fraction in reduced form.
25. Write the following proportion: 4 is to 7 as 28 is to x.
26. There is a law stating that "the ratio of width to length for the American flag should be 10 to 19." Is a flag measuring 40 by 76 feet of the correct ratio?
27. Solve the proportion: $\frac{x}{2} = \frac{2}{8}$
28. Solve the proportion: $\frac{10}{p} = \frac{2}{3}$
29. A student traveled 500 miles on 19 gallons of gas. How many miles per gallon did the student get? (Round to the nearest whole number.)
30. A 20-ounce jar of popcorn costs \$2.49. What is the unit price in cents per ounce? (Answer to the nearest cent.)
31. A pound of fertilizer covers 1000 square feet of lawn. How many pounds are needed to cover a lawn measuring 80 by 50 feet (4000 square feet)?
32. The protein RDA for females is 52 grams per day. Three ounces of a certain product provide four grams of protein. How many ounces of the product are needed to provide 52 grams of protein?
33. Write 67% as a decimal.
34. Write $3\frac{3}{4}\%$ as a decimal.
35. Write 0.09 as a percent.
36. 80% of 60 is what number?
37. What is $66\frac{2}{3}\%$ of 54?
38. What percent of 32 is 16?
39. 12 is 40% of what number?
40. The sales tax rate in a certain state is 4%. Find the total price paid for a pair of shoes that costs \$18.

41. Find the simple interest earned on $300 invested at 4.5% for 5 years.

42. The following table shows the distribution of families by income in Racine, Wisconsin.

Income Level	Percent of Families
$0–9999	3
10,000–14,999	10
15,000–19,999	27
20,000–24,999	33
25,000–34,999	10
35,000–49,999	9
50,000–79,999	4
80,000–119,000	3
120,000 and over	1

What percent of the families in Racine have incomes between $15,000 and $19,999?

43. The following graph represents the yearly rainfall in inches in Sagamore County for 2001–2006. Find the rainfall for 2001.

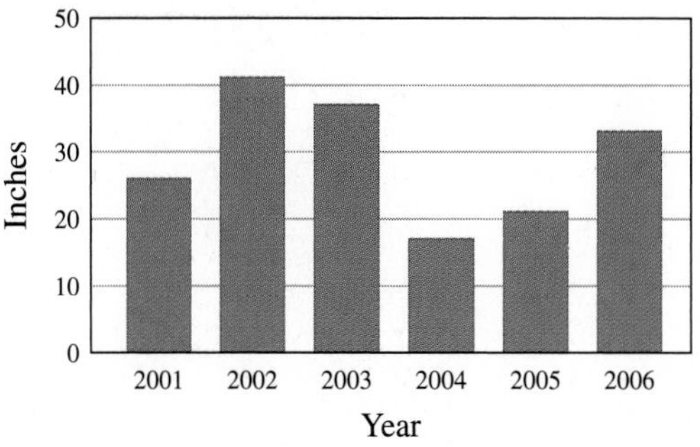

44. The following graph represents the monthly average temperature for seven months of the year. How much higher is the average temperature in May than it is in March?

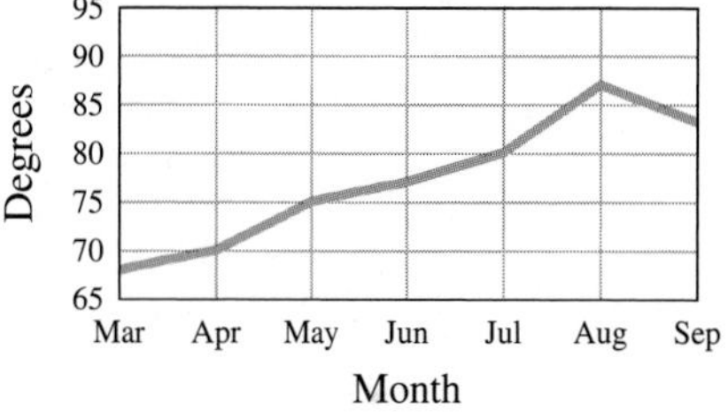

45. The number of hours required in each discipline of a college core curriculum is represented by the following circle graph. What percent of these hours is in P.E. and art combined?

History 15%
P.E. 30%
English 15%
Math 20%
Art 20%

46. What is the mode of the following set of numbers?
2, 23, 16, 21, 12, 2, 9, 2, 6, 2

47. What is the mean of the following set of numbers?
3, 2, 30, 17, 2, 2, 2, 18, 3, 4, 16

48. What is the median of the following set of numbers?
4, 7, 24, 4, 18, 14, 23, 6, 29, 29, 29

Section

Chapter 7

seven

Measurement and the Metric System

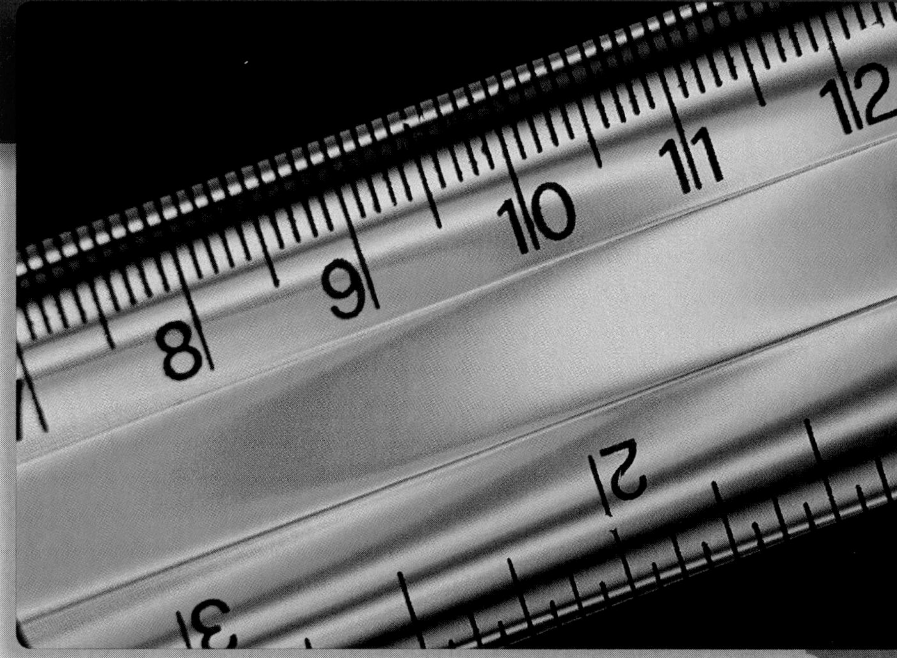

The Human Side of Mathematics

Here is a quiz for you: name one country that uses the metric system. Almost any country will do *except* the United States. Take a romantic vacation in France. It is a cool day; so you take a 30-minute walk to a nearby village, get some cheese and some wine. The temperature (in degrees Celsius, not Fahrenheit), the distance (kilometers, not miles), the cheese (kilograms, not pounds), and the wine (liters, not quarts) are all in the metric system. How did this happen?

Weights and measures are among the earliest tools invented by humans. Babylonians and Egyptians measured length with the forearm, hand, or finger, but the system used in the United States is based on the English system. The need for a single, worldwide, coordinated system of measurement was recognized over 300 years ago by Gabriel Mouton, vicar of St. Paul's Church in Lyons, France. In 1790, the National Assembly of France requested the French Academy of Sciences to "deduce an invariable standard for all measures and weights." The result was the metric system, made compulsory in France in 1840. What about the United States? As early as 1789, Thomas Jefferson submitted a report proposing a decimal-based system of measurement. Congress took no action on his proposal but later, in 1832, directed the Treasury Department to standardize the measures used by customs officials at U.S. ports. Congress allowed this report to stand without taking any formal action. Finally, in 1866, Congress legalized the use of the metric system and in 1875, the United States became one of the original signers of the Treaty of the Meter.

7.1 Length: U.S. System

Objectives

You should be able to:

A › Convert a given measure in the U.S. system to an equivalent one also in the U.S. system.

B › Solve applications involving U.S. units of length.

To Succeed, Review How To . . .

Multiply and divide fractions and decimals. (pp. 130–136, 212–216)

Getting Started

In the cartoon, Hugo is **measuring** the board and finds it to be $31\frac{1}{2}$ inches. When we measure an object, we assign a **number** to it that indicates the size of the object, in this case, $31\frac{1}{2}$.

Reprinted with permission of King Features Syndicate.

We also use **units** (such as inches) comparing the quantity being measured to some standard unit. Thus, any measurement involves two things:

1. A *number* (whole, fraction, or decimal)
2. A *unit*

Suppose $\overline{AB}$ A ——1—— B is a unit segment. Using $\overline{AB}$ as our standard, we can measure other segments such as $\overline{CD}$.

C ——1——1——1—— D

Since the segment $\overline{CD}$ contains three of the $\overline{AB}$ segments end to end, we say that the segment $\overline{CD}$ is three units long. When the length of a segment is not a whole number, we can use fractional parts to indicate the length. Thus, the length of segment $\overline{EF}$ is $2\frac{1}{2}$ units.

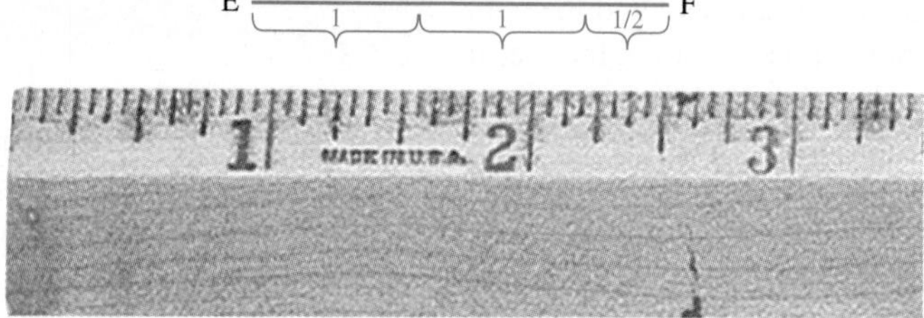

A standard ruler shows the units we use in everyday life.

A › Converting U.S. Units of Length

There are two major systems of measurement, the **U.S. customary** or **American system** and the **metric** system, which will be covered in Section 7.2.* In ancient times most units were based on measurements related to the body. For example, the **inch** originated

*The U.S. (American) system used to be called the English system, but the English have changed to the metric system, so we call this system the U.S. customary or American system.

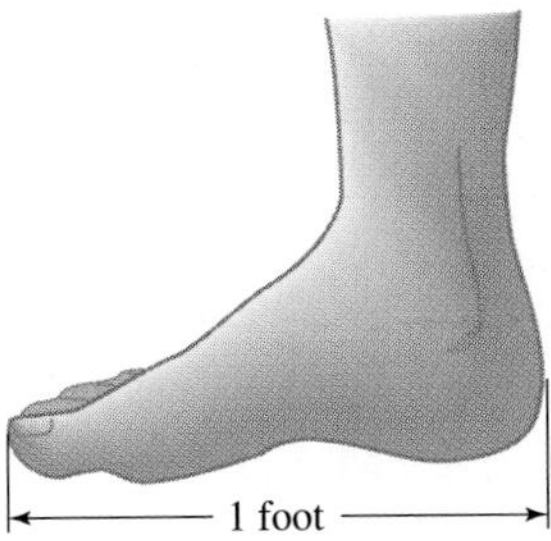

with the Greeks, who based it on the breadth of the thumb. (The word for *inch* in Spanish is *pulgada,* a word derived from the word *pulgar,* meaning thumb. The Latin word for inch is *uncia,* which means $\frac{1}{12}$, later evolving into inch.) The **foot** was the length of a man's foot, and a **yard** was the circumference of a person's waist (*gird* in Saxon, later evolving into *yard*). It is said that Henry I decreed that the yard should be the distance from the tip of his nose to the end of his thumb. The problem with these units is that their size depends on whose finger, foot, or arm you use. To alleviate this problem, the yard was defined to be the distance between marks on a brass bar kept in London. The foot was exactly $\frac{1}{3}$ of this standard yard and the inch, exactly $\frac{1}{36}$ of the yard.

These relationships and the abbreviations used are given in the table.

U.S. Units of Length
1 foot (ft) = 12 inches (in.)
1 yard (yd) = 3 ft
1 mile (mi) = 5280 ft

To change from one unit to another, we can make *substitutions,* treating the names of the units as if they were numbers. For example, to find how many inches are in a yard, we write

$$1 \text{ yard} = 3 \text{ ft} = 3 \cdot (12 \text{ in.}) = 36 \text{ in.}$$

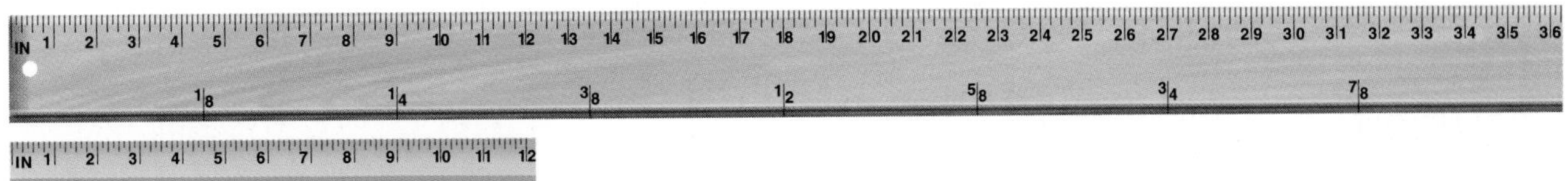

EXAMPLE 1 Converting yards to inches

2 yd = ________ in.

SOLUTION

$$1 \text{ yd} = 3 \text{ ft}$$

$$2 \text{ yd} = 6 \text{ ft} = 6 \cdot 12 \text{ in.} = 72 \text{ in.}$$

Thus, 2 yd = 72 in.

PROBLEM 1

3 yd = ________ in.

How do we convert from a smaller unit to a larger one? For example,

60 in. = ______ ft

First, note that there are 12 inches in a foot; thus, there is $\frac{1}{12}$ of a foot in an inch. Also, 3 ft = 1 yd, so 1 ft = $\frac{1}{3}$ yd. Finally, 1 mi = 5280 ft, so 1 ft = $\frac{1}{5280}$ mi. The information is summarized as follows:

$$1 \text{ in.} = \frac{1}{12} \text{ ft}$$

$$1 \text{ ft} = \frac{1}{3} \text{ yd}$$

$$1 \text{ ft} = \frac{1}{5280} \text{ mi}$$

To solve the problem 60 in. = ________ ft, we can now use a procedure similar to that used in Example 1. Thus,

$$60 \text{ in.} = 60 \cdot \left(\frac{1}{12} \text{ ft}\right) = \frac{60}{12} \text{ ft} = 5 \text{ ft}$$

Another method of conversions involves **unit fractions**—that is, fractions that equal 1. For example, since

$$1 \text{ ft} = 12 \text{ in.},$$

$$1 = \frac{12 \text{ in.}}{\text{ft}} \quad \text{and} \quad 1 = \frac{\text{ft}}{12 \text{ in.}}$$

Answers to PROBLEMS

1. 108 in.

are both unit fractions. To use unit fractions to convert 60 inches to feet, we use the second of these fractions (because it has the desired units in the numerator). Thus,

$$60 \text{ in.} = 60 \text{ in.} \cdot 1 = 60 \text{ in.} \cdot \frac{\text{ft}}{12 \text{ in.}} = \frac{60}{12} \text{ ft} = 5 \text{ ft}$$

We shall work the examples using both the substitution and the unit fraction methods.

EXAMPLE 2 Converting inches to feet

76 in. = _______ ft

SOLUTION

Method 1. $76 \text{ in.} = 76 \cdot \frac{1}{12} \text{ ft} = \frac{76}{12} \text{ ft} = 6\frac{1}{3} \text{ ft}$

Thus, 76 in. = $6\frac{1}{3}$ ft.

Method 2. $76 \text{ in.} = 76 \text{ in.} \cdot 1 = 76 \text{ in.} \cdot \frac{\text{ft}}{12 \text{ in.}} = \frac{76}{12} \text{ ft} = 6\frac{1}{3} \text{ ft}$

PROBLEM 2

40 in. = _______ ft

EXAMPLE 3 Converting inches to feet

11 in. = _______ ft

SOLUTION

Method 1. $11 \text{ in.} = 11 \cdot \frac{1}{12} \text{ ft} = \frac{11}{12} \text{ ft}$

Thus, 11 in. = $\frac{11}{12}$ ft.

Method 2. $11 \text{ in.} = 11 \text{ in.} \cdot 1 = 11 \text{ in.} \cdot \frac{\text{ft}}{12 \text{ in.}} = \frac{11}{12} \text{ ft}$

PROBLEM 3

7 in. = _______ ft

EXAMPLE 4 Converting feet to yards

26 ft = _______ yd

SOLUTION

Method 1. $26 \text{ ft} = 26 \cdot \frac{1}{3} \text{ yd} = \frac{26}{3} \text{ yd} = 8\frac{2}{3} \text{ yd}$

Thus, 26 ft = $8\frac{2}{3}$ yd.

Method 2. Since 3 ft = 1 yd, $1 = \frac{\text{yd}}{3 \text{ ft}}$. Thus,

$26 \text{ ft} = 26 \text{ ft} \cdot 1 = 26 \text{ ft} \cdot \frac{\text{yd}}{3 \text{ ft}} = \frac{26}{3} \text{ yd} = 8\frac{2}{3} \text{ yd}$

PROBLEM 4

38 ft = _______ yd

EXAMPLE 5 Converting feet to miles

15,840 ft = _______ mi

SOLUTION

Method 1. $15{,}840 \text{ ft} = 15{,}840 \cdot \frac{1}{5280} \text{ mi} = \frac{15{,}840}{5280} \text{ mi} = 3 \text{ mi}$

Thus, 15,840 ft = 3 mi.

Method 2. $15{,}840 \text{ ft} = 15{,}840 \text{ ft} \cdot 1 = 15{,}840 \text{ ft} \cdot \frac{\text{mi}}{5280 \text{ ft}}$

$= \frac{15{,}840}{5280} \text{ mi} = 3 \text{ mi}$

PROBLEM 5

10,560 ft = _______ mi

EXAMPLE 6 Converting feet to inches

2 ft = _______ in.

SOLUTION

Method 1. $2 \text{ ft} = 2 \cdot 12 \text{ in.} = 24 \text{ in.}$

Thus, 2 ft = 24 in.

Method 2. $2 \text{ ft} = 2 \text{ ft} \cdot 1 = 2 \text{ ft} \cdot \frac{12 \text{ in.}}{\text{ft}} = 24 \text{ in.}$

PROBLEM 6

4 ft = _______ in.

Answers to PROBLEMS

2. $3\frac{1}{3}$ ft **3.** $\frac{7}{12}$ ft **4.** $12\frac{2}{3}$ yd

5. 2 mi **6.** 48 in.

B › Applications Involving U.S. Units of Length

EXAMPLE 7 Converting height from feet to inches

How tall are you? Probably not as tall as Angus MacAskill, the tallest "true" (nonpathological) giant, who measured 7.75 feet. How many inches is that?

SOLUTION

Method 1. $7.75\text{ ft} = 7.75 \cdot 12\text{ in.} = 93\text{ in.}$

Method 2. $7.75 = 7.75\text{ ft} \cdot 1$

$= 7.75\text{ ft} \cdot \frac{12\text{ in.}}{\text{ft}} = 93\text{ in.}$

$$\begin{array}{r} 7.75 \\ \times\ 12 \\ \hline 15\ 50 \\ 77\ 5\ \ \\ \hline 93.00 \end{array}$$

PROBLEM 7

One of the tallest living humans was John Carroll, who measured 8.75 feet tall. How many inches is that?

Before you try the exercises, let us talk about estimation. Look at Example 1, where we are converting 2 yards to inches. Since the yard contains many inches (36 to be exact), your estimated answer must be much larger than 2. As a matter of fact, if you remembered that one yard is 36 inches, it would be exactly twice the 36 or $2 \cdot 36 = 72$ inches as in the example.

On the other hand, in Example 5 we are converting 15,840 feet to miles. Since one foot is much smaller than a mile ($\frac{1}{5280}$ of a mile, to be exact), your estimated answer must be much smaller than 15,840, a fraction of it, $\frac{1}{5280}$ of the 15,840, to be exact. Now, try the exercises and remember to estimate before you give the final answer!

› Exercises 7.1

‹ A › Converting U.S. Units of Length

In Problems 1–30, fill in the blank.

1. 4 yd = ________ in.
2. 5 yd = ________ in.
3. 2.5 yd = ________ in.
4. 1.5 yd = ________ in.
5. $2\frac{1}{3}$ yd = ________ in.
6. $1\frac{1}{4}$ yd = ________ in.
7. 50 in. = ________ ft
8. 72 in. = ________ ft
9. 84 in. = ________ ft
10. 30 in. = ________ ft
11. 3 in. = ________ ft
12. 4 in. = ________ ft
13. 9 in. = ________ ft
14. 5 in. = ________ ft
15. 30 ft = ________ yd
16. 33 ft = ________ yd
17. 37 ft = ________ yd
18. 32 ft = ________ yd
19. 5280 ft = ________ mi
20. 26,400 ft = ________ mi
21. 4 ft = ________ in.
22. 5 ft = ________ in.
23. 1 yd = ________ ft
24. 3 yd = ________ ft
25. 1 mi = ________ ft
26. 2 mi = ________ ft
27. 1 mi = ________ yd
28. 3 mi = ________ yd
29. 1760 yd = ________ mi
30. 3520 yd = ________ mi

‹ B › Applications Involving U.S. Units of Length

31. *Refrigerator* The opening for the refrigerator in a new house is 3 ft wide.

a. How many inches is that?

b. If the refrigerator is 33 in. wide, will it fit?

32. *Football* The longest field goal in NFL competition was 189 ft. How many yards is that? By the way, the field goal was kicked by Tom Dempsey of the New Orleans Saints, and it beat the Detroit Lions 19–17 on the last play of the game on November 8, 1970. This record was tied by Jason Elam of the Denver Broncos on October 25, 1998.

Answers to PROBLEMS

7. 105 in.

33. *Stove* The opening for the stove in a new home is $2\frac{1}{2}$ ft wide.

a. How many inches is that?

b. Would a 28 in. wide stove fit?

34. *Baseball* Mickey Mantle hit the longest officially measured home run in a major league game, 565 ft. How many yards is that?

35. *Tibet* Lhasa, the capital of Tibet, is 12,087 ft above sea level. How many yards is that?

36. *Battleships* The U.S.S. *New Jersey* is the longest battleship, 888 ft. How many yards is that?

37. *Cars* The longest production car was the Bugatti Royale Type 41. It was 22 feet 1 in. long. How many inches is that?

38. *Bananas* A banana must be 8 in. long to qualify for the Chiquita label. How many feet is that?

39. *Horse races* Horse races are measured in furlongs. If a furlong is 220 yd, what is the length in miles of an 8-furlong race?

40. *Height* A woman is 5 ft 8 in. in height. How many inches is that?

41. *Dog jumps* One of the highest dog jumps was made by Young Sabre, a German shepherd, who scaled an 11.75 ft wall. How many inches is that?

42. *Birds* The largest flying bird (prehistoric) had a wing span of 24 ft. How many yards is that?

43. *Height* The average woman is 5 ft 4 in. tall. How many inches is that?

44. *Height* The average man is 5 ft 9 in. tall. How many inches is that?

45. *Snakes* A snake's average speed is 2 mi/hr. How many feet per hour is that?

46. *Plants* Water rises inside a plant's stem an average of 4 ft every hour.

a. How many inches is that?

b. If a plant is 72 in. tall, how long would it take for water to travel from the bottom to the top of the plant?

47. *Walking speed* The average urbanite walks about 6 ft per second.

a. How many yards per second is that?

b. How many yards would an urbanite go in one minute?

48. *Walking speed* The average urbanite walks about 2 yd per second.

a. How many seconds would it take to go 100 yd?

b. How many seconds would it take to go 440 yd?

49. *Dessert* The longest banana split ever made was 4 mi 686 yd long. How many feet is that? However, a 4.55-mile banana split was made in Selinsgrove, Pennsylvania, in 1988. Which is longer? (*Hint:* 0.55 miles is 0.55×5280 ft)

50. *Food* One of the longest sausages recorded was 9.89 mi long. How many feet is that? (Answer to the nearest foot.) The Nowicki Sausage Show made a sausage 8773 ft long. How many miles is that? Is it longer than the 9.89 mi sausage?

51. *Cheetah speed* The cheetah is believed to be the fastest animal on land. It can run as fast as 102 ft per second. How many yards per second is that?

52. *Antelope speed* An antelope can run at 87 ft per second. How many yards per second is that?

53. *100-yard dash* Houston McTear, who ran for Baker High School in Florida during the early 1970s, is credited with being the only man to ever run the 100-yard dash in 9 seconds. How many feet per second is that? Write the answer as a mixed number.

54. *Travel* You are traveling along and you see this sign. How many yards away is the road closure? Write the answer as a mixed number.

55. *Travel* You are traveling on the Kentucky Parkway but it ends in 2500 ft.

a. How many yards is that? Write the answer as a mixed number.

b. If you are traveling at 60 miles per hour, you are moving at about 29 yd/sec. Can you prove that?

c. If you are traveling at 60 miles per hour and the Parkway ends in 2500 ft, in how many seconds will the Parkway end? Answer to the nearest second.

Using Your Knowledge

World Records When the measurement of an item is used in setting world records, the results and claims can be varied. For example, if you research the "longest sausage," several claims surface. We report some of the claims; you decide which one is the longest! Convert each measurement to feet first.

56. A 28-mi, 1354-yd sausage in Ontario, Canada (1995).

57. The Sheffield, England, sausage, 36.75 mi long (Oct. 2000).

58. A super-sausage reported on the Internet, 13 mi long http://website.lineone.net

Write On

59. Write in your own words how you think the names of the units for the American system (inches, yards, feet, and miles) originated.

60. Write in your own words how you use estimation when converting from one American unit to another.

61. Write in your own words the two methods used in this text to convert from one American unit to another one.

62. Which of the two methods used to convert from one American unit to another do you prefer? Explain why.

Concept Checker

Fill in the blank(s) with the correct word(s), phrase or mathematical statement.

63. The **two** major **systems** of measurement are the ______________ system and the ______________ system.

64. The ______________ was the **circumference** of a person's waist.

metric

foot

meter

U.S.

yard

Mastery Test

65. Tom is 6.5 inches tall. Convert his height to inches.

In Problems 66–72, fill in the blank.

66. 6 ft = __________ in.

67. 31,680 ft = __________ mi

68. 3 mi = __________ ft

69. 25 ft = __________ yd

70. 11 in. = __________ ft

71. 75 in. = __________ ft

72. 31 ft = __________ yd

Skill Checker

Perform the indicated operations.

73. $4232 \div 1000$

74. $3 \div 10$

75. $83.5 \cdot 100$

76. $340 \cdot 10$

77. $100 \cdot 0.465$

7.2 Length: The Metric System

Objective

You should be able to:

A › Convert a metric unit of length to an equivalent metric unit of length.

To Succeed, Review How To . . .

1. Multiply by powers of 10 (10, 100, 1000, etc.). (p. 213)
2. Divide by powers of 10. (p. 217)

Getting Started

In the cartoon the girl is very upset about thinking metric. Actually, the metric system is used by most nations in the world, with one notable exception—the United States. The system was invented many years ago, and it divides the distance from the Earth's equator to the North Pole into 10 million equal parts, each called a **meter.** (The official definition of a meter is 1,650,763.73 wavelengths of the orange-red light of the element krypton.) The meter is about 39.37 in., that is, about $3\frac{1}{2}$ in. longer than a yard.

A › Converting Metric Units of Length

The number 10 is very important in the metric system, because each unit of length in the system is 10 times longer or 10 times shorter than the next measuring unit. The following table shows the relationship between some metric units of length. You should memorize these names and abbreviations.

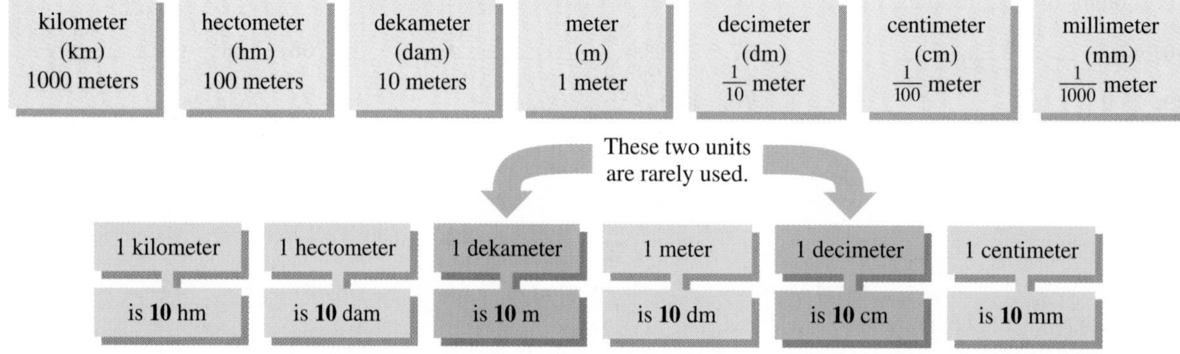

How would the metric system relate to you? When stating the distance between two cities, for example, you now use miles. In the metric system you would use kilometers. Smaller linear dimensions, such as tool sizes, would be measured in centimeters or millimeters. As in the American system, measurements in the metric system involve a

number (whole, fraction, or decimal) and a **unit.** The ruler shows one of the units used in the metric system: the centimeter. For comparison purposes, 1 inch = 2.54 centimeters.

As before, to change from one unit to another, we simply substitute the correct equivalence. For example, to find out how many meters there are in 2 km, that is, to find 2 km = ________ m, we proceed as follows:

$$2\text{ km} = 2 \cdot 1000\text{ m} = 2000\text{ m}$$

We can do the problem using unit fractions. Since there are 1000 m in 1 km,

$$1 = \frac{1000\text{ m}}{\text{km}}$$

so

$$2\text{ km} = 2\text{ km} \cdot 1 = 2\cancel{\text{km}} \cdot \frac{1000\text{ m}}{\cancel{\text{km}}} = 2000\text{ m}$$

As before, we shall work the examples using the substitution and the unit fraction methods.

EXAMPLE 1 Converting kilometers to meters

3 km = ________ m

SOLUTION

Method 1. $3\text{ km} = 3 \cdot 1000\text{ m} = 3000\text{ m}$

Thus, 3 kilometers = 3000 meters.

Method 2. $3\text{ km} = 3\text{ km} \cdot 1 = 3\cancel{\text{km}} \cdot \frac{1000\text{ m}}{\cancel{\text{km}}} = 3000\text{ m}$

PROBLEM 1

4 km = ________ m

To convert from a smaller unit to a larger unit in the metric system, we use the following table:

$$1\text{ m} = \frac{1}{1000}\text{ km}$$

$$1\text{ m} = \frac{1}{100}\text{ hm}$$

$$1\text{ m} = \frac{1}{10}\text{ dam}$$

Now we can convert meters to kilometers. For example, to find out how many kilometers there are in 3254 meters, we have to solve

3254 m = ________ km

Method 1. $3254\text{ m} = 3254 \cdot \left(\frac{1}{1000}\text{ km}\right)$

$$= \frac{3254}{1000}\text{ km} = 3.254\text{ km}$$

Thus, 3254 meters = 3.254 kilometers.

Remember that to divide by 1000, we simply move the decimal point three places to the left. Thus

$$\frac{3254}{1000} = 3.254$$

3 zeros 3 places

Answers to PROBLEMS

1. 4000 m

Method 2. Since

$$1 = \frac{\text{km}}{1000\text{ m}}$$

$$3254\text{ m} = 3254\text{ m} \cdot 1 = 3254\text{ m} \cdot \frac{\text{km}}{1000\text{ m}} = 3.254\text{ km}$$

EXAMPLE 2 Converting meters to dekameters

2 m = ________ dam

SOLUTION

Method 1. $2\text{ m} = 2 \cdot \frac{1}{10}\text{ dam} = \frac{2}{10}\text{ dam} = 0.2\text{ dam}$

Thus, 2 meters = 0.2 dekameters.

Method 2. $2\text{ m} = 2\text{ m} \cdot 1 = 2\text{ m} \cdot \frac{\text{dam}}{10\text{ m}} = \frac{2}{10}\text{ dam} = 0.2\text{ dam}$

PROBLEM 2

3 m = ________ dam

EXAMPLE 3 Converting dekameters to meters

340 dam = ________ m

SOLUTION

Method 1. $340\text{ dam} = 340 \cdot 10\text{ m} = 3400\text{ m}$

Thus, 340 dekameters = 3400 meters.

Method 2. $340\text{ dam} = 340\text{ dam} \cdot 1 = 340\text{ dam} \cdot \frac{10\text{ m}}{\text{dam}} = 3400\text{ m}$

PROBLEM 3

4 dam = ________ m

EXAMPLE 4 Converting meters to centimeters

83.5 m = ________ cm

SOLUTION

Method 1. $83.5\text{ m} = 83.5 \cdot 100\text{ cm} = 8350\text{ cm}$

Thus, 83.5 meters = 8350 centimeters.

Method 2. $83.5\text{ m} = 83.5\text{ m} \cdot 1 = 83.5\text{ m} \cdot \frac{100\text{ cm}}{\text{m}} = 8350\text{ cm}$

PROBLEM 4

215 m = ________ cm

EXAMPLE 5 Converting centimeters to meters

397 cm = ________ m

SOLUTION

Method 1. $397\text{ cm} = 397 \cdot \frac{1}{100}\text{ m} = 3.97\text{ m}$

Thus, 397 centimeters = 3.97 meters.

Method 2. $397\text{ cm} = 397\text{ cm} \cdot 1 = 397\text{ cm} \cdot \frac{\text{m}}{100\text{ cm}} = 3.97\text{ m}$

PROBLEM 5

581 cm = ________ m

After all these examples you probably have noticed that changing from one unit to another in the metric system can be done simply by moving the decimal point. Thus, if we keep the following table in mind, we can do these conversions mentally.

km	hm	dam	m	dm	cm	mm
1000 m	100 m	10 m	1 m	$\frac{1}{10}$ m	$\frac{1}{100}$ m	$\frac{1}{1000}$ m

Answers to PROBLEMS

2. 0.3 dam **3.** 40 m
4. 21,500 cm **5.** 5.81 m

Thus, to solve 315 cm = ________ km, we think, "To go from centimeters to kilometers in the table, we must move five places left." So we move the decimal point five places to the *left,* like this:

315 cm = .00315 km = 0.00315 km

Similarly, to solve 58.2 dam = ________ mm, we notice that to go from dekameters to millimeters, we must move four places to the *right*. Thus,

58.2 dam = 58 2000 mm = 582,000 mm

If you need practice on this, you can go back and rework all the examples using this method.

› Exercises 7.2

‹ A › Converting Metric Units of Length In Problems 1–20, use the table to fill in the blank.

km	hm	dam	m	dm	cm	mm
1000 m	100 m	10 m	1 m	$\frac{1}{10}$ m	$\frac{1}{100}$ m	$\frac{1}{1000}$ m

1. 5 km = ______ m **2.** 4.2 km = ______ m **3.** 1877 m = ______ km **4.** 157 m = ______ km

5. 4 dm = ______ m **6.** 8 dm = ______ m **7.** 49 m = ______ dm **8.** 1.6 m = ______ dm

9. 182 cm = ______ m **10.** 43 cm = ______ m **11.** 22 m = ______ cm **12.** 4.5 m = ______ cm

13. 3 m = ______ mm **14.** 0.35 m = ______ mm **15.** 2358 mm= ______ m **16.** 425 mm = ______ cm

17. 30 cm = ______ mm **18.** 2.4 cm = ______ mm **19.** 67 mm = ______ cm **20.** 184 mm = ______ cm

› Web IT go to mhhe.com/bello for more lessons

››› *Applications*

In Problems 21–25, select the answers that are most nearly correct.

21. *Height of a basketball player* The height of a professional basketball player is:

a. 200 mm. **b.** 200 m. **c.** 200 cm.

22. *Living room dimensions* The dimensions of the living room in an ordinary home are:

a. 4 m by 5 m. **b.** 4 cm by 5 cm. **c.** 4 mm by 5 mm.

23. *Diameter of aspirin* The diameter of an aspirin tablet is:

a. 1 cm. **b.** 1 mm. **c.** 1 m.

24. *Length of 100-yd dash* The length of the 100-yd dash is about:

a. 100 cm. **b.** 100 mm. **c.** 100 m.

25. *Length of a pencil* The length of an ordinary lead pencil is:

a. 19 mm. **b.** 19 cm. **c.** 19 m.

26. *Bed length* A bed is 210 cm long. How many meters is that?

27. *Diameter of a tablet* The diameter of a vitamin C tablet is 6 mm. How many centimeters is that?

28. *Race length* The length of a certain race is 1.5 km. How many meters is that?

29. *Swimming pool depth* The depth of a swimming pool is 1.6 m. How many centimeters is that?

30. *Noah's ark* Dr. James Strange of the University of South Florida wishes to explore Mount Ararat in Turkey, searching for Noah's ark. According to the book of Genesis, the dimensions of the ark are as follows: length, 300 cubits; breadth, 50 cubits; height, 30 cubits. If a cubit is 52.5 cm, give each dimension in meters.

31. *Stride length* The average stride (step) of a runner is about 1 meter in length. If a runner has entered the 5 km (usually written as the 5K) race, how many steps does the runner take to complete the race?

32. *Length of a race* A runner took 10,000 steps to complete a race. If the average stride (step) for the runner was about 1 meter in length, how long was the race?

33. *Height of animals* Here are the average heights of three animals in Sedgwick County zoo, Wichita, Kansas:

a. 1.1 meters, 3.2 meters and 1.5 meters. Write the heights in cm.

b. The animals are an African elephant, a chimp and a zebra. What is the height of each in centimeters?

Source: http://tinyurl.com

34. *Heights of birds* Here are the average heights of three birds in Sedgwick County zoo in Wichita, Kansas:

a. 88 cm, 150 cm, and 60 cm. Write the heights in meters.

b. The birds are a bald eagle, a trumpeter swan, and a brush turkey. What is the height of each in meters?

Source: http://tinyurl.com

35. *Lengths of reptiles* Here are the lengths of three reptiles in the Sedgwick County zoo in Wichita, Kansas:

a. 240 cm, 76 cm, and 17 cm. Write the lengths in meters.

b. The reptiles are a copperhead, a Madagascar boa, and a barred tiger salamander. What is the length of each in meters?

Using Your Knowledge

36. *Human Bones* The *humerus* is the bone in a person's upper arm. With this bone as a clue, an anthropologist can tell about how tall a person was. If the bone is that of a female, then the height of the person is about

$$(2.75 \times \text{humerus length}) + 71.48 \text{ cm}$$

Suppose the humerus of a female was found to be 31 cm long. About how tall was the person in centimeters?

In Problem 37, match each item in the first column with an appropriate measure in the second column.

37.

i. A letter-sized sheet of paper	**a.** 20 mm × 25 mm
ii. A newspaper	**b.** 54 mm × 86 mm
iii. A credit card	**c.** 70 mm × 150 mm
iv. A regular bank check	**d.** 21.5 cm × 28 cm
v. A postage stamp	**e.** 35 cm × 56 cm

Write On

38. Explain in your own words why it is easier to change from one unit to another in the metric system than it is in the U.S. system.

39. Name three different metric units of length you have seen in your daily life and what they were measuring.

Concept Checker

Fill in the blank(s) with the correct word(s), phrase, or mathematical statement.

40. **Measurements** in the **metric system** involve a ________ and a ________.

41. To **change** from **one unit to another** simply ________ the correct equivalency.

variable	**abbreviation**
substitute	**prefix**
number	**unit**

Mastery Test

In Problems 42–47, fill in the blank.

42. 457 cm = ______ m

43. 92.4 m = ______ cm

44. 380 dam = ______ m

45. 3 m = ______ dam

46. 9 km = ______ m

47. 2300 m = ______ km

Skill Checker

In Problems 48–53, find the product.

48. $60 \cdot 2.54$

49. $72 \cdot 2.54$

50. $100 \cdot 0.914$

51. $50 \cdot 1.6$

52. $70 \cdot 0.62$

53. $10 \cdot 0.4$

7.3 Length: U.S. to Metric and Metric to U.S. Conversions

Objectives

You should be able to:

A Convert a U.S. (American) unit of length to a metric unit of length and vice versa.

B Solve applications involving changing units from U.S. (American) to metric and vice versa.

To Succeed, Review How To . . .

1. Multiply and divide decimals. (pp. 212–216)
2. Perform basic operations with fractions and decimals. (pp. 130–136, 151–156, 203–206, 212–216)

Getting Started

Metric Sayings What do you think would happen if the United States adopted the metric system right this minute? Many people believe that they would have to be making conversions from American to metric and from metric to American constantly. This is just not true! For example, you probably buy soda or spring water in one- and two-liter bottles, but this does not require that you convert liters to quarts. You may have a foreign car that is totally metric, but you do not need to convert tire or engine sizes to the American system. However, if you insist on converting, here are a few sayings you may want to convert to the metric system:

1. A miss is as good as ______ km.
   (a mile)
2. I wouldn't touch it with a ______ m pole.
   (10-foot)
3. He was so stubborn he wouldn't give ______ cm.
   (an inch)
4. Walk ______ km in my moccasins.
   (a mile)

How would you do that? We give you a table of conversions next, so you know!

A › Converting between U.S. System and Metric System

Here is the table you need to convert length measurements between the U.S. system and the metric system and vice versa.

U.S. System to Metric Conversions

1 in. = 2.54 cm*	1 cm = 0.4 in.
1 yd = 0.914 m	1 m = 1.1 yd
1 mi = 1.6 km	1 km = 0.62 mi

To make the conversions, you can use the same methods that we discussed in Sections 7.1 and 7.2.

EXAMPLE 1 Converting feet to centimeters

5 ft = ______ cm

SOLUTION Remember, 1 ft = 12 in.

Method 1. 5 ft = 60 in. = 60 · 2.54 cm = 152.4 cm

Method 2. $5 \text{ ft} = 60 \text{ in.} = 60 \text{ in.} \cdot 1 = 60 \cancel{\text{in.}} \cdot \frac{2.54 \text{ cm}}{\cancel{\text{in.}}} = 152.4 \text{ cm}$

Thus, 5 ft = 152.4 cm.

PROBLEM 1

7 ft = ______ cm

EXAMPLE 2 Converting yards to meters

200 yd = ______ m

SOLUTION From the table, 1 yd = 0.914 m

Method 1. 200 yd = 200 yd = 200 · 0.914 m = 182.8 m

Method 2. $200 \text{ yd} = 200 \text{ yd} = 200 \text{ yd} \cdot 1 = 200 \cancel{\text{yd}} \cdot \frac{0.914 \text{ m}}{\cancel{\text{yd}}} = 182.8 \text{ m}$

Thus, 200 yd = 182.8 m.

PROBLEM 2

300 yd = ______ m

EXAMPLE 3 Converting miles to kilometers

60 mi = ______ km

SOLUTION From the table, 1 mi = 1.6 km

Method 1. 60 mi = 60 mi = 60 · 1.6 km = 96 km

Method 2. $60 \text{ mi} = 60 \text{ mi} = 60 \text{ mi} \cdot 1 = 60 \cancel{\text{mi}} \cdot \frac{1.6 \text{ km}}{\cancel{\text{mi}}} = 96 \text{ km}$

Thus, 60 mi = 96 km.

PROBLEM 3

70 mi = ______ km

EXAMPLE 4 Converting centimeters to inches

300 cm = ______ in.

SOLUTION From the table, 1 cm = 0.4 in.

Method 1. 300 cm = 300 cm = 300 · 0.4 in. = 120 in.

Method 2. $300 \text{ cm} = 300 \text{ cm} = 300 \text{ cm} \cdot 1 = 300 \cancel{\text{cm}} \cdot \frac{0.4 \text{ in.}}{\cancel{\text{cm}}} = 120 \text{ in.}$

Thus, 300 cm = 120 in.

PROBLEM 4

200 cm = ______ in.

*The inch is defined as 2.54 cm. The other measurements are approximate.

Answers to PROBLEMS

1. 213.36 cm **2.** 274.2 m
3. 112 km **4.** 80 in.

EXAMPLE 5 **Converting meters to yards**

200 m = ______ yd

SOLUTION From the table, 1 m = 1.1 yd

Method 1. 200 m = 200 m = 200 · 1.1 yd = 220 yd

Method 2. 200 m = 200 m = 200 m · 1 = 200 m · $\frac{1.1 \text{ yd}}{\text{m}}$ = 220 yd

Thus, 200 m = 220 yd.

PROBLEM 5

300 m = ______ yd

EXAMPLE 6 **Converting kilometers to miles**

90 km = ______ mi

SOLUTION From the table, 1 km = 0.62 mi

Method 1. 90 km = 90 km = 90 · 0.62 mi = 55.8 mi

Method 2. 90 km = 90 km = 90 km · 1 = 90 km · $\frac{0.62 \text{ mi}}{\text{km}}$ = 55.8 mi

Thus, 90 km = 55.8 mi.

PROBLEM 6

70 km = ______ mi

Before you go on to the applications, remember your estimation:

Inches are longer than centimeters, so when you convert inches to centimeters, the answer should be **larger** than the given number. Yards are shorter than meters, so when you convert yards to meters, the answer should be **smaller** than the given number. Miles are longer than kilometers, so when you convert miles to kilometers, the answer should be **larger** than the given number.

B › Applications Involving U.S. and Metric Units of Length

EXAMPLE 7 **Converting feet to centimeters**

A man is 6 ft tall. How many centimeters is that?

SOLUTION 1 ft = 12 in.

Method 1. 6 ft = 72 in. = 72 · 2.54 cm = 182.88 cm

Method 2. 6 ft = 72 in. = 72 in. · 1 = 72 in. · $\frac{2.54 \text{ cm}}{\text{in.}}$ = 182.88 cm

PROBLEM 7

A woman is 5 feet tall. How many centimeters is that?

EXAMPLE 8 **Converting yards to meters**

A football field is 100 yd long. How many meters is that?

SOLUTION

Method 1. 100 yd = 100 · 0.914 m = 91.4 m

Method 2. 100 yd = 100 yd · 1 = 100 yd · $\frac{0.914 \text{ m}}{\text{yd}}$ = 91.4 m

PROBLEM 8

A pool is 50 yards long. How many meters is that?

EXAMPLE 9 **Converting miles to kilometers**

A car is traveling at 50 mi/hr. How many kilometers per hour is that?

SOLUTION

Method 1. 50 mi = 50 · 1.6 km = 80 km

Method 2. 50 mi = 50 mi · 1 = 50 mi · $\frac{1.6 \text{ km}}{\text{mi}}$ = 80 km

Thus 50 mi/hr is equivalent to 80 km/hr.

PROBLEM 9

A car is traveling at 30 miles per hour. How many kilometers per hour is that?

Answers to PROBLEMS

5. 330 yd **6.** 43.4 mi **7.** 152.40 **8.** 45.7 **9.** 48

EXAMPLE 10 Converting meters to yards

An Olympic pool is 100 m long. How many yards is that?

SOLUTION

Method 1. $100 \text{ m} = 100 \cdot 1.1 \text{ yd} = 110 \text{ yd}$

Method 2. $100 \text{ m} = 100 \text{ m} \cdot 1 = 100 \text{ m} \cdot \frac{1.1 \text{ yd}}{\text{m}} = 110 \text{ yd}$

PROBLEM 10

A track race in the Olympics is the 200-meter event. How many yards is that?

EXAMPLE 11 Converting kilometers to miles

A car is traveling at 70 km/hr. How many miles per hour is that?

SOLUTION

Method 1. $70 \text{ km} = 70 \cdot 0.62 \text{ mi} = 43.40 \text{ mi}$

Method 2. $70 \text{ km} = 70 \text{ km} \cdot 1 = 70 \text{ km} \cdot \frac{0.62 \text{ mi}}{\text{km}} = 43.40 \text{ mi}$

Thus, the car is traveling at 43.40 mi/hr.

PROBLEM 11

A car is traveling 80 kilometers per hour. How many miles per hour is that?

EXAMPLE 12 Converting centimeters to inches

A cigarette is 10 cm long. How many inches is that?

SOLUTION

Method 1. $10 \text{ cm} = 10 \cdot 0.4 \text{ in.} = 4 \text{ in.}$

Method 2. $10 \text{ cm} = 10 \text{ cm} \cdot 1 = 10 \text{ cm} \cdot \frac{0.4 \text{ in.}}{\text{cm}} = 4 \text{ in.}$

PROBLEM 12

A ruler is 30 centimeters long. How many inches is that?

Exercises 7.3

Web IT go to mhhe.com/bello for more lessons

⟨A⟩ Converting between U.S. System and Metric System

In Problems 1–12, fill in the blanks.

1. 4 ft = ______ cm
2. 3 ft = ______ cm
3. 30 yd = ______ m
4. 20 yd = ______ m
5. 90 mi = ______ km
6. 100 mi = ______ km
7. 20 cm = ______ in.
8. 50 cm = ______ in.
9. 600 m = ______ yd
10. 500 m = ______ yd
11. 10 km = ______ mi
12. 400 km = ______ mi

⟨B⟩ Applications Involving U.S. and Metric Units of Length

13. *Football* Steve O'Neal of the New York Jets kicked a 98-yd punt on September 21, 1969. How many meters is that? (Give answer to one decimal place.)

14. *Largest swimming pool* The largest swimming pool in Casablanca, Morocco, is 82 yd wide. How many meters is that? (Give answer to one decimal place.)

15. *Mount Everest* Mount Everest is 8848 m high. How many yards is that? (Give answer to the nearest hundred.)

16. *Speed limit* The speed limit on a highway is 55 mi/hr. How many kilometers per hour is that?

17. *Speed limit in Europe* The speed limit on European highways is 100 km/hr. How many miles per hour is that?

18. *Author's dimensions* The author of this book once had measurements of 46-30-36 (in inches). How would you write that in centimeters? (Use 1 in. = 2.5 cm.)

19. *Miss World contest* In a Miss World contest the winner had measurements of 90-60-90 (in centimeters). What is that in inches?

20. *TV screen dimensions* The screen of a TV set measures 24 in. diagonally. How many centimeters is that?

Answers to PROBLEMS

10. 220 **11.** 49.60 **12.** 12

21. *Venice* The city of Venice is built on wooden planks that are sinking at the rate of seven centimeters per century. How many inches per century is that?

22. *Venice* A recent study by an American group, however, estimated that Venice sank a whopping 24 centimeters in the last century alone! How many inches is that?

Source: http://www.classbrain.com.

23. *Shanghai* Parts of Shanghai are now sinking at a rate of 1 inch a year, largely as a result of a massive building boom there over the last 10 years. How many centimeters per year is that?

24. *Louisiana* The sinking in Louisiana has run anywhere from 6 to 20 inches over the past 20 years, according to Roy Dokka, a professor at Louisiana State University's Center for Geoinformatics. How much has Louisiana been sinking in centimeters over the last 20 years?

> > > Using Your Knowledge

25. *Airplanes* A Boeing 747 requires about 1900 m for a takeoff runway. About how many miles is that?

26. *Egg toss* One of the longest recorded distances for throwing (and catching) a raw hen's egg without breaking it is 316 ft 51 in. About how many meters is that?

27. *Glider* You have probably heard the expression, "Hang in there!" Well, Rudy Kishazy did just that. He hung onto a glider that took off from Mount Blanc and landed 35 min later at Servoz, France, a distance of 15 mi. How many kilometers is that?

> > > Write On

28. Write in your own words how you use estimation to do the calculations in this section.

29. If you are converting miles to kilometers, will the result be larger or smaller than the number of miles originally given? Explain.

> > > Concept Checker

Fill in the blank(s) with the correct word(s), phrase, or mathematical statement.

30. When **converting inches to centimeters,** the answer should be ______________ **(larger, smaller)** than the original number.

31. When **converting yards to meters,** the answer should be ______________ **(larger, smaller)** than the original number.

equal

larger

smaller

> > > Mastery Test

32. Cuban José Castelar Cairo was officially recorded in the *Guinness Book of World Records* for the longest-ever hand-rolled Havana cigar: about 13.5 meters long. How many inches is that?

33. A car is traveling at 40 km/hr. How many miles per hour is that?

34. A pool is 50 m long. How many yards is that?

35. A car is traveling at 60 mph. How many kilometers per hour is that?

36. A soccer field can be as long as 130 yards. How many meters is that?

37. High schooler Ha Seung-jin of Korea is 7.15 ft tall. How many meters is that? (Answer to three decimal places.)

In Problems 38–43, fill in the blank.

38. 15 ft = ________ cm

39. 800 yd = ________ m

40. 30 cm = ________ ft

41. 600 cm = ________ in.

42. 800 m = ________ yd

43. 20 km = ________ mi

Skill Checker

Perform the indicated operations.

44. $(1000)^2$

45. $27 \cdot \frac{1}{9}$

46. $5 \cdot 4840$

47. $4 \cdot 2.47$

7.4 Area: U.S., Metric, and Conversions

Objectives

You should be able to:

A Convert from one metric unit of area to another.

B Convert from one U.S. unit of area to another.

C Convert units of area from metric to U.S. (American) and vice versa.

D Solve applications involving U.S. and metric units of area.

To Succeed, Review How To . . .

1. Multiply a whole number by a fraction. (p. 132)
2. Multiply and divide decimals. (pp. 212–216)

Getting Started

Suppose there was a parade and *everybody* came. How can you estimate the crowd? Robert Gillette, a reporter for the *Los Angeles Times,* estimates attendance at the Tournament of Roses Parade by using areas. First, he measures the depth of the standing room area at 23 ft. (This depth is bounded by the blue line behind which spectators must stand and by the buildings at the back of the crowd.) Then he multiplies by the 5.5-mi parade route and doubles the amount, since there are spectators on both sides of the street. So far, the calculation is

$$23 \text{ ft} \cdot 5.5 \text{ mi} \cdot 2$$

Unfortunately, the answer is in feet × miles. We can convert this to square feet if we remember that 1 mi = 5280 ft. Substituting,

$$\begin{aligned} 23 \text{ ft} \cdot 5.5 \text{ mi} \cdot 2 &= 23 \text{ ft} \cdot 5.5 \cdot 5280 \text{ ft} \cdot 2 \\ &= 1{,}335{,}840 \text{ ft}^2 \end{aligned}$$

That is, an area of 1,335,840 ft^2 is available along the parade route. Mr. Gillette finishes his calculations by assuming that each spectator occupies 2 ft^2 (2 ft thick and 1 ft wide). Dividing 2 ft^2 into 1,335,840 ft^2 gives the estimated attendance as

$$\frac{1{,}335{,}840 \text{ ft}^2}{2 \text{ ft}^2} = 667{,}920 \text{ persons}$$

A Converting from One Metric Unit of Area to Another

Sometimes we wish to convert one metric unit of area to another. To do this we substitute the correct equivalent unit as shown in the examples.

EXAMPLE 1 **Converting square meters to square centimeters**

$1\ m^2 =$ ______ cm^2

SOLUTION We proceed as follows:

$$1\ m^2 = 1 \cdot (100\ cm)^2 = 1 \cdot (100\ cm) \cdot (100\ cm) = 10{,}000\ cm^2$$

PROBLEM 1

$5\ m^2 =$ ______ cm^2

EXAMPLE 2 **Converting square kilometers to square meters**

$1\ km^2 =$ ______ m^2

SOLUTION

$$1\ km^2 = 1 \cdot (1000\ m)^2 = 1 \cdot (1000\ m) \cdot (1000\ m) = 1{,}000{,}000\ m^2$$

PROBLEM 2

$2\ km^2 =$ ______ m^2

B > Converting from One U.S. Unit of Area to Another

A similar procedure can be used when changing units in the customary U.S. system.

EXAMPLE 3 **Converting square feet to square inches**

$1\ ft^2 =$ ______ $in.^2$

SOLUTION Substituting,

$$1\ ft^2 = 1 \cdot (12\ in.)^2 = 144\ in.^2$$

PROBLEM 3

$5\ ft^2 =$ ______ $in.^2$

When converting from smaller to larger units in the customary system, it is advantageous to remember the following facts. Since 12 in. = 1 ft, 1 in. = $\frac{1}{12}$ ft. Thus, $(1\ in.)^2 = (\frac{1}{12}\ ft)^2 = (\frac{1}{12}\ ft) \cdot (\frac{1}{12}\ ft) = \frac{1}{144}\ ft^2$. That is,

$$1\ in.^2 = \frac{1}{144}\ ft^2$$

Similarly, 3 ft = 1 yd and 1 ft = $\frac{1}{3}$ yd. Thus,

$$(1\ ft)^2 = \left(\frac{1}{3}\ yd\right)^2 = \left(\frac{1}{3}\ yd\right) \cdot \left(\frac{1}{3}\ yd\right) = \frac{1}{9}\ yd^2$$

That is,

$$1\ ft^2 = \frac{1}{9}\ yd^2$$

We summarize these two facts in the table.

$1\ in.^2 = \frac{1}{144}\ ft^2$
$1\ ft^2 = \frac{1}{9}\ yd^2$

EXAMPLE 4 **Converting square inches to square feet**

$288\ in.^2 =$ ______ ft^2

SOLUTION Substituting,

$$288\ in.^2 = 288 \cdot \left(\frac{1}{144}\ ft^2\right) = 288 \cdot \frac{1}{144}\ ft^2 = 2\ ft^2$$

PROBLEM 4

$432\ in.^2 =$ ______ ft^2

Answers to PROBLEMS

1. $50{,}000\ cm^2$ **2.** $2{,}000{,}000\ m^2$

3. $720\ in.^2$ **4.** $3\ ft^2$

EXAMPLE 5 **Converting square feet to square yards**

$27 \text{ ft}^2 =$ ______ yd^2

SOLUTION

$$27 \text{ ft}^2 = 27 \cdot \frac{1}{9} \text{ yd}^2 = 3 \text{ yd}^2$$

PROBLEM 5

$36 \text{ ft}^2 =$ ______ yd^2

In the U.S. (American) system, large areas are measured in **acres.** An acre is 4840 yd^2, that is,

$$1 \text{ acre} = 4840 \text{ yd}^2$$

EXAMPLE 6 **Converting acres to square yards**

What is the area in square yards of a five-acre lot?

SOLUTION We know that

$$1 \text{ acre} = 4840 \text{ yd}^2$$

Thus, $5 \text{ acres} = 5 \cdot 4840 \text{ yd}^2 = 24{,}200 \text{ yd}^2$

PROBLEM 6

What is the area in square yards of a 20-acre lot?

C › Converting Units of Area from Metric to U.S. (American) and Vice Versa

When using metric units, we measure large areas in hectares. A **hectare** is the area of a square 100 meters on each side. Thus,

$$1 \text{ hectare} = 10{,}000 \text{ m}^2$$

EXAMPLE 7 **Converting hectares to square meters**

7 hectares = ______ m^2

SOLUTION We know that

$$1 \text{ hectare} = 10{,}000 \text{ m}^2$$

Thus, $7 \text{ hectares} = 7 \cdot 10{,}000 \text{ m}^2 = 70{,}000 \text{ m}^2$

PROBLEM 7

12 hectares = ______ m^2

The relationship between hectares and acres is as follows:

$$1 \text{ hectare} = 2.47 \text{ acres}$$

EXAMPLE 8 **Converting hectares to acres**

2 hectares = ______ acres

SOLUTION Since

$$1 \text{ hectare} = 2.47 \text{ acres}$$

$$2 \text{ hectares} = 2 \cdot 2.47 \text{ acres} = 4.94 \text{ acres}$$

PROBLEM 8

6 hectares = ______ acres

Answers to PROBLEMS

5. 4 yd^2 **6.** 96,800 yd^2

7. 120,000 m^2 **8.** 14.82 acres

D > Applications Involving U.S. and Metric Units of Area

EXAMPLE 9 **Converting acres to hectares: Deck space on cruise ships**

Freedom of the Seas, one of the world's largest cruise ships, has 40.755 acres of deck space. How many hectares is that?

SOLUTION Since

$$1 \text{ hectare} = 2.47 \text{ acres}$$
$$1 \text{ acre} = \frac{1 \text{ hectare}}{2.47}$$

and

$$\mathbf{40.755} \text{ acres} = \mathbf{40.755} \cdot \frac{1 \text{ hectare}}{2.47}$$
$$= \frac{\mathbf{40.755}}{2.47} \text{ hectares}$$
$$= 16.5 \text{ hectares}$$

PROBLEM 9

The Miami Terminal of Carnival Corporation has grown from 76 acres to 132 acres to accommodate its ships. How many hectares is that? Answer to the nearest hectare.

> Exercises 7.4

< A > Converting from One Metric Unit of Area to Another
< B > Converting from One U.S. Unit of Area to Another

In Problems 1–10, fill in the blanks.

1. $3 \text{ km}^2 = \_\_\_\_\_ \text{ m}^2$ **2.** $4 \text{ km}^2 = \_\_\_\_\_ \text{ m}^2$ **3.** $2 \text{ ft}^2 = \_\_\_\_\_ \text{ in.}^2$ **4.** $8 \text{ ft}^2 = \_\_\_\_\_ \text{ in.}^2$

5. $432 \text{ in.}^2 = \_\_\_\_\_ \text{ ft}^2$ **6.** $720 \text{ in.}^2 = \_\_\_\_\_ \text{ ft}^2$ **7.** $54 \text{ ft}^2 = \_\_\_\_\_ \text{ yd}^2$ **8.** $162 \text{ ft}^2 = \_\_\_\_\_ \text{ yd}^2$

9. $2 \text{ acres} = \_\_\_\_\_ \text{ yd}^2$ **10.** $3 \text{ acres} = \_\_\_\_\_ \text{ yd}^2$

< C > Converting Units of Area from Metric to U.S. (American) and Vice Versa

In Problems 11–14, fill in the blanks.

11. $3 \text{ hectares} = \_\_\_\_\_ \text{ acres}$ **12.** $5 \text{ hectares} = \_\_\_\_\_ \text{ acres}$ **13.** $2 \text{ hectares} = \_\_\_\_\_ \text{ m}^2$ **14.** $5 \text{ hectares} = \_\_\_\_\_ \text{ m}^2$

< D > Applications Involving U.S. and Metric Units of Area

15. *Bookstores* One of the largest bookstores is that of Barnes and Noble in New York, with 154,250 ft^2 of space. To the nearest square yard, how many square yards is that?

16. *Department stores* The world's largest store, R. H. Macy and Co., occupies 46 acres. How many square yards is that?

17. *Volkswagen plant* The Volkswagen Wolfsburg plant in Germany occupies 1730 acres. How many square yards is that?

18. *Area of Monaco* Monaco, on the south coast of France, has an area of 370 acres. How many square yards is that?

19. *Area of a house* The carpeted area of a house is 50 ft long and 30 ft wide.

a. How many square feet is that?

b. How many square yards of carpet is that?

20. *Area of Vehicle Assembly Building* The Vehicle Assembly Building (VAB) near Cape Canaveral has a floor area of 343,500 ft^2. How many acres is that? (Answer to two decimal places.)

21. *Convention space* The convention space at the Hilton Hotel in Las Vegas covers 125,000 ft^2. How many acres is that? (Answer to two decimal places.)

22. *Area of a recreation deck* The recreation deck in the roof of the Hilton Hotel in Las Vegas covers 10 acres. How many square yards is that?

Answers to PROBLEMS

9. 31 to 53

> Web IT go to mhhe.com/bello for more lessons

23. *Casinos* Circus Circus in Las Vegas covers an area of 129,000 ft². How many acres is that? (Answer to two decimal places.)

All about *Freedom* (the ship)

24. *Cruise ships* At the present time, the biggest cruise ship is the *Freedom of the Seas,* which has more than 100 acres of outdoor park, recreation, exercise, and community space! How many square yards is that?

25. *Cruise ships* You can actually buy a 474 m² "residency" on board. How many square feet in your residency? (*Hint:* $1 \text{ m}^2 \approx 11 \text{ ft}^2$.) By the way, the cost of the residency was $9,340,600.

26. *Cruise ships* You can have a smaller 125 m² place with a water view for a mere $1,154,500. How many square feet will you get for your money?

Using Your Knowledge

Area Many practical problems around the house require some knowledge of the material we have studied. For example, let us say your living room is 12 ft by 10 ft. Its area is $12 \text{ ft} \times 10 \text{ ft} = 120 \text{ ft}^2$. Since carpets are sold by the square yard, to carpet this area we need to know how many square yards we have. Here is how we do it.

From one of the tables, $1 \text{ ft}^2 = \frac{1}{9} \text{ yd}^2$. Thus,

$$120 \text{ ft}^2 = 120 \cdot \frac{1}{9} \text{ yd}^2 = \frac{120}{9} \text{ yd}^2 = 13\frac{1}{3} \text{ yd}^2$$

Hence, we need $13\frac{1}{3}$ square yards of carpet.

27. How many square yards of carpet do we need to carpet a room 12 ft by 11 ft?

28. How many square yards of carpet do we need to carpet a room 12 ft by 15 ft?

We can use these ideas outdoors, too. Suppose your lawn is 30 yd by 20 yd. Its area is

$$30 \text{ yd} \cdot 20 \text{ yd} = 600 \text{ yd}^2$$

If you wish to plant new grass in this lawn, you can buy sod—squares of grass that can be simply laid on the ground. Each sod square is approximately 1 ft². How many squares do you need? You first convert 600 yd² to square feet. Thus,

$$600 \text{ yd}^2 = 600 \cdot (3 \text{ ft})^2 = 600 \cdot (9 \text{ ft}^2) = 5400 \text{ ft}^2$$

Hence, you need 5400 squares of sod.

29. How many squares of sod do we need to cover a piece of land 50 yd by 20 yd?

30. How many squares of sod do we need to cover a piece of land 40 yd by 15 yd?

31. Wallpaper comes in a roll containing 36 ft² of paper. A wall is 12 ft by 9 ft. How many rolls do we need to paper this wall?

Write On

32. A hectometer is 100 meters. What is a hectare in terms of hectometers?

33. Which is larger and why: A square yard or a square meter?

Mastery Test

In Problems 34–40, fill in the blank.

34. 2 m² = ________ cm²

35. 54 ft² = _____ yd²

36. 576 in.² = ______ ft²

37. 3 ft² = ________ in.²

38. 5 km² = ___________ m²

39. 4 hectares = ________ acres

40. 6 hectares = ________ m²

41. What is the area in square yards of a 10-acre lot?

Skill Checker

42. Multiply $\frac{1}{2} \cdot 4$

43. Write $\frac{500}{240}$ as a mixed number.

44. Write $\frac{400}{240}$ as a mixed number.

7.5 Volume (Capacity): U.S., Metric, and Conversions

Objectives

You should be able to:

A Convert units of volume in the U.S. system to metric and vice versa.

B Convert units of volume from one metric unit to another.

C Solve applications using U.S. and metric units of volume.

To Succeed, Review How To . . .

1. Divide by powers of 10. (p. 217)
2. Multiply fractions. (pp. 130–133)

Getting Started

The Tropicana pack in the photograph holds 2 quarts, or 1892 milliliters (mL). (A **milliliter** is defined to be the volume of a cube 1 cm on each edge.) In the metric system the basic unit of volume is the *liter*. A **liter** is the volume of a cube 10 cm on each edge. The U.S. customary system uses the **quart** to measure volume. From the Tropicana pack we know that

$$2 \text{ quarts} = 1892 \text{ milliliters}$$

Thus

$$1 \text{ quart} = 946 \text{ milliliters}$$

Since a milliliter is $\frac{1}{1000}$ of a liter,

$$\boxed{1 \text{ quart} = 0.946 \text{ liter}}$$

Thus, a quart is slightly less than a liter.

A > U.S. (American) to Metric Conversion

As usual, the units used to measure volume in the U.S. (American) system are more complicated. Here they are for comparison.

U.S. (American) System	Metric System
8 ounces (oz) = 1 cup (c)	1 liter (L) = 1000 milliliters (mL)
2 cups (c) = 1 pint (pt)	1 cubic centimeter (cm^3) = 1 mL
16 ounces (oz) = 1 pint (pt)	
2 pt = 1 quart (qt) = 32 (oz)	
4 qt = 1 gallon (gal)	
1 qt = 0.946 L 1 L = 1.06 qt	

To make some conversions from the customary to the metric system, you can capitalize on the fact that 1 qt is approximately equal to 1 L (1 qt $\approx$ 1 L). For example, to solve

$$1 \text{ gal} = \_\_\_\_\_ \text{ L}$$

you can think like this:

$$1 \text{ gal} = 4 \text{ qt}$$

and since 1 qt $\approx$ 1 L, then

$$1 \text{ gal} \approx 4 \text{ L}$$

Note that $\approx$ means "is approximately equal to."

EXAMPLE 1 Converting U.S. units of volume to metric units

Fill in the blanks.

a. $\frac{1}{2}$ gal $\approx$ _____ L **b.** 8 oz $\approx$ _____ L **c.** 20 gal $\approx$ _____ L

SOLUTION

a. Since

1 gal = 4 qt

$\frac{1}{2}$ gal $= \frac{1}{2}$(4 qt) = 2 qt

Thus, $\frac{1}{2}$ gal $\approx$ 2L.

b. Since

32 oz = 1 qt

$\frac{1}{4}$(32 oz) $= \frac{1}{4}$ qt $\approx \frac{1}{4}$ L

Thus, 8 oz $\approx \frac{1}{4}$ L.

c. Since

1 gal = 4 qt

20 gal = 20 (4 qt) = 80 qt $\approx$ 80 L

Thus, 20 gal $\approx$ 80 L.

PROBLEM 1

Fill in the blanks.

a. $\frac{1}{4}$ gal $\approx$ _____ L

b. 16 oz $\approx$ _____ L

c. 10 gal $\approx$ _____ L

B > Converting from One Metric Measure to Another

As before, volume conversions in the metric system are just a matter of moving the decimal point. Here is a table to aid in the process.

Kiloliter (kL)	hectoliter (hL)	dekaliter (daL)	liter (L)	deciliter (dL)	centiliter (cL)	milliliter (mL)
1000 L	100 L	10 L	1 L	$\frac{1}{10}$ L	$\frac{1}{100}$ L	$\frac{1}{1000}$ L

Thus, to solve 5 hL = _____ L, we see that to go from hectoliters to liters in the table, we must move *two* places to the right. So we move the decimal point in 5 *two* places to the right, obtaining

$$5 \text{ hL} = 500 \text{ L}$$

That is, 5 hL = 500 L.

EXAMPLE 2 Converting kiloliters to liters

8 kL = _____ L

SOLUTION To go from kiloliters to liters in the table, we must move three places to the right. Thus

$$8 \text{ kL} = 8000 \text{ L}$$

PROBLEM 2

9 kL = _____ L

Answers to PROBLEMS

1. a. 1 L **b.** $\frac{1}{2}$ L **c.** 40 L **2.** 9000 L

EXAMPLE 3 **Converting milliliters to liters**

481 mL = ______ L

SOLUTION To go from milliliters to liters in the table, we must move three places to the left. Thus

$$481 \text{ mL} = 0.481 \text{ L}$$

PROBLEM 3

247 mL = ______ L

C › Applications Involving U.S. and Metric Units of Volume

Capacity can also be measured using the *household method,* a method using common measures such as *teaspoons, tablespoons, fluid ounces,* and *cups.*

Here are the equivalencies between common capacity measures and metric system measures. By the way, a milliliter (mL) is the capacity of an eyedropper or $\frac{1}{5}$ of a teaspoon.

Household and Metric Measures

1 teaspoon = 5 mL	1 fluid oz = 30 mL
1 tablespoon = 15 mL	1 cup = 240 mL

EXAMPLE 4 **Converting units for a recipe**

Do you want to make your own cleaner? *The Old Farmer's Almanac* gives the following recipe for an oven cleaner:

Oven Cleaner

2 tablespoons dishwashing liquid	$\frac{1}{4}$ cup ammonia
2 teaspoons borax	$1\frac{1}{2}$ cups warm water

a. How many milliliters of dishwashing liquid do you need?

b. How many milliliters of borax do you need?
(See http://www.almanac.com/home/cleaners.html.)

SOLUTION

a. As usual, we use substitution to solve the problem.
We need two tablespoons of dishwashing liquid.
We know that 1 tablespoon = 15 mL
Thus, 2 tablespoons = 2 · 15 mL = 30 mL
This means that we need 30 mL of dishwashing liquid.

b. We need two teaspoons of borax.
We know that 1 teaspoon = 5 mL
Thus, 2 teaspoons = 2 · 5 mL = 10 mL
This means that we need 10 mL of borax.

PROBLEM 4

a. How many milliliters of ammonia do you need?

b. How many milliliters of warm water do you need?

Answers to PROBLEMS

3. 0.247 L

4. a. 60 mL **b.** 360 mL

EXAMPLE 5 Converting units for a recipe

A recipe for Sopa Azteca calls for 500 mL of cream of tomato soup. How many cups is that?

SOLUTION

We know that $240 \text{ mL} = 1 \text{ cup}$ (See table.)

This means that $1 \text{ mL} = \frac{1 \text{ cup}}{240}$

Thus, $500 \text{ mL} = \frac{500 \text{ cup}}{240} = 2\frac{1}{12} \text{ cups}$

This means that we need $2\frac{1}{12}$ cups of tomato soup. Probably two cups will do!

PROBLEM 5

Sopa Azteca also calls for 400 mL of consommé. How many cups is that?

Many medical applications require knowledge of the metric system. For example, most liquids in your local pharmacy are labeled in liters (L) or milliliters (mL). Drug dosages are usually given in milliliters or cubic centimeters (cc) or cm^3. Here is the relationship between milliliters (mL), cc, and cubic centimeters.

$$1 \text{ mL} = 1 \text{ cm}^3 = 1 \text{ cc}$$

EXAMPLE 6 Unit conversions and medicine

A doctor orders 20 ounces of IV (intravenous) fluid for a patient.

a. How many mL is that?

b. How many cc is that?

SOLUTION

a. We first have to convert fluid ounces to mL.

We know that 1 fluid ounce = 30 mL

Thus, $20 \text{ fluid ounces} = 20 \cdot 30 \text{ mL} = 600 \text{ mL}$

This means that the doctor ordered 600 mL of IV fluid.

b. Now, we have to convert mL to cc.

We know that 1 mL = 1 cc

Thus, $600 \text{ mL} = 600 \cdot 1 \text{ cc} = 600 \text{ cc}$

This means the doctor ordered 600 cc of IV fluid.

PROBLEM 6

A doctor orders 30 fluid ounces of acetaminophen pediatric elixir.

a. How many mL is that?

b. How many cc is that?

› Exercises 7.5

‹A› U.S. (American) to Metric Conversion In Problems 1–10, use the fact that 1 qt ≈ 1 L to fill in the blanks.

1. $\frac{3}{4}$ gal ≈ ________ L
2. $\frac{1}{8}$ gal ≈ ________ L
3. 2 L ≈ ________ qt
4. 5 L ≈ ________ qt
5. 8 L ≈ ________ gal
6. 12 L ≈ ________ gal
7. 5 gal ≈ ________ L
8. 6 gal ≈ ________ L
9. 8 qt ≈ ________ L
10. 12 qt ≈ ________ L

Answers to PROBLEMS

5. $1\frac{2}{3}$ cups
6. a. 900 mL **b.** 900 cc

‹ B › Converting from One Metric Measure to Another

Use this table to solve Problems 11–30.

kiloliter (kL)	hectoliter (hL)	dekaliter (daL)	liter (L)	deciliter (dL)	centiliter (cL)	milliliter (mL)
1000 L	100 L	10 L	1 L	$\frac{1}{10}$ L	$\frac{1}{100}$ L	$\frac{1}{1000}$ L

11. 177 mL = ________ L
12. 781 mL = ________ L
13. 3847 mL = ________ L
14. 9381 mL = ________ L
15. 205 cL = ________ L
16. 804 cL = ________ L
17. 55 dL = ________ L
18. 49 dL = ________ L
19. 6 daL = ________ L
20. 11 daL = ________ L
21. 7 hL = ________ L
22. 13 hL = ________ L
23. 6 kL = ________ L
24. 5 kL = ________ L
25. 5 kL = ________ daL
26. 8 kL = ________ daL
27. 4 daL = ________ cL
28. 8 daL = ________ cL
29. 9 daL = ________ dL
30. 7 daL = ________ dL

31. A permanganate solution has 200 mL of permanganate and 1800 mL of water.

a. Change 200 mL to liters.

b. Change 1800 mL to liters.

32. A Lysol solution has 160 mL of Lysol and 3840 mL of water.

a. Change 160 mL to liters.

b. Change 3840 mL to liters.

33. A 44-lb patient needs to receive 1500 mL of fluids to meet his fluid maintenance needs. How many liters is that?

34. A 30-kg patient needs 1700 mL of fluids to meet her fluid maintenance needs. How many liters is that?

‹ C › Applications Involving U.S. and Metric Units of Volume

Cleaning products The following information for mixing a general-purpose cleaner will be used in Problems 35–40.

General-Purpose Cleaner

1 teaspoon borax
$\frac{1}{2}$ teaspoon washing soda (found in laundry section of stores)
2 teaspoons vinegar
$\frac{1}{4}$ teaspoon dishwashing liquid
2 cups hot water

Source: Data from *The Old Farmer's Almanac*.

35. How many mL of borax are there in the cleaner?

36. How many mL of washing soda are there in the cleaner?

37. How many mL of vinegar are there in the cleaner?

38. How many mL of dishwashing liquid are there in the cleaner?

39. How many mL of hot water are there in the cleaner?

40. How many liters of cleaner will result when you mix all ingredients?

41. *Pain reliever* Biofreeze, a pain-relieving gel, is sold in a 15-fluid-ounce size. How many mL is that?

42. *Medicine* Stopain spray comes in an 11.25-fluid-ounce bottle. How many mL is that?

43. *Eye drops* Clear Eyes eye drops come in a 0.5-fluid-ounce bottle. How many mL is that?

44. *Mouth wash* The dosage for Periocheck mouth rinse is $\frac{1}{8}$ fluid ounce per rinse. How many mL is that?

45. *Vitamins* Centrum liquid vitamin comes in a 240-mL bottle. How many fluid ounces is that?

46. *Medicine* Mylanta comes in a 720-mL bottle. How many fluid ounces is that?

47. *Fluid per day* A child weighing 15 kg requires 1500 mL of fluid per day.

a. How many fluid ounces per day is that?

b. How many cups of fluid per day is that?

48. *Fluid per day* A child who weighs 8 kg requires 800 mL of fluid per day.

a. How many fluid ounces per day is that?

b. How many cups of fluid per day is that?

Web IT go to mhhe.com/bello for more lessons

49. *Daily water* How much water should a patient drink each day? The Institute of Medicine advises that men consume about 13 cups of total beverages a day.

a. How many mL is that?

b. How many fluid ounces is that?

50. *Daily water* What about women? They should consume about 9 cups of total beverages a day.

a. How many mL is that?

b. How many fluid ounces is that?

Using Your Knowledge

51. *Eye Drops* The dosage for Clear Eyes eye drops is two drops per eye, four times a day.

a. How many drops per day is that?

b. If one drop is 0.2 mL, how many mL are used each day?

c. If Clear Eyes comes in a 15 mL bottle, about how many days will the bottle last? Answer to four decimal places.

52. *Mouthwash* The dosage for Periocheck mouth rinse is $\frac{1}{8}$ fluid ounces two times a day.

a. How many fluid ounces per day is that?

b. If the bottle of Periocheck contains 10 fluid ounces of liquid, how many days will it last?

Write On

53. Write in your own words the advantages of the metric over the American system when measuring capacities.

54. Can you think of other household measures that were not mentioned in the text? What are they and how are they used?

Concept Checker

Fill in the blank(s) with the correct word(s), phrase, or mathematical statement.

55. A **milliliter** is defined as the volume of a cube one ________ on each edge.

56. A ________ is the **volume of a cube** 10 cm on each edge.

liter **cm**

in **quart**

m

Mastery Test

57. A doctor orders 25 fluid ounces of IV fluid for a patient.

a. How many mL is that?

b. How many cc is that?

58. A recipe calls for 600 mL of cream of mushroom soup. How many cups is that?

59. A rug-cleaning solution is made by mixing $\frac{1}{4}$ teaspoon dishwashing liquid and 1 cup lukewarm water. How many mL of dishwashing liquid and how many mL of water do you need?

Fill in the blanks.

60. a. 3 gal ≈ ________ L

b. 16 oz ≈ ________ L

61. 10 kL = ________ L

62. 347 mL = ________ L

Skill Checker

In Problems 63–66, perform the indicated operations.

63. $80 \cdot \frac{1}{16}$

64. $5000 \cdot \frac{1}{2000}$

65. $6 \cdot 0.45$

66. $\frac{5(41 - 32)}{9}$

7.6 Weight and Temperature: U.S., Metric, and Conversions

Objectives

You should be able to:

A › Convert a U.S. unit of weight to an equivalent U.S. unit of weight.

B › Convert a metric unit of mass to an equivalent metric unit of mass.

C › Convert a U.S. unit of weight to an equivalent metric unit of weight and vice versa.

D › Convert temperatures from Celsius to Fahrenheit and vice versa.

To Succeed, Review How To . . .

1. Multiply fractions. (pp. 130–136)
2. Multiply decimals. (pp. 212–216)

Getting Started

A › Converting U.S. Units of Weight

In the cartoon, the prisoner is getting three glops of food a day. To discuss how much food that is, we need a system of **unit weights.** The customary system (also called the avoirdupois system) uses the following units of weight:

U.S. Units of Weight

1 ton = 2000 pounds (lb)	1 pound = $\frac{1}{2000}$ ton
1 pound = 16 ounces (oz)	1 oz = $\frac{1}{16}$ lb

We can change from one unit to another in the customary system by using *substitution.* For example, to solve

$$3 \text{ lb} = \underline{\qquad} \text{ oz}$$

we write

$$3 \text{ lb} = 3 \cdot (16 \text{ oz}) = 48 \text{ oz}$$

Similarly, to solve the problem

$$80 \text{ oz} = \underline{\qquad} \text{ lb}$$

we write

$$80 \text{ oz} = 80 \cdot \left(\frac{1}{16} \text{ lb}\right) = \frac{80}{16} \text{ lb} = 5 \text{ lb}$$

We can also use **unit fractions.** Since 1 lb contains 16 oz,

$$1 = \frac{\text{lb}}{16 \text{ oz}}$$

and

$$80 \text{ oz} = 80 \text{ oz} \cdot 1 = 80 \cancel{\text{oz}} \cdot \frac{\text{lb}}{16 \cancel{\text{oz}}}$$
$$= 5 \text{ lb}$$

EXAMPLE 1 **Converting pounds to ounces**

2 lb = _______ oz

SOLUTION

$$2 \text{ lb} = 2 \cdot (16 \text{ oz}) = 32 \text{ oz}$$

PROBLEM 1

4 lb = _______ oz

EXAMPLE 2 **Converting ounces to pounds**

48 oz = _______ lb

SOLUTION

$$48 \text{ oz} = 48 \cdot \left(\frac{1}{16} \text{ lb}\right) = 48 \cdot \frac{1}{16} \text{ lb} = 3 \text{ lb}$$

PROBLEM 2

32 oz = _______ lb

EXAMPLE 3 **Converting tons to pounds**

3 tons = _______ lb

SOLUTION

$$3 \text{ tons} = 3 \cdot (2000 \text{ lb}) = 6000 \text{ lb}$$

PROBLEM 3

2 tons = _______ lb

EXAMPLE 4 **Converting pounds to tons**

5000 lb = _______ tons

SOLUTION

$$5000 \text{ lb} = 5000 \cdot \left(\frac{1}{2000} \text{ ton}\right) = 5000 \cdot \frac{1}{2000} \text{ ton} = 2\frac{1}{2} \text{ tons}$$

PROBLEM 4

7000 lb = _______ tons

B > Converting Metric Units of Mass

As usual, the metric system is easier. In this system the unit of **mass** is called the gram. (Note that we wrote *mass* and not *weight*. There is a difference, but the terms are used interchangeably.)* A **gram** is the mass of 1 cm^3 (1 mL) of water. Here is the table giving the information used to convert from one unit to another in the metric system:

kilogram (kg)	hectogram (hg)	dekagram (dag)	gram (g)	decigram (dg)	centigram (cg)	milligram (mg)
1000 g	100 g	10 g	1 g	$\frac{1}{10}$ g	$\frac{1}{100}$ g	$\frac{1}{1000}$ g

Answers to PROBLEMS

1. 64 oz 2. 2 lb
3. 4000 lb 4. $3\frac{1}{2}$ tons

*Weight is related to the force of gravity and mass is not. If you weigh 150 pounds on Earth you would only weigh 25 pounds on the moon (gravity is less there), but your mass would be the same as on Earth.

To convert from one unit to another is just a matter of moving the decimal point the correct number of places. Thus, to solve

3 hg = _______ cg

we have to move from hectograms to centigrams in the table; that is, we have to move four places to the right. Thus, we move the decimal point in 3 four places to the right, obtaining

3 hg = 30,000 cg

EXAMPLE 5 **Converting dekagrams to decigrams**

4 dag = _______ dg

SOLUTION To move from dekagrams to decigrams in the table, we have to move *two* places to the right. Thus we move the decimal point in 4 *two* places to the right, obtaining.

4 dag = 400 dg

PROBLEM 5

3 dag = _______ dg

EXAMPLE 6 **Converting milligrams to grams**

401 mg = _______ g

SOLUTION To move from milligrams to grams in the table, we have to move *three* places to the left. Thus we move the decimal point in 401 *three* places to the left, obtaining

401 mg = 0.401 g

PROBLEM 6

103 mg = _______ g

C › Converting Units of Weight between Metric and U.S. (American) Customary

We are now ready to convert weights from the U.S. customary system to the metric system, or vice versa. To do this we need the following table:

U.S. System to Metric Conversions	
1 kg = 2.2 lb	1 lb = 0.45 kg

EXAMPLE 7 **Converting kilograms to pounds**

3 kg = _______ lb

SOLUTION

$$3 \text{ kg} = 3 \cdot (2.2 \text{ lb}) = 6.6 \text{ lb}$$

PROBLEM 7

2 kg = _______ lb

EXAMPLE 8 **Converting pounds to kilograms**

6 lb = _______ kg

SOLUTION

$$6 \text{ lb} = 6 \cdot (0.45 \text{ kg}) = 2.70 \text{ kg}$$

PROBLEM 8

4 lb = _______ kg

Answers to PROBLEMS

5. 300 dg **6.** 0.103 g

7. 4.4 lb **8.** 1.80 kg

D > Converting from Fahrenheit to Celsius and Vice Versa

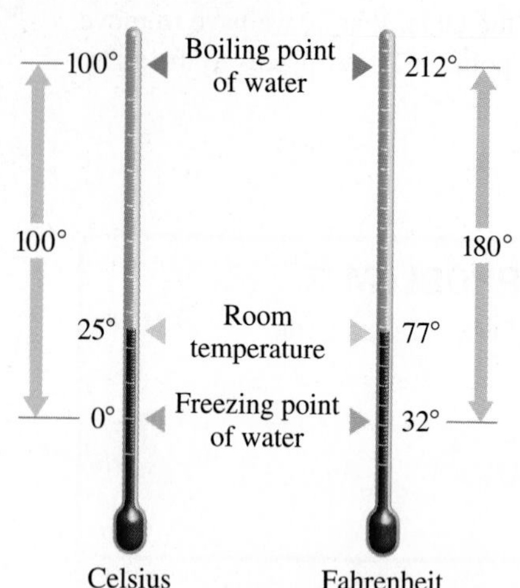

Comparison of the Celsius and Fahrenheit thermometer scales

Finally, we study the measurement of temperatures in both the customary and the metric systems. The customary temperature scale was invented in 1714 by Gabriel Daniel Fahrenheit, a German scientist. In this scale, called the **Fahrenheit** scale, the point at which water boils was labeled 212°F (read "212 degrees Fahrenheit"), and the point at which water freezes was labeled 32°F. The Fahrenheit scale was later modified by Anders Celsius. This new scale, called the **Celsius** scale, was divided into 100 units. (The Celsius scale used to be called the centigrade scale because it has 100 units.) The boiling point occurs at 100°C (read "100 degrees Celsius") and the freezing point at 0°C. You can see how the temperatures in these two scales are compared by looking at the drawing.

Here are the formulas for converting from one scale to the other. In these formulas C stands for the Celsius temperature and F for the Fahrenheit temperature.

Temperature Conversions

$$C = \frac{5(F - 32)}{9} \qquad F = \frac{9C}{5} + 32$$

Note that $5(F - 32)$ means $5 \cdot (F - 32)$, $9C$ means $9 \cdot C$, and that $F - 32$ is in *parentheses.* This means that when F is known, $F - 32$ is calculated *first.*

At the freezing point of water, $F = 32$ and

$$C = \frac{5(32 - 32)}{9} = \frac{5(0)}{9} = 0$$

Similarly, at the boiling point of water, $C = 100$ and

$$F = \frac{9 \cdot 100}{5} + 32 = \frac{900}{5} + 32$$
$$= 180 + 32 = 212$$

EXAMPLE 9 **Converting from Fahrenheit to Celsius**

41°F = _______ °C

SOLUTION From the table,

$$C = \frac{5(F - 32)}{9} = \frac{5(41 - 32)}{9} = \frac{5 \cdot 9}{9} = 5$$

Thus, 41°F = 5°C.

PROBLEM 9

50°F = _______ °C

EXAMPLE 10 **Converting from Celsius to Fahrenheit**

15°C = _______ °F

SOLUTION From the table,

$$F = \frac{9C}{5} + 32 = \frac{9 \cdot 15}{5} + 32 = 27 + 32 = 59$$

Thus, 15°C = 59°F.

PROBLEM 10

20°C = _______ °F

Answers to PROBLEMS

9. 10°C **10.** 68°F

Calculator Corner

The formula to convert degrees Fahrenheit to Celsius is especially suitable for a calculator with parentheses ( ) keys. But the calculator does not do it all. You must know that to convert 41°F to degrees Celsius, you must multiply 5 by $(F - 32)$ and then divide by 9. The computation in Example 9 is like this:

$$F = \frac{5(41 - 32)}{9}$$

With a calculator, press 5 × (4 1 − 3 2) ÷ 9 ENTER and the correct answer, 5, will be given.

If your calculator does not have parentheses keys, you must know even more! To find the answer using a calculator without parentheses keys, find 41 − 32 first, as explained in the text. The calculations look like this:

4 1 − 3 2 ENTER × 5 ÷ 9 ENTER

Exercises 7.6

⟨A⟩ Converting U.S. Units of Weight

In Problems 1–16, fill in the blanks.

1. 3 lb = ________ oz
2. 5 lb = ________ oz
3. 4.5 lb = ________ oz
4. $3\frac{1}{2}$ lb = ________ oz
5. 64 oz = ________ lb
6. 96 oz = ________ lb
7. 72 oz = ________ lb
8. 40 oz = ________ lb
9. 4 tons = ________ lb
10. 5 tons = ________ lb
11. $2\frac{1}{2}$ tons = ________ lb
12. 4.5 tons = ________ lb
13. 3000 lb = ________ tons
14. 9000 lb = ________ tons
15. 4000 lb = ________ tons
16. 6000 lb = ________ tons

⟨B⟩ Converting Metric Units of Mass

Use this table to solve Problems 17–30.

kilogram (kg)	hectogram (hg)	dekagram (dag)	gram (g)	decigram (dg)	centigram (cg)	milligram (mg)
1000 g	100 g	10 g	1 g	$\frac{1}{10}$ g	$\frac{1}{100}$ g	$\frac{1}{1000}$ g

17. 2 kg = ________ cg
18. 4 kg = ________ cg
19. 2 dag = ________ dg
20. 5 dag = ________ dg
21. 2 kg = ________ dag
22. 6 kg = ________ dag
23. 5 dag = ________ g
24. 4 dag = ________ g
25. 899 mg = ________ g
26. 301 mg = ________ g
27. 30 mg = ________ g
28. 51 mg = ________ g
29. 57 cg = ________ g
30. 64 cg = ________ g

⟨C⟩ Converting Units of Weight between Metric and U.S. (American) Customary

In Problems 31–40, fill in the blanks.

31. 4 kg = ________ lb
32. 5 kg = ________ lb
33. 10 kg = ________ lb
34. 12 kg = ________ lb
35. 2 lb = ________ kg
36. 8 lb = ________ kg
37. 10 kg = ________ hg
38. 20 kg = ________ hg
39. 16 oz = ________ kg
40. 64 oz = ________ kg

Web IT go to mhhe.com/bello for more lessons

‹ D › Converting from Fahrenheit to Celsius and Vice Versa In Problems 41–50, fill in the blanks.

41. 59°F = ______________ °C **42.** 68°F = ______________ °C **43.** 77°F = ______________ °C

44. 86°F = ______________ °C **45.** 113°F = ______________ °C **46.** 122°F = ______________ °C

47. 10°C = ______________ °F **48.** 25°C = ______________ °F **49.** 35°C = ______________ °F

50. 30°C = ______________ °F

› › › Applications

51. *Melting point of gold* The melting point of gold is 1000°C. How many degrees Fahrenheit is that?

52. *Normal body temperature* The normal body temperature is 98.6°F. How many degrees Celsius is that?

53. *High fever* 104°F is considered a high fever. How many degrees Celsius is that?

54. *Weight* A woman weighs 48 kg. How many pounds is that?

55. *Weight* A man weighs 160 lb. How many kilograms is that?

56. *Air temperature* The highest dry-air temperature endured by heavily clothed men in an Air Force experiment was 500°F. How many degrees Celsius is that?

57. *Coimbra temperature* In September 1933, the temperature in Coimbra, Portugal, rose to 70°C. How many degrees Fahrenheit is that?

58. *Death Valley temperature* On July 10, 1913, the temperature in Death Valley was 134°F. How many degrees Celsius is that?

59. *Maximum weight for flyweight* The maximum weight for a flyweight wrestler is 52 kg. How many pounds is that?

60. *Maximum weight for bantamweight* The maximum weight for a bantamweight wrestler is $123\frac{1}{4}$ lb. How many kilograms is that?

Did you know that there are two kinds of elephants, African and Asian? Let us look at some facts about each of them. *Source:* http://www.sandiegozoo.org

61. *Elephants* In general, wild elephants eat all types of vegetation, from grass and fruit to leaves and bark—about 220 to 440 pounds each day. How many kilograms is that?

62. *Elephants* The elephants at the San Diego Zoo eat less: about 125 pounds of food each day. How many kilograms is that? Round your answer to the nearest kilogram.

63. *Elephants* The largest elephant on record was an adult male African elephant. It weighed about 10,886 kilograms. How many pounds is that? Round your answer to the nearest pound.

64. *Elephants* At birth, an elephant weights 50 to 113 kg. How many pounds is that? Round your answer to the nearest pound. By the way, the gestation (pregnancy) period for elephants is 20 to 22 months.

› › › Using Your Knowledge

Desirable Weights The tables show the desirable weight for certain heights at age 25 or over. Find these weights to the nearest kilogram.

Men

	Height (1 in. heels)		Weight	
	Feet	Inches	Pounds	Kilograms
65.	5	2	130	____
66.	5	6	142	____
67.	5	10	158	____
68.	6	2	178	____
69.	6	3	184	____

Women

	Feet	Inches	Pounds	Kilograms
70.	5	2	120	____
71.	5	6	135	____
72.	5	10	145	____
73.	5	11	155	____

Write On

74. Write in your own words the advantages of the metric system over the U.S. (American) system when measuring weight.

75. Do you know of any other measures for weight? What are they and how are they used?

76. Write in your own words how to use estimation to make the calculations in this section.

Concept Checker

Fill in the blank(s) with the correct word(s), phrase, or mathematical statement.

77. A ________ is the **mass** of **1 cm³** (cubic centimeter) of water.

78. In the **customary** (U.S.) system, **temperature** is measured in degrees ____________ while the **metric system** is measured in degrees ____________.

kilogram **pound**
Fahrenheit **Celsius**
gram

Mastery Test

79. 40°C = ______ F

80. 68°F = ______ C

81. 8 kg = ______ lb

82. 10 lb = ______ kg

83. 384 mg = ______ g

84. 8 dag = ______ dg

85. 12,000 lb = ______ tons

86. 7 tons = ________ lb

87. 160 oz = ______ lb

88. 6 lb = ______ oz

Skill Checker

Evaluate:

89. $180 - x$, when $x = 50$

90. $180 - x$, when $x = 160$

91. $180 - x - 59$, when $x = 47$

92. $180 - x - 48$, when $x = 45$

Collaborative Learning

The object of this Collaborative Learning is to propose a method for converting the United States to the metric system. Form three groups of students:

1. Commercial users
2. General users
3. Scientific users

Questions to be answered by each group:

Group 1 (Farmers, manufacturers, producers)

1. Why does the United States need to change to the metric system?
2. Why doesn't the United States accept the metric system, even though it is now used in other countries?
3. What are the advantages and disadvantages of the metric system?
4. What would your plan be to convert the United States to metric?

Group 2 (Household users, mechanics, builders, students)

1. What are the benefits of having all measurements in metric rather than in a combination of standard and metric?
2. Name some items measured in standard measurement, some items measured in metric, and some items measured in both.
3. Why is the government interested in changing to metric?

<CONTINUED>

4. What would your plan be to convert the United States to metric?

Group 3 (Researchers, students in medical professions)

1. What are the benefits of a total metric system adoption in the United States?
2. How difficult would it be for the general public to learn the metric system? How can you make it easier?
3. Why is the federal government making the metric transition?
4. What would your plan be to convert the United States to metric?

Adapted from: Metric System: A WebQuest for 7–10 Grade Science by Deborah L. Folis

Research Questions

1. When was the metric system made legal (but not mandatory) in the United States?
2. Write a paragraph about Gabriel Mouton and his contribution to the creation of the metric system.
3. Where are the words *inch, foot,* and *yard* derived from?
4. When was the Metric Conversion Act passed by Congress?
5. Find the official definition of:
   a. The unit of length (meter)
   b. The unit of mass (kilogram)
   c. The unit of time (second)
6. What is the "iron ulna," what distance did it represent, and what king used it as measurement?

Summary Chapter 7

Section	Item	Meaning	Example
7.1A, B	Inch	The breadth of a thumb, defined as $\frac{1}{12}$ of a foot	
	Foot	Length of a man's foot (12 in.)	12 in. = 1 ft
	Yard	The circumference of a person's waist, defined as 36 inches	
	Unit fractions	Fractions that equal 1	$\frac{12\text{ in.}}{\text{ft}}$
7.2A	Kilometer	1 kilometer = 1000 meters	5 km = 5000 m
	Hectometer	1 hectometer = 100 meters	3 hm = 300 m
	Dekameter	1 dekameter = 10 meters	4 dam = 40 m
	Meter	The basic unit of length in the metric system	
	Decimeter	1 decimeter = $\frac{1}{10}$ meter	7 dm = $\frac{7}{10}$ m
	Centimeter	1 centimeter = $\frac{1}{100}$ meter	9 cm = $\frac{9}{100}$ m
	Millimeter	1 millimeter = $\frac{1}{1000}$ meter	3 mm = $\frac{3}{1000}$ m

Section	Item	Meaning	Example
7.3A	Inch	1 in. = 2.54 cm	2 in. = 5.08 cm
	Yard	1 yd = 0.914 m	3 yd = 2.742 m
	Mile	1 mi = 1.6 km	5 mi = 8.0 km
	Centimeter	1 cm = 0.4 in.	5 cm = 2.0 in.
	Meter	1 m = 1.1 yd	10 m = 11 yd
	Kilometer	1 km = 0.62 mi	10 km = 6.2 mi
7.4A, B, C	1 in.^2	$\frac{1}{144}\ \text{ft}^2$	
	1 ft^2	$\frac{1}{9}\ \text{yd}^2$	
	1 hectare	10,000 m^2 = 2.47 acres	20,000 m^2 = 4.94 acres
	1 acre	4840 yd^2	
7.5	Liter	Volume of a cube 10 cm on each edge	
	Quart	0.946 L	2 qt = 1.892 L
	Liter	1.06 qt	4 L = 4.24 qt
7.5B	Kiloliter	1 kL = 1000 L	5 kL = 5000 L
	Hectoliter	1 hL = 100 L	3 hL = 300 L
	Dekaliter	1 daL = 10 L	10 daL = 100 L
	Liter	The basic unit of volume in the metric system	
	Deciliter	1 dL = $\frac{1}{10}$ L	7 dL = $\frac{7}{10}$ L
	Centiliter	1 cL = $\frac{1}{100}$ L	9 cL = $\frac{9}{100}$ L
	Milliliter	1 mL = $\frac{1}{1000}$ L	3 mL = $\frac{3}{1000}$ L
7.5C	Teaspoon	5 mL	2 Teaspoons = 10 mL
	Tablespoon	15 mL	2 Tablespoons = 30 mL
	Fluid ounce	30 mL	3 Fluid ounces = 90 mL
	Cup	240 mL	2 Cups = 480 mL
	cm^3	cc (cubic centimeter)	
7.6A	1 Ton	2000 lb	2 tons = 4000 lb
	1 Pound	$\frac{1}{2000}$ ton	4000 lb = 2 tons
	1 Pound	16 oz	3 pounds = 48 oz
	1 Ounce	$\frac{1}{16}$ lb	32 ounces = 2 lb
7.6B	Kilogram	1 kg = 1000 g	10 kg = 10,000 g
	Hectogram	1 hg = 100 g	5 hg = 500 g
	Dekagram	1 dag = 10 g	7 dag = 70 g
	Gram	The basic unit of mass in the metric system	
	Decigram	1 dg = $\frac{1}{10}$ g	7 dg = $\frac{7}{10}$ g
	Centigram	1 cg = $\frac{1}{100}$ g	3 cg = $\frac{3}{100}$ g
	Milligram	1 mg = $\frac{1}{1000}$ g	9 mg = $\frac{9}{1000}$ g

(continued)

Section	Item	Meaning	Example
7.6C	1 Kilogram 1 Pound	2.2 lb 0.45 kg	2 kg = 4.4 lb 10 lb = 4.5 kg
7.6D	Degree Celsius Degree Fahrenheit	$C = \frac{5(F-32)}{9}$ (F = the Fahrenheit temperature) $F = \frac{9C}{5} + 32$ (C = the Celsius temperature)	$41°F = \frac{5(41-32)}{9} = 5°C$ $10°C = \frac{9 \cdot 10}{5} + 32 = 50°C$

Review Exercises Chapter 7

(If you need help with these exercises, look in the section indicated in brackets.)

1. ⟨7.1A⟩ *Find how many inches there are in:*
   a. 2 yd
   b. 3 yd
   c. 4 yd
   d. 6 yd
   e. 7 yd

2. ⟨7.1A⟩ *Find how many feet there are in:*
   a. 12 in.
   b. 24 in.
   c. 36 in.
   d. 40 in.
   e. 65 in.

3. ⟨7.1A⟩ *Find how many feet there are in:*
   a. 2 in.
   b. 3 in.
   c. 5 in.
   d. 7 in.
   e. 14 in.

4. ⟨7.1A⟩ *Find how many yards there are in:*
   a. 6 ft
   b. 12 ft
   c. 18 ft
   d. 20 ft
   e. 29 ft

5. ⟨7.1A⟩ *Find how many miles there are in:*
   a. 5280 ft
   b. 10,560 ft
   c. 26,400 ft
   d. 21,120 ft
   e. 15,840 ft

6. ⟨7.2A⟩ *Find how many meters there are in:*
   a. 2 km
   b. 7 km
   c. 4.6 km
   d. 0.45 km
   e. 45 km

7. ⟨7.2A⟩ *Find how many meters there are in:*
   a. 2 dam
   b. 3 dam
   c. 7 dam
   d. 9 dam
   e. 10 dam

8. ⟨7.2A⟩ *Find how many decimeters there are in:*
   a. 100 m
   b. 300 m
   c. 350 m
   d. 450 m
   e. 600 m

9. ⟨7.2A⟩ *Find how many meters there are in:*
   a. 200 cm
   b. 395 cm
   c. 405 cm
   d. 234 cm
   e. 499 cm

10. ⟨7.3A⟩ *Find how many cm there are in:*
    a. 30 in.
    b. 40 in.
    c. 50 ft
    d. 60 ft
    e. 70 ft

11. ⟨7.3A⟩ *Find how many meters there are in:*
    a. 100 yd
    b. 200 yd
    c. 350 yd
    d. 450 yd
    e. 500 yd

12. ⟨7.3A⟩ *Find how many kilometers there are in:*
    a. 30 mi
    b. 50 mi
    c. 90 mi
    d. 100 mi
    e. 250 mi

13. ⟨7.3A⟩ *Find how many miles there are in:*
    a. 40 km
    b. 50 km
    c. 60 km
    d. 70 km
    e. 80 km

14. ⟨7.4A⟩ *Find how many square centimeters there are in:*
    a. 2 m^2
    b. 3 m^2
    c. 4 m^2
    d. 5 m^2
    e. 6 m^2

15. ⟨7.4A⟩ *Find how many square meters there are in:*
    a. 2 km^2
    b. 3 km^2
    c. 4 km^2
    d. 5 km^2
    e. 6 km^2

16. ⟨7.4B⟩ *Find how many square inches there are in:*
    a. 2 ft^2
    b. 3 ft^2
    c. 4 ft^2
    d. 5 ft^2
    e. 6 ft^2

17. ⟨7.4B⟩ *Find how many square feet there are in:*
    a. 144 in.^{2}
    b. 288 in.^{2}
    c. 360 in.^{2}
    d. 432 in.^{2}
    e. 504 in.^{2}

18. ⟨7.4B⟩ *Find the area, in square yards, of:*
    a. A 1-acre lot
    b. A 3-acre lot
    c. A 2-acre lot
    d. A 1.5-acre lot
    e. A 4-acre lot

19. ⟨7.4C⟩ *Find how many acres there are in:*
    a. 5 hectares
    b. 4 hectares
    c. 1 hectare
    d. 3 hectares
    e. 2 hectares

20. ⟨7.5A⟩ *Find (approximately) how many liters there are in:*
    a. 7 qt
    b. 9 qt
    c. 10 qt
    d. 13 qt
    e. 2.2 qt

21. ⟨7.5B⟩ *Find how many liters there are in:*
    a. 4 kL
    b. 7 kL
    c. 9 kL
    d. 2.3 kL
    e. 5.97 kL

22. ⟨7.5B⟩ *Find how many liters there are in:*
    a. 452 mL
    b. 48 mL
    c. 3 mL
    d. 1657 mL
    e. 456 mL

23. ⟨7.5C⟩ *Find how many mL there are in:*
    a. 4 cups
    b. 5 cups
    c. 6 cups
    d. 7 cups
    e. 9 cups

24. ⟨7.5C⟩ *Find how many cups there are in:*
    a. 120 mL
    b. 360 mL
    c. 480 mL
    d. 600 mL
    e. 720 mL

25. ⟨7.5C⟩ *Find how many mL there are in:*
    a. 120 fluid ounces
    b. 180 fluid ounces
    c. 240 fluid ounces
    d. 300 fluid ounces
    e. 360 fluid ounces

26. ⟨7.5C⟩ *Find how many mL there are in:*
    a. 7 tablespoons
    b. 8 tablespoons
    c. 10 teaspoons
    d. 11 teaspoons
    e. 13 teaspoons

27. ⟨7.6A⟩ *Find how many ounces there are in:*
    a. 3 lb
    b. 4 lb
    c. 5 lb
    d. 6 lb
    e. 7 lb

28. ⟨7.6A⟩ *Find how many pounds there are in:*
    a. 16 oz
    b. 24 oz
    c. 32 oz
    d. 40 oz
    e. 48 oz

29. ⟨7.6A⟩ *Find how many pounds there are in:*
    a. 2 tons
    b. 3 tons
    c. 4 tons
    d. 5 tons
    e. 6 tons

30. ⟨7.6A⟩ *Find how many tons there are in:*
    a. 3000 lb
    b. 5000 lb
    c. 7000 lb
    d. 9000 lb
    e. 18,000 lb

31. ‹ **7.6B** › *Find how many decigrams there are in:*
   a. 1 dag
   b. 3 dag
   c. 5 dag
   d. 4 dag
   e. 2 dag

32. ‹ **7.6B** › *Find how many grams there are in:*
   a. 307 mg
   b. 40 mg
   c. 3245 mg
   d. 2 mg
   e. 10,342 mg

33. ‹ **7.6C** › *Find how many pounds there are in:*
   a. 1 kg
   b. 7 kg
   c. 6 kg
   d. 4 kg
   e. 8 kg

34. ‹ **7.6C** › *Find how many kilograms there are in:*
   a. 1 lb
   b. 3 lb
   c. 6 lb
   d. 4 lb
   e. 10 lb

35. ‹ **7.6D** › *Find how many degrees Celsius there are in:*
   a. 32°F
   b. 41°F
   c. 50°F
   d. 59°F
   e. 212°F

36. ‹ **7.6D** › *Find how many degrees Fahrenheit there are in:*
   a. 10°C
   b. 15°C
   c. 20°C
   d. 25°C
   e. 30°C

› Practice Test Chapter 7

(Answers on pages 450)

Visit www.mhhe.com/bello to view helpful videos that provide step-by-step solutions to several of the problems below.

1. 5 yd = ______ in.

2. 4 in. = ______ ft

3. 25 ft = ______ yd

4. 10,560 ft = ______ mi

5. 3 ft = ______ in.

6. 5 km = ______ m

7. 5 dam = ______ m

8. 250 m = ______ dm

9. 482 cm = ______ m

10. 20 ft = ______ cm

11. 300 yd = ______ m

12. 40 mi = ______ km

13. 80 km = ______ mi

14. 6 m^2 = ______ cm^2

15. 432 $in.^2$ = ______ ft^2

16. Find the area in square yards of a 2-acre lot.

17. 3 hectares = ______ acres

18. 3 qt ≈ ______ L

19. 3 kL = ______ L

20. 393 mL = ______ L

21. A cleaning solution contains $\frac{3}{4}$ cup of ammonia. How many mL is that?

22. A recipe calls for 400 mL of broth. How many cups is that?

23. A doctor orders 20 fluid oz of cough medicine. How many mL is that?

24. 6 lb = ______ oz

25. 64 oz = ______ lb

26. 4 tons = ______ lb

27. 7000 lb = ______ tons

28. 6 dag = ______ dg

29. 401 mg = ______ g

30. 5 kg = ______ lb

31. 5 lb = ______ kg

32. 59°F = ______ °C

33. 10°C = ______ °F

› Answers to Practice Test Chapter 7

Answer	If You Missed	Review		
	Question	Section	Examples	Page
1. 180	1	7.1	1	411
2. $\frac{1}{3}$	2	7.1	2, 3	412
3. $8\frac{1}{3}$	3	7.1	4	412
4. 2	4	7.1	5	412
5. 36	5	7.1	6	412
6. 5000	6	7.2	1	417
7. 50	7	7.2	3	418
8. 2500	8	7.2	2	418
9. 4.82	9	7.2	5	418
10. 609.6	10	7.3	1	422
11. 274.2	11	7.3	2	422
12. 64	12	7.3	3	422
13. 49.60	13	7.3	4–6	422–423
14. 60,000	14	7.4	1	427
15. 3	15	7.4	4	427
16. 9680 yd^2	16	7.4	6	428
17. 7.41	17	7.4	8	428
18. 3	18	7.5	1	432
19. 3000	19	7.5	2	432
20. 0.393	20	7.5	3	433
21. 180	21	7.5	4	433
22. $1\frac{2}{3}$	22	7.5	5	434
23. 600	23	7.5	6	434
24. 96	24	7.6	1	438
25. 4	25	7.6	2	438
26. 8000	26	7.6	3	438
27. $3\frac{1}{2}$	27	7.6	4	438
28. 600	28	7.6	5	439
29. 0.401	29	7.6	6	439
30. 11	30	7.6	7	439
31. 2.25	31	7.6	8	439
32. 15	32	7.6	9	440
33. 50	33	7.6	10	440

Cumulative Review Chapters 1–7

1. Write two thousand, nine hundred ten in standard form.

2. Simplify: $4 \div 2 \cdot 2 + 9 - 7$

3. Write $\frac{31}{6}$ as a mixed number.

4. Write $4\frac{1}{3}$ as an improper fraction.

5. Subtract: $745.42 - 17.5$

6. Multiply: $0.503 \cdot 0.16$

7. Round 549.851 to the nearest tenth.

8. Divide: $50 \div 0.13$ (Round answer to two decimal digits.)

9. What decimal part of 12 is 3?

10. Solve for y: $1.6 = 0.4y$

11. Solve for z: $9 = \frac{z}{3.9}$

12. There is a law stating that "the ratio of width to length for the American flag should be 10 to 19." Is a flag measuring 50 by 97 feet of the correct ratio?

13. Solve the proportion: $\frac{s}{3} = \frac{3}{27}$

14. The protein RDA for males is 60 grams per day. Three ounces of a certain product provide 4 grams of protein. How many ounces of the product are needed to provide 60 grams of protein?

15. Write 12% as a decimal.

16. Write $7\frac{1}{4}\%$ as a decimal.

17. 40% of 50 is what number?

18. What is $33\frac{1}{3}\%$ of 6?

19. What percent of 28 is 14?

20. 6 is 30% of what number?

21. Find the simple interest earned on $500 invested at 8.5% for 5 years.

22. Referring to the circle graph, which is the main source of pollution?

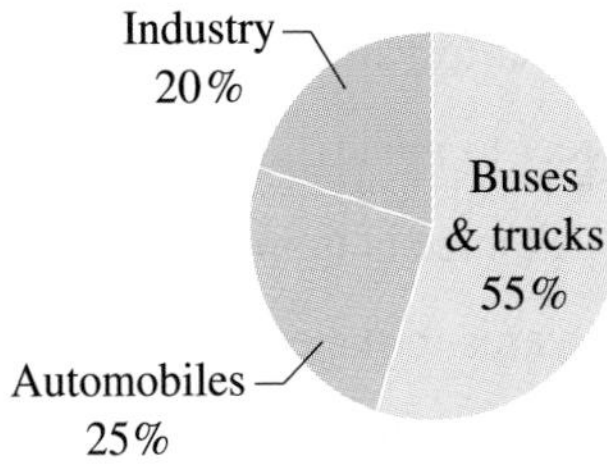

23. Make a circle graph for these data:

Family	Budget	(Monthly)
Savings	(S)	$500
Housing	(H)	$700
Food	(F)	$400
Clothing	(C)	$400

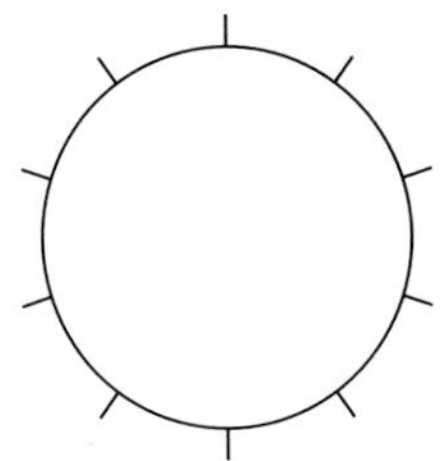

24. The following table shows the distribution of families by income in Tampa, Florida.

Income Level	Percent of Families
$0–9,999	3
10,000–14,999	8
15,000–19,999	19
20,000–24,999	43
25,000–34,999	11
35,000–49,999	7
50,000–79,999	5
80,000–119,000	3
120,000 and over	1

What percent of the families in Tampa have incomes between $20,000 and $24,999?

25. The following graph represents the yearly rainfall in inches in Sagamore County for 2000–2005. Find the rainfall for 2005.

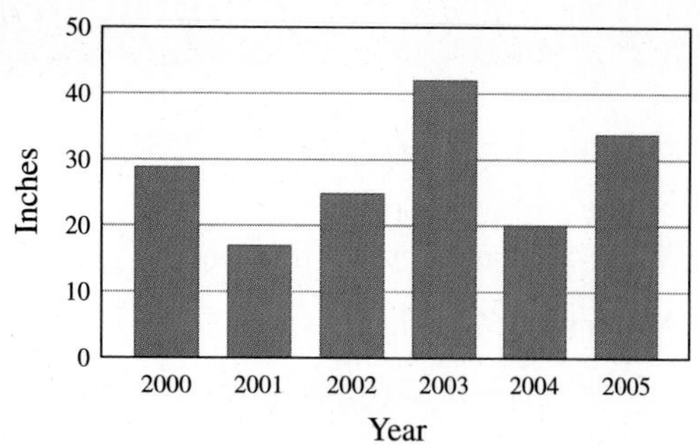

26. The following graph represents the monthly average temperature for 7 months of the year. How much higher is the average temperature in July than it is in May?

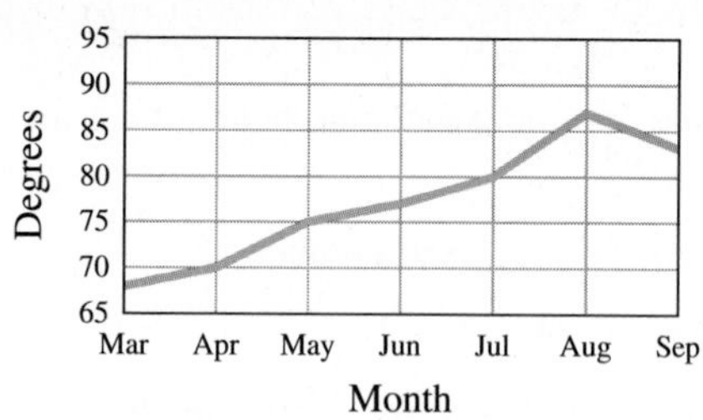

27. The number of hours required in each discipline of a college core curriculum is represented by the following circle graph. What percent of these hours is in math and English combined?

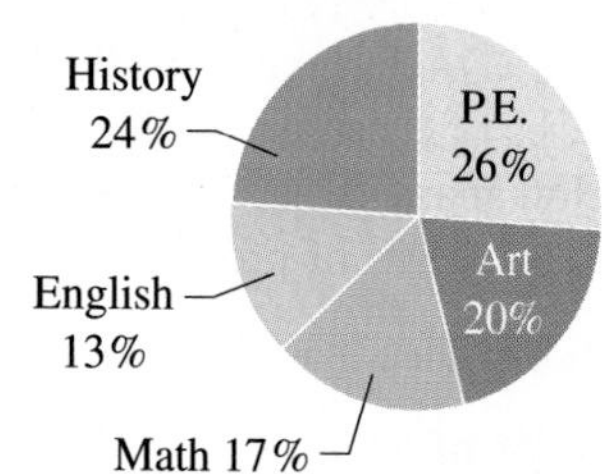

28. What is the mode of the following set of numbers? 8, 11, 6, 12, 6, 10, 6, 24, 20, 23, 6

29. What is the mean of the following numbers? 11, 3, 4, 12, 1, 7, 1, 25, 1, 22, 1

30. What is the median of the following numbers? 6, 27, 25, 16, 27, 13, 27, 12, 27

31. Convert 12 yards to inches.

32. Convert 17 inches to feet.

33. Convert 47 feet to yards.

34. Convert 26,400 feet to miles.

35. Convert 8 feet to inches.

36. Convert 2 kilometers to meters.

37. Convert 5 dekameters to meters.

38. Convert 150 meters to decimeters.

39. Convert 50 yards to meters.

40. Convert 66 miles to kilometers.

41. Convert 100 kilometers to miles.

42. Find the area in square yards of a 4-acre lot.

43. Convert 6 hectares to acres.

Section

Chapter 8 eight

Geometry

The Human Side of Mathematics

One of the most famous mathematicians of all time is Euclid, who taught in about 300 B.C. at the university in Alexandria, the main Egyptian seaport. Unfortunately, very little is known about Euclid personally; even his birthdate and birthplace are unknown. However, two stories about him have survived. One concerns the Emperor Ptolemy, who asked if there was no easy way to learn geometry and received Euclid's reply, "There is no royal road to geometry." The other story is about a student who studied geometry under Euclid and, when he had mastered the first theorem, asked, "But what shall I get by learning these things?" Euclid called a slave and said "Give him a penny, since he must make gain from what he learns."

Geometry evolved from the more or less rudimentary ideas of the ancient Egyptians (about 1500 B.C.), who were concerned with practical problems involving the measurement of areas and volumes. The Egyptians were satisfied with the geometry that was needed to construct buildings and pyramids; they cared little about mathematical derivations or proofs of formulas. We shall follow that model in this chapter.

Euclid's greatest contribution was his collection and systemization of most of the Greek mathematics of his time. His reputation rests mainly on his work titled *The Elements,* which contains geometry, number theory, and some algebra. Most U.S. textbooks on plane and solid geometry contain essentially the material in the geometry portions of Euclid's *The Elements.* No work, except the Bible, has been so widely used or studied, and probably no work has influenced scientific thinking more than this one. Over a thousand editions of *The Elements* have been published since the first printed edition appeared in 1482, and for more than 2000 years, this work has dominated the teaching of geometry.

8.1 Lines, Angles, and Triangles

Objectives

You should be able to:

A › Identify and name points, lines (parallel, intersecting, and perpendicular), segments, and rays.

B › Name an angle three different ways.

C › Classify an angle.

D › Identify complementary and supplementary angles and find the complement or supplement of a given angle.

E › Classify a triangle.

F › Find the measure of the third angle, given the measures of two angles in a triangle.

To Succeed, Review How To . . .

1. Add, subtract, multiply, and divide whole numbers. (pp. 24, 37, 49, 62)
2. Solve equations. (pp. 89–91, 240–241)

Getting Started

The painting, Blanc et Rouge/*Composition in Red, Blue, and Yellow,* was painted by Piet Mondrian, a founder of the De Stijl art movement, which he left in 1923. Mondrian's own artistic theory was "Neoplasticism" a search for harmony and balance, which he pursued for the rest of his career. Neoplasticism used horizontal or vertical black lines of varied thickness to create a grid of planar rectangles, some of which were filled in with black or white, or vivid red, blue, or yellow. This characteristic abstract style expunged all reference to the real world ("Trees! How ghastly!" Mondrian said). In this section we shall study lines (parallel, perpendicular, and intersecting) like the ones in the painting. If you want to make your own painting using lines, squares, and rectangles, go to the paint machine at http://www.ptank.commondrian.

Mondrian, Piet (1872–1944)

II: Blanc et Rouge (white and red), 1937/Composition in Red, Blue, and Yellow, 1937–42.

Oil on canvas, $23\frac{3}{4} \times 21\frac{7}{8}$ (60.3 × 55.4 cm).

A › Points, Lines, and Rays

The word **geometry** is derived from the Greek words *geo* (earth) and *metron* (measure). The basic elements of geometry are **points, lines,** and **planes.**

A **point** can be regarded as a *location* in space. A point has no breadth, no width, and no length. We represent a *point* as a *dot* and label it with capital letters such as A, B, and C or P, Q, and R like this:

A	B	Q	R
•	•	•	•
Point A	Point B	Point Q	Point R

We can use points to make **lines.** A **line** is a set of points that extends *infinitely* in both directions. A line has no width or breadth, but it does have *length.* Lines are named with lowercase letters such as *l*, *m*, or *n* or by using two of the points on the line.

Line *l* $\quad$ $\overline{AB}$ or $\overleftrightarrow{AB}$

The arrowheads at each end of these lines signify that the line extends *infinitely* in each direction.

A piece of a line using *A* and *B* as endpoints is a **line segment** and is denoted by $\overline{AB}$.

Line segment $\overline{AB}$

Lines that are on the same flat surface (*plane*) but never intersect (cross) are called **parallel lines,** while lines that intersect (cross) are called **intersecting lines.** The symbol || is used to indicate that two lines are **parallel.**

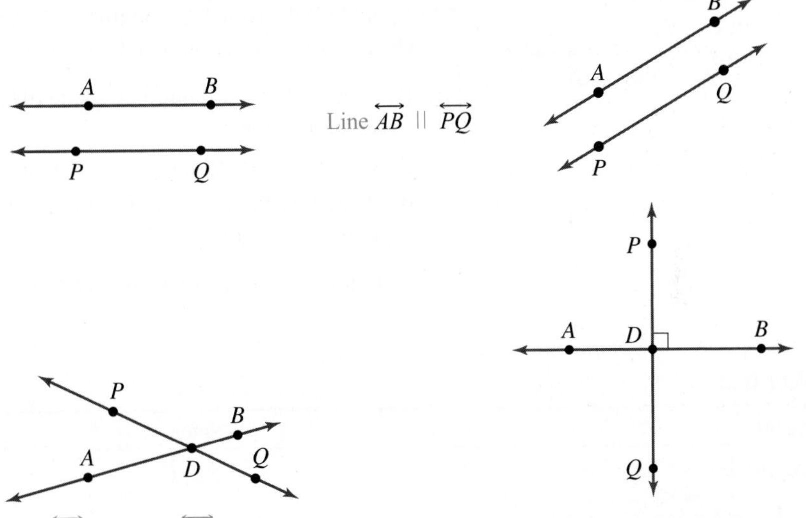

Line $\overleftrightarrow{AB} \parallel \overleftrightarrow{PQ}$

Line $\overleftrightarrow{AB}$ intersects $\overleftrightarrow{PQ}$ at point *D*.

If the intersection is at right angles, the lines are perpendicular ($\perp$). Line $\overleftrightarrow{AB} \perp \overleftrightarrow{PQ}$.

A **ray** is part of a line with one *endpoint* and extending *infinitely* in one direction.

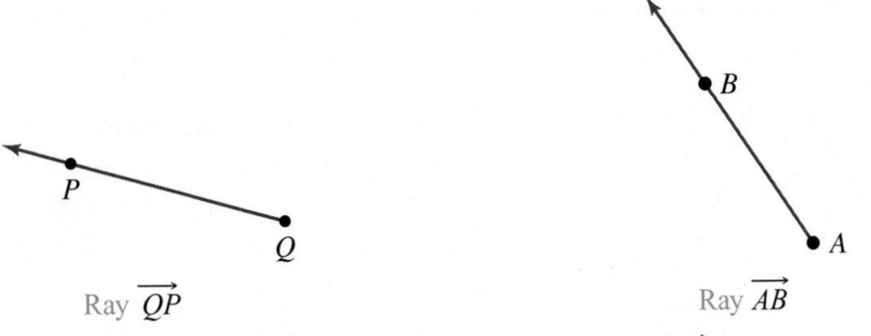

Ray $\overrightarrow{QP}$ $\quad$ Ray $\overrightarrow{AB}$

Note that the endpoint of ray $\overrightarrow{QP}$ is *Q* and the endpoint of ray $\overrightarrow{AB}$ is *A*.

B > Naming Angles

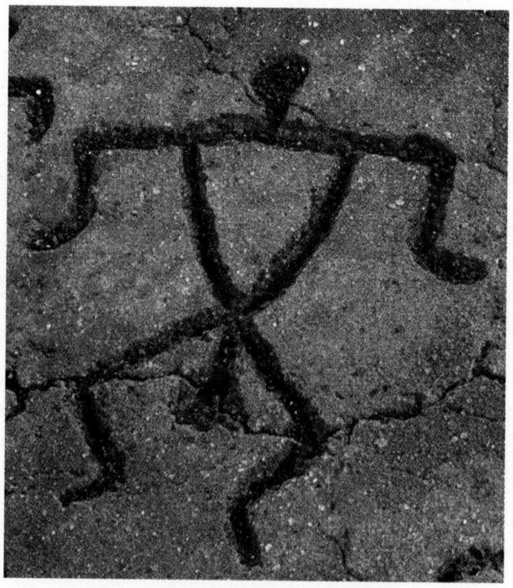

The photo shows one of the many examples of ancient Hawaiian rock carvings (petroglyphs) at Waikoloa. These carvings contain images of the human figure varying in complexity from simple angular figures to triangular figures to muscular figures. How many angles do you see in the photo? How would you measure these angles?

Have you heard the expression "I am looking at it from a new angle"? In geometry, an angle is the figure formed by two sides with a common point called the vertex. Here is the definition.

ANGLE

An angle is the figure formed by two rays (sides) with a common endpoint called the vertex.

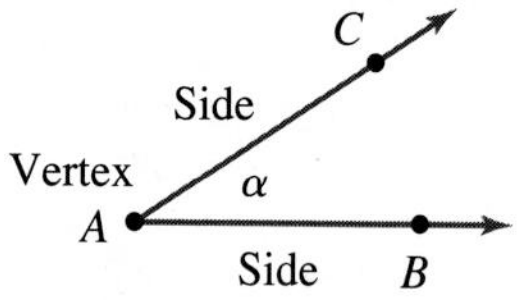

The two rays $\overleftrightarrow{AB}$ and $\overleftrightarrow{AC}$ are called the **sides** of the angle. We use the symbol $\angle$ (read "angle") in naming angles. Thus, the angle in the figure can be named $\angle\alpha$ (angle alpha), $\angle BAC$, or $\angle CAB$. (Note that the middle letter designates the vertex.)

The angle in the margin can be named in the following three ways:

1. By using a letter or a number inside the angle. Thus, we would name the angle $\angle\alpha$ (read "angle alpha").
2. By using the vertex letter only, such as $\angle A$.
3. By using three letters, one from each ray, with the vertex letter in the middle. The angle would be named $\angle BAC$ or $\angle CAB$.

Table 8.1 summarizes the concepts we have discussed.

Table 8.1

Concept	Name	Notation
A **point** shows a location. • P	Point ***P***	P
A **line** extends infinitely in both directions. 	Line ***AB***	$\overleftrightarrow{AB}$
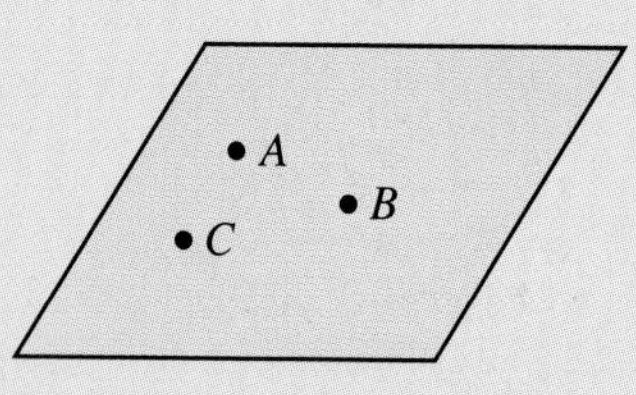 Three points determine a **plane.**	Plane ***ABC***	ABC

Table 8.1 *(continued)*

Concept	Name	Notation
The **line segment** *AB* includes the endpoints *A* and *B* and all points between *A* and *B*. A B	Segment ***AB***	$\overline{AB}$
The line *AB* is **parallel** to the line *PQ*. 	Line ***AB*** is **parallel** to line ***PQ*.**	Line $\overleftrightarrow{AB} \parallel \overleftrightarrow{PQ}$
The line *AB* is **perpendicular** to the line *PQ*. The green symbol indicates that the lines *AB* and *PQ* intersect at a 90° angle.	Line ***AB*** is **perpendicular** to line ***PQ*.**	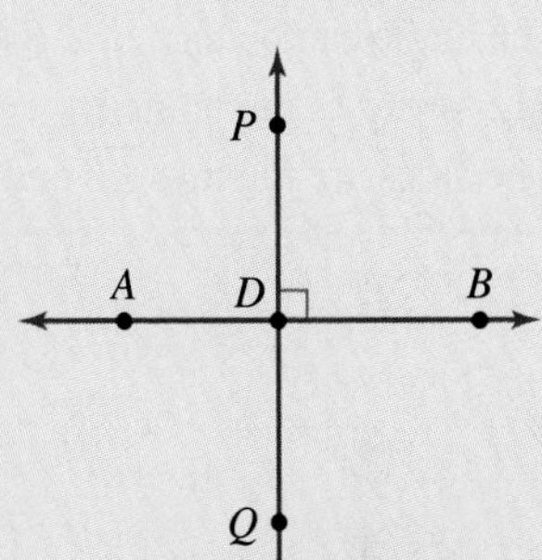 Line $\overleftrightarrow{AB} \perp \overleftrightarrow{PQ}$
The **ray** *AB* extends infinitely in one direction starting at endpoint *A* and going through point *B*. 	Ray ***AB***	$\overrightarrow{AB}$

EXAMPLE 1 Naming geometric figures

Refer to the figure and identify

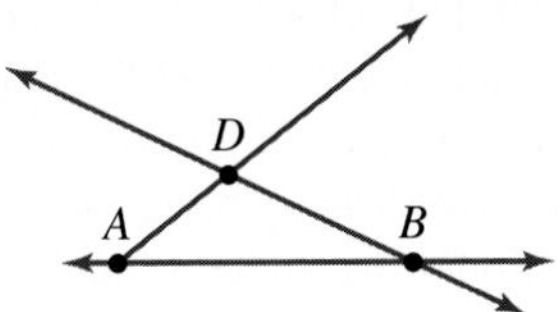

a. Three points
b. Two lines
c. Three line segments
d. A pair of intersecting lines
e. A ray

SOLUTION

a. The three points in the figure are *A*, *B*, and *D*.
b. There are two lines in the figure: line $\overleftrightarrow{AB}$ and line $\overleftrightarrow{BD}$. (You can also call $\overleftrightarrow{BD}$ line $\overleftrightarrow{DB}$.)

(continued)

PROBLEM 1

Refer to the figure and identify

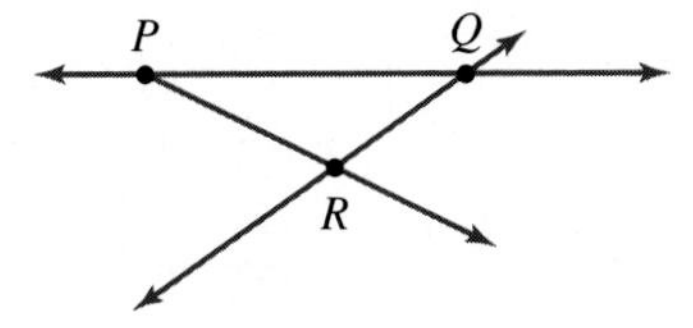

a. Three points
b. Two lines
c. Three line segments
d. A pair of intersecting lines
e. A ray

Answers to PROBLEMS

1. a. *P, Q, R* **b.** $\overleftrightarrow{PQ}$ and $\overleftrightarrow{RQ} = \overleftrightarrow{QR}$ **c.** $\overline{PQ}$, $\overline{PR}$, and $\overline{RQ}$ **d.** $\overleftrightarrow{PQ}$ and $\overleftrightarrow{RQ}$ intersect **e.** $\overrightarrow{PR}$

c. $\overleftrightarrow{AB}$, $\overleftrightarrow{AD}$, and $\overleftrightarrow{DB}$

d. Lines $\overleftrightarrow{AB}$ and $\overleftrightarrow{BD}$ intersect.

e. There is only one ray in the figure: $\overrightarrow{AD}$ (its endpoint is at A and it goes through point D).

EXAMPLE 2 Naming angles and vertices

Consider the angle in the figure shown here.

a. Name the angle in three different ways.

b. Name the vertex of the angle.

c. Name the sides of the angle.

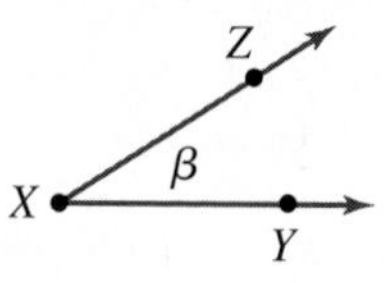

SOLUTION

a. The angle can be named $\angle\beta$ (Greek letter beta), $\angle X$, or $\angle YXZ$ (or $\angle ZXY$).

b. The vertex is the point X.

c. The sides are the rays $\overrightarrow{XZ}$ and $\overrightarrow{XY}$.

PROBLEM 2

Name the angle shown in three different ways.

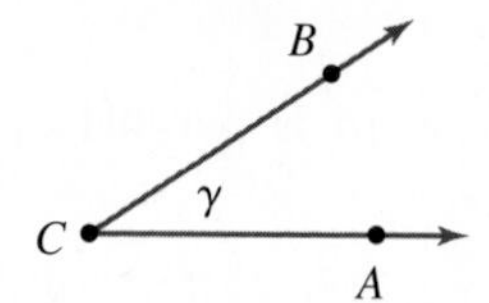

C > Classifying Angles

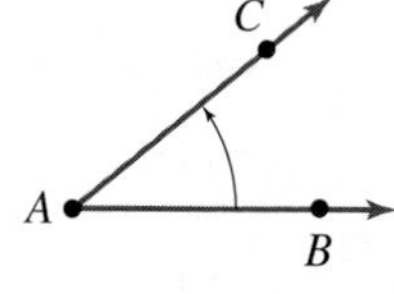

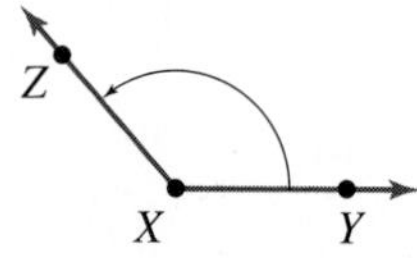

>Figure 8.1

For most practical purposes, we need to have a way of measuring angles. We first consider the amount of **rotation** needed to turn one side of an angle so that it *coincides with* (falls exactly on top of) the other side. Figure 8.1 shows two angles, $\angle CAB$ and $\angle ZXY$, with curved arrows to indicate the rotation needed to turn the rays $\overrightarrow{AB}$ and $\overrightarrow{XY}$ so that they coincide with the rays $\overrightarrow{AC}$ and $\overrightarrow{XZ}$, respectively. Clearly, the amount needed for $\angle ZXY$ is greater than that for $\angle CAB$. To find how much greater, we have to measure the amounts of rotation.

The most common unit of measure for an angle is the *degree*. We can trace the degree system back to the ancient Babylonians, who used a base 60 system of numeration. The Babylonians considered a *complete revolution* of a ray as indicated in Figure 8.2, and divided that into 360 equal parts. Each part is **1 degree,** denoted by **1°**. Thus, a complete revolution is equal to 360°. One-half of a complete revolution is 180° and gives us an angle that is called a **straight angle** (see Figure 8.3). One-quarter of a complete revolution is 90° and gives a **right angle** (see Figure 8.4). Notice the small square at Y to denote that it is a right angle.

In practice, the size of an angle is measured with a protractor (see Figure 8.5). The protractor is placed with its center at a vertex of the angle and the straight side of the protractor along one side of the angle, as in Figure 8.6. The measure of $\angle BAC$ is then read as 70° (because it is obviously less than 90°) and the measure of $\angle DAC$ is read as 110°. Surveying and navigational instruments, such as the sextant, use the idea of a protractor to measure angles very precisely. The measure of $\angle BAC$ is written $m\angle BAC$.

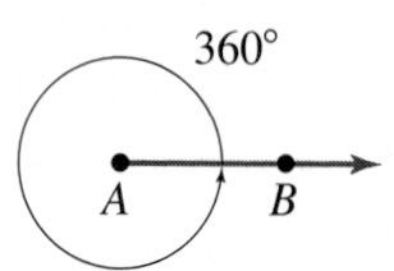

>Figure 8.2 A complete revolution.

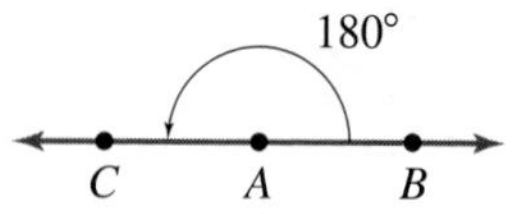

>Figure 8.3 The **straight** angle CAB.

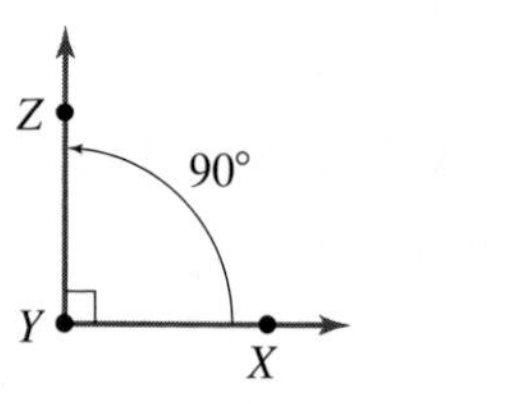

>Figure 8.4 The **right** angle XYZ.

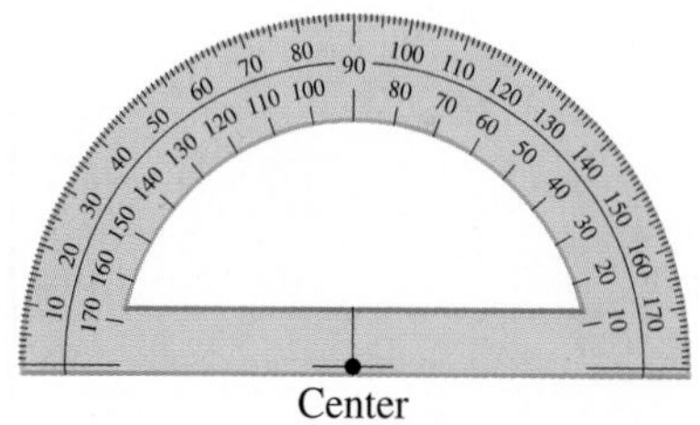

>Figure 8.5 A protractor.

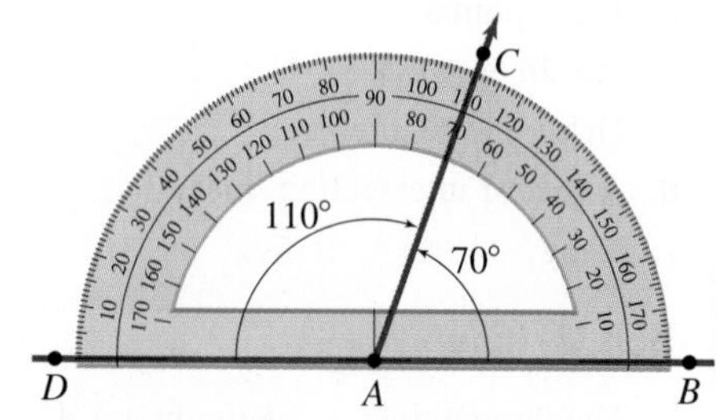

>Figure 8.6 Measuring an angle.

Answers to PROBLEMS

2. $\angle\gamma$, $\angle C$, $\angle ACB$ (or $\angle BCA$)

We have already named two angles: a *straight angle* (180°) and a *right angle* (90°). Certain other angles are classified as follows:

TYPES OF ANGLES

An **acute angle** is an angle of measure *greater* than 0° and *less* than 90°.
An **obtuse angle** is an angle of measure *greater* than 90° and *less* than 180°.

We can summarize this discussion by classifying angles according to their measurement. Here is the way we do it:

TYPES OF ANGLES

Right angle: An angle whose measure is 90°.
Straight angle: An angle whose measure is 180°.
Acute angle: An angle whose measure is greater than 0° and less than 90°.
Obtuse angle: An angle whose measure is greater than 90° and less than 180°.

Here are some examples:

The following angles are all acute angles (between 0° and 90°):

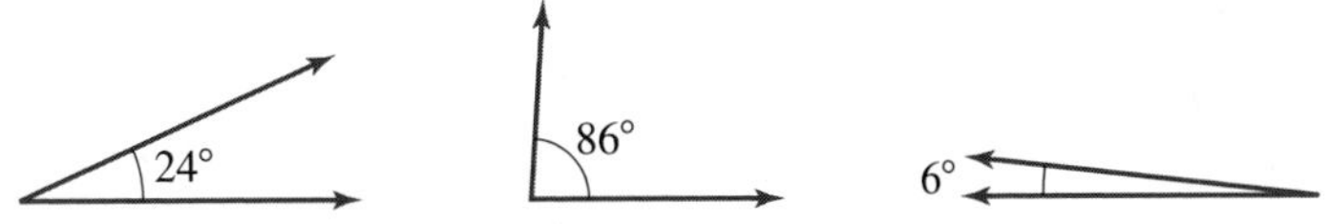

The following angles are all obtuse angles (between 90° and 180°):

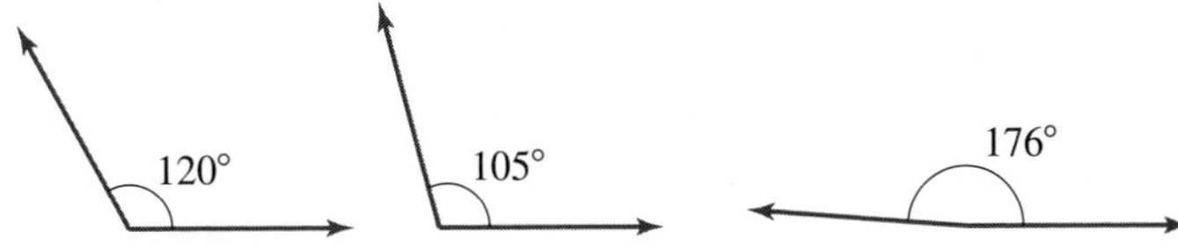

EXAMPLE 3 Classifying angles

Classify the given angles.

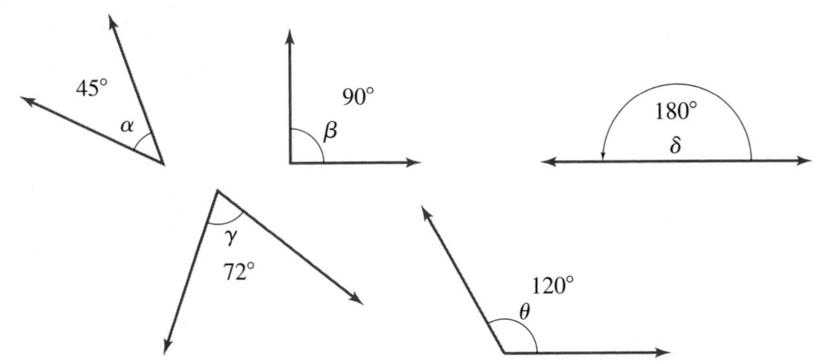

SOLUTION

α is between 0° and 90°; it is an acute angle.
β is exactly 90°; it is a right angle.
δ (delta) is exactly 180°; it is a straight angle.
γ (gamma) is between 0° and 90°; it is an acute angle.
θ (theta) is between 90° and 180°; it is an obtuse angle.

PROBLEM 3

Classify the given angles.

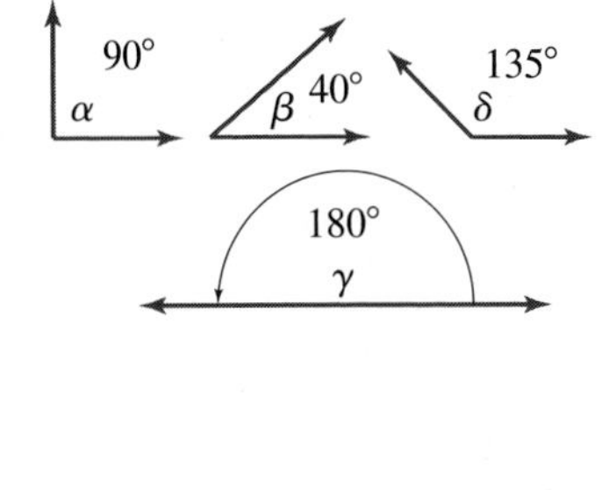

Answers to PROBLEMS

3. α is a right angle, β is an acute angle, δ is an obtuse angle, and γ is a straight angle.

EXAMPLE 4 Classifying angles: Solar eclipse angles

Why do we have solar eclipses (when the moon blocks the sun from view)? It is because the angle subtended (taken up) by the sun (0.52° to 0.54°) and the moon (0.49° to 0.55°) are almost identical. Classify these angles.

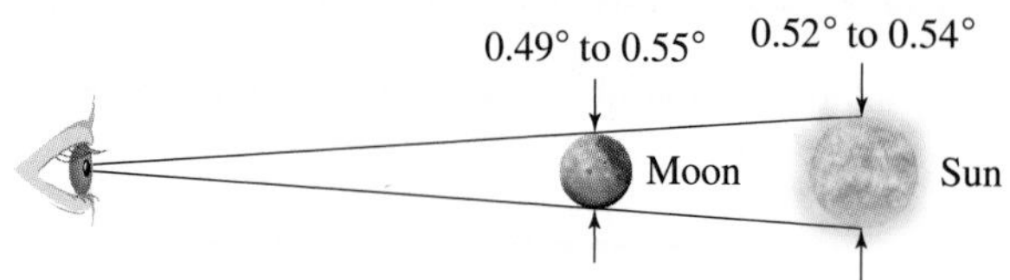

Source: National Aeronautics and Space Administration.

SOLUTION All the angles are actually smaller than one degree, so all of them are acute angles.

PROBLEM 4

The diagram shows the path of a satellite from noon to 12:03 P.M. Classify angles γ, β, and δ.

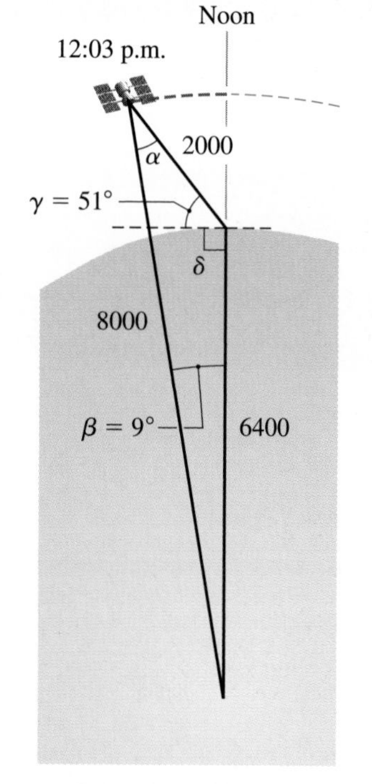

D > Complementary and Supplementary Angles

How far is the leg from being vertical, that is, how much is it bending? The answer can be obtained by measuring angle α. As you can see the sum of the measures of angles θ and α is 90°. This makes angles θ and α complementary angles. Here is the definition.

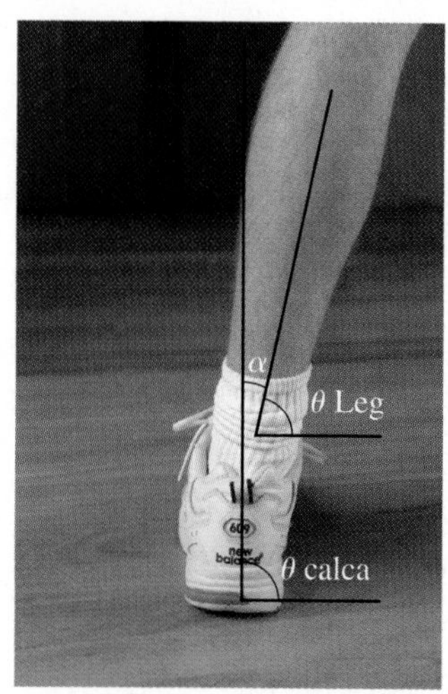

COMPLEMENTARY ANGLES

Complementary angles are two angles whose sum is 90°.* If $\angle A$ and $\angle B$ are complementary $m\angle A + m\angle B = 90°$ or, equivalently, $m\angle A = 90° - m\angle B$.

In the following figure, $m\angle A = 60°$ and $m\angle B = 30°$, so $m\angle A + m\angle B = 30° + 60° = 90°$. Thus, $\angle A$ and $\angle B$ are *complementary* angles.

Note that $m\angle A = 90° - m\angle B$.

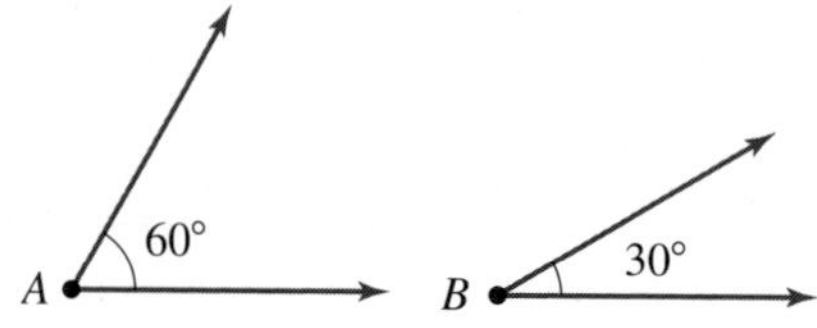

* Technically, the **sum** of their **measures** is 90°.

Answers to PROBLEMS

4. γ is an acute angle, β is an acute angle, and δ is a right angle.

SUPPLEMENTARY ANGLES

Supplementary angles are two angles whose sum is 180°.* If $\angle A$ and $\angle B$ are supplementary $m\angle A + m\angle B = 180°$ or, equivalently, $m\angle A = 180° - m\angle B$.

In the next figure, $m\angle A = 130°$ and $m\angle B = 50°$, so $m\angle A + m\angle B = 180°$. Thus, $\angle A$ and $\angle B$ are *supplementary* angles.

Note that $m\angle A = 180° - m\angle B$.

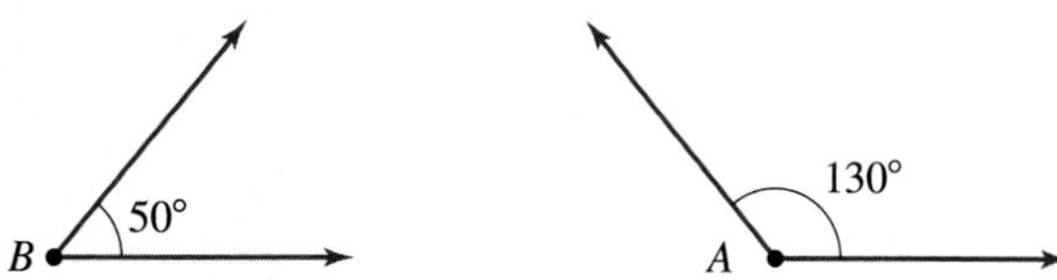

EXAMPLE 5 Identifying complementary angles

Identify all complementary angles in the figure.

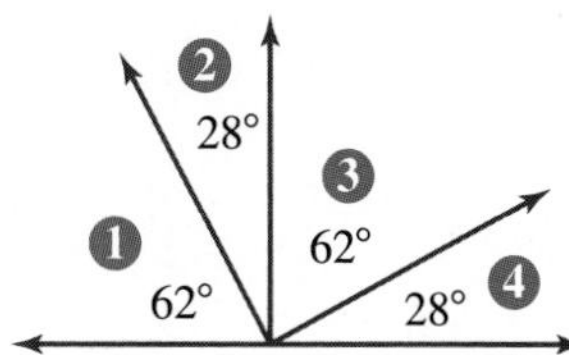

SOLUTION ∠❶ and ∠❷; ∠❷ and ∠❸; ∠❸ and ∠❹; ∠❶ and ❹.

PROBLEM 5

Identify all complementary angles in the figure.

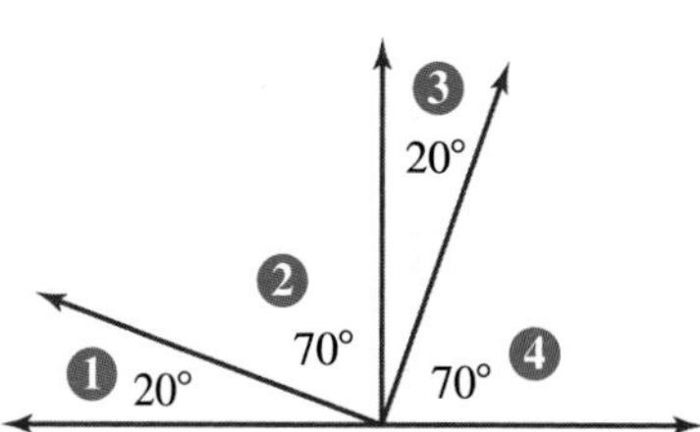

EXAMPLE 6 Finding the complement of an angle

Find the measure of the complement of a 50° angle.

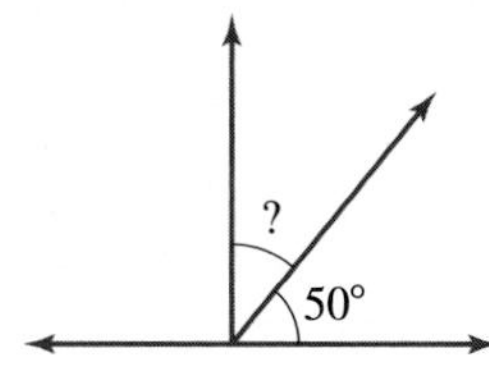

SOLUTION The measure of the complement of a 50° angle is $90° - 50° = 40°$.

CHECK $50° + 40° = 90°$.

PROBLEM 6

Find the measure of the complement of a 30° angle.

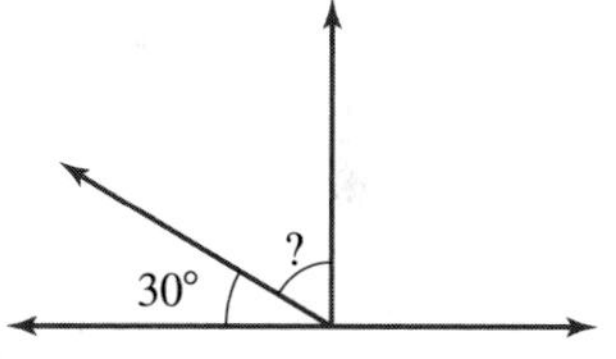

EXAMPLE 7 Finding the supplement of an angle

Find the measure of the supplement of angle ❶.

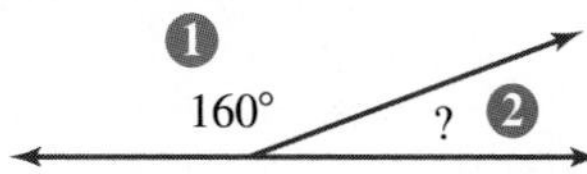

SOLUTION Since ❶ and ❷ are supplementary and $m\angle$❶ $= 160°$, $m\angle$❷ $= 180° - 160° = 20°$.

CHECK $160° + 20° = 180°$.

PROBLEM 7

Find the measure of the supplement of angle ❷.

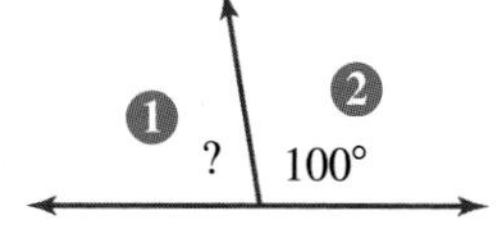

* Technically, the **sum** of their **measures** is 180°.

Answers to PROBLEMS

5. ∠❶ and ∠❷; ∠❷ and ∠❸; ∠❸ and ∠❹; ∠❶ and ∠❹ **6.** 60° **7.** 80°

E > Triangles

Now that we know how to work with angles, we expand our studies to triangles (literally, three angles). What is the most famous triangle you know? It is probably the Bermuda Triangle, a 1.5-million-mile triangular area of open sea with vertices at Miami, Bermuda, and Puerto Rico. What kind of a triangle is the Bermuda triangle? Certainly a mysterious

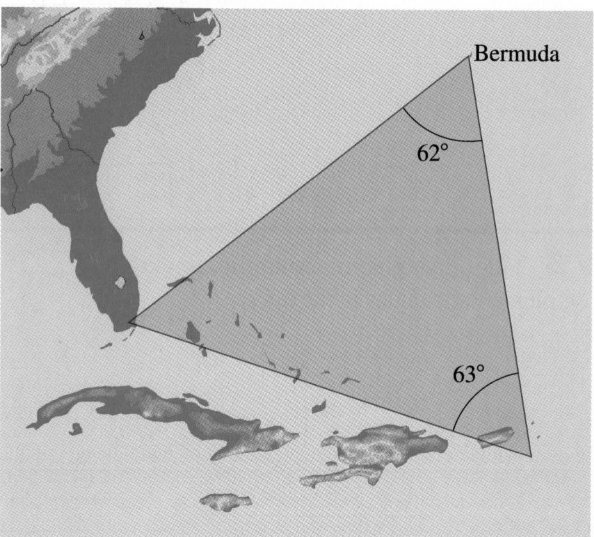

one, but as far as geometry is concerned, triangles are classified according to their angles or the number of equal sides. Here are the classifications:

CLASSIFYING TRIANGLES BY THEIR ANGLES

Right triangle: A triangle containing a *right* angle

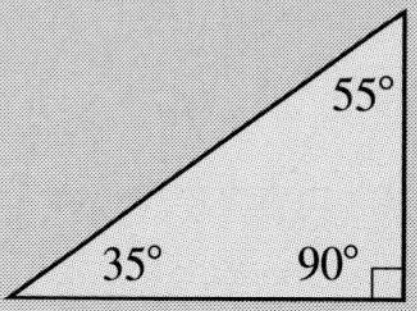

Acute triangle: A triangle in which all the angles are *acute*

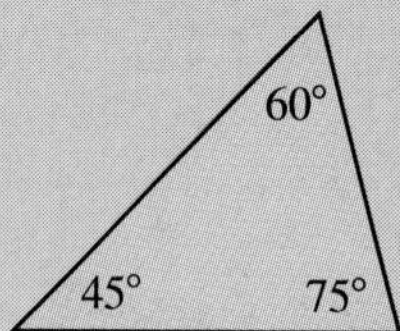

Obtuse triangle: A triangle containing an *obtuse* angle

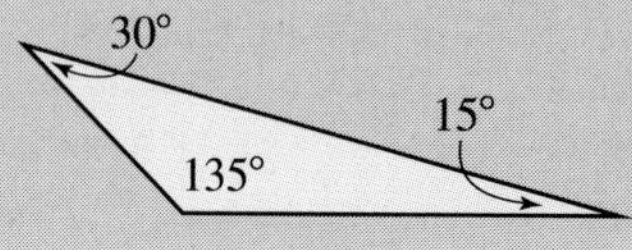

CLASSIFYING TRIANGLES BY THEIR SIDES

Scalene triangle: A triangle with *no* equal sides (Note that the sides are labeled I, II, and III to show that the lengths of the sides are different.)

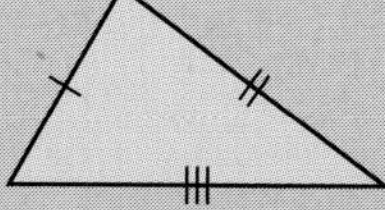

Isosceles triangle: A triangle with *two* equal sides

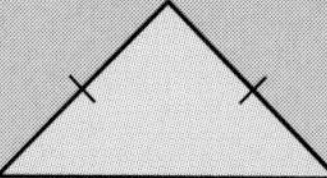

Equilateral triangle: A triangle with *all* three sides equal

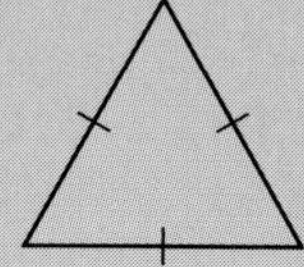

As you can see, all three angles in the Bermuda triangle are acute, so the triangle is an acute triangle. In addition, no two sides have the same length, that is, there are no equal sides, which makes it a scalene triangle. We can then say that the Bermuda triangle is an acute, scalene triangle. Triangles are named using their vertices. Thus, the triangle with vertices A, B, and C is named $\triangle ABC$, and the triangle with vertices P, Q, and R is named $\triangle PQR$.

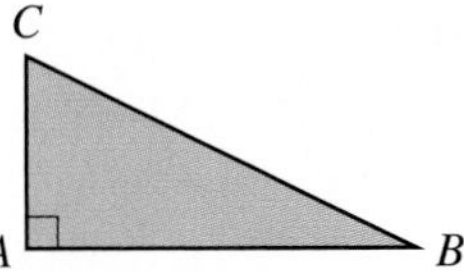

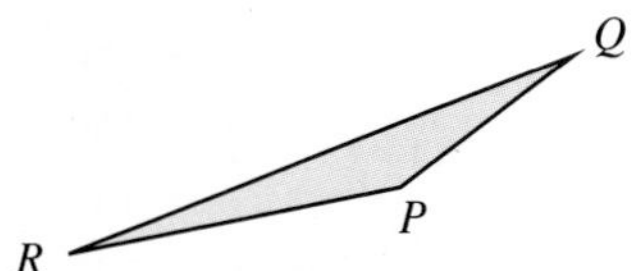

EXAMPLE 8 **Classifying triangles**

Classify the given triangles according to their angles and their sides.

a.

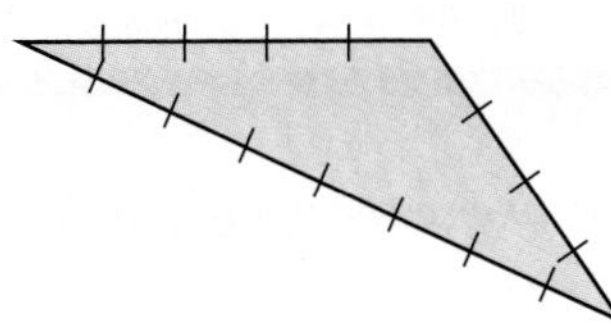

b.

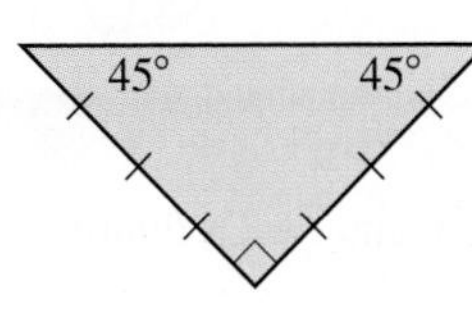

c.

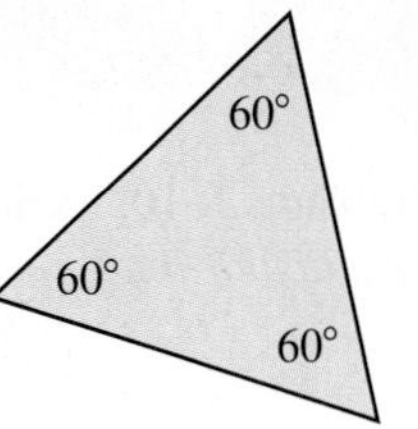

SOLUTION

a. The triangle has an *obtuse* angle and *no* equal sides; it is an *obtuse, scalene* triangle.

b. The triangle has *two* equal sides and a *right* angle; it is an *isosceles, right* triangle.

c. The triangle has three 60° angles; it is an *equilateral* triangle, which is also *equiangular*.

PROBLEM 8

Classify the given triangles according to their angles and their sides.

a.

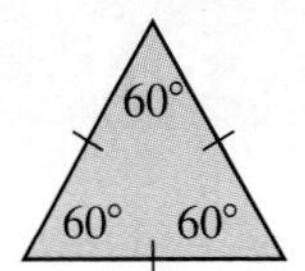

b.

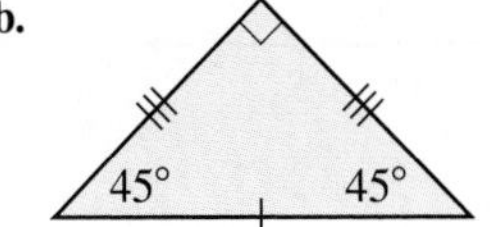

c.

F > Sum of the Angles of a Triangle

What do you notice about the sum of the measure of the angles of the triangles in Example 8, $45° + 45° + 90°$ and $60° + 60° + 60°$? The sum is 180°. This is always the case!

Here is a way to show this. Cut a triangle out of a sheet of paper as shown in the following figure. Label the angles 1, 2, and 3 and cut them off the triangle. Place the vertices of angles 1, 2, and 3 together. Now two of the sides form a straight line!

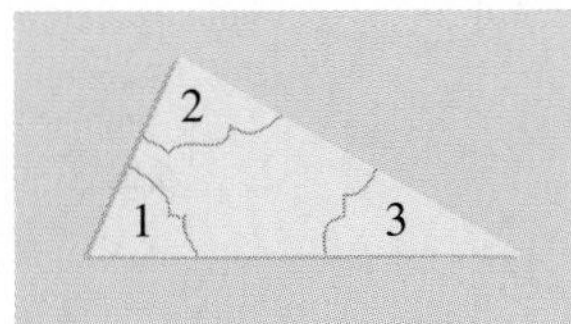

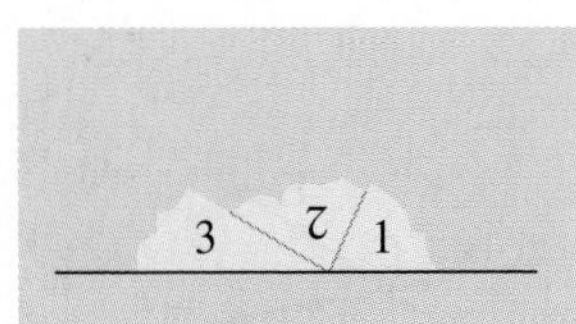

Thus, we have shown the following:

SUM OF THE MEASURES OF THE ANGLES IN A TRIANGLE

The sum of the measure of the three angles in any triangle is 180°.

Answers to PROBLEMS

8. a. Equilateral, equiangular triangle **b.** Isosceles, right triangle **c.** Obtuse, scalene triangle

EXAMPLE 9 **Finding the measure of an angle in a triangle**

In a triangle ABC, $m\angle A = 47°$, $m\angle B = 59°$. Find $m\angle C$.

SOLUTION

Since $m\angle A + m\angle B + m\angle C = 180°$

$$\begin{aligned} m\angle C &= 180° - m\angle A - m\angle B \\ &= 180° - 47° - 59° \\ &= 180° - 106° \\ &= 74° \end{aligned}$$

Thus, $m\angle C = 74°$.

PROBLEM 9

Find the third angle for the Bermuda Triangle shown on page 462.

Exercises 8.1

Web IT go to mhhe.com/bello for more lessons

⟨A⟩ Points, Lines, and Rays In Problems 1–10, identify and name each figure as a *line, line segment, ray, parallel lines, intersecting lines,* or *perpendicular lines*

1. R S

2. 

3. 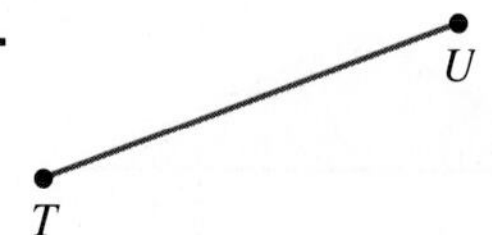

4. 

5. 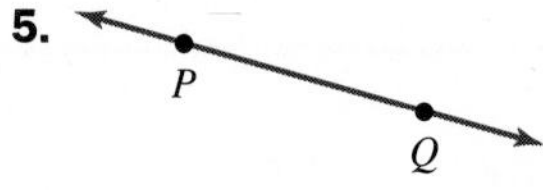

6. 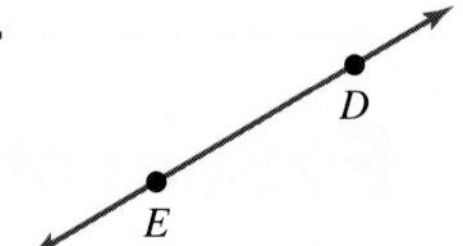

7. 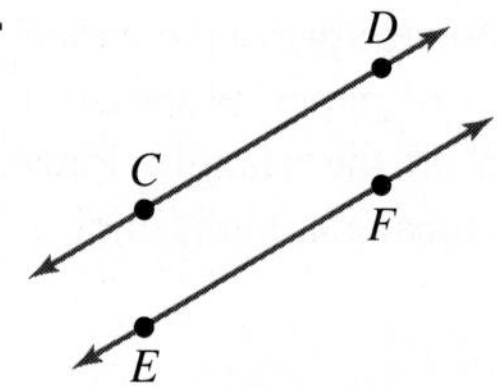

8. 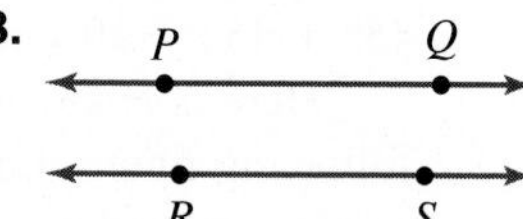

9. 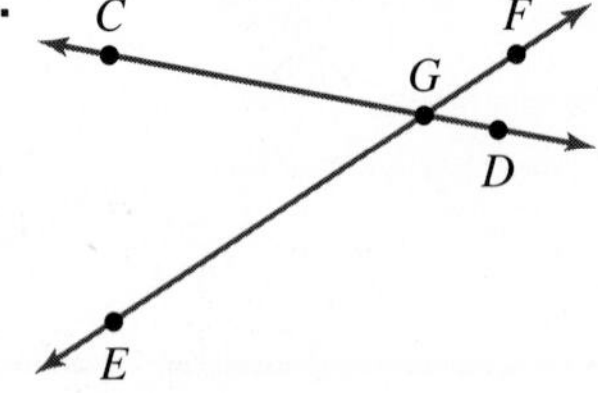

10. 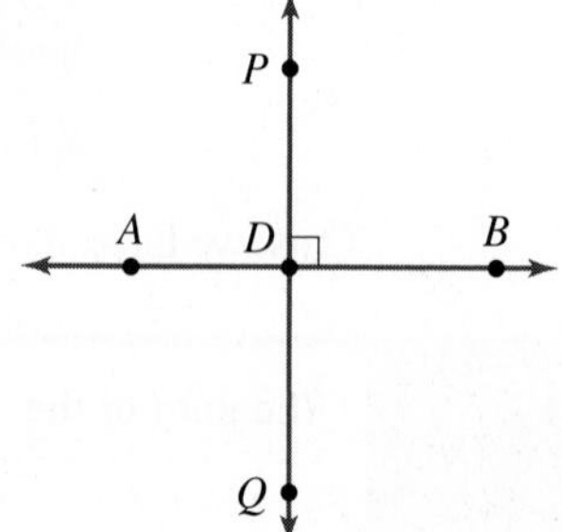

Answers to PROBLEMS

9. $180° - 62° - 63° = 55°$

⟨B⟩ Naming Angles In Problems 11 and 12, name the angle in three different ways.

11.

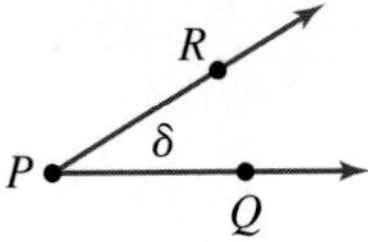

12.

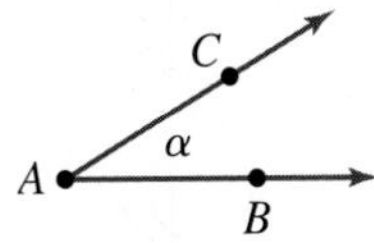

For Problems 13–16, refer to Figure 8.7.

13. Name $\angle\alpha$ in another way.

14. Name $\angle EAF$ in another way.

15. Name $\angle\beta$ in another way.

16. Name $\angle CAB$ in another way.

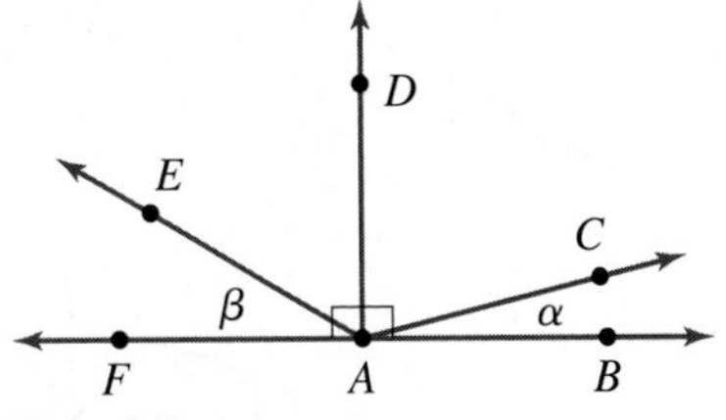

Figure 8.7

⟨C⟩ Classifying Angles In Problems 17–22, classify the given angle in Figure 8.7.

17. $\angle\alpha$

18. $\angle\beta$

19. $\angle DAB$

20. $\angle DAF$

21. $\angle FAB$

22. $\angle BAF$

23. List all the right angles in Figure 8.7.

24. List all the obtuse angles in Figure 8.7.

25. Name one straight angle in Figure 8.7.

⟨D⟩ Complementary and Supplementary Angles For Problems 26–36, refer to Figure 8.7.

26. Name the complement of $\angle\alpha$.

27. Name the complement of $\angle\beta$.

28. Name an angle that is the complement of $\angle EAF$.

29. Name an angle that is the complement of $\angle BAC$.

30. Name an angle that is supplementary to $\angle\alpha$.

31. Name an angle that is supplementary to $\angle\beta$.

32. Name an angle that is supplementary to $\angle BAD$.

33. If $m\ \angle\alpha = 15°$, find $m\ \angle CAD$.

34. If $m\ \angle\beta = 55°$, find $m\ \angle DAE$.

35. If $m\ \angle DAE = 35°$, find $m\ \angle\beta$.

36. If $m\ \angle CAD = 75°$, find $m\ \angle\alpha$.

⟨E⟩ Triangles In Problems 37–44, classify the triangle as scalene, isosceles, or equilateral and acute, right, or obtuse.

37.

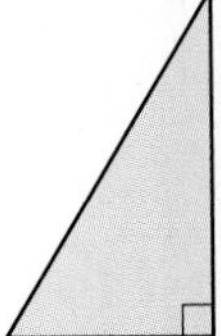

38.

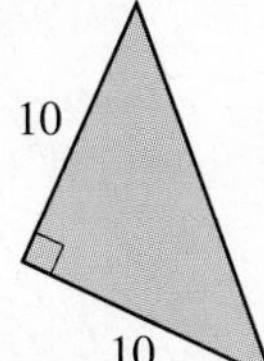

39.

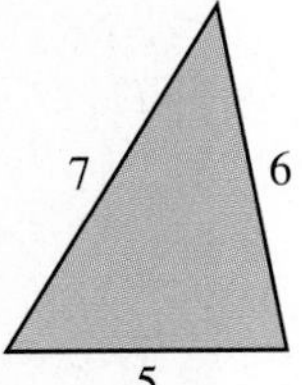

40.

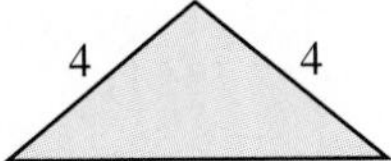

41.

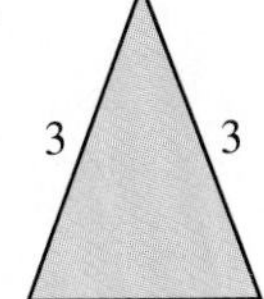

42.

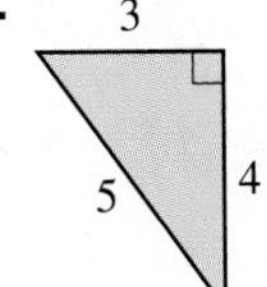

❯ Web IT go to **mhhe.com/bello** for more lessons

43.

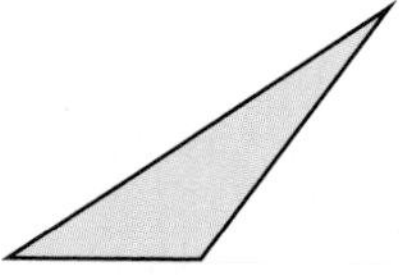

44.

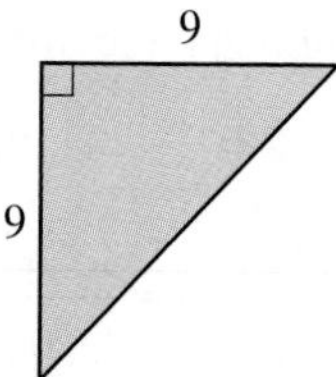

‹ F › Sum of the Angles of a Triangle In Problems 45–50, find the measure of the missing angle.

45.

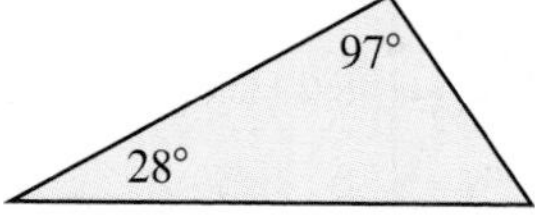

46.

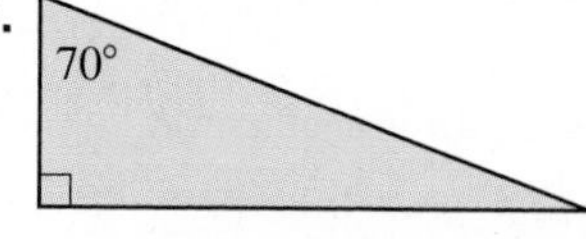

47.

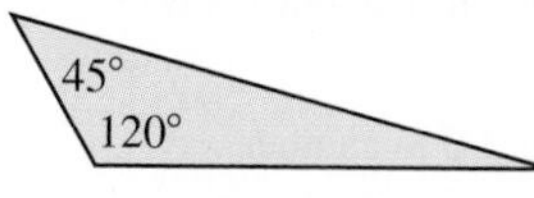

48.

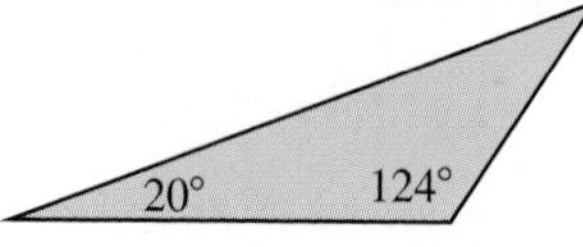

49.

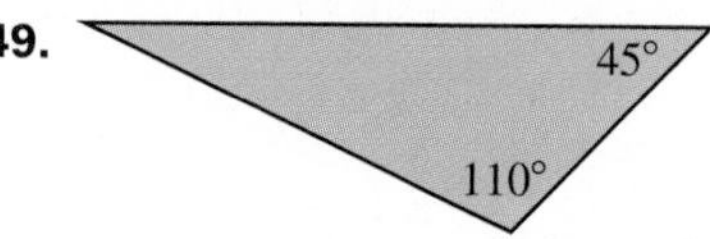

50.

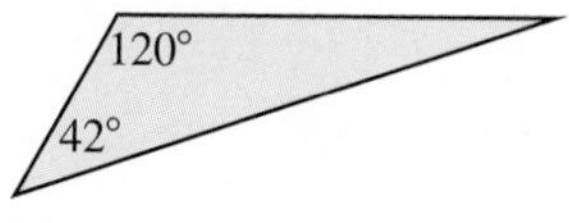

51. In a triangle ABC, $m\angle A = 37°$ and $m\angle C = 53°$. Find $m\angle B$.

52. In a triangle ABC, $m\angle B = 67°$ and $m\angle C = 105°$. Find $m\angle A$.

››› Applications

Angles Classify the angles shown in the photos as acute, obtuse, or right.

53. Classify $\angle\alpha$

54. Classify $\angle\beta$

55. Classify $\angle\gamma$

56. Classify $\angle\delta$

57. Classify $\angle\phi$

58. Classify $\angle\lambda$

59. Classify $\angle\mu$

60. Classify $\angle\theta$

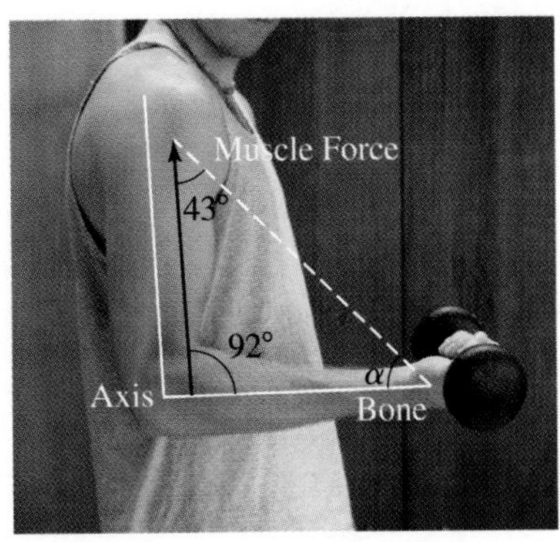

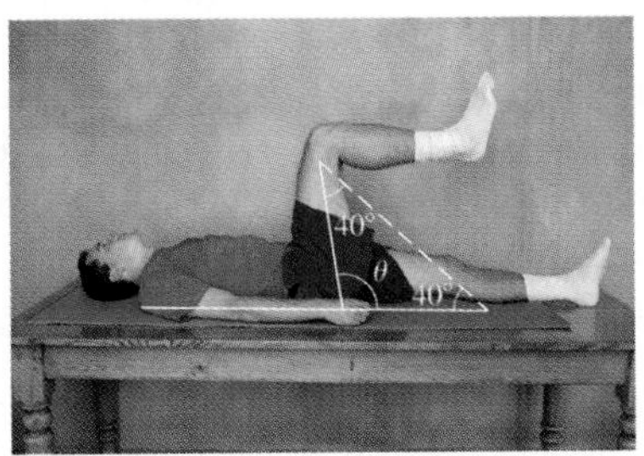

61. What is the measure of angle α? (left figure above)

62. What is the measure of angle θ? (right figure above)

Angles and triangles For Problems 63–68 refer to the rock art drawing.

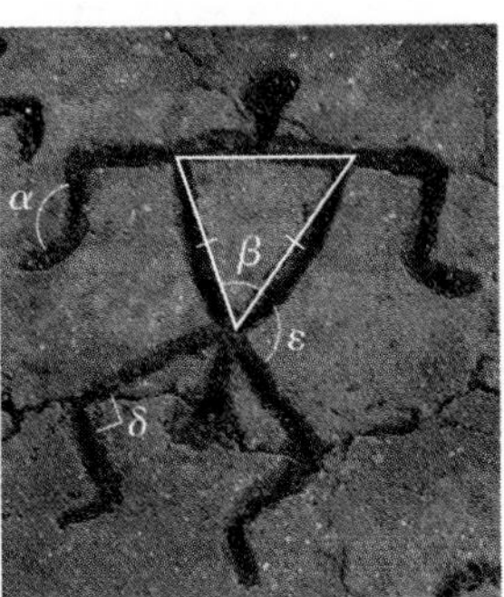

63. Classify angle α.

64. Classify angle β.

65. Classify angle δ.

66. Classify angle ε.

67. The body of the rock carving is in the shape of a triangle. Classify the triangle.

68. If the triangle is an isosceles triangle and the two equal angles are 62°, what is the measure of β?

Using Your Knowledge

Parallel Lines and a Transversal Line We have studied parallel lines, intersecting lines, and angles. Let us use our knowledge to generalize these ideas. What happens when a line intersects a pair of parallel lines?

A line that intersects a pair of parallel lines is called a **transversal.** In the figure, lines L_1 and L_2 are parallel and line L_3 is a transversal.

Think of $\angle 1$ as a **"small angle"** and $\angle 2$ as a **"big angle"**

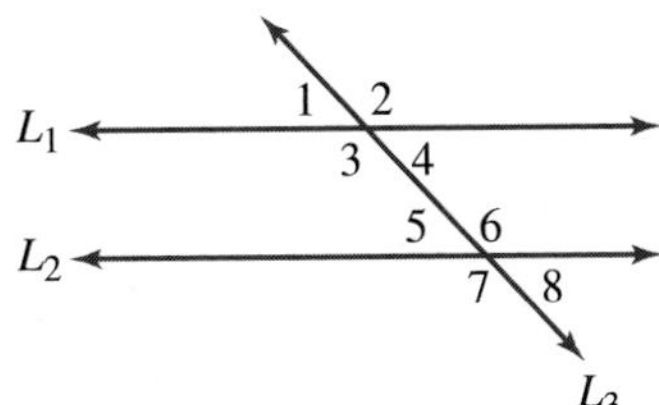

There is a special rule in geometry (the *Transversal Postulate*) that involves angles and transversals. It says that if two parallel lines are intersected by a transversal, then the corresponding angles are congruent (equal), denoted by using the symbol $\cong$.

Corresponding angles (angles on the same side of the transversal) are **congruent.**

$$m\angle 1 \cong \angle 5, \quad \angle 3 \cong \angle 7, \quad \angle 2 \cong \angle 6, \quad \text{and} \quad \angle 4 \cong \angle 8$$

Alternate interior angles (pairs of angles between the parallel lines and on opposite sides of the transversal) are **congruent.**

$$m\angle 3 \cong \angle 6, \quad \text{and} \quad \angle 4 \cong \angle 5.$$

Vertical angles are congruent.

$$m\angle 1 \cong \angle 4, \quad \angle 2 \cong \angle 3, \quad \angle 5 \cong \angle 8, \quad \text{and} \quad \angle 6 \cong \angle 7$$

Do you have to memorize all this? Absolutely not!

Think of $\angle 1$ as "a small angle" and $\angle 2$ as "a big angle." Here is the summary:

All small angles are equal.
All big angles are equal.

Now, you remember adding apples and apples and bananas and bananas? Let's apply the postulate to the fruits shown in the picture. Which angles are congruent (equal)?

If the two horizontal lines are parallel, all "small" angles: cherry, lemon, and strawberry are congruent (have the same measure). By the way, if the angle represented by the cherry is 60°, then the angle represented by the cantaloupe must be 120°, since those two angles (cherry and cantaloupe) are supplementary angles. Can you name the rest of the angles that are congruent (equal)? (*Hint:* All the "big" angles are equal.)

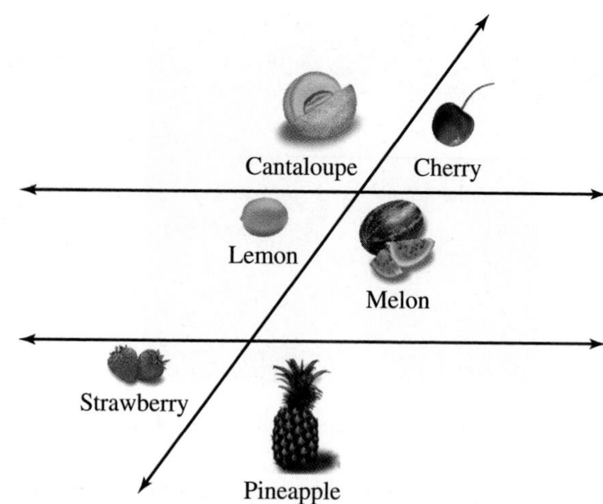

The figure will be used in Problems 69–70.

Lines L_1 and L_2 are parallel and L_3 is a transversal

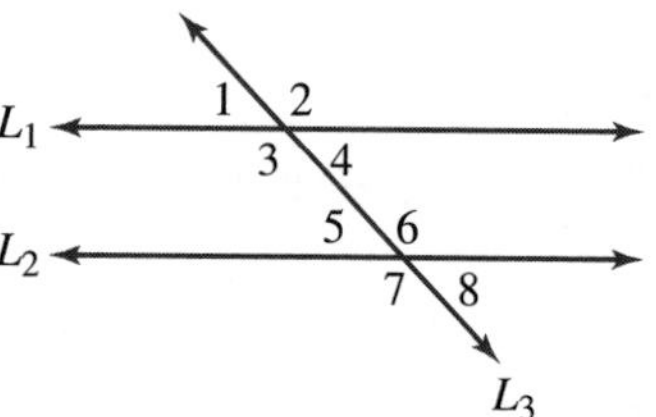

69. Assume that $m\angle 2 = 135°$. What is $m\angle 1$?

70. If $m\angle 2$ is a "big" angle, list all the angles that are congruent to $\angle 2$.

The figure will be used in Problems 71–72.

Lines L_1 and L_2 are parallel and L_3 is a transversal

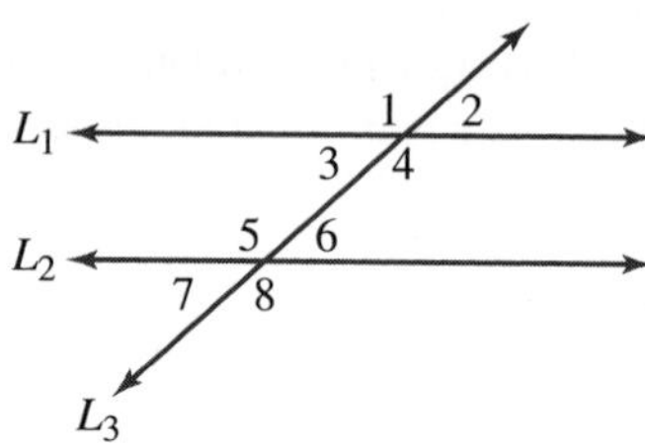

71. Assume that $m\angle 2 = 42°$. What is $m\angle 1$?

72. If $m\angle 2$ is a "small" angle, list all the angles that are congruent to $\angle 2$.

Music The graph shows the music preferences of young adults 14–19. Each of the sections in the circle (pie) graph shows an angle.

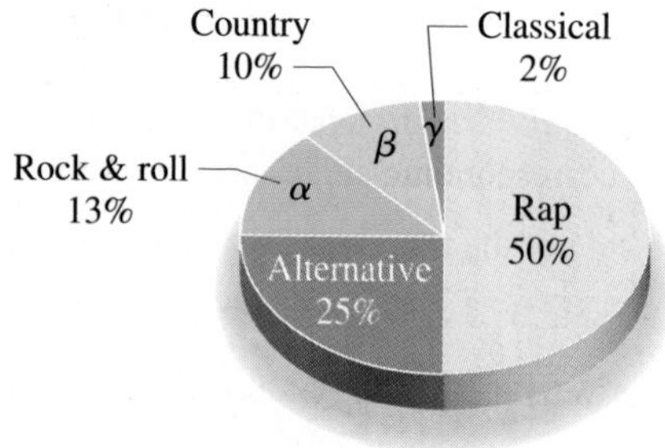

Source: Statistics Canada.

73. Which section shows a straight angle?

74. Which section shows a right angle?

75. How many sections show acute angles?

76. A complete circle covers 360°. The straight angle is 50% of 360°. How many degrees is that?

77. The right angle covers 25% of 360°. How many degrees is that?

78. Use the percents given in the graph to find $m\angle\alpha$.

79. Use the percents given in the graph to find $m\angle\beta$.

80. Use the percents given in the graph to find $m\angle\gamma$.

Write On

81. Write in your own words the difference between a line and a ray.

82. You can actually measure the length of a line segment. Can you measure the length of a line? Explain.

83. Can you measure the length of a ray? Explain

84. If the length of a ray is given by the letter r and the length of a line is given by the letter L, what symbol would you write ($=, <, >$) to make this statement true r ___ L? Explain your reasoning.

85. Write in your own words the type of angle you need when you want to use the phone on the right.

86. How many degrees would that angle be when the phone is not being used.

87. What angle do you think is best for viewing the screen on the phone? Explain.

Concept Checker

Fill in the blank(s) with the correct word(s), phrase, or mathematical statement.

88. A ______________ shows a **location.**

89. A ______________ extends **infinitely** in both directions.

90. A line ______________ includes its **endpoints.**

91. To denote that the line *AB* is **parallel** to the line *CD*, we use the symbol ______________.

92. A ______________ extends **infinitely** in one direction.

93. An ______________ is the figure formed by **two rays** with a **common endpoint.**

94. A ______________ angle is an angle whose **measure is 90°.**

95. A ______________ angle is an angle whose **measure is 180°.**

96. An ______________ angle is an angle whose measure is **greater than 0° and less than 90°.**

97. An ______________ angle is an angle whose measure is **greater than 90° and less than 180°.**

98. ______________ angles are two angles whose **sum is 90°.**

99. ______________ angles are two angles whose **sum is 180°.**

100. The **sum** of the measure of the **three angles** in a triangle is ______________.

straight
line
Supplementary
angle
ray
point
acute
segment
$\parallel$
90°
Complementary
180°
right
obtuse
$\perp$

Mastery Test

The figure will be used in Problems 101–105.

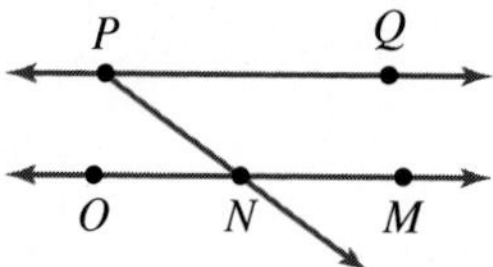

101. Name three points in the figure.

102. Name two lines in the figure.

103. Name three line segments in the figure.

104. Which pair of lines appears to be parallel lines in the figure.

105. Name a ray in the figure.

106. In triangle ABC, $m\angle A = 50°$, $m\angle B = 30°$. Find $m\angle C$.

107. Classify each of the triangles according to their angles and their sides.

a. 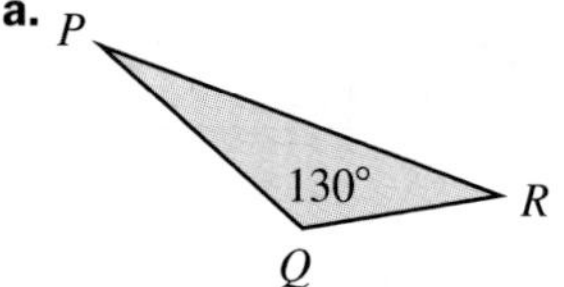

b. 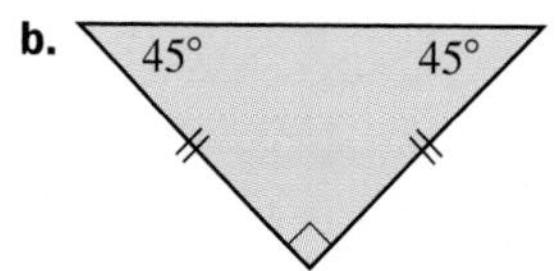

c. 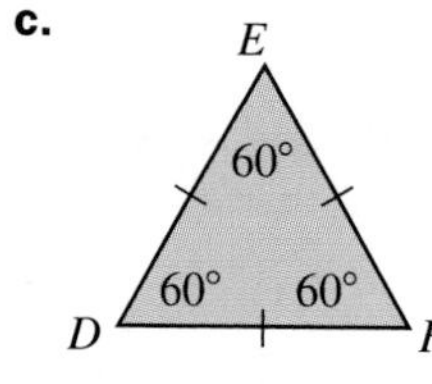

d. 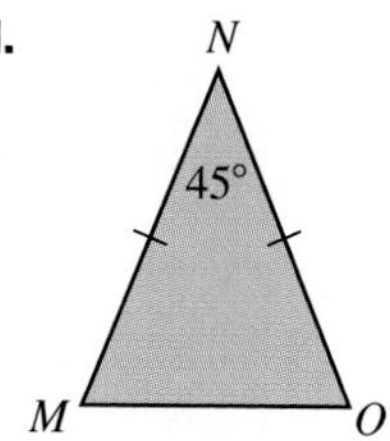

108. Find the measure of the complement of the 70° angle shown.

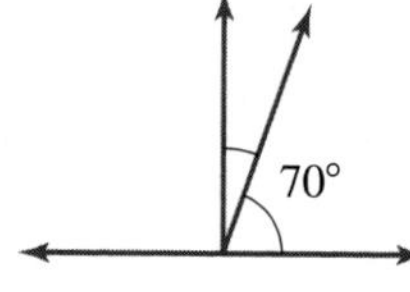

109. Find the measure of the supplement of the 35° angle shown.

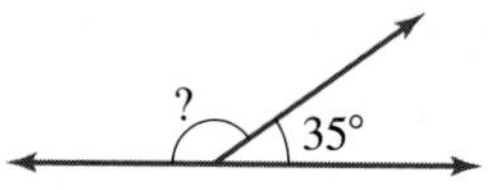

110. Classify the given angles.

a. 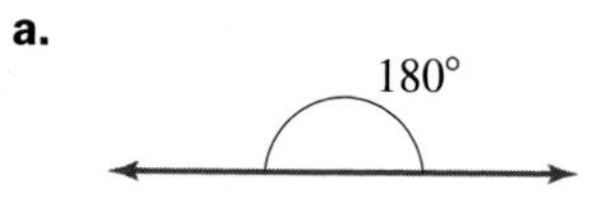

b. 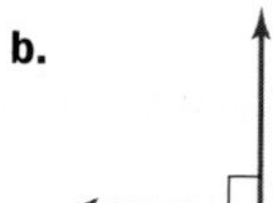

c.

d. 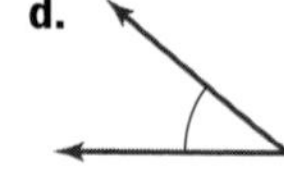

111. Name the angle in three different ways.

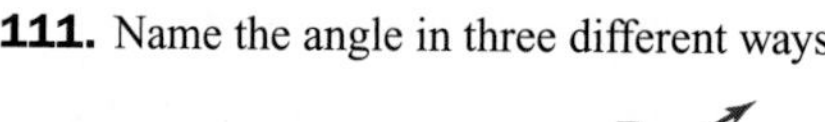

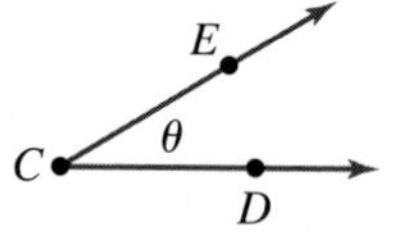

Skill Checker

In Problems 112–116, evaluate the expression.

112. $3.14 \cdot 100$ **113.** $2 \cdot 5.1 + 2 \cdot 3.2$ **114.** $2 \cdot 400 + 2 \cdot 500$ **115.** $4 \cdot 3\frac{1}{4}$ **116.** $5 \cdot 921$

8.2 Finding Perimeters

Objectives

You should be able to:

A > Find the perimeter of a polygon.

B > Find the circumference of a circle.

C > Solve applications involving the concepts studied.

To Succeed, Review How To . . .

1. Multiply decimals and fractions. (pp. 130–133, 212–214)
2. Add decimals and fractions. (pp. 151–155, 203–204)

Getting Started

Suppose you want to fence the lot. How many linear feet of fencing do you need? To answer this question we need to find the *perimeter* (distance around) of the lot by adding the lengths of the sides. (The symbol ['] means feet and ["] means inches.) The perimeter is:

(439 + 180 + 534 + 293) ft, or 1446 ft

A > Finding Perimeters

In general, the **perimeter** is the distance around an object.

PERIMETER OF A POLYGON

The **perimeter of a polygon** is the sum of the lengths of the sides.

Note: In a **regular** polygon all sides are of **equal** length.

Table 8.2 will give you an idea of the shapes and names of some of the polygons we will study.

Table 8.2

A **regular polygon** is a polygon with all *sides of equal length* and *all angles of equal measure*. They are usually named by using the *number* of sides. For example,

A **tri**angle is a **3**-sided polygon.
Quadrilateral is a **4**-sided polygon.
Pentagon is a **5**-sided polygon.
Hexagon is a **6**-sided polygon.
Heptagon is a **7**-sided polygon.
Octagon is an **8**-sided polygon.
Nonagon is a **9**-sided polygon.
Decagon is a **10**-sided polygon.

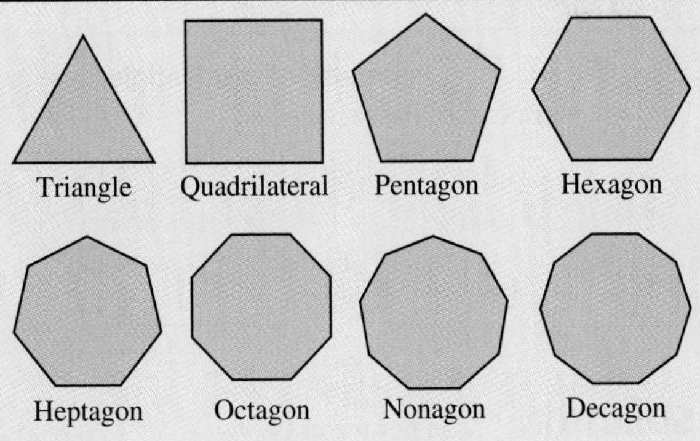

(continued)

Table 8.2 ***(continued)***

A **trapezoid** is *a quadrilateral* with *exactly one pair of opposite sides parallel.*

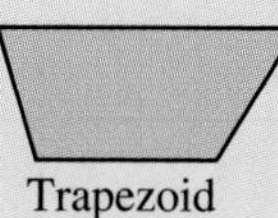
Trapezoid

A **parallelogram** is *a quadrilateral* in which *the opposite sides are parallel.*

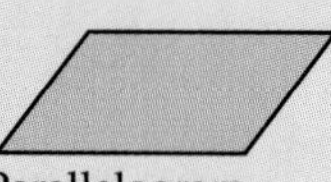
Parallelogram

A **rhombus** is *a parallelogram* with *all sides equal in length.*

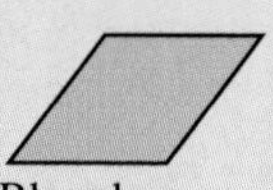
Rhombus

A **rectangle** is *a parallelogram* with *four right angles.*

Rectangle

A **square** is *a rectangle* with *all sides equal in length.*

Square

Now, we are ready to find the perimeter of some of these polygons.

EXAMPLE 1 **Perimeter of a polygon**

Find the perimeter of the polygon.

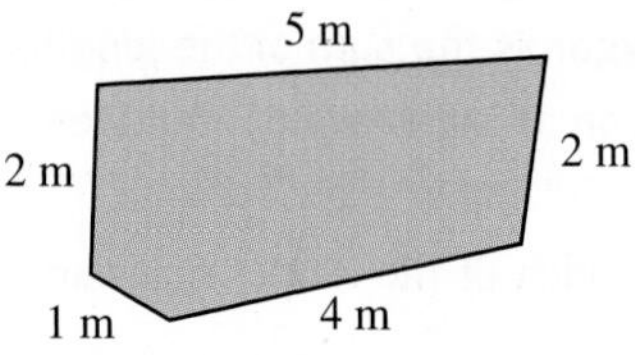

SOLUTION The perimeter is

$$(1 + 2 + 5 + 2 + 4)\text{ m} = 14\text{ m}$$

PROBLEM 1

Find the perimeter of the polygon.

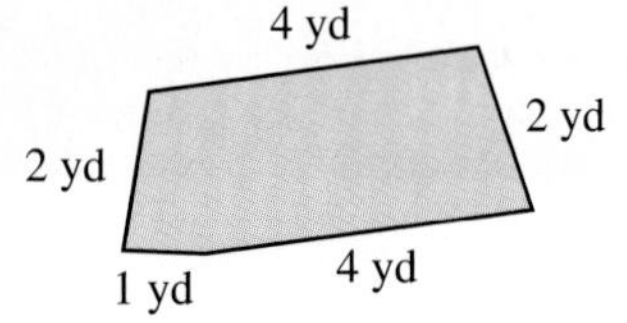

EXAMPLE 2 **Perimeter of a rectangle**

Find the perimeter of the rectangle.

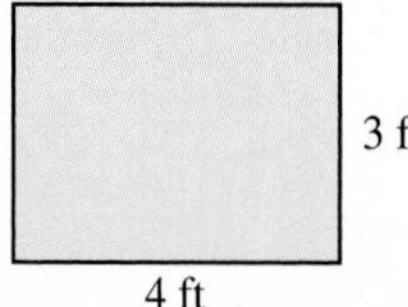

SOLUTION The perimeter is

$$(4 + 3 + 4 + 3)\text{ ft} = 14\text{ ft}$$

PROBLEM 2

Find the perimeter of the rectangle.

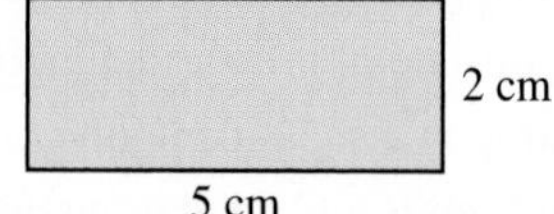

Answers to PROBLEMS

1. 13 yd **2.** 14 cm

Note that in Example 2 we added the length twice and the width twice because a rectangle has two pairs of sides with equal length. Here is the formula:

PERIMETER OF A RECTANGLE

The perimeter *P* of a rectangle is twice the length *L* plus twice the width *W*, that is,

$$P = 2 \cdot L + 2 \cdot W$$

EXAMPLE 3 Perimeter of a rectangle

Find the perimeter of a rectangle 5.1 centimeters long by 3.2 centimeters wide.

SOLUTION

$$\begin{aligned} P &= (2 \cdot 5.1 + 2 \cdot 3.2) \text{ cm} \\ &= (10.2 + 6.4) \text{ cm} \\ &= 16.6 \text{ cm} \end{aligned}$$

PROBLEM 3

Find the perimeter of a rectangle 6.3 inches long by 3.4 inches wide.

EXAMPLE 4 Perimeter of a square

Find the perimeter of the square.

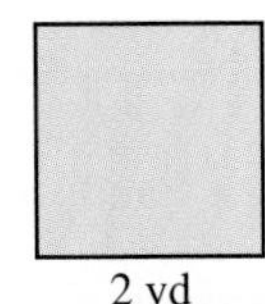

SOLUTION The perimeter is

$$(2 + 2 + 2 + 2) \text{ yd} = 8 \text{ yd}$$

PROBLEM 4

Find the perimeter of the square.

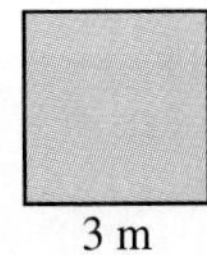

To find the perimeter of the square in Example 4 we added the length of the side four times because all four sides of a square have equal length. Here is the formula:

PERIMETER OF A SQUARE

The perimeter *P* of a square is four times the length of its side *S*, that is,

$$P = 4 \cdot S$$

EXAMPLE 5 Perimeter of a square

Find the perimeter of the square.

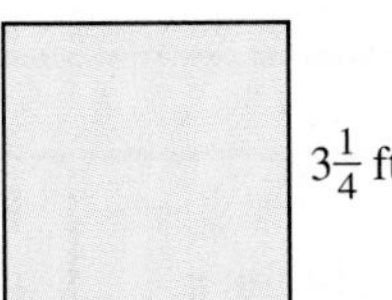

SOLUTION The perimeter is

$$\begin{aligned} P &= 4 \cdot 3\frac{1}{4} \text{ ft} \\ &= 4 \cdot \frac{13}{4} \text{ ft} \\ &= 13 \text{ ft} \end{aligned}$$

PROBLEM 5

Find the perimeter of the square.

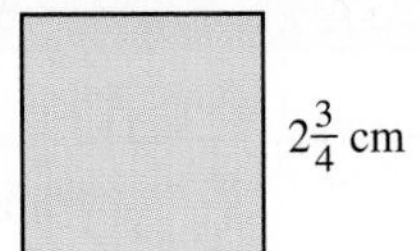

Answers to PROBLEMS

3. 19.4 in. **4.** 12 m
5. 11 cm

EXAMPLE 6 Perimeter of the Pentagon

Each of the outermost sides of the Pentagon building in Arlington, Virginia, is 921 feet long. What is the perimeter of this building?

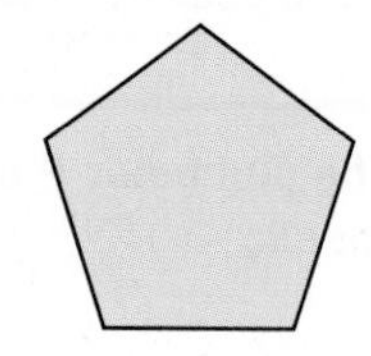

SOLUTION A pentagon has five sides, as shown. Thus, the perimeter of the building is $5 \cdot 921$ ft $= 4605$ ft.

PROBLEM 6

Find the perimeter of the pentagon.

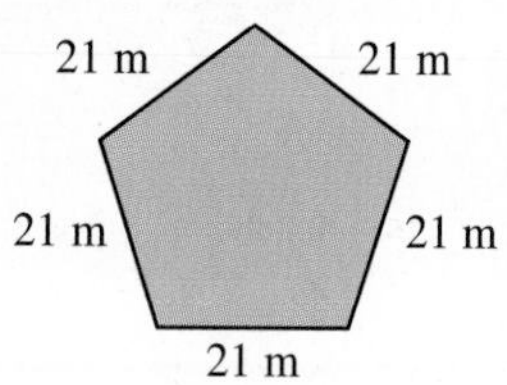

B > Finding Circumferences

You are probably familiar with the geometric figure called a circle. The distance around a circle is called the *circumference*. The figure shows a circle with center O. The line segment AB is called the **diameter,** d, and OC is called the **radius,** r. Note that, in general,

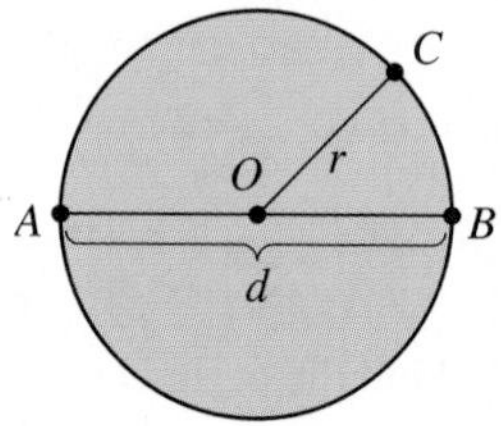

RELATIONSHIP BETWEEN DIAMETER d AND RADIUS r

$$d = 2 \cdot r \quad \text{and} \quad r = \frac{d}{2}$$

This means that the diameter d is always twice the radius r, and the radius r is half the diameter. Thus, if the diameter of a circle is 8 inches, the radius is half of that, or 4 inches. Also, if the radius of a circle is 3 centimeters, its diameter is twice that, or 6 centimeters. Now, suppose you measure the circumference C and the diameter d of a soda can. If you divide C by d, the ratio $\frac{C}{d}$ is very close to 3.14159 . . . . (The three dots mean the decimal continues.) If we find the ratio $\frac{C}{d}$ for different sized cans, the answer continues to be close to 3.14159 . . . . As a matter of fact, for any circle, we symbolize the ratio of $\frac{C}{d}$ by the letter π (pi, read "pie")—that is, $\frac{C}{d} = \pi$. Multiplying on both sides by d gives $d \cdot \frac{C}{d} = d \cdot \pi$, or $C = \pi \cdot d$. Thus,

CIRCUMFERENCE OF A CIRCLE

The circumference C of a circle of radius r equals two times π times the radius r or π times the diameter d. In symbols,

$$C = 2\pi r \quad \text{or} \quad C = \pi d$$

Note: π is approximately 3.14 or $\frac{22}{7}$.

EXAMPLE 7 Finding the circumference

Find the circumference of a circle whose radius is 4 centimeters. Use 3.14 for π.

SOLUTION Since the radius is 4 centimeters, the diameter is twice that, or 8 centimeters. The circumference is

$$\begin{aligned} C &= \pi \cdot d \\ &= 3.14 \cdot 8 \text{ cm} \\ &= 25.12 \text{ cm} \end{aligned}$$

PROBLEM 7

Find the circumference of a circle whose radius is 5 feet. Use 3.14 for π.

Answers to PROBLEMS

6. 105 m **7.** 31.4 ft

C > Applications Involving Perimeter and Circumference

EXAMPLE 8 Distance traveled by Venus

The planet Venus revolves around the sun in a nearly circular orbit whose diameter is about 100 million kilometers. Find the distance traveled by Venus in one revolution around the sun. Use 3.14 for π.

SOLUTION The distance traveled is the circumference C.

$$C = \pi \cdot d$$
$$= 3.14 \cdot 100 \text{ million kilometers}$$
$$= 314 \text{ million kilometers}$$

PROBLEM 8

Mars's orbit about the sun is approximately a circle 200 million kilometers in diameter. Find the distance traveled by Mars in one revolution around the sun. Use 3.14 for π.

EXAMPLE 9 Perimeter of exercise area

Have you heard the words "secure the perimeter"? One application of this concept is to establish security for a zone of a given perimeter, in this case, a corrections exercise area measuring 500 feet by 400 feet. What is the length of the security cable surrounding the exercise area?

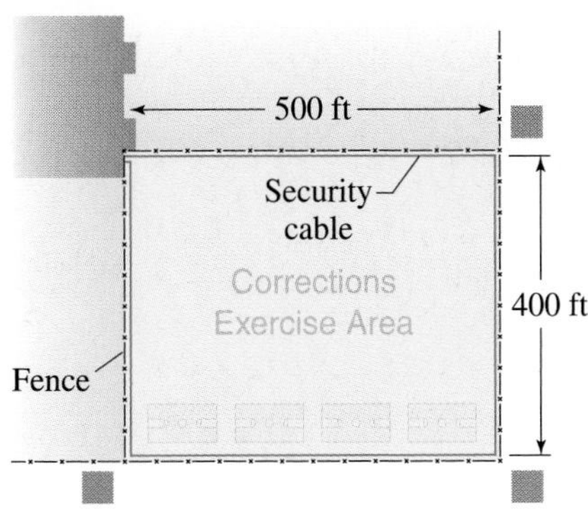

Source: Data from Fiber SenSys.

SOLUTION The cable is as long as the perimeter of a rectangle measuring 500 feet by 400 feet. Since the perimeter P of a rectangle is $P = 2L + 2W$, the perimeter of the area is

$$P = 2(500 \text{ ft}) + 2(400 \text{ ft}) = 800 \text{ ft} + 1000 \text{ ft} = 1800 \text{ ft}$$

Thus, the length of the security cable surrounding the exercise area is 1800 feet. (*Note:* Sometimes more than one security cable is used around the perimeter.)

PROBLEM 9

What would be the length of a security cable surrounding a smaller exercise area measuring 400 feet by 300 feet?

Answers to PROBLEMS

8. 628 million kilometers
9. 1400 ft

Exercises 8.2

Boost your grade at mathzone.com!
- Practice Problems
- NetTutor
- Self-Tests
- e-Professors
- Videos

Web IT go to mhhe.com/bello for more lessons

A **Finding Perimeters** In Problems 1–20, find the perimeters of the polygons.

1.

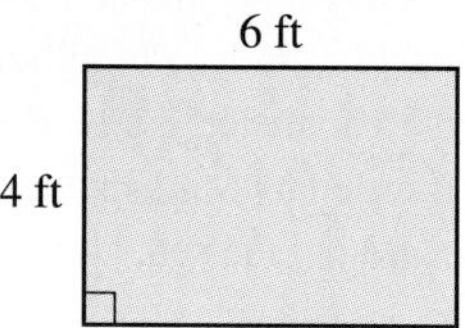

2.

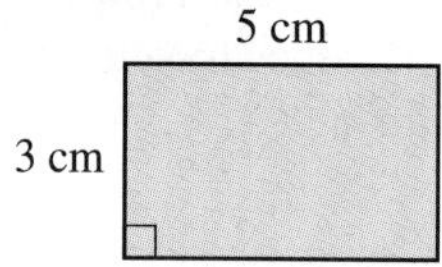

3.

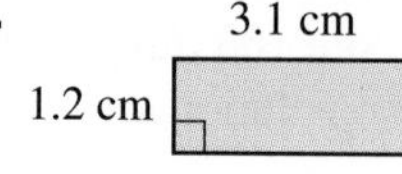

4.

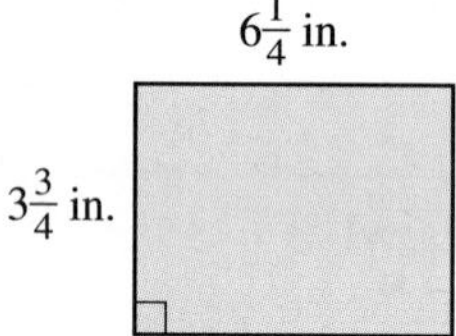

5.

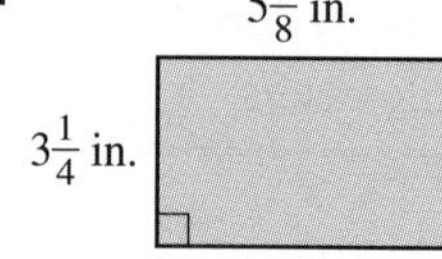

6.

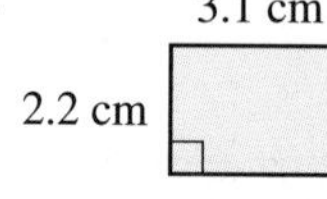

7.

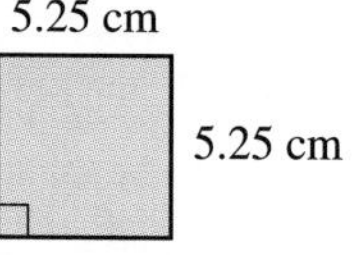

8.

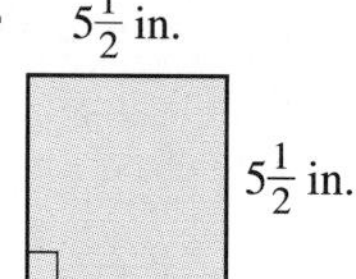

9.

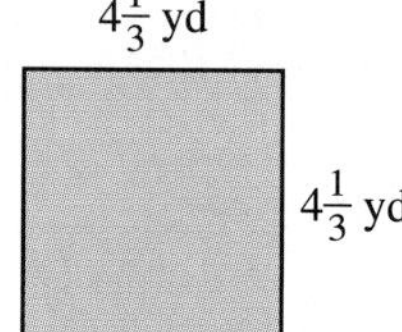

10.

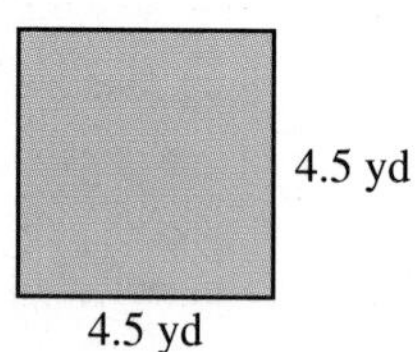

11.

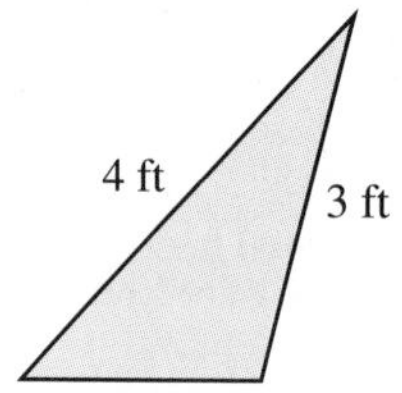

12.

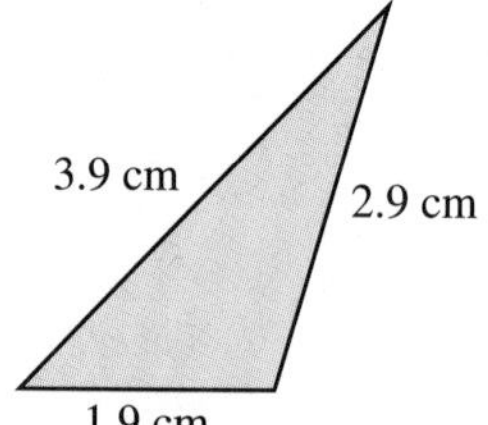

13.

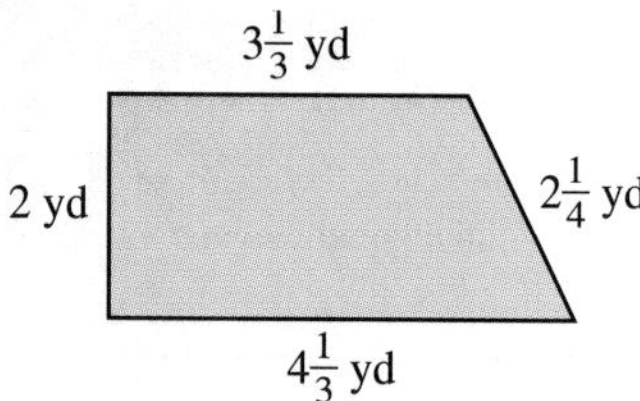

14.

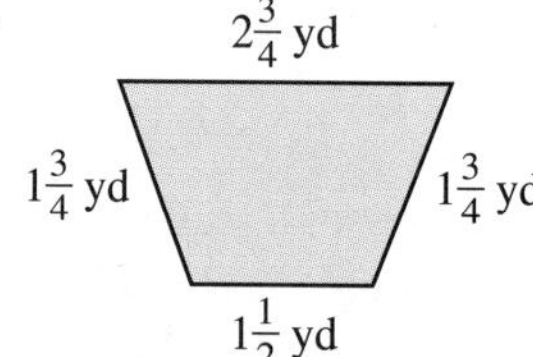

15.

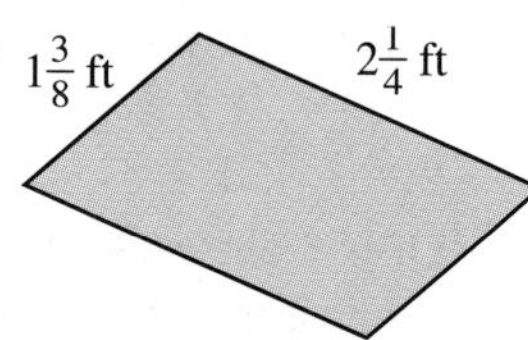

16.

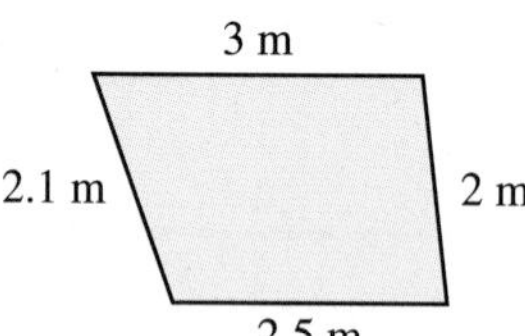

17.

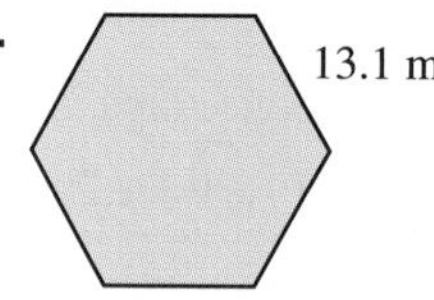

18.

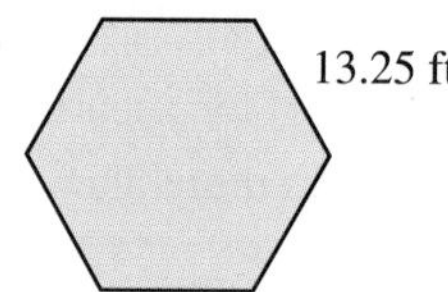

19.

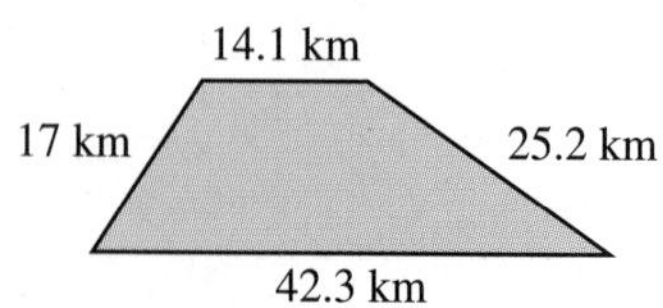

20.

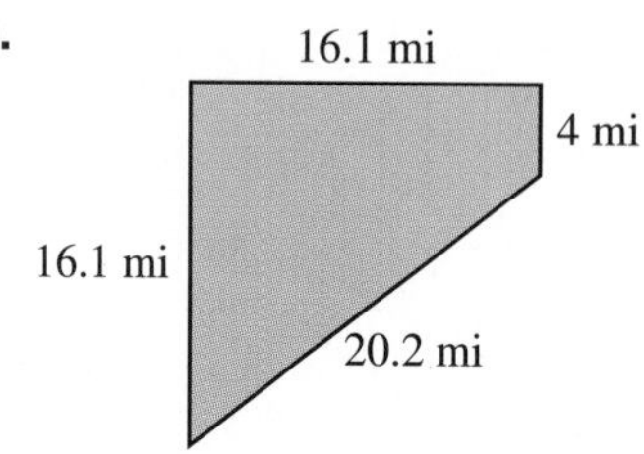

⟨ B ⟩ Finding Circumferences In Problems 21–28, find the circumference (use 3.14 for π).

21.

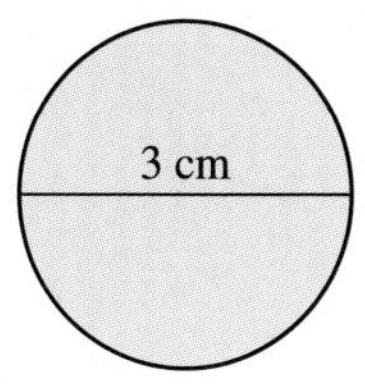

22.

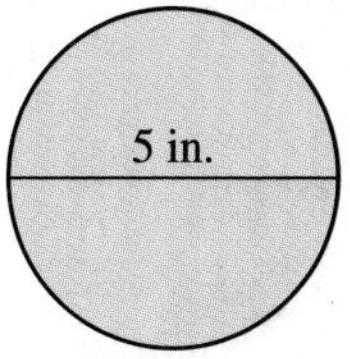

23.

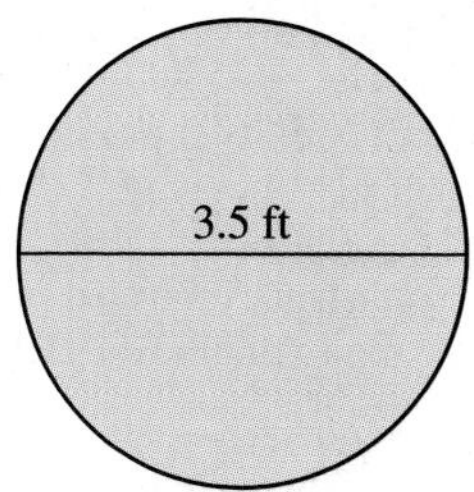

24.

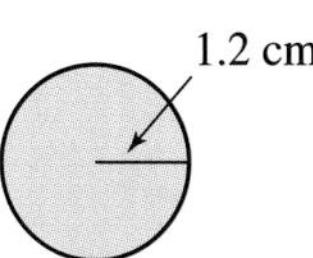

25.

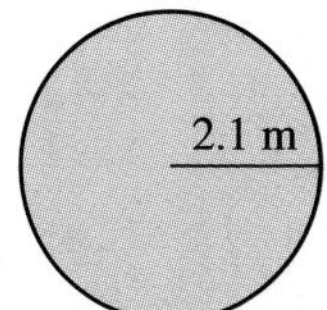

26.

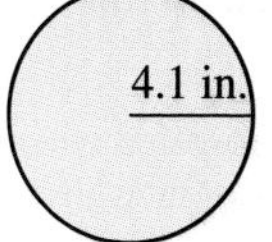

27.

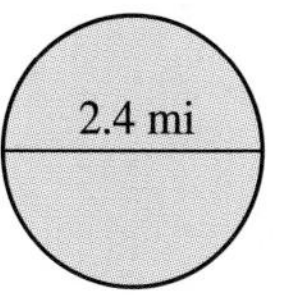

28.

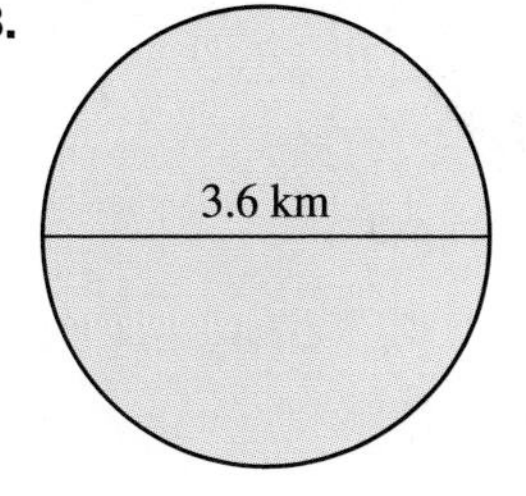

⟨ C ⟩ Applications Involving Perimeter and Circumference

29. *Perimeter of largest pool* The largest pool in the world is in Casablanca, Morocco, and it is 480 meters long and 75 meters wide. What is the perimeter of this pool?

30. *Perimeter of VAB* The rectangular Vehicle Assembly Building (VAB) at Cape Canaveral is 716 feet long and 518 feet wide. What is the perimeter of this building?

31. *Perimeter of largest drydock* The largest drydock is Okopo No. 1 in South Korea. The rectangular structure is 1772.4 feet long by 430 feet wide. How many feet do you have to walk to cover the perimeter of this drydock?

32. *Poster perimeter* One of the largest posters was made in Japan and measured 328 feet by 328 feet. How many feet of frame were required to frame this poster?

33. *Framing the Mona Lisa* The *Mona Lisa,* a painting by Leonardo da Vinci, measures 30.5 inches by 20.9 inches. How many inches of frame (measured on the inside) were needed to frame the painting?

34. *Baseball diamond* A baseball diamond is actually a square. If the distance to first base is 90 feet, how far do you have to run to cover all bases?

35. *Circumference of bicycle tire* The diameter of a bicycle tire is 60 centimeters. What is the circumference of the tire? Use 3.14 for π.

36. *Clock hand movement* The minute hand of a clock is 3 inches long. How far does the tip of the hand move in 1 hour? Use 3.14 for π.

37. *Soccer* The regulation soccer field shown is a rectangle 60 yards wide and 100 yards long. What is the perimeter of the field?

38. *Soccer* The penalty box is a rectangle 44 yards wide and 18 yards long. What is the perimeter of the penalty box?

39. *Soccer* The goal area is a smaller rectangle inside the "penalty area," centered on the goal. The measurements of this box are 20 yards wide by 6 yards deep. What is the perimeter of this goal area?

40. *Soccer* The center circle is a circle with a 10-yard radius. Defenders must stay outside this circle at the start of a kickoff. What is the circumference of the circle? (Give an exact answer.)

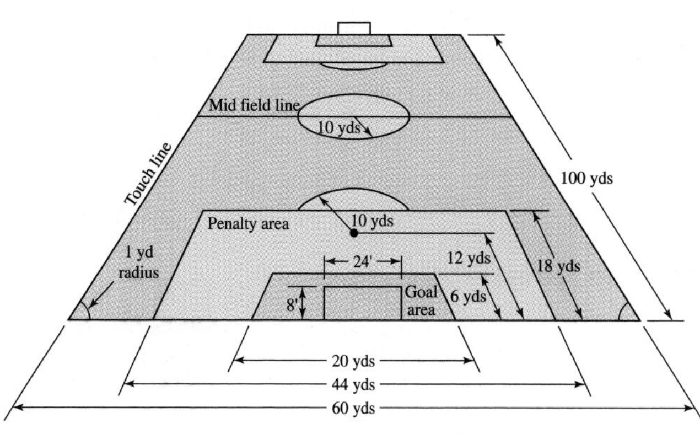

Source: http://www.sportsknowhow.com.

Using Your Knowledge

Distance Traveled If you are traveling by car, the distance between two points is shown on travel maps.

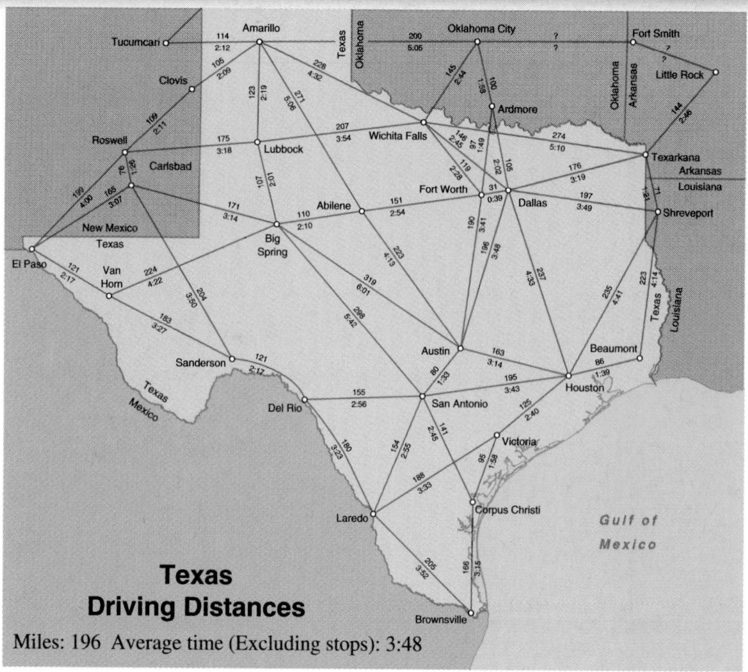

41. Find the distance traveled when going from Laredo to San Antonio to Del Rio and back to Laredo.

42. Find the distance traveled when going from Dallas to Houston to Shreveport and back to Dallas.

43. Find the distance traveled when starting at Austin, driving to Ft. Worth, then to Abilene, and back to Austin.

Write On

44. Do you remember the Distributive Property from Chapter 1? In this section, we mentioned that the perimeter P of a rectangle is $P = 2L + 2W$. Use the Distributive Property to write this formula in a different way.

45. A rectangle is 20 meters by 10 meters. Discuss three different ways in which you can find the perimeter. Which is the fastest?

Concept Checker

Fill in the blank(s) with the correct word(s), phrase, or mathematical statement.

46. The ____________ of a polygon is the **sum of the lengths** of its sides.

47. The ____________ of a **circle** of diameter d is πd.

perimeter	**volume**
area	**circumference**

Mastery Test

48. Find the perimeter of a rectangular recreation area measuring 200 feet by 100 feet.

49. Jupiter's orbit around the sun is approximately a circle with an 800-million-kilometer diameter. Find the distance traveled by Jupiter in one revolution around the sun. Use 3.14 for π.

50. A pentagonal building is constructed so that each of its sides is 700 feet long. What is the perimeter of this building?

51. Find the perimeter of a square whose side is $5\frac{1}{4}$ centimeters.

52. Find the perimeter of the square shown.

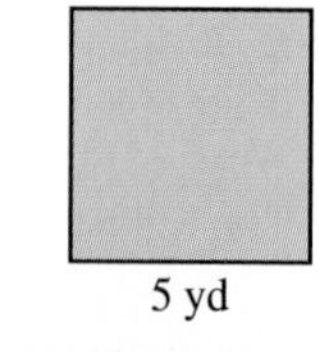

5 yd

53. Find the perimeter of a rectangle 3.4 centimeters long by 1.2 centimeters wide.

54. Find the perimeter of the rectangle.

5 ft

18 ft

55. Find the perimeter of the polygon.

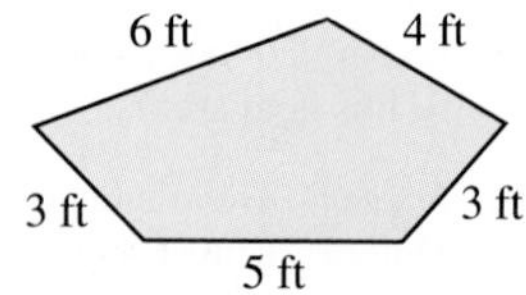

Skill Checker

Multiply.

56. $\frac{1}{2} \cdot 15 \cdot 10$

57. $\frac{1}{2} \cdot 15 \cdot 20$

58. $\frac{1}{2} \cdot 13 \cdot 8$

59. $\frac{1}{2} \cdot 3 \cdot 5$

60. $\frac{1}{2} \cdot 9 \cdot 7$

8.3 Finding Areas

Objectives

You should be able to:

A Find the area of a rectangle or square.

B Find the area of a triangle.

C Find the area of a parallelogram.

D Find the area of a trapezoid.

E Find the area of a circle.

To Succeed, Review How To . . .

Multiply fractions and decimals. (pp. 130–133, 212–214)

Getting Started

The picture shows some fish in an aquarium. Do you know how many fish you can have living in an aquarium? The approximate number can be found by estimating the *area* of the water on the surface. If the aquarium is 20 inches long by 10 inches wide, as shown below, the area of the water surface is

$$10 \text{ in.} \cdot 20 \text{ in.} = 200 \text{ in.}^2$$ (read "200 square inches")

10 in.

20 in.

A › Areas of Rectangles and Squares

In the metric system the **area** of a square 1 centimeter on each side is

$$1 \text{ cm} \cdot 1 \text{ cm} = 1 \text{ cm}^2$$ (read "one square centimeter")

Note that we treat the dimension symbol, the centimeter, as if it were a number. Thus,

$$\text{cm} \cdot \text{cm} = \text{cm}^2$$
$$\text{in.} \cdot \text{in.} = \text{in.}^2$$

Similarly, in the customary system we can define the area of a square 1 in. on each side as

$$1 \text{ in.} \cdot 1 \text{ in.} = 1 \text{ in.}^2$$

These two areas are shown in the following illustration.

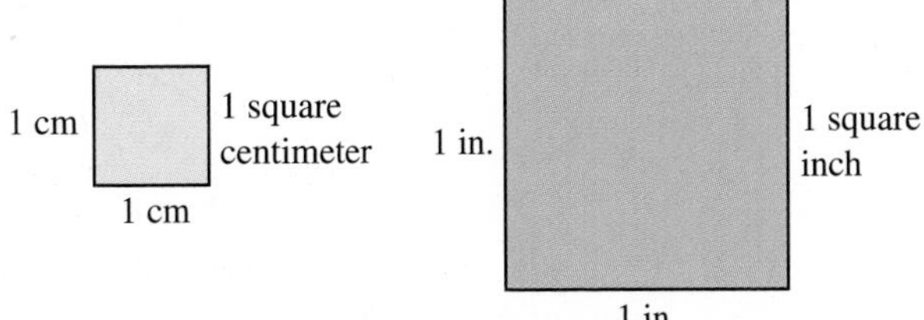

To find the area of a figure, we must find the number of square units it contains. For example, the area of a rectangle* 3 centimeters by 4 centimeters is $3 \text{ cm} \cdot 4 \text{ cm} = 12 \text{ cm}^2$. This is because such a rectangle contains 12 squares, each 1 square centimeter.

In general, you can find the area of any rectangle by multiplying its length L by its width W, as shown.

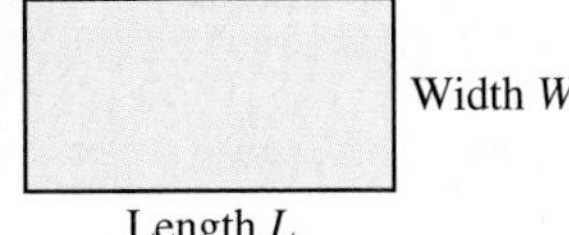

*A rectangle is a four-sided polygon in which the opposite sides are parallel and which has four 90° angles.

AREA OF A RECTANGLE

The area *A* of a rectangle is the product of its length *L* and its width *W*. In symbols,

$$A = L \cdot W$$

EXAMPLE 1 **Area of a rectangle**

Find the area of a rectangle 6 meters by 4 meters.

SOLUTION $A = L \cdot W = (6 \text{ m}) \cdot (4 \text{ m}) = 24 \text{ m}^2$

Thus the area of the rectangle is 24 square meters.

PROBLEM 1

Find the area of a rectangle 5 yards by 8 yards.

The area of a square is easier to find because length and width are the same. For example, if a square has a side of length S, its area is $S \cdot S = S^2$.

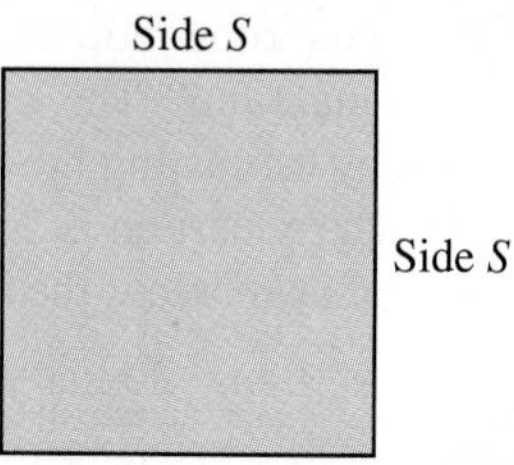

AREA OF A SQUARE

The area *A* of a square is the square of its side *S*. In symbols,

$$A = S \cdot S = S^2$$

EXAMPLE 2 **Area of a square**

Find the area of a square whose side is 6 inches long.

SOLUTION $A = S^2 = S \cdot S$

$= (6 \text{ in.}) \cdot (6 \text{ in.}) = 36 \text{ in.}^2$

PROBLEM 2

Find the area of a square whose side is 8 centimeters long.

B › Area of a Triangle

If we know the area of a rectangle, we can always find the area of a triangle. Look at the following figure and see if you can find the formula for the area of the triangle.

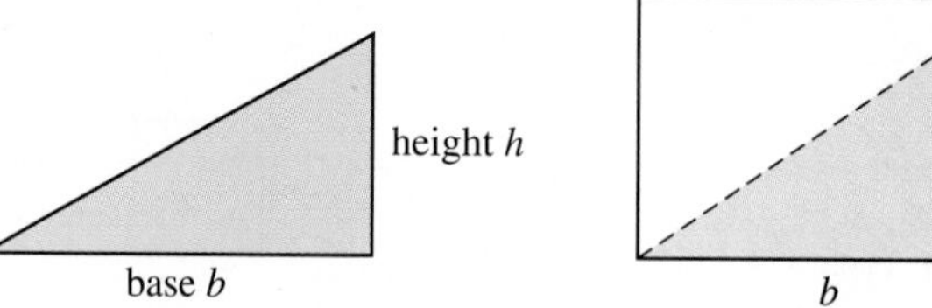

Since the area of the triangle (shaded) is $\frac{1}{2}$ the area of the rectangle, which is bh, the area of the triangle is $\frac{1}{2}bh$.

AREA OF A TRIANGLE

The area *A* of a triangle is the product of $\frac{1}{2}$ its base *b* by its height *h*. In symbols,

$$A = \frac{1}{2} \cdot b \cdot h$$

Answers to PROBLEMS

1. 40 yd^2 **2.** 64 cm^2

This formula holds true for any type of triangle.

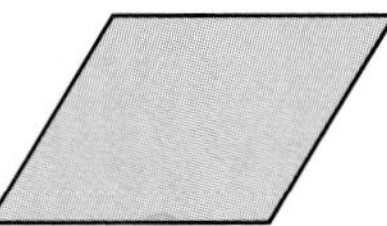

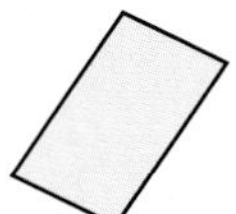

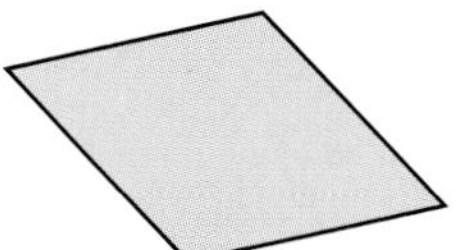

EXAMPLE 3 **Area of a triangle**

Find the area of a triangular piece of cloth 15 centimeters long and 10 centimeters high.

SOLUTION

$$A = \frac{1}{2} \cdot 15 \text{ cm} \cdot 10 \text{ cm} = \frac{1}{\not{2}} \cdot 15 \text{ cm} \cdot \overset{5}{\not{10}} \text{ cm} = 75 \text{ cm}^2$$

PROBLEM 3

Find the area of a triangular piece of metal 12 inches high and 10 inches long.

C › Area of a Parallelogram

If you know the area of a rectangle, you can find the area of a **parallelogram,** a four-sided figure with two pairs of parallel sides. Here are some parallelograms.

To find the area of the following parallelogram, we cut the triangular piece and move it to the other side, obtaining a rectangle. Since the area of the rectangle is the length times the width, the area A of the parallelogram is $b \cdot h$.

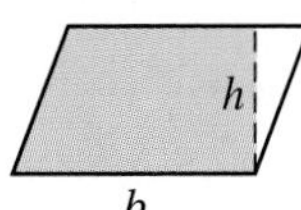

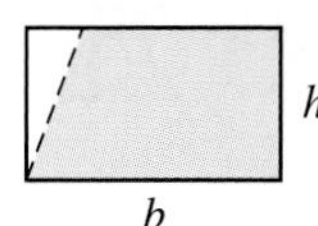

AREA OF A PARALLELOGRAM

The area A of a parallelogram is the product of its base b times its height h. In symbols,

$$A = b \cdot h$$

EXAMPLE 4 **Area of a parallelogram**

Find the area of the parallelogram.

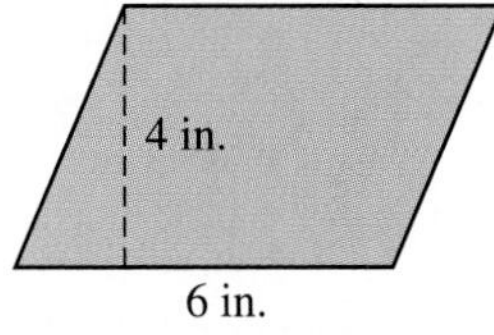

SOLUTION

$$A = (6 \text{ in.}) \cdot (4 \text{ in.})$$
$$= 24 \text{ in.}^2$$

PROBLEM 4

Find the area of the parallelogram.

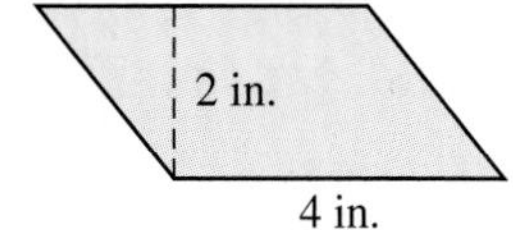

Answers to PROBLEMS

3. 60 in.^{2} **4.** 8 in.^{2}

D > Area of a Trapezoid

A **trapezoid** is a four-sided figure that has exactly one pair of parallel sides. Here are some trapezoids.

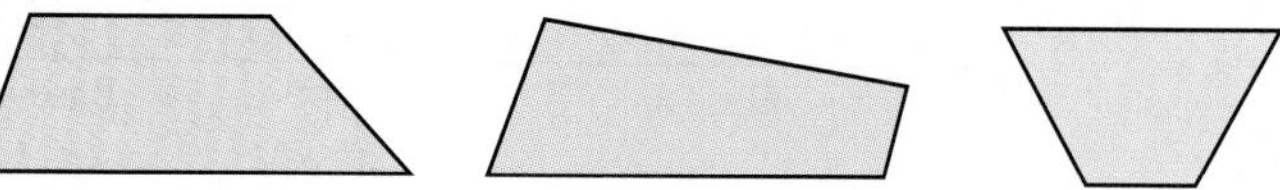

To find the area of a trapezoid, we construct another just like it and place the two together to form a parallelogram.

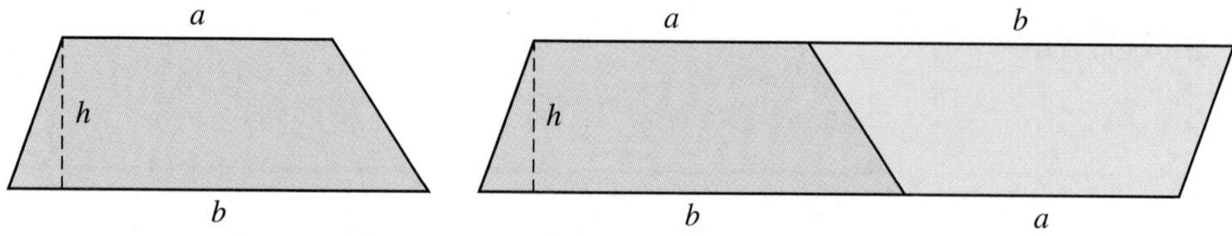

The area of the parallelogram is the length of the base $(a + b)$ times the height (h), or $h \cdot (a + b)$; however, the area of the trapezoid is one-half the area of the parallelogram or $\frac{1}{2}(a + b)h$.

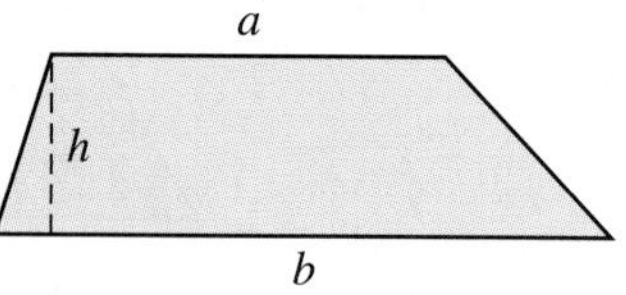

AREA OF A TRAPEZOID

The area A of a trapezoid is $\frac{1}{2}$ times the sum of the length of the bases $(a + b)$ and its height h. In symbols,

$$A = \frac{1}{2} \cdot (a + b) \cdot h$$

EXAMPLE 5 **Area of a trapezoid**

Find the area of the trapezoid.

SOLUTION

$$\begin{aligned} A &= \frac{1}{2} \cdot (4 \text{ cm}) \cdot (5 \text{ cm} + 6 \text{ cm}) \\ &= \frac{1}{2} \cdot (4 \text{ cm}) \cdot (11 \text{ cm}) \\ &= \frac{4 \cdot 11}{2} \text{ cm}^2 \\ &= 22 \text{ cm}^2 \end{aligned}$$

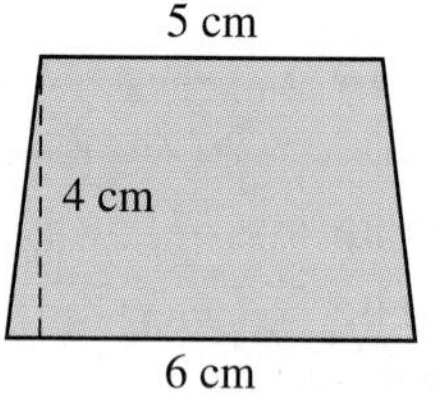

PROBLEM 5

Find the area of the trapezoid

3 cm

4 cm

5 cm

E > Area of a Circle

The area of a circle can be found if we know the distance from the center of the circle to its edge, the radius r, of the circle. The formula for finding the area of a circle also involves the number π, which we approximate by 3.14 or $\frac{22}{7}$.

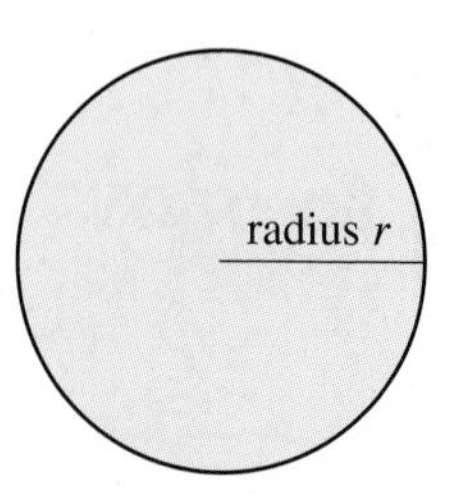

Answers to PROBLEMS

5. 16 cm^2

AREA OF A CIRCLE

The area A of a circle is the product of π and the square of the radius r. In symbols,

$$A = \pi \cdot r^2$$

Note: $\pi \approx 3.14$ **or** $\frac{22}{7}$.

EXAMPLE 6 Area of a circle

Find the area of a circle with a radius of 3 centimeters. Use 3.14 for π.

SOLUTION

$$A = (3.14) \cdot (3\text{ cm})^2 = 3.14 \cdot (3\text{ cm}) \cdot (3\text{ cm})$$
$$= (3.14) \cdot (9\text{ cm}^2) = 28.26\text{ cm}^2$$

PROBLEM 6

Find the area of a circle with a radius of 2 inches. Use 3.14 for π.

EXAMPLE 7 Area of a circle

The Fermi National Accelerator Lab has the atom smasher shown in the photo. The smasher has a radius of 0.60 miles. What area does it cover ? Use 3.14 for π.

SOLUTION The area A of a circle is $A = \pi r^2$, where $r = 0.60$. Thus, the area is $A = \pi(0.60\text{ mi})^2 \approx 3.14(0.36\text{ mi}^2) = 1.1304\text{ mi}^2$. That is, the accelerator covers about 1.1304 square miles.

PROBLEM 7

The photo shows several of the mysterious wheat circles appearing in Rockville, California. The bigest circle is claimed to be 140 feet in diameter. What is the area of this circle? Use 3.14 for π.

Calculator Corner

The squaring key x^2 is especially useful when working with area problems. For example, to find the area of a circle with a radius of 3 centimeters, you need to find $(3.14) \times (3)^2$, as in Example 6. This can be done by pressing 3 . 1 4 × 3 x^2 ENTER. The correct result, 28.26, will be displayed.

In other mathematics courses, a more precise approximation for π might be necessary. If your calculator has a π key, this approximation is built into the calculator. Thus pressing the π key may give 3.1415927 as an approximation.

Answers to PROBLEMS

6. 12.56 in.^{2} **7.** 15,386 ft^2

Exercises 8.3

⟨ A ⟩ Areas of Rectangles and Squares In Problems 1–9, find the area of the rectangle or square.

1. A rectangle 10 feet by 15 feet

2. A rectangle 5 inches by 3 inches

3. A rectangle 2 inches by 3 inches

4. A rectangle 3 yards by 2 yards

5. A rectangle 8 centimeters by 9 centimeters

6. A square 5 centimeters on each side

7. A square 9 inches on each side

8. A square 15 meters on each side

9. A square 9 yards on each side

⟨ B ⟩ Area of a Triangle In Problems 10–20, find the area of the triangle.

10. A triangle whose base is 6 inches and whose height is 10 inches

11. A triangle of base 8 centimeters and height 7 centimeters

12. A triangle 20 millimeters high and with a 10-millimeter base

13. A triangle with a 50-millimeter base and a 7-millimeter height

14.

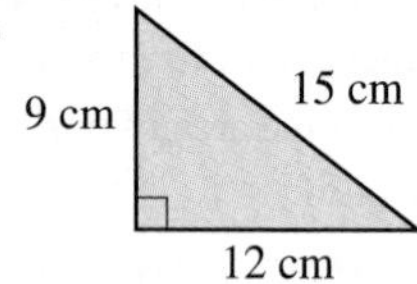

15.

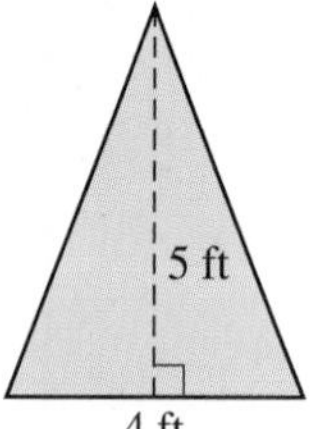

16.

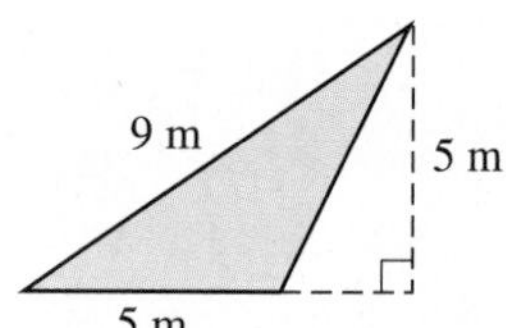

17.

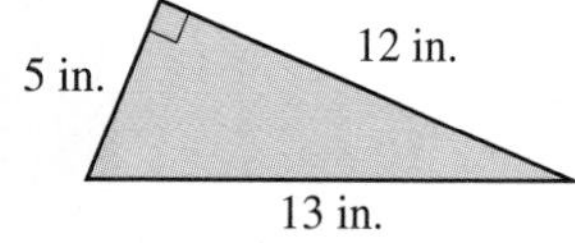

18.

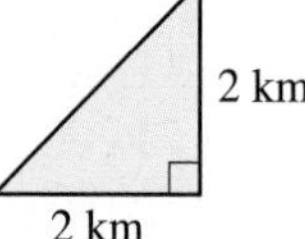

19.

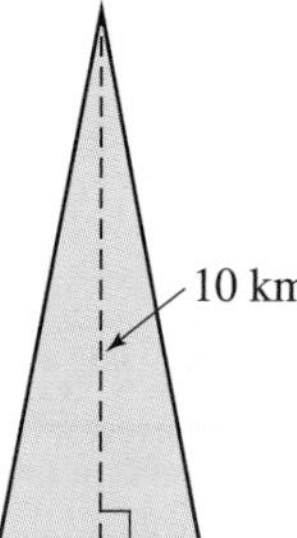

20.

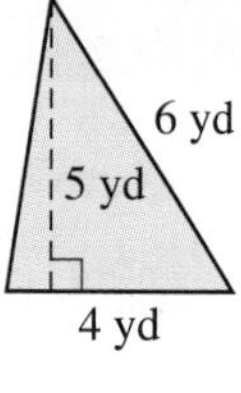

⟨ C ⟩ Area of a Parallelogram In Problems 21–28, find the area of the parallelogram.

21.

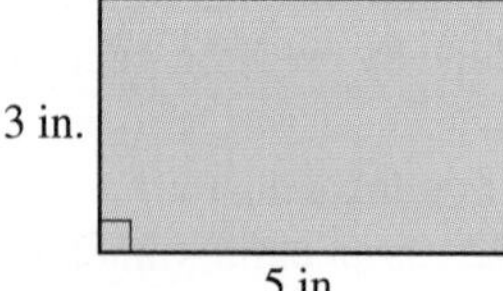

22.

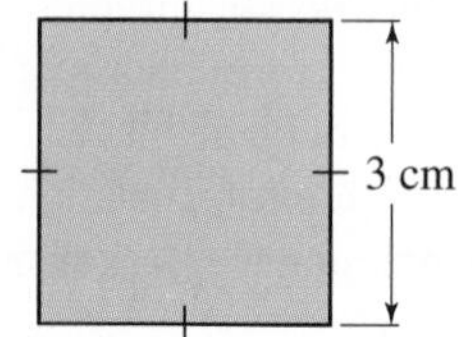

23.

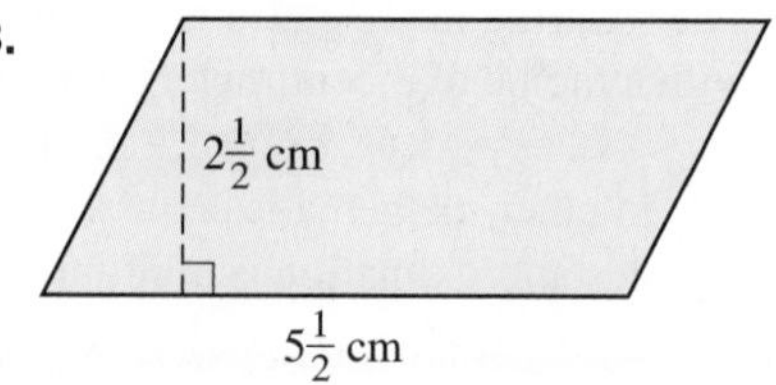

24. $3\frac{1}{2}$ ft 4 ft

25. 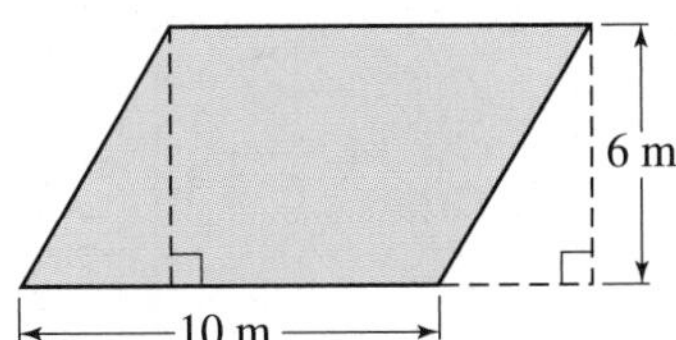

26. 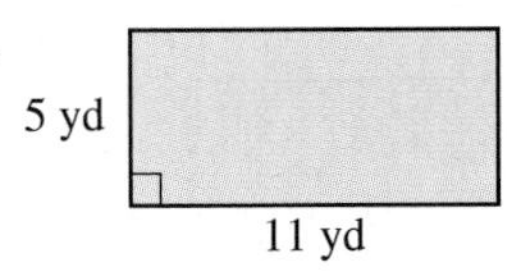

27. 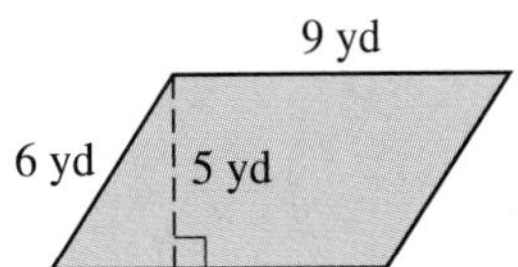

28. 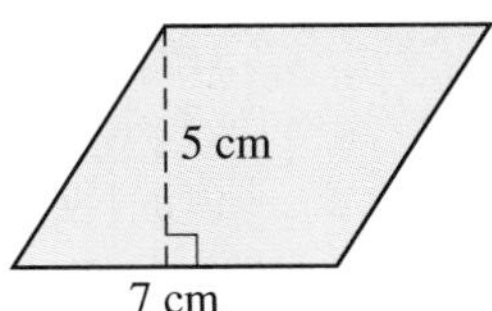

⟨D⟩ Area of a Trapezoid In Problems 29–33, find the area of the trapezoid.

29. 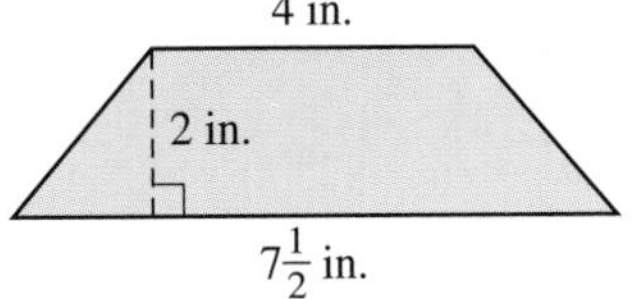

30. 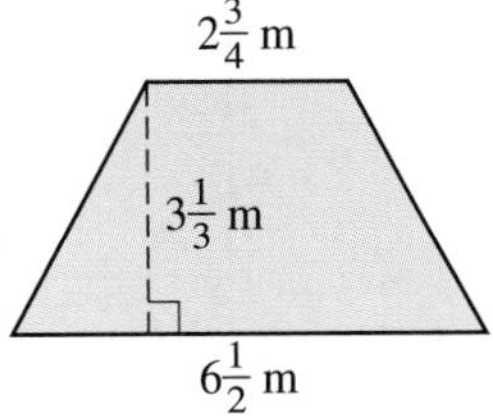

31. Find the area of the lot.

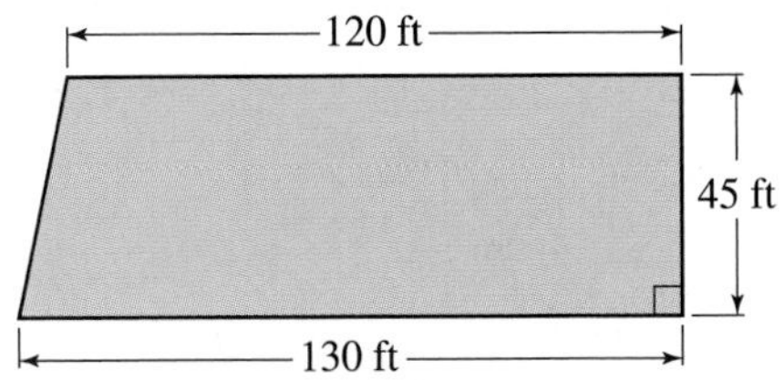

32. What is the area of the missile wing in square inches?

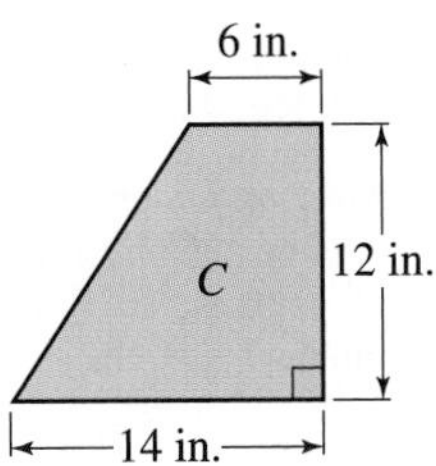

33. What is the area of the yard shown?

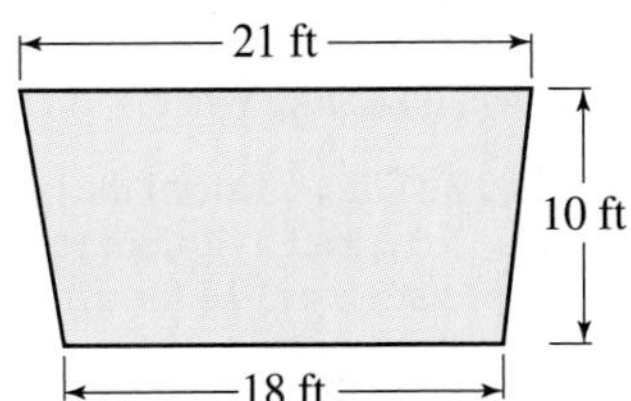

⟨E⟩ Area of a Circle In Problems 34–40, find the area of the circle. Use 3.14 for π.

34. A circle with a 5-inch radius

35. A circle with a 4-inch radius

36. A circle with a 2-foot radius

37. A circle with a 7-centimeter radius

38. A circle with a 10-millimeter radius

39. A circle with a 1-meter radius

40. A circle with a 3-yard radius

In Problems 41–50, find the area of the shaded region by adding or subtracting the individual regions.

41. 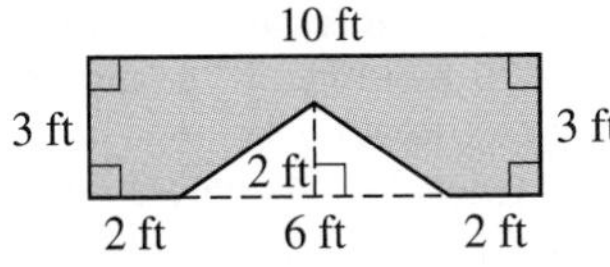

42. 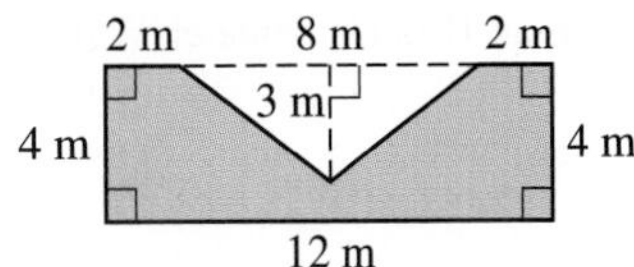

43. 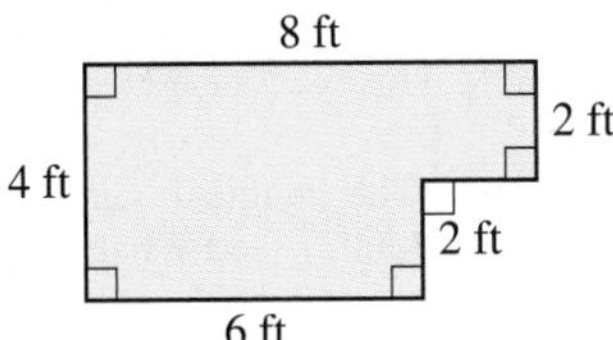

44. 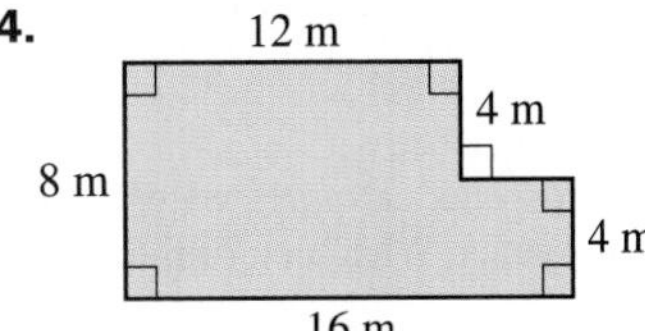

45.

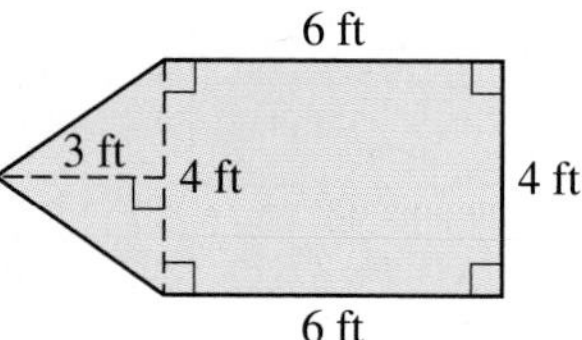

46.

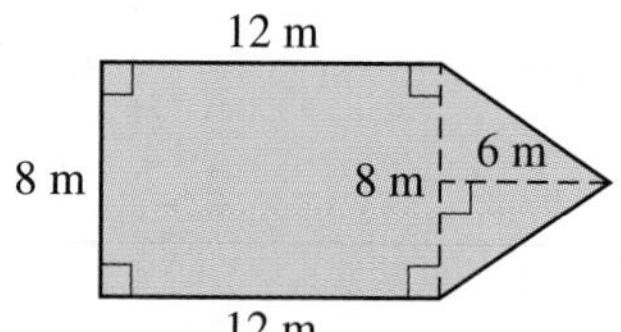

47. Use $\pi \approx 3.14$.

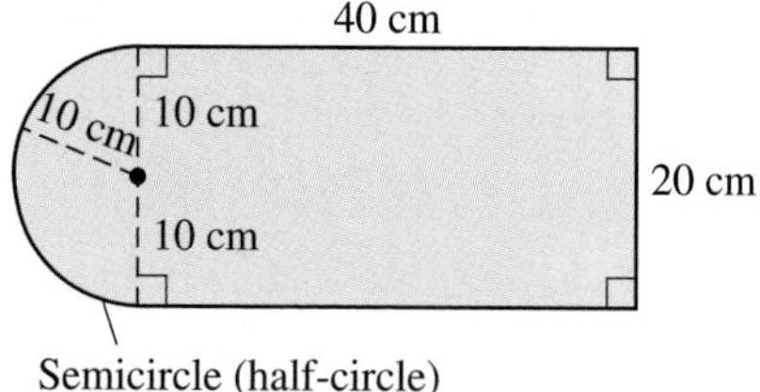

48. Use $\pi \approx 3.14$.

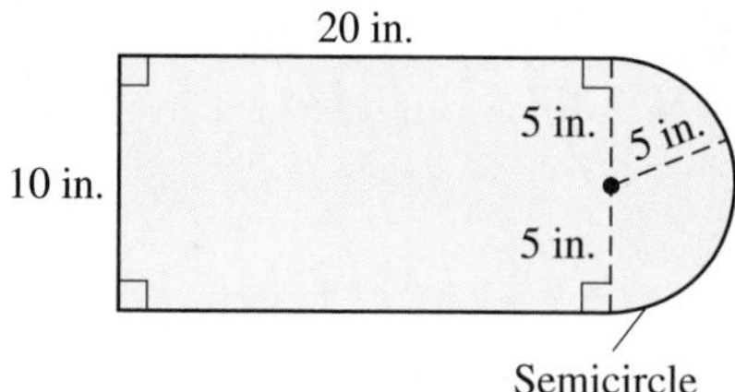

49. Use $\pi \approx 3.14$.

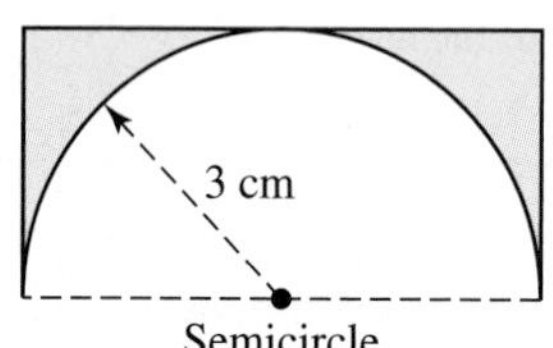

50. Use $\pi \approx 3.14$.

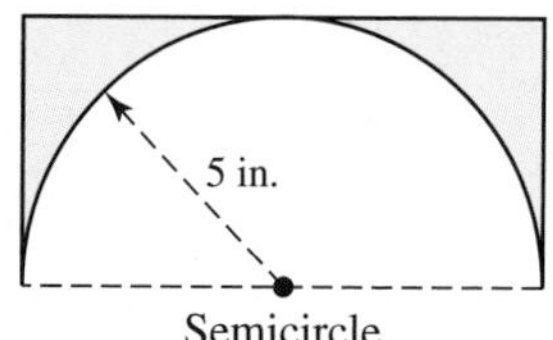

Applications

51. *Area of a baseball diamond* A baseball diamond is a square 90 feet on a side. What is its area?

52. *Area of football field* A football field is 120 yards long and 160 feet wide. What is its area in square yards?

53. *Area of a pizza* One of the largest pizzas ever made (in Glens Falls, N.Y.) had a 40-foot radius. What area was covered by this pizza? (Use 3.14 for π.)

54. *Area of an omelet* One of the largest omelets made was cooked at Conestoga College, in Ontario, Canada. It measured 30 feet by 10 feet. What was the area of this rectangular omelet?

55. *Area of a crater* The largest proven circular crater, which is in northern Arizona, is about 5200 feet wide. What is the area of this crater? (Use 3.14 for π.)

In Problems 56–65, use 3.14 for π and give your answer to two decimal places.

56. *Area of dinner plate* Find the area of a 6-inch dinner plate.

57. *Area of Dutch crop circle* The diameter of the largest reported Dutch crop circle is 12 meters. What is its area?

58. *Area of largest crop circle* The largest recorded crop circle appeared in Wiltshire, England, and is 240 meters in diameter. What is the area of the circle?

59. *Area of bicycle wheel* The diameter of a bicycle wheel is 20 inches. What is the area of the wheel?

60. *Area of a pizza* A pizza has an 18-inch diameter. What is the area of this pizza?

61. *Mowed lawn area* A self-propelled lawnmower is tied to a pole in the backyard with a 20-foot rope. If the mower goes around in decreasing circles (because the rope is getting tied around the pole), what is the area the mower can mow?

62. *Storm area* A summer storm is circular in shape with a 50-mile-long diameter. What area does this storm cover?

63. *Area of sectors in wheel of fortune* A wheel of fortune has six sectors, half of which are red and half of which are yellow. If the radius of the wheel is 3 feet, what area is covered by the red sectors?

64. *Sprinkled lawn area* A lawn sprinkler sprays water 10 feet away in every direction as it rotates. What area of the lawn is being sprinkled?

65. *Area of Disneyland Carousel* The magical Carousel in Disneyland Paris has a radius of about 27 feet. What area does it cover?

66. *Soccer* The regulation soccer field shown is a rectangle 60 yards wide and 100 yards long. What is the area of the field?

67. *Soccer* The penalty box is a rectangle 44 yards wide and 18 yards long. What is the area of the penalty box?

68. *Soccer* The goal area is a smaller rectangle inside the "penalty area," centered on the goal. The measurements of this box are 20 yards wide by 6 yards deep. What is the area of this rectangular box?

69. *Soccer* The field dimensions for international matches in soccer are given in meters (m) and are as follows:

Length: 100 m to 110 m

Width: 64 m to 75 m

What is the area of the smallest field allowed?

70. *Soccer* Using the dimensions in Problem 69, what is the area of the largest field allowed?

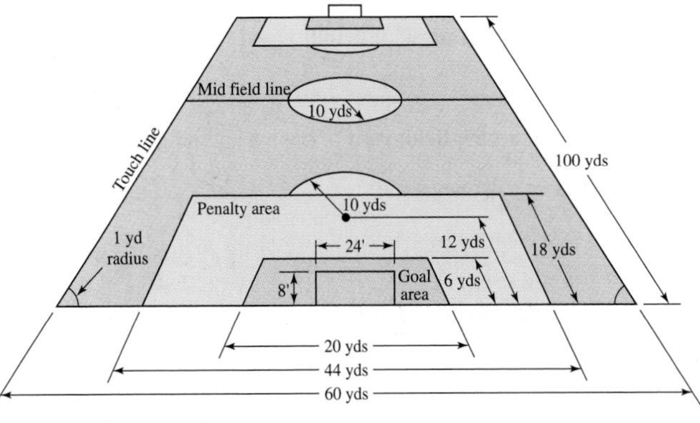

Source: http://www.sportsknowhow.com.

Using Your Knowledge

Areas of Irregular Figures You can use your knowledge of the formulas studied to find the areas of more complicated figures. For example, the area of the following sheet metal blank can be found by adding the area of the rectangle to the area of the two half-circles (which add up to the area of a circle).

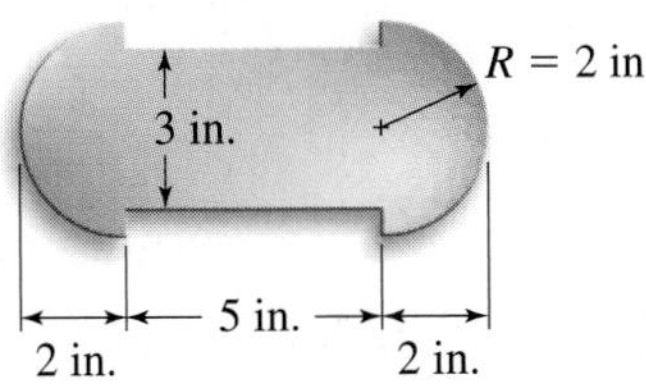

71. Find the area of the rectangle.

72. Find the area of each half-circle. Use 3.14 for π.

73. Find the total area.

Sometimes we *subtract* to find areas. The area of the following plate is the area of the trapezoid *minus* the area of the rectangle.

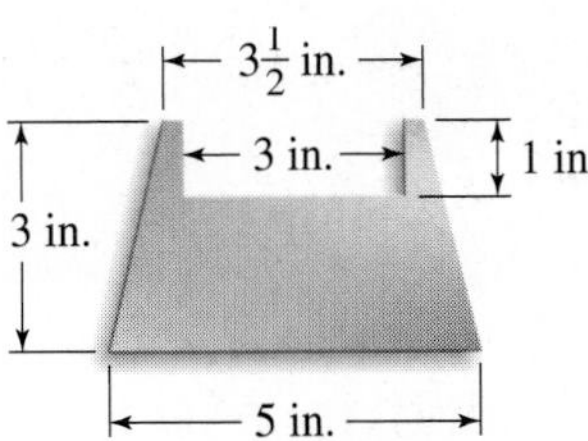

74. Find the area of the trapezoid.

75. Find the area of the rectangle.

76. Find the total area of the plate.

Write On

77. A small pizza 11 inches in diameter cost $8 and a large pizza 15 inches in diameter cost $15. Which is the better deal, to buy two small pizzas or one large one? *Hint:* Find out how many square inches you get per dollar.

78. Which do you think is a better illustration of a circle: a perfectly round penny or a bicycle tire? Explain.

Concept Checker

Fill in the blank(s) with the correct word(s), phrase, or mathematical statement.

79. The **area** of a **rectangle** of length L and width W is ______________.

80. The **area** of a ______________ of base b and height h is ______________.

81. The **area** of a **parallelogram** of base b and height h is ______________.

82. The **area** of a **circle** of radius r is ______________.

πr^2	πr
triangle	$2\pi r$
$\frac{1}{2}bh$	bh
$L + W$	LW
rectangle	

Mastery Test

83. A wheat circle is 120 feet in diameter. What is the area of this circle? (Use 3.14 for π.)

84. Find the area of a circle with a radius of 10 centimeters. (Use 3.14 for π.)

85. Find the area of the trapezoid.

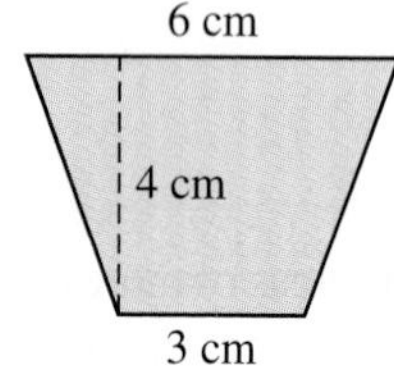

86. Find the area of the parallelogram.

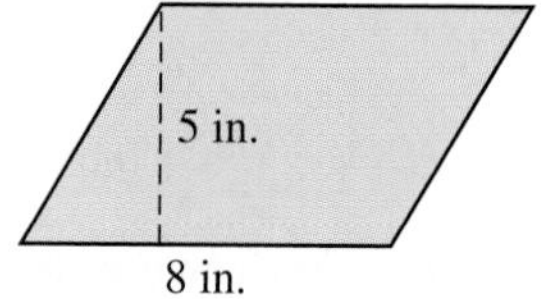

87. Find the area of a triangular piece of cloth 20 centimeters long and 10 centimeters high.

88. Find the area of a square whose side is 10 inches long.

89. Find the area of a rectangle 8 meters by 3 meters.

Skill Checker

In Problems 90–94, perform the indicated operations.

90. $3 \cdot 2 \cdot \frac{1}{3}$

91. $\frac{10}{3} \cdot 2 \cdot \frac{1}{2}$

92. (2)(18)(0.333)

93. (0.5)(22)(20)

94. (0.62)(0.62)(0.62) (Answer to two decimal places.)

8.4 Volume of Solids

Objectives

You should be able to:

A Find the volume of a rectangular solid.

B Find the volume of a cylinder.

C Find the volume of a sphere.

To Succeed, Review How To . . .

1. Perform the four fundamental operations using fractions. (pp. 130–136, 151–156)
2. Evaluate an expression containing exponents. (pp. 77, 82–84)

Getting Started

Which bread has more volume, the Cuban bread (cylindrical) or the Greek bread (doughnut shaped)? To answer the question, you need to know the formula for finding the volume of a **cylinder.** Before we do that, let us learn how to calculate the volume of a simpler **rectangular solid** shaped like a box.

D › Find the volume of a circular cone.

E › Find the volume of a pyramid.

F › Solve applications involving the volume of solids.

A › Volume of Rectangular Solids

The volume of a rectangular solid is the number of unit cubes it takes to fill it. In the metric system a solid cube 1 centimeter on each side is defined to be the *unit* of volume, $1 \text{ cm} \times 1 \text{ cm} \times 1 \text{ cm} = 1 \text{ cm}^3$ (read "one cubic centimeter"). Other units of volume are the cubic meter and, in the customary system, the cubic foot and cubic yard. (Note that volume is measured in *cubic* units.)

As in the case of areas, the volume of a rectangular solid object equals the number of units of volume it contains. Suppose you have a block 3 centimeters long, 2 centimeters wide, and 2 centimeters high. You can see that the top layer has 3 rows each containing 2 unit volumes, that is, 6 cubic centimeters (cm^3) in all. Since the bottom layer is identical, the volume of this cube is 12 cubic centimeters. This volume can also be found by multiplying $2 \text{ cm} \times 3 \text{ cm} \times 2 \text{ cm} = 12 \text{ cm}^3$. In general, the volume V of a rectangular solid is found by multiplying the length L times the width W times the height H. This is also the area of the base $(L \cdot W)$ times the height (H) of the solid.

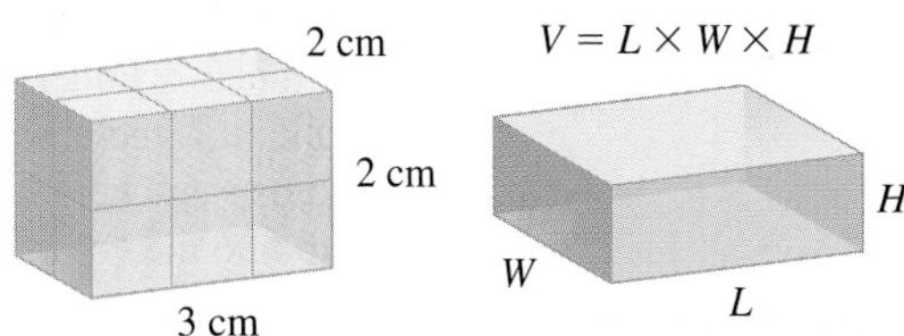

VOLUME OF A RECTANGULAR SOLID

The volume V of a rectangular solid is the product of its length L, its width W, and its height H. In symbols,

$$V = \underbrace{L \cdot W}_{\text{Base area}} \cdot H$$

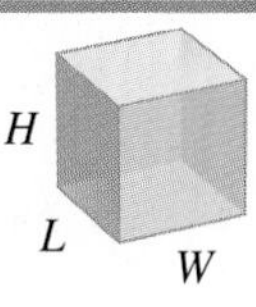

EXAMPLE 1 Volume of a rectangular solid

Find the volume of this solid.

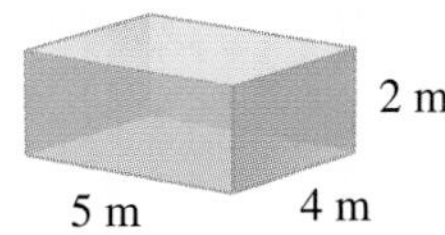

SOLUTION The volume is $5 \text{ m} \cdot 4 \text{ m} \cdot 2 \text{ m} = 40 \text{ m}^3$.

PROBLEM 1

Find the volume of this solid.

4 m
6 m
2 m

Answers to PROBLEMS

1. 48 m^3

In some instances, we need to do some conversions within a system to obtain the desired units in the answer. For example, concrete is sold by the cubic yard (yd^3). Thus, to estimate how much concrete is needed to fill a wooden box form 9 feet long, 6 feet wide, and 12 inches thick, we first have to write all the measurements in yards. Thus,

$$9 \text{ ft} = 3 \text{ yd}$$
$$6 \text{ ft} = 2 \text{ yd}$$
$$12 \text{ in.} = \frac{1}{3} \text{ yd}$$

The volume of the box is

$$V = L \cdot W \cdot H = 3 \text{ yd} \cdot 2 \text{ yd} \cdot \frac{1}{3} \text{ yd} = 2 \text{ yd}^3$$

EXAMPLE 2 Volume of a rectangular space

How many cubic yards of concrete are needed to fill a hole 10 feet long, 6 feet wide, and 18 inches thick?

SOLUTION Recall that 3 ft = 1 yd and 1 ft = 12 inches.

$10 \text{ ft} = 3\frac{1}{3} \text{ yd}$, $6 \text{ ft} = 2 \text{ yd}$, and $18 \text{ in.} = \frac{1}{2} \text{ yd}$

Thus, the volume is

$$V = L \cdot W \cdot H = 3\frac{1}{3} \text{ yd} \cdot \cancel{2} \text{ yd} \cdot \frac{1}{\cancel{2}} \text{ yd}$$
$$= 3\frac{1}{3} \text{ yd}^3$$

PROBLEM 2

How many cubic yards of concrete are needed to fill a hole 18 feet long, 4 feet wide, and 9 inches thick?

B > Volume of Cylinders

Most soda cans are made of aluminum and are in the shape of a cylinder about 5 inches in height and 2 inches in diameter. How do we find the volume of such a can? The volume of a circular cylinder like the can is found in the same way as the volume of a rectangular solid: it is the product of the area of the base times the height. For a circular cylinder of radius r and height h, the area of the base will be the area of a circle, that is, πr^2. In order to find the volume we multiply the base area (πr^2) by the height h.

VOLUME OF A CIRCULAR CYLINDER

The volume V of a circular cylinder of radius r and height h is the product of π, the square of the radius r, and its height h. In symbols,

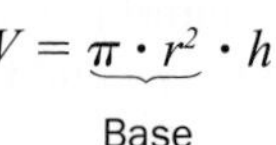

$$V = \underbrace{\pi \cdot r^2}_{\text{Base area}} \cdot h$$

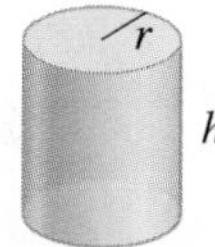

EXAMPLE 3 Volume of a cylinder

Find the volume of the can. Use 3.14 for π.

SOLUTION The can is in the shape of a circular cylinder, and its volume is

$$V = \pi \cdot r^2 \cdot h$$
$$= 3.14 \cdot (1 \text{ in.})^2 \cdot (5 \text{ in.})$$
$$= 15.7 \text{ in.}^3$$

$r = 1$ in.

$h = 5$ in.

PROBLEM 3

Find the volume of a can 6 inches high with a radius of 1 inch. Use 3.14 for π.

Answers to PROBLEMS

2. 2 yd^3 **3.** 18.84 $in.^3$

C > Volume of Spheres

The two bowling balls are shaped like **spheres.** A sphere is a three-dimensional solid that is defined as the set of all points in space that are a given distance (the radius r) from its center. The volume of a sphere of radius r is given by the following formula.

VOLUME OF A SPHERE

The volume V of a sphere of radius r is the product of $\frac{4}{3}\pi$ and the radius cubed. In symbols,

$$V = \frac{4}{3}\pi \cdot r^3$$

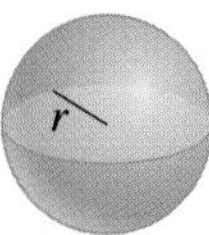

EXAMPLE 4 **Volume of bowling ball**

The radius of a bowling ball is 10.9 centimeters. Find the volume of the ball. Use 3.14 for π and round the answer to the nearest hundredth of a cubic centimeter.

SOLUTION The volume of a sphere is $V = \frac{4}{3} \cdot \pi \cdot r^3$

$$= \frac{4}{3} \cdot 3.14 \cdot (10.9 \text{ cm})^3$$

$$= \frac{4 \cdot 3.14 \cdot 1295.029}{3} \text{ cm}^3$$

$$\approx 5421.85 \text{ cm}^3$$

PROBLEM 4

Find the volume of a bowling ball that is about 4.3 inches in diameter. Use 3.14 for π and round the answer to the nearest hundredth of a cubic inch.

D > Volume of Circular Cones

Do you know the name of the geometric shape of the hat the boy is wearing? In geometry, it is called a circular cone, but you probably know it as a "dunce cap," a term originating from the name of the 18th century scholastic theologian, John Duns Scotus. The volume of a circular cone of radius r and height h is given by

VOLUME OF A CIRCULAR CONE

The volume V of a cone of radius r is $\frac{1}{3}\pi$ times the product of the radius squared and the height h. In symbols,

$$V = \frac{1}{3}\underbrace{\pi \cdot r^2}_{\text{Base area}} \cdot h$$

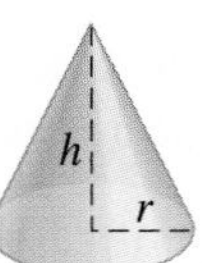

Answers to PROBLEMS

4. 41.61 in.^{3}

EXAMPLE 5 Volume of a cone hat

If the cone on the boy's head is 10 centimeters high and has a radius of 5 centimeters, what is its volume? Use 3.14 for π and round the answer to the nearest hundredth of a cubic centimeter.

SOLUTION The volume of a circular cone is $V = \frac{1}{3}\pi \cdot r^2 \cdot h$

$$= \frac{1}{3} \cdot 3.14 \cdot (5 \text{ cm})^2 \cdot (10 \text{ cm})$$

$$\approx 261.67 \text{ cm}^3$$

PROBLEM 5

Find the volume of the pink princess party and dressup hat, which is 14 inches tall and has a diameter of 4 inches. Use 3.14 for π and round the answer to the nearest hundredth of a cubic inch.

E > Volume of Pyramids

Do you recognize the shape of the structure in the photo? It is a **pyramid** (from the Greek word *pyra* meaning fire, light, or visible, and the word *midos* meaning measures, even though other scholars claim that the origin of the word is the Greek word *pyramis*, meaning wheat cake!). The pyramid in the photo is the Great Pyramid at Giza built by King Khufus, also known as Cheops, from 2589 to 2566 B.C. What is the volume of this pyramid?

VOLUME OF A PYRAMID

The volume V of a pyramid is the product of $\frac{1}{3}$ of its base area B and its height h, In symbols,

$$V = \frac{1}{3}\underbrace{B}_{\text{Base area}} \cdot h$$

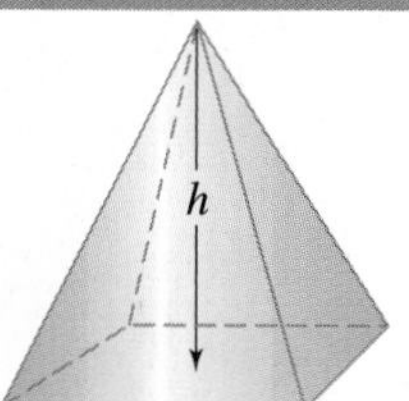

EXAMPLE 6 Volume of the original Great Pyramid

The base of the Great Pyramid is a square with each side measuring 230 meters. If the height of the pyramid is 147 meters, what is its volume?

SOLUTION The volume of a pyramid is $V = \frac{1}{3}Bh$.

The area of the square base B is 230 m · 230 m = 52,900 m^2 and the height h is 147 meters. Substituting for B and h,

PROBLEM 6

Did you know that the Great Pyramid has shrunk? Actually, due to the loss of its outer casing stones, the base is now 227 meters and its height is 137 meters (having lost 10 meters in height). To the nearest cubic meter, what is the volume of the pyramid now?

Answers to PROBLEMS

5. 58.61 in.^{3} **6.** 2,353,158 m^3

$$\begin{aligned} V &= \frac{1}{3}Bh \\ &= \frac{1}{3}(52{,}900\text{ m}^2)(147\text{ m}) \\ &= (52{,}900\text{ m}^2)\left(\frac{147\text{ m}}{3}\right) \\ &= (52{,}900\text{ m}^2)(49\text{ m}) \\ &= 2{,}592{,}100\text{ m}^3 \end{aligned}$$

Thus, the volume of the original pyramid is 2,592,100 cubic meters. But the pyramid has shrunk; see Problem 6 to discover by how much!
Source: http://www.pbs.org.

F > Applications Involving Volume

So many applications and so little space! The formulas for the volume of the solids we have discussed can be used to find the volume of a Pet Taxi (carrier for animals), the amount of soil to be removed from a contaminated landfill, the amount of medication in a cylindrical pill, a microwave or toaster oven's capacity, the volume of boxes or trucks used for moving, and many other applications. Here are some samples.

EXAMPLE 7 Volume of soil removed

How much soil was removed from the portion of the hole shown? Give the answer in cubic yards.

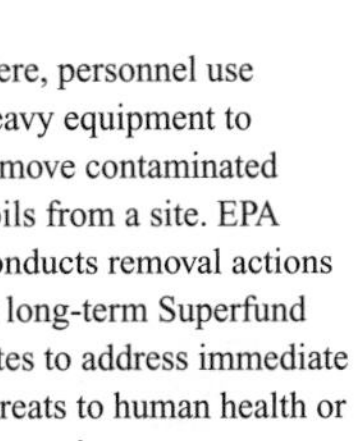
Here, personnel use heavy equipment to remove contaminated soils from a site. EPA conducts removal actions at long-term Superfund sites to address immediate threats to human health or the environment.

SOLUTION Since the answer has to be in cubic yards, we assume that the hole is a rectangular box 20 yards long, 2 yards wide, and 1 yard deep.
(*Note:* 6 ft = 2 yd and 3 ft = 1 yd.)

The volume of a rectangular solid is

$$\begin{aligned} V &= L \cdot W \cdot H \\ &= (20\text{ yd}) \cdot (2\text{ yd}) \cdot (1\text{ yd}) \\ &= 40\text{ yd}^3 \end{aligned}$$

This is the capacity of two large dump trucks!

PROBLEM 7

A much larger trench has been dug in the background. If the dimensions of that trench are 120 feet long by 6 feet wide by 3 feet deep, how many cubic yards of soil were removed from that trench?

Answers to PROBLEMS

7. 80 yd^3

EXAMPLE 8 Volume of a capsule

The capsules are in the shape of a cylinder 12 millimeters long with two half spheres with a diameter of 2 millimeters at each end. What is the volume of the capsule? Use 3.14 for π and round the answer to the nearest hundredth of a cubic millimeter.

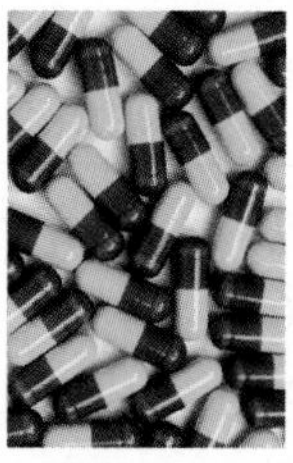

SOLUTION We have to find the volume V of the cylinder plus the volume of the two half spheres (which make up one whole sphere).

The volume of the cylinder is $V = \pi \cdot r^2 \cdot h$

Here,

$r = 1$ mm and $h = 12$ mm, so $V = 3.14 \cdot (1 \text{ mm})^2 \cdot (12 \text{ mm})$

$= 37.68 \text{ mm}^3$

The volume of the sphere is $V = \frac{4}{3} \cdot \pi \cdot r^3$

$= \frac{4}{3} \cdot 3.14 \cdot (1 \text{ mm})^3$

$= \frac{4 \cdot 3.14 \cdot 1}{3} \text{ mm}^3$

$\approx 4.19 \text{ mm}^3$

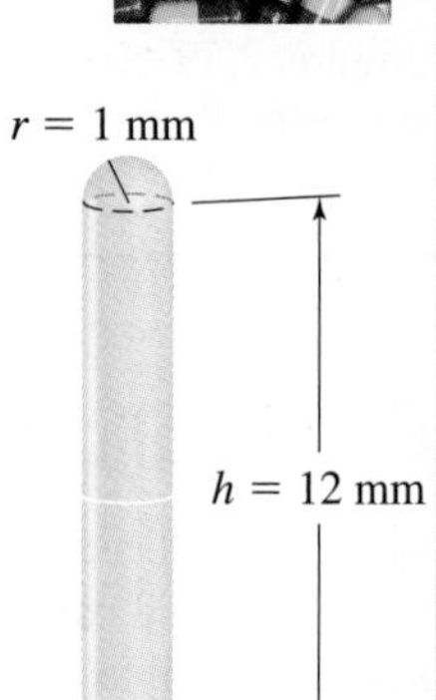

The volume of the whole capsule is $37.68 + 4.19 = 41.87 \text{ mm}^3$.

PROBLEM 8

Suppose the capsules are in the shape of a cylinder 10 millimeters long with two half spheres of diameter 2 millimeters at each end. What is the volume of each capsule? Use 3.14 for π and round the answer to the nearest hundredth of a cubic millimeter.

Answers to PROBLEMS

8. 35.59 mm^3

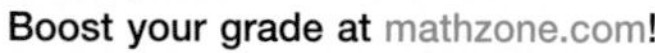

Exercises 8.4

Web IT go to mhhe.com/bello for more lessons

⟨A⟩ Volume of Rectangular Solids In Problems 1–4, find the volume of the rectangular solid with the indicated dimensions.

	Length	Width	Height
1.	8 cm	5 cm	6 cm
2.	15 cm	20 cm	9 cm

	Length	Width	Height
3.	$9\frac{1}{2}$ in.	3 in.	$4\frac{1}{2}$ in.
4.	$5\frac{1}{4}$ in.	$1\frac{3}{8}$ in.	$\frac{5}{8}$ in.

5. The Pet Kennel measures 26 inches long, 19 inches high, and 16 inches wide. What is the volume of the Pet Kennel?

6. The Pet Taxi is 16 inches long, 10 inches high, and 11 inches wide. What is the volume of the Pet Taxi?

7. The inside dimensions of the toaster are 11 inches wide, 8 inches deep, and 5 inches high. What is the volume of the inside of the toaster?

8. The inside dimensions of the microwave are 16 inches by 11 inches by 13 inches. What is the volume of the inside of the microwave?

9. A swimming pool measuring 30 feet by 18 feet by 6 feet is to be excavated. How many cubic yards of dirt must be removed? (*Hint:* Change the feet to yards first.)

10. How many cubic yards of sand are being transported in a dump truck whose loading area measures 9 feet by 4 feet by 3 feet?

11. A rectangular tank measures 30 feet by 3 feet by 2 feet. How many gallons of water will it hold if 1 ft^3 = 7.5 gallons?

12. A hot tub measures 4 feet by 6 feet and is 3 feet deep. How many gallons of water will it hold if 1 ft^3 = 7.5 gallons?

13. Where do you use the most water at home? In the toilet! If the toilet tank is 18 inches long, 8 inches wide, and the water is 12 inches deep, how many gallons of water does the toilet tank contain? (1 ft^3 = 7.5 gal; answer to the nearest gallon.)

14. How much concrete is needed to build a roadway 45 feet wide, 1 foot thick, and 10 miles long (1 mile = 5280 ft)?

⟨ B ⟩ Volume of Cylinders In Problems 15–20, find the volume of a cylinder with the given radius and height. (Use 3.14 for π and round the answer to the nearest tenth.)

	Radius	Height
15.	10 in.	8 in.
16.	4 in.	18 in.

	Radius	Height
17.	10 cm	20 cm
18.	3.5 cm	2.5 cm

	Radius	Height
19.	1.5 m	4.5 m
20.	0.8 m	3.2 m

21. Find the volume of the large 4-foot high, 1-foot diameter cylinder. Use 3.14 for π and round the answer to the nearest hundredth.

22. Find the volume of the small 2-foot high, 1-foot diameter cylinder. Use 3.14 for π and round the answer to the nearest hundredth.

23. Each sugar container is 5 inches high and 3 inches in diameter. Find the volume of one container. Use 3.14 for π and round the answer to the nearest hundredth.

24. What is the volume of the sugar in the three containers? Note that one of the containers is half-full. Use 3.14 for π and round your answer to the nearest whole number.

In Problems 25–28, use 3.14 for π, and round the answer to the nearest tenth.

25. A cylindrical tank has a 20-foot diameter and is 40 feet high. What is its volume?

26. A steel rod is $\frac{1}{2}$ inch in diameter and 18 feet long. What is its volume?

27. A coffee cup has a 3-inch diameter and is 4 inches high. What is its volume? If 1 $in.^3$ = 0.6 fl oz, how many fluid ounces does the cup hold?

28. A coffee cup is 3 inches in diameter and 3.5 inches high. Find the volume and determine if a 12-fluid-ounce soft drink will fit in the cup (1 $in.^3$ = 0.6 fl oz).

⟨ C ⟩ Volume of Spheres In Problems 29–32, use 3.14 for π and round the answer to the nearest tenth.

29. A spherical water tank has a 24-foot radius. How many gallons of water will it hold if 1 cubic foot holds about 7.5 gallons?

30. A spherical water tank has a 7.2-meter radius. How many liters of water will it hold if 1 cubic meter holds 1000 liters?

31. The fuel tanks on some ships are spheres, of which only the top halves are above deck. If one of these tanks is 120 feet in diameter, how many gallons of fuel does it hold (1 ft^3 = 7.5 gal)?

32. The Christmas ornament has a 2-inch diameter. What is its volume?

Web IT go to mhhe.com/bello for more lessons

Web IT go to mhhe.com/bello for more lessons

‹D› Volume of Circular Cones

In Problems 33–37, find the volume of a circular cone with the given radius and height. Use 3.14 for π and round the answer to the nearest tenth.

	Radius	Height
33.	10 in.	6 in.
34.	18 in.	12 in.
35.	50 ft	20 ft

	Radius	Height
36.	10 ft	50 ft
37.	0.6 m	1.2 m

38. *Volume of funnel* The inside of the funnel is 8 centimeters in diameter and 7 centimeters high. What is the volume of the funnel? Use 3.14 for π and round the answer to the nearest tenth.

39. *Volume of smaller funnel* The inside of the funnel is 7 centimeters in diameter and 6 centimeters high. What is the volume of the funnel? Use 3.14 for π and round the answer to the nearest tenth.

‹E› Volume of Pyramids

40. *Volume of a pyramid* Most people associate pyramids with Egypt (see Example 6), but did you know that there are ancient pyramids in Teotihuacan, Mexico? The Pyramid of the Sun has a height of 233.5 feet (about half that of the Great Pyramid) but its square base is 733 feet on each side (the Great Pyramid is 756 feet on each side). What is the volume of the Pyramid of the Sun? Answer to the nearest cubic foot.

The Great Pyramid superimposed over the Pyramid of the Sun.

41. *Volume of a pyramid* The second photo shows the Pyramid of the Moon also built at Teotihuacan, Mexico, between A.D. 150 and 225. Its base measures 492 feet on each side and its height is 138 feet. What is the volume of the Pyramid of the Moon?

Pyramid of the Moon

‹F› Applications Involving Volume

42. *Volume of boxes* Have you moved to the dorm or an apartment lately? You probably needed some boxes with some of the dimensions shown. Give the answer in both cubic inches and cubic feet. (1 cubic foot = 1728 cubic inches)

a. Find the volume of the small box.

b. Find the volume of the medium box.

c. Find the volume of the large box.

d. Which of the three volumes shown for the boxes (1.5 cu ft, 3.0 cu ft, 4.5 cu ft, and 6.0 cu ft) agrees exactly with your answer, the box in **a**, **b**, or **c**?

- **Small Box**
  16″ × 12″ × 12″ 1.5 cu/ft
- **Medium Box**
  18″ × 18″ × 16″ 3.0 cu/ft
- **Large Box**
  18″ × 18″ × 24″ 4.5 cu/ft
- **Extra-Large Box**
  24″ × 18″ × 24″ 6.0 cu/ft

43. *Volume of a truck* You probably needed a truck to move. The U-Haul truck has the dimensions shown.

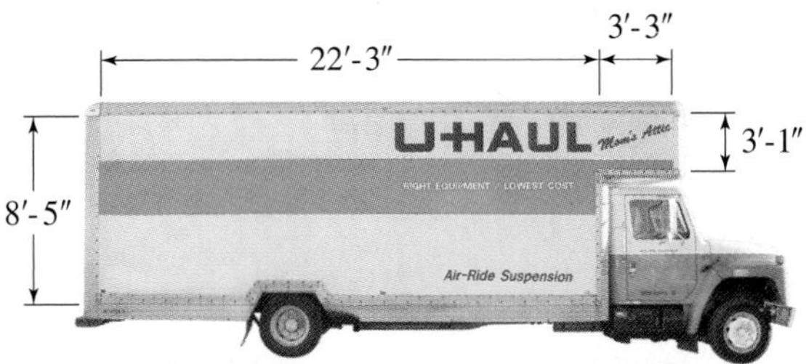

Inside dimensions: 22′-3″ × 7′-7″ × 8′-5″ (*L* × *W* × *H*)
Dimensions over the truck cab: 3′-3″ × 7′-7″ × 3′-1″ (*L* × *W* × *H*)

a. Approximate the inside dimensions of the truck to 22 feet by 7 feet by 8 feet. What is the volume?

b. Approximate the dimensions over the truck cab to 3 feet by 7 feet by 3 feet. What is the volume?

c. You estimate that you have 1300 cubic feet of stuff to be moved. Does your stuff theoretically fit in the truck? Explain (*Hint:* Don't forget the space in over the truck cab.)

44. *U-Haul recommendations* If you have a two- to three-room apartment, U-Haul recommends a 1200- to 1600-cubic foot truck.

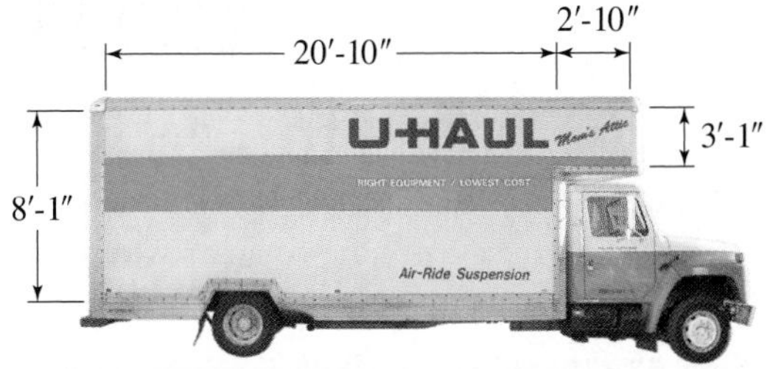

Inside dimensions: 20′-10″ × 7′-6″ × 8′-1″ (*L* × *W* × *H*)
Dimensions over the truck cab: 2′-10″ × 7′-6″ × 3′-1″ (*L* × *W* × *H*)

a. Approximate the inside dimensions of the truck to 20 feet by 7 feet by 8 feet. What is the volume?

b. Approximate the dimensions over the truck cab to 3 feet by 7 feet by 3 feet. What is the volume?

c. If you have a two-room apartment, U-Haul recommends a truck with a minimum of 1200 cubic feet of space. Does this truck meet that recommendation?

45. *Bench volume* Some benches at the University of South Florida are in the form of a hemisphere (half of a sphere) with a 34-inch diameter. Use 3.14 for π and give the answer (to the nearest tenth) in both cubic inches and cubic feet. ($1 \text{ ft}^3 = 1728 \text{ in.}^3$)

a. If the bench were a sphere (instead of a hemisphere), what would be its volume?

b. Since the bench is a hemisphere, its volume is $\frac{1}{2}$ the volume of a sphere. What is this volume?

46. *Pendulum volume* The pendulum at USF has a diameter of 2 feet. If we assume that the pendulum is a sphere, how much metal was used to make it? Use 3.14 for π and give the answer to the nearest tenth.

47. *UPS box volume* The UPS box is approximately 36 inches high by 24 inches wide by 20 inches deep. Give the answer in both cubic inches and cubic feet. ($1 \text{ ft}^3 = 1728 \text{ in.}^3$)

a. What is the volume of the box?

b. You cannot fill the box to the top. The actual effective area where you can place your mail is only 24 inches high. What is the volume of the effective area for the box?

48. *Volume of a funnel cloud* Funnel clouds can be 300 to 2000 feet in diameter. Assume the funnel cloud in the photo is 1000 feet in diameter and 2000 feet high. What is the volume of this funnel cloud? Use 3.14 for π and give the answer to the nearest tenth.

49. *Volume of a funnel cloud* Assume a funnel cloud has a 300-foot diameter and it is 600 feet high. What is the volume of the funnel cloud? Use 3.14 for π.

Web IT go to mhhe.com/bello for more lessons

Web IT go to mhhe.com/bello for more lessons

Volume of garbage cans What is the volume (capacity) of the garbage cans they use at your school? The volume of the can consists of two parts: the cylindrical bottom (2.5 feet high, 1.5 feet diameter) and the hemisphere (half of a sphere). (Use 3.14 for π and round the answer to the nearest tenth.)

50. What is the volume of the cylindrical part?

51. What is the volume of the hemispherical top?

52. What is the volume of the entire trash can?

53. The chemistry building has three floors and there are four trash cans on each floor. What is the volume of the trash cans on the three floors of the chemistry building?

54. *Trash Capacity* The USF garbage truck carries about 500 cubic feet of trash. How many cans of trash can it carry? (Use the volume you found for the trash can in Problem 52.)

55. *Trucks needed for the garbage* If we assume that every building has three floors and four trash cans on each floor, how many buildings can the truck service? (*Hint:* How many 12-can loads fit in the truck?)

Using Your Knowledge

Volume of Bread In the *Getting Started,* we raised the question: which bread has more volume? We now have the knowledge to answer. Use 3.14 for π and round the answer to the nearest tenth.

56. The Cuban bread loaf is almost a cylinder 20 inches long with a 3-inch diameter. What is the approximate volume of the Cuban loaf?

57. The Greek bread is almost shaped like a cylinder 8 inches in diameter and 2 inches high. What is the approximate volume of the cylinder?

58. The answer in Exercise 57 does not really give the volume of the Greek bread because the bread has a hole 3 inches in diameter in the middle. What is the volume of this hole? Actually, 0 because there is no bread there! Find the volume of the empty space in the middle of the bread and subtract it from the answer you got in Exercise 57; that is the volume of the Greek bread! Is it more or less than the volume of the Cuban bread of Exercise 56?

59. Here is another way to find the volume of the Greek bread. Suppose you make a vertical cut perpendicular to the ground on the bread and "stretch" the bread into a loaf. How long would that loaf be? Remember, the outside diameter is 8 inches.

60. Since the original Greek bread was 2 inches high, we can assume that the new loaf of Greek bread has a 2-inch diameter. Now, use your answer to Exercise 59 (that would give you the length of the new loaf) and find the volume of the Greek loaf. Is it approximately the same answer as the one you obtained in Exercise 58?

Write On

61. Burger King, McDonald's, and other hamburger stores compare the amount of beef they have in their hamburgers. When doing so, should you compare the circumference, the area, or the volume of the hamburgers? Explain.

62. "In men's play, a basketball is 29.5 to 30 inches (74.9 to 76.2 centimeters) in diameter." What is wrong with that statement? Write in your own words how the statement should be stated.

Concept Checker

Fill in the blank(s) with the correct word(s), phrase, or mathematical statement.

63. The volume of a **rectangular solid** of length L, width W, and height H is ______________.

64. The **volume** of a **cylinder** of radius r and height h is ______________.

$\frac{1}{3}\pi r^2 h$ LWH

πr^3 $\frac{1}{3}\pi r^3$

65. The **volume** of a **sphere** of radius r is ______________.

66. The **volume** of a **cone** of radius r and height h is ______________.

$\pi r^2 h$ $\frac{4}{3}\pi r^3$

πrh

Mastery Test

In Problems 67–71, use 3.14 for π when necessary and round the answer to the nearest tenth.

67. Find the volume of a circular cone 10 centimeters high and with a 6-centimeter radius.

68. How many cubic yards of sand are needed to fill a sandbox 9 feet long by 3 feet wide and 4 inches deep?

69. Find the volume of a bowling ball with a 4.2-inch diameter.

70. Find the volume of the solid.

2 m

3 m

1 m

71. Find the volume of a can 8 inches high with a 1-inch radius.

Skill Checker

In Problems 72–73 evaluate:

72. $9^2 + 12^2$

73. $8^2 + 11^2$

8.5 Square Roots and the Pythagorean Theorem

Objectives

You should be able to:

A Find the exact and approximate square root of a number.

B Use the Pythagorean theorem to find the length of one of the sides of a right triangle when the two other sides are given.

C Solve applied problems using the Pythagorean theorem.

To Succeed, Review How To . . .

1. Add, subtract, multiply, and divide whole numbers. (pp. 24, 37, 49, 62)
2. Solve equations. (pp. 89–91, 240–241)

Getting Started

In the photo at the beginning of the chapter (shown here reduced), the length c of the upper arm of the player is $\sqrt{0.0884}$ meter (read "the square root of 0.0884") and was obtained by using the Pythagorean theorem. What does "the square root of a number" mean? What is the Pythagorean theorem? We will discuss both topics next.

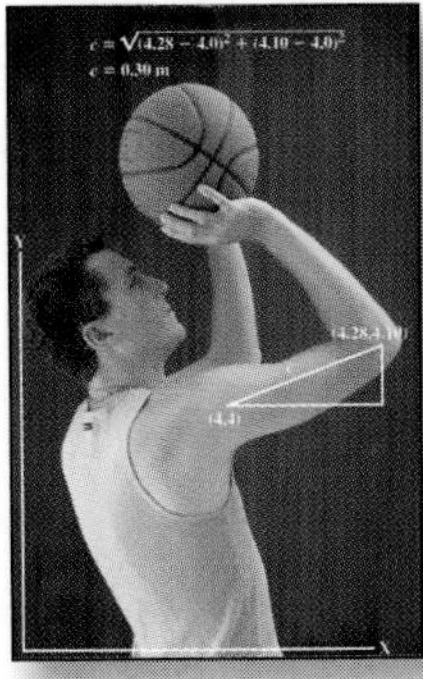

A › Square Roots

We have already studied how to square a number. Look at the table.

x	x^2
1	$1^2 = 1$
2	$2^2 = 4$
3	$3^2 = 9$
4	$4^2 = 16$

Numbers such as 1, 4, 9, 16, $\frac{9}{16}$, and $\frac{4}{25}$ are called **perfect squares** because they are squares of whole numbers or fractions.

SQUARE ROOT

The square root of a number a, denoted by $\sqrt{a}$, is one of the equal factors b of the number a, that is,

$$\sqrt{a} = b \text{ means that } a = b^2$$

Note that $\sqrt{a}$ is always positive.

Remember, to find the square root of a ($\sqrt{a}$), simply find a number whose square is a. Thus, $\sqrt{16} = 4$ because $16 = 4^2$, $\sqrt{25} = 5$ because $5^2 = 25$, and $\sqrt{36} = 6$ because $6^2 = 36$.

EXAMPLE 1 Finding square roots

Find: **a.** $\sqrt{64}$ **b.** $\sqrt{121}$

SOLUTION

a. We need a number whose square is 64. The number is 8.

Thus, $\sqrt{64} = 8$. **CHECK** $8^2 = 64$

b. This time we need a number whose square is 121. The number is 11.

Thus, $\sqrt{121} = 11$. **CHECK** $11^2 = 121$

PROBLEM 1

Find: **a.** $\sqrt{49}$ **b.** $\sqrt{81}$

Now, suppose you want to find $\sqrt{122}$. What number can you square so that you get 122? Unfortunately, 122 is not a perfect square, so there is no whole number whose square is 122. There are many square roots that are not whole numbers or fractions. For example, $\sqrt{2}$, $\sqrt{3}$, $\sqrt{5}$, $\sqrt{6}$, and $\sqrt{7}$ are not whole numbers. The best we can do for these numbers is approximate the answer with a calculator.

To find $\sqrt{2}$ with your calculator, press 2nd x^2 2) ENTER or 2 $\sqrt{\ }$ ENTER. Check your calculator manual to determine which method you should use.

Thus,

$$\sqrt{2} \approx 1.4142136 \approx 1.414 \text{ (to three decimal places)}$$

$$\sqrt{3} \approx 1.7320508 \approx 1.732 \text{ (to three decimal places)}$$

$$\sqrt{5} \approx 2.236068 \approx 2.236 \text{ (to three decimal places)}$$

EXAMPLE 2 Approximating square roots

Find $\sqrt{6}$ to three decimal places.

SOLUTION Using a calculator, $\sqrt{6} \approx 2.4494897 \approx 2.449$.

CHECK $(2.449)^2 = 5.997601$, very close to 6!

PROBLEM 2

Find $\sqrt{7}$ to three decimal places.

Answers to PROBLEMS

1. a. 7 **b.** 9 **2.** 2.646

B › The Pythagorean Theorem

In the photo in the *Getting Started,* the length of the arm of the player is c, the longest side (the hypotenuse) of a right triangle whose sides (legs) measure 0.28 meter and 0.10 meter as shown:

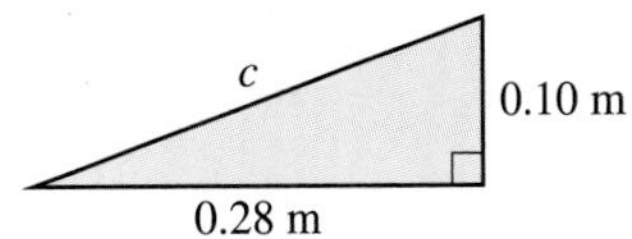

The relationship between the length of the legs a and b and the length of the hypotenuse c is given by $a^2 + b^2 = c^2$.

This fact, proved by the Greek mathematician Pythagoras over 2500 years ago, is true for all right triangles and is stated next.

THE PYTHAGOREAN THEOREM

In any right triangle with legs of length a and b and hypotenuse c,

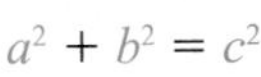

$$a^2 + b^2 = c^2$$

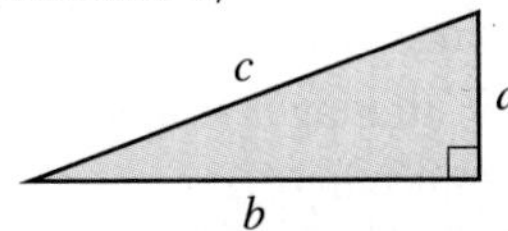

Thus, if we are given the lengths of any two sides of a right triangle, we can always find the length of the third side using the Pythagorean theorem. Note that the converse of the theorem is also true.

CONVERSE OF THE PYTHAGOREAN THEOREM

If $a^2 + b^2 = c^2$, then the triangle is a right triangle.

EXAMPLE 3 Finding the length of the hypotenuse

Find the length c of the hypotenuse of the given right triangle.

SOLUTION

Using the Pythagorean theorem,	$a^2 + b^2 = c^2$
Substituting $a = 3$, $b = 4$	$3^2 + 4^2 = c^2$
Simplifying	$9 + 16 = c^2$
Adding	$25 = c^2$
Solving for c	$c = \sqrt{25} = 5$

Thus, the length of the hypotenuse c is 5.

CHECK $3^2 + 4^2 = 5^2$

PROBLEM 3

Find the length c of the hypotenuse of the triangle.

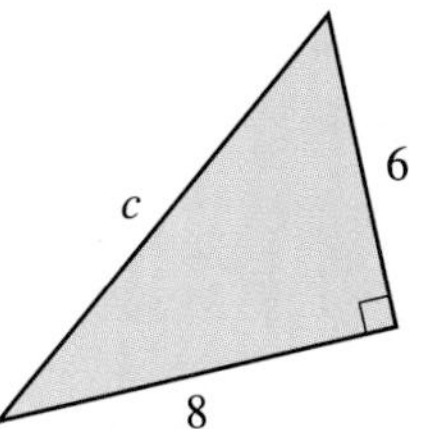

EXAMPLE 4 Finding the length of a leg

Find the length a for the given right triangle.

SOLUTION

Using the Pythagorean theorem, with $b = 5$ and $c = 6$	$a^2 + 5^2 = 6^2$
Simplifying	$a^2 + 25 = 36$
Subtracting 25 from both sides	$a^2 = 11$
Solving for a	$a = \sqrt{11}$

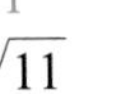

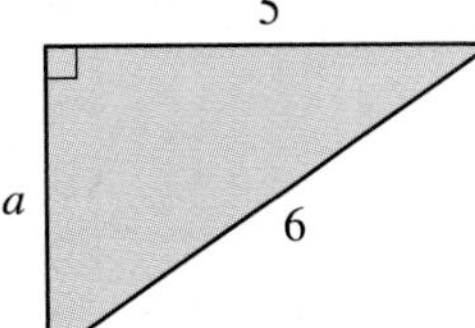

Thus, the exact length of a is $\sqrt{11}$.

The approximate length of a, using a calculator, is 3.317.

PROBLEM 4

Find the length a for the given triangle.

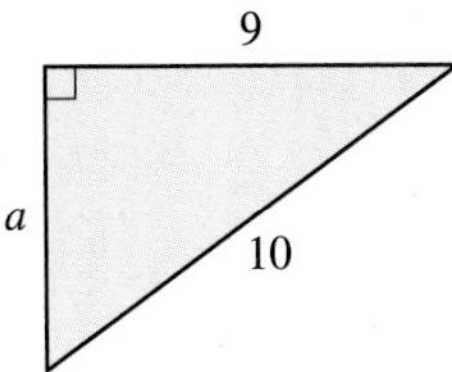

Answers to PROBLEMS

3. $c = 10$ **4.** $a = \sqrt{19} \approx 4.359$

C > Applications of the Pythagorean Theorem

Television sets and computer monitors are classified according to the length of the diagonal (hypotenuse) of the set or screen. Thus, a 15-inch screen means that the diagonal of the set or monitor is 15 inches. For example, the screen of the laptop is about 12 inches wide by 9 inches high. Is it really a 15-inch monitor?

We shall see next.

EXAMPLE 5 Finding the length of the hypotenuse

Find the length of the diagonal of the screen with the given dimensions.

SOLUTION

9

12

Using the Pythagorean theorem,	$a^2 + b^2 = c^2$
Substituting $a = 9$, $b = 12$	$9^2 + 12^2 = c^2$
Simplifying	$81 + 144 = c^2$
Adding	$225 = c^2$
Solving for c	$c = \sqrt{225} = 15$

Thus, the hypotenuse is 15, and we do have a 15-inch screen.

PROBLEM 5

Find the length of the diagonal of a television screen 16 inches wide and 12 inches high.

EXAMPLE 6 Finding the length of the hypotenuse

A baseball diamond is really a square with sides of 90 feet. To the nearest whole number, how far is it from first base to third base?

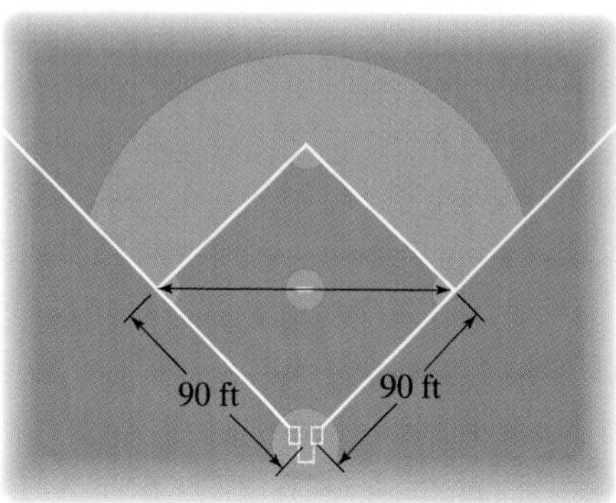

SOLUTION In this case $a = b = 90$.

Using the Pythagorean theorem,	$a^2 + b^2 = c^2$
Substituting $a = 90$, $b = 90$	$90^2 + 90^2 = c^2$
Simplifying	$8100 + 8100 = c^2$
Adding	$16{,}200 = c^2$
Solving for c	$c = \sqrt{16{,}200} \approx 127$ ft

PROBLEM 6

A softball diamond is a square with sides of 60 feet. To the nearest whole number, how far is it from home plate to second base?

Answers to PROBLEMS

5. 20 in. **6.** $\sqrt{7200} \approx 85$ ft

> Exercises 8.5

⟨ A ⟩ Square Roots In Problems 1–10, find the square root of the number.

1. $\sqrt{100}$ **2.** $\sqrt{144}$ **3.** $\sqrt{196}$ **4.** $\sqrt{289}$ **5.** $\sqrt{361}$

6. $\sqrt{324}$ **7.** $\sqrt{400}$ **8.** $\sqrt{900}$ **9.** $\sqrt{169}$ **10.** $\sqrt{10{,}000}$

In Problems 11–20, approximate the answer to three decimal places.

11. $\sqrt{8}$ **12.** $\sqrt{10}$ **13.** $\sqrt{11}$ **14.** $\sqrt{17}$ **15.** $\sqrt{23}$

16. $\sqrt{27}$ **17.** $\sqrt{29}$ **18.** $\sqrt{33}$ **19.** $\sqrt{108}$ **20.** $\sqrt{405}$

⟨ B ⟩ The Pythagorean Theorem In Problems 21–30, find the hypotenuse c. Give the answer to three decimal places when appropriate.

21.
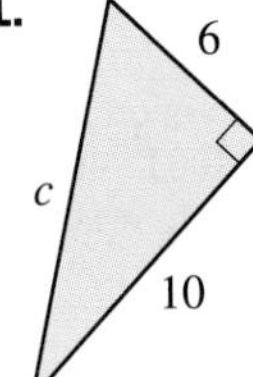

22.
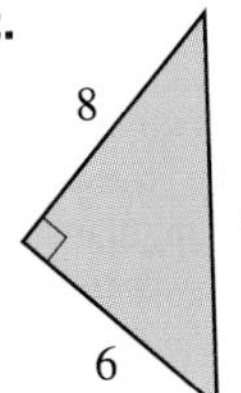

23.
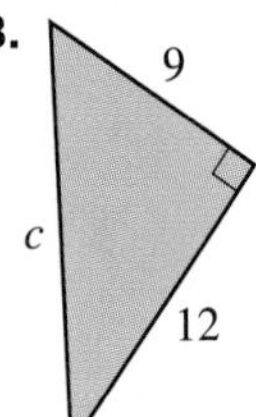

24.
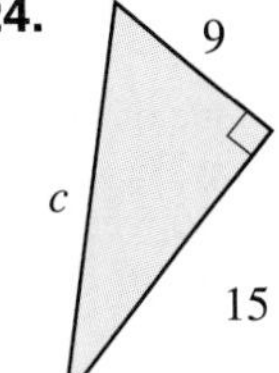

25.
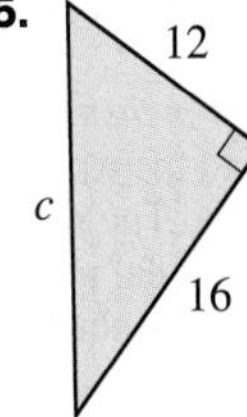

26.
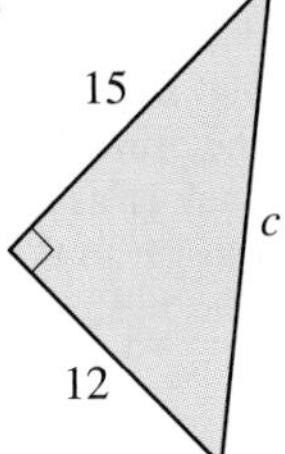

27.
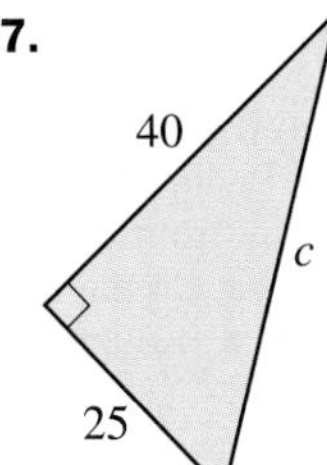

28.
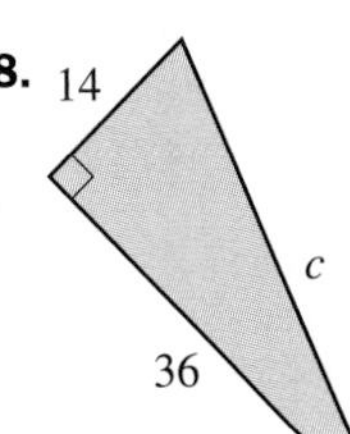

29.
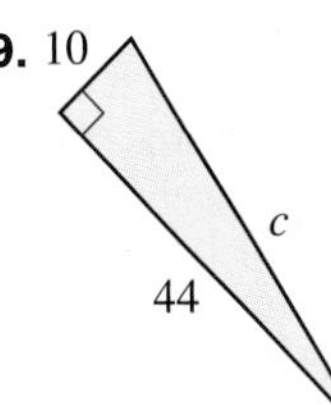

30.
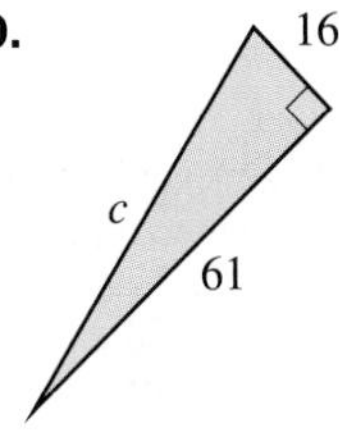

In Problems 31–38, find the missing side. Give the answer to three decimal places when appropriate.

31.
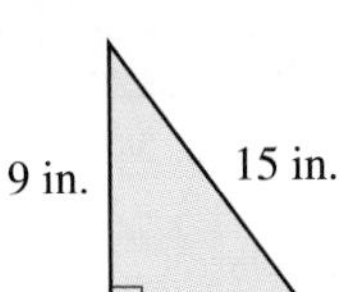

32.
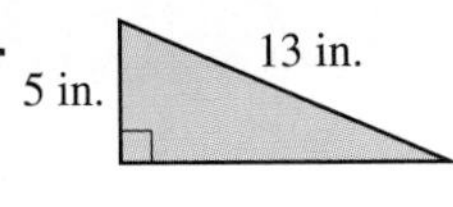

33.
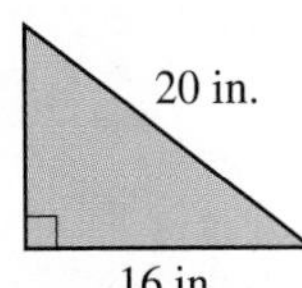

34.
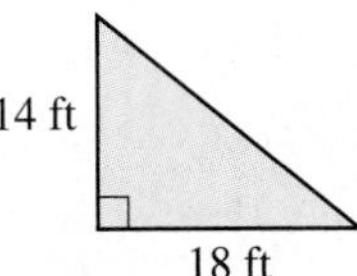

35.
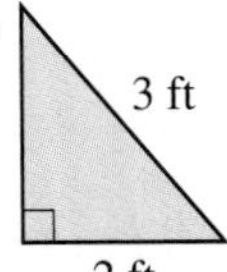

36.
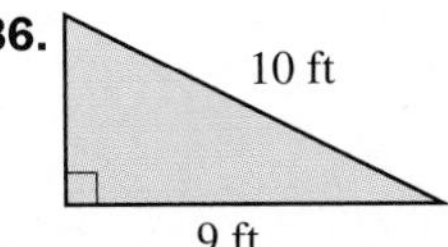

37.
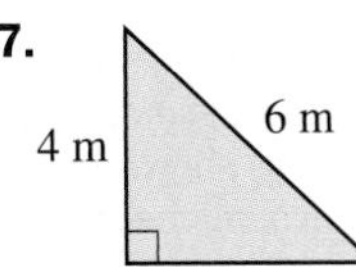

38.
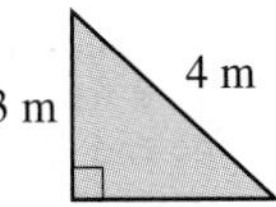

‹ C › Applications of the Pythagorean Theorem

39. *Body measurements* The diagram shows the distance from the hip to the knee ($b \approx 0.30$ m) and the distance from the knee to the ankle ($a \approx 0.44$ m). Use the Pythagorean theorem to find the distance (in meters m) from the hip to the ankle if the leg is bent at a right angle. Round the answer to the nearest thousandth.

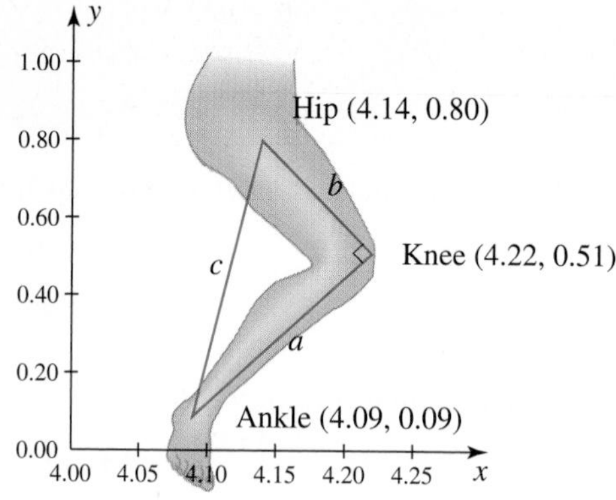

Source: Indiana University–Purdue University Indianapolis.

40. *Triangle dimensions* *The Human Side of Mathematics* on page 453 shows a triangle like this one. If the dimensions of the triangle are as shown, find c and approximate it to two decimal places. Do you get the same answer as was shown in the *Getting Started?*

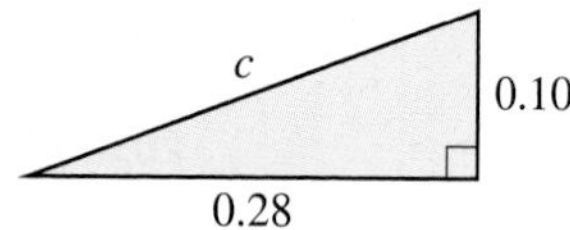

41. *Height of large television* One of the largest television sets ever built was the 289 Sony Jumbo Tron (289 feet diagonally!). If the length of the screen was 150 feet, what was the height? Round the answer to the nearest thousandth.

42. *Height of television* A 10-inch television screen (measured diagonally) is 8 inches wide. How high is it?

43. *Diagonal measurement of screen* One of the largest cinema screens in the world is at the Pictorium Theater in Santa Clara, California. The rectangular screen is 70 feet by 96 feet. What is the diagonal measurement of the screen? Round the answer to the nearest thousandth.

44. *Width of LCD TV* One of the largest LCD television screens in the world is the Samsung SyncMaster 400T 40 inch (diagonal). If the rectangular screen is 20 inches high, what is the width? Round the answer to the nearest thousandth.

45. *Length of jewel case* A rectangular DVD box is 14 centimeters wide and 11 centimeters high. What is the diagonal length of the box? Round the answer to the nearest thousandth.

46. *Length of a conveyor belt* Pedro Mendieta operates the conveyor belt machine used to haul materials to the top of the concrete mixing container. If the belt ends 40 feet above the ground, and starts 120 feet away from the base of the container, as shown in the photo, how long is the belt? Round the answer to the nearest foot.

47. *Length of a conveyor belt* Another conveyor belt machine operated by John Taylor ends 45 feet above the ground, but it starts 118 feet away from the base of the container as shown in the photo. How far do the materials travel to get to the top of the conveyor belt? Round the answer to the nearest foot.

48. *Airplane travel* The plane shown is ascending at a 10° angle. If the plane is 40 feet high, and the horizontal distance from the plane to the end of the runway is 220 feet as shown in the photo, how many feet has the plane traveled? Round the answer to the nearest foot.

49. *Bridges* The Dames Point Bridge spans the St. John River in Jacksonville, Florida. The longest cable supporting the bridge is 720 feet long and is secured 650 feet from the pole holding the cable as shown in the photo. What is the height h of the pole? Round the answer to the nearest whole number.

50. *Bridges* If in Problem 49 the cable was 700 feet long, how high would the pole be? Round the answer to the nearest whole number.

Using Your Knowledge

Approximation Suppose you want to approximate $\sqrt{18}$ *without* using your calculator. Since $\sqrt{16} = 4$ and $\sqrt{25} = 5$, you have $\sqrt{16} = 4 < \sqrt{18} < \sqrt{25} = 5$. This means that $\sqrt{18}$ is between 4 and 5. To find a better approximation of $\sqrt{18}$, you can use a method that mathematicians call **interpolation.** Don't be scared by this name! The process is *really* simple. If you want to find an approximation for $\sqrt{18}$, follow the steps in the diagram; $\sqrt{20}$ and $\sqrt{22}$ are also approximated.

$18 - 16 = 2$	$20 - 16 = 4$	$22 - 16 = 6$
$\sqrt{16} = 4$	$\sqrt{16} = 4$	$\sqrt{16} = 4$
$\sqrt{18} \approx 4 + \frac{2}{9}$	$\sqrt{20} \approx 4 + \frac{4}{9}$	$\sqrt{22} \approx 4 + \frac{6}{9}$
$\sqrt{25} = 5$	$\sqrt{25} = 5$	$\sqrt{25} = 5$
$25 - 16 = 9$	$25 - 16 = 9$	$25 - 16 = 9$

As you can see,

$$\sqrt{18} \approx 4\tfrac{2}{9} \quad \sqrt{20} \approx 4\tfrac{4}{9} \quad \text{and} \quad \sqrt{22} \approx 4\tfrac{6}{9} \approx 4\tfrac{2}{3}$$

By the way, do you see a pattern? What would $\sqrt{24}$ be?

Using a calculator,

$$\sqrt{18} \approx 4.2 \qquad \sqrt{20} \approx 4.5 \quad \text{and} \quad \sqrt{22} \approx 4.7$$

$$\sqrt{18} = 4\tfrac{2}{9} \approx 4.2, \quad \sqrt{20} = 4\tfrac{4}{9} \approx 4.4 \quad \sqrt{22} = 4\tfrac{6}{9} \approx 4.7$$

As you can see, we can get very close approximations!

Use this knowledge to approximate the following roots, giving the answer as a mixed number.

51. **a.** $\sqrt{26}$ **b.** $\sqrt{28}$ **c.** $\sqrt{30}$

52. **a.** $\sqrt{67}$ **b.** $\sqrt{70}$ **c.** $\sqrt{73}$

Write On

53. Squaring a number and finding the square root of a number are inverse operations. Can you find two other inverse operations?

54. Explain how you would prove or disprove that a given triangle is a right triangle.

Concept Checker

Fill in the blank(s) with the correct word(s), phrase, or mathematical statement.

55. In any **right triangle** with legs of length a and b and hypotenuse c, ________.

56. If a and b are the lengths of the legs of a triangle with hypotenuse c and $a^2 + b^2 = c^2$ then **the triangle must be** a ______ triangle.

$a^2 + b^2 = c^2$
scalene
right
isosceles

Mastery Test

57. Find the length a for the given right triangle. Round the answer to the nearest thousandth.

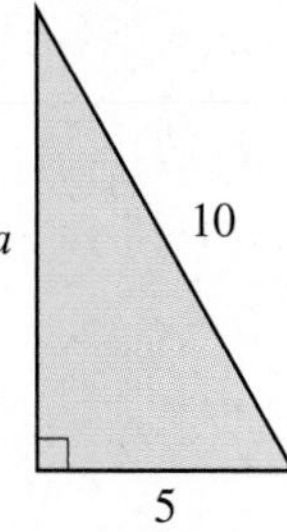

58. Find the hypotenuse for the given right triangle.

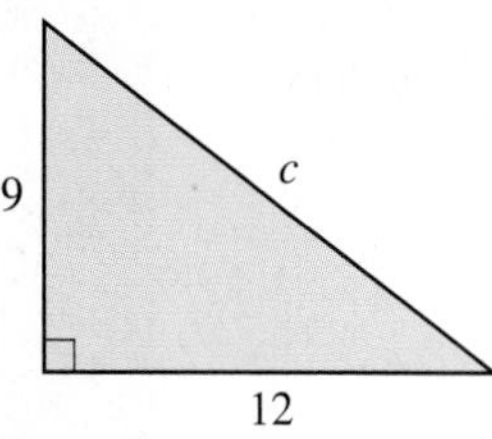

59. Find the hypotenuse for the given right triangle. Round the answer to the nearest thousandth.

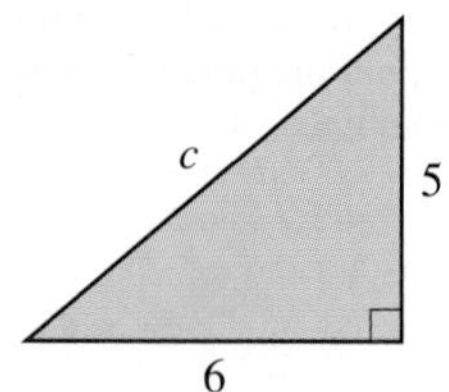

60. Find $\sqrt{1600}$.

61. Find $\sqrt{21}$ to three decimal places.

62. The diagonal of a computer monitor is 22 inches. If the monitor is 12 inches high, how wide is it? Round the answer to the nearest thousandth.

Skill Checker

In Problems 63–67, subtract and check by addition.

63. $50 - 37$ **64.** $38 - 23$ **65.** $23 - 19$ **66.** $53 - 27$ **67.** $92 - 78$

Collaborative Learning

Did you know that your hat size and your ring size are almost identical? By the way, there is a little flaw here. This is true only of men's hat sizes. Women's hat sizes are measured differently and more logically! Women's hat sizes are simply the circumference of the inner band of the hat. Nevertheless, we would like to see if there is a relationship between men's hat sizes—for men and women—and ring sizes. Form several groups of men and women.

1. Measure the circumference of the heads $C(H)$ of the participants.
2. Measure the circumference of their ring fingers $C(f)$.
3. Complete the following table, where $C(H)$ is the circumference of the head, $C(f)$ the circumference of the finger, d the diameter of the finger, H the men's hat size, and s the ring size.

$C(H)$	$\frac{C(H)}{\pi} = H$	$C(f)$	$\frac{C(f)}{\pi} = d$	$\frac{d - 0.458}{0.032} = s$

4. Is there a relationship between H and s?
5. Is there a relationship between hat size and ring size for women? If so, what is it?

Research Questions

1. Many scholars ascribe the Pythagorean theorem to, surprise, Pythagoras! However, other versions of the theorem like the following exist:

   The ancient Chinese proof   Bhaskara's proof   Euclid's proof

   Garfield's proof   Pappus's generalization

   Select three of these versions and write a paper giving details, if possible, about where they appeared, who authored them, and what was stated in each theorem.
2. There are different versions of the death of Pythagoras. Write a short paper detailing the circumstances of his death, where it occurred, and how.
3. Write a report about Pythagorean triples.
4. The Pythagoreans studied arithmetic, music, geometry, and astronomy. Write a report about the Pythagoreans' theory of music.
5. Write a report about the Pythagoreans' theory of astronomy.

Summary Chapter 8

Section	Item	Meaning	Example
8.1A	Point	A location in space	P•, Q•, and R• are points.
8.1A	Line	A set of points that extends *infinitely* in both directions	The line $\overleftrightarrow{AB}$
8.1A	Line segment	A piece of a line using A and B as endpoints	Line segment $\overleftrightarrow{AB}$
8.1A	Parallel lines	Two lines in the same plane that extend *infinitely* in both directions and never intersect	Line $\overleftrightarrow{AB} \parallel \overleftrightarrow{PQ}$
8.1A	Intersecting lines	Two lines with one common point	Line $\overleftrightarrow{AB}$ intersects $\overleftrightarrow{PQ}$ at point D.

(continued)

Section	Item	Meaning	Example
8.1A	Perpendicular lines	Two lines that intersect at right angles	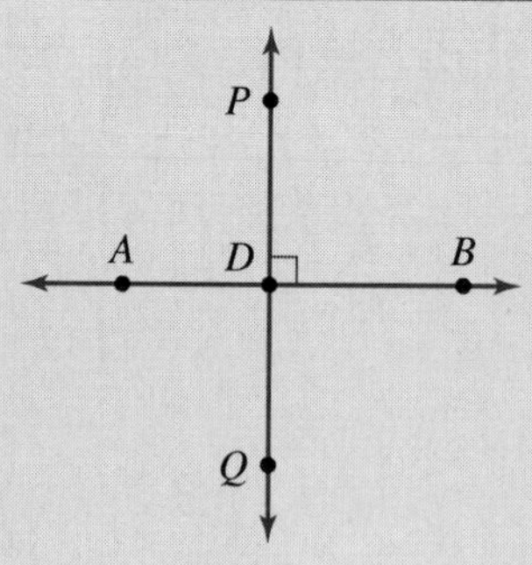 Line $\overleftrightarrow{AB} \perp \overleftrightarrow{PQ}$
8.1A	Ray	A ray is part of a line with one *endpoint* and extending *infinitely* in one direction.	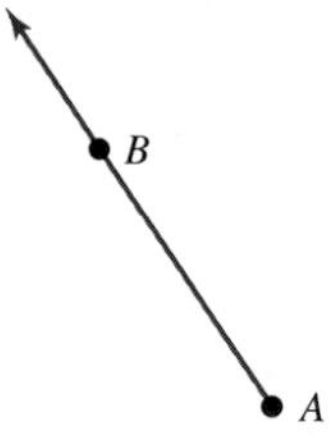 Ray $\overrightarrow{AB}$
8.1B	Angle Vertex of an angle Sides of an angle	The figure formed by two rays with a common endpoint called the vertex The common point of the two rays The rays forming the angle	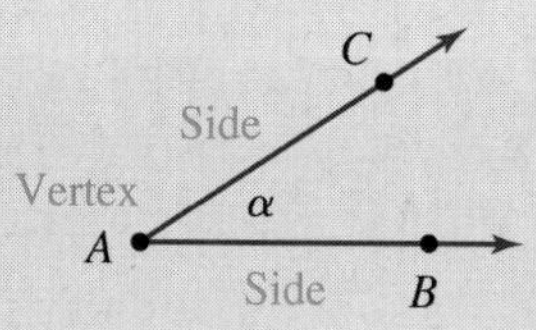 Vertex A, Side, Side, C, B, α
8.1C	Degree	$\frac{1}{360}$ of a complete revolution	
	Straight angle	One-half of a complete revolution	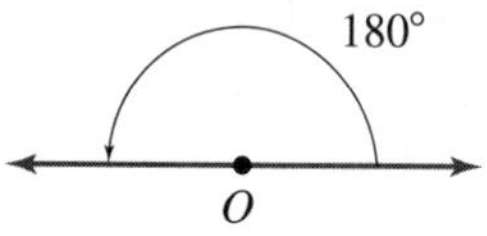 180°, O
	Right angle	One-quarter of a complete revolution	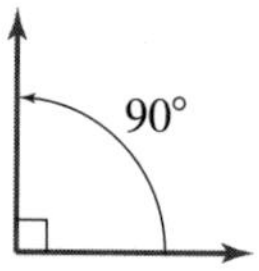 90°
	Acute angle	An angle greater than 0° and less than 90°	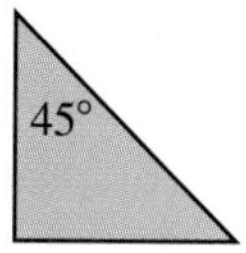 45°
	Obtuse angle	An angle greater than 90° and less than 180°	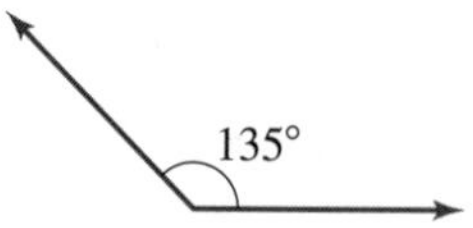 135°

Section	Item	Meaning	Example
8.1D	Supplementary angles	Angles whose measures add to 180°	
	Complementary angles	Angles whose measures add to 90°	
8.1E	Acute triangle	All three angles are acute.	
	Right triangle	One of the angles is a right angle.	
	Obtuse triangle	One of the angles is an obtuse angle.	
	Scalene triangle	No equal sides	15, 9, 20
	Isosceles triangle	Two equal sides	3, 3, 2
8.1E	Equilateral triangle	A triangle with three equal sides	
8.1F	Sum of the measures in a triangle	The sum of the measures in any triangle is 180°.	If in triangle ABC, $m\angle A = 50°$ and $m\angle B = 60°$, then $m\angle C = 180° - 50° - 60° = 70°$
8.2A	Perimeter Perimeter of a rectangle	Distance around $P = 2L + 2W$ (L = length, W = width)	The perimeter P of a rectangle 5 inches long and 3 inches wide is $P = 2 \cdot 5$ in. $+ 2 \cdot 3$ in. $= 16$ in.
	Perimeter of a square	$P = 4S$ (S = length of side)	The perimeter P of a square 3 centimeters on a side is $P = 4 \cdot 3$ cm $= 12$ cm.
8.2B	Circumference	$C = \pi d = 2\pi r$	The circumference of a circle of radius 6 centimeters is $2 \cdot \pi \cdot 6$ cm $= 12\pi$ cm ≈ 37.68 cm.
8.3A	Area of a rectangle	$A = LW$ (L = length, W = width)	The area A of a rectangle 4 inches long and 2 inches wide is $A = 4$ in. $\cdot$ 2 in. $= 8$ in.^{2}.
	Area of a square	$A = S^2$	The area A of a square 5 centimeters on a side is $A = 5^2$ cm^2 $= 25$ cm^2.
8.3B	Area of a triangle	$A = \frac{1}{2}bh$ (b = base, h = height)	The area A of a triangle of base 8 inches and height 10 inches is $A = \frac{1}{2} \cdot 8$ in. $\cdot$ 10 in. $= 40$ in.^{2}.

(continued)

Section	Item	Meaning	Example
8.3C	Area of a parallelogram	$A = bh$ (b = base, h = height)	The area A of a parallelogram of base 10 centimeters and height 15 centimeters is $A = 10 \text{ cm} \cdot 15 \text{ cm} = 150 \text{ cm}^2$.
8.3D	Area of a trapezoid	$A = \frac{1}{2}h(a + b)$(h = height, a and b are bases)	The area A of a trapezoid of height 8 centimeters and bases 6 centimeters and 10 centimeters, respectively, is $A = \frac{1}{2} \cdot 8 \text{ cm}(6 \text{ cm} + 10 \text{ cm}) = 64 \text{ cm}^2$.
8.3E	Area of a circle	$A = \pi r^2$ (r = radius)	The area of a circle of radius 6 centimeters is $\pi(6 \text{ cm})^2$, or $36\pi \text{ cm}^2 \approx 113 \text{ cm}^2$.
8.4A	Volume of a rectangular solid	$V = L \cdot W \cdot H$ (L = length, W = width, H = height)	The volume V of a rectangular box of length 5 meters, width 6 meters, and height 2 meters is $V = 5 \text{ m} \cdot 6 \text{ m} \cdot 2 \text{ m} = 60 \text{ m}^3$.
8.4B	Volume of a circular cylinder	$V = \pi r^2 h$ (r = radius, h = height)	The volume V of a circular cylinder of radius 3 inches and height 2 inches is $V = \pi \cdot (3 \text{ in.})^2 \cdot 2 \text{ in.} = 18\pi \text{ in.}^3 \approx 57 \text{ in.}^3$.
8.4C	Volume of a sphere	$V = \frac{4}{3}\pi r^3$ (r = radius)	The volume V of a sphere of radius 3 feet is $V = \frac{4}{3}\pi \cdot (3 \text{ ft})^3 = 36\pi \text{ ft}^3 \approx 113 \text{ ft}^3$.
8.4D	Volume of a circular cone	$V = \frac{1}{3}\pi r^2 h$ (r = radius, h = height)	The volume V of a circular cone of radius 2 centimeters and height 3 centimeters is $V = \frac{1}{3}\pi \cdot (2 \text{ cm})^2 \cdot 3 \text{ cm} = 4\pi \text{ cm}^3 \approx 13 \text{ cm}^3$.
8.4E	Volume of a pyramid	$V = \frac{1}{3}Bh$ (B = base area, h = height)	The volume V of a pyramid with each side measuring 3 feet with a height of 9 feet is $V = \frac{1}{3} \cdot 3 \text{ ft} \cdot 3 \text{ ft} \cdot 9 \text{ ft} = \frac{1}{3} \cdot 81 \text{ ft}^3 = 27 \text{ ft}$.
8.5A	$\sqrt{a} = b$	The square root of a is a number b so that $a = b^2$.	$\sqrt{9} = 3$ because $3^2 = 9$ $\sqrt{121} = 11$ because $11^2 = 121$
8.5B	Pythagorean theorem $c^2 = a^2 + b^2$	The square of the hypotenuse c of a right triangle equals the sum of the squares of the other two sides, a and b.	b c a

› Review Exercises Chapter 8

The figure will be used in Problems 1–5.

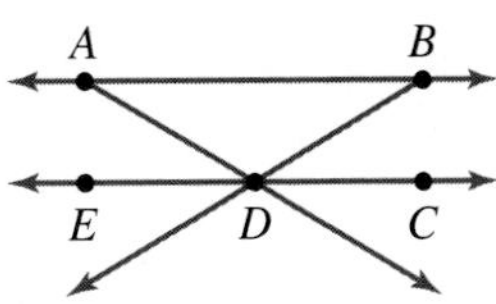

1. ‹ **8.1A** › *Name a point in the figure.*

a. **b.**

c. **d.**

e.

2. ‹ **8.1A** › *Name a line in the figure.*

a.

b.

3. ‹ **8.1A** › *Name a line segment in the figure.*

a. **b.**

c. **d.**

e.

4. ‹ **8.1A** › *Name a pair of parallel lines in the figure.*

5. ‹ **8.1A** › *Name a ray in the figure.*

a.

b.

6. ‹ **8.1B** › *Name the angle in three different ways:*

a.

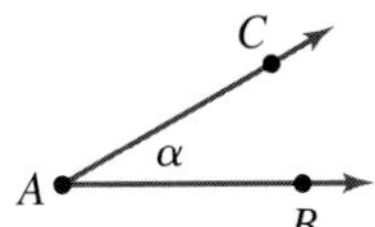

b.

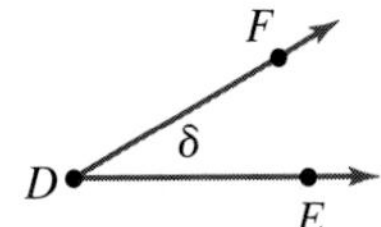

c.

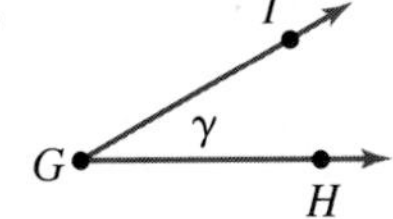

d.

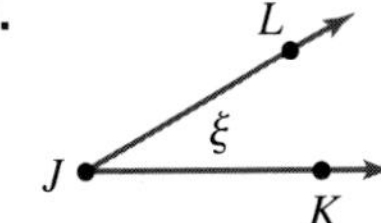

e.

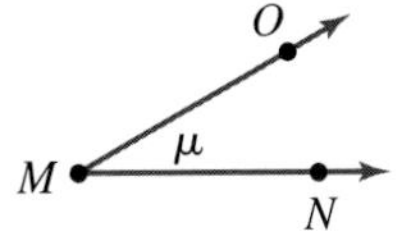

7. ⟨ **8.1C** ⟩ *Classify the angles as acute, right, straight, or obtuse.*

a. 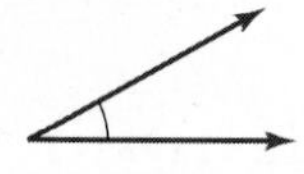

b. 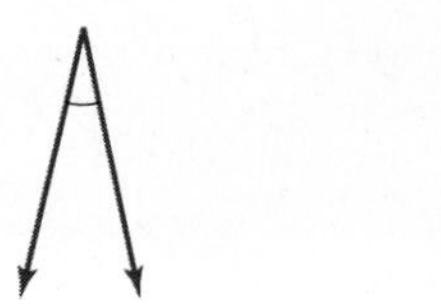

c. 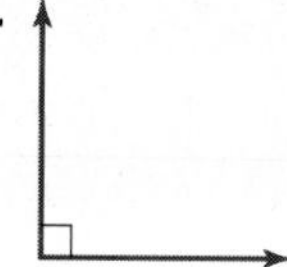

d.

e.

8. ⟨ **8.1D** ⟩ *Find the measure of the complement of a:*

a. 10° angle

b. 15° angle

c. 20° angle

d. 30° angle

e. 80° angle

9. ⟨ **8.1D** ⟩ *Find the measure of the supplement of a:*

a. 10° angle

b. 15° angle

c. 20° angle

d. 30° angle

e. 80° angle

10. ⟨ **8.1E** ⟩ *Classify the triangle according to the angles and sides.*

a. 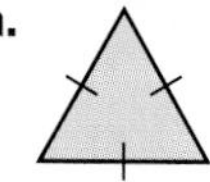

b. 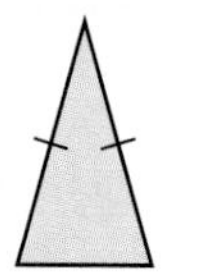

c. 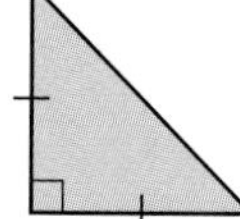

d.

e.

11. ⟨ **8.1F** ⟩ *Find* $m\angle C$ *in triangle ABC if:*

a. $m\angle A = 30°$ and $m\angle B = 40°$

b. $m\angle A = 40°$ and $m\angle B = 50°$

c. $m\angle A = 50°$ and $m\angle B = 60°$

d. $m\angle A = 60°$ and $m\angle B = 70°$

e. $m\angle A = 70°$ and $m\angle B = 80°$

12. ⟨ **8.2A** ⟩ *Find the perimeter of a rectangle:*

a. 5.1 meters long and 3.2 meters wide

b. 4.1 centimeters long and 3.2 centimeters wide

c. 3.5 inches long and 4.1 inches wide

d. 5.2 yards wide and 4.1 yards long

e. 4.1 feet wide and 6.2 feet long

13. ⟨ **8.2A** ⟩ *Find the perimeter of the polygon.*

a. 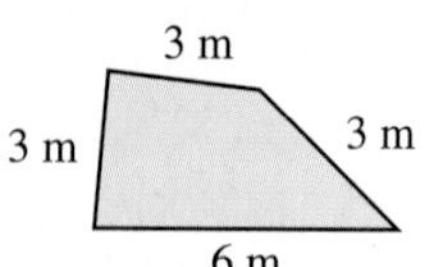

b. 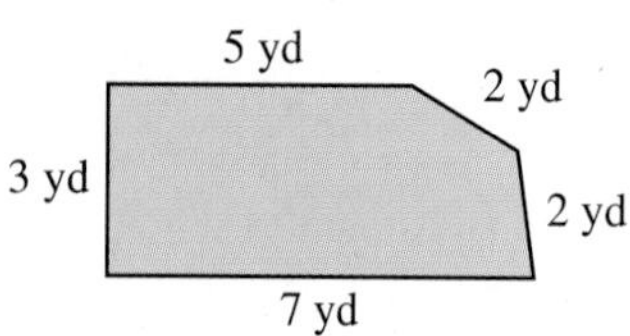

c. 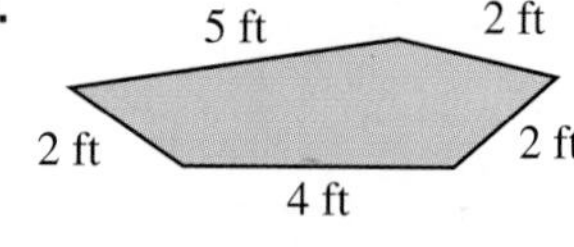

d. 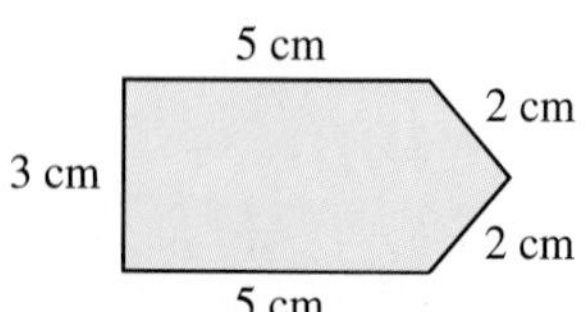

e. 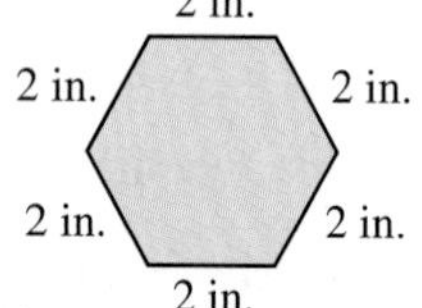

14. **⟨8.2B⟩** *Use 3.14 for π to find the circumference of a circle with radius:*

a. 6 cm

b. 8 cm

c. 10 in.

d. 12 in.

e. 14 ft

15. **⟨8.3A⟩** *Find the area of a rectangle:*

a. 5 m by 6 m

b. 7 m by 8 m

c. 4 in. by 6 in.

d. 3 in. by 7 in.

e. 9 in. by 12 in.

16. **⟨8.3B⟩** *Find the area of a triangle:*

a. With a 10-inch base and a 12-inch height

b. With an 8-inch base and a 10-inch height

c. With a 6-centimeter base and an 8-centimeter height

d. With a 4-meter base and a 6-meter height

e. With a 2-meter base and a 4-meter height

17. **⟨8.3C⟩** *Find the area of the parallelogram:*

a.

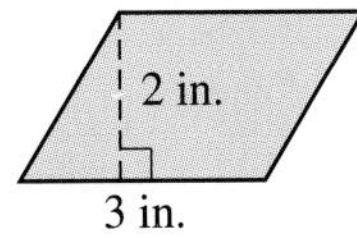

b.

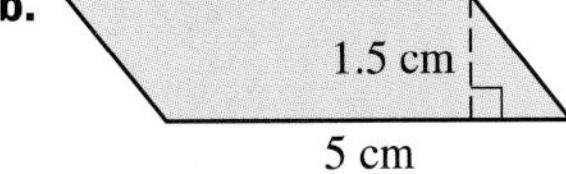

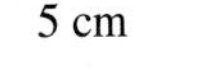

c.
2 m
4 m

d.

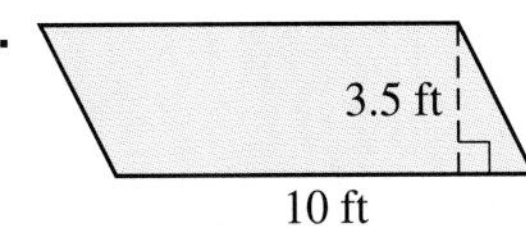

e.

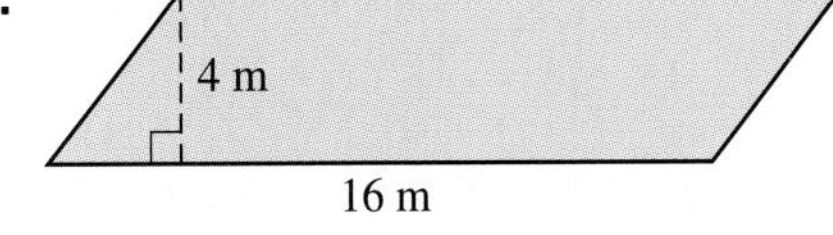

18. **⟨8.3D⟩** *Find the area of the trapezoid:*

a.

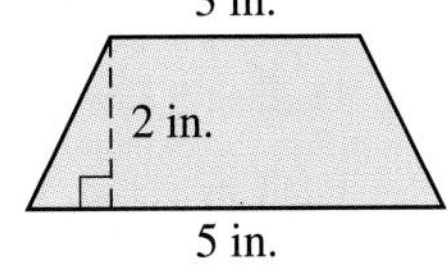

b.

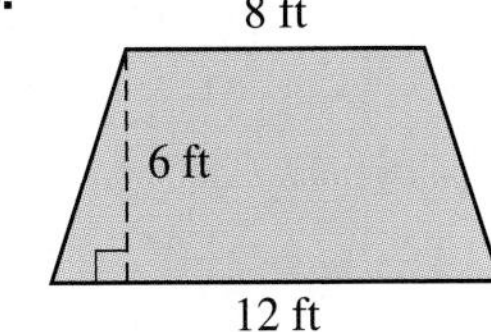

c.

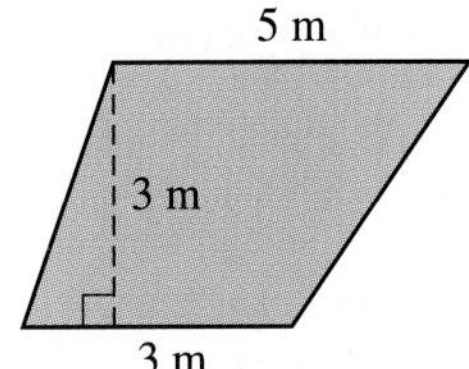

d.

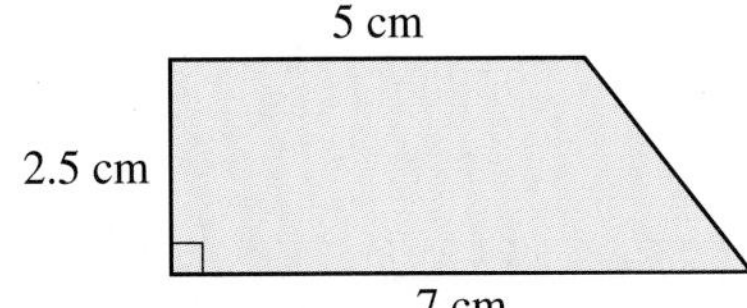

e.

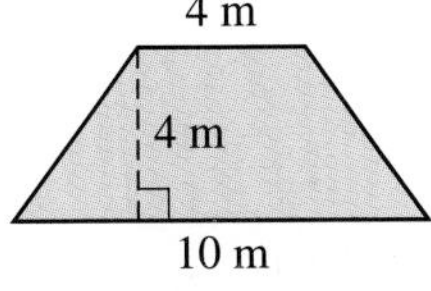

19. **⟨8.3E⟩** *Find the area of a circle with the given radius. Use 3.14 for π.*

a. 7 inches

b. 1 centimeter

c. 3 inches

d. 2 feet

e. 5 yards

20. **⟨8.4A⟩** *Find the volume of a box:*

a. 2 centimeters wide, 6 centimeters long, and 7 centimeters high

b. 3 centimeters wide, 4 centimeters long, and 5 centimeters high

c. 2 centimeters wide, 5 centimeters long, and 4 centimeters high

d. 3 centimeters wide, 6 centimeters long, and 5 centimeters high

e. 2 centimeters wide, 7 centimeters long, and 5 centimeters high

21. **⟨8.4B⟩** *Use 3.14 for π to find the volume (to the nearest hundredth) of a can in the shape of a circular cylinder with:*

a. A 1-inch radius and 6-inch height

b. A 1-inch radius and 7-inch height

c. A 1-inch radius and 8-inch height

d. A 2-inch radius and 9-inch height

e. A 2-inch radius and 10-inch height

22. **⟨8.4C⟩** *Use 3.14 for π to find the volume (to the nearest hundredth) of a sphere with a radius of:*

a. 6 inches

b. 7 inches

c. 8 inches

d. 9 inches

e. 10 inches

23. **⟨8.4D⟩** *Use 3.14 for π to find the volume (to the nearest hundredth) of a circular cone with:*

a. A 10-inch radius and 15-inch height

b. An 8-inch radius and 12-inch height

c. A 6-inch radius and 9-inch height

d. A 4-inch radius and 6-inch height

e. A 2-inch radius and 3-inch height

24. **⟨8.4E⟩** *Find the volume of a pyramid 120 meters high and with a square base measuring:*

a. 210 meters on each side

b. 216 meters on each side

c. 220 meters on each side

d. 221 meters on each side

e. 225 meters on each side

25. **⟨8.5A⟩** *Find:*

a. $\sqrt{25}$

b. $\sqrt{9}$

c. $\sqrt{121}$

d. $\sqrt{225}$

e. $\sqrt{169}$

26. **⟨8.5A⟩** *Find to three decimal places:*

a. $\sqrt{2}$

b. $\sqrt{8}$

c. $\sqrt{12}$

d. $\sqrt{22}$

e. $\sqrt{17}$

27. **⟨8.5B⟩** *Find the length of the hypotenuse of a right triangle whose sides measure:*

a. 5 cm and 12 cm

b. 4 cm and 3 cm

c. 9 cm and 12 cm

d. 12 cm and 16 cm

e. 15 cm and 20 cm

28. **⟨8.5B⟩** *Find (to three decimal places) the length of the hypotenuse of a right triangle whose sides measure:*

a. 2 cm and 3 cm

b. 4 cm and 2 cm

c. 5 cm and 4 cm

d. 3 cm and 6 cm

e. 7 cm and 1 cm

29. **⟨8.5B⟩** *Find (to three decimal places where necessary) the length of side a for a triangle:*

a. With hypotenuse 5 and side $b = 3$

b. With hypotenuse 8 and side $b = 4$

c. With hypotenuse 9 and side $b = 5$

d. With hypotenuse 10 and side $b = 6$

e. With hypotenuse 12 and side $b = 8$

› Practice Test Chapter 8

(Answers on page 517)

Visit mhhe.com/bello to view helpful videos that provide step-by-step solutions to several of the problems below.

The figure will be used in Problems 1–5.

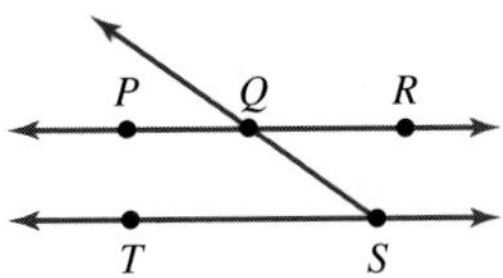

1. Name three points in the figure.

2. Name two lines in the figure.

3. Name three line segments in the figure.

4. Name a pair of lines that appear to be parallel in the figure.

5. Name a ray in the figure.

6. Name the angle three different ways.

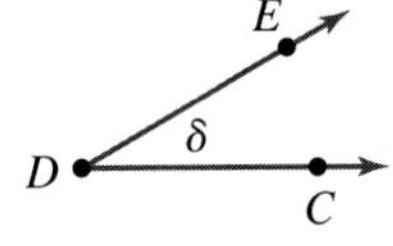

7. Classify the angle:

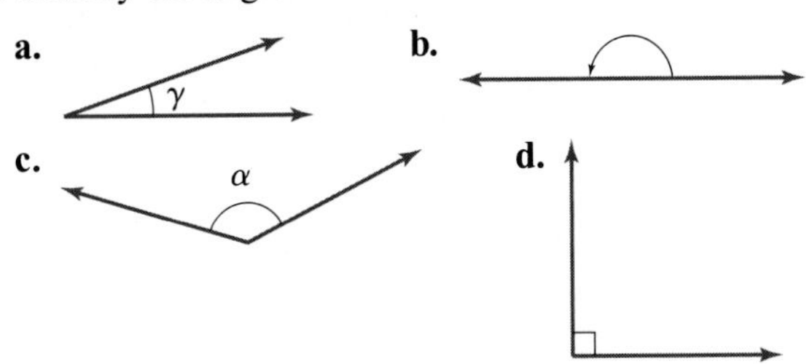

8.

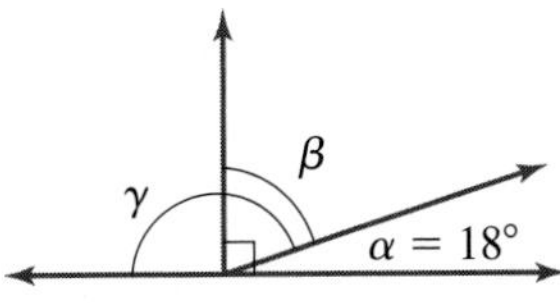

If $m\angle\alpha = 18°$

a. What is the measure of the complement of α? Name the complement of α.

b. What is the measure of the supplement of α? Name the supplement of α.

9. Classify the triangles according to their angles and sides.

a. 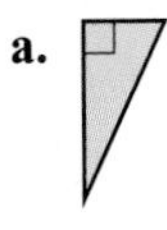  **b.** 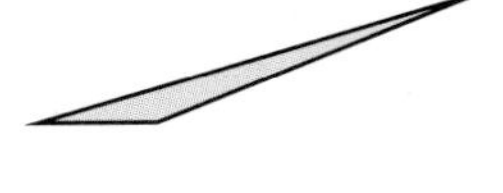 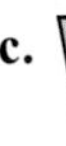 **c.** 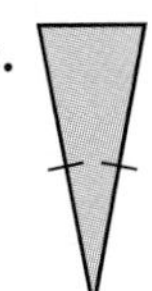

10. In a triangle ABC, $m\angle A = 47°$, $m\angle B = 53°$. Find $m\angle C$.

11. Find the perimeter of a rectangle 4.1 centimeters long by 2.3 centimeters wide.

12. Find the circumference of a circle whose radius is 5 inches. (Use 3.14 for π.)

13. Find the area of a rectangle 3 meters by 5 meters.

14. Find the area of a triangular piece of cloth that has a 10-centimeter base and is 10 centimeters high.

15. Find the area of the trapezoid.

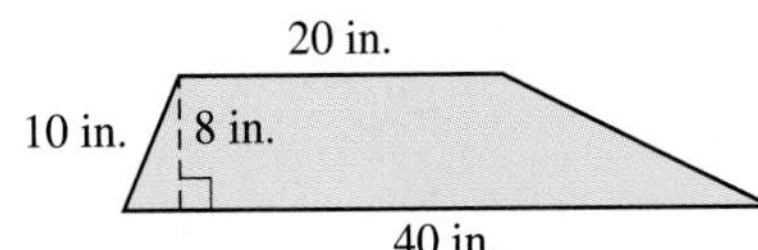

16. Find the area of a circle with a radius of 4 centimeters. (Use 3.14 for π.)

17. Find the volume of the box.

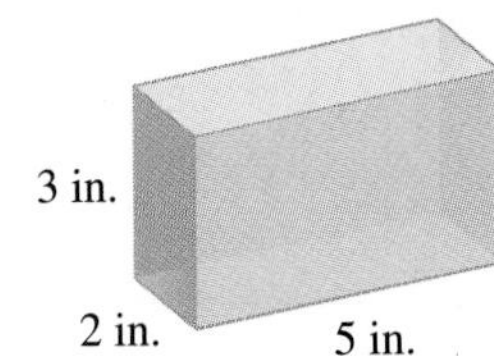

18. Find the volume of a cylinder 6 inches high and with a 3-inch radius. Use 3.14 for π and give the answer to two decimal places.

19. Find the volume of a sphere with a 6-inch radius. Use 3.14 for π and give the answer to two decimal places.

20. Find the volume of a cone 9 inches high and with a 6-inch radius. Use 3.14 for π and give the answer to two decimal places.

21. Find $\sqrt{3600}$.

22. Find $\sqrt{6}$ (give the answer to three decimal places).

23. Find the length of the hypotenuse c.

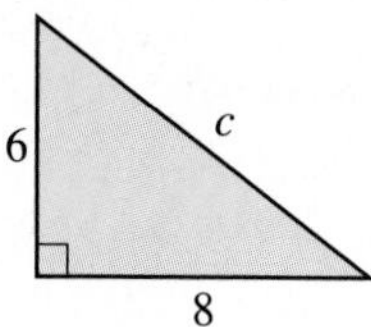

24. Find the length of a.

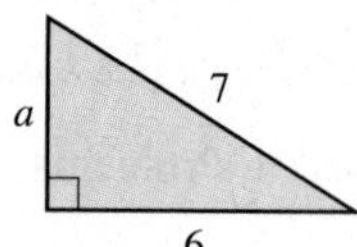

25. The diagonal length of a television screen is 15 inches. If the width is 12 inches, how high is it?

Answers to Practice Test Chapter 8

Answer	If You Missed	Review		
	Question	Section	Examples	Page
1. Pick any three: P, Q, and R for example.	1	8.1	1a	457
2. $\overleftrightarrow{PR}$ and $\overleftrightarrow{TS}$	2	8.1	1b	457
3. $\overline{PQ}$, $\overline{QR}$, and $\overline{TS}$	3	8.1	1c	457–458
4. $\overleftrightarrow{PR}$ and $\overleftrightarrow{TS}$ appear to be parallel	4	8.1	1d	457–458
5. $\overrightarrow{SQ}$	5	8.1	1e	457–458
6. $\angle\delta$, $\angle D$, $\angle CDE$ (or $\angle EDC$)	6	8.1	2	458
7. **a.** Acute **b.** Straight **c.** Obtuse **d.** Right	7	8.1	3, 4	459–460
8. **a.** 72°; β **b.** 162°; γ	8	8.1	6, 7	461
9. **a.** Right scalene **b.** Obtuse scalene **c.** Acute isosceles	9	8.1	8	463
10. 80°	10	8.1	9	464
11. 12.8 cm	11	8.2	1, 2, 3, 4	472–473
12. 31.4 in.	12	8.2	7, 8	474–475
13. 15 m^2	13	8.3	1, 2	480
14. 50 cm^2	14	8.3	3	481
15. 240 $in.^2$	15	8.3	5	482
16. 50.24 cm^2	16	8.3	6, 7	483
17. 30 $in.^3$	17	8.4	1, 2	489–490
18. 169.56 $in.^3$	18	8.4	3	490
19. 904.32 $in.^3$	19	8.4	4	491
20. 339.12 $in.^3$	20	8.4	5	492
21. 60	21	8.5	1	500
22. 2.449	22	8.5	2	500
23. $c = 10$	23	8.5	3	501
24. $a = \sqrt{13} \approx 3.606$	24	8.5	4	501
25. 9 in.	25	8.5	5, 6	502

Cumulative Review Chapters 1–8

1. Simplify: $9 \div 3 \cdot 3 + 5 - 4$

2. Subtract: $745.42 - 17.5$

3. Round 749.851 to the nearest ten.

4. Divide: $40 \div 0.13$ (Round answer to two decimal places.)

5. Solve for y: $3.6 = 0.6y$

6. Solve for z: $4 = \frac{z}{4.4}$

7. Solve the proportion: $\frac{j}{5} = \frac{5}{125}$

8. Write 89% as a decimal.

9. 50% of 90 is what number?

10. What percent of 24 is 6?

11. 18 is 30% of what number?

12. Find the simple interest earned on $600 invested at 6.5% for 2 years.

13. Referring to the circle graph, which is the main source of pollution?

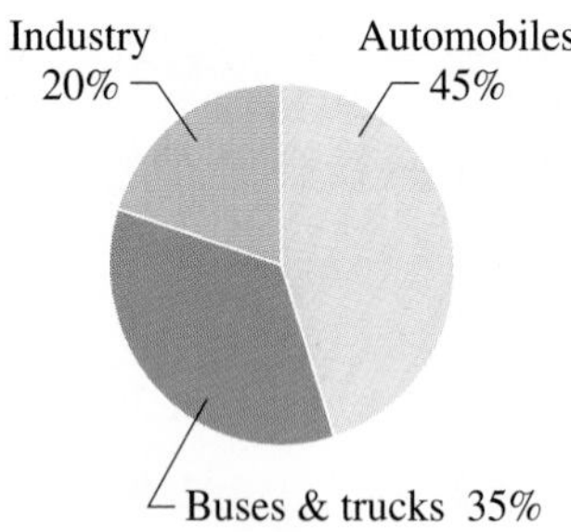

14. Make a circle graph for this data:

Family Budget (Monthly)

Savings	(S)	$200
Housing	(H)	$900
Food	(F)	$400
Clothing	(C)	$500

15. The following table shows the distribution of families by income in Portland, Oregon.

Income Level	Percent of Families
$0–9,999	3
10,000–14,999	10
15,000–19,999	27
20,000–24,999	34
25,000–34,999	12
35,000–49,999	6
50,000–79,999	4
80,000–119,000	3
120,000 and over	1

What percent of the families in Portland have incomes between $10,000 and $14,999?

16. The following graph represents the monthly average temperature for seven months of the year. How much higher is the average temperature in July than it is in May?

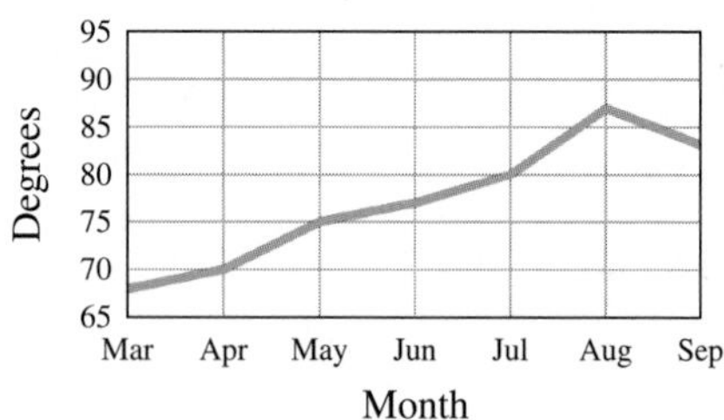

17. The number of hours required in each discipline of a college core curriculum is represented by the following circle graph. What percent of these hours is in math and English combined?

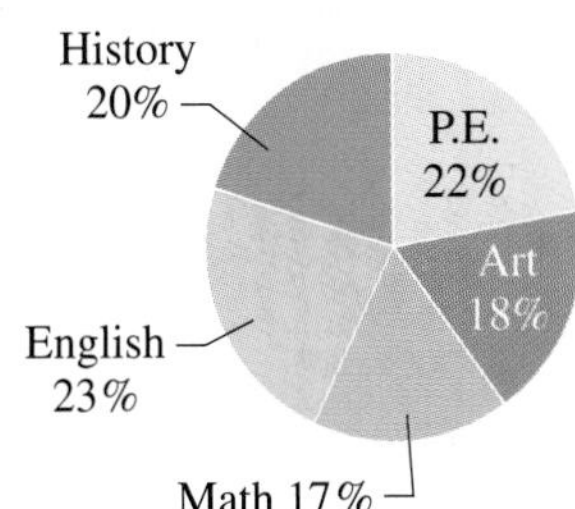

18. What is the mode of the following numbers?

1, 18, 5, 18, 9, 8, 26, 28, 18, 7

19. What is the mean of the following numbers?

5, 7, 27, 24, 27, 27, 27, 18, 25, 6, 27

20. What is the median of the following numbers?

4, 21, 22, 23, 4, 15, 23, 11, 25, 23, 16

21. Convert 11 yards to inches.

22. Convert 9 inches to feet.

23. Convert 7 feet to yards.

24. Convert 5 feet to inches.

25. Convert 4 kilometers to meters.

26. Convert 2 dekameters to meters.

27. Convert 140 meters to decimeters.

28. Find the area in square yards of a 10-acre lot.

29. Find the perimeter of a rectangle 5.1 inches long by 2.8 inches wide.

30. Find the circumference of a circle whose radius is 6 centimeters. (Use 3.14 for π.)

31. Find the area of a triangular piece of cloth with a 20-centimeter base and 12-centimeter height.

32. Find the area of a circle with a radius of 8 centimeters. (Use 3.14 for π.)

33. Find the area of the trapezoid.

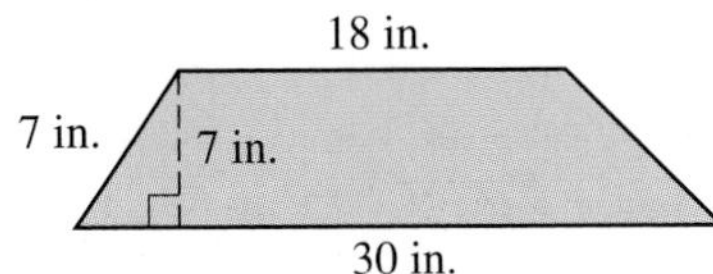

34. Find the volume of a box 2 centimeters wide, 4 centimeters long, and 9 centimeters high.

35. Find the volume of a sphere with a 6-inch radius. Use 3.14 for π and give the answer to two decimal places.

36. Classify the angle:

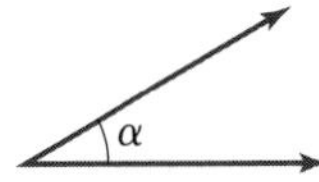

37. If in triangle ABC, $m\ \angle A = 20°$, and $m\ \angle B = 35°$, what is $m\ \angle C$?

38. Find $\sqrt{3600}$.

39. Find the hypotenuse c of the given triangle:

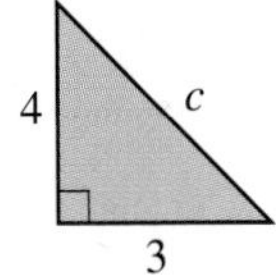

Section

Chapter

9 nine

The Real Numbers

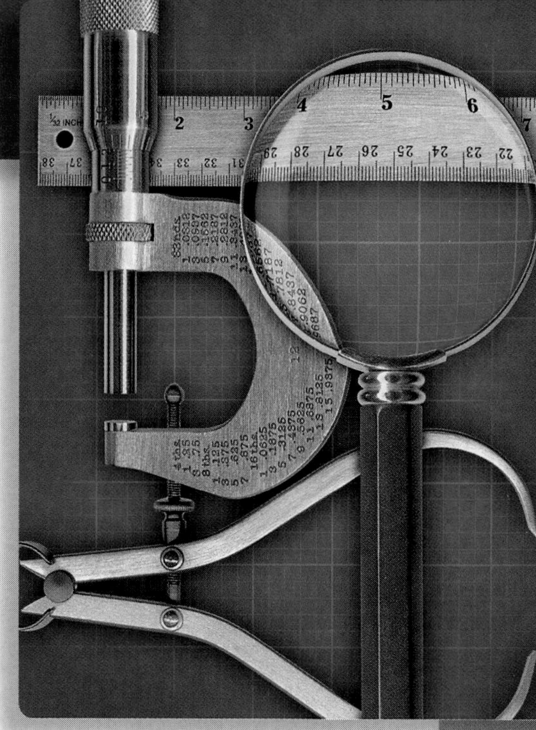

The Human Side of Mathematics

We have already studied the natural or counting numbers 1, 2, 3, and so on, but these numbers are not enough for the needs of everyday life. We need to measure subzero temperatures, yardage losses in football games, and financial losses in the stock market. We need **negative numbers!** Who invented and used these numbers and when? Here is a partial answer.

China: Negative numbers were used since the first century. On their calculation tables, black rods represented negative numbers and red rods positive ones.

India: Negative numbers were used by the **Indian mathematicians** (Hindu) of the sixth and seventh centuries. For example, **Bramagupta** (seventh century) taught the way of making additions and subtractions on goods, debts, and nothingness. "*A debt cut off from nothingness becomes a good, a good cut off from nothingness becomes a debt.*"

Italy: Negative numbers were used at the end of the fifteenth century as solutions for equations. For example, the writings of the Italian mathematician Girolamo Cardan (1501–1576) included negative numbers.

France: Unlike Cardan, the French mathematician Francois Viete (1540–1603) only gave the positive solutions of the equations.

But the story does not end there. Dividing one integer by another yields a new number, a **rational** number. These numbers (which we also call fractions) can also be written as decimals. We might imagine that our numerical journey ends with the rational numbers. However, the Pythagoreans, a secret society of scholars in ancient Greece (ca. 540–500 B.C.), made a stunning discovery: the **irrational** numbers ($\sqrt{2}$ and π are irrational). These numbers defied their knowledge of number properties because they could not be written as the ratio of two whole numbers. We use the rationals and irrationals in a new set of numbers called the **real** numbers.

9.1 Addition and Subtraction of Integers

Objectives

You should be able to:

A › Classify, use, represent, graph, and compare integers.

B › Find the additive inverse (opposite) of an integer.

C › Find the absolute value of an integer.

D › Add two integers using a number line.

E › Add two integers using the rules given in the text.

F › Subtract integers by adding additive inverses.

G › Solve applications involving integers.

To Succeed, Review How To . . .

1. Construct a number line containing the whole numbers. (p. 14)
2. Check a subtraction problem by addition. (p. 37)

Getting Started

In the cartoon below, Linus is trying to subtract six from four. He wants to find

$$4 - 6 = \square$$

As you recall, $4 - 6 = \square$ can be written as

$$4 = 6 + \square$$

We cannot find a *whole* number $\square$ that when added to 6 will give us 4. Let us think about it this way. Say you are in a poker game. You win \$6 and then added to this you lose \$2. You are still winning \$4. Therefore,

$$6 + \boxed{-2} = 4$$

Thus a *loss* of \$2 is written as -2, and a *gain* of \$6 is written as 6 (or $+6$ if you prefer). The number -2 is called a **negative** number. Here are some other quantities that can be measured using negative numbers:

10 degrees below zero: $-10°$

a loss of \$5: $-\$5$

15 feet below see level: -15 feet

A › Introduction to the Integers

In the *Getting Started* we have added an important concept to the idea of a number: the concept of direction. To visualize this we draw a straight line, choose a point on this line, and label it 0, the **origin.** (See Figure 9.1.) We then measure successive equal intervals to the

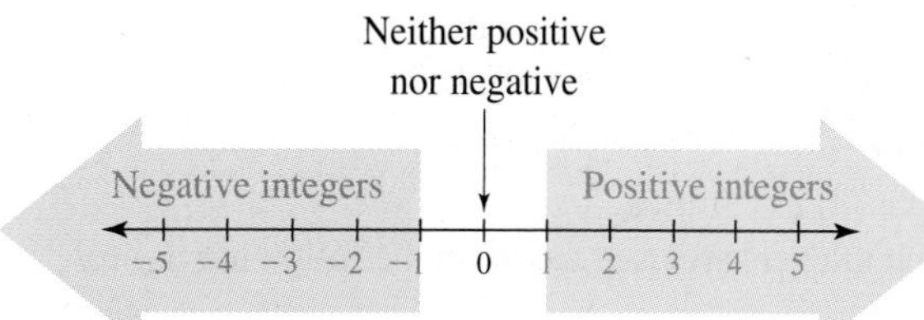

>Figure 9.1

right of 0 and label the division points with the **positive integers** (natural or **counting** numbers) in their order, 1, 2, 3, ... Note that the positive integers can be written as 1, 2, 3, . . . , or as +1, +2, +3, ... We usually write 1 instead of +1, 2 instead of +2, and so on. Similarly, the points to the *left* of 0, the **negative integers,** are labeled −1, −2, −3, . . . (read "negative one, negative two, negative three, and so on"). Note that 0 is neither positive nor negative.

We have drawn our line a little over five units long on each side. The arrows at either end indicate that the line could be drawn to any desired length.

The **positive** and the **negative** numbers are collectively called **signed** numbers. Why? Because we identify the **positive** numbers with the + **(plus)** sign and **negative** numbers with the − **(negative)** sign. Note that we **do not** use the minus sign (−), which is reserved for the operation of *subtraction*. What about 0? The number 0 is neither positive nor negative. Based on this discussion we introduce a new set of numbers called the **integers.** The set of integers consists of

1. The positive numbers {1, 2, 3, . . . }.
2. The number 0.
3. The negative numbers {. . . −3, −2, −1}.

Thus, the set of integers is I = **{. . . −3, −2, −1, 0, 1, 2, 3 . . .}.**

To do Example 1, recall that the set of natural numbers is N = {1, 2, 3, . . .} and the set of whole numbers is W = {0, 1, 2, 3, . . .}.

EXAMPLE 1 Classifying numbers

Which numbers in the set {−5, −4, 0, 1, 2, 3} are

a. Natural numbers?
b. Whole numbers?
c. Positive integers?
d. Negative integers?
e. Integers?

SOLUTION

a. The natural numbers in the set are 1, 2, and 3.
b. The whole numbers in the set are 0, 1, 2, and 3.
c. The positive integers in the set are 1, 2, and 3.
d. The negative integers in the set are −5 and −4.
e. The integers in the set are −5, −4, 0, 1, 2, and 3.

PROBLEM 1

Which numbers in the set {−3, −1, 0, 4, 5, 7} are

a. Natural numbers?
b. Whole numbers?
c. Positive integers?
d. Negative integers?
e. Integers?

Answers to PROBLEMS

1. a. 4, 5, 7 **b.** 0, 4, 5, 7
c. 4, 5, 7 **d.** −3, −1
e. −3, −1, 0, 4, 5, 7

The integers can be used in many situations. Some of these situations are illustrated in Example 2.

EXAMPLE 2 Uses of integers

Describe how the integers are used based on the following figures.

a.

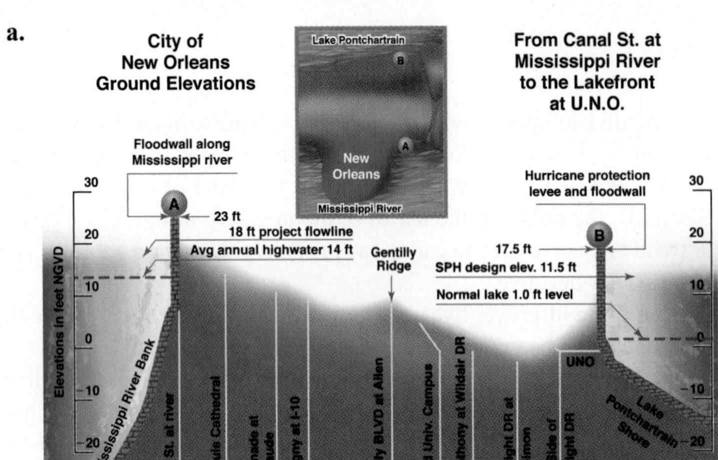

Source: http://www.publichealth.hurricane.lsu.edu.

b.

Golf: Best-Ball Stroke Play

Pos	Name	Thru	Total
1	Sue/Fred	5	−2
2	Al/Bob	2	−1
3	Cal/Don	2	E
4	Mike/Tami	5	+1

Source: http://www.motherhensw.com.

SOLUTION

a. Indicates elevations in New Orleans

b. The integers indicate golf scores.

PROBLEM 2

Describe how the integers are used in the following diagram.

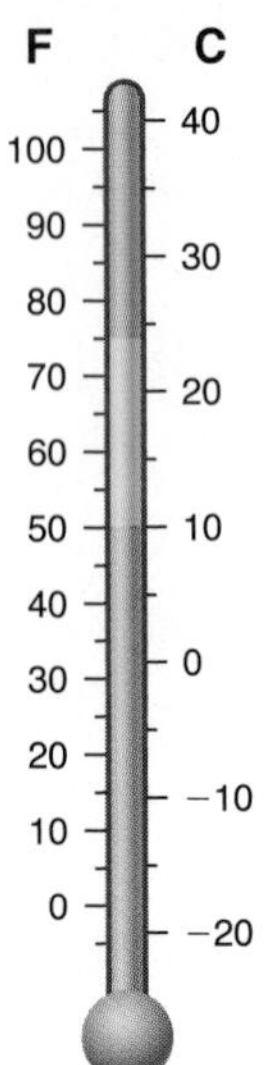

EXAMPLE 3 Uses of integers

Represent a real-world situation using integers.

a. A bank account is overdrawn by \$50 and a \$30 deposit is made.

b. The height of the Eiffel tower and its antenna is 1052 feet.

c. The elevation of Death Valley, which is 282 feet below sea level.

SOLUTION

a. −\$50 and +\$30 (or simply \$30)

b. +1052 feet (or simply 1052 feet)

c. −282 feet

PROBLEM 3

Represent a real-world situation using integers.

a. The national debt is \$8,391,014,420,537.

b. The height of the Empire State Building and its lightning rod is 1454 feet.

c. The elevation of the Dead Sea is 408 meters below sea level.

In Figure 9.1, on the preceding page, we have a picture of a number line. We can graph individual integers on a number line as we did in Section 1.2. We illustrate how in Example 4.

Answers to PROBLEMS

2. The integers indicate temperatures in degrees Celsius or Fahrenheit.

3. a. −\$8,391,014,420,537 (this is more than 8 trillion dollars in debt!) **b.** +1454 ft (or 1454 ft) **c.** −408 m

EXAMPLE 4 Graphing integers

Graph the integers -3, 0, and 5 on a number line.

SOLUTION

Draw a number line with equally spaced points, as in Figure 9.1.
To graph -3 place a heavy dot on the number line above -3.

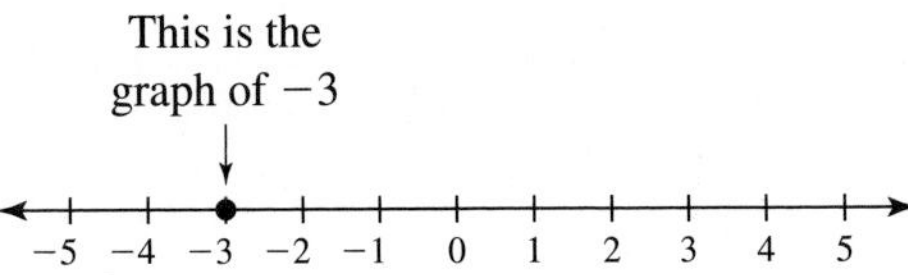

Do the same for 0 and 5.

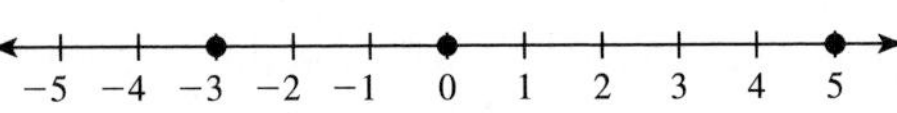

PROBLEM 4

Graph the integers -4, 1, and 3 on a number line.

In Chapter 1 we used the symbols $<$ (less than) and $>$ (greater than) to compare numbers. We can use the same symbols to compare integers using a number line.

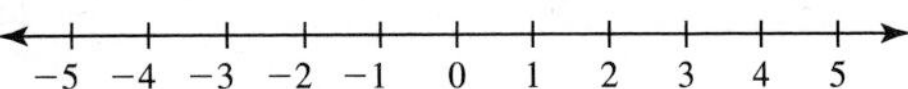

As you can see, the integers are written *in order* and they increase as you move from left to right on the number line, so any integer is always **greater than** ($>$) any other integer to its left and is **less than** ($<$) any integer to its right. Here is the definition we need:

LESS THAN ($<$) AND GREATER THAN ($>$)

If a and b are two integers, a is less than b if a lies to the left of b on the number line. In symbols $a < b$.

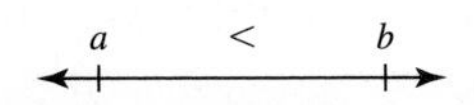

This also means that b is to the right of a. That is, b is greater than a. In symbols, $b > a$.

EXAMPLE 5 Comparing integers

Graph the integers and insert the symbol $<$ or $>$ to make the statement true.

a. -2 ____ -5 **b.** 1 ____ 3

SOLUTION

a. Draw a number line and graph -2 and -5, as shown.

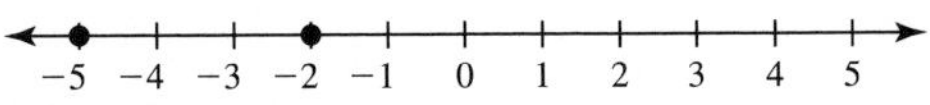

$-2 > -5$, since -2 is to the right of -5. Thus, -2 __$>$__ -5.

b. Draw a number line and graph 1 and 3.

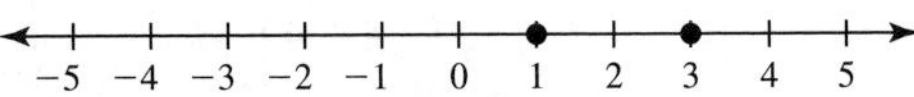

$1 < 3$, since 1 is to the left of 3. Thus, 1 __$<$__ 3.

PROBLEM 5

Graph the integers and insert the symbol $<$ or $>$ to make the statement true.

a. -4 ____ -1 **b.** 2 ____ 0

Answers to PROBLEMS

4. 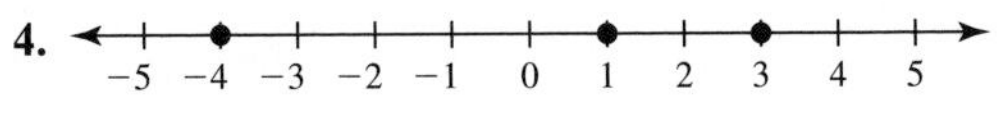

5. 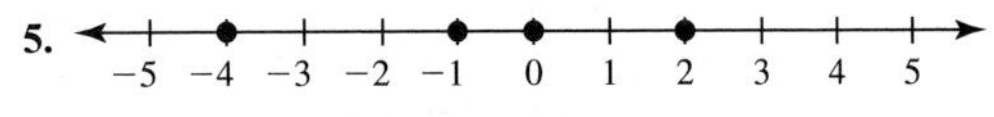

a. -4 __$<$__ -1 **b.** 2 __$>$__ 0

B › Additive Inverses

For every *positive* integer, there is a corresponding *negative* integer. Thus, associated with the positive integer 3, we have the negative integer -3. Since -3 and 3 are the same distance from the origin but in opposite directions, 3 and -3 are called **opposites, additive inverses,** or simply **inverses.** Similarly, the additive inverse (opposite) of 2 is -2, and the additive inverse (opposite) of -1 is 1, as shown in Figure 9.2

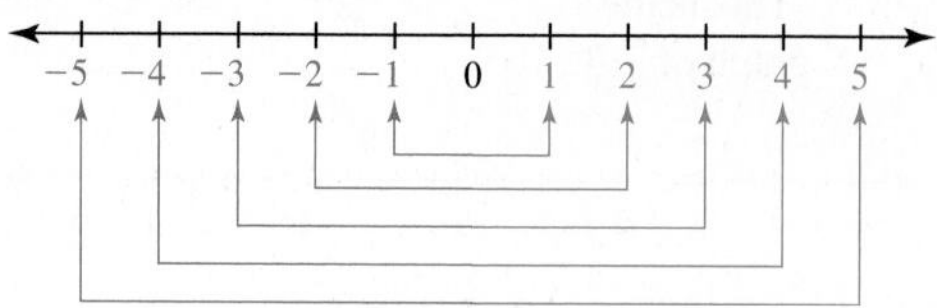

>**Figure 9.2**

Thus for every positive integer there is a corresponding negative integer.

3 and -3 are additive inverses (opposites).

-1 and 1 are additive inverses (opposites).

If we follow a pattern to find inverses (opposites) we have

Integer	Additive Inverse
3	-3
2	-2
-1	$-(-1) = 1$
-2	$-(-2) = 2$

In general,

ADDITIVE INVERSE OF A NUMBER

The inverse of a nonzero number a is $-a$ and the inverse of $-a$ is $-(-a) = a$.
***Note:* $0 = -0$**

EXAMPLE 6 **Finding the additive inverse**

Find the additive inverse.

a. 9 **b.** -7

SOLUTION

a. The additive inverse of 9 is -9.

b. The additive inverse of -7 is $-(-7) = 7$.

PROBLEM 6

Find the additive inverse.

a. 17 **b.** -17

C › Absolute Value

Now, let us go back to the number line. What is the distance between 3 and 0? The answer is 3 units.

$|-3| = 3$ units $|3| = 3$ units

−5 −4 −3 −2 −1 0 1 2 3 4 5

What is the distance between -3 and 0? The answer is still 3 units.

Answers to PROBLEMS

6. a. -17 **b.** 17

The distance between any number n and 0 is called the **absolute value** of the number and is denoted by $|n|$. Thus, $|-3| = 3$ and $|3| = 3$. Note that since the absolute value represents a *distance,* the absolute value of a number is *never* negative.

ABSOLUTE VALUE OF A NUMBER

The absolute value of a number n is its distance from 0 and is denoted by $|n|$. In general,

$$|n| = \begin{cases} n, & \text{if } n > 0 \\ 0, & \text{if } n = 0 \\ -n, & \text{if } n < 0 \end{cases}$$

EXAMPLE 7 Absolute value of an integer

Find the absolute value.

a. $|-6|$ **b.** $|7|$ **c.** $|0|$

SOLUTION

a. $|-6| = 6$ −6 is 6 units from 0.

$|-6| = 6$

−7 −6 −5 −4 −3 −2 −1 0 1

b. $|7| = 7$ 7 is 7 units from 0.

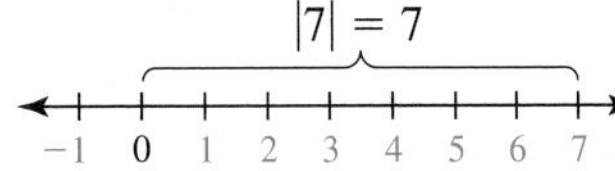
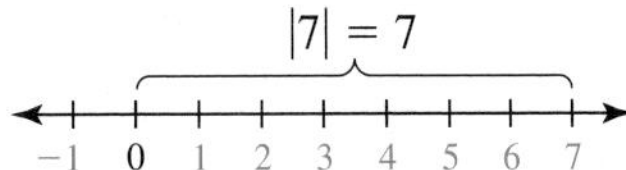

c. $|0| = 0$ 0 is 0 units from 0.

$|0| = 0$

−4 −3 −2 −1 0 1 2 3 4

PROBLEM 7

Find the absolute value.

a. $|-18|$ **b.** $|8|$ **c.** $|-0|$

D > Addition of Integers Using a Number Line

We are now ready to do addition using the number line. Here is the procedure to do it.

TO ADD $a + b$ ON THE NUMBER LINE

1. Start at 0 and move to a (to the right if a is positive, to the left if negative).
2. **a.** If b is positive, move right b units.
   b. If b is negative, move left b units.
   c. If b is zero, stay at a.

For example, the sum 2 + 3, or (+2) + (+3), is found by moving 2 units to the right, followed by 3 more units to the right. Thus, 2 + 3 = 5, as shown in Figure 9.3.

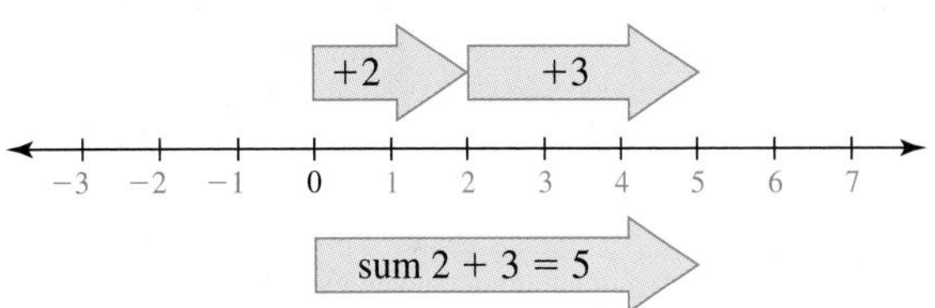

>**Figure 9.3**

Answers to PROBLEMS

7. a. 18 **b.** 8 **c.** 0

EXAMPLE 8 Adding integers

Add: $5 + (-2)$

SOLUTION First move 5 units to the right, followed by 2 units to the left. The end result is 3. Thus $5 + (-2) = 3$, as shown in Figure 9.4.

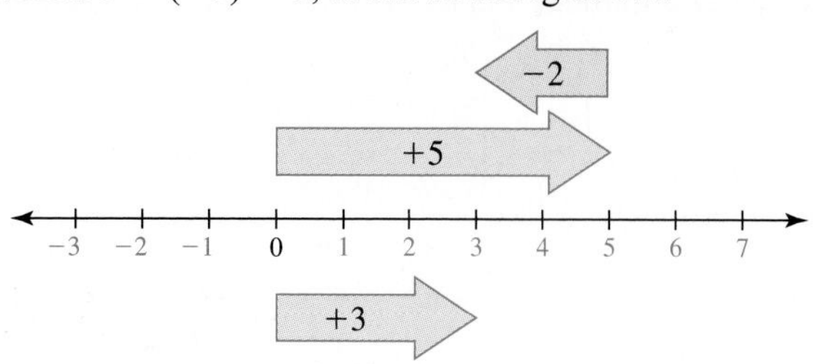

>Figure 9.4

PROBLEM 8

Add: $4 + (-3)$

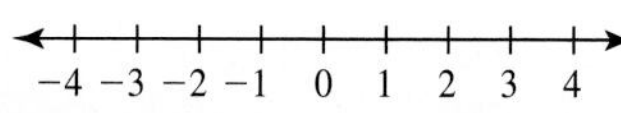

EXAMPLE 9 Adding integers

Add: $-5 + 3$

SOLUTION First move 5 units to the left, followed by 3 units to the right. The end result is -2 as shown in Figure 9.5. Thus, $-5 + 3 = -2$.

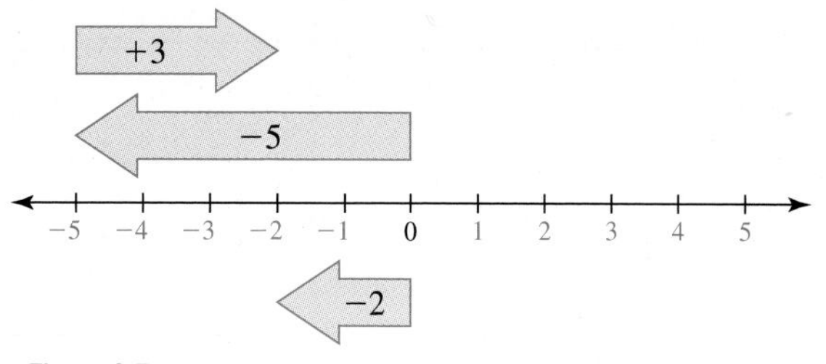

>Figure 9.5

PROBLEM 9

Add: $-3 + 1$

−4 −3 −2 −1 0 1 2 3 4

Can we add two negative integers? Of course. However, we have to be careful when writing such problems. For example, to add -3 and -2 we should write

$$-3 + (-2)$$

The parentheses are needed because

$$-3 + -2$$

is confusing.

CAUTION

Never use two signs together without parentheses.

EXAMPLE 10 Adding integers

Add: $-3 + (-2)$

PROBLEM 10

Add: $-2 + (-1)$

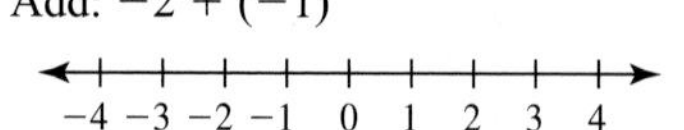

Answers to PROBLEMS

8.

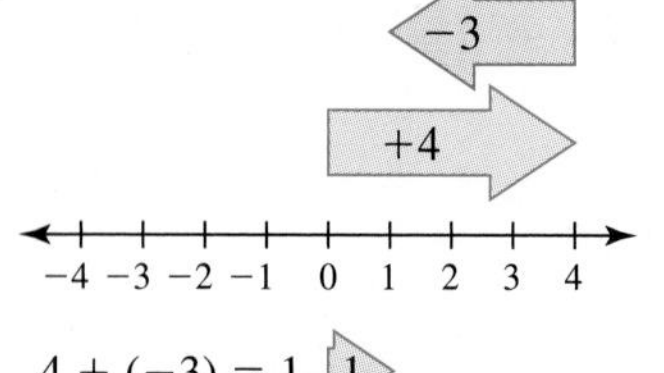

9.

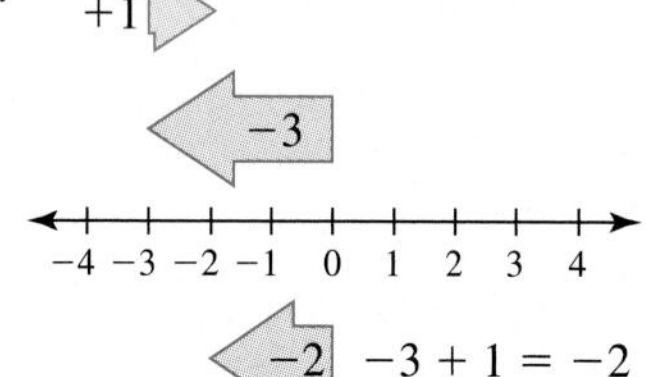

10.

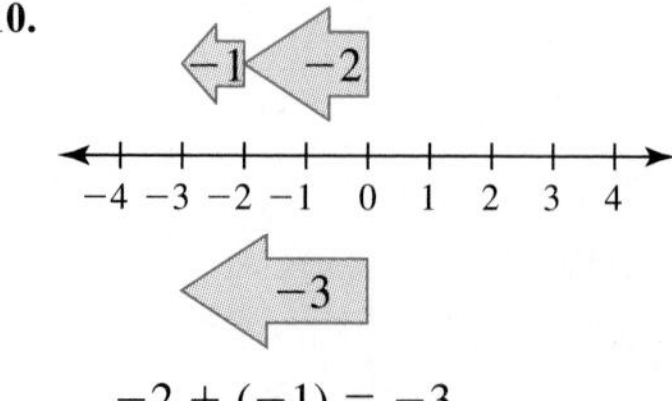

SOLUTION First move 3 units left, followed by 2 more units left. The result is 5 units to the left; that is, $-3 + (-2) = -5$.

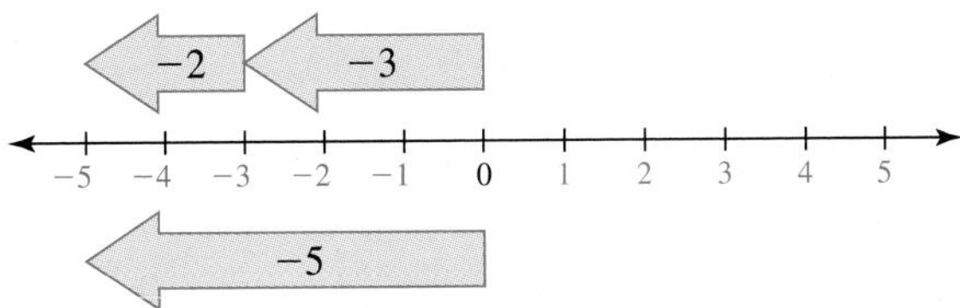

E > Adding Integers Using Addition Rules

As you have seen from these examples, if we add numbers with the *same* sign, the result is a number with this sign. Thus, $2 + 3 = 5$ and $-3 + (-2) = -5$. If we add numbers with *opposite* signs, the answer carries the sign of the number with the larger *absolute value*; thus

$$5 + (-2) = 3 \quad \text{but} \quad -5 + 2 = -3$$

Here is the rule summarizing this discussion:

TO ADD INTEGERS

1. If both numbers have the *same* sign, add their absolute values and give the sum the common sign.
2. If the numbers have *opposite* signs, subtract their absolute values and give the difference the sign of the number with the greater absolute value.

Thus, $3 + 7 = 10$, $4 + 9 = 13$, $-3 + (-7) = -10$, and $-4 + (-9) = -13$. To add $-8 + 5$ we first notice that the numbers have *opposite* signs. So we subtract their absolute values and give the difference the sign of the number with the greater absolute value. Thus,

$$-8 + 5 = -(8 - 5) = -3$$

Use the sign of the larger integer. Subtract the smaller integer from the larger one.

(You can think about this problem in another way. You are adding a negative to a positive, but you have more negatives, so the answer is negative. How many more negatives do you have? 3, so the answer is -3.) Similarly,

$$+8 + (-5) = +(8 - 5) = 3$$

Use the sign of the larger integer. Subtract the smaller integer from the larger one.

EXAMPLE 11 **Adding integers**

Add:

a. $-8 + (-6)$

b. $-10 + 6$

c. $10 + (-6)$

d. $10 + (-3) + 8 + (-2)$

PROBLEM 11

Add:

a. $-9 + (-4)$

b. $-17 + 8$

c. $17 + (-8)$

d. $12 + (-4) + 5 + (-3)$

(continued)

Answers to PROBLEMS

11. a. -13 **b.** -9 **c.** 9 **d.** 10

SOLUTION

a. $-8 + (-6) = -(8 + 6) = -14$

b. $-10 + 6 = -(10 - 6) = -4$

c. $10 + (-6) = 10 - 6 = 4$

d. We add the positives 10 and 8 and the negatives (-3) and (-2), then add the results like this

$$10 + (-3) + 8 + (-2)$$

$$18 + (-5) = 13$$

F › Subtracting Integers

We are now ready to subtract integers. Suppose a bank uses positive integers to indicate deposits and negative integers to indicate withdrawals. If you deposit \$4 and then withdraw \$6, you owe \$2. Thus

$$4 - 6 = 4 + (-6) = -2$$

Also,

$$-4 - 3 = -4 + (-3) = -7$$

because if you owe \$4 and then withdraw \$3 more, you now owe \$7. These results can be checked using the definition of subtraction of integers.

DEFINITION OF SUBTRACTION

The difference $a - b = c$ means $a = b + c$.

Thus $4 - 6 = -2$ because $4 = 6 + (-2)$, and $-4 - 3 = -7$ because $-4 = 3 + (-7)$.

What about $5 - (-2)$? We claim that $5 - (-2) = 7$ because if you deposit \$5 and the bank cancels (subtracts) a \$2 debt (represented by -2), your account *increases* by \$2. Thus $5 - (-2) = 5 + 2 = 7$. Note that

$$4 - 6 = 4 + (-6)$$
$$-4 - 3 = -4 + (-3)$$
$$5 - (-2) = 5 + 2$$

In general, we have the following rule for subtracting integers.

TO SUBTRACT INTEGERS

For any real numbers a and b,

$$a - b = a + (-b)$$

To *subtract* an integer b, *add* its opposite $(-b)$.

Note that the opposite of $-a$ is a, that is, $-(-a) = a$.

Thus

$$3 - 7 = 3 + (-7) = -4$$
$$9 - 3 = 9 + (-3) = 6$$
$$-8 - 3 = -8 + (-3) = -11$$
$$-8 - (-4) = -8 + 4 = -4$$

EXAMPLE 12 Subtracting integers

Subtract:

a. $15 - 6$

b. $-20 - 3$

c. $-10 - (-8)$

d. $8 - 13$

e. $4 - (-5) - 3 + 7$

SOLUTION

a. $15 - 6 = 15 + (-6) = +(15 - 6) = 9$

b. $-20 - 3 = -20 + (-3) = -(20 + 3) = -23$

c. $-10 - (-8) = -10 + 8 = -(10 - 8) = -2$

d. $8 - 13 = 8 + (-13) = -(13 - 8) = -5$

e. Use the definition of subtraction to write as an addition problem, then add the positives together and the negatives together:

$$4 - (-5) - 3 + 7$$
$$= 4 + 5 + (-3) + 7$$
$$= 16 + (-3)$$
$$= 13$$

PROBLEM 12

Subtract:

a. $13 - 7$

b. $-18 - 4$

c. $-3 - (-9)$

d. $7 - 15$

e. $5 - (-8) - 6 + 9$

G > Applications Involving Integers

Is there a relationship between how much education your parents received (educational attainment) and your scores on the SAT? We shall explore that question in the next example.

EXAMPLE 13 Interpreting a graph involving integers

The chart shows that the verbal scores for test takers with parents who had no high school (H.S.) diploma is −95, that is, 95 points *below* average. (By the way, the average score in math was 514 and the average verbal score was 506.)

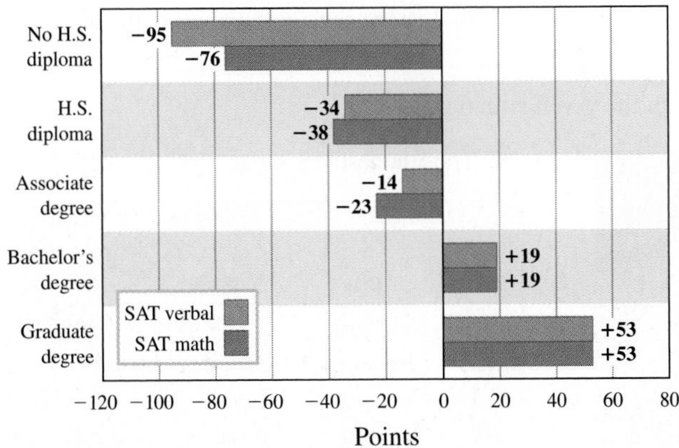

a. What integer corresponds to the verbal scores of test takers whose parents had a high school diploma? What does the integer mean?

b. What integer corresponds to the math scores of test takers whose parents had a graduate degree? What does the integer mean?

PROBLEM 13

What integer corresponds to the verbal scores of test takers whose parents earned an associate degree? What does the integer mean?

(continued)

Answers to PROBLEMS

12. a. 6 **b.** −22 **c.** 6 **d.** −8 **e.** 16 **13.** −14. The verbal scores were 14 points below 506, that is, $506 - 14 = 492$.

SOLUTION

a. The verbal scores for test takers whose parents had a high school diploma appear to the left of the red bar labeled H.S. diploma. The score is −34. It means that the verbal scores of the test takers were 34 points *below* the verbal average (506). Thus, the verbal scores for the test takers were 506 − 34 = 472.

b. The math scores for test takers whose parents had a graduate degree appear to the right of the blue bar labeled Graduate degree. The score is +53. It means that the math scores of the test takers were 53 points *above* the math average (514). Thus, the math scores for the test takers were 514 + 53 = 567.

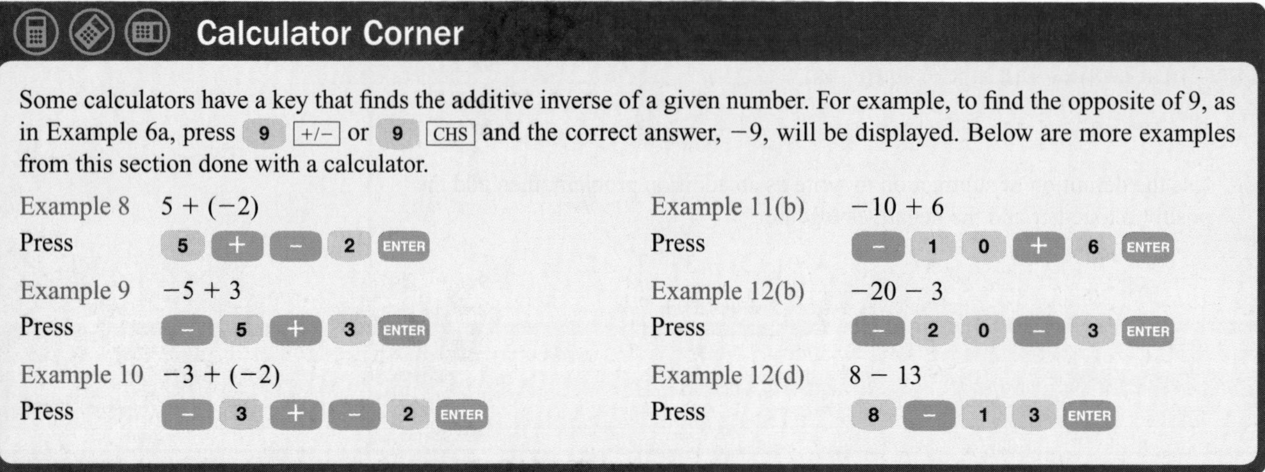

Calculator Corner

Some calculators have a key that finds the additive inverse of a given number. For example, to find the opposite of 9, as in Example 6a, press 9 +/− or 9 CHS and the correct answer, −9, will be displayed. Below are more examples from this section done with a calculator.

Example 8 5 + (−2)
Press 5 + − 2 ENTER

Example 9 −5 + 3
Press − 5 + 3 ENTER

Example 10 −3 + (−2)
Press − 3 + − 2 ENTER

Example 11(b) −10 + 6
Press − 1 0 + 6 ENTER

Example 12(b) −20 − 3
Press − 2 0 − 3 ENTER

Example 12(d) 8 − 13
Press 8 − 1 3 ENTER

› Exercises 9.1

Boost your grade at mathzone.com!
> Practice Problems
> NetTutor
> Self-Tests
> e-Professors
> Videos

‹ A › Introduction to the Integers The set {−5, −2, 0, 1, 3} will be used in Problems 1–3.

1. a. List the whole numbers in the set.
b. List the positive numbers in the set.

2. a. List the natural numbers in the set.
b. List the negative numbers in the set.

3. List the integers in the set.

In problems 4–5, indicate how the integers are used in the given situation.

4.

5. The 5 largest U.S. cities

Rank	City	State	Population in 2003	Change since 2000
1	New York	New York	8,085,742	77,464
2	Los Angeles	California	3,819,951	125,131
3	Chicago	Illinois	2,869,121	−26,895
4	Houston	Texas	2,009,690	56,059
5	Philadelphia	Pennsylvania	1,479,339	−38,211

Source: http://www.citymayors.com.

› Web IT go to **mhhe.com/bello** for more lessons

In Problems 6–7, write the integer that represents the situation.

6. a. A $100 withdrawal from your bank account.

b. A $40 deposit to your bank account.

7. a. A $50 million increase in the budget.

b. A $30 million deficit in the budget.

In Problems 8–10, graph the integers on the number line then fill in the blank with < or > to make the statement true.

8. a. $-1, 2, 4$

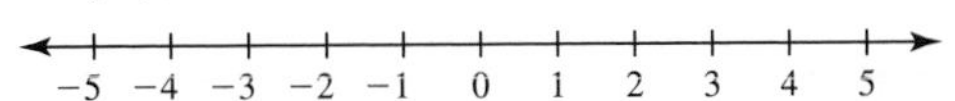

b. -1 ________ 2 $\quad 4$ ________ 2

9. a. $-5, 0, -2$

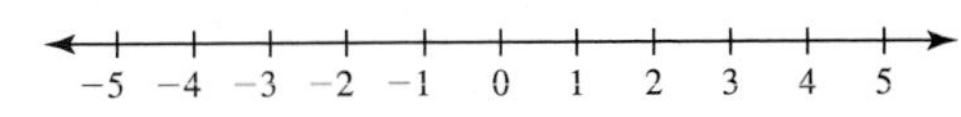

b. -2 ________ -5 $\quad 0$ ________ -2

10. a. $-4, -1, 5$

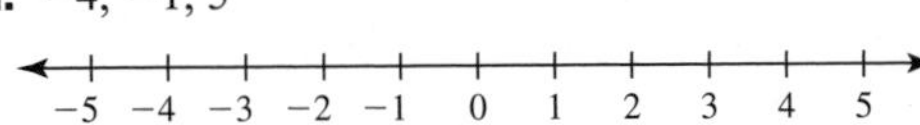

b. -1 ________ -4 $\quad -1$ ________ 5

⟨B⟩ Additive Inverses In Problems 11–14, find the additive inverse (opposite).

11. 0 **12.** -11 **13.** -17 **14.** 10

⟨C⟩ Absolute Value In Problems 15–20, find each value.

15. $|2|$ **16.** $|21|$ **17.** $|-11|$ **18.** $|-15|$ **19.** $|-30|$ **20.** $|-16|$

⟨D⟩ Addition of Integers Using a Number Line In Problems 21–30, add (verify your answer using a number line).

21. $2 + 1$ **22.** $3 + 3$ **23.** $-4 + 3$ **24.** $-5 + 1$ **25.** $5 + (-1)$

26. $6 + (-5)$ **27.** $-3 + (-3)$ **28.** $-2 + (-5)$ **29.** $-4 + 4$ **30.** $3 + (-3)$

⟨E⟩ Adding Integers Using Addition Rules In Problems 31–40, add.

31. $-3 + 5$ **32.** $-18 + 21$ **33.** $8 + (-1)$ **34.** $19 + (-6)$

35. $-8 + 13$ **36.** $-9 + 11$ **37.** $-17 + 4$ **38.** $-18 + 9$

39. $-4 + (+8) + 6 + (-2)$ **40.** $-17 + (+5) + (-6) + 7$

⟨F⟩ Subtracting Integers In Problems 41–50, subtract.

41. $-4 - 7$ **42.** $-5 - 11$ **43.** $-9 - 11$ **44.** $-4 - 16$

45. $8 - 12$ **46.** $7 - 13$ **47.** $8 - (-4)$ **48.** $9 - (-7)$

49. $0 - (-4) - 3 - (-3)$ **50.** $0 - 4 - (-5) - 5$

⟨G⟩ Applications Involving Integers

51. *Ocean floor depth* Mount Pico in the Azores is 7615 feet above sea level. The distance from the ocean floor to the crest of Mount Pico is 23,615 feet. How many feet below sea level is the ocean floor?

52. *Distance between high and low* The highest point on Earth is Mount Everest, 9 kilometers above sea level. The lowest point is the Mariannas trench in the Pacific, 11 kilometers below sea level. What is the distance between these two extremes?

53. *Temperature differences* The temperature in Death Valley is 53°C, and on Mount McKinley it is −20°C. What is the difference in temperature between Death Valley and Mount McKinley?

54. *Temperature differences* The temperature in the troposphere (the air layer around the Earth) is −56°C, and on Mount McKinley it is −20°C. Find the temperature difference between the troposphere and Mount McKinley.

55. *Temperature differences* The temperature in the center core of the Earth reaches +5000°C. In the thermosphere (a region in the upper atmosphere) it is +1500°C. Find the difference in temperature between the center of the Earth and the thermosphere.

56. *Water temperature differences* The water temperature of the White Sea is −2°C. In the Persian Gulf it is 36°C. Find the difference in temperature between the Persian Gulf and the White Sea.

57. *Temperature differences* The record high temperature in Calgary, Alberta, is +99°F. The lowest is −46°F. Find the difference between these extremes.

58. *Stock prices* The price of a certain stock at the beginning of a week is $42. Here are the changes in price during the week: +1, +2, −1, −2, −1. What is the price of the stock at the end of the week?

59. *Elevator rides* Joe was on the tenth floor of a building. He got on the elevator and he went three floors up, two down, five up, seven down. Representing these trips as +3, −2, +5, and −7, on what floor is Joe now?

60. *Temperature changes* Here are the temperature changes by the hour in a certain city:

1 PM	+2
2 PM	+1
3 PM	−1
4 PM	−3

If the temperature was initially 15°C, what was it at 4 PM?

For Problems 61–68, use the chart in Example 13.

61. *SAT scores* What integer corresponds to the verbal scores of test takers whose parents earned a bachelor's degree? What was their score?

62. *SAT scores* What integer corresponds to the math scores of test takers whose parents earned a bachelor's degree? What was their score?

63. *SAT scores* What integer corresponds to the verbal scores of test takers whose parents earned a graduate degree? What was their score?

64. *SAT scores* Find the difference between the integers corresponding to the verbal (−95) and math (−76) scores of test takers whose parents did not earn a high school diploma.

65. *SAT scores* Find the difference between the integers corresponding to the verbal and math scores of test takers whose parents earned a high school diploma.

66. *SAT scores* Find the difference between the integers corresponding to the verbal and math scores of test takers whose parents earned an associate degree.

67. *SAT scores* Find the difference between the integers corresponding to the verbal and math scores of test takers whose parents earned a bachelor's degree.

68. *SAT scores* Find the difference between the integers corresponding to the verbal and math scores of test takers whose parents earned a graduate degree.

The following graph will be used in problems 69–74.

Net farm income The red line shows the graph of the net farm income for the Farm Management Association. Is there a relationship between age bracket and net farm income? Let us see.

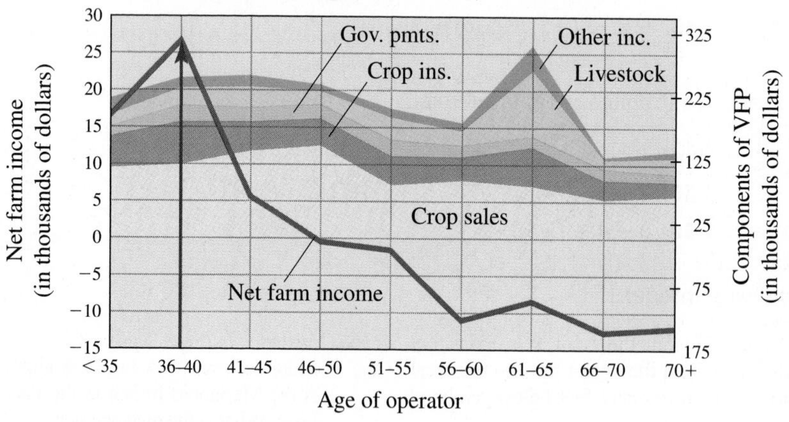

Source: Data from Northwest Kansas Farm Management Association.

69. The blue arrow shows the highest net farm income (in thousands) for the Farm Management Association (FMA). What was the net farm income and what was the corresponding age bracket?

70. What was the lowest net farm income produced and to what age bracket does it correspond?

71. At what age bracket do the farms break even?

72. If you are in the 56–60 age bracket, what is the corresponding net farm income?

73. What is the net farm income for farmers over 70?

74. At what age bracket would you expect to have a $5000 net farm income?

Surpluses and deficits The following graph compares estimated budget surpluses and deficits (in blue) to actual results (in red) and will be used in Problems 75–78.

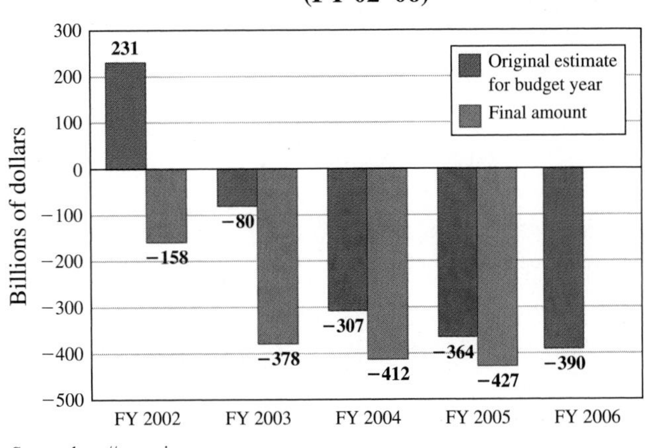

Source: http://newsaic.com.

75. a. What is the original estimate for Fiscal Year (FY) 2002?

b. What is the final amount for FY 2002?

c. What is the difference between the original estimate and the final amount?

76. a. What is the original estimate for Fiscal Year (FY) 2003?

b. What is the final amount for FY 2003?

c. What is the difference between the original estimate and the final amount?

77. a. What is the original estimate for Fiscal Year (FY) 2004?

b. What is the final amount for FY 2004?

c. What is the difference between the original estimate and the final amount?

78. a. What is the original estimate for Fiscal Year (FY) 2005?

b. What is the final amount for FY 2005?

c. What is the difference between the original estimate and the final amount?

Web IT go to mhhe.com/bello for more lessons

Using Your Knowledge

A Little History The accompanying chart contains some important historical dates.

Important Historical Dates

323 B.C.	Alexander the Great dies
216 B.C.	Hannibal defeats the Romans
A.D. 476	Fall of the Roman Empire
A.D. 1492	Columbus lands in America
A.D. 1776	The Declaration of Independence signed
A.D. 1939	World War II starts
A.D. 1988	Reagan and Gorbachev hold summit

We can use negative integers to represent years b.c. For example, the year Alexander the Great died can be written as −323, while the fall of the Roman Empire occurred in +476 (or simply 476). To find the number of years that elapsed between the fall of the Roman Empire and the defeat by Hannibal, we write

$$476 - (-216) = 476 + 216 = 692$$

Fall of the Roman Empire (A.D. 476) — Hannibal defeats the Romans (216 B.C.) — Years elapsed

Use these ideas to find the number of years elapsed between the following time periods.

79. The fall of the Roman Empire and the death of Alexander the Great.

80. Columbus's landing in America and Hannibal's defeat of the Romans.

81. The landing in America and the signing of the Declaration of Independence.

82. The year Reagan and Gorbachev held a summit conference and the signing of the Declaration of Independence.

83. The start of World War II and the death of Alexander the Great.

››› Write On

84. Write in your own words what you mean by the additive inverse of an integer.

85. Write in your own words why a and $-a$ are called opposites.

86. Write in your own words the procedure you use to add two integers with different signs.

87. Write in your own words the procedure you use to subtract two integers with different signs.

››› Concept Checker

Fill in the blank(s) with the correct word(s), phrase, or mathematical statement.

88. The **positive** integers are ______________.

89. The **negative** integers are ______________.

90. The integer a is less than the integer b ($a < b$) if a lies to the ______________ of b on the number line.

91. **b is greater than a** is written as ______________.

92. The **additive inverse** of a is ______________.

93. The **absolute value** of a number n is its ______________ from 0.

94. $a - b = c$ means ______________.

95. To **subtract** an integer ______________ its opposite.

$-1, -2, -3$	**0**
right	**$a = b + c$**
left	**$b = a + c$**
$0, 1, 2, 3$	**add**
$1, 2, 3, \ldots$	**subtract**
$b < a$	**distance**
$b > a$	**difference**
$-a$	**$-1, -2, -3, \ldots$**

››› Mastery Test

96. Find the additive inverse of each number.

a. 19 **b.** -23

97. Find the absolute value of each number.

a. $|-8|$ **b.** $|4|$ **c.** $|0|$

98. Use the number line to add $5 + (-3)$.

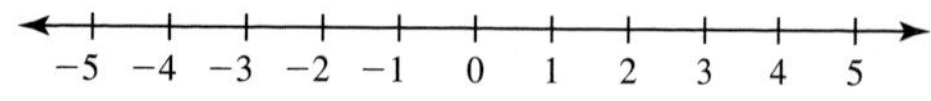

99. Use the number line to add $-4 + 2$.

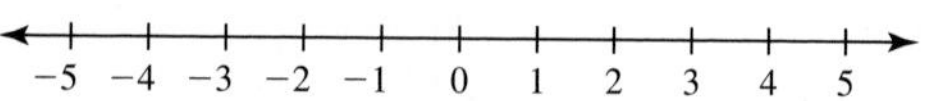

100. Use the number line to add $-2 + (-3)$.

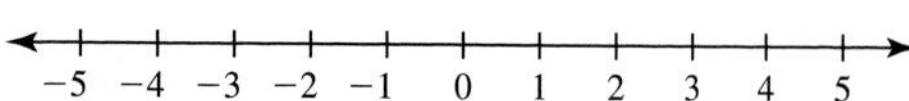

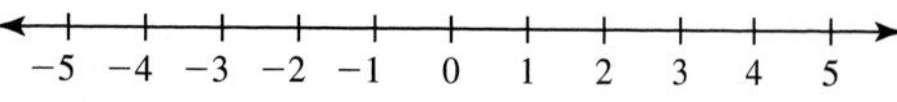

101. Add.

a. $-7 + (-5)$ **b.** $-9 + 5$

c. $9 + (-5)$

102. Subtract.

a. $14 - 7$ **b.** $-21 - 4$

c. $-8 - (-5)$ **d.** $7 - 10$

103. The chart shows the percent change for employment growth in Oklahoma (red) by quarters. When was the percent change the lowest and what was the percent change?

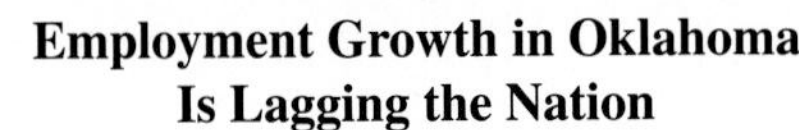

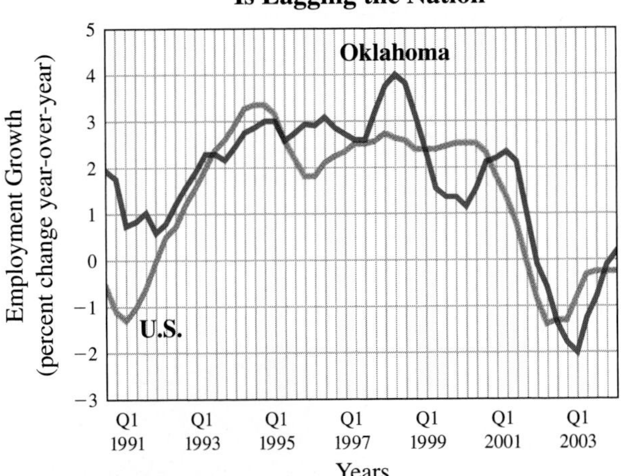

Source: Federal Deposit Insurance Corporation.

104. Consider the set of numbers $\{-3, -1, 0, 2, 6\}$

a. List the natural numbers in the set.

b. List the whole numbers in the set.

c. List the positive integers in the set.

d. List the negative integers in the set.

105. Give an **integer** that represents the following situation:

a. The stock market gained 80 points.

b. An investor lost $100,000.

106. Describe how the integers are being used in the given situation.

2006 Master's Golf Tournament

NAME	SCORE
Phil Mickelson	−7
Tim Clark	−5
Retief Goosen	−4
Jose Maria Olazabal	−4
Tiger Woods	−4

107. Graph the integers -4, 0, -1, and 5.

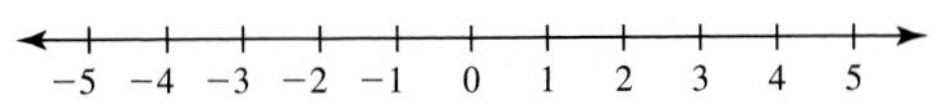

108. Insert the symbol $>$ or $<$ to make the statement true.
-4 ________ -1

109. Insert the symbol $>$ or $<$ to make the statement true.
0 ________ -4

110. Insert the symbol $>$ or $<$ to make the statement true.
0 ________ -1

Skill Checker

Find:

111. 5^2

112. 3^2

113. $2^3 \cdot 3^2$

114. $5^2 \cdot 3^3$

9.2 Multiplication and Division of Integers

Objectives

You should be able to:

A Multiply or divide two given integers.

B Raise an integer to a given power.

C Multiply more than two given integers.

To Succeed, Review How To . . .

1. Multiply and divide whole numbers. (pp. 49, 62)
2. Raise a whole number to a given power. (pp. 77, 82–84)

Getting Started

The scientists in the cartoon have forgotten how to multiply whole numbers. Products and quotients of integers can be found in the same way as products and quotients of whole numbers. The only difference is that we have to determine if the result is positive or negative. Look at the pattern that follows and see if you can discover a rule for multiplying integers.

"Say, I think I see where we went off. Isn't eight times seven fifty six"

This number decreases by 1. This number decreases by 3. This number decreases by 1. This number increases by 3.

$3 \cdot 3 = +9$	$3 \cdot (-3) = -9$
$2 \cdot 3 = +6$	$2 \cdot (-3) = -6$
$1 \cdot 3 = +3$	$1 \cdot (-3) = -3$
$0 \cdot 3 = 0$	$0 \cdot (-3) = 0$
$-1 \cdot 3 = -3$	$-1 \cdot (-3) = +3$
$-2 \cdot 3 = -6$	$-2 \cdot (-3) = +6$
$-3 \cdot 3 = -9$	$-3 \cdot (-3) = +9$

A > Multiplication and Division of Integers

The numbers shown in red represent the products of integers with *opposite* (*unlike*) signs; their product is *negative*. The numbers in blue have the *same* sign; their product is *positive*. Recall that 0 is neither positive nor negative.

Since the same rule applies to division (because, as you recall $a \div b = c$ means $a = b \times c$), we can summarize this discussion with the following rule:

Sign Rules for × and ÷

When Multiplying or Dividing Numbers with	The Answer Is
Same (like) signs	+
Opposite (unlike) signs	−

We can explain why the multiplication of integers with *unlike* signs is negative. Take $(3) \cdot (-4)$. Since multiplication is a process of repeated addition,

$$\begin{aligned}(3) \cdot (-4) &= (-4) + (-4) + (-4) \\ &= -12\end{aligned}$$

For example:

$$\left.\begin{aligned}7 \cdot 8 &= +56 \\ -3 \cdot (-8) &= +24\end{aligned}\right\}$$ Same signs, answer is +.

$$\left.\begin{aligned}16 \div 2 &= +8 \\ -24 \div (-6) &= +4\end{aligned}\right\}$$ Same signs, answer is +.

$$\left.\begin{aligned}-7 \cdot 8 &= -56 \\ -3 \cdot 5 &= -15\end{aligned}\right\}$$ Opposite (unlike) signs, answer is −.

$$\left.\begin{aligned}-30 \div 6 &= -5 \\ 27 \div (-3) &= -9\end{aligned}\right\}$$ Opposite (unlike) signs, answer is −.

When multiplying numbers you should remember the special case of multiplying any number by zero: The multiplication property of zero.

MULTIPLICATION PROPERTY OF ZERO

For any real number *a*,

$$a \cdot 0 = 0 \cdot a = 0$$

This means that the product of any real number and 0 is 0.

EXAMPLE 1 **Multiplying integers**

Multiply:

a. $9 \cdot 8$ **b.** $-5 \cdot 3$

c. $-5 \cdot (-4)$ **d.** $7 \cdot (-6)$

SOLUTION

a. $9 \cdot 8 = 72$

b. $\underbrace{-5 \cdot 3}_{\text{Unlike signs}} = -15$ ← Negative answer

c. $\underbrace{-5 \cdot (-4)}_{\text{Like signs}} = 20$ ← Positive answer

d. $\underbrace{7 \cdot (-6)}_{\text{Unlike signs}} = -42$ ← Negative answer

PROBLEM 1

Multiply:

a. $6 \cdot 7$ **b.** $-4 \cdot 2$

c. $-3 \cdot (-8)$ **d.** $-2 \cdot (8)$

Answers to PROBLEMS

1. a. 42 **b.** −8 **c.** 24 **d.** −16

We are now ready to do division, but we should note that there are some division problems that do not have integer answers (try $\frac{17}{5}$ or $-18 \div 7$). We discuss numbers such as $\frac{17}{5}$ and $-18 \div 7$ in Section 9.3. Moreover, division by 0 is *not* defined as indicated next.

DIVISION BY ZERO

For all real numbers *a*,

$$a \div 0 \quad \text{or} \quad \frac{a}{0} \text{ is } \textit{not} \text{ defined.}$$

Even though division by 0 is not defined, we can always divide 0 by any nonzero real number a. The result is 0, as shown next.

DIVIDENDS OF ZERO

For all real nonzero numbers *a*,

$$\frac{0}{a} = 0; a \neq 0$$

EXAMPLE 2 Dividing integers

Divide (if possible):

a. $35 \div 7$ **b.** $\frac{-12}{4}$

c. $24 \div (-12)$ **d.** $\frac{-10}{-5}$

e. $\frac{0}{10}$ **f.** $\frac{10}{0}$

SOLUTION

a. $35 \div 7 = 5$ **b.** $\frac{-12}{4} = -3$

c. $24 \div (-12) = -2$ **d.** $\frac{-10}{-5} = 2$

e. $\frac{0}{10} = 0$ **f.** $\frac{10}{0}$ is not defined

PROBLEM 2

Divide (if possible):

a. $42 \div 6$ **b.** $\frac{-18}{9}$

c. $49 \div (-7)$ **d.** $\frac{-15}{-3}$

e. $\frac{0}{15}$ **f.** $\frac{15}{0}$

B > Integers and Exponents

As you will recall, an exponent is a number that indicates how many times a number, called the base, is used as a factor. Thus,

$$2^2 = 2 \cdot 2 = 4$$
$$2^3 = 2 \cdot 2 \cdot 2 = 8$$

What about $(-2)^2$? Using the definition of exponent, we have

$$(-2)^2 = \underbrace{(-2) \cdot (-2)}_{\text{Like signs}} = 4$$

Note that $(-2)^2 = 4$ but $-2^2 = -(2 \cdot 2) = -4$. (Remember that -2^2 means the inverse of 2^2, that is, $-2^2 = -4$.) Thus, $-2^2 \neq (-2)^2$ (not equal). The placing of the parentheses in the expression $(-2)^2$ is very important!

Answers to PROBLEMS

2. a. 7 **b.** −2 **c.** −7 **d.** 5 **e.** 0 **f.** Not defined

EXAMPLE 3 Integers raised to a power

Evaluate:

a. $(-3)^2$ b. -3^2

SOLUTION

a. $(-3)^2 = (-3) \cdot (-3) = 9$

b. $-3^2 = -(3 \cdot 3) = -9$

PROBLEM 3

Evaluate:

a. $(-4)^2$ b. -4^2

EXAMPLE 4 Integers raised to a power

Evaluate:

a. $(-2)^3$ b. -2^3

SOLUTION

a. $(-2)^3 = (-2) \cdot (-2) \cdot (-2) = 4 \cdot (-2) = -8$

b. $-2^3 = -(2 \cdot 2 \cdot 2) = -8$

PROBLEM 4

Evaluate:

a. $(-3)^3$ b. -3^3

C › Multiplying More than Two Integers

When multiplying more than two signed numbers, we can use some examples to determine the sign of the final product.

a. $(-3)(2)(-5)$ — two negative factors
$= -6(-5)$
$= +30$ — + product

b. $2(-4)(-3)(-1)(-2)$ — four negative factors
$= -8(-3)(-1)(-2)$
$= 24(-1)(-2)$

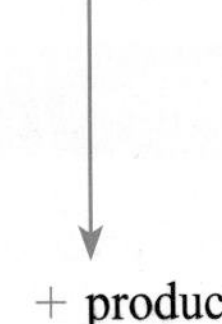

$= -24(-2)$
$= +48$ — + product

c. $-1(4)(-3)(-2)$ — three negative factors
$= -4(-3)(-2)$
$= 12(-2)$

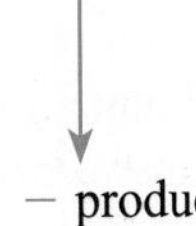

$= -24$ — − product

d. $3(-2)(-1)(-5)(-3)(-1)$ — five negative factors
$= -6(-1)(-5)(-3)(-1)$
$= 6(-5)(-3)(-1)$
$= -30(-3)(-1)$
$= 90(-1)$
$= -90$ — − product

As you can see from the pattern, the sign of the product is determined by the number of **negative** factors in the multiplication. This result is summarized next.

Answers to PROBLEMS

3. a. 16 b. −16

4. a. −27 b. −27

PROCEDURE TO MULTIPLY MORE THAN TWO SIGNED NUMBERS

First, multiply their absolute values. The sign of the final product is

1. Positive $(+)$ if there are an *even* number of negative factors.
2. Negative $(-)$ if there are an *odd* number of negative factors.

EXAMPLE 5 Multiplying signed numbers

Multiply:

a. $4(-7)(-1)$ **b.** $(-6)(-2)(-3)$ **c.** $(-5)(-3)(2)(-4)$

SOLUTION

a. There are **two** negative factors (-7) and (-1), which is an *even* number of negative factors, so the final product is *positive*. Multiply the absolute value of the numbers and remember that the final product is *positive*.

$$\begin{aligned} 4(-7)(-1) &= (+)4(7)(1) \\ &= 28 \end{aligned}$$

b. There are **three** negative factors, which is an *odd* number of negative factors, so the final product is *negative*. Multiply the absolute value of the numbers and assign the final product a *negative* sign.

$$\begin{aligned} (-6)(-2)(-3) &= (-)(6)(2)(3) \\ &= -36 \end{aligned}$$

c. We have **three** negative factors, so the final product is *negative*. Multiply the absolute value of the numbers and assign the final product a *negative* sign.

$$\begin{aligned} (-5)(-3)(2)(-4) &= -(5)(3)(2)(4) \\ &= -120 \end{aligned}$$

PROBLEM 5

Multiply:

a. $-2(-5)(-3)$

b. $(-3)(-6)(2)$

c. $(2)(-4)(-3)(-5)$

Calculator Corner

The +/− or CHS key is essential when multiplying or dividing integers, as shown for Examples 1 and 2.

Example 1

a. 9×8 Press 9 × 8 ENTER.

b. -5×3 Press − 5 × 3 ENTER.

c. $-5 \times (-4)$ presents a problem. If you press − 5 × − 4 ENTER, the calculator gives -9 for an answer. To obtain the correct answer you must key in − 5 × 4 +/− ENTER. This time the calculator multiplies -5 by the inverse of 4, or -4, to obtain the correct answer, 20.

d. $7 \times (-6)$ Press 7 × 6 +/− ENTER.

Example 2

a. $35 \div 7$ Key in 3 5 ÷ 7 ENTER.

b. $\frac{-12}{4}$ Key in − 1 2 ÷ 4 ENTER.

c. $24 \div (-12)$ presents the same problem as part **c** of Example 1. If you key in 2 4 ÷ − 1 2, you do not get the correct answer. As before, the proper procedure is to key in 2 4 ÷ 1 2 +/− ENTER, which gives the correct answer, -2.

d. $-10 \div (-5)$ has to be done like part **c.** The correct entries are − 1 0 ÷ 5 +/− ENTER.

e. $10 \div 0$ Key in 1 0 ÷ 0 ENTER. The calculator will show ERROR.

The examples involving decimals may be done similarly.

Answers to PROBLEMS

5. a. -30 **b.** 36 **c.** -120

MathZone

Boost your grade at mathzone.com!
- Practice Problems
- NetTutor
- Self-Tests
- e-Professors
- Videos

Exercises 9.2

Web IT go to mhhe.com/bello for more lessons

⟨A⟩ Multiplication and Division of Integers In Problems 1–30, multiply or divide.

1. $16 \cdot 2$
2. $9 \cdot 4$
3. $-7 \cdot 8$
4. $-10 \cdot 4$
5. $2 \cdot (-5)$
6. $9 \cdot (-9)$
7. $-4 \cdot (-5)$
8. $-6 \cdot (-3)$
9. $-7 \cdot (-10)$
10. $-9 \cdot (-2)$
11. $10 \div 2$
12. $\frac{14}{2}$
13. $\frac{-20}{5}$
14. $-50 \div 10$
15. $-40 \div 8$
16. $\frac{-30}{10}$
17. $150 \div (-15)$
18. $96 \div (-6)$
19. $\frac{140}{-7}$
20. $\frac{91}{-13}$
21. $-98 \div (-14)$
22. $-120 \div (-30)$
23. $\frac{-98}{-7}$
24. $\frac{-92}{-4}$
25. $\frac{0}{-10}$
26. $\frac{0}{-15}$
27. $-0 \div 8$
28. $\frac{-0}{3}$
29. $\frac{-8}{0}$
30. $-5 \div 0$

⟨B⟩ Integers and Exponents In Problems 31–40, find the value.

31. $(-4)^2$
32. -4^2
33. -5^2
34. $(-5)^2$
35. -6^3
36. $(-6)^3$
37. $(-3)^4$
38. -3^4
39. $(-5)^3$
40. -5^3

⟨C⟩ Multiplying More than Two Integers In Problems 41–50, multiply.

41. $-3(4)(-5)$
42. $-5(2)(-3)$
43. $-4(-2)(5)$
44. $-2(-5)(9)$
45. $-3(-5)(-2)$
46. $-3(-10)(-2)$
47. $-4(-5)(2)(3)$
48. $-10(-3)(6)(2)$
49. $-2(4)(-3)(-2)$
50. $3(-2)(-1)(-5)$

Applications

51. *Stockholders' losses* The net loss of a company was \$6400. If the company has 3200 stockholders and the losses are equally distributed among them, how much money did each stockholder lose?

52. *Profit or loss* A rock concert promoter sold 9000 tickets at \$5 each and 3000 tickets at \$8 each. If the performers charged \$50,000, what was the promoters' profit or loss?

Use the following tables for Problems 53–56.

1 all-beef frank: +45 calories
1 slice of bread: +65 calories

Running: −15 calories (1 minute)	Swimming: −7 calories (1 minute)

53. *Calories* If a person eats 2 beef franks and runs for 5 minutes, what is the caloric gain or loss?

54. *Calories* If the person in Problem 53 also swims for 10 minutes, what is the caloric gain or loss?

55. *Calories* If a person eats 2 beef franks in 2 slices of bread and then runs for 8 minutes, what is the caloric gain or loss?

56. *Calories* If the person of Problem 55 also swims for 15 minutes, what is the caloric gain or loss?

57. *Money* A child opened a savings account and made four deposits of $15 each and three withdrawals of $10 each. How much money was in the account after these transactions?

58. *Stock* A man bought 10 shares of a certain stock. Here are the changes in price for the stock during the following week: $+1, -2, -3, +1, -2$. How much money did he gain or lose on his 10 shares?

59. *Investments* An investment increased in value by $\frac{5}{8}$. A woman had invested $400. How much did the value of her investment increase?

60. *Investments* An investment went down in value by $\frac{1}{8}$. A woman had invested $1600. How much did the value of her investment decrease?

The environmental lapse Wait, wait, don't worry: we are not forgetting the environment! The *Environmental Lapse* is the rate of decrease of temperature with altitude (elevation): the higher you are, the lower the temperature. As a matter of fact, the temperature drops about **4°F** (−4°F) for each **1000** feet of altitude. *Source:* www.answers.com.

Fill in the blanks in the chart:

	Altitude in (1000 ft)	Altitude × Rate of Change	Temperature Change
	1	1(−4)	−4°F
	2	2(−4)	−8°F
61.	5	______	______
62.	10	______	______
63.	15	______	______

The metric environmental lapse In the metric system, the environmental lapse is about −7°C (negative 7 degrees Celsius) for each kilometer of altitude.

Fill in the blanks in the chart:

	Altitude in kilometers	Altitude × Rate of Change	Temperature Change
	1	1(−7)	−7°C
	2	2(−7)	−14°C
64.	3	______	______
65.	5	______	______
66.	6	______	______

67. *Pikes Peak* You are at the base of Pikes Peak and the temperature is about 70°F. You climb to the top of the peak, about 14,000 feet. Give an expression that would tell you the temperature at the top. (*Hint:* Use the facts given for Problems 61–63.) What is that temperature in degrees Fahrenheit?

68. *Mount McKinley* Mt. McKinley is about 20,000 feet high. If the temperature at the base is 70°F, what is the temperature at the top in degrees Fahrenheit?

69. *Mount McKinley* Mt. McKinley is about 6000 meters high and the temperature at the base is 20°C. What is the temperature at the top in degrees Celsius? (*Hint:* Use the facts given for Problems 64–67.)

70. *Mount Everest* You are at the base of Mt. Everest and the temperature is about 20°C. If you could climb to the top of Mt. Everest, an elevation of more than 8000 meters, what expression would tell you the temperature at the top? What is that temperature in degrees Celsius?

Using Your Knowledge

Splitting the Atom The valence (or oxidation number) of a compound is found by using the sum of the valences of each individual atom present in the compound. For example, the valence of hydrogen is +1, the valence of sulphur is +6, and that of oxygen is −2. Thus the valence of sulphuric acid is

$$H_2SO_4$$

$$\begin{aligned} &2(\text{valence of H}) + (\text{valence of S}) + 4(\text{valence of O}) \\ &= 2(+1) + (+6) + 4(-2) \\ &= 2 + 6 + (-8) = 0 \end{aligned}$$

Use this idea to solve Problems 71–75.

71. Find the valence of phosphate, PO_4, if the valence of phosphorus (P) is +5 and that of oxygen (O) is −2.

72. Find the valence of nitrate, NO_3, If the valence of nitrogen (N) is +5 and that of oxygen (O) is −2.

73. Find the valence of sodium bromate, $NaBrO_3$, if the valence of sodium (Na) is +1, the valence of bromine (Br) is +5, and the valence of oxygen (O) is −2.

74. Find the valence of sodium dichromate, $Na_2Cr_2O_7$, if the valence of sodium (Na) is +1, the valence of chromium (Cr) is +6, and that of oxygen (O) is −2.

75. Find the valence of water, H_2O, if the valence of hydrogen (H) is +1 and that of oxygen (O) is −2.

Write On

76. Write in your own words why the product of two negative numbers should be positive.

77. Write in your own words why the product of two integers with different signs should be negative.

78. Write in your own words why division by 0 is not defined.

Concept Checker

Fill in the blank(s) with the correct word(s), phrase, or mathematical statement.

79. When **multiplying** or **dividing** numbers with the **same** sign the answer is ______________.

80. **Division** by **zero** is not ______________.

81. For $a \neq 0$, $\frac{0}{a} =$ ______________.

82. An **exponent** is a number that indicates how many times the **base** is used as a ______________.

product	***a***
factor	**negative**
defined	**positive**
0	

Mastery Test

83. Find: **a.** $(-10)^2$ **b.** -10^2

84. Find: **a.** $(-10)^3$ **b.** -10^3

85. Find: **a.** $42 \div 7$ **b.** $\frac{-16}{4}$ **c.** $36 \div (-12)$ **d.** $\frac{-16}{-4}$

86. Find: **a.** $4 \cdot 7$ **b.** $-8 \cdot 4$ **c.** $-8 \cdot (-4)$ **d.** $8 \cdot (-9)$

87. Find: **a.** $(-3)(4)(-5)$ **b.** $(-5)(-2)(-3)$

88. Find: **a.** $(2)(-5)(3)(-4)$ **b.** $(-3)(2)(-5)(-6)$

Skill Checker

Perform the indicated operations.

89. $\frac{3}{4} \times \frac{12}{5}$

90. $\frac{5}{8} \times \frac{6}{7}$

91. $\frac{5}{8} \div \frac{7}{16}$

92. $\frac{4}{9} \div \frac{11}{18}$

93. 0.32×8

94. 0.82×9

95. $0.49 \div 7$

96. $0.54 \div 9$

97. $\frac{0.64}{0.8}$

98. $\frac{0.72}{0.09}$

9.3 The Rational Numbers

Objectives

You should be able to:

A Find the additive inverse of a rational number.

B Find the absolute value of a rational number.

C Add two rational numbers.

D Subtract two rational numbers.

E Multiply two rational numbers.

F Divide two rational numbers.

G Classify real numbers.

To Succeed, Review How To . . .

1. Find the additive inverse and the absolute value of an integer. (pp. 526, 527)
2. Perform the four fundamental operations with fractions and decimals. (pp. 130–136, 151–156, 203–206, 212–216)

Getting Started

The ad for Burlington PrintWorks multipurpose paper uses whole numbers (500), mixed numbers ($8\frac{1}{2}$), and decimals (0.008/sh.). We have studied:

The natural (counting) numbers: 1, 2, 3, . . .
The whole numbers: 0, 1, 2, 3, . . .
The integers: . . . , $-3, -2, -1, 0, 1, 2, 3,$ . . .

We are now ready to study the *rational numbers*.

Burlington
Multipurpose Paper
$8\frac{1}{2} \times 11''$
500 sh.
Price/Unit: $0.008/sh.
Price: $3.99

RATIONAL NUMBERS

The rational numbers consist of all numbers that can be written in the form $\frac{a}{b}$, where a and b are integers and $b \neq 0$.
In symbols, $\{r | r = \frac{a}{b}$, a and b integers and $b \neq 0\}$
The "|" bar is read as "such that".

Since any number of the form $\frac{a}{b}$ can be written as a decimal (by dividing a by b), the rational numbers include all the corresponding *decimals* as well. Moreover, *integers* are rational numbers, since any integer a can be written as $\frac{a}{1}$. Fortunately, everything we have said about the integers works for the rational numbers. We now discuss some properties that apply to these rational numbers.

A > Additive Inverses (Opposites)

As with the integers, every rational number has an *additive inverse* (*opposite*) and an *absolute value*. The table shows some rational numbers and their additive inverses.

Rational Number	Additive Inverse (Opposite)
$\frac{9}{8}$	$-\frac{9}{8}$
$-\frac{3}{4}$	$\frac{3}{4}$
5.9	−5.9
−6.8	6.8

Note that the sum (addition) of additive inverses is 0. For example, $\frac{9}{8} + \left(-\frac{9}{8}\right) = 0$ and $-6.8 + 6.8 = 0$.

EXAMPLE 1 Additive inverse of a rational number

Find the additive inverse (opposite).

a. $\frac{5}{2}$ **b.** -4.8 **c.** $-3\frac{1}{3}$ **d.** 7.2

SOLUTION

a. The additive inverse of $\frac{5}{2}$ is $-\frac{5}{2}$.
b. The additive inverse of -4.8 is 4.8.
c. The additive inverse of $-3\frac{1}{3}$ is $3\frac{1}{3}$.
d. The additive inverse of 7.2 is -7.2.

PROBLEM 1

Find the additive inverse.

a. $\frac{3}{4}$ **b.** -5.1 **c.** $-9\frac{1}{4}$ **d.** 3.9

B > Absolute Value

The absolute value of a rational number is obtained in the same way as the absolute value of an integer, as shown next.

EXAMPLE 2 Absolute value of a rational number

Find the absolute value of:

a. $-\frac{3}{7}$ **b.** 2.1 **c.** $-2\frac{1}{2}$ **d.** -4.1

SOLUTION

a. $\left|-\frac{3}{7}\right| = \frac{3}{7}$ **b.** $|2.1| = 2.1$

c. $\left|-2\frac{1}{2}\right| = 2\frac{1}{2}$ **d.** $|-4.1| = 4.1$

PROBLEM 2

Find the absolute value of:

a. $\frac{-1}{8}$ **b.** 3.4
c. $-3\frac{1}{4}$ **d.** -8.2

Answers to PROBLEMS

1. a. $-\frac{3}{4}$ **b.** 5.1 **c.** $9\frac{1}{4}$ **d.** -3.9 **2. a.** $\frac{1}{8}$ **b.** 3.4 **c.** $3\frac{1}{4}$ **d.** 8.2

C > Addition of Rational Numbers

Rational numbers are added in the same way as integers. Here is the rule.

ADDING RATIONAL NUMBERS

1. If both numbers have the *same* sign, add their absolute values and give the sum the common sign.
2. If the numbers have *opposite* signs, subtract their absolute values and give the difference the sign of the number with the greater absolute value.

This is illustrated in Examples 3 and 4.

EXAMPLE 3 Adding rational numbers

Add:

a. $-8.6 + 3.4$ **b.** $6.7 + (-9.8)$

c. $-2.3 + (-4.1)$

SOLUTION

a. $-8.6 + 3.4 = -(8.6 - 3.4) = -5.2$

b. $6.7 + (-9.8) = -(9.8 - 6.7) = -3.1$

c. $-2.3 + (-4.1) = -(2.3 + 4.1) = -6.4$

PROBLEM 3

Add:

a. $-7.5 + 2.1$

b. $8.3 + (-9.7)$

c. $-1.4 + (-6.1)$

EXAMPLE 4 Adding rational numbers

Add:

a. $-\frac{3}{7} + \frac{5}{7}$ **b.** $\frac{2}{5} + \left(-\frac{5}{8}\right)$

SOLUTION

a. $-\frac{3}{7} + \frac{5}{7} = +\left(\frac{5}{7} - \frac{3}{7}\right) = \frac{2}{7}$

b. As usual, we first find the LCD of 5 and 8, which is 40. We then write

$$\frac{2}{5} = \frac{16}{40} \text{ and } -\frac{5}{8} = -\frac{25}{40}$$

Thus,

$$\frac{2}{5} + \left(-\frac{5}{8}\right) = \frac{16}{40} + \left(-\frac{25}{40}\right) = -\left(\frac{25}{40} - \frac{16}{40}\right) = -\frac{9}{40}$$

PROBLEM 4

Add:

a. $-\frac{5}{9} + \frac{7}{9}$

b. $\frac{3}{4} + \left(-\frac{5}{3}\right)$

As with the whole numbers, the addition of rational numbers is *associative* and *commutative.*

D > Subtraction of Rational Numbers

As with the integers, we define subtraction as follows:

SUBTRACTION

For any rational numbers *a* and *b*,

$$a - b = a + (-b)$$

Remember, this means that to subtract b, we add the inverse (opposite) of b. Then use the rule for adding rational numbers.

Answers to PROBLEMS

3. a. -5.4 **b.** -1.4 **c.** -7.5

4. a. $\frac{2}{9}$ **b.** $-\frac{11}{12}$

EXAMPLE 5 **Subtracting rational numbers**

Subtract:

a. $-4.2 - (-3.1)$ **b.** $-2.5 - (-7.8)$

c. $\frac{2}{9} - \left(-\frac{4}{9}\right)$ **d.** $-\frac{5}{6} - \frac{7}{4}$

SOLUTION First, write the problem as an addition, then use the rules to add rational numbers.

a. $-4.2 - (-3.1) = -4.2 + 3.1 = -(4.2 - 3.1) = -1.1$

b. $-2.5 - (-7.8) = -2.5 + 7.8 = +(7.8 - 2.5) = 5.3$

c. $\frac{2}{9} - \left(-\frac{4}{9}\right) = \frac{2}{9} + \frac{4}{9} = \frac{6}{9} = \frac{2}{3}$

d. The LCD is 12:

$$-\frac{5}{6} = -\frac{10}{12} \text{ (}\times 2\text{)} \quad \text{and} \quad \frac{7}{4} = \frac{21}{12} \text{ (}\times 3\text{)}$$

Thus,

$$-\frac{5}{6} - \frac{7}{4} = -\frac{5}{6} + \left(-\frac{7}{4}\right) = -\frac{10}{12} + \left(-\frac{21}{12}\right) = -\left(\frac{10}{12} + \frac{21}{12}\right) = -\frac{31}{12}$$

PROBLEM 5

Subtract:

a. $-3.8 - (-2.5)$

b. $-4.7 - (-6.9)$

c. $\frac{3}{8} - \left(-\frac{1}{8}\right)$

d. $-\frac{7}{8} - \frac{5}{6}$

E > Multiplication of Rational Numbers

The multiplication of rational numbers uses the same rules of signs as the multiplication of integers.

Sign Rules for Multiplication and Division

When Multiplying Two Rational Numbers with	The Product Is
Like (same) signs	Positive (+)
Unlike (opposite) signs	Negative (−)

EXAMPLE 6 **Multiplying rational numbers**

Multiply:

a. $-3.1 \cdot 4.2$ **b.** $-1.2 \cdot (-3.4)$

c. $-\frac{3}{4} \cdot \left(-\frac{5}{2}\right)$ **d.** $\frac{5}{6} \cdot \left(-\frac{4}{7}\right)$

SOLUTION

a. -3.1 and 4.2 have *unlike* signs. The result is *negative*. Thus,

$$-3.1 \cdot 4.2 = -13.02$$

b. -1.2 and -3.4 have *like* signs. The result is *positive*. Thus,

$$-1.2 \cdot (-3.4) = 4.08$$

c. $-\frac{3}{4}$ and $-\frac{5}{2}$ have *like* signs. The result is *positive*. Thus,

$$-\frac{3}{4} \cdot \left(-\frac{5}{2}\right) = \frac{15}{8}$$

d. $\frac{5}{6}$ and $-\frac{4}{7}$ have *unlike* signs. The result is *negative*. Thus,

$$\frac{5}{6} \cdot \left(-\frac{4}{7}\right) = -\frac{20}{42} = -\frac{10}{21}$$

PROBLEM 6

Multiply:

a. $-2.2 \cdot 3.2$ **b.** $-1.3 \cdot (-4)$

c. $-\frac{3}{7} \cdot \left(-\frac{4}{5}\right)$ **d.** $\frac{6}{7} \cdot \left(-\frac{2}{3}\right)$

Note that the multiplication of rational numbers is *associative* and *commutative*.

Answers to PROBLEMS

5. a. -1.3 **b.** 2.2 **c.** $\frac{1}{2}$ **d.** $-\frac{41}{24}$ **6. a.** -7.04 **b.** 5.2 **c.** $\frac{12}{35}$ **d.** $-\frac{4}{7}$

F › Reciprocals and Division of Rational Numbers

The division of rational numbers is related to the *reciprocal* of a number. As you recall, two nonzero numbers a and $\frac{1}{a}$ are *reciprocals* (or *multiplicative inverses*) and their product is 1. Here are some numbers and their reciprocals.

Number	Multiplicative Inverse (Reciprocal)	Check
$\frac{3}{4}$	$\frac{4}{3}$	$\frac{3}{4} \cdot \frac{4}{3} = 1$
$-\frac{5}{2}$	$-\frac{2}{5}$	$-\frac{5}{2} \cdot \left(-\frac{2}{5}\right) = 1$

RECIPROCAL OF A NUMBER

The reciprocal of any nonzero number $\frac{a}{b}$ is $\frac{b}{a}$ and the reciprocal of $-\frac{a}{b}$ is $-\frac{b}{a}$.

Note that $\frac{a}{b} \cdot \frac{b}{a} = 1$ and $-\frac{a}{b} \cdot \left(-\frac{b}{a}\right) = 1$.

EXAMPLE 7 Finding the reciprocal

Find the reciprocal.

a. $\frac{3}{7}$ **b.** $-\frac{8}{3}$

SOLUTION

a. The reciprocal of $\frac{3}{7}$ is $\frac{7}{3}$ because $\frac{3}{7} \cdot \frac{7}{3} = 1$

b. The reciprocal of $-\frac{8}{3}$ is $-\frac{3}{8}$ because $-\frac{8}{3} \cdot \left(-\frac{3}{8}\right) = 1$

PROBLEM 7

Find the reciprocal.

a. $\frac{4}{5}$ **b.** $-\frac{7}{9}$

As in Section 2.3, the idea of a reciprocal can be used in division as follows:

DIVISION OF RATIONALS

To divide $\frac{a}{b}$ by $\frac{c}{d}$, multiply $\frac{a}{b}$ by the reciprocal of $\frac{c}{d}$; that is, $\frac{a}{b} \div \frac{c}{d} = \frac{a}{b} \cdot \frac{d}{c}$, where b, c, and $d \neq 0$.

EXAMPLE 8 Dividing rational numbers

Divide:

a. $\frac{2}{5} \div \left(-\frac{3}{4}\right)$ **b.** $-\frac{5}{6} \div \left(-\frac{7}{2}\right)$

c. $-\frac{3}{7} \div \frac{6}{7}$

SOLUTION

a. $\frac{2}{5} \div \left(-\frac{3}{4}\right) = \frac{2}{5} \cdot \left(-\frac{4}{3}\right) = -\frac{8}{15}$

b. $-\frac{5}{6} \div \left(-\frac{7}{2}\right) = -\frac{5}{6} \cdot \left(-\frac{2}{7}\right) = \frac{10}{42} = \frac{5}{21}$

c. $-\frac{3}{7} \div \frac{6}{7} = -\frac{3}{7} \cdot \frac{7}{6} = -\frac{21}{42} = -\frac{1}{2}$

PROBLEM 8

Divide:

a. $\frac{3}{5} \div \left(-\frac{4}{7}\right)$

b. $-\frac{6}{7} \div \left(-\frac{3}{5}\right)$

c. $-\frac{4}{5} \div \frac{8}{5}$

Answers to PROBLEMS

7. a. $\frac{5}{4}$ **b.** $-\frac{9}{7}$ **8. a.** $-\frac{21}{20}$ **b.** $\frac{10}{7}$ **c.** $-\frac{1}{2}$

Exercises 9.3

⟨A⟩ Additive Inverses (Opposites) In Problems 1–6, find the additive inverse (opposite).

1. $\frac{7}{3}$ **2.** $-\frac{8}{9}$ **3.** -6.4

4. -2.3 **5.** $3\frac{1}{7}$ **6.** $4\frac{1}{8}$

⟨B⟩ Absolute Value In Problems 7–12, find each value.

7. $\left|-\frac{4}{5}\right|$ **8.** $\left|-\frac{9}{2}\right|$ **9.** $|-3.4|$

10. $|-2.1|$ **11.** $\left|1\frac{1}{2}\right|$ **12.** $\left|3\frac{1}{4}\right|$

⟨C⟩ Addition of Rational Numbers In Problems 13–30, find each value.

13. $-7.8 + 3.1$ **14.** $-6.7 + 2.5$ **15.** $3.2 + (-8.6)$

16. $4.1 + (-7.9)$ **17.** $-3.4 + (-5.2)$ **18.** $-7.1 + (-2.6)$

19. $-\frac{2}{7} + \frac{5}{7}$ **20.** $-\frac{5}{11} + \frac{7}{11}$ **21.** $-\frac{3}{4} + \frac{1}{4}$

22. $-\frac{5}{6} + \frac{1}{6}$ **23.** $\frac{3}{4} + \left(-\frac{5}{6}\right)$ **24.** $\frac{5}{6} + \left(-\frac{7}{8}\right)$

25. $-\frac{1}{6} + \frac{3}{4}$ **26.** $-\frac{1}{8} + \frac{7}{6}$ **27.** $-\frac{1}{3} + \left(-\frac{2}{7}\right)$

28. $-\frac{4}{7} + \left(-\frac{3}{8}\right)$ **29.** $-\frac{5}{6} + \left(-\frac{8}{9}\right)$ **30.** $-\frac{4}{5} + \left(-\frac{7}{8}\right)$

⟨D⟩ Subtraction of Rational Numbers In Problems 31–40, find each value.

31. $-3.8 - (-1.2)$ **32.** $-6.7 - (-4.3)$ **33.** $-3.5 - (-8.7)$

34. $-6.5 - (-9.9)$ **35.** $4.5 - 8.2$ **36.** $3.7 - 7.9$

37. $\frac{3}{7} - \left(-\frac{1}{7}\right)$ **38.** $\frac{5}{6} - \left(-\frac{1}{6}\right)$ **39.** $-\frac{5}{4} - \frac{7}{6}$

40. $-\frac{2}{3} - \frac{3}{4}$

⟨E⟩ Multiplication of Rational Numbers In Problems 41–50, find each value.

41. $-2.2 \cdot 3.3$ **42.** $-1.4 \cdot 3.1$ **43.** $-1.3 \cdot (-2.2)$

44. $-1.5 \cdot (-1.1)$ **45.** $\frac{5}{6} \cdot \left(-\frac{5}{7}\right)$ **46.** $\frac{3}{8} \cdot \left(-\frac{5}{7}\right)$

47. $-\frac{3}{5} \cdot \left(-\frac{5}{12}\right)$ **48.** $-\frac{4}{7} \cdot \left(-\frac{21}{8}\right)$ **49.** $-\frac{6}{7} \cdot \frac{35}{8}$

50. $-\frac{7}{5} \cdot \frac{15}{28}$

⟨F⟩ Reciprocals and Division of Rational Numbers In Problems 51–60, find each value.

51. $\frac{3}{5} \div \left(-\frac{4}{7}\right)$ **52.** $\frac{4}{9} \div \left(-\frac{1}{7}\right)$ **53.** $-\frac{2}{3} \div \left(-\frac{7}{6}\right)$

54. $-\frac{5}{6} \div \left(-\frac{25}{18}\right)$ **55.** $-\frac{5}{8} \div \frac{7}{8}$ **56.** $-\frac{4}{5} \div \frac{8}{15}$

57. $\frac{-3.1}{6.2}$ **58.** $\frac{1.2}{-4.8}$ **59.** $\frac{-1.6}{-9.6}$

60. $\frac{-9.8}{-1.4}$

Web IT go to mhhe.com/bello for more lessons

⟨ G ⟩ Classifying Real Numbers In Problems 61–68, classify each number by making a check mark in the appropriate row.

	61. $\sqrt{16}$	62. $-1\frac{7}{9}$	63. 0	64. 9.2	65. 3	66. $\sqrt{11}$	67. $0.\overline{68}$	68. $-1\frac{3}{4}$
Natural number								
Whole number								
Integer								
Rational number								
Irrational number								
Real number								

Sets and numbers The natural numbers (N), the whole numbers (W), the integers (I), and the rational numbers (Q) can be symbolized using set notation.

$$N = \{1, 2, 3, \ldots\}$$
$$W = \{0, 1, 2, \ldots\}$$
$$I = \{\ldots, -2, -1, 0, 1, 2, \ldots\}$$
$$Q = \{r \mid r = \tfrac{a}{b}, \text{ where } a \text{ and } b \text{ are integers and } b \neq 0\}$$

Moreover, if all of the elements of set A are also members of set B, we say that A is contained in B (or A is a proper subset of B) and write $\boldsymbol{A \subset B}$.

In Problems 69–72 fill in the blank with the $\subset$ symbol and write the meaning where indicated.

69. N _____ W means that every __________ is also a __________.

70. W _____ I means that every __________ is also an __________.

71. I _____ Q means that every __________ is also a __________.

72. N _____ W _____ I _____ Q.

Web IT go to mhhe.com/bello for more lessons

⟩⟩⟩ Using Your Knowledge

Moods Have you met anybody nice today, or did you have an unpleasant experience? Perhaps the person you met was *very nice* or your experience was *very unpleasant*. Psychologists and linguists have a numerical way to indicate the difference between nice and very nice or between unpleasant and very unpleasant. Suppose you assign a positive number (+2, for example) to the adjective *nice*, and a negative number (say, −2) to *unpleasant,* and a positive number greater than 1 (say +1.75) to *very*. Then, very nice means

Very nice

$$(1.75) \cdot (2) = 3.50 \quad (1.75)(2) \text{ are multiplied.}$$

and very unpleasant means

Very unpleasant

$$(1.75) \cdot (-2) = -3.50$$

Here are some adverbs and adjectives and their numerical values.

Adverbs		Adjectives	
Slightly	0.54	Wicked	−2.5
Rather	0.84	Disgusting	−2.1
Decidedly	0.16	Average	−0.8
Very	1.75	Good	3.1
Extremely	1.45	Lovable	2.4

Find the value of each mood.

73. Slightly wicked

74. Decidedly average

75. Extremely disgusting

76. Rather lovable

77. Very good

By the way, if you got all the answers correct, you are 4.495!

Write On

78. Write in your own words why $-a^2$ is always negative ($a \neq 0$).

79. Write in your own words why $(-a)^2$ is always positive ($a \neq 0$).

80. Why do you think that $\frac{a}{0}$ is not defined?

Concept Checker

Fill in the blank(s) with the correct word(s), phrase, or mathematical statement.

81. The **additive inverse** (opposite) of ***a*** is ________.

82. By the definition of subtraction, ***a*** − ***b*** = ________ when *a* and *b* are rational numbers.

83. When **multiplying** two rational numbers with **like** (same) signs the **product** is ________.

84. When **multiplying** two rational numbers with **unlike** (opposite) signs the **product** is ________.

85. The **reciprocal** of $\frac{a}{b}$ is ________ ($b \neq 0$).

86. The **reciprocal** of $-\frac{a}{b}$ is ________ ($b \neq 0$).

$\frac{b}{a}$, $a \neq 0$	$\frac{a}{b}$, $a \neq 0$
$\frac{1}{a}$	negative
$-a$	0
$-\frac{b}{a}$, $a \neq 0$	$-\frac{a}{b}$, $b \neq 0$
positive	$a + (-b)$

Mastery Test

87. Find the additive inverse of $\frac{13}{10}$.

88. Find the additive inverse of -4.7.

89. Find the additive inverse of $-1\frac{1}{5}$.

90. Find the additive inverse of 5.6.

91. Find: $\left|-\frac{6}{17}\right|$

92. Find: $|5.9|$

93. Find: $\left|-7\frac{9}{11}\right|$

94. Find: $|-3.5|$

95. Add: $-4.6 + 4.3$

96. Add: $5.5 + (-2.8)$

97. Add: $-1.1 + (-2.4)$

98. Add: $-\frac{5}{11} + \frac{10}{11}$

99. Add: $\frac{1}{5} + \left(-\frac{1}{6}\right)$

100. Subtract: $-5.3 - (-2.3)$

101. Subtract: $-7.8 - 1.7$

102. Subtract: $\frac{5}{11} - \left(-\frac{6}{11}\right)$

103. Subtract: $-\frac{1}{8} - \frac{1}{6}$

104. Multiply: $-1.4 \cdot 6.3$

105. Multiply: $-2.4 \cdot (-2.6)$

106. Multiply: $-\frac{3}{8} \cdot \frac{5}{9}$

107. Multiply: $\frac{5}{4} \cdot \left(-\frac{2}{5}\right)$

108. Find the reciprocal of $-\frac{1}{3}$.

109. Divide: $\frac{3}{5} \div \frac{7}{15}$

110. Divide: $-\frac{7}{8} \div \left(-\frac{5}{24}\right)$

111. Divide: $-\frac{8}{7} \div \frac{4}{7}$

112. Divide: $\frac{-6.4}{1.6}$

113. Divide: $\frac{-1.5}{-6}$

114. Divide: $\frac{3.6}{-1.2}$

115. Classify the given numbers.

	a. 3.2	b. $-\frac{8}{9}$	c. 9	d. $\sqrt{21}$
Natural number				
Whole number				
Integer				
Rational number				
Irrational number				
Real number				

116. Classify the given numbers.

	a. 0	b. $\sqrt{49}$	c. $0.\overline{34}$	d. $-2\frac{1}{4}$
Natural number				
Whole number				
Integer				
Rational number				
Irrational number				
Real number				

Skill Checker

In Problems 117–120, perform the indicated operations.

117. $7 \cdot 9 - 5$

118. $36 \div 3 \cdot 2$

119. $6 \div 3 - (3 - 5)$

120. $8 \div 4 \cdot 2 + 2 \cdot (3 - 5)$

9.4 Order of Operations

Objectives

You should be able to:

A Evaluate expressions using the order of operations.

B Evaluate expressions using fraction bars as grouping symbols.

To Succeed, Review How To . . .

1. Add, subtract, multiply, and divide rational numbers. (pp. 548–550)
2. Use the order of operations for fractions. (pp. 82–84, 235)

Getting Started

Do you know your handicap in bowling? No, it is not the fact that your ball goes in the gutter all the time! In bowling, "Handicapping is a means of placing bowlers and teams with varying degrees of skill on as equitable a basis as possible for their competition against each other" (*source:* Bowl.com). This handicap is defined as *180 minus your average, multiplied by 9/10.* Suppose your average is 130, what is your handicap? Is it

$$180 - 130 \cdot \frac{9}{10}?$$

or

$$(180 - 130) \cdot \frac{9}{10}?$$

Note that

$$180 - 130 \cdot \frac{9}{10} = 180 - 13 \cdot 9 = 180 - 117 = 63$$

but

$$(180 - 130) \cdot \frac{9}{10} = 50 \cdot \frac{9}{10} = 45$$

To make the meaning precise, we should use **parentheses** to indicate that your average, 130, must be subtracted from 180 first, so your handicap is 45.

A › Using the Order of Operations

Here is the order of operations for real numbers.

ORDER OF OPERATIONS (PEMDAS)

1. Do all calculations inside *parentheses* and other grouping symbols ( ), [ ], { } **first.**
2. Evaluate all *exponential* expressions.
3. Do *multiplications* and *divisions* in order from left to right.
4. Do *additions* and *subtractions* in order from left to right.

P E M D A S

Remember PEMDAS so you can do the operations in the correct order!

Before going to the examples, let us explain what each of the lines mean.

1. Do all calculations inside *parentheses* and other grouping symbols ( ), [ ], { } **first.**

This means that when evaluating $\left\{\left(5+\frac{1}{2}\right)+\left[1-\frac{1}{2}\right]\right\}$

you follow steps ❶, ❷, and ❸ (❶ is the parentheses, ❷ the brackets, ❸ the braces)

❶ $\left(5+\frac{1}{2}\right)=\left(5\frac{1}{2}\right)$ $\qquad = \left\{\left(5\frac{1}{2}\right)+\left[1-\frac{1}{2}\right]\right\}$

Add inside parentheses

❷ $\left[1-\frac{1}{2}\right]=\left[\frac{1}{2}\right]$ $\qquad = \left\{\left(5\frac{1}{2}\right)+\left[\frac{1}{2}\right]\right\}$

Subtract inside brackets

❸ $\left(5\frac{1}{2}\right)+\left[\frac{1}{2}\right]=6$ $\qquad = \{\quad 6 \quad\}$

Add inside braces

2. Evaluate all *exponential* expressions.

This step is easy to follow, just look for the exponents! When you see 2^3 or 4^2, evaluate it: $2^3 = 8$ and $4^2 = 16$

3. Do *multiplications* and *divisions* **in order** from left to right.

The key words are **in order.** Of course, you must proceed from left to right.
Thus, if you have $12 \div 4 \cdot 2$ go from left to right and do the division $12 \div 4$ **first.**

$$
\begin{aligned}
& 12 \div 4 \cdot 2 & \\
&= 3 \cdot 2 & \text{Divide } 12 \div 4 = 3 \text{ first} \\
&= 6 & \text{Multiply } 3 \cdot 2 = 6 \text{ next}
\end{aligned}
$$

However, if you have $12 \cdot 2 \div 4$, multiply $12 \cdot 2 = 24$ then divide by 4, like this:

$$
\begin{aligned}
& 12 \cdot 2 \div 4 & \\
& 24 \div 4 & \text{Multiply } 12 \cdot 2 = 24 \\
& 6 & \text{Divide } 24 \div 4 = 6
\end{aligned}
$$

4. Do *additions* and *subtractions* in order from left to right.

Again, go in order from left to right.

$$
\begin{aligned}
& 3 - 4 + 5 & \\
&= -1 + 5 & \text{Subtract 4 from 3 first} \\
&= 4 & \text{Do the addition } -1 + 5 = 4 \text{ next}
\end{aligned}
$$

Similarly,

$$
\begin{aligned}
& 3 + 4 - 5 & \\
& 7 - 5 & \text{Do the addition } 3 + 4 = 7 \text{ first} \\
& 2 & \text{Then subtract 5 from 7}
\end{aligned}
$$

EXAMPLE 1 Evaluating expressions

Find the value of:

a. $-\frac{8}{9}\cdot 9 - 3$

b. $-27 + \frac{3}{5}\cdot 5$

SOLUTION

a. $-\frac{8}{9}\cdot 9 - 3$

$= -8 - 3$ — **M, D:** Do multiplications and divisions in order from left to right ($-\frac{8}{9}\cdot 9 = -8$).

$= -11$ — **A, S:** Then do additions and subtractions in order from left to right ($-8 - 3 = -11$).

b. $-27 + \frac{3}{5}\cdot 5$

$= -27 + 3$ — **M, D:** Do multiplications and divisions in order from left to right ($\frac{3}{5}\cdot 5 = 3$).

$= -24$ — **A, S:** Then do additions and subtractions in order from left to right ($-27 + 3 = -24$).

PROBLEM 1

Find the value of:

a. $-\frac{5}{7}\cdot 7 - 3$

b. $-20 + \frac{4}{9}\cdot 9$

EXAMPLE 2 Evaluating expressions

Find the value of:

a. $-6 \div 2 \cdot 5$

b. $-6 \cdot 2 \div \frac{4}{7}$

SOLUTION

a. $-6 \div 2 \cdot 5$

$= -3 \cdot 5$ — **D:** Do multiplications and divisions in **order** from left to right. ($-6 \div 2 = -3$, the division occurred first, so it was done first.)

$= -15$ — **M:** Next, do the multiplication ($-3 \cdot 5 = -15$).

b. $-6 \cdot 2 \div \frac{4}{7}$

$= -12 \div \frac{4}{7}$ — **M:** Do multiplications and divisions in **order** from left to right. ($-6 \cdot 2 = -12$, the multiplication occurred first, so it was done first.)

$= -12 \cdot \frac{7}{4}$ — To divide by $\frac{4}{7}$: multiply by the reciprocal $\frac{7}{4}$.

$= -\overset{3}{\cancel{12}} \cdot \frac{7}{\cancel{4}}$ — Simplify by dividing -12 by 4.

$= -21$ — **M:** Do multiplications and divisions in **order** from left to right ($-3 \cdot 7 = -21$).

PROBLEM 2

Find the value of:

a. $-8 \div 2 \cdot 5$

b. $-8 \cdot 2 \div \frac{4}{7}$

EXAMPLE 3 Evaluating expressions

Find the value of $8 \div (-2) \cdot \frac{5}{4} - 3 + \frac{1}{2}$:

SOLUTION

$8 \div (-2) \cdot \frac{5}{4} - 3 + \frac{1}{2}$

$= -4 \cdot \frac{5}{4} - 3 + \frac{1}{2}$ — **M, D:** Do multiplications and divisions in **order** from left to right ($8 \div (-2) = -4$).

$= -5 - 3 + \frac{1}{2}$ — **M, D:** Do multiplications and divisions in **order** from left to right ($-4 \cdot \frac{5}{4} = -5$).

$= -8 + \frac{1}{2}$ — **A, S:** Do the additions and subtractions in order from left to right ($-5 - 3 = -8$).

$= -7\frac{1}{2}$ — **A, S:** Do the additions and subtractions in order from left to right ($-8 + \frac{1}{2} = -7\frac{1}{2}$).

PROBLEM 3

Find the value of

$16 \div (-2) \cdot \frac{5}{4} - 7 + \frac{1}{2}$:

Answers to PROBLEMS

1. a. -8 **b.** -16 **2. a.** -20 **b.** -28 **3.** $-16\frac{1}{2}$

EXAMPLE 4 Expressions with grouping symbols and exponents

Find the value of:

a. $-63 \div \frac{7}{9} - (2 + 3)$ **b.** $-8 \div 2^3 + 3 - 1$

SOLUTION

a. $-63 \div \frac{7}{9} - (2 + 3)$

$= -63 \div \frac{7}{9} - 5$ **P:** First do the operation inside the parentheses $(2 + 3 = 5)$.

$= -81 - 5$ **D:** Do the division $(-63 \div \frac{7}{9} = -63 \cdot \frac{9}{7} = -81)$.

$= -86$ **S:** Then do the subtraction $(-81 - 5 = -86)$.

b. $-8 \div 2^3 + 3 - 1$

$= -8 \div 8 + 3 - 1$ **E:** First do the exponentiation $(2^3 = 8)$.

$= -1 + 3 - 1$ **D:** Next do the division $(-8 \div 8 = -1)$.

$= 2 - 1$ **A, S:** Then do the additions and subtractions in order from left to right $(-1 + 3 = 2)$.

$= 1$ **S:** Do the final subtraction.

PROBLEM 4

Find the value of:

a. $-64 \div \frac{8}{3} - (4 + 1)$

b. $-27 \div 3^3 + 5 - 2$

EXAMPLE 5 Expressions with grouping symbols

Find the value of $-8 \div 4 \cdot 2 + 3(5 - 2) - 3 \cdot \frac{2}{3}$:

SOLUTION

$-8 \div 4 \cdot 2 + 3(\underbrace{5 - 2}) - 3 \cdot \frac{2}{3}$

$= \underbrace{-8 \div 4} \cdot 2 + 3\ (3) - 3 \cdot \frac{2}{3}$ **P:** First do the operations inside the parentheses, **M, D:** Now do the multiplications and divisions, in order from left to right:

$= \underbrace{-2 \cdot 2} + 3(3) - 3 \cdot \frac{2}{3}$ **D:** This means do $-8 \div 4 = -2$ first.

$= -4 + \underbrace{3(3)} - 3 \cdot \frac{2}{3}$ **M:** Then do $-2 \cdot 2 = -4$.

$= -4 + 9 \underbrace{- 3 \cdot \frac{2}{3}}$ **M:** Next do $3(3) = 9$.

$= \underbrace{-4 + 9} - 2$ **M:** And finally, do $-3 \cdot \frac{2}{3} = -2$.

$= \underbrace{5 - 2}$ **A:** We are through with multiplications and divisions. Now do the addition of -4 and 9.

$= 3$ **S:** The final operation is a subtraction, $5 - 2 = 3$.

PROBLEM 5

Find the value of

$-6 \div 3 \cdot 4 + 4(7 - 5) - 5 \cdot \frac{4}{5}$:

EXAMPLE 6 Expressions with grouping symbols and exponents

Find the value of $-8 \div 4 \cdot 5 - 3(5 - 2)^2 + 5 \cdot 3\frac{1}{5}$:

SOLUTION

$-8 \div 4 \cdot 5 - 3(5 - 2)^2 + 5 \cdot 3\frac{1}{5}$

$= -8 \div 4 \cdot 5 - 3(3)^2 + 5 \cdot 3\frac{1}{5}$ **P:** $[(5 - 2) = (3)]$

$= -8 \div 4 \cdot 5 - 3(9) + 5 \cdot 3\frac{1}{5}$ **E:** $(3^2 = 9)$

$= -2 \cdot 5 - 3(9) + 5 \cdot 3\frac{1}{5}$ **D:** $(-8 \div 4 = -2)$

$= -10 - 27 + 16$ **M:** $(-2 \cdot 5 = -10, -3(9) = -27, 5 \cdot 3\frac{1}{5} = 16)$

$= -21$ **S, A:** $(-10 - 27 = -37, -37 + 16 = -21)$

PROBLEM 6

Find the value of:

$9 \div 3 \cdot 2 - 3(4 - 1)^2 + 5 \cdot 2\frac{1}{5}$

Answers to PROBLEMS

4. a. -29 **b.** 2

5. -4

6. -10

B > Using Fraction Bars as Grouping Symbols

Fraction bars are sometimes used as grouping symbols to indicate an expression representing a single number. To find the value of such expressions, simplify above and below the fraction bars following the order of operations. Thus,

$$\frac{-2(3+8)+4}{2(4)-10}$$

$$=\frac{-2(11)+4}{2(4)-10}$$ **A:** Add inside the parentheses in the numerator $(3+8=11)$.

$$=\frac{-22+4}{8-10}$$ **M:** Multiply in the numerator $[-2(11)=-22]$ and in the denominator $[2(4)=8]$.

$$=\frac{-18}{-2}$$ **A, S:** Add in the numerator $(-22+4=-18)$, subtract in the denominator $(8-10=-2)$.

$$=9$$ **D:** Do the final division. (Remember to use the rules of signs.)

EXAMPLE 7 Using the fraction bar as a grouping symbol

Find the value of $-5^2+\frac{3(4-8)}{2}+10\div 5$:

SOLUTION

As usual, we use the order of operations.

$=-5^2+\frac{3(4-8)}{2}+10\div 5$ Given.

$=-5^2+\frac{3(-4)}{2}+10\div 5$ **P:** Subtract inside the parentheses $(4-8=-4)$.

$=-25+\frac{3(-4)}{2}+10\div 5$ **E:** Do the exponentiation $(5^2=25$, so $-5^2=-25)$.

$=-25+\frac{-12}{2}+10\div 5$ **M:** Multiply above the division bar $[3(-4)=-12]$.

$=-25+(-6)+10\div 5$ **D:** Divide $\left(\frac{-12}{2}=-6\right)$.

$=-25+(-6)+2$ **D:** Divide $(10\div 5=2)$.

$=-31+2$ **A:** Add $[-25+(-6)=-31]$.

$=-29$ Do the final addition.

PROBLEM 7

Find the value of:

$-6^2+\frac{2(6-2)}{2}+15\div 3$

Calculator Corner

Most calculators contain a set of parentheses on the keyboard. These keys will allow you to specify the exact order in which you wish operations to be performed. Thus to find the value of $(4\cdot 5)+6$, key in

(4 × 5) + 6 ENTER

to obtain 26. Similarly, if you key in

4 × (5 + 6) ENTER

you obtain 44. However, in many calculators you won't be able to find the answer to the problem $4(5+6)$ unless you key in the multiplication sign × between the 4 and the parentheses.

It's important to note that some calculators follow the order of operations automatically and others do not. To find out whether yours does, enter 2 + 3 × 4 ENTER. If your calculator follows the order of operations, you should get 14 for an answer.

Answers to PROBLEMS

7. -27

Exercises 9.4

‹A› Using the Order of Operations In Problems 1–20, find the value of the expression.

1. $\frac{-4}{5} \cdot 5 + 6$

2. $\frac{-3}{4} \cdot 4 + 6$

3. $-7 + \frac{3}{2} \cdot 2$

4. $-6 + \frac{9}{2} \cdot 2$

5. $\frac{-7}{4} \cdot 8 - 3$

6. $\frac{-3}{4} \cdot 8 - 9$

7. $20 - \frac{3}{5} \cdot 5$

8. $30 - \frac{6}{5} \cdot 5$

9. $48 \div \frac{3}{4} - (3 + 2)$

10. $81 \div \frac{9}{2} - (4 + 5)$

11. $3 \cdot 4 \div \frac{2}{3} + (6 - 2)$

12. $3 \cdot 6 \div \frac{2}{3} + (5 - 2)$

13. $-36 \div 3^2 + 4 - 1$

14. $-16 \div 2^3 + 3 - 2$

15. $-8 \div 2^3 - 3 + 5$

16. $-9 \div 3^2 - 8 + 5$

17. $-10 \div 5 \cdot 2 + 8 \cdot (6 - 4) - 3 \cdot 4$

18. $-15 \div 3 \cdot 3 + 2 \cdot (5 - 2) + 8 \div 4$

19. $-4 \cdot 8 \div 2 - 3(4 - 1) + 9 \div 3$

20. $-6 \cdot 3 \div 3 - 2(3 - 2) - 8 \div 2$

‹B› Using Fraction Bars as Grouping Symbols In Problems 21–34, find the value of the expression.

21. $\frac{4 \cdot (6 - 2)}{-8} - \frac{6}{-2}$

22. $\frac{5 \cdot (6 - 2)}{-4} - \frac{16}{-4}$

23. $-5^2 + \frac{2 - 10}{4} + 12 \div 4$

24. $-4^2 + \frac{3 - 7}{2} + 18 \div 9$

25. $-3^3 + 4 - 6 \cdot 8 \div 4 - \frac{8 - 2}{-3}$

26. $-2^3 + 6 - 6 \div 3 \cdot 2 - \frac{9 - 3}{-6}$

27. $4 \cdot 9 \div 3 \cdot 10^3 - 2 \cdot (6 + 4)^2$

28. $5 \cdot 8 \div 4 \cdot (7 + 3)^3 - 2 \cdot (9 + 1)^2$

29. $(4 - 6)^2 \div 4 - \frac{2(7 - 9)}{4} - 4 \cdot 3 \div 2^2$

30. $(5 - 10)^2 \div 5 \cdot 2 - \frac{4(8 - 10)}{2} - 6 \cdot 4 \div 2^3$

31. $-7^2 + \frac{3(8 - 4)}{4} + 10 \div 2 \cdot 3$

32. $-5^2 + \frac{6(3 - 7)}{4} + 9 \div 3 \cdot 2$

33. $(-6)^2 \cdot 4 \div 4 - \frac{3(7 - 9)}{2} - 4 \cdot 3 \div 2^2$

34. $(-4)^2 \cdot 3 \div 8 - \frac{4(6 - 10)}{2} - 3 \cdot 8 \div 2^3$

Web IT go to mhhe.com/bello for more lessons

Applications

35. *Octane rating* Have you noticed the octane rating of gasoline at the gas pump? This octane rating is given by

$$\frac{R + M}{2}$$

where R is a number measuring the performance of gasoline using the Research Method and M is a number measuring the performance of gasoline using the Motor Method. If a certain gasoline has $R = 92$ and $M = 82$, what is its octane rating?

36. *Octane rating* If a gasoline has $R = 97$ and $M = 89$, what is its octane rating (see Problem 35)?

37. *Pulse rate* If A is your age, the minimum pulse rate you should maintain during aerobic activities is $0.72(220 - A)$. What is the minimum pulse rate you should maintain if you are:

a. 20 years old? **b.** 45 years old?

38. *Pulse rate* If A is your age, the maximum pulse rate you should maintain during aerobic activities is $0.88(220 - A)$. What is the maximum pulse rate you should maintain if you are

a. 20 years old? **b.** 45 years old?

Pizza Remember the measurement chapter: length, area, and capacity? What units of measurement would you use to buy pizza? Square units, of course!

The illustration shows the price of four popular pizza sizes: 10 inch, 12 inch, 14 inch, and 16 inch. For convenience, we round prices to \$8, \$10, \$12.50, and \$15, use $\pi \approx 3.14$, and round the final answers (in cents) to one decimal place.

THIN CRUST PIZZA (Non Veg Momose)				
Giordano's FAMOUS STUFFED PIZZA	10" Small	12" Medium	14" Large	16" E-Large ....
CHEESE with	\$7.85	\$9.95	\$12.45	\$14.95

39. **a.** What is the area of the 10-inch pizza?

b. What is the price per square inch for the 10-inch pizza?

40. **a.** What is the area of the 12-inch pizza?

b. What is the price per square inch for the 12-inch pizza?

41. **a.** What is the area of the 14-inch pizza?

b. What is the price per square inch for the 14-inch pizza?

42. **a.** What is the area of the 16-inch pizza?

b. What is the price per square inch for the 16-inch pizza?

Using Your Knowledge

43. Insert any operation signs (parentheses, +, −, ·, or ÷) so that the string of numbers equals the given answer.

$$3\ 2\ 5\ 6 = 19$$

44. Repeat the procedure of Problem 43 for the numbers

$$9\ 5\ 6\ 7 = 14$$

45. Have you heard of Einstein's problem? The idea is to use the digits 1, 2, 3, 4, 5, 6, 7, 8, 9 and any combination of the operation signs (+, −, ·, ÷) to write an expression that equals 100. Oh, yes, the numbers have to be in consecutive order from left to right!

Write On

46. Write in your own words why you think the order of operations is needed. Give examples.

47. When evaluating an expression, do you *always* have to do multiplications before divisions? Give examples to support your answer.

48. When evaluating expressions, do you *always* have to do additions before subtractions? Give examples to support your answer.

49. Write in your own words why the parentheses are needed when evaluating $2 + (3 \cdot 4)$ or $(2 \cdot 3) + 4$.

Concept Checker

Fill in the blank(s) with the correct word(s), phrase, or mathematical statement.

In Problems 50–55, we will refer to the abbreviation **PEMDAS.**

50. The **P** in PEMDAS means ________.

51. The **E** in PEMDAS means ________.

52. The **M** in PEMDAS means ____________.

53. The **D** in PEMDAS means ________.

54. The **A** in PEMDAS means ________.

55. The **S** in PEMDAS means _________.

exponent	**parentheses**
multiplication	**subtraction**
simplify	**addition**
division	**exclude**

Mastery Test

Find the value of each expression.

56. $63 \div 9 - (2 + 5)$

57. $-64 \div 8 - (4 - 2)^2$

58. $-16 + 2^3 + 3 - 9$

59. $\frac{-3}{4} \cdot 4 - 18$

60. $-18 + \frac{4}{5} \cdot 5$

61. $-12 \div 4 \cdot 2 + 2(5 - 3) - \frac{3}{4} \cdot 4$

62. $-6^2 + \frac{4(6 - 12)}{3} + 8 \div 2$

63. The ideal heart rate while exercising for a person A years old is $[(205 - A) \cdot 7] \div 10$. What is the ideal heart rate for a 35-year-old person?

64. $4 \div (-2) \cdot \frac{3}{2} + (3 - 1)^2 + \frac{1}{2}$

Skill Checker

In Problems 65–68, multiply two different ways: first by using the distributive property, then by adding inside the parentheses before multiplying.

65. $5(7 + 2)$

66. $2(4 + 3)$

67. $3(4 + 6)$

68. $8(4 + 5)$

Collaborative Learning

Form two or more groups.

1. Let each group select **three** integers from 0 to 9 and make a three digit number using the three integers, then subtract its *reversal* (the *reversal* of 856 is 658). Then, if the difference is *positive,* add its reversal. If the difference is *negative,* subtract its reversal. Here are three possibilities for three numbers.

$$\begin{array}{r} 856 \\ -\ 658 \\ \hline 198 \\ +\ 891 \\ \hline \end{array} \qquad \begin{array}{r} 159 \\ -\ 951 \\ \hline -792 \\ -\ 297 \\ \hline \end{array} \qquad \begin{array}{r} 872 \\ -\ 278 \\ \hline 594 \\ +\ 495 \\ \hline \end{array}$$

   What answer did your group get? Did all groups get the same answer? Why do you think this works?

2. Look at these patterns.

$$99 + 100 + 101 + 102 + 103 = 505$$
$$99^2 + 100^2 + 101^2 + 102^2 + 103^2 = 51{,}015$$

   The answers 505 and 51,015 are *palindromic* numbers. Do not get alarmed; it only means that 505 and 51,015 read the same from left to right as from right to left. Will the pattern continue if six numbers are used instead of five? Let the members of your group decide.

3. Look at the next patterns.

$$10{,}099 + 10{,}100 + 10{,}101 + 10{,}102 + 10{,}103 = 50{,}505$$
$$10{,}099^2 + 10{,}100^2 + 10{,}101^2 + 10{,}102^2 + 10{,}103^2 = 510{,}151{,}015$$

   Will the pattern continue if six numbers are used instead of five? Let the members of your group decide.

Research Questions

1. In previous chapters we have used the customary symbols to denote the operations of addition, subtraction, multiplication, and division. Who invented the +, −, ×, and ÷ symbols, and in what publications did they first appear? *Note:* Answers may not be unique; try http://www.roma.unisa.edu.au/07305/symbols.htm#Plus for starters.

2. Write a report about Johan Widmann's *Mercantile Arithmetic* (1489), indicating which symbols of operation were found in the book for the first time and the manner in which they were used.

(continued)

3. We have also used the horizontal bar to denote fractions. Who invented the horizontal bar to write common fractions, and who was the first European mathematician to use it?

4. We have also used the diagonal fraction bar to write fractions as *a/b*. Who were the first users of the diagonal fraction bar? From what did the diagonal bar evolve?

Summary Chapter 9

Section	Item	Meaning	Example
9.1A	Positive integers Negative integers Integers	1, 2, 3, and so on $-1, -2, -3$, and so on $\ldots, -2, -1, 0, 1, 2, \ldots$	19 and 28 are positive integers. -41 and -56 are negative integers.
9.1B	Additive inverse	The additive inverse of a is $-a$.	-7 and 7 are additive inverses.
9.1C	Absolute value	The distance of a number from 0	$\lvert -7 \rvert = 7$ and $\lvert 4 \rvert = 4$
9.1E	Adding integers	If both integers have the *same* sign, add their absolute values and give the sum the common sign. If the integers have *opposite* signs, subtract their absolute values and give the difference the sign of the number with the larger absolute value.	$-2 + (-7) = -(2 + 7) = -9$ $-7 + 2 = -(7 - 2) = -5$
9.1F	Subtracting integers	To subtract one integer from another, add the opposite of the number to be subtracted.	$3 - 9 = 3 + (-9) = -6$
9.2A	Multiplication and division of integers	The product or quotient of integers with *like* signs is + and with *unlike* signs is −.	$-2 \cdot (-7) = 14$ $-24 \div (-4) = 6$ $-2 \cdot 7 = -14$ $24 \div (-4) = -6$
9.3	Rational numbers	Numbers that can be written in the form $\frac{a}{b}$, where a and b are integers and $b \neq 0$	$\frac{3}{4}$, $-5\frac{1}{2}$, $-\frac{7}{8}$, 0.23, and $0.\overline{6}$ are rational.
9.3A	Additive inverse	The additive inverse of a is $-a$.	The additive inverse of $\frac{3}{4}$ is $-\frac{3}{4}$ and the additive inverse of -0.53 is 0.53.
9.3B	Absolute value	The distance of a number from 0	$\left\lvert \frac{3}{4} \right\rvert = \frac{3}{4}$ and $\lvert -0.7 \rvert = 0.7$
9.3F	Reciprocal Division	The reciprocal of any nonzero number $\frac{a}{b}$ is $\frac{b}{a}$. $\frac{a}{b} \div \frac{c}{d} = \frac{a}{b} \cdot \frac{d}{c}$	$\frac{2}{3}$ and $\frac{3}{2}$ are reciprocals. $\frac{3}{4} \div \left(-\frac{1}{2}\right) = \frac{3}{4} \cdot \left(-\frac{2}{1}\right) = -\frac{3}{2}$
9.3G	Irrational numbers	Numbers that cannot be written in the form $\frac{a}{b}$, where a and b are integers and $b \neq 0$. When written as decimals, irrationals are nonrepeating and nonterminating.	$\sqrt{2}$ and $\sqrt{5}$ are irrational numbers. 32.01234 . . . and 0.101001000 . . . are irrational (nonrepeating, nonterminating).

Section	Item	Meaning	Example
9.4	Order of operations (PEMDAS)	1. Do all calculations inside *parentheses* and other grouping symbols ( ), [ ], { } **first.** 2. Evaluate all *exponential* expressions. 3. Do *multiplications* and *divisions* in order from left to right. 4. Do *additions* and *subtractions* in order from left to right.	

› Review Exercises Chapter 9

(If you need help with these exercises, look in the section indicated in brackets.)

1. ‹ **9.1A** › *Consider the set* $\{-6, -4, -1, 0, 2, 5\}$.

a. List the negative numbers in the set.

b. List the positive numbers in the set.

c. List the whole numbers in the set.

d. List the natural (counting) numbers in the set.

e. List the integers in the set.

2. ‹ **9.1A** › *Indicate how the integers are used in each situation.*

a. Financial markets

Symbol	Price	Change
AMEX	1866	+7
Nasdaq	2121	−1
Dow	10989	−30
NYSE	7924	+4

b. Outdoor thermometer

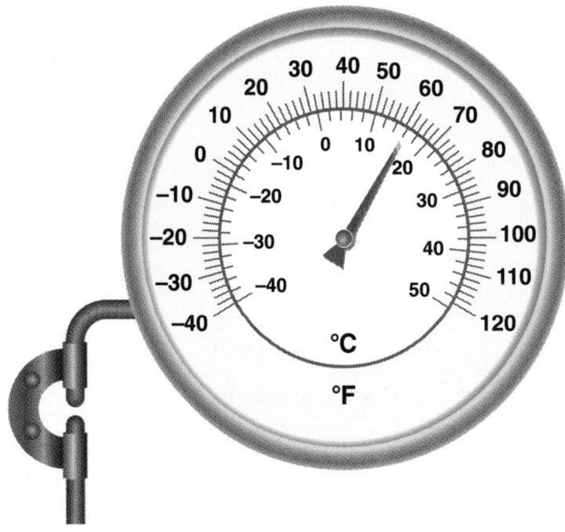

c. North America

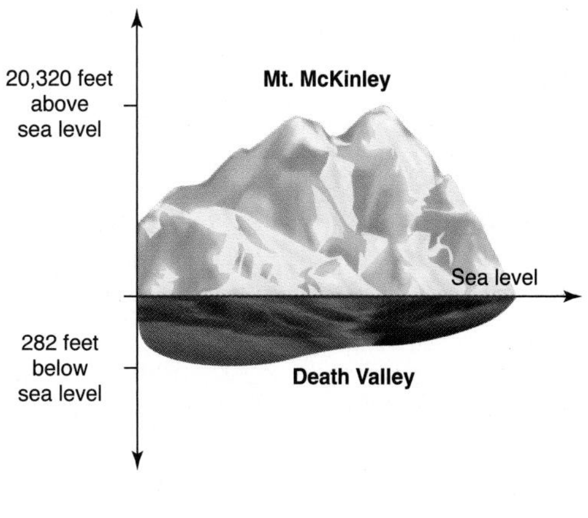

d. New Orleans elevations

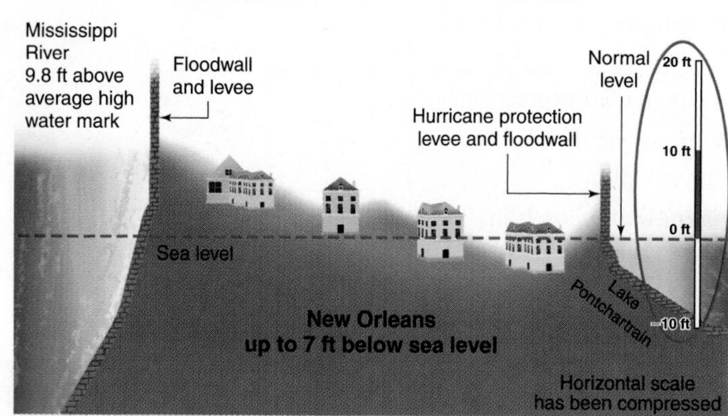

e. Population changes

2000–2050 Population Change (millions)	
Poland	−12
Portugal	−18
Romania	−24
Russia	−29
Saudi Arabia	152

3. ⟨**9.1A**⟩ *Write the integer that represents the situation.*

a. The highest recorded temperature in Israel, 129°F

b. The lowest recorded temperature in Canada, −63°C

c. A $53 withdrawal from your bank account

d. An $8 trillion deficit in the budget

e. Submarine Kaiko descent to the lowest part of the Marianas Trench, 10,897 meters below the ocean surface.

4. ⟨**9.1A**⟩ *Graph the given integers on the number line.*

a. −3, 0, 4

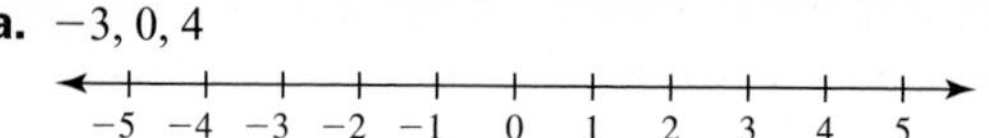

b. −4, −2, 5

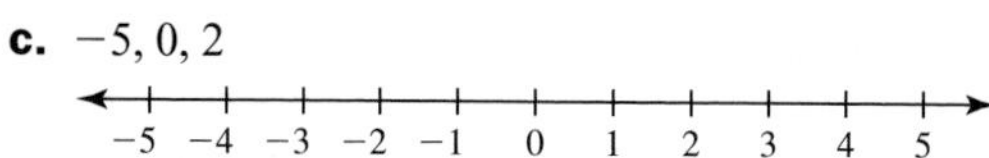

c. −5, 0, 2

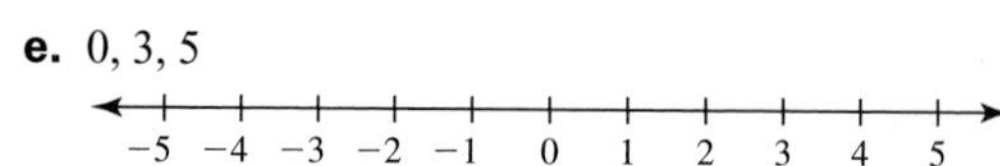

d. −4, −3, −1

−5 −4 −3 −2 −1 0 1 2 3 4 5

e. 0, 3, 5

−5 −4 −3 −2 −1 0 1 2 3 4 5

5. ⟨**9.1A**⟩ *Fill in the blank with $<$ or $>$ to make the statement true.*

a. −5 ________ −1

b. −1 ________ −3

c. −4 ________ −2

d. 1 ________ 5

e. 0 ________ −2

6. ⟨**9.1B**⟩ *Find the additive inverse (opposite).*

a. 10

b. 11

c. 12

d. 13

e. 14

7. ⟨**9.1B**⟩ *Find the additive inverse (opposite).*

a. −9

b. −8

c. −7

d. −6

e. −5

8. ⟨**9.1C**⟩ *Find each value.*

a. $|-7|$

b. $|8|$

c. $|-9|$

d. $|10|$

e. $|-11|$

9. ⟨**9.1E**⟩ *Add.*

a. $-8 + (-5)$

b. $-8 + (-4)$

c. $-8 + (-3)$

d. $-8 + (-2)$

e. $-8 + (-1)$

10. ⟨**9.1E**⟩ *Add.*

a. $-12 + 5$

b. $-12 + 6$

c. $-12 + 7$

d. $-12 + 8$

e. $-12 + 9$

11. ⟨**9.1E**⟩ *Add.*

a. $14 + (-4)$

b. $14 + (-5)$

c. $14 + (-6)$

d. $14 + (-7)$

e. $14 + (-8)$

12. ⟨**9.1F**⟩ *Subtract.*

a. $-15 - 3$

b. $-15 - 4$

c. $-15 - 5$

d. $-15 - 6$

e. $-15 - 7$

13. ⟨**9.1F**⟩ *Subtract.*

a. $-10 - (-2)$

b. $-10 - (-3)$

c. $-10 - (-1)$

d. $-10 - (-4)$

e. $-10 - (-5)$

14. ⟨**9.1F**⟩ *Subtract.*

a. $9 - 14$

b. $9 - 15$

c. $9 - 16$

d. $9 - 17$

e. $9 - 18$

15. ⟨**9.2A**⟩ *Multiply.*

a. $-6 \cdot 4$

b. $-6 \cdot 5$

c. $-6 \cdot 6$

d. $-6 \cdot 7$

e. $-6 \cdot 8$

16. ⟨**9.2A**⟩ *Multiply.*

a. $-8 \cdot (-5)$

b. $-8 \cdot (-6)$

c. $-8 \cdot (-7)$

d. $-8 \cdot (-8)$

e. $-8 \cdot (-9)$

17. ⟨**9.2A**⟩ *Multiply.*

a. $5 \cdot (-5)$

b. $6 \cdot (-5)$

c. $7 \cdot (-5)$

d. $8 \cdot (-5)$

e. $9 \cdot (-5)$

18. ⟨**9.2A**⟩ *Divide.*

a. $\frac{-58}{2}$

b. $\frac{-48}{2}$

c. $\frac{-38}{2}$

d. $\frac{-28}{2}$

e. $\frac{-18}{2}$

19. ⟨**9.2A**⟩ *Divide.*

a. $72 \div (-12)$

b. $72 \div (-18)$

c. $72 \div (-24)$

d. $72 \div (-36)$

e. $72 \div (-72)$

20. ⟨**9.2A**⟩ *Divide.*

a. $\frac{-15}{-5}$

b. $\frac{-25}{-5}$

c. $\frac{-35}{-5}$

d. $\frac{-45}{-5}$

e. $\frac{-55}{-5}$

21. ⟨**9.2B**⟩ *Find each value.*

a. $(-4)^2 =$ ________

b. $(-5)^2 =$ ________

c. $(-6)^2 =$ ________

d. $(-7)^2 =$ ________

e. $(-8)^2 =$ ________

22. ⟨**9.2B**⟩ *Find each value.*

a. $-4^2 =$ ________

b. $-5^2 =$ ________

c. $-6^2 =$ ________

d. $-7^2 =$ ________

e. $-8^2 =$ ________

23. ⟨**9.3A**⟩ *Find the additive inverse.*

a. $\frac{3}{11}$

b. $\frac{4}{11}$

c. $\frac{5}{11}$

d. $\frac{6}{11}$

e. $\frac{7}{11}$

24. ⟨**9.3A**⟩ *Find the additive inverse.*

a. -3.4

b. -4.5

c. -5.6

d. -6.7

e. -7.8

25. ⟨9.3A⟩ *Find the additive inverse.*
   a. $-3\frac{1}{2}$
   b. $-4\frac{1}{2}$
   c. $-5\frac{1}{2}$
   d. $-6\frac{1}{2}$
   e. $-7\frac{1}{2}$

26. ⟨9.3B⟩ *Find each value.*
   a. $\left|-\frac{2}{11}\right|$
   b. $\left|-\frac{3}{11}\right|$
   c. $\left|-\frac{4}{11}\right|$
   d. $\left|\frac{5}{11}\right|$
   e. $\left|\frac{6}{11}\right|$

27. ⟨9.3B⟩ *Find each value.*
   a. $\left|-3\frac{1}{4}\right|$
   b. $\left|4\frac{1}{4}\right|$
   c. $\left|5\frac{1}{4}\right|$
   d. $\left|-6\frac{1}{4}\right|$
   e. $\left|-7\frac{1}{4}\right|$

28. ⟨9.3B⟩ *Find each value.*
   a. $|-5.1|$
   b. $|6.2|$
   c. $|-7.3|$
   d. $|8.4|$
   e. $|-9.5|$

29. ⟨9.3C⟩ *Add.*
   a. $-8.7 + 3.1$
   b. $-8.7 + 3.2$
   c. $-8.7 + 3.3$
   d. $-8.7 + 3.4$
   e. $-8.7 + 3.5$

30. ⟨9.3C⟩ *Add.*
   a. $6.2 + (-9.3)$
   b. $6.2 + (-9.4)$
   c. $6.2 + (-9.5)$
   d. $6.2 + (-9.6)$
   e. $6.2 + (-9.7)$

31. ⟨9.3C⟩ *Add.*
   a. $-2.1 + (-3.2)$
   b. $-2.1 + (-3.3)$
   c. $-2.1 + (-3.4)$
   d. $-2.1 + (-3.5)$
   e. $-2.1 + (-3.6)$

32. ⟨9.3C⟩ *Add.*
   a. $-\frac{3}{11} + \frac{5}{11}$
   b. $-\frac{3}{11} + \frac{6}{11}$
   c. $-\frac{3}{11} + \frac{7}{11}$
   d. $-\frac{3}{11} + \frac{8}{11}$
   e. $-\frac{3}{11} + \frac{9}{11}$

33. ⟨9.3C⟩ *Add.*
   a. $\frac{1}{5} + \left(-\frac{2}{9}\right)$
   b. $\frac{1}{5} + \left(-\frac{4}{9}\right)$
   c. $\frac{1}{5} + \left(-\frac{5}{9}\right)$
   d. $\frac{1}{5} + \left(-\frac{7}{9}\right)$
   e. $\frac{1}{5} + \left(-\frac{8}{9}\right)$

34. ⟨9.3D⟩ *Subtract.*
   a. $-5.9 - (-3.1)$
   b. $-5.9 - (-3.2)$
   c. $-5.9 - (-3.3)$
   d. $-5.9 - (-3.4)$
   e. $-5.9 - (-3.5)$

35. ⟨9.3D⟩ *Subtract.*
   a. $-3.2 - (-7.5)$
   b. $-3.2 - (-7.6)$
   c. $-3.2 - (-7.7)$
   d. $-3.2 - (-7.8)$
   e. $-3.2 - (-7.9)$

36. ⟨9.3D⟩ *Subtract.*
   a. $\frac{2}{11} - \left(-\frac{3}{11}\right)$
   b. $\frac{2}{11} - \left(-\frac{4}{11}\right)$
   c. $\frac{2}{11} - \left(-\frac{5}{11}\right)$
   d. $\frac{2}{11} - \left(-\frac{6}{11}\right)$
   e. $\frac{2}{11} - \left(-\frac{7}{11}\right)$

37. ⟨9.3D⟩ *Subtract.*
   a. $-\frac{5}{6} - \frac{4}{3}$
   b. $-\frac{5}{6} - \frac{5}{3}$
   c. $-\frac{5}{6} - \frac{7}{3}$
   d. $-\frac{5}{6} - \frac{8}{3}$
   e. $-\frac{5}{6} - \frac{2}{3}$

38. ⟨9.3E⟩ *Multiply.*
   a. $-3.1 \cdot 4.2$
   b. $-3.1 \cdot 4.3$
   c. $-3.1 \cdot 4.4$
   d. $-3.1 \cdot 4.5$
   e. $-3.1 \cdot 4.6$

39. ⟨9.3E⟩ *Multiply.*
   a. $-3.1 \cdot (-2.1)$
   b. $-3.1 \cdot (-2.2)$
   c. $-3.1 \cdot (-2.3)$
   d. $-3.1 \cdot (-2.4)$
   e. $-3.1 \cdot (-2.5)$

40. ⟨**9.3E**⟩ *Multiply.*

a. $-\frac{2}{3}\cdot\left(-\frac{2}{3}\right)$

b. $-\frac{2}{3}\cdot\left(-\frac{4}{3}\right)$

c. $-\frac{2}{3}\cdot\left(-\frac{5}{3}\right)$

d. $-\frac{2}{3}\cdot\left(-\frac{7}{3}\right)$

e. $-\frac{2}{3}\cdot\left(-\frac{8}{3}\right)$

41. ⟨**9.3E**⟩ *Multiply.*

a. $\frac{5}{2}\cdot\left(-\frac{2}{7}\right)$

b. $\frac{5}{2}\cdot\left(-\frac{3}{7}\right)$

c. $\frac{5}{2}\cdot\left(-\frac{4}{7}\right)$

d. $\frac{5}{2}\cdot\left(-\frac{6}{7}\right)$

e. $\frac{5}{2}\cdot\left(-\frac{8}{7}\right)$

42. ⟨**9.3F**⟩ *Find the reciprocal.*

a. $-\frac{2}{11}$

b. $-\frac{3}{11}$

c. $-\frac{4}{11}$

d. $-\frac{5}{11}$

e. $-\frac{6}{11}$

43. ⟨**9.3F**⟩ *Divide.*

a. $\frac{1}{5}\div\left(-\frac{1}{7}\right)$

b. $\frac{1}{5}\div\left(-\frac{2}{7}\right)$

c. $\frac{1}{5}\div\left(-\frac{3}{7}\right)$

d. $\frac{1}{5}\div\left(-\frac{4}{7}\right)$

e. $\frac{1}{5}\div\left(-\frac{5}{7}\right)$

44. ⟨**9.3F**⟩ *Divide.*

a. $-\frac{5}{2}\div\left(-\frac{1}{4}\right)$

b. $-\frac{5}{2}\div\left(-\frac{1}{6}\right)$

c. $-\frac{5}{2}\div\left(-\frac{1}{8}\right)$

d. $-\frac{5}{2}\div\left(-\frac{1}{10}\right)$

e. $-\frac{5}{2}\div\left(-\frac{1}{12}\right)$

45. ⟨**9.3F**⟩ *Divide.*

a. $-\frac{2}{11}\div\frac{3}{11}$

b. $-\frac{2}{11}\div\frac{4}{11}$

c. $-\frac{2}{11}\div\frac{5}{11}$

d. $-\frac{2}{11}\div\frac{6}{11}$

e. $-\frac{2}{11}\div\frac{7}{11}$

46. ⟨**9.3F**⟩ *Divide.*

a. $\frac{-2.2}{1.1}$

b. $\frac{-3.3}{1.1}$

c. $\frac{-4.4}{1.1}$

d. $\frac{-5.5}{1.1}$

e. $\frac{-6.6}{1.1}$

47. ⟨**9.3F**⟩ *Divide.*

a. $\frac{-1.1}{-2.2}$

b. $\frac{-1.1}{-3.3}$

c. $\frac{-1.1}{-4.4}$

d. $\frac{-1.1}{-5.5}$

e. $\frac{-1.1}{-6.6}$

48. ⟨**9.3F**⟩ *Divide.*

a. $\frac{2.2}{-1.1}$

b. $\frac{3.3}{-1.1}$

c. $\frac{4.4}{-1.1}$

d. $\frac{5.5}{-1.1}$

e. $\frac{6.6}{-1.1}$

49. ⟨**9.3G**⟩ *Classify the given numbers by placing a check mark in the appropriate row(s).*

	a.	b.	c.	d.	e.
	3.7	$\sqrt{121}$	$0.\overline{56}$	$-3\frac{1}{5}$	$\sqrt{21}$
Natural number					
Whole number					
Integer					
Rational number					
Irrational number					
Real number					

50. ⟨**9.4A**⟩ *Find the value of:*

a. $-\frac{7}{2}\cdot 2-5$

b. $-\frac{7}{3}\cdot 3-6$

c. $-\frac{7}{4}\cdot 4-7$

d. $-\frac{7}{5}\cdot 5-8$

e. $-\frac{7}{6}\cdot 6-9$

51. ⟨**9.4A**⟩ *Find the value of:*

a. $-30\div\frac{3}{4}-(3+4)$

b. $-30\div\frac{3}{5}-(3+4)$

c. $-30\div\frac{3}{7}-(3+4)$

d. $-30\div\frac{3}{8}-(3+4)$

e. $-30\div\frac{3}{9}-(3+4)$

52. ⟨**9.4A**⟩ *Find the value of:*

a. $-64\div 4\cdot 2+3(5-3)-4\cdot\frac{3}{4}$

b. $-64\div 8\cdot 2+3(5-3)-8\cdot\frac{3}{8}$

c. $-64\div 16\cdot 2+3(5-3)-16\cdot\frac{3}{16}$

d. $-64\div 32\cdot 2+3(5-3)-32\cdot\frac{3}{32}$

e. $-64\div 64\cdot 2+3(5-3)-64\cdot\frac{3}{64}$

53. ⟨**9.4B**⟩ *Find the value of:*

a. $-6^2+\frac{3(4-8)}{2}+10\div 5$

b. $-7^2+\frac{3(4-8)}{2}+20\div 5$

c. $-7^2+\frac{3(4-8)}{2}+30\div 5$

d. $-8^2+\frac{3(4-8)}{2}+40\div 5$

e. $-9^2+\frac{3(4-8)}{2}+50\div 5$

Practice Test Chapter 9

(Answers on page 571)

Visit www.mhhe.com/bello to view helpful videos that provide step-by-step solutions to several of the problems below.

Consider the set $\{-4, -3, 0, 3, 4\}$.

1. **a.** List the negative numbers in the set.
b. List the positive numbers in the set.
c. List the integers in the set.

2. Describe how the integers are used in the given situation.

Average High Temperature MINNEAPOLIS-ST. PAUL

Jan.	Feb.	Mar.	Apr.	May
-6	-3	4	13	20

Source: National Weather Service, San Francisco. Data collected through 1993.

3. Write the integer that represents the given situation.
a. A scuba diver is 30 feet below the surface.
b. A bird is flying at 500 feet.

4. Graph the integers -4, 0, and 3 on the number line.

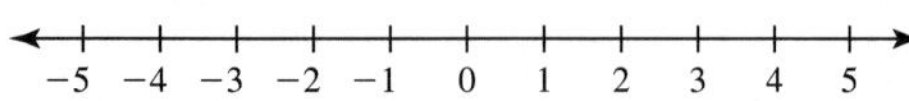

5. Fill in the blank with < or > to make the statement true.
a. -4 ________ -2 **b.** -2 ________ -5

Find the additive inverse.

6. **a.** $\frac{9}{2}$ **b.** -3.8

7. **a.** $-4\frac{1}{4}$ **b.** 9.2

Find each value.

8. **a.** $\left|-\frac{4}{9}\right|$ **b.** $|3.7|$

9. **a.** $\left|-5\frac{1}{2}\right|$ **b.** $|-9.2|$

Add.

10. **a.** $-8.9 + 4.7 =$ ____ **b.** $7.3 + (-9.9) =$ ____

11. $-3.5 + (-5.1) =$ ________

12. **a.** $-\frac{2}{9} + \frac{8}{9} =$ ______ **b.** $\frac{2}{5} + \left(-\frac{5}{8}\right) =$ ______

Subtract.

13. **a.** $-5.2 - (-4.1) =$ ________
b. $-1.5 - (-7.8) =$ ________

14. **a.** $\frac{2}{9} - \left(-\frac{1}{9}\right) =$ ____ **b.** $-\frac{5}{6} - \frac{7}{4} =$ ____

Multiply.

15. **a.** $-3.1 \cdot 4.4$ **b.** $-1.2 \cdot (-2.3)$

16. **a.** $-\frac{3}{5} \cdot \frac{9}{2}$ **b.** $\frac{5}{7} \cdot \left(-\frac{5}{8}\right)$

Find the reciprocal.

17. $-\frac{9}{4}$

Divide.

18. **a.** $\frac{2}{5} \div \left(-\frac{3}{4}\right)$ **b.** $-\frac{5}{6} \div \left(-\frac{7}{2}\right)$

19. **a.** $-\frac{3}{11} \div \frac{5}{11}$ **b.** $\frac{-3.2}{1.6}$

20. **a.** $\frac{-3.2}{-19.2}$ **b.** $\frac{4.2}{-1.4}$

Find each value.

21. **a.** $(-7)^2$ **b.** -7^2

Find the value of:

23. $-90 \div \frac{3}{4} - (6 + 7)$

24. $-32 \div 4 \cdot 2 + 5(5 - 3) - 7 \cdot \frac{3}{7}$

25. $-3^2 + \frac{5(4 - 8)}{2} + 54 \div 6$

22. Classify the given numbers by placing a check mark in the appropriate row(s).

	a.	b.	c.	d.	e.
	7.9	$\sqrt{169}$	$0.\overline{34}$	$-9\frac{1}{5}$	$\sqrt{23}$
Natural number					
Whole number					
Integer					
Rational number					
Irrational number					
Real number					

Answers to Practice Test Chapter 9

Answer		If You Missed	Review		
		Question	Section	Examples	Page
1. a. $-4, -3$ c. $-4, -3, 0, 3, 4$	b. $3, 4$	1	9.1	1	523
2. Listing the average high temperature in Minneapolis-St. Paul.		2	9.1	2	524
3. a. -30 ft	b. $+500$ ft	3	9.1	3	524
4. $-5 \quad -4 \quad -3 \quad -2 \quad -1 \quad 0 \quad 1 \quad 2 \quad 3 \quad 4 \quad 5$		4	9.1	4	525
5. a. $<$	b. $>$	5	9.1	5	525
6. a. $-\frac{9}{2}$	b. 3.8	6	9.1, 9.3	6, 1	526, 547
7. a. $4\frac{1}{4}$	b. -9.2	7	9.1, 9.3	6, 1	526, 547
8. a. $\frac{4}{9}$	b. 3.7	8	9.1, 9.3	7, 2	527, 547
9. a. $5\frac{1}{2}$	b. 9.2	9	9.1, 9.3	7, 2	527, 547
10. a. -4.2	b. -2.6	10	9.1, 9.3	11, 3	529–530, 548
11. -8.6		11	9.1, 9.3	11, 3	529–530, 548
12. a. $\frac{2}{3}$	b. $-\frac{9}{40}$	12	9.1, 9.3	11, 4	529–530, 548
13. a. -1.1	b. 6.3	13	9.1, 9.3	12, 5	531, 549
14. a. $\frac{1}{3}$	b. $-\frac{31}{12}$	14	9.1, 9.3	12, 5	531, 549
15. a. -13.64	b. 2.76	15	9.2, 9.3	1, 6	539, 549
16. a. $-\frac{27}{10}$	b. $-\frac{25}{56}$	16	9.2, 9.3	1, 6	539, 549
17. $-\frac{4}{9}$		17	9.3	7	550
18. a. $-\frac{8}{15}$	b. $\frac{5}{21}$	18	9.2, 9.3	2, 8	540, 550
19. a. $-\frac{3}{5}$	b. -2	19	9.2, 9.3	2, 9	540, 551
20. a. $\frac{1}{6}$	b. -3	20	9.2, 9.3	2, 9	540, 551
21. a. 49	b. -49	21	9.2	3	541
22. (see table below)		22	9.3	10	552
23. -133		23	9.4	1, 4	558, 559
24. -9		24	9.4	5	559
25. -10		25	9.4	7	560

22.

a. 7.9	b. $\sqrt{169}$	c. $0.\overline{34}$	d. $-9\frac{1}{5}$	e. $\sqrt{23}$
	✓			
	✓			
	✓			
✓	✓	✓	✓	
				✓
✓	✓	✓	✓	✓

Cumulative Review Chapters 1–9

1. Simplify: $9 \div 3 \cdot 3 + 5 - 4$

2. Round 449.851 to the nearest ten.

3. Divide: $90 \div 0.13$. (Round answer to two decimal places.)

4. Solve for y: $7.2 = 0.9y$

5. Solve for z: $6 = \frac{z}{6.6}$

6. Solve the proportion: $\frac{c}{4} = \frac{4}{64}$

7. Write 23% as a decimal.

8. 80% of 40 is what number?

9. What percent of 4 is 2?

10. 30 is 40% of what number?

11. Find the simple interest earned on $900 invested at 4.5% for 4 years.

12. Referring to the circle graph, which is the main source of pollution?

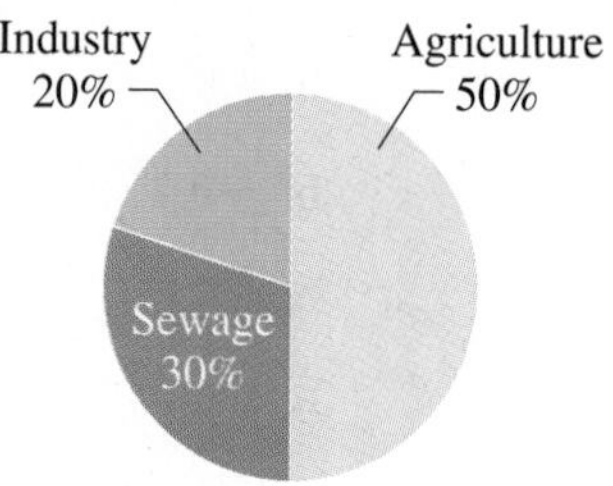

13. Make a circle graph for this data.

Family Budget	(Monthly)
Savings (S)	$200
Housing (H)	$900
Food (F)	$400
Clothing (C)	$500

14. The following table shows the distribution of families by income in Chicago, Illinois.

Income Level	Percent of Families
$0–9,999	3
10,000–14,999	6
15,000–19,999	20
20,000–24,999	45
25,000–34,999	12
35,000–49,999	6
50,000–79,999	4
80,000–119,999	3
120,000 and over	1

What percent of the families in Chicago have incomes between $50,000 and $79,999?

15. The following graph represents the monthly average temperature for 7 months of the year. How much higher is the average temperature in June than it is in May?

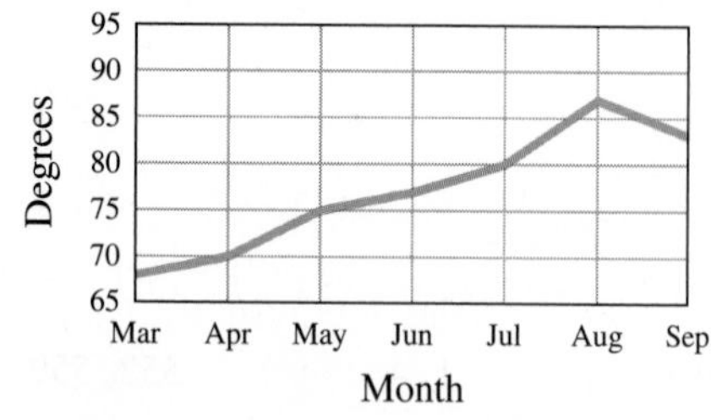

16. The number of hours required in each discipline of a college core curriculum is represented by the following circle graph. What percent of these hours is in art and English combined?

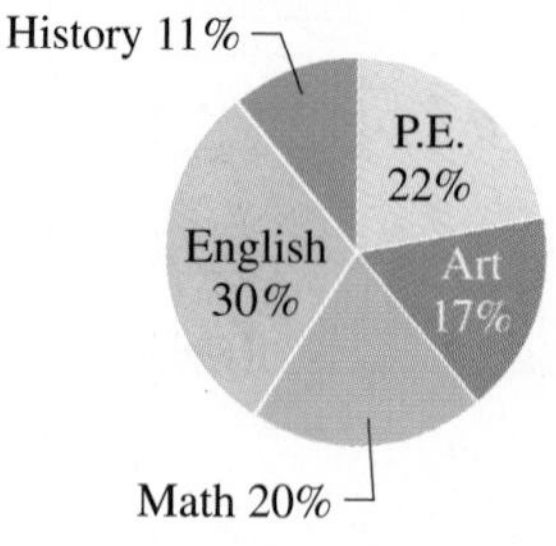

17. What is the mode for the following numbers?
8, 30, 6, 1, 26, 24, 6, 10, 6

18. What is the mean for the following numbers?
11, 21, 7, 21, 18, 21, 4, 27, 16, 21, 20

19. What is the median for the following numbers?
4, 2, 23, 19, 3, 2, 29, 10, 2, 2, 14

20. Convert 9 yards to inches.

21. Convert 16 inches to feet.

22. Convert 17 feet to yards.

23. Convert 6 feet to inches.

24. Convert 170 meters to decimeters.

25. Find the perimeter of a rectangle 4.4 inches long by 2.3 inches wide.

26. Find the circumference of a circle whose radius is 8 feet. Use 3.14 for π and round the answer to two decimal places.

27. Find the area of a triangular piece of cloth whose base is 16 centimeters and whose height is 20 centimeters.

28. Find the area of a circle with a radius of 6 centimeters. Use 3.14 for π and round the answer to two decimal places.

29. Find the area of the trapezoid.

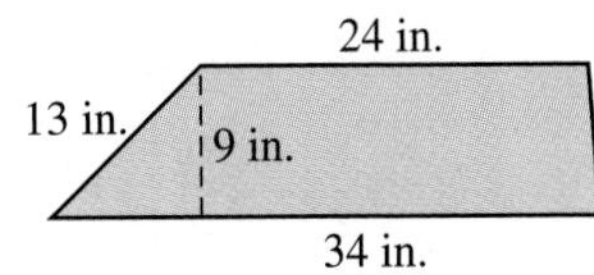

30. Find the volume of a box 5 centimeters wide, 4 centimeters long, and 10 centimeters high.

31. Find the volume of a sphere with a 9-inch radius. Use 3.14 for π and round the answer to two decimal places.

32. Classify the angle:

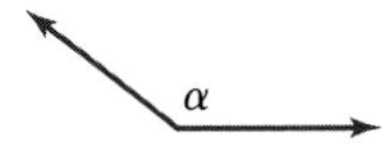

33. If in triangle ABC, $m\ \angle A = 30°$ and $m\ \angle B = 25°$, what is $m\ \angle C$?

34. Find $\sqrt{4900}$.

35. Find the hypotenuse c of the given triangle:

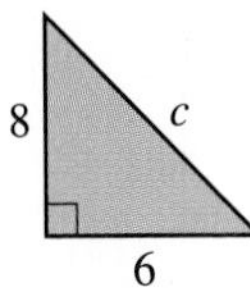

36. Find the additive inverse of $-4\frac{3}{4}$.

37. Find: $|-9.8|$

38. Add: $-7.2 + (-2.6)$

39. Add: $\frac{1}{5} + \left(-\frac{1}{7}\right)$

40. Subtract: $-4.2 - (-8.5)$

41. Subtract: $-\frac{1}{5} - \frac{1}{6}$

42. Multiply: $-\frac{3}{8} \cdot \frac{8}{9}$

43. Divide: $-\frac{9}{8} \div \frac{5}{8}$

44. Evaluate: $(-9)^2$

Section

Chapter

10 ten

Introduction to Algebra

The Human Side of Mathematics

What is algebra and where did it come from? The word *algebra* is the European derivation of *al-jabr,* part of the title of a book written by Abu Ja'far Muhammad ibn Musa al-Khowarizmi, born in Baghdad probably in 820. Some historians suggest that al-Khowarizmi means Mohammed the son of Moses of Khowarismi and was really born in the city of Khwarizm, south of the Aral Sea in central Asia. What is indeed certain is that he wrote a book *Hisab al-jabr w'al-muqabala,* "The Science of Reunion and Reduction" or "Restoring and Simplification," from which the name "algebra" is derived. What about the "reunion" and the "reduction"? The reduction is carried out using two operations: *al-jabr* and *al-muqabala,* where *al-jabr* means "completion": the process of removing negative terms from an equation. Using one of al-Khwarizmi's examples, *al-jabr* changes $x^2 = 40x - 4x^2$ into $5x^2 = 40x$. (We do this now by simply *adding* $4x^2$ to both sides of the equation.) The term *al-muqabala* means "balancing." Thus, *al-muqabala* balances

$$50 + 3x + x^2 = 29 + 10x \quad (1)$$

to

$$21 + x^2 = 7x \quad (2)$$

In modern terms, we simply say that we *subtract* 29 and *subtract* $3x$ from both sides of equation (1) to obtain equation (2).

10.1 Introduction to Algebra

Objectives

You should be able to:

A › Identify the terms in an expression.

B › Use the distributive property to simplify an expression.

C › Use the distributive property to factor an expression.

D › Combine like terms in an expression.

E › Remove parentheses in an expression.

F › Evaluate expressions.

To Succeed, Review How To . . .

Add, subtract, and multiply rational numbers. (pp. 548, 549)

Getting Started

The poster uses the language of algebra to tell you how to be successful! The letters X, Y, and Z are used as *placeholders*. In algebra, we use the letters of the alphabet as **placeholders, unknowns,** or **variables.** By tradition, the letters t, u, v, w, x, y, and z are frequently used for unknowns. Of course, we use the same symbols as in arithmetic to denote the operations of addition (+) and subtraction (−). In algebra, multiplication is indicated by using the raised dot, $\cdot$, using parentheses, ( ), or simply writing the variables next to each other. Thus, the *product* of a and b can be written as:

$$a \cdot b, \quad (a)(b), \quad a(b), \quad (a)b, \quad \text{or} \quad ab$$

A › Expressions and Terms

How do we express mathematical ideas in the language of algebra? By using expressions, of course! As stated in the *Getting Started,* in algebra we use letters of the alphabet as placeholders (variables) for unknown value(s). An **expression** is a collection of numbers and letters connected by operation signs. Thus, xy^2, $x + y$, and $3x^2 - 2y + 9$ are expressions. In these expressions, the parts that are to be added or subtracted are called **terms.**

1. The expression xy^2 has *one* term (it is a monomial), xy^2.
2. The expression $x + y$ has *two* terms (a binomial), x and y.
3. The expression $3x^2 - 2y + 9$ has *three* terms (a trinomial), $3x^2$, $-2y$, and $+9$ (or simply 9, without the + sign).

EXAMPLE 1 Identifying the terms of an expression

What are the terms in $5y^3 + 2y - 3$?

SOLUTION The terms are $5y^3$, $2y$, and -3.

PROBLEM 1

What are the terms in $8z^2 - 5z + 2$?

Answers to PROBLEMS

1. $8z^2$, $-5z$, and 2

If the division involves rational numbers written as decimals, the rules of signs are, of course, unchanged. Just remember that it is easier to divide by a whole number than by a decimal. Thus to find $\frac{-3.6}{1.8}$, multiply the numerator and denominator by $10(1.8 \times 10 = 18$, a whole number), obtaining:

$$\frac{-3.6}{1.8} = \frac{-3.6 \times 10}{1.8 \times 10} = \frac{-36}{18} = -2$$

EXAMPLE 9 Dividing rational numbers

Divide:

a. $\frac{-4.2}{2.1}$ **b.** $\frac{-8.1}{-16.2}$ **c.** $\frac{9.6}{-3.2}$

SOLUTION

a. $\frac{-4.2}{2.1} = \frac{-4.2 \cdot 10}{2.1 \cdot 10} = \frac{-42}{21} = -2$

b. $\frac{-8.1}{-16.2} = \frac{-8.1 \cdot 10}{-16.2 \cdot 10} = \frac{-81}{-162} = \frac{1}{2}$

c. $\frac{9.6}{-3.2} = \frac{9.6 \cdot 10}{-3.2 \cdot 10} = \frac{96}{-32} = -3$

PROBLEM 9

Divide:

a. $\frac{-3.6}{1.2}$ **b.** $\frac{-3.1}{-12.4}$

c. $\frac{6.5}{-1.3}$

G > Classifying Real Numbers

Rational numbers have two very important properties:

1. They can be expressed as a fraction or ratio (rational) of the form $\frac{a}{b}$, where a and b are integers and $b \neq 0$.
2. When the numerator a is divided by the denominator b, the rational number $\frac{a}{b}$ becomes a terminating or a repeating decimal.

For example, $\frac{1}{2} = 0.5$ and $\frac{2}{5} = 0.4$ are terminating decimals, but $\frac{1}{3} = 0.\overline{3}$ and $\frac{1}{6} = 0.1\overline{6}$ are repeating decimals.

It can be shown that only decimals that terminate or repeat can be written in the form $\frac{a}{b}$. A decimal number that is not a rational number is nonterminating and nonrepeating. It is an **irrational** number.

IRRATIONAL NUMBERS

An irrational number cannot be written in the form $\frac{a}{b}$ and will never terminate or repeat when written as a decimal.

Here are some irrational numbers:

$$0.12345\ldots$$
$$0.101001000\ldots$$
$$\sqrt{2} = 1.414213562\ldots$$
$$\pi = 3.141592653\ldots$$

What about $\sqrt{9}$? Here you must remember that $\sqrt{9} = 3$, which is a natural number (not irrational). Before classifying real numbers, let us look at the relationship among numbers we have studied.

Real numbers

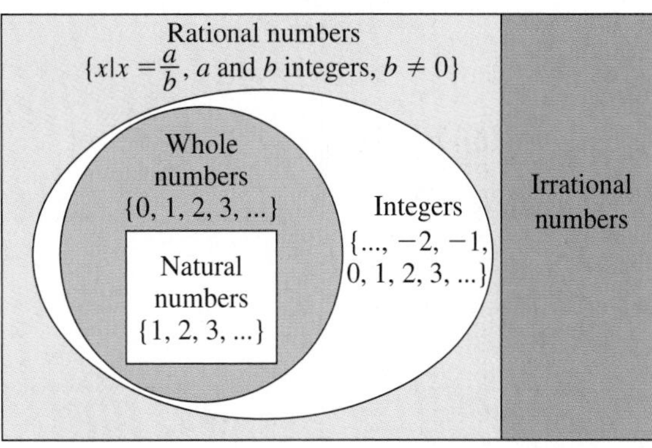

Answers to PROBLEMS

9. a. -3 **b.** $\frac{1}{4}$ **c.** -5

Note that, in particular, the real numbers are composed of all the rational and all the irrational numbers. To see the relationship among these numbers, see Problems 69 to 72.

For now, note that:

All natural numbers are whole numbers.
Example: 3 is a natural number *and* a whole number.

All whole numbers are integers.
Example: 0 and 5 are whole numbers so 0 and 5 are integers.

All integers are rational numbers.
Example: -3 and 5 are integers so -3 and 5 are rational numbers.

All rational numbers are real numbers.
Example: $-\frac{5}{6}$ and 0.3 are rational numbers, so $-\frac{5}{6}$ and 0.3 are real numbers.

All irrational numbers are real numbers.
Example: $\sqrt{2}$ is an irrational number, so $\sqrt{2}$ is a real number.

EXAMPLE 10 Classifying numbers

Classify each number by making a check mark in the appropriate row.

	a.	b.	c.	d.	e.	f.	g.	h.
	5.7	$-\frac{7}{9}$	0	$\sqrt{8}$	3	$\sqrt{25}$	$0.\overline{66}$	$-1\frac{1}{4}$
Natural number					✓	✓		
Whole number			✓		✓	✓		
Integer			✓		✓	✓		
Rational number	✓	✓	✓		✓	✓	✓	✓
Irrational number				✓				
Real number	✓	✓	✓	✓	✓	✓	✓	✓

SOLUTION

a. 5.7 is a terminating decimal, so it is rational and real.

b. $-\frac{7}{9}$ is of the form $\frac{a}{b}$, so it is rational and real.

c. 0 is a whole number and thus an integer, a rational, and real number.

d. $\sqrt{8}$ is nonterminating and nonrepeating, so it is irrational and real.

e. 3 is a natural number and thus a whole number, an integer, a rational number, and a real number.

f. Do not be fooled! $\sqrt{25} = 5$, so it is a natural number, a whole number, an integer, a rational number, and a real number.

g. $0.\overline{66}$ is a repeating decimal, so it is rational and real.

h. $-1\frac{1}{4}$ can be written as $-\frac{5}{4}$, so it is rational and real.

PROBLEM 10

Classify each number by making a check mark in the appropriate row.

	a.	b.	c.	d.	e.	f.	g.	h.
	$-\frac{2}{7}$	3.4	8	$\sqrt{81}$	$-2\frac{3}{4}$	$\sqrt{2}$	$0.\overline{36}$	0
N								
W								
I								
Rat.								
Irr.								
Real								

Answers to PROBLEMS

10.

	a.	b.	c.	d.	e.	f.	g.	h.
	$-\frac{2}{7}$	3.4	8	$\sqrt{81}$	$-2\frac{3}{4}$	$\sqrt{2}$	$0.\overline{36}$	0
N			✓	✓				
W			✓	✓				✓
I			✓	✓				✓
Rat.	✓	✓	✓	✓	✓		✓	✓
Irr.						✓		
Real	✓	✓	✓	✓	✓	✓	✓	✓

B › The Distributive Property

If you wish to multiply a real number by a *sum,* you can either add and then multiply or multiply and then add. For example, suppose you wish to multiply a number, say 7, by the sum of 4 and 5. The product $7 \cdot (4 + 5)$ can be obtained in two ways:

$$7 \cdot (4+5) \quad 7 \cdot 9 \quad 63$$ Adding within the parentheses first

$$(7 \cdot 4) + (7 \cdot 5) \quad 28 + 35 \quad 63$$ Multiplying and then adding

Thus,

$$7 \cdot (4 + 5) = (7 \cdot 4) + (7 \cdot 5)$$

The parentheses in $(7 \cdot 4) + (7 \cdot 5)$ can be omitted as long as we agree that *multiplications must be done first.* The fact that $a(b + c) = ab + ac$ is called the **distributive property of multiplication over addition.** There is also a distributive property of multiplication over subtraction. When we multiply a real number by a *difference,* we can either subtract and then multiply or multiply and then subtract. These two properties are stated next.

DISTRIBUTIVE PROPERTY FOR MULTIPLICATION

For any real numbers *a*, *b*, and *c*,

$$a(b + c) = ab + ac \quad \text{and} \quad a(b - c) = ab - ac$$

EXAMPLE 2 Using the distributive property

Multiply.

a. $2(x + 2y)$ **b.** $5(3x - 4y)$ **c.** $-3(x - 2y + z)$

SOLUTION

a. $2(x + 2y) = 2 \cdot x + 2 \cdot 2y = 2x + 4y$

b. $5(3x - 4y) = 5 \cdot 3x - 5 \cdot 4y = 15x - 20y$

c. $-3(x - 2y + z) = -3 \cdot x - (-3) \cdot (2y) - 3 \cdot z = -3x + 6y - 3z$

PROBLEM 2

Multiply.

a. $3(a + 5b)$

b. $6(4a - 5b)$

c. $-2(a - 3b + c)$

C › Factoring Out the Greatest Common Factor

Do you remember how to factor 15? To *factor* 15, we write it as the product $5 \cdot 3$. Thus, factoring is the reverse of multiplying.

FACTOR

To factor an expression is to find an equivalent expression that is a product.

Now, look at Example 2(a). To factor $2x + 4y$, we would write it as the product $2(x + 2y)$. Similarly, to factor $15x - 20y$ we write it as $5(3x - 4y)$. Note that in each case, we removed the **largest** (greatest) common factor (GCF for short).

EXAMPLE 3 Factoring out the GCF

Factor.

a. $3x - 6y$ **b.** $ax + ay - az$ **c.** $10x - 20y - 30z$

PROBLEM 3

Factor.

a. $5a - 15b$

b. $ab + ac - ad$

c. $6a - 12b + 18c$

(continued)

Answers to PROBLEMS

2. a. $3a + 15b$ **b.** $24a - 30b$ **c.** $-2a + 6b - 2c$ **3. a.** $5(a - 3b)$ **b.** $a(b + c - d)$ **c.** $6(a - 2b + 3c)$

SOLUTION

a. $3x - 6y = 3 \cdot x - 3 \cdot 2y = 3(x - 2y)$
b. $ax + ay - az = a \cdot x + a \cdot y - a \cdot z = a(x + y - z)$
c. $10x - 20y - 30z = 10 \cdot x - 10 \cdot 2y - 10 \cdot 3z = 10(x - 2y - 3z)$

Note that $5(2x - 4y - 6z)$ is also a factored form for $10x - 20y - 30z$ but $5(2x - 4y - 6z)$ is not **completely factored.** When we say *factor,* we must factor out the *greatest* common factor. Moreover, $-5x + 30$ is factored as $-5(x - 6)$ **not** $5(-x + 6)$.

D › Combining Like Terms*

One of the fundamental operations in algebra is **combining like terms.** Combining like terms is really simple; you only have to be sure that the variable parts of the terms to be combined are **identical** and then add (or subtract) the numerical factors involved. Thus,

$$2a + 3a = (2 + 3)a = 5a$$

and

$$2b + 3b = (2 + 3)b = 5b$$

Similarly,

$$5a - 2a = (5 - 2)a = 3a$$

and

$$5b - 2b = (5 - 2)b = 3b$$

However, $2a^2 + 3a$ and $5b^2 - 2b$ *cannot* be simplified any further, since the terms in $2a^2 + 3a$ and $5b^2 - 2b$ are **not** like terms.

EXAMPLE 4 **Combine like terms (simplify)**

Simplify.

a. $7x - 2x$

b. $3x + 5y - x + 4y$

c. $0.2x + 0.31y - 0.6x + 0.23y$

d. $\frac{1}{7}x + \frac{3}{5}y + \frac{4}{7}x - \frac{2}{5}y$

SOLUTION

a. $7x - 2x = (7 - 2)x = 5x$

b. $3x + 5 \quad - x + 4 \quad = 3x - x + 5 \quad + 4$ Since $5y - x = -x + 5y$
$= (3 - 1)x + (5 + 4) \quad = 2x + 9$

c. $0.2x + 0.31 \quad - 0.6x + 0.23 \quad = 0.2x - 0.6x + 0.31 \quad + 0.23$
$= (0.2 - 0.6)x + (0.31 + 0.23)$
$= -0.4x + 0.54$

d. $\frac{1}{7}x + \frac{3}{5} \quad + \frac{4}{7}x - \frac{2}{5} \quad = \frac{1}{7}x + \frac{4}{7}x + \frac{3}{5} \quad - \frac{2}{5}$
$= \left(\frac{1}{7} + \frac{4}{7}\right)x + \left(\frac{3}{5} - \frac{2}{5}\right) \quad = \frac{5}{7}x + \frac{1}{5}$

PROBLEM 4

Simplify.

a. $6a - 2a$

b. $5a + 7b - a + 2b$

c. $0.3a + 0.21b - 0.8a + 0.32b$

d. $\frac{1}{5}a + \frac{4}{7}b + \frac{2}{5}a - \frac{1}{7}b$

E › Removing Parentheses

Sometimes it is necessary to remove parentheses before combining like terms. Thus, to combine like terms in

$$(4x + 2) + (5x + 1)$$

we have to remove the parentheses first. As long as the parentheses are preceded by a *plus* sign (or no sign at all), we can simply remove the parentheses. That is,

Answers to PROBLEMS

4. a. $4a$ **b.** $4a + 9b$
c. $-0.5a + 0.53b$ **d.** $\frac{3}{5}a + \frac{3}{7}b$

* For a more rigorous development, see the *Using Your Knowledge* section.

REMOVING PARENTHESES PRECEDED BY + OR NO SIGN

If a and b are real numbers, then

$$(a + b) = a + b$$

Thus,

$$(4x + 2) + (5x + 1) = 4x + 2 + 5x + 1$$

(like terms: $4x$ and $5x$; like terms: 2 and 1)

$$= 9x + 3$$

What about $-(a + b)$? Since $-1 \cdot a = -a$, then, by the distributive property, $-(a + b) = -1 \cdot (a + b) = -1 \cdot a + (-1) \cdot b = -a - b$. That is,

$$\boxed{-(a + b) = -a - b}$$

Thus,

$$(5a + 3b) - (4a + 7b) = 5a + 3b - 4a - 7b$$

$$= a - 4b$$

Note that when a minus sign precedes the parentheses, we change the sign of each of the terms inside using the property. You can think of this as multiplying each term by -1.

REMOVING PARENTHESES PRECEDED BY –

If a and b are real numbers, then

$$-(a - b) = -a + b$$

These two rules can be summarized as follows:

REMOVING PARENTHESES

1. If there is a *plus* sign (or no sign) in *front* of the parentheses, simply remove the parentheses.
2. If there is a *minus* sign in *front* of the parentheses, the parentheses can be removed by *changing the sign* of *each* of the terms *inside* the parentheses.

For example,

$$-(a + 7b) = -a - 7b$$
$$-(2a + 4b) = -2a - 4b$$
$$-(a - 3b) = -a + 3b$$
$$-(4a - 7b) = -4a + 7b$$

EXAMPLE 5 Removing parentheses

Simplify.

a. $(3a + 5b) - (a + 2b)$ **b.** $(5a + 2b) - (6a - 3b)$

SOLUTION

a. $(3a + 5b) - (a + 2b) = 3a + 5b - a - 2b$

$$= 2a + 3b$$

b. $(5a + 2b) - (6a - 3b) = 5a + 2b - 6a + 3b$

$$= -a + 5b$$

PROBLEM 5

Simplify.

a. $(4a + 5b) - (a + 3b)$

b. $(6a + 5b) - (8a - 2b)$

Answers to PROBLEMS

5. a. $3a + 2b$ **b.** $-2a + 7b$

EXAMPLE 6 **Removing parentheses**

Simplify.

a. $-(-a + 2b) - 5a$

b. $-(-a^2 - 3b) + a$

SOLUTION

a. $-(-a + 2b) - 5a = a - 2b - 5a$
$= -4a - 2b$

b. $-(-a^2 - 3b) + a = a^2 + 3b + a$

(*Note:* a^2 and a are not combined, since they *are not* like terms.)

PROBLEM 6

Simplify.

a. $-(-2x + 5y) - 6x$

b. $-(-2x^2 - 3y) + 3x$

F › Evaluating Expressions

Sometimes in algebra, we have to substitute a number for a variable in an expression. This process is called **evaluating** the expression. For example, the number of students in your class is $M + W$, where M is the number of men and W is the number of women in your class. If $M = 20$ and $W = 18$, then the number of students in your class is 20 + 18 or 38. To find the answer 38 we evaluated the expression $M + W$ using $M = 20$ and $W = 18$.

EXAMPLE 7 **Evaluating expressions**

Evaluate:

a. $x + y$ when $x = 3$ and $y = -7$

b. $4y + x$ when $x = 2$ and $y = 5$

c. $-(y + z)$ when $y = 7$ and $z = 4$

SOLUTION

a. Substituting $x = 3$ and $y = -7$, $x + y = 3 + (-7) = -4$.

b. Substituting $x = 2$ and $y = 5$, $4y + x = 4(5) + 2 = 22$.

c. Substituting $y = 7$ and $z = 4$, $-(y + z) = -(7 + 4) = -11$.

PROBLEM 7

Evaluate:

a. $a + b$ when $a = 4$ and $b = -9$

b. $5a + b$ when $a = 3$ and $b = 5$

c. $-(a + b)$ when $a = 6$ and $b = 3$

EXAMPLE 8 **Vehicle insurance for persons under 25**

According to the Bureau of Labor Statistics the amount of money I (in billions) spent annually on vehicle insurance by persons under 25 years of age is as shown in the graph in ***blue*** and can be approximated by

$$I = 22.2x + 390 \text{ (billion)}$$

where x is the year number (1, 2, 3, 4, or 5).

a. Find the amount of money (in billions) spent by persons under 25 years of age during year **1** by evaluating ***I*** when $x = 1$.

b. How much money is spent annually on vehicle insurance by persons under 25 years of age during year 5?

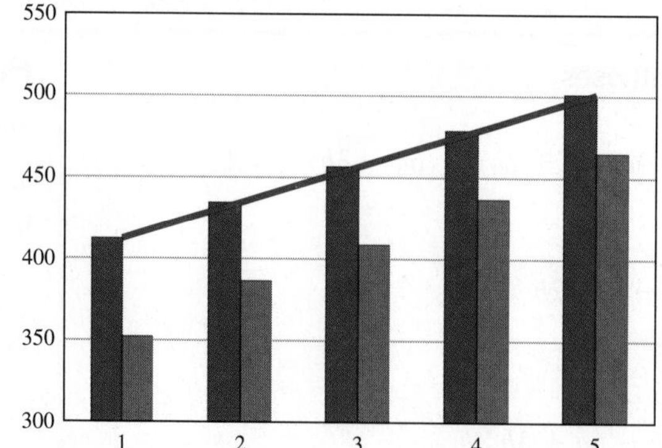

PROBLEM 8

The amount A spent annually on alcoholic beverages is in **red** and can be approximated by

$A = 28.2x + 324$ **(billion).**

a. Find the amount of money (in billions) spent on alcoholic beverages by persons under 25 years of age during year **1** by evaluating A when $x = 1$.

b. How much money is spent annually on alcoholic beverages by persons under 25 years of age during year 5?

Answers to PROBLEMS

6. a. $-4x - 5y$ **b.** $2x^2 + 3x + 3y$

7. a. -5 **b.** 20 **c.** -9

8. a. \$352.2 (billion)

b. \$465 (billion)

SOLUTION

a. When $x = 1$, $I = 22.2x + 390$ becomes

$$I = 22.2(1) + 390$$
$$= \$412.2 \text{ (billion)}$$

b. To find the amount spent during year **5**, substitute $x = 5$ in I.

$$I = 22.2(5) + 390 \text{ (billion)}$$
$$= 111 + 390 \text{ (billion)}$$
$$= \$501 \text{ (billion)}$$

Exercises 10.1

⟨A⟩ Expressions and Terms In Problems 1–6, list the terms in the expression.

1. $2x - 3$ **2.** $3y - 6$ **3.** $-3x^2 + 0.5x - 6$

4. $-5y^2 - 0.2y + 8$ **5.** $\frac{1}{5}x - \frac{3}{4}y + \frac{1}{8}z$ **6.** $\frac{3}{4}x^2 - \frac{1}{5}y + \frac{4}{9}z^3$

⟨B⟩ The Distributive Property In Problems 7–30, multiply.

7. $3(2x + y)$ **8.** $5(3x + y)$ **9.** $3(a - b)$

10. $4(a - b)$ **11.** $-5(2x - y)$ **12.** $-6(3x - y)$

13. $8(3x^2 + 2)$ **14.** $7(2x^2 + 5)$ **15.** $-6(2x^2 - 3)$

16. $-2(3x^2 - 4)$ **17.** $3(2x^2 + 3x + 5)$ **18.** $5(3x^2 + 2x + 2)$

19. $-5(3x^2 - 2x - 3)$ **20.** $-6(3x^2 - 5x - 4)$ **21.** $0.5(x + y - 2)$

22. $0.8(a + b - 6)$ **23.** $\frac{6}{5}(a - b + 5)$ **24.** $\frac{2}{3}(x - y + 4)$

25. $-2(x - y + 4)$ **26.** $-4(a - b + 8)$ **27.** $-0.3(x + y - 6)$

28. $-0.2(a + b - 3)$ **29.** $-\frac{5}{2}(a - 2b + c - 1)$ **30.** $-\frac{4}{7}(2a - b + 3c - 5)$

⟨C⟩ Factoring the Greatest Common Factor In Problems 31–40, factor.

31. $3x + 15$ **32.** $5x + 45$ **33.** $9y - 18$

34. $11y - 33$ **35.** $-5y + 20$ **36.** $-4y + 28$

37. $-3x - 27$ **38.** $-6x - 36$ **39.** $bx - by + bz$

40. $cx + cy - cz$

⟨D⟩ Combining Like Terms In Problems 41–60, combine like terms (simplify).

41. $8a + 2a$ **42.** $3b + 9b$ **43.** $4x + 2x$

44. $5y + 9y$ **45.** $8a^2 + 2a^2$ **46.** $8a^3 + 3a^3$

Web IT go to mhhe.com/bello for more lessons

47. $13x - 2x$

48. $5y - 5y$

49. $17y^2 - 12y^2$

50. $4z^3 - 3z^3$

51. $7x + 3y - 2x$

52. $8x - 3y - 4x$

53. $13x + 5 - 2y - 3x - 9 - y$

54. $8a - 2b + 4 - 3a - b - 7$

55. $3.9a + 4.5b - 3 - 1.5a - 7.5b - 4$

56. $-2.8a + 5 - 4.2b - 2 + 3.8a - 1.3b$

57. $\frac{1}{7}a - \frac{1}{5}b + \frac{3}{7}a - \frac{3}{5}b$

58. $\frac{3}{5}a - \frac{4}{11}b + \frac{1}{5}a - \frac{3}{11}b$

59. $-\frac{1}{8}x - \frac{3}{11}b + \frac{3}{8}x - \frac{2}{11}b$

60. $-\frac{3}{7}x - \frac{2}{9}a + \frac{5}{7}x - \frac{1}{9}a$

⟨ E ⟩ Removing Parentheses In Problems 61–70, simplify.

61. $-(-a + 5b) - (7a + 8b)$

62. $-(-a + 3b) - (9b + 2a)$

63. $-(-2a + b) - (4b + a)$

64. $-(-3b + a) - (5a + 2b)$

65. $-(-b^2 - 2a) + (3a - 4b^2)$

66. $-(-a^2 - 3b) + (4b - 3a^2)$

67. $-(-a^2 - 3b) + (2a^2 - 5b)$

68. $-(-b^2 - 3a) + (4b^2 - 7a)$

69. $-(-2b^2 - 3a) - (6b^2 - 5a)$

70. $-(-3a^2 - 5a) - (7a^2 - 8a)$

⟨ F ⟩ Evaluating Expressions In Problems 71–80, evaluate the expression.

71. $m + n$ when $m = 7$ and $n = -9$

72. $5r + s$ when $r = 4$ and $s = 5$

73. $-(u + v)$ when $u = 5$ and $v = 3$

74. $4y^2$ when $y = 5$

75. $3x^2y$ when $x = 3$ and $y = 5$

76. $3a - b$ when $a = -1$ and $b = 4$

77. $\frac{x}{3} + x$ when $x = 6$

78. $\frac{x - y}{3}$ when $x = 4$ and $y = -5$

79. LWH when $L = 2$, $W = 3.2$, and $H = 5$

80. $\frac{5}{9}(F - C)$ when $F = 212$ and $C = 32$

Books, maps, magazines, newspapers, and sheet music The table shows the amount of money spent annually (in billions) over a 5-year period on books and maps (column 2) and can be approximated by

$$B = 1.7x + 27.8 \text{ (in billions)}$$

The table also shows the amount spent on magazines, newspapers, and sheet music and can be approximated by

$$M = 0.52x + 32.3 \text{ (billion)}$$

where x is the year number.

	Books, Maps	Mag, News, Sheet Music
1	28.8	32.1
2	31.6	33.5
3	33.7	35.0
4	34.6	34.5
5	35.8	34.2

81. Evaluate B for $x = 1, 2, 3, 4$, and 5. How close are the results to the values given in the table?

82. Predict the amount of money B that will be spent on books and maps in 10 years.

83. Evaluate M for $x = 1, 2, 3, 4$, and 5. How close are the results to the values given in the table?

84. Predict the amount of money M that will be spent on magazines, newspapers, and sheet music in 10 years.

Source: U.S. Bureau of Economic Analysis, *Survey of Current Business*, January 2004; http://www.census.gov/prod/2004pubs/04statab/arts.pdf.

Vehicle insurance According to the Bureau of Labor Statistics the amount of money I spent annually (in billions) on vehicle insurance by persons between 25 and 34 years of age is as shown in the table and can be approximated by

$$I = 50.5x + 665.7 \text{ (billion)}$$

where x is the year number (1, 2, 3, 4, or 5).

	Ins	Alc Bev
1	705	365
2	774	431
3	822	393
4	872	395
5	910	446

85. Evaluate I for $x = 1, 2, 3, 4$, and 5. How close are the results to the values given in the table?

86. Predict the amount of money I that will be spent on vehicle insurance for persons between 25 and 34 years of age in 10 years.

Alcohol The annual amount A spent on alcoholic beverages by persons between 25 and 34 years of age can be approximated by

$$A = 12.6x + 368.2 \text{ (billion)}$$

87. Evaluate A for $x = 1, 2, 3, 4$, and 5. How close are the results to the values given in the table?

88. Predict the amount A that will be spent on alcoholic beverages by persons between 25 and 34 years of age in 10 years (see Exercise 87).

Total enrollment The graph shows the total enrollment of 18-year- to 24-year-old students in degree-granting institutions and can be approximated by

$$E = 0.15N + 7.77 \text{ (millions)}$$

where N represents the number of years after 1989.

89. Use the equation to find the projected number E of students enrolled in 2002 ($N = 13$). How close is your answer to the value 9.9 given in the graph?

90. Use the equation to find the projected number E of students enrolled in 2014 ($N = 25$). How close is your answer to the value 11.5 given in the graph?

91. The graph does not show a projected value for the year 2010. Use the equation to predict the enrollment for the year 2010 ($N = 21$).

Enrollment by Age of Student

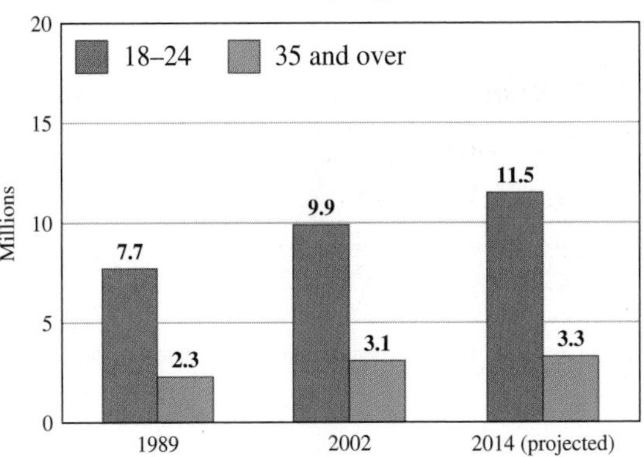

Source: U.S. Dept. of Education, NCES: Integrated Post-secondary Education Data System (IPEDS), "Fall Enrollment Survey"; http://nces.ed.gov/pubs2005/2005074_1.pdf.

Total enrollment The graph shows the total student enrollment (in millions) in degree-granting institutions and can be approximated by

$$E = 0.24N + 13.5$$

where N represents the number of years after 1989.

92. Use the equation to find the projected number E of students enrolled in 2002 ($N = 13$). How close is your answer to the value 16.7 given in the graph?

93. Use the equation to find the projected number E of students enrolled in 2014 ($N = 25$). How close is your answer to the value 19.5 given in the graph?

Enrollment in Millions

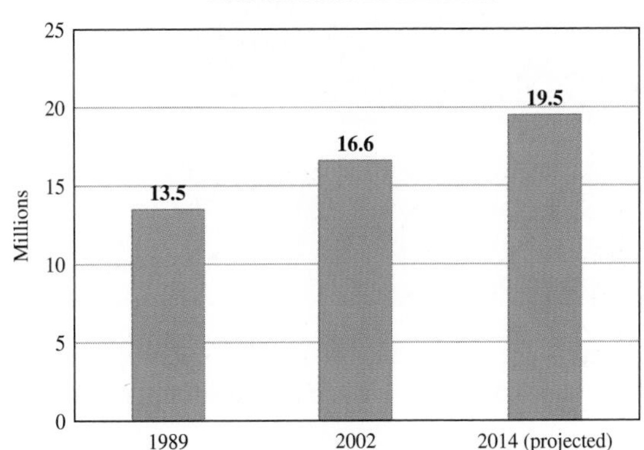

Using Your Knowledge

Properties of Numbers In this section, we will study properties of addition and multiplication that will be used to explain *why* we can simplify expressions. Some of these properties were mentioned in Chapter 1, and they also apply to the real numbers. The first of these properties is the commutative property, which we have mentioned previously and which states the following for any real numbers a and b.

COMMUTATIVE PROPERTY OF ADDITION

$$a + b = b + a$$

Changing the order of two addends does not change their sum.

This property is used to add two numbers without regard to the order in which the numbers are added. Thus,

$$3 + 7 = 7 + 3$$
$$8 + 4 = 4 + 8$$
$$101 + 17 = 17 + 101$$

Sometimes we have to add three or more numbers. For example, to add $4 + 8 + 2$ we can *first* add 4 and 8 and then add 2 to this result; that is,

$$\underbrace{(4 + 8)} + 2 = \underbrace{12} + 2 = 14$$

To find $4 + 8 + 2$ we could also add $8 + 2$ *first* and then add 4 to this result; that is,

$$4 + \underbrace{(8 + 2)} = 4 + \underbrace{10} = 14$$

Because the answers are the same, we see that it makes *no difference* which numbers we add first. For example,

$$3 + (4 + 6) = (3 + 4) + 6$$

and

$$7 + (3 + 6) = (7 + 3) + 6$$

Thus, for any three real numbers a, b, and c, we have the following property.

ASSOCIATIVE PROPERTY OF ADDITION

$$a + (b + c) = (a + b) + c$$

The grouping of numbers (addends) does not change their sum.

Here is how we use these two properties to simplify expressions such as $(3 + a) + 5$.

$(3 + a) + 5 = (a + 3) + 5$ By the commutative property of addition
$= a + (3 + 5)$ By the associative property of addition
$= a + 8$

We also have commutative and associative properties for multiplication. These properties are stated next.

COMMUTATIVE PROPERTY OF MULTIPLICATION

$$a \cdot b = b \cdot a$$

Changing the *order* of two factors does not change their product.

ASSOCIATIVE PROPERTY OF MULTIPLICATION

$$a(bc) = (ab)c$$

Changing the *grouping* of two factors does not change the product.

Thus, $4 \cdot 5 = 5 \cdot 4$ and $3 \cdot 7 = 7 \cdot 3$. Remember, to indicate the product of 4 and 5, we write $4 \cdot 5$. To indicate the product of a and b, we simply write ab. To simplify $3(4a)$ we proceed like this:

$3 \cdot (4a) = (3 \cdot 4)a$ By the associative property of multiplication
$= 12a$

Similarly, to simplify $(4a) \cdot 5$ we write

$(4a) \cdot 5 = 5 \cdot (4a)$ By the commutative property of multiplication
$= (5 \cdot 4)a$ By the associative property of multiplication
$= 20a$

How do we simplify $3a + 5a$? We need the distributive property.

DISTRIBUTIVE PROPERTY OF MULTIPLICATION

$$a(b + c) = ab + ac$$

When multiplying a number by a sum, multiply each addend by the number and then add.

By the commutative property of multiplication, we can rewrite $ab + ac$ as shown next.

DISTRIBUTIVE PROPERTY (FOR FACTORING)

$$ba + ca = (b + c)a$$

Factor the *greatest common factor* out from each term in the expression.

Now,

$$\begin{aligned} 3a + 5a &= (3 + 5)a && \text{By the distributive property for factoring} \\ &= 8a \end{aligned}$$

Similarly, to simplify $5x + (3x + 2)$, we write

$$\begin{aligned} 5x + (3x + 2) &= (5x + 3x) + 2 && \text{By the associative property of addition} \\ &= (5 + 3)x + 2 && \text{By the distributive property for factoring} \\ &= 8x + 2 \end{aligned}$$

To simplify expressions involving a negative sign in front of the parentheses, we need the following property:

MULTIPLICATIVE PROPERTY OF −1

$$(-1)a = a(-1) = -a$$

The product of any number and −1 is the opposite of the number.

Thus

$$\begin{aligned} -(x + 3) &= -1(x + 3) = -1 \cdot x + (-1) \cdot 3 && \text{By the distributive property for multiplication} \\ &= -x - 3 \end{aligned}$$

We are now ready to simplify $10 - (2x + 5)$.

$$\begin{aligned} 10 - (2x + 5) &= 10 - 1 \cdot (2x + 5) && \text{By the multiplicative property of } -1 \\ &= 10 - 1 \cdot 2x - 1 \cdot 5 && \text{By the distributive property for multiplication} \\ &= 10 - 2x - 5 && \text{Multiplying} \\ &= 5 - 2x \end{aligned}$$

Now you know the reasons why we can simplify expressions.

> > > Write On

94. Write in your own words the difference between a factor and a term.

95. Write in your own words the procedure you use to combine like terms.

96. Write in your own words the procedure you use to remove parentheses when no sign or a plus sign precedes an expression within parentheses.

97. Write in your own words the procedure you use to remove parentheses when a minus sign precedes an expression within parentheses.

Concept Checker

Fill in the blank(s) with the correct word(s), phrase, or mathematical statement.

98. An __________ is a **collection of numbers and letters** connected by operation signs.

99. The **parts that are to be added or subtracted** in an expression are called the __________ of the expression.

100. An **expression** with **two terms** is called a __________.

101. An **expression** with **three** terms is called a __________.

102. According to the distributive property, $a(b + c) =$ __________.

103. $-(a - b) =$ __________.

104. $-(a + b) =$ __________.

105. If we **substitute a number** for a **variable** in an expression, we have __________ the expression.

evaluated	**variables**
$ab + ac$	**terms**
binomial	**monomial**
trinomial	$-a + b$
$a - b$	$a - c$
sentence	$-a - b$
expression	**multiplied**

Mastery Test

106. Evaluate the expressions when $x = 4$ and $y = 7$.

a. $x + y$

b. $5y + x$

c. $-(x + y)$

107. Simplify.

a. $(4x + 3y) - (x + 2y)$

b. $(5x + 3y) - (7x - 2y)$

108. Simplify.

a. $-(-2x + 3y) - 6x$

b. $-(-x^2 - 5y) + x$

109. Combine like terms (simplify).

a. $7a - 3a$

b. $5a + 3b - a + 3b$

c. $\frac{2}{9}a + \frac{5}{7}b + \frac{3}{9}a - \frac{4}{7}b$

110. Factor.

a. $4a - 8b$

b. $bx + by - bz$

c. $5a - 10b - 15c$

111. Multiply.

a. $3(a + 2b)$

b. $5(2a - 3b)$

c. $-4(a - 3b + c)$

112. What are the terms in $-5x^3 + 2x - 3$?

Skill Checker

Write as a product of factors without exponents and multiply.

113. $3^2 \cdot 5^2 \cdot 7^0$

114. $2^3 \cdot 5^2 \cdot 7^1$

115. $10^3 \cdot 3^0 \cdot 5^2 \cdot 8^1$

10.2 The Algebra of Exponents

Objectives

You should be able to:

A › Write a number with a negative exponent as a fraction and vice versa.

B › Multiply expressions involving exponents.

C › Divide expressions involving exponents.

D › Raise a power to a power.

E › Solve applications involving the concepts studied.

To Succeed, Review How To . . .

1. Understand the meaning of the word *exponent.* (p. 77)
2. Add, subtract, and multiply numbers. (pp. 548–550)

Getting Started

As we mentioned in Section 1.7, exponential notation is used to indicate how many times a quantity is to be used as a *factor*. For example, the area A of the square in the figure can be written as

$$A = x \cdot x = x^2$$ Read "x squared or x to the second power."

The exponent 2 indicates that the x is used as a factor *twice*. Similarly, the volume V of the cube is

$$V = x \cdot x \cdot x = x^3$$ Read "x cubed or x to the third power."

This time, the exponent 3 indicates that the x is used as a factor *three* times.

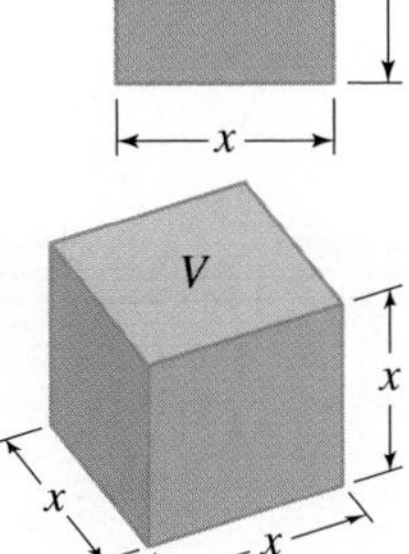

A › Integer Exponents

If a symbol x (called the *base*) is to be used n times as a factor, we use the following definition:

EXPONENT

$$\underbrace{x \cdot x \cdot x \cdot \ldots \cdot x}_{n \text{ factors}} = x^n$$

← exponent; x = base

Sometimes x^n is called a **power** of x, or x to the nth power. Note that if the base carries no exponent, the exponent is assumed to be 1; that is,

$$\boxed{a = a^1,\ b = b^1, \text{ and } c = c^1.}$$

We can use integers as exponents. Look at the pattern obtained by dividing by 10 in each step.

$10^3 = 1000$

$10^2 = 100$

$10^1 = 10$ Note that the exponents decrease by 1 at each step.

$10^0 = 1$

As you can see, this procedure yields $10^0 = 1$. In general, we make the following definition.

ZERO EXPONENT

Any nonzero number raised to the 0 power is 1.

$$x^0 = 1 \quad \text{for } x \neq 0$$

FOLLOW THE PATTERN!

$$10^{-1} = \frac{1}{10} = \frac{1}{10^1}$$
$$10^{-2} = \frac{1}{100} = \frac{1}{10^2}$$
$$10^{-3} = \frac{1}{1000} = \frac{1}{10^3}$$

Now look at the numbers in the box above. Note that we obtained

$$10^{-1} = \frac{1}{10}, \quad 10^{-2} = \frac{1}{10^2}, \quad \text{and} \quad 10^{-3} = \frac{1}{10^3}$$

Thus, we make the following definition:

NEGATIVE EXPONENT

If n is a positive integer, any nonzero number x raised to the $-n$ power is the reciprocal of x^n.

$$x^{-n} = \frac{1}{x^n}, \quad x \neq 0$$

In general, negative exponents should *not* be included in the final answer (except for scientific notation), so this definition helps you to simplify them. Since $x^n \cdot x^{-n} = x^n \cdot \frac{1}{x^n} = \frac{x^n}{x^n} = 1$, x^{-n} and x^n are reciprocals.

The definition of the negative exponent $-n$ means that

$$6^{-2} = \frac{1}{6^2} = \frac{1}{6 \cdot 6} = \frac{1}{36}$$
$$2^{-3} = \frac{1}{2^3} = \frac{1}{2 \cdot 2 \cdot 2} = \frac{1}{8}$$
$$\frac{1}{4^2} = 4^{-2}$$
$$\frac{1}{3^4} = 3^{-4}$$

EXAMPLE 1 **Simplifying negative exponents**

Write as a fraction and simplify.

a. 6^{-2}

b. 4^{-3}

SOLUTION

a. $6^{-2} = \frac{1}{6^2} = \frac{1}{6 \cdot 6} = \frac{1}{36}$

b. $4^{-3} = \frac{1}{4^3} = \frac{1}{4 \cdot 4 \cdot 4} = \frac{1}{64}$

PROBLEM 1

Write as a fraction and simplify.

a. 5^{-2}

b. 3^{-3}

EXAMPLE 2

Write using negative exponents.

a. $\frac{1}{5^4}$

b. $\frac{1}{7^5}$

SOLUTION

a. Using the definition, we have $\frac{1}{5^4} = 5^{-4}$.

b. $\frac{1}{7^5} = 7^{-5}$

PROBLEM 2

Write using negative exponents.

a. $\frac{1}{7^4}$

b. $\frac{1}{6^5}$

Answers to PROBLEMS

1. a. $\frac{1}{25}$ **b.** $\frac{1}{27}$ **2. a.** 7^{-4} **b.** 6^{-5}

B › Multiplying Exponential Expressions

We are now ready to multiply expressions involving exponents. For example, to multiply x^2 by x^3, we first write

$$\begin{aligned} &\overbrace{x^2}\cdot\overbrace{x^3} \\ &= \underbrace{x\cdot x\cdot x\cdot x\cdot x} \\ &= x^{2+3} \\ &= x^5 \end{aligned}$$

Clearly, we have simply *added* the exponents of x^2 and x^3 to find the exponent of the result. Similarly,

$$a^3\cdot a^4 = a^{3+4} = a^7$$
$$b^2\cdot b^4 = b^{2+4} = b^6$$

What about multiplying x^5 by x^{-2}? We have:

$$\begin{aligned} x^5\cdot x^{-2} &= (x\cdot x\cdot x\cdot x\cdot x)\cdot\left(\frac{1}{x\cdot x}\right) \\ &= x\cdot x\cdot x\cdot\left(\frac{x\cdot x}{1}\right)\cdot\left(\frac{1}{x\cdot x}\right) = x\cdot x\cdot x\cdot\frac{x\cdot x}{x\cdot x} = x^3 \end{aligned}$$

Adding exponents gives

$$x^5\cdot x^{-2} = x^{5+(-2)} = x^3 \quad \text{Same answer!}$$

Similarly,

$$x^{-3}\cdot x^{-2} = \frac{1}{x^3}\cdot\frac{1}{x^2} = \frac{1}{x^{3+2}} = \frac{1}{x^5} = x^{-5}$$

Adding exponents gives

$$x^{-3}\cdot x^{-2} = x^{-3+(-2)} = x^{-5} \quad \text{Same answer again!}$$

We state the resulting rule here for your convenience.

THE PRODUCT RULE OF EXPONENTS

For any nonzero number x and any integers m and n,

$$x^m\cdot x^n = x^{m+n};\quad x\neq 0$$

This rule states that when *multiplying* exponential expressions with the *same* base, *add* the exponents.

EXAMPLE 3 Multiplying exponential expressions

Multiply and simplify.

a. $2^6\cdot 2^2$ **b.** $x^3\cdot x^4$ **c.** $4^3\cdot 4^{-5}$
d. $y^2\cdot y^{-3}$ **e.** $a^{-5}\cdot a^5$

SOLUTION

a. $2^6\cdot 2^2 = 2^{6+2} = 2^8 = 256$

b. $x^3\cdot x^4 = x^{3+4} = x^7$

c. $4^3\cdot 4^{-5} = 4^{3+(-5)} = 4^{-2} = \frac{1}{4^2} = \frac{1}{16}$

Note that we wrote the answer **without** using negative exponents.

d. $y^2\cdot y^{-3} = y^{2+(-3)} = y^{-1} = \frac{1}{y}$

Again, we wrote the answer without negative exponents.

e. $a^{-5}\cdot a^5 = a^{-5+5} = a^0 = 1$

PROBLEM 3

Multiply.

a. $2^5\cdot 2^{-3}$ **b.** $x^5\cdot x^3$
c. $3^4\cdot 3^{-7}$ **d.** $x^3\cdot x^{-4}$
e. $y^3\cdot y^{-3}$

Answers to PROBLEMS

3. a. 4 **b.** x^8 **c.** $\frac{1}{27}$ **d.** $\frac{1}{x}$ **e.** 1

C > Dividing Exponential Expressions

To divide expressions involving exponents, we need to develop a rule to handle these exponents. For example, to divide x^5 by x^3, we first write

$$\frac{x^5}{x^3} = \frac{x \cdot x \cdot x \cdot x \cdot x}{x \cdot x \cdot x}, \qquad x \neq 0$$

Since $(x \cdot x \cdot x)$ is common to the numerator and denominator, we have

$$\frac{x^5}{x^3} = \frac{(\mathbf{x} \cdot \mathbf{x} \cdot \mathbf{x}) \cdot x \cdot x}{(\mathbf{x} \cdot \mathbf{x} \cdot \mathbf{x})} = x^{5-3} = x \cdot x = x^2$$

Here the bold x's mean that we divided the numerator and denominator by the common factor $(\mathbf{x} \cdot \mathbf{x} \cdot \mathbf{x})$. Of course, you can immediately see that the exponent 2 in the answer is simply the *difference* of the original two exponents, that is,

$$\frac{x^5}{x^3} = x^{5-3} = x^2$$

Similarly,

$$\frac{x^7}{x^4} = x^{7-4} = x^3$$

$$\frac{y^4}{y} = y^{4-1} = y^3$$

The rule used then can be extended to any integers.

THE QUOTIENT RULE OF EXPONENTS

For any nonzero number x and any integers m and n,

$$\frac{x^m}{x^n} = x^{m-n}; \qquad x \neq 0$$

This rule states that when *dividing* exponential expressions with the *same* base, *subtract* the exponents.

Thus

$$\frac{2^4}{2^{-1}} = 2^{4-(-1)} = 2^{4+1} = 2^5 = 32$$

$$\frac{x^3}{x^5} = x^{3-5} = x^{-2} = \frac{1}{x^2}$$

(We write the answer without negative exponents.)

EXAMPLE 4 Dividing exponential expressions

Divide and simplify.

a. $\frac{6^3}{6^{-2}}$ b. $\frac{x}{x^5}$ c. $\frac{y^{-2}}{y^{-2}}$ d. $\frac{z^{-3}}{z^{-4}}$

SOLUTION

a. $\frac{6^3}{6^{-2}} = 6^{3-(-2)} = 6^{3+2} = 6^5 = 7776$

b. $\frac{x}{x^5} = x^{1-5} = x^{-4} = \frac{1}{x^4}$

c. $\frac{y^{-2}}{y^{-2}} = y^{-2-(-2)} = y^{-2+2} = y^0 = 1$

d. $\frac{z^{-3}}{z^{-4}} = z^{-3-(-4)} = z^{-3+4} = z^1 = z$

PROBLEM 4

Divide and simplify.

a. $\frac{7^2}{7^{-3}}$ b. $\frac{y}{y^6}$

c. $\frac{x^{-3}}{x^{-3}}$ d. $\frac{z^{-4}}{z^{-5}}$

Answers to PROBLEMS

4. a. $7^5 = 16{,}807$ **b.** $\frac{1}{y^5}$
c. 1 **d.** z

D › Raising a Power to a Power

Suppose we wish to find $(5^3)^2$. By definition,

$$(5^3)^2 = 5^3 \cdot 5^3 = 5^{3+3}, \qquad \text{or} \qquad 5^6$$

We could get this answer by multiplying exponents in $(5^3)^2$, obtaining $5^{3\cdot 2} = 5^6$. Similarly,

$$(4^{-2})^3 = \frac{1}{4^2} \cdot \frac{1}{4^2} \cdot \frac{1}{4^2} = \frac{1}{4^6} = 4^{-6}$$

Again, we could have multiplied exponents in $(4^{-2})^3$, obtaining $4^{-2\cdot 3} = 4^{-6}$. We use these ideas to state the following rule.

THE POWER RULE OF EXPONENTS

For any nonzero number x and any integers m and n,

$$(x^m)^n = x^{mn}; \qquad x \neq 0$$

This rule states that when raising an exponential expression to a *power, multiply* the exponents.

EXAMPLE 5 **Raising a power to a power**

Simplify.

a. $(2^3)^2$ **b.** $(x^{-2})^3$ **c.** $(y^4)^{-5}$ **d.** $(z^{-2})^{-3}$

SOLUTION

a. $(2^3)^2 = 2^{3\cdot 2} = 2^6 = 64$

b. $(x^{-2})^3 = x^{-2\cdot 3} = x^{-6} = \frac{1}{x^6}$

c. $(y^4)^{-5} = y^{4(-5)} = y^{-20} = \frac{1}{y^{20}}$

d. $(z^{-2})^{-3} = z^{-2(-3)} = z^6$

PROBLEM 5

Simplify.

a. $(5^3)^2$ **b.** $(x^{-3})^4$

c. $(y^3)^{-6}$ **d.** $(z^{-3})^{-5}$

Sometimes we need to raise several factors inside parentheses to a power, as in $(x^2y^3)^3$. We use the definition of cubing and write

$$\begin{aligned}(x^2y^3)^3 &= x^2y^3 \cdot x^2y^3 \cdot x^2y^3 \\ &= (x^2 \cdot x^2 \cdot x^2)(y^3 \cdot y^3 \cdot y^3) \\ &= (x^2)^3(y^3)^3 \\ &= x^6y^9\end{aligned}$$

We could get the same answer by multiplying each of the exponents in x^2y^3 by 3, obtaining $x^{2\cdot 3}y^{3\cdot 3} = x^6y^9$.

Thus, to raise several factors inside parentheses to a power, we raise each factor to the given power, as shown next.

THE PRODUCT TO A POWER RULE OF EXPONENTS

For any real numbers x and y and any integers m, n, and k,

$$(x^my^n)^k = (x^m)^k(y^n)^k = x^{mk}y^{nk}$$

This rule states that when raising *several factors* inside parentheses to a *power*, raise each factor to the given *power*.

Answers to PROBLEMS

5. a. $5^6 = 15{,}625$ **b.** $\frac{1}{x^{12}}$

c. $\frac{1}{y^{18}}$ **d.** z^{15}

EXAMPLE 6 **Raising a product to a power**

Simplify.

a. $(x^2y^{-2})^3$ **b.** $(x^{-2}y^3)^3$ **c.** $(x^{-2}y^3)^{-2}$

SOLUTION

a. $(x^2y^{-2})^3 = (x^2)^3(y^{-2})^3$
$= x^6y^{-6}$
$= \dfrac{x^6}{y^6}$

b. $(x^{-2}y^3)^3 = (x^{-2})^3(y^3)^3$
$= x^{-6}y^9$
$= \dfrac{y^9}{x^6}$

c. $(x^{-2}y^3)^{-2} = (x^{-2})^{-2}(y^3)^{-2}$
$= x^4y^{-6}$
$= \dfrac{x^4}{y^6}$

PROBLEM 6

Simplify.

a. $(x^3y^{-2})^3$

b. $(x^{-3}y^2)^3$

c. $(x^3y^2)^{-2}$

E > Applications: Compound Interest

Suppose you invest P dollars at 10% compounded annually. At the end of 1 year you will have your original principal P plus the interest, $10\% \cdot P$, that is, $P + 0.10P = (1 + 0.10)P = 1.10P$. At the end of 2 years, you will have $1.10P$ plus the interest earned on $1.10P$, that is, $1.10P + 0.10(1.10P) = 1.10P(1 + 0.10) = 1.10P(1.10)$, or $(1.10)^2P$. If you follow this pattern, at the end of 3 years you will have $(1.10)^3P$, and so on. Here is the general formula:

COMPOUND INTEREST

If the principal *P* is invested at rate *r*, compounded annually, in *n* years the compound amount *A* will be

$$A = P(1 + r)^n$$

EXAMPLE 7 **Calculating a compounded amount**

If \$1000 is invested at 10% compounded annually, how much will be in the account at the end of 3 years?

SOLUTION Here $P = \$1000$, $r = 10\% = 0.10$, and $n = 3$. We have

$$A = 1000(1 + 0.10)^3$$
$$= 1000(1.10)^3 = 1000(1.331) = \$1331$$

PROBLEM 7

If \$500 is invested at 10% compounded annually, how much will be in the account at the end of 2 years?

Calculator Corner

If you have a scientific calculator with a y^x key (some students call y^x the power key), the numerical calculations in this section become simple. For example, here is the way to find 5^3: Press 5 y^x 3 ENTER. The answer, 125, will be displayed. Similarly, to find 6^{-2}: Press 6 y^x 2 ± ENTER. Note that you *do not* enter -2, but enter 2 and ±, which will change the sign of 2 to -2. The answer is given as a decimal.

Answers to PROBLEMS

6. a. $\dfrac{x^9}{y^6}$ **b.** $\dfrac{y^6}{x^9}$ **c.** $\dfrac{1}{x^6y^4}$

7. \$605

> Exercises 10.2

Boost your grade at mathzone.com!
> Practice Problems > Self-Tests
> NetTutor > e-Professors
> Videos

⟨A⟩ Integer Exponents
In Problems 1–6, write as a fraction and simplify.

1. a. 4^{-2} b. x^{-2}
2. a. 2^{-3} b. x^{-3}
3. a. 5^{-3} b. y^{-3}
4. a. 7^{-2} b. y^{-2}
5. a. 3^{-4} b. z^{-4}
6. a. 6^{-3} b. a^{-3}

In Problems 7–12, write using negative exponents.

7. a. $\frac{1}{2^3}$ b. $\frac{1}{x^3}$
8. a. $\frac{1}{3^4}$ b. $\frac{1}{x^4}$
9. a. $\frac{1}{4^5}$ b. $\frac{1}{y^5}$
10. a. $\frac{1}{5^6}$ b. $\frac{1}{y^6}$
11. a. $\frac{1}{3^5}$ b. $\frac{1}{z^5}$
12. a. $\frac{1}{7^4}$ b. $\frac{1}{z^4}$

⟨B⟩ Multiplying Exponential Expressions
In Problems 13–36, multiply and simplify.

13. a. $3^2 \cdot 3^3$ b. $x^5 \cdot x^8$
14. a. $4^2 \cdot 4^2$ b. $x^2 \cdot x^2$
15. a. $2^{-5} \cdot 2^7$ b. $y^{-3} \cdot y^8$
16. a. $3^8 \cdot 3^{-5}$ b. $y^7 \cdot y^{-3}$
17. a. $4^{-6} \cdot 4^4$ b. $x^{-7} \cdot x^3$
18. a. $5^{-4} \cdot 5^2$ b. $x^{-5} \cdot x^2$
19. a. $6^{-1} \cdot 6^{-2}$ b. $y^{-3} \cdot y^{-4}$
20. a. $3^{-2} \cdot 3^{-1}$ b. $y^{-8} \cdot y^{-4}$
21. a. $2^{-4} \cdot 2^{-2}$ b. $x^{-3} \cdot x^{-7}$
22. a. $4^{-1} \cdot 4^{-2}$ b. $x^{-2} \cdot x^{-6}$
23. $x^6 \cdot x^{-4}$
24. $y^7 \cdot y^{-2}$
25. $y^{-3} \cdot y^5$
26. $x^{-7} \cdot x^8$
27. $a^3 \cdot a^{-8}$
28. $b^4 \cdot b^{-7}$
29. $x^{-5} \cdot x^3$
30. $y^{-6} \cdot y^2$
31. $x \cdot x^{-3}$
32. $y \cdot y^{-5}$
33. $a^{-2} \cdot a^{-3}$
34. $b^{-5} \cdot b^{-2}$
35. $b^{-3} \cdot b^3$
36. $a^6 \cdot a^{-6}$

⟨C⟩ Dividing Exponential Expressions
In Problems 37–50, divide and simplify.

37. $\frac{3^4}{3^{-1}}$
38. $\frac{2^2}{2^{-2}}$
39. $\frac{4^{-1}}{4^2}$
40. $\frac{3^{-2}}{3^3}$
41. $\frac{y}{y^3}$
42. $\frac{x}{x^4}$
43. $\frac{x}{x^{-2}}$
44. $\frac{y}{y^{-3}}$
45. $\frac{x^{-3}}{x^{-1}}$
46. $\frac{x^{-4}}{x^{-2}}$
47. $\frac{x^{-3}}{x^4}$
48. $\frac{y^{-4}}{y^5}$
49. $\frac{x^{-2}}{x^{-5}}$
50. $\frac{y^{-3}}{y^{-6}}$

⟨D⟩ Raising a Power to a Power
In Problems 51–74, simplify.

51. $(3^2)^2$
52. $(2^3)^2$
53. $(3^{-1})^2$
54. $(2^{-2})^2$
55. $(2^{-2})^{-3}$
56. $(3^{-1})^{-2}$
57. $(3^2)^{-1}$
58. $(2^3)^{-2}$
59. $(x^3)^{-3}$
60. $(y^2)^{-4}$
61. $(y^{-3})^2$
62. $(x^{-4})^3$
63. $(a^{-2})^{-3}$
64. $(b^{-3})^{-5}$
65. $(x^3y^{-2})^3$
66. $(x^2y^{-3})^2$
67. $(x^{-2}y^3)^2$
68. $(x^{-4}y^4)^3$
69. $(x^3y^2)^{-3}$
70. $(x^5y^4)^{-4}$
71. $(x^{-6}y^{-3})^2$
72. $(y^{-4}z^{-3})^5$
73. $(x^{-4}y^{-4})^{-3}$
74. $(y^{-5}z^{-3})^{-4}$

⟨E⟩ Applications: Compound Interest
Round to the nearest cent.

75. *Investment problem* If \$1000 is invested at 8% compounded annually, how much will be in the account at the end of 3 years?

76. *Investment problem* If \$500 is invested at 10% compounded annually, how much will be in the account at the end of 3 years?

77. *Investment problem* Suppose \$1000 is invested at 10% compounded annually. How much will be in the account at the end of 4 years?

Web IT go to mhhe.com/bello for more lessons

Applications

78. *Distance traveled by* Pioneer *in 10 months* The first manufactured object to leave the solar system was *Pioneer 10*, which attained a velocity of 5.1×10^4 kilometers per hour when leaving Earth on its way to Jupiter. After 10 months, it had traveled

$$(5.1 \times 10^4) \times (7.2 \times 10^3), \text{ or}$$

$$(5.1 \times 7.2) \times (10^4 \times 10^3) \text{ km.}$$

How many kilometers is that?

79. *Distance traveled by* Pioneer *in 10 years* The top speed of *Pioneer 10* was 1.31×10^5 kilometers per hour. At this rate, in 8.7×10^4 hours (about 10 years), *Pioneer 10* would go

$$(1.31 \times 10^5) \times (8.7 \times 10^4), \text{ or}$$

$$(1.31 \times 8.7) \times (10^5 \times 10^4) \text{ km.}$$

How many kilometers is that?

80. *Petroleum and gas reserves* The estimated petroleum and gas reserves for the United States are about 2.8×10^{17} kilocalories. If we consume these reserves at the rate of 1.4×10^{16} kilocalories per year, they will last $\frac{2.8 \times 10^{17}}{1.4 \times 10^{16}}$, or $\frac{2.8}{1.4} \times \frac{10^{17}}{10^{16}}$ years. How many years is that?

81. *Internet sites* During January 2006 almost 3×10^7 people visited flower, greeting, and gift sites on the Internet. On average, each visitor spent about \$9 from January 1 to February 9 on these sites. How many million dollars were spent on flowers, greetings, and gift sites during this period?

Source: clickz.com; http://www.clickz.com/stats/sectors/traffic_patterns/article.php/3585186#table2.

82. *Travel expenses* A global panel of 2×10^6 people spent about $\$3.1 \times 10^4$ annually on travel expenses. What was the total amount spent during the year?

Source: clickz.com; http://www.clickz.com/stats/sectors/retailing/article.php/3575456#table1.

83. *U.S. national debt* At the present time, the U.S. national debt amounts to $\$8.43 \times 10^{12}$. Since the population of the United States is approximately $3 \cdot 10^8$, the share of each person amounts to $\frac{\$8.43 \times 10^{12}}{3 \times 10^8}$. How much is that?

84. *U.S. national debt* If the U.S. national debt is rounded to $\$8.4 \cdot 10^{12}$, the average monthly increase on the debt is $\frac{\$8.4 \times 10^{12}}{1.2 \times 10}$. How many billions per month is that?

Using Your Knowledge

Patterns There are many interesting patterns involving exponents. Use your knowledge to find the answers to the following problems.

85.

$$1^2 = 1$$
$$(11)^2 = 121$$
$$(111)^2 = 12{,}321$$
$$(1111)^2 = 1{,}234{,}321$$

a. Find $(11{,}111)^2$.

b. Find $(111{,}111)^2$.

86.

$$1^2 = 1$$
$$2^2 = 1 + 2 + 1$$
$$3^2 = 1 + 2 + 3 + 2 + 1$$
$$4^2 = 1 + 2 + 3 + 4 + 3 + 2 + 1$$

a. Use this pattern to write 5^2.

b. Use this pattern to write 6^2.

87.

$$1 + 3 = 2^2$$
$$1 + 3 + 5 = 3^2$$
$$1 + 3 + 5 + 7 = 4^2$$

a. Find $1 + 3 + 5 + 7 + 9$.

b. Find $1 + 3 + 5 + 7 + 9 + 11 + 13$.

88. In this problem, discover your own pattern. What is the largest number you can construct by using the number 9 three times? (It is *not* 999!)

Write On

89. Explain why the product rule does not apply to the expression $x^n \cdot y^n$.

90. Write in your own words the difference between the product rule and the power rule.

91. Write in your own words three different reasons for the fact that $x^0 = 1$ $(x \neq 0)$.

Concept Checker

Fill in the blank(s) with the correct word(s), phrase, or mathematical statement.

92. **Exponential notation** is used to indicate how many times a quantity is to be used as a ______________.

93. For $x \neq 0$, $x^0 =$ ______________.

94. If n is a positive integer, $x^{-n} =$ ______________.

95. If m and n are integers, $x^m \cdot x^n =$ ______________.

96. If m and n are integers, $\frac{x^m}{x^n} =$ ______________.

97. If m and n are integers, $(x^m)^n =$ ______________.

98. If m, n, and k are integers, $(x^m \cdot y^n)^k =$ ______________.

99. If a principal P is invested at a rate r compounded annually for n years, the compound amount A is ______________.

x^{m+n}	$P(1+n)^r$
$x^{m \cdot n}$	factor
x^{m-n}	term
$x^{-m \cdot n}$	1
$P(1+r)^n$	$\frac{1}{x^n}$
$r(1+P)^n$	$x^{m \cdot k}y^{n \cdot k}$

Mastery Test

100. If \$1000 is invested at 4% compounded annually, what will be the compound amount A at the end of 2 years?

101. Simplify.

a. $(2^2)^3$ **b.** $(x^{-3})^2$

c. $(a^3)^{-5}$ **d.** $(b^{-2})^{-5}$

102. Divide and simplify.

a. $\frac{3^4}{3^{-2}}$ **b.** $\frac{a}{a^4}$

c. $\frac{x^{-4}}{x^{-4}}$ **d.** $\frac{c^{-4}}{c^{-5}}$

103. Simplify.

a. $(a^3b^{-2})^4$ **b.** $(a^{-4}b^4)^3$

c. $(a^{-3}b^4)^{-2}$

104. Multiply and simplify.

a. $3^3 \cdot 3^2$ **b.** $a^3 \cdot a^4$

c. $3^4 \cdot 3^{-5}$ **d.** $a^4 \cdot a^{-5}$

e. $a^{-7} \cdot a^7$

105. Write using negative exponents.

a. $\frac{1}{6^3}$ **b.** $\frac{1}{c^4}$

106. Write as a fraction and simplify.

a. 3^{-2} **b.** 4^{-3}

Skill Checker

107. Find 7.31×10^1.

108. Find 7.31×10^2.

109. Find 7.31×10^4.

110. Find 7.31×10^5.

10.3 Scientific Notation

Objectives

You should be able to:

A Convert between ordinary decimal notation and scientific notation.

B Multiply and divide numbers in scientific notation.

C Solve applications involving the concepts studied.

To Succeed, Review How To . . .

1. Multiply and divide a number by a power of 10 (10, 100, 1000, etc.). (pp. 54, 213, 217)
2. Use the properties of exponents. (pp. 77, 82–84)

Getting Started

How many facts do you know about the sun? Here is some information taken from an encyclopedia article.

Mass: 2.19×10^{27} tons

Temperature: 1.8×10^{6} degrees Fahrenheit

Energy per minute: 2.4×10^{4} hp

Each number involved is written as a *product* of a number between 1 and 10 and an appropriate power of 10. This form is called **scientific notation,** which we use to write numbers that are either very large or very small.

A › Scientific Notation

SCIENTIFIC NOTATION

A number in scientific notation is written as

$M \times 10^{n}$ M = a number between 1 and 10

n = an integer

How do we change a whole number to scientific notation? First, recall that when we *multiply* a number by a power of 10 ($10^1 = 10$, $10^2 = 100$, and so on) we simply *move* the decimal point as many places to the *right* as indicated by the *exponent* of 10. Thus

$7.31 \times 10^{1} = 73.1$ (Exponent 1, move 1 place right.)

$72.813 \times 10^{2} = 7281.3$ (Exponent 2, move 2 places right.)

$160.7234 \times 10^{3} = 160723.4$ (Exponent 3, move 3 places right.)

On the other hand, if we *divide* a number by a power of 10, we move the decimal point as many places to the *left* as indicated by the exponent of 10. Thus,

$$\frac{7}{10} = 0.7 = 7 \times 10^{-1}$$

$$\frac{8}{100} = 0.08 = 8 \times 10^{-2}$$

and

$$\frac{4.7}{10{,}000} = 00.00047 = 4.7 \times 10^{-4}$$

TO WRITE A NUMBER IN SCIENTIFIC NOTATION ($M \times 10^n$)

1. Move the decimal point in the given number so that there is only one digit to its left. The resulting number is M.
2. Count how many places you have to move the decimal point in step 1. If the decimal point must be moved to the *left*, n is *positive;* if it must be moved to the *right*, n is *negative*.
3. Write $M \times 10^n$.

For example,

$5.3 = 5.3 \times 10^0$ — The decimal point in 5.3 must be moved **0** places.

$87 = 8.7 \times 10^1 = 8.7 \times 10$ (1) — The decimal point in 87 must be moved **1** place to the left to get 8.7.

$68{,}000 = 6.8 \times 10^4$ (4) — The decimal point in 68,000 must be moved **4** places to the left to get 6.8.

$0.49 = 4.9 \times 10^{-1}$ (1) — The decimal point in 0.49 must be moved **1** place to the right to get 4.9.

$0.072 = 7.2 \times 10^{-2}$ (2) — The decimal point in 0.072 must be moved **2** places to the right to get 7.2.

Note that if the standard notation of a number (like 68,000) is large the exponent of 10 in scientific notation is *positive*. If the standard notation of a number (like 0.072) is small the exponent of 10 is *negative*.

EXAMPLE 1 Writing a number in scientific notation

The approximate distance to the sun is 93,000,000 miles and the wavelength of its ultraviolet light is 0.000035 centimeters. Write 93,000,000 and 0.000035 in scientific notation.

SOLUTION

$$93{,}000{,}000 = 9.3 \times 10^7$$

$$0.000035 = 3.5 \times 10^{-5}$$

PROBLEM 1

The distance to the moon is about 239,000 miles and its mass is 0.0123456 times that of the earth. Write 239,000 and 0.0123456 in scientific notation.

EXAMPLE 2 Writing a number in standard notation

A jumbo jet weighs 7.75×10^5 pounds, whereas a house spider weighs 2.2×10^{-4} pounds. Write 7.75×10^5 and 2.2×10^{-4} in standard notation.

SOLUTION

$$7.75 \times 10^5 = 775{,}000$$

$$2.2 \times 10^{-4} = 0.00022$$

PROBLEM 2

The Concorde weighs 4.08×10^5 pounds and a cricket weighs 3.125×10^{-4} pounds. Write 4.08×10^5 and 3.125×10^{-4} in standard notation.

B > Multiplying and Dividing Using Scientific Notation

Consider the product $300 \cdot 2000 = 600{,}000$. In scientific notation, we would write

$$(3 \times 10^2) \cdot (2 \times 10^3) = 6 \times 10^5$$

Answers to PROBLEMS

1. 2.39×10^5; 1.23456×10^{-2}

2. 408,000; 0.0003125

To find the answer, we can multiply 3 by 2 to obtain 6 and 10^2 by 10^3, obtaining 10^5. To multiply numbers in scientific notation, we proceed in a similar manner:

1. Multiply the decimal parts first and write the result in scientific notation.
2. Multiply the powers of 10.
3. The answer, which should be simplified, is the product of steps 1 and 2.

EXAMPLE 3 Multiplying numbers in scientific notation

Multiply.

a. $(5 \times 10^3) \times (8.1 \times 10^4)$ **b.** $(3.2 \times 10^2) \times (4 \times 10^{-5})$

SOLUTION

a. We multiply the decimal parts first.
$5 \times 8.1 = 40.5 = 4.05 \times 10$
Then multiply the powers of 10.
$10^3 \times 10^4 = 10^7$ (adding exponents)
The answer is $(4.05 \times 10) \times 10^7$, or 4.05×10^8.

b. Multiply the decimals.
$3.2 \times 4 = 12.8 = 1.28 \times 10$
Multiply the powers of 10.
$10^2 \times 10^{-5} = 10^{2-5} = 10^{-3}$
The answer is $(1.28 \times 10) \times 10^{-3}$, or $1.28 \times 10^{1+(-3)} = 1.28 \times 10^{-2}$.

PROBLEM 3

Multiply.

a. $(6 \times 10^4) \times (2.2 \times 10^3)$

b. $(4.1 \times 10^2) \times (3 \times 10^{-5})$

Division is done in the same manner. For example, $\frac{3.2 \times 10^5}{1.6 \times 10^2}$ is found by dividing 3.2 by 1.6 (yielding 2) and 10^5 by 10^2, which gives 10^3. The answer is 2×10^3.

EXAMPLE 4 Dividing numbers in scientific notation

Divide: $(1.24 \times 10^{-2}) \div (3.1 \times 10^{-3})$.

SOLUTION First divide 1.24 by 3.1, obtaining $0.4 = 4 \times 10^{-1}$.
Now divide powers of 10:

$$10^{-2} \div 10^{-3} = 10^{-2-(-3)} = 10^{-2+3} = 10^1$$

The answer is $(4 \times 10^{-1}) \times 10^1 = 4 \times 10^0$ or 4.

PROBLEM 4

Divide.

$(2.52 \times 10^{-2}) \div (4.2 \times 10^{-3})$

C › Applications Involving Scientific Notation

EXAMPLE 5 Dividing numbers in scientific notation

The total energy received from the sun each minute is 1.02×10^{19} calories. Since the area of Earth is 5.1×10^{18} cm², the amount of energy received per square centimeter of Earth surface every minute (the solar constant) is:

$$\frac{1.02 \times 10^{19}}{5.1 \times 10^{18}}$$

Simplify this expression.

SOLUTION Dividing 1.02 by 5.1, we obtain $0.2 = 2 \times 10^{-1}$. Now, $10^{19} \div 10^{18} = 10^{19-18} = 10^1$. Thus, the final answer is $(2 \times 10^{-1}) \times 10^1 = 2 \times 10^0 = 2$. This means that Earth receives about 2 calories of energy per square centimeter each minute.

PROBLEM 5

The width of the asteroid belt is 2.8×10^8 km. The speed of *Pioneer 10* in passing through this belt was 1.4×10^5 km/h. Thus, *Pioneer 10* took $\frac{2.8 \times 10^8}{1.4 \times 10^5}$ h to go through the belt.

Answers to PROBLEMS

3. a 1.32×10^8 **b.** 1.23×10^{-2} **4.** 6×10^0 or 6 **5.** 2×10^3

Calculator Corner

If you have a scientific calculator and you multiply 9,800,000 by 4,500,000, the display will show: 4.41   13
This means that the answer is 4.41×10^{13}.

1. The display on a calculator shows: 3.34   5
Write this number in scientific notation.

2. The display on a calculator shows: −9.97   −6
Write this number in scientific notation.

To enter large or small numbers in a calculator with scientific notation, you must first write the number using scientific notation. Thus, to enter the number 8,700,000,000 in the calculator you must know that 8,700,000,000 is 8.7×10^9; then you can key in 
The calculator displays: 8.7   09

3. What would the display read when you enter the number 73,000,000,000?

4. What would the display read when you enter the number 0.000000123?

Exercises 10.3

⟨A⟩ Scientific Notation In Problems 1–10, write in scientific notation.

1. 68,000,000 (working women in the United States)

2. 78,000,000 (working men in the United States)

3. 293,000,000 (U.S. population now)

4. 281,000,000 (U.S. population in the year 2000)

5. 1,900,000,000 (dollars spent on waterbeds and accessories in one year)

6. 0.035 (ounces in a gram)

7. 0.00024 (probability of four of a kind in poker)

8. 0.000005 (the gram-weight of an amoeba)

9. 0.000000002 (the gram-weight of one liver cell)

10. 0.00000009 (wavelength of an X ray in centimeters)

In Problems 11–20, write in standard notation.

11. 2.35×10^2 (pounds of meat consumed per person per year in the United States)

12. 2.87×10^2 (pounds of fresh fruit consumed per person per year in the United States)

13. 8×10^6 (bagels eaten per day in the United States)

14. 22×10^6 (jobs created in service industries between now and the year 2010)

15. 6.8×10^9 (estimated worth of the five wealthiest women)

16. 1.68×10^{10} (estimated worth of the five wealthiest men)

17. 2.3×10^{-1} (kilowatts per hour used by your TV)

18. 4×10^{-2} in. (1 mm)

19. 2.5×10^{-4} (thermal conductivity of glass)

20. 4×10^{-11} joules (energy released by splitting one uranium atom)

⟨B⟩ Multiplying and Dividing Using Scientific Notation In Problems 21–30, perform the indicated operations (give your answer in scientific notation).

21. $(3 \times 10^4) \times (5 \times 10^5)$

22. $(5 \times 10^2) \times (3.5 \times 10^3)$

23. $(6 \times 10^{-3}) \times (5.1 \times 10^6)$

24. $(3 \times 10^{-2}) \times (8.2 \times 10^5)$

25. $(4 \times 10^{-2}) \times (3.1 \times 10^{-3})$

26. $(3.1 \times 10^{-3}) \times (4.2 \times 10^{-2})$

27. $\frac{4.2 \times 10^5}{2.1 \times 10^2}$

28. $\frac{5 \times 10^6}{2 \times 10^3}$

29. $\frac{2.2 \times 10^4}{8.8 \times 10^6}$

30. $\frac{2.1 \times 10^3}{8.4 \times 10^5}$

Web IT go to mhhe.com/bello for more lessons

‹ C › Applications Involving Scientific Notation

31. *Annual vegetable consumption* The average American eats 160 pounds of vegetables each year. Since there are about 300 million Americans, the number of pounds of vegetables consumed each year should be:

$$(1.6 \times 10^2) \times (3 \times 10^8)$$

a. Write this number in scientific notation.

b. Write this number in standard notation.

32. *Average soft drink consumption* The average American drinks 54.2 gallons of soft drinks each year. Since there are about 300 million Americans, the number of gallons of soft drinks consumed each year should be:

$$(5.42 \times 10^1) \times (3 \times 10^8)$$

a. Write this number in scientific notation.

b. Write this number in standard notation.

33. *Garbage per person* America produces 250 million tons of garbage each year. Since a ton is 2000 pounds and there are about 360 days in a year and 300 million Americans, the number of pounds of garbage produced each day of the year for each man, woman, and child in America is:

$$\frac{(2.5 \times 10^8) \times (2 \times 10^3)}{(3 \times 10^8) \times (3.6 \times 10^2)}$$

Write this number in standard notation to two decimal places.

34. *Velocity of light* The velocity of light can be measured by dividing the distance from the sun to Earth (1.47×10^{11} meters) by the time it takes for sunlight to reach Earth (4.9×10^2 seconds). Thus, the velocity of light is

$$\frac{1.47 \times 10^{11}}{4.9 \times 10^2}$$

How many meters per second is that? Write the number in standard notation.

35. *Energy in a gram of uranium* Nuclear fission is used as an energy source. The amount of energy a gram of uranium 235 releases is:

$$\frac{4.7 \times 10^9}{235} \text{ kilocalories}$$

Write this number in scientific notation.

Is it crowded where you live? How many people per square mile live in your town? The *population density* is a measurement of population per unit area.

36. *Population density* The population density of New York State is 3.5×10^2. If all those people live in an area of 5.5×10^3 square miles, this makes the population of New York State $(3.5 \times 10^2) \times (5.5 \times 10^4)$. How many people live in New York State? Write the answer in scientific and in standard notation.

37. *Population density* The population density of Paris is 9.2×10^3. If they live in an area of 1.05×10^3 square miles, this makes the population of Paris $(9.2 \times 10^3) \times (1.05 \times 10^3)$. How many people live in Paris? Write the answer in scientific and in standard notation.

38. *Population density* What about the population in the entire world? The population density of the world is a mere 43 persons per square kilometer. Since the land area of the world is 1.5×10^8, this means the world population is $(4.3 \times 10) \times (1.5 \times 10^8)$. How many people is that? Write the answer in scientific and in standard notation.

39. *Population density* One of the most densely populated places on Earth is Macau, in China. Their population density is $\frac{4.68 \times 10^5}{2.6 \times 10 \text{ km}^2}$, which means 4.68×10^5 inhabitants live in an area of 2.6×10 square kilometers. What is the population density of Macau? Write the answer in scientific and in standard notation.

40. *Population density* Another densely populated place on Earth is Singapore. Their population density is $\frac{4.416 \times 10^6}{6.9 \times 10^2 \text{ km}^2}$, which means 4.416×10^6 inhabitants live in an area of 6.9×10^2 square kilometers. What is the population density of Singapore? Write the answer in scientific and in standard notation.

41. *Sunlight* Since the sun is 1.5×10^{11} meters from Earth, it takes light traveling at 3×10^8 meters per second $\frac{1.5 \times 10^{11}}{3 \times 10^8}$ seconds to reach the Earth. How many seconds is that? Write the answer in scientific and in standard notation.

42. *Moon* The moon is about 3.9×10^8 meters from Earth. It takes light traveling at 3×10^8 meters per second $\frac{3.9 \times 10^8}{3 \times 10^8}$ seconds to reach the Earth. How many seconds is that?

›› Using Your Knowledge

Astronomy

43. Scientific notation is especially useful when very large quantities are involved. For example, in astronomy we find that the speed of light is 299,792,458 meters per second. Write 299,792,458 in scientific notation.

44. Astronomical distances are so large that they are measured in astronomical units (A.U. for short). An astronomical unit is defined as the average separation (distance) of Earth and the sun, that is, 150,000,000 kilometers. Write 150,000,000 in scientific notation.

45. Distances in astronomy are also measured in *parsecs:* 1 parsec = 2.06×10^5 A.U. Thus 1 parsec = $(2.06 \times 10^5) \times (1.5 \times 10^8)$ kilometers. Written in scientific notation, how many kilometers is that?

46. Astronomers also measure distances in light-years, the distance light travels in 1 year: 1 light-year = 9.46×10^{12} kilometers. The closest star, Proxima Centauri, is 4.22 light-years away. In scientific notation, rounded to two decimal places, how many kilometers is that?

47. Since 1 parsec = 3.09×10^{13} km (see Problem 45) and 1 light-year = 9.46×10^{12} km, the number of light-years in a parsec is

$$\frac{3.09 \times 10^{13}}{9.46 \times 10^{12}}$$

Write this number in standard notation rounded to two decimal places.

Write On

48. Write in your own words the procedure you use to write a number in scientific notation.

49. What are the advantages and disadvantages of writing numbers in scientific notation?

Concept Checker

Fill in the blank(s) with the correct word(s), phrase, or mathematical statement.

50. A number in **scientific notation** is written as ________________, where M is a number between 1 and 10 and n is an integer.

51. To write a number in the scientific notation $M \times 10^n$, **move** the decimal point in the number so that there is only **one** digit to its ________________; the result is M.

left

$n \times 10^M$

$M \times 10^n$

right

Mastery Test

52. The Earth is approximately 9.3×10^7 miles from the sun, and light travels at a speed of 1.86×10^5 miles per second. Thus, it takes

$$\frac{9.3 \times 10^7}{1.86 \times 10^5}$$

seconds for the light from the sun to reach Earth. Written in standard notation, how many seconds is that?

53. Divide 2.48×10^{-2} by 6.2×10^{-4}. Write the answer in standard notation.

54. Multiply and write the answer in standard notation:

a. $(5 \times 10^2) \times (6.1 \times 10^4)$

b. $(6.4 \times 10^2) \times (2 \times 10^{-5})$

55. The half-life of uranium 234 is 250,000 years. Write 250,000 in scientific notation.

56. One of the fastest computers in the world, at the Lawrence Livermore National Laboratory, could perform a single calculation in 0.000 000 000 000 26 second. Write this number in scientific notation.

57. The half-life of uranium 238 is 4.5×10^9 years. Write this number in standard notation.

58. The wavelength of ultraviolet light is 3.5×10^{-5} cm. Write this number in standard notation.

Skill Checker

Solve the following equations.

59. $x - 1 = -4$

60. $x - \frac{1}{4} = \frac{1}{2}$

61. $x + \frac{3}{5} = \frac{1}{2}$

62. $x - 0.3 = 0.9$

63. $x + 0.7 = 0.2$

64. $0.4x = -0.8$

65. $-0.3x = 0.9$

66. $-0.4x = -0.8$

10.4 Solving Linear Equations

Objective

You should be able to:

A › Solve equations using the addition, subtraction, multiplication, or division principle.

B › Solve linear equations using the five-step procedure given in the text.

To Succeed, Review How To . . .

1. Do addition, subtraction, multiplication, and division using rational numbers. (pp. 548–550)
2. Solve equations. (pp. 89–91, 240–241)

Getting Started

Do you know the man in the picture? He is Albert Einstein, who discovered the equation $E = mc^2$. In this section we shall study a type of equation called a *linear* equation. An **equation** is a sentence having a variable in which the verb is *equals* (or an equivalent verb). Here are some equations.

English Language	Algebra
One added to x gives 7.	$x + 1 = 7$
One subtracted from z equals 9	$z - 1 = 9$
Twice a number n is 8.	$2n = 8$
Half of y is 3.	$\frac{y}{2} = 3$

A › Solving Equations Using the Addition, Subtraction, Multiplication, or Division Principle

The equations shown in the *Getting Started* are also called **linear,** or **first-degree.** The variables x, z, n, and y show no exponent, which means that the exponent 1 is *understood* ($x = x^1$, $z = z^1$, $n = n^1$, and $y = y^1$); this is why they are called *first-degree* equations.

We can solve (find the solution of) these equations by using the same principles we studied before.

PRINCIPLES FOR SOLVING EQUATIONS

The equation $a = b$ is equivalent to

$a + c = b + c$ (Addition principle)

$a - c = b - c$ (Subtraction principle)

$a \cdot c = b \cdot c$ (Multiplication principle, $c \neq 0$)

$a \div c = b \div c$ Division principle, $c \neq 0$)

or $\frac{a}{c} = \frac{b}{c}$

Thus, to *solve* an equation, we may add or subtract the same number on *both* sides and multiply or divide *both* sides by the same nonzero number. (Exactly the same operations *must* be done on both sides of the equation to obtain equivalent equations.) The idea is to obtain an equation with the variable (letter) by itself on one side. Let us practice with these principles.

EXAMPLE 1 Solving linear equations

Solve.

a. $x - 1 = -3$ **b.** $y + \frac{1}{3} = \frac{1}{2}$

SOLUTION

a.

$x - 1 = -3$ Given.

$x - 1 + 1 = -3 + 1$ Add 1 to both sides so the *x* is by itself.

$x = -2$ Simplify.

CHECK We substitute $x = -2$ in the original equation:

$x - 1 \stackrel{?}{=} -3$	
$-2 - 1$	-3
-3	

Thus, our solution $x = -2$ is correct.

b.

$y + \frac{1}{3} = \frac{1}{2}$ Given.

$y + \frac{1}{3} - \frac{1}{3} = \frac{1}{2} - \frac{1}{3}$ Subtract $\frac{1}{3}$ from both sides to have *y* by itself.

$y = \frac{1 \cdot 3}{2 \cdot 3} - \frac{1 \cdot 2}{3 \cdot 2}$ Simplify (we need to use the LCD for $\frac{1}{2}$ and $\frac{1}{3}$,

$= \frac{3}{6} - \frac{2}{6}$ which is 6, write both fractions with that denominator, and subtract).

$= \frac{1}{6}$

CHECK Replacing *y* with $\frac{1}{6}$ in the original equation,

$y + \frac{1}{3} \stackrel{?}{=} \frac{1}{2}$	
$\frac{1}{6} + \frac{1}{3}$	$\frac{1}{2}$
$\frac{3}{6}$	
$\frac{1}{2}$	

Thus, our solution $y = \frac{1}{6}$ is correct.

PROBLEM 1

Solve.

a. $x - 6 = -8$

b. $y + \frac{1}{5} = \frac{1}{2}$

EXAMPLE 2 Solving linear equations

Solve.

a. $-x = 6$ **b.** $\frac{-y}{3} = 4$ **c.** $-3.2z = 6.4$

SOLUTION

a.

$-x = 6$ Given.

$-1x = 6$ Since $-x = -1x$.

$-1 \cdot (-1x) = -1 \cdot 6$ Multiply both sides by -1.

$x = -6$ Simplify ($-1 \cdot (-1x) = 1x = x$).

PROBLEM 2

Solve.

a. $-x = 8$ **b.** $\frac{-y}{4} = 6$

c. $-1.2z = 4.8$

(continued)

Answers to PROBLEMS

1. a. $x = -2$ **b.** $y = \frac{3}{10}$

2. a. $x = -8$ **b.** $y = -24$

c. $z = -4$

CHECK Replace x with -6 in the equation.

$$\begin{array}{c|c} -x & \stackrel{?}{=} 6 \\ \hline -(-6) & 6 \\ 6 & \end{array}$$

Hence, $x = -6$ is the correct solution.

Note that the equation $-1x = 6$ can also be solved by dividing both sides by -1 (using the division principle). The answer is still $x = -6$.

b.

$\frac{-y}{3} = 4$	Given.
$3 \cdot \frac{-y}{3} = 3 \cdot 4$	Multiply both sides by 3.
$-y = 12$	
$-1 \cdot y = 12$	Rewrite the equation.
$-1 \cdot (-1 \cdot y) = -1 \cdot 12$	Multiply both sides by -1.
$y = -12$	Simplify.

CHECK

$$\begin{array}{c|c} \frac{-y}{3} & \stackrel{?}{=} 4 \\ \hline \frac{-(-12)}{3} & 4 \\ \frac{12}{3} & \\ 4 & \end{array}$$

The solution $y = -12$ is correct.

c.

$-3.2z = 6.4$	Given.
$\frac{-3.2z}{-3.2} = \frac{6.4}{-3.2}$	Divide both sides by -3.2.
$z = -2$	Simplify.

CHECK Replace z with -2 in the original equation.

$$\begin{array}{c|c} -3.2z & \stackrel{?}{=} 6.4 \\ \hline -3.2(-2) & 6.4 \\ 6.4 & \end{array}$$

Thus, the solution is $z = -2$.

B › Solving Linear Equations

Unfortunately, not all equations are as simple as the ones discussed. Moreover, we must know what procedure to follow when confronted with any linear equation. Here is the way to do it:

PROCEDURE FOR SOLVING LINEAR EQUATIONS

1. Simplify both sides of the equations, if necessary.
2. Add or subtract the same numbers to or from both sides of the equation so that one side contains variables only.
3. Add or subtract the same terms to or from both sides of the equation so that the other side contains numbers only.
4. If the coefficient of the variable (the number multiplied by the variable) is not 1, divide both sides of the equation by this number.
5. The resulting number is the solution of the equation. Check in the original equation to be sure it is correct.

Remember, the goal when solving an equation is to have the variable by itself (isolated) on one side of the equation.

We will follow these steps to solve the equation $3(x + 1) = x + 7$.

Step 1. $3(x + 1) = x + 7$

$3x + 3 = x + 7$ Simplify. $3(x + 1) = 3x + 3$

Step 2. $3x + 3 - 3 = x + 7 - 3$ Subtract 3.

$3x = x + 4$

Step 3. $3x - x = x - x + 4$ Subtract x.

$2x = 4$

Step 4. $\frac{2x}{2} = \frac{4}{2}$ Divide by 2.

Step 5. $x = 2$

CHECK

$3(x + 1) \stackrel{?}{=} x + 7$	
$3(2 + 1)$	$2 + 7$
$3(3)$	9
9	

Of course, not all equations involve every step. We will give some examples so that you can see how it is done. A general rule to follow is that we wish to have the variables on one side of the equation and the numbers on the other side.

EXAMPLE 3 **Solving a linear equation**

Solve: $-3x + 6 = 10$.

SOLUTION We use the given steps.

Step 1. The equation is already simplified. $-3x + 6 = 10$

Step 2. Subract 6 from both sides. $-3x + 6 - 6 = 10 - 6$

Step 3. The right-hand side has numbers only. $-3x = 4$

Step 4. Divide by the coefficient of x, that is, by -3. $\frac{-3x}{-3} = \frac{4}{-3}$

Step 5. $x = -\frac{4}{3}$

CHECK

$-3x + 6 \stackrel{?}{=} 10$	
$-3 \cdot \left(-\frac{4}{3}\right) + 6$	10
$4 + 6$	
10	

The solution is $x = -\frac{4}{3}$.

PROBLEM 3

Solve: $-2x + 6 = 9$.

EXAMPLE 4 **Solving a linear equation**

Solve: $2x - 5 = 7$.

SOLUTION

Step 1. The equation is already simplified. $2x - 5 = 7$

Step 2. Add 5 to both sides. $2x - 5 + 5 = 7 + 5$

Step 3. The right-hand side has numbers only. $2x = 12$

Step 4. Divide by the coefficient of x, that is, by 2. $\frac{2x}{2} = \frac{12}{2}$

Step 5. $x = 6$

(continued)

PROBLEM 4

Solve: $3x - 5 = 7$.

Answers to PROBLEMS

3. $x = -\frac{3}{2}$ **4.** $x = 4$

CHECK

$2x - 5 \stackrel{?}{=} 7$	
$2(6) - 5$	7
$12 - 5$	
7	

The solution is $x = 6$.

EXAMPLE 5 Solving a linear equation involving decimals

Solve: $3x + 5.6 = x + 7.8$.

SOLUTION

Step 1. The equation is already simplified. $3x + 5.6 = x + 7.8$

Step 2. Subtract 5.6 from both sides.

$$3x + 5.6 - 5.6 = x + 7.8 - 5.6$$
$$3x = x + 2.2$$

Step 3. Subtract x from both sides so that there is no variable on the right-hand side of the equation.

$$3x - x = x - x + 2.2$$
$$2x = 2.2$$

Step 4. Divide by the coefficient of the variable, that is, by 2.

$$\frac{2x}{2} = \frac{2.2}{2}$$
$$x = 1.1$$

Step 5.

CHECK

$3x + 5.6 \stackrel{?}{=} x + 7.8$	
$3(1.1) + 5.6$	$1.1 + 7.8$
$3.3 + 5.6$	8.9
8.9	

The solution is $x = 1.1$.

PROBLEM 5

Solve: $3x + 2.4 = x + 8.6$.

EXAMPLE 6 Solving a linear equation containing parentheses

Solve: $4(x + 1) = 2x + 12$.

SOLUTION

Step 1. Simplify.

$$4(x + 1) = 2x + 12$$
$$4x + 4 = 2x + 12$$

Step 2. Subtract 4 from both sides.

$$4x + 4 - 4 = 2x + 12 - 4$$
$$4x = 2x + 8$$

Step 3. Subtract $2x$ from both sides.

$$4x - 2x = 2x - 2x + 8$$
$$2x = 8$$

Step 4. Divide by the coefficient of x, that is, by 2.

$$\frac{2x}{2} = \frac{8}{2}$$

Step 5.

$$x = 4$$

CHECK

$4(x + 1) \stackrel{?}{=} 2x + 12$	
$4(4 + 1)$	$2(4) + 12$
$4(5)$	$8 + 12$
20	20

The solution is $x = 4$.

PROBLEM 6

Solve: $5(x + 2) = 2x + 19$.

EXAMPLE 7 Solving a linear equation involving parentheses

Solve: $20 = 5(x + 2)$.

SOLUTION Before solving this equation, remember that we wish to have the variables on one side of the equation and the numbers on the other side.

PROBLEM 7

Solve: $10 = 2(x + 3)$.

Answers to PROBLEMS

5. $x = 3.1$ **6.** $x = 3$ **7.** $x = 2$

Step 1. Simplify. $20 = 5(x + 2)$

$20 = 5x + 10$

Step 2. Subtract 10 from both sides. $20 - 10 = 5x + 10 - 10$

$10 = 5x$

Step 3. There is no variable on the left-hand side.

Step 4. Divide both sides by 5. $\frac{10}{5} = \frac{5x}{5}$

Step 5. $2 = x$

$x = 2$

CHECK

$20 = 5(x + 2)$	
20	$5(2 + 2)$
	$5(4)$
	20

The solution is $x = 2$.

EXAMPLE 8 Solving linear equations involving fractions

Solve.

a. $\frac{2}{3}x = -5$ **b.** $\frac{1}{8}x - \frac{3}{8} + \frac{1}{2}x = \frac{1}{2} + x$

SOLUTION

a. Since all the x's are alone on the left side, we can skip steps 1, 2, and 3. The coefficient of x is $\frac{2}{3}$, so we can *divide* both sides by $\frac{2}{3}$. By the definition of division of fractions, dividing by $\frac{2}{3}$ is the same as *multiplying* by the *reciprocal* of $\frac{2}{3}$—that is multiplying by $\frac{3}{2}$. Since it is easier to multiply both sides by $\frac{3}{2}$, we proceed in that fashion.

$\frac{2}{3}x = -5$ Given.

$\frac{3}{2} \cdot \frac{2}{3}x = \frac{3}{2} \cdot (-5)$ Multiply both sides by $\frac{3}{2}$.

$x = -\frac{15}{2}$ Simplify.

CHECK

$\frac{2}{3}x \stackrel{?}{=} -5$	
$\frac{2}{3}\left(-\frac{15}{2}\right)$	-5
-5	

Thus, the solution is $x = -\frac{15}{2}$.

b. We can start by combining like terms, but we can make the problem easier if we multiply both sides by the LCD of all terms (8). This is an *optional* step, but it will save work. Here is the procedure.

$\frac{1}{8}x - \frac{3}{8} + \frac{1}{2}x = \frac{1}{2} + x$ Given.

$8 \cdot \left(\frac{1}{8}x - \frac{3}{8} + \frac{1}{2}x\right) = 8 \cdot \left(\frac{1}{2} + x\right)$ Multiply both sides by 8.

$8 \cdot \frac{1}{8}x - 8 \cdot \frac{3}{8} + 8 \cdot \frac{1}{2}x = 8 \cdot \frac{1}{2} + 8 \cdot x$ Simplify.

$x - 3 + 4x = 4 + 8x$

$5x - 3 = 4 + 8x$

Now, on which side of the equation should we leave the x's? It is easier (because it avoids expressions with a negative sign in front) to leave the x's on the right. (As a rule of thumb, leave the x's on the side that has more of them.) Here is the rest of the problem.

(continued)

PROBLEM 8

Solve:

a. $\frac{3}{4}x = -2$

b. $\frac{1}{3}x - \frac{1}{4} + \frac{3}{4}x = \frac{1}{3} + 2x$

Answers to PROBLEMS

8. a. $x = -\frac{8}{3}$ **b.** $x = -\frac{7}{11}$

Step 1. The equation is simplified.	$5x - 3 = 4 + 8x$
Step 2. Subtract 4 from both sides.	$5x - 3 - 4 = 4 - 4 + 8x$
	$5x - 7 = 8x$
Step 3. Subtract $5x$ from both sides.	$5x - 5x - 7 = 8x - 5x$
	$-7 = 3x$
Step 4. Divide both sides by 3.	$-\frac{7}{3} = \frac{3x}{3}$
Step 5.	$x = -\frac{7}{3}$

The check is left for you. The solution is $x = -\frac{7}{3}$.

When solving Example 8b, we introduced two new steps that have to be done when solving a linear equation involving fractions: **C**learing the fractions (by multiplying both sides of the equation by the LCD) and **R**emoving parentheses. You can remember the steps to use if you remember to **CRAM** as shown next.

PROCEDURE FOR SOLVING LINEAR EQUATIONS (CRAM)

Clear fractions by multiplying both sides of the equation by the LCD.
Remove parentheses (simplify).
Add or subtract numbers and expressions so the variables are on one side (isolated).
Multiply or divide by the coefficient of the variable.

› Exercises 10.4

› Web IT go to mhhe.com/bello for more lessons

‹ A › Solving Equations Using the Addition, Subtraction, Multiplication, or Division Principle In Problems 1–26, solve the equation.

1. $x - 2 = -4$
2. $-7 = y - 5$
3. $-\frac{3}{4} = y - \frac{1}{4}$
4. $-\frac{4}{5} = x - \frac{1}{5}$
5. $x - 1.9 = -8.9$
6. $y - 3.7 = -9.7$
7. $\frac{1}{2} = y + \frac{1}{5}$
8. $\frac{1}{3} = x + \frac{1}{4}$
9. $x + 3.8 = 9.9$
10. $y + 2.5 = 6.9$
11. $-x = 8$
12. $-y = 3$
13. $-y = -\frac{1}{5}$
14. $-x = -\frac{1}{7}$
15. $-x = 2.3$
16. $-y = 5.7$
17. $\frac{-x}{2} = 8$
18. $\frac{-y}{3} = 7$
19. $3.1 = \frac{-y}{4}$
20. $2.2 = \frac{-x}{5}$
21. $-2.1z = 4.2$
22. $-3.8z = 1.9$
23. $-1.2 = 2.4x$
24. $-1.8 = 3.6x$
25. $3.6y = 4.8$
26. $6.4y = 3.2$

‹ B › Solving Linear Equations In Problems 27–60, solve the equation.

27. $2x + 7.1 = 9.3$
28. $3y + 7.2 = 13.8$
29. $6.5 = 3y + 3.2$
30. $21 = 2.5x + 1$
31. $\frac{3}{5}x - \frac{6}{5} = \frac{4}{5}$
32. $\frac{3}{5}y - \frac{2}{5} = \frac{4}{5}$
33. $\frac{1}{3} = \frac{5}{2}x - \frac{8}{3}$
34. $\frac{1}{2} = \frac{3}{4}x - \frac{3}{2}$
35. $3.5x + 5 = 1.5x + 7$
36. $20 - 3.3x = 5.7x + 2$
37. $21 - 1.5y = 6.5y + 5$
38. $2(x + 5) = -12$

39. $3(y + 2) = -24$

40. $8(y - 1) = y + 2$

41. $6(x - 1) = x + 6$

42. $x + 5 = -3(x + 1)$

43. $y + 6 = -2(y + 2)$

44. $11 - x = 4(x - 1)$

45. $1 - y = 5(y - 1)$

46. $2x + 1 = 3(x + 1)$

47. $3x + 1 = 2(x + 1)$

48. $5x + 1 = -5x + 1$

49. $6x + 2 = -6x + 2$

50. $7x + 3 = -7x + 3$

51. $\frac{3}{5}x = -4$

52. $\frac{2}{7}y = -3$

53. $-\frac{3}{8}y = 2$

54. $-\frac{4}{5}x = 3$

55. $-\frac{2}{5}x = -3$

56. $-\frac{3}{7}y = -2$

57. $\frac{5}{6}y - \frac{1}{4} = \frac{1}{2} - \frac{2}{3}y$

58. $\frac{3}{5}x + \frac{2}{3} = \frac{4}{3} + \frac{9}{5}x$

59. $\frac{3}{2}y - \frac{1}{3} = \frac{5}{4}y + \frac{1}{8}$

60. $\frac{4}{9}x - \frac{3}{2} = \frac{5}{6}x - \frac{3}{2}$

61. *Fuel* The 2.8 billion gallons of ethanol (a type of alcohol used as a gas additive) produced in 2003 increased by I to reach 3.4 billion gallons in 2004.

a. Write an equation for this situation.

b. What principle do you use to solve for I?

c. Find the value of I.

62. *Fuel* In 2004, production of ethanol reached 3.4 billion gallons, double the production P for the year 2000.

a. Write an equation for this situation.

b. What principle do you use to solve for P?

c. Find the value of P.

63. *Fuel* The 2.8 billion gallons of ethanol produced in 2003 was only $\frac{1}{50}$ of the total gas T used in the United States that year.

a. Write an equation for this situation.

b. What principle do you use to solve for T?

c. Find the value of T.

64. *Fuel* In the meantime, oil production P in the United States declined by 6 million barrels per day to 5 million barrels per day from 1980 to 2005.

a. Write an equation for this situation.

b. What principle do you use to solve for P?

c. Find the value of P.

> > > Using Your Knowledge

The weekly salary of a salesperson is given by

Salary		Commission		Total pay
S	$+$	C	$=$	T

65. If the person made \$66 on commissions and the total pay was \$176, what was the person's salary?

66. If a person's salary was \$133 and the total pay was \$171, how much did the person earn in commissions?

Your bank balance is given by

Deposits		Withdrawals		Balance
D	$-$	W	$=$	B

67. A person deposited \$308. The balance was \$186. How much money did the person withdraw from her account?

68. The balance in a checking account was \$147. The deposits were \$208. How much money was withdrawn from the account?

The weight of a man is related to his height by

Weight (pounds)		Height (in inches)	
W	$=$	$5H$	$- 190$

69. A man weighs 160 pounds. What is his height?

> Web IT go to mhhe.com/bello for more lessons

Write On

70. Write in your own words the difference between an expression and an equation.

71. When solving the equation $-\frac{3}{4}x = 6$, would it be easier to multiply by the reciprocal $-\frac{4}{3}$ or to divide both sides by $-\frac{3}{4}$?

72. Describe in your own words the solution of an equation. Can you make an equation with no solution?

Concept Checker

Fill in the blank(s) with the correct word(s), phrase, or mathematical statement.

73. An ______________ is a **sentence having a variable** in which the verb is *equals*.

74. By the **addition principle,** the equation $a = b$ is equivalent to ______________.

75. By the **subtraction principle,** the equation $a = b$ is equivalent to ______________.

76. By the **multiplication principle,** the equation $a = b$ is equivalent to ______________.

77. By the **division principle,** the equation $a = b$ is equivalent to ______________.

78. The **first step** in the procedure for solving an equation is to ______________ both sides of the equation.

simplify

$\frac{a}{c} = \frac{b}{c}, c \neq 0$

$a \cdot c = b \cdot c$

$a - c = b - c$

$a + c = b + c$

multiply

equation

Mastery Test

Solve the equation.

79. $x - 3 = -4$

80. $y + \frac{1}{4} = \frac{1}{2}$

81. $-y = 7$

82. $\frac{-x}{4} = 3$

83. $-3.3z = 9.9$

84. $-4x + 6 = 9$

85. $3x - 6 = 9$

86. $4x + 6.6 = x + 7.8$

87. $3(x + 1) = 2x + 7$

88. $30 = 6(x + 2)$

89. $\frac{2}{3}x = -6$

90. $\frac{1}{8}x - \frac{1}{8} + \frac{1}{2}x = \frac{3}{4} + x$

Skill Checker

In Problems 91–92, multiply:

91. $\frac{2}{3} \cdot 1000$

92. $\frac{3}{4} \cdot 500$

In Problems 93–94, divide:

93. $67.50 \div 0.25$

94. $90.3 \div 0.15$

10.5 Applications: Word Problems

Objective

A > You should be able to solve word problems using the RSTUV procedure.

To Succeed, Review How To . . .

1. Use the RSTUV method presented in Sections 2.8 and 3.5. (p. 92)
2. Solve equations. (pp. 89–91, 240–241, 602–608)

Getting Started

Do you believe the United States sent a mission to the moon? Some people don't, but the illustration shows the Lunar Rover used by the astronauts of *Apollo 15*. The weight of this vehicle on Earth (450 pounds) is 6 times the weight of the vehicle on the moon. What is the weight of the vehicle on the moon?

A > Solving Word Problems

In this section we are going to use the language of algebra and translate word sentences into equations. We will then solve these equations using the methods of the previous section and the **RSTUV** procedure we studied in Section 2.8. We repeat the procedure here for your convenience.

RSTUV PROCEDURE TO SOLVE WORD PROBLEMS

1. **R**ead the problem carefully and decide what is asked for (the *unknown*).
2. **S**elect □ or a letter to represent the unknown.
3. **T**ranslate the problem into an equation.
4. **U**se the rules we have studied to solve the equation.
5. **V**erify the answer.

We now use these five steps to solve the problem in the *Getting Started*.

Step 1. **R**ead the problem. Decide what number is asked for. In our problem we are asked for the weight of the vehicle on the moon.

Step 2. **S**elect w to represent the weight of the vehicle on the moon.

Step 3. Translate. According to the problem,

The weight of the vehicle on Earth | is | 6 times the weight of the vehicle on the moon.

$$450 = 6w$$

Step 4. Use algebra to solve the equation $450 = 6w$:

$$450 = 6w$$

$$\frac{450}{6} = \frac{6w}{6} \quad \text{Divide both sides by 6.}$$

$$w = 75$$

Step 5. Verify the requirements of the problem. Is the weight of the vehicle on Earth (450 pounds) 6 times the weight of the vehicle on the moon? Since $450 = 6 \cdot 75$, our answer is correct. Thus the weight of the vehicle on the moon is 75 pounds.

EXAMPLE 1 Problem solving: Astronauts

One of the moon astronauts, Eugene "Buzz" Aldrin, weighed 180 pounds on Earth. His Earth weight was 6 times his weight on the moon. How much did he weigh on the moon?

SOLUTION We proceed by steps.

Step 1. Read the problem carefully and decide what number it asks for. It asks for Aldrin's weight on the moon.

Step 2. Select w to be this weight.

Step 3. Translate. According to the problem,

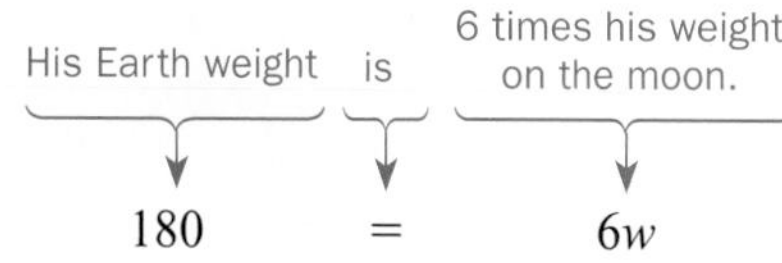

Step 4. Use algebra to solve $180 = 6w$:

$$180 = 6w$$

$$\frac{180}{6} = \frac{6w}{6} \quad \text{Divide by 6.}$$

$$30 = w$$

$$w = 30$$

Step 5. Verify. Is it true that $180 = 6 \cdot 30$? Yes, our answer is correct. Thus, Aldrin's weight on the moon was 30 pounds.

PROBLEM 1

Another astronaut, Neil Armstrong, weighed 162 pounds on Earth. This weight was 6 times his weight on the moon. How much did he weigh on the moon?

EXAMPLE 2 Problem solving: Dieting

Celesta Geyer, alias Dolly Dimples, decreased her weight by 401 pounds, to 152, in 14 months. How much did she weigh at the beginning of her diet?

SOLUTION

Step 1. Read the problem carefully and find out what we are asked for. In this problem we want to know her weight at the beginning of the diet.

Step 2. Select w to be this weight.

Step 3. Translate. According to the problem,

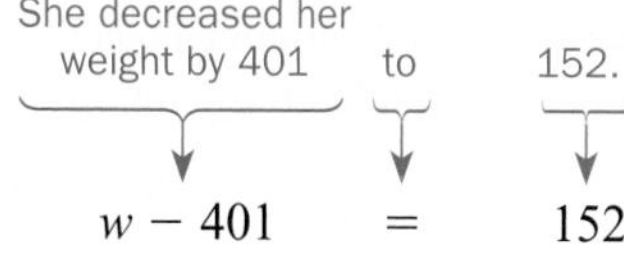

PROBLEM 2

Paul Kimmelman holds the speed record for weight reducing. His weight decreased by 357 pounds, to 130, in about eight months. How much did he weigh at the beginning of his diet?

Answers to PROBLEMS

1. 27 lb **2.** 487 lb

Step 4. Use algebra to solve $w - 401 = 152$:

$$w - 401 = 152$$

$$w - 401 + 401 = 152 + 401 \quad \text{Add 401 to both sides.}$$

$$w = 553$$

Step 5. Verify the answer. Since $553 - 401 = 152$, our result is correct. Thus, her weight was 553 pounds.

EXAMPLE 3 Problem solving: Anthropology

Do you know how anthropologists classify the different forms of humans? By measuring the brain in cubic centimeters. *Homo erectus,* of which two subspecies have been identified, had a brain that was 1000 cubic centimeters, or about $\frac{2}{3}$ that of a modern person. How many cubic centimeters of brain does a modern person have?

SOLUTION

Step 1. Read the problem. It asks for the number of cubic centimeters of brain in a modern person.

Step 2. Select c to be this number.

Step 3. Translate. According to the problem,

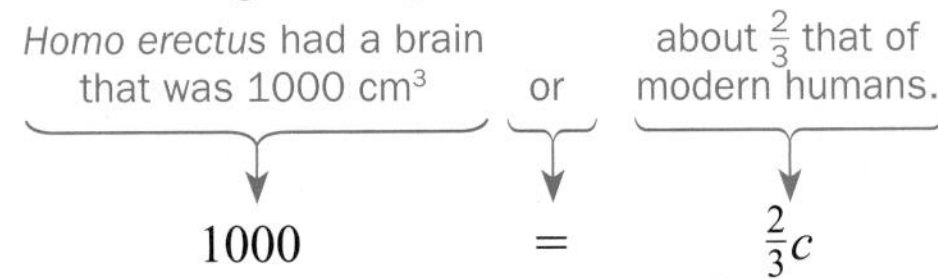

$$1000 = \frac{2}{3}c$$

Step 4. Use algebra to solve the equation $1000 = \frac{2}{3}c$:

$$1000 = \frac{2}{3}c$$

$$\frac{3}{2} \cdot 1000 = \frac{3}{2} \cdot \frac{2}{3}c \quad \text{Multiply both sides by the reciprocal of } \frac{2}{3}, \frac{3}{2}.$$

$$\frac{3000}{2} = c$$

$$c = 1500$$

Step 5. Verify the answer. Since

$$1000 = \frac{2}{\cancel{3}} \cdot \overset{500}{\cancel{1500}} = 1000$$

our result is correct. Thus, a modern person has 1500 cubic centimeters of brain capacity.

PROBLEM 3

Water is made up of eight parts of oxygen and one of hydrogen by weight, which means that $\frac{8}{9}$ of water is oxygen. If the amount of oxygen in a bucket of water is 368 grams, how much water is there in the bucket?

EXAMPLE 4 Problem solving: Renting a car

The daily cost C (in dollars) of renting a car is \$20 plus \$0.25 per mile traveled. If m is the number of miles traveled, then

$$C = 20 + 0.25m$$

An executive rents a car for a day and pays \$87.50. How many miles did the executive travel?

SOLUTION

Step 1. Read the problem. It asks for the number of miles traveled.

Step 2. Select m to be this number.

Step 3. Translate. According to the problem,

$$C = 20 + 0.25m$$

or

$$87.50 = 20 + 0.25m$$

PROBLEM 4

The daily cost of renting a car is given by

$$C = 15 + 0.20m$$

where C is the daily cost (in dollars) and m the number of miles traveled. A person rented a car for a day and paid \$65. How many miles did the person travel?

(continued)

Answers to PROBLEMS

3. 414 g **4.** 250 miles

Step 4. Use algebra to solve the equation.

$$87.50 = 20 + 0.25m$$
$$87.50 - 20 = 20 - 20 + 0.25m \quad \text{Subtract 20.}$$
$$67.50 = 0.25m \quad \text{Simplify.}$$
$$\frac{67.50}{0.25} = m \quad \text{Divide by 0.25.}$$

```
        2 70
0.25 )67.50
       50
       17 5
       17 5
          00
```

Thus, $m = 270$.

Step 5. Verify the answer. Since $0.25 \cdot 270 + 20 = 67.50 + 20 = 87.50$, our result is correct. Thus the executive traveled 270 miles.

EXAMPLE 5 Problem solving: Rate of interest

Angie bought a 6-month, $10,000 certificate of deposit. At the end of the 6 months, she received $650 simple interest. What rate of interest did the certificate pay?

SOLUTION

Step 1. Read the problem. It asks for the rate of simple interest.

Step 2. Select the variable r to represent this rate.

Step 3. Translate the problem. Here, we need to know that the formula for simple interest is

$$I = Prt$$

where I is the amount of interest, P is the principal, r is the interest rate, and t is the time in years. For our problem, $I = \$650$, $P = \$10{,}000$, r is unknown, and $t = \frac{1}{2}$ year. Thus, we have

$$650 = (10{,}000)(r)\left(\frac{1}{2}\right)$$
$$650 = 5000r$$

Step 4. Use algebra to solve the equation.

$$650 = 5000r$$
$$\frac{650}{5000} = r \quad \text{Divide by 5000.}$$
$$r = 0.13 = 13\%$$

Step 5. Verify the answer. Is the interest earned on a $10,000, 6-month certificate at a 13% rate $650? Evaluating Prt, we have

$$(10{,}000)(0.13)\left(\frac{1}{2}\right) = 650$$

Since the answer is yes, 13% is correct. The certificate paid 13% simple interest.

PROBLEM 5

Angel bought a 6-month, $10,000 certificate of deposit. At the end of the 6 months, he received $600 simple interest. What rate of interest did the certificate pay?

You have probably noticed the frequent occurrence of certain words in the statements of word problems. Because these words are used often, here is a small mathematics dictionary to help you translate them properly.

Answers to PROBLEMS

5. 12%

Mathematics Dictionary

Words	Translation	Example	Translation
Add, more than, sum, increased by, added to	$+$	Add n to 7 7 more than n The sum of n and 7 n increased by 7 7 added to n	$n + 7$
Subtract, less than, minus, difference, decreased by, subtracted from	$-$	Subtract 9 from x 9 less than x x minus 9 Difference of x and 9 x decreased by 9 9 subtracted from x	$x - 9$
Of, the product, times, multiply by	$\times$	$\frac{1}{2}$ of a number x The product of $\frac{1}{2}$ and x $\frac{1}{2}$ times a number x Multiply $\frac{1}{2}$ by x.	$\frac{1}{2}x$
Divide, divided by, the quotient	$\div$	Divide 10 by x. 10 divided by x The quotient of 10 and x	$\frac{10}{x}$
The same, yields, gives, is	$=$	$\frac{6}{3}$ is the same as 2. 6 divided by 3 yields 2. 6 divided by 3 gives 2. The quotient of 6 and 3 is 2.	$\frac{6}{3} = 2$

Before we practice solving some more word problems, let us see how the words contained in the mathematics dictionary are used in real life. Example 6 shows actual excerpts from an article entitled *Biofuels for Transportation: Selected Trends and Facts* published by the Worldwatch Institute. We are going to use the dictionary to translate each of the lines.

Source: http://www.worldwatch.org/node/4081.

EXAMPLE 6 Translating words into mathematical expressions

Translate the underlined words into a mathematical expression.

a. From 2002 to 2004, world oil demand D increased by 5.3%

b. Consumption C in the U.S. rose by 4.9%

c. Canada 10.2%

SOLUTION

a. The key word is *increased,* which means +. The translation is $D + 0.053D$.

b. The key word is *rose,* which means *increased* and +. The translation is $C + 4.9\%$.

c. Here we have to assume that Canada's consumption *rose* by 10.2%, which is translated as $C + 10.2\%$.

PROBLEM 6

Translate the underlined words into mathematical expressions.

a. The United Kingdom's consumption C increased by 6.3%.

b. Demand D in Germany and Japan, meanwhile, dropped by 1 percent and

c. 2.6%, respectively.

EXAMPLE 7 Problem solving: Integers

If 7 is added to twice a number n, the result is 9 times the number. Find the number.

PROBLEM 7

If 6 is added to 3 times a number, the result is 6 times the number. Find the number.

(continued)

Answers to PROBLEMS

6. a. $C + 0.063C$ **b.** $D - 1\%$ **c.** $D - 2.6\%$ **7.** 2

SOLUTION

Step 1. Read the problem.

Step 2. Select n to be the number.

Step 3. Translate the problem.

If 7 is	added to	twice a number n	the result is	9 times the number.
↓	↓	↓	↓	↓
7	+	$2n$	=	$9n$

Step 4. Use algebra to solve the equation.

$$7 + 2n = 9n$$

$$7 + 2n - 2n = 9n - 2n \quad \text{Subtract } 2n.$$

$$7 = 7n \quad \text{Simplify.}$$

$$\frac{7}{7} = \frac{7n}{7} \quad \text{Divide by 7.}$$

$$1 = n$$

Step 5. Verify the answer. Is $7 + 2(1) = 9(1)$? Since $7 + 2 = 9$, our result is correct. The answer is 1.

EXAMPLE 8 Problem solving: Hamburger calories

A McDonald's hamburger and large fries contain 820 calories. Which has more calories, the burger or the fries? The fries, of course! In fact, the fries have almost *double* the number of calories of the burger, but not quite. The exact relationship is *if you double the calories in the hamburger and reduce the result by 20, you will have the same number of calories as are in the fries.* How many calories are in the hamburger and how many are in the fries?

SOLUTION

Step 1. Read the problem.

Step 2. Select h to be the number of calories in the hamburger and f to be the number of calories in the fries.

Step 3. Translate the problem.

If you double the calories in the hamburger ($2h$) and reduce the result by 20, you will have the same number of calories as are in the fries (f), that is, $2h - 20 = f$.

However, $2h - 20 = f$ has two variables. Anything missing?
Yes, the 820. It does say at the beginning of the problem that a McDonald's hamburger and large fries contain 820 calories.

That is, $h + f = 820$

Substituting for f $\quad h + (2h - 20) = 820$

Step 4. Use algebra to solve.

$$3h - 20 = 820 \quad \text{Simplify.}$$

$$3h = 840 \quad \text{Add 20.}$$

$$h = 280 \quad \text{Divide by 3.}$$

To find the calories in the fries, double 280 and subtract 20 to get 540 calories ($2 \cdot 280 - 20 = 540$) for the fries.

Step 5. Verify the answer. Since $h = 280$ and $f = 540$, we substitute those numbers in the original equation:

$$h + f = 820$$

$$280 + 540 = 820$$

$$820 = 820$$

So the results for h and f are correct.

PROBLEM 8

Believe it or not, if you eat a Burger King hamburger with large fries, you will also get 820 calories. However, if you double the calories in the burger, you will have 140 more calories than are in the fries. How many calories are in each?

Answers to PROBLEMS

8. $h = 320$; $f = 500$

Before you attempt to solve the exercises, practice translating!

TRANSLATE THIS

The third step in the RSTUV procedure is to **TRANSLATE** the information into an equation. In Problems 1–10, **TRANSLATE** the sentence and match the correct translation with one of the equations A–O.

1. The amount of financial aid F received by graduate and undergraduate students in a recent year was \$122 billion plus 11% over the amount P received the preceding year.
2. The \$23 billion distributed by colleges and universities in grant aid represents 19% of total student aid funds F.

   Source: The College Board.
3. In a recent year, the average federal grant to students was \$2421, which falls \$24,000 short of Tuition and Fees TF at Harvard.
4. A survey of T employed American adults indicated that $\frac{1}{2}$ of them pay off debt when they receive extra money. This represents 1365 of the people surveyed.
5. The speed D (in centimeters per second) of an ant is the product of $\frac{1}{6}$ and the difference of C and 4 (C the temperature in degrees Celsius).
6. The number of chirps N a cricket makes in 1 minute is the product of 4 and the difference of F and 40 (F the temperature in degrees Fahrenheit).
7. The charges C from a plumbing company amounted to \$35 for the service call plus \$40 for each hour h spent on the job.
8. The equation y of a line is the sum of b and the product of its slope m and the variable x.
9. The number of 20-year-old females enrolled in the Student Center exceeds the number of males M by 10,000. Write an equation representing the total membership of 72,000 students enrolled.
10. A survey of W young women, ages 15 to 24, found that 60% or 300, of them were currently involved in an ongoing abusive relationship.

A. $2421 = TF - 24,000$
B. $D = \frac{1}{6}(C - 4)$
C. $2421 - TF = 24,000$
D. $23 = 0.19F$
E. $D = \frac{1}{6}C - 4$
F. $C = 35 + 40h$
G. $y = b + mx$
H. $N = 4(F - 40)$
I. $F = 122 + 0.11P$
J. $72,000 = M + (M + 10,000)$
K. $F = 122 + 1.11P$
L. $C = 40 + 35h$
M. $N = 4F - 40$
N. $\frac{1}{2}T = 1365$
O. $0.60W = 300$

Boost your grade at mathzone.com!
> Practice Problems
> NetTutor
> Self-Tests
> e-Professors
> Videos

› Exercises 10.5

‹ A › Solving Word Problems For Problems 1–30, try to estimate the answer, then follow the RSTUV procedure given in the text to solve the problems.

1. A woman deposited \$300 in her savings account. At the end of the year, her balance was \$318. If this amount represents her \$300 plus interest, how much interest did she earn during the year?
2. A man borrowed \$600 from a loan company. He made payments totaling \$672. If this amount is the original \$600 plus interest, how much interest did he pay?
3. In a recent election the winning candidate received 10,839 votes, 632 more votes than the loser. How many votes did the loser get?
4. The total cost of a used car is \$3675. You can buy the car with \$500 down and the rest to be financed. How much do you have to finance?
5. In a recent election the losing candidate received 9347 votes, 849 votes less than the winning candidate. How many votes did the winner get?
6. A woman's weekly salary after deductions is \$257. If her deductions are \$106, what is her salary before the deductions?
7. The cost of an article is \$57. If the profit when selling it is \$11, what is the selling price of the article?
8. In accounting,

   assets owner's equity = liabilities

   If the owner's equity is \$4787 and the liabilities \$1688, what are the assets?
9. During strenuous exercise the flow of blood pumped by the heart is increased to four times its normal rate, reaching 40 pints per minute. How many pints of blood per minute does the heart pump when at rest?
10. It takes 9 times as much steel as nickel to make stainless steel. If 81 tons of stainless steel are used in constructing a bridge, how many tons of nickel are needed?

› Web IT go to mhhe.com/bello for more lessons

11. During strenuous exercise the number of breaths a person takes each minute increases to 5 times the normal rate, reaching 60 breaths per minute. How many breaths per minute should a person take when at rest?

12. Nutritionists have discovered that a cup of coffee has 144 mg of caffeine, 3 times the amount contained in an average cola drink. How many milligrams are there in an average cola drink?

13. When a gas tank is $\frac{2}{3}$ full, it contains 18 gallons of gas. What is the capacity of this tank?

14. A man is entitled to $\frac{3}{4}$ his pay when he retires. If he receives \$570 per month after retirement, what was his regular salary?

15. The dome of the village church of Equord, Germany, is $\frac{1}{16}$ as high as the dome of St. Peter's Basilica. If the dome of the village church is 43.75 feet high, how high is the dome of St. Peter's Basilica?

16. The density of a substance is given by the equation

$$d = \frac{m}{v}$$

where m is the mass and v is the volume. If the density of sulfur is 2 grams per milliliter and the volume is 80 milliliters, what is the mass of the sulfur?

17. The price-earnings (P/E) ratio of a stock is given by the equation

$$\text{P/E} = \frac{M}{E}$$

where M is the market price of the stock and E is the earnings. If the P/E ratio of a stock earning \$3 per share is 12, what should the market value of the stock be?

18. Supporting tissues, like bones, ligaments, tendons, and fats make up $\frac{1}{3}$ of the body. If a woman has 40 pounds of supporting tissue in her body, what is her total weight?

19. The best linguist of all times is Cardinal Giuseppe Caspor Mezzofanti, who can translate 186 languages and dialects. If he can translate 42 more languages than dialects, how many dialects can he translate?

20. Mezzofanti can speak 6 more than 3 times the number of languages he uses in interviews. If he can speak 39 languages, how many languages does he use in interviews? (See Problem 19.)

21. Brenda bought a \$1000 certificate of deposit. At the end of the year, she received \$80 simple interest. What rate of interest did the certificate pay?

22. Hadish deposited \$800 in a savings account. At the end of the year, he received \$40 simple interest? What rate of interest did the account pay?

23. Tonya invested \$3000 in municipal bonds. At the end of the year, she received \$240 simple interest. What was the rate of interest paid by the bonds?

24. Pedro buys a \$10,000, 3-month certificate of deposit. At the end of the 3 months, he receives simple interest of \$350. What was the rate of interest on the certificate?

25. A loan company charges \$588 for a 2-year loan of \$1400. What simple interest rate is the loan company charging?

26. If 4 times a number is increased by 5, the result is 29. Find the number.

27. Eleven more than twice a number is 19. Find the number.

28. The sum of 3 times a number and 8 is 29. Find the number.

29. If 6 is added to 7 times a number, the result is 69. Find the number.

30. If the product of 3 and a number is decreased by 2, the result is 16. Find the number.

Problems 31–38 are based on a Websense Inc. survey of Information Technology (IT) decision-makers conducted by Harris Interactive.

Source: http://www.websense.com.

31. *Surfing at work* IT decision-makers estimate employees spend about 5.7 hours a week on personal surfing at work. Employees deny this claim. They say the 5.7 hours is 1.9 times as many hours as they are willing to admit. How many hours are employees willing to admit to personal surfing at work?

32. *Surfing for the weather* When employees were asked what types of non-work-related websites they visited, 375 of them (75%), said they visited weather sites. How many employees were surveyed?

33. *Unauthorized disk space* IT decision-makers estimate that 7.5% of their organization's total disk space is filled with non-work-related files (like music and photo files). If this 7.5% amounts to 600 gigs of disk space, what is the total disk space (in gigs) for the company?

34. *Unauthorized disk space* Mid-sized companies tend to have a greater percentage of their disk space taken up by non-work-related files: 10.1%. If this 10.1% of the total disk space amounts to 505 gigs of space, what is the total disk space (in gigs) for this mid-sized company?

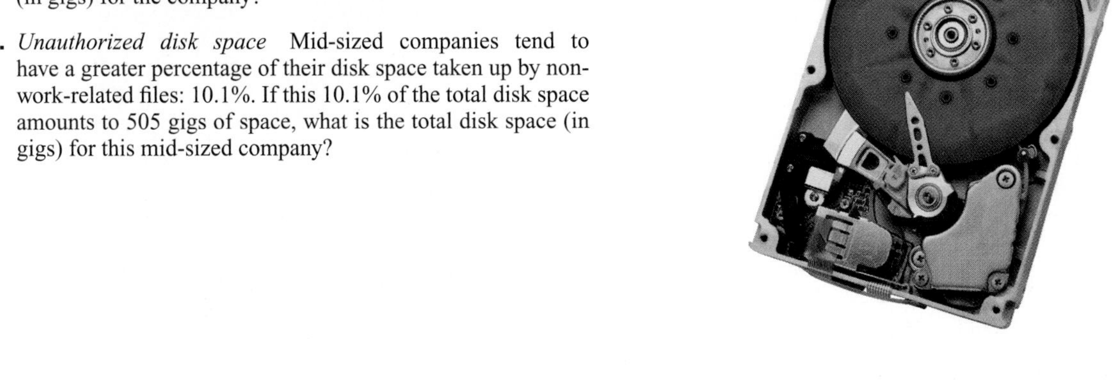

35. *Total people* The survey covered exactly 851 people: 149 more employees than IT decision-makers. How many employees and how many IT decision-makers were surveyed?

36. *Phishing* Do you know what phishing is? It is a type of fraud in which the perpetrator entices people to share passwords or credit card numbers so they can be used illegally at a later time. In the survey, about 4 out of 5 or 280 IT decision-makers reported that their employees had received a phishing attack. About how many IT decision-makers were surveyed?

Source: www.plus.es.

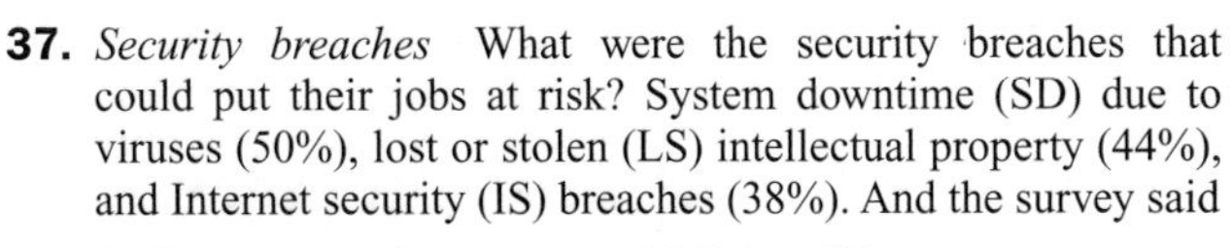

37. *Security breaches* What were the security breaches that could put their jobs at risk? System downtime (SD) due to viruses (50%), lost or stolen (LS) intellectual property (44%), and Internet security (IS) breaches (38%). And the survey said

20 fewer respondents answered LS than SD

42 fewer respondents answered IS than SD

If the total number of responses was 463, how many responded SD, LS, and IS, respectively?

38. *Intellectual loss* The loss of intellectual property is a main concern with IT decision-makers. Forty percent said they are very or extremely concerned (VE) with it, 35% say they are somewhat concerned (SC) with it, and 30% said they are not very (NV) or not at all concerned with it. If there are 35 fewer respondents answering NV than VE, 17 fewer respondents answering SC than VE, and the total number of respondents is 368, how many respondents are in each category? The percents add up to 132%, so multiple responses were allowed, presumably.

Problems 39–42 refer to another survey about viruses and the illnesses they cause conducted by Harris Interactive.
Source: http://www.harrisinteractive.com.

39. *Survey coverage* The survey covered 1373 children and teenagers. The number of teens (13–18) surveyed exceed the number of tweens (8–12) by 227. How many teens and how many tweens were surveyed?

40. *HIV/AIDS* Kids were asked if it was likely that they or one of their friends would catch HIV/AIDS. Here are the results: the number of teens that answered "yes" exceeded the number of tweens answering "yes" by 56. If the total number of "yes" answers was 216, how many teens and how many tweens answered "yes"?

41. *Hepatitis C* Thirty-five more teens than tweens answered "yes" when asked if it was likely that they or one of their friends would catch hepatitis C. If the total number of "yes" answers was 173, how many teens and how many tweens answered "yes"?

42. *Antibiotics* This time 61% of the tweens (349.53 of them), said "yes," while 59% of the teens (472 of them) agreed. How many tweens and how many teens were surveyed?

Problems 43–50 are based on the article: How Many Calories Does Your Body Need?
Source: http://www.primusweb.com.

43. *Basal metabolic rate* One of the ways you can answer that question is by finding your basal metabolic rate (BMR); this is the number of calories you need per day to maintain weight. For males, the BMR is defined as follows: Multiply your body weight W (in pounds) by 10 and add double the body weight to the value.

a. Write a formula for the BMR of a male.

b. A male needed 1800 calories per day to maintain his weight. What was his weight?

44. *BMR for adult females* The BMR for females is different than for males. To find the BMR for a female multiply her body weight W (in pounds) by 10 and add the body weight to the value.

a. Write a formula for the BMR of a female.

b. A female needed 1320 calories per day to maintain her weight. What was her weight?

45. *Exercise and weight loss* One way to lose weight is to exercise! Physical activity contributes 20% to 30% of the body's total energy output. If you burn 700 calories through exercise and this represents 20% of your caloric output O for a week:

a. Write an equation for the caloric output O for the week.

b. What is your caloric output O for the week?

c. To lose one pound, you need to burn 3500 calories. If your caloric output is as in part **b,** how many pounds would you lose that week?

46. *Caloric output for different activities and weights* Suppose you weigh 150 pounds.

Activity (one hour)	100 lb	150 lb	200 lb
Bicycling, 6 mph	160	240	312
Bicycling, 12 mph	270	410	534
Jogging, 7 mph	610	920	1230

a. Write a formula for the number of calories C you will use in h hours of bicycling at 6 mph.

b. Use the formula to find the number of calories C you will use in 3 hours of bicycling at 6 mph.

c. A person used 1230 calories bicycling at 12 mph. Consult the chart, write an equation for the number of calories used in bicycling for h hours, and find out how many hours the person bicycled.

47. *Caloric output for jogging* Suppose you weigh 150 pounds.

a. Write a formula for the number of calories C you will use in h hours of jogging at 7 mph.

b. Use the formula to find the number of calories C you will use in 3 hours of jogging at 7 mph. Relax; you don't have to do it all at once!

c. Who will burn more calories jogging at 7 mph: a 100-pound person jogging for 3 hours or a 150-pound person jogging at 7 mph for 2 hours?

48. *Caloric requirements* A simpler way to estimate the daily caloric requirements you need to maintain your weight is to use these formulas:

For sedentary people: $S = 14W$, where W is your weight in pounds

For moderatively active people: $M = 17W$, where W is your weight in pounds

For active people: $A = 20W$, where W is your weight in pounds

Find the caloric requirements for a 140-pound person that is

a. Sedentary

b. Moderatively active

c. Active

49. *Weight* A person is consuming 2380 calories a day. Refer to the formulas in Problem 48 and find the person's weight if the person is

a. Sedentary

b. Moderatively active

c. Active

50. *Weight* The recommended number of daily calories to maintain weight for a 50+-year-old male is 2300. The recommended weight for an average height 50-year-old is 150 pounds. Which of the three categories of Problem 48 (S, M, or A) can this 50+-year-old be and still maintain his weight?

Problems 51 and 52 are based on "Your Weight on Other Worlds."
Source: http://www.exploratorium.edu.

51. *Weight* Do you want to lose weight instantly and without dieting? Move to Pluto! Your weight P on Pluto is only 0.067 of your weight E on Earth.

a. Write a formula for your weight P on Pluto.

b. If you weigh 150 pounds on Earth, how much do you weigh on Pluto?

c. If you weigh 9.38 pounds on Pluto, how much do you weigh on Earth?

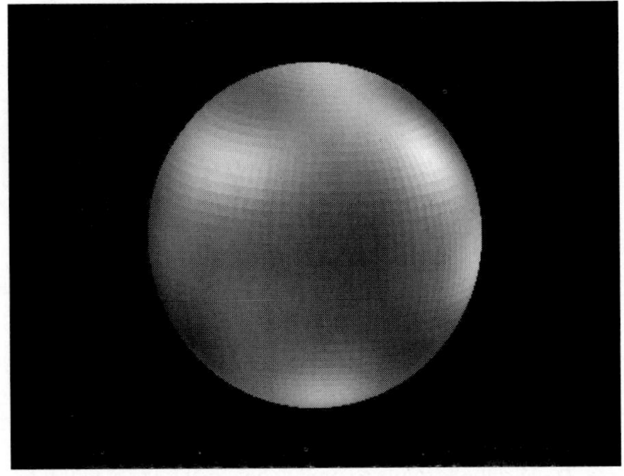

52. *Weight* If you want to *gain* weight instantly move to Jupiter. Your weight J on Jupiter will be 253.3 times your weight E on Earth.

a. Write a formula for your weight J on Jupiter.

b. If you weigh 150 pounds on Earth, how much would you weigh on Jupiter?

c. If you weigh 40,528 pounds on Jupiter, how much do you weigh on Earth?

To check the answers to Problems 51 and 52, or if you want to know how much you weigh on many other planets, you can go to the website: http://www.exploratorium.edu.

Web IT go to mhhe.com/bello for more lessons

Web IT go to mhhe.com/bello for more lessons

Problems 53–56 are based on the Agriculture Fact Book: Profiling Food Consumption in America.
Source: http://www.usda.gov.

53. *Available calories per person per day* How many calories C are available per person per day? According to the U.S. Department of Agriculture the answer is 3800 but 1100 of these calories are lost to spoilage, place waste, and cooking, leaving 2700 calories daily for each American. The approximate number of calories C is given by the formula:

$C = 18t + 2170$, where t is the number of years after 1970.

a. How many calories per person per day were available in 1970 ($t = 0$)?

b. How many calories per person per day were available in 2000 ($t = 30$)?

c. Use the formula to predict how many calories per person per day would be available in 2020 ($t = 50$).

54. *Predicting available calories* Use the formula $C = 18t + 2170$ of Problem 53 to find:

a. How many years will it take to reach $C = 3000$ calories per person/day? Answer to the nearest whole number.

b. In what year will this happen?

56. *Food consumption* In Problem 55 we used the formula $P = -0.07t + 3.3$, where t is the number of years after 1970 to predict the consumption of veal and lamb.

a. Predict (to the nearest whole number) how many years it would take for P to be 0.

b. Based on your answer to part **a**, can you use the formula for $t \geq 47$?

Try it for $t = 50$!

55. *Food consumption* Which types of food consumption are declining? Veal and lamb, for example. The formula for the annual number of pounds P of veal and lamb consumed per capita is given by the equation

$P = -0.07t + 3.3$, where t is the number of years after 1970

a. What was the annual per capita consumption of veal and lamb in 1970?

b. What would be the annual consumption of veal and lamb in 2010?

See the decline!

Problems 57 and 58 are based on *Vital Signs 2006–2007,* Worldwatch Institute.

57. *Biodiesels* Biodiesel is the name of a clean burning alternative fuel, produced from domestic, renewable resources. The production P (in millions of liters) can be approximated by

$P = 526.2t + 893$, where t is the number of years after 2000

a. Use the formula to find the biodiesel production in 2000 ($t = 0$).

b. Use the formula to find the biodiesel production in 2005 ($t = 5$).

c. Use the formula to predict the biodiesel production in 2010 ($t = 10$).

58. *World population* Do you know the population of the world? You can go to the website and find out, but it is certainly over 6 billion and growing. Here is an approximation of the world population P (in billions):

The Population of the Earth is
6,528,777,630

Source: http://opr.princeton.edu.

$P = 0.073t + 6.08$, where t is the number of years after 2000

a. Use the formula to predict the world population in 2010 ($t = 10$).

b. Use the formula to predict how many years will it take for the world population to reach 7 billion. Answer to the nearest whole number.

Problems 59 and 60 are based on an Associated Press Report.

Source: The Associated Press; http://www.msnbc.msn.com.

59. *U.S. population* The formula to estimate the U.S. population (in millions) is

$P = 2.7t + 300$, where t is the number of years after 2006

a. Use the formula to predict the U.S. population (in millions) in 2010 ($t = 4$).

b. In how many years (after 2006) will the U.S. population reach 500 million?

Answer to the nearest whole number.

c. In what year will the population reach 500 million?

60. *Ukraine population* Is there a place in the world where the population is decreasing? The answer is yes! Ukraine. (Look it up on a map and maybe you will see why!) The formula for the population P of Ukraine (in millions) is

$P = -0.46t + 46$, where t is the number of years after 2006

a. Use the formula to predict the population of Ukraine in 2006 ($t = 0$).

b. Use the formula to predict the number of years it will take for the population of Ukraine to be 0.

c. Is this a realistic answer?

››› Using Your Knowledge

Oxidation Number Do you know why water is so common on Earth? It is because hydrogen and oxygen, the elements that compose water, are so common and they combine very easily. The number that indicates how an atom combines with other atoms in a compound is called the *valence*. (In modern chemistry books the term *valence* has been replaced by the term **oxidation number.**) For example, the formula for water is

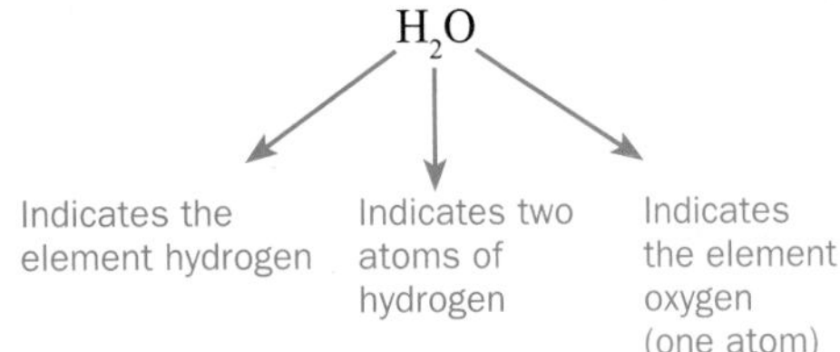

The valence of hydrogen is H and that of oxygen is -2. Thus, if we have two hydrogen atoms and one oxygen atom, the valence of the compound is

H_2O

$$2(\text{H}) + 1(-2) = 0$$

The valence of a stable compound is always 0.

Solving for H,

$$\text{H} = +1$$

In Problems 61–64, assume that all compounds are stable.

61. Find the valence of chromium (Cr) in sodium dichromate, $Na_2Cr_2O_7$, if the other valences are Na $= +1$ and O $= -2$.

62. Find the valence of bromine (Br) in sodium bromate, $NaBrO_3$, if the other valences are Na $= +1$ and O $= -2$.

63. Find the valence of phosphorus (P) in phosphate, PO_4, if the valence of oxygen is O $= -2$.

64. Find the valence of nitrogen (N) in nitrate, NO_3, if the valence of oxygen is O $= -2$.

››› Write On

65. When reading a word problem, what is the first thing you try to determine?

66. How do you verify your answer in a word problem?

67. In Example 8, the information needed to solve the problem was there, but you had to dig for it. In real life, irrelevant information (called *red herrings*) may also be present. Find some problems with *red herrings* and point them out.

Concept Checker

Fill in the blank(s) with the correct word(s), phrase, or mathematical statement.

68. The **procedure** used to **solve word problems** is called the __________ procedure.

69. In the **RSTUV** procedure, the **R** stands for __________.

70. In the **RSTUV** procedure, the **S** stands for __________.

71. In the **RSTUV** procedure, the **T** stands for __________.

72. In the **RSTUV** procedure, the **U** stands for __________.

73. In the **RSTUV** procedure, the **V** stands for __________.

Translate	**RSTUV**
Solving	**Read**
Recite	**Select**
Verify	**Transpose**
See	**Use**

Mastery Test

74. A McDonald's cheeseburger and small fries contain 600 calories. If you double the number of calories in the fries, you get 90 more calories than in the cheeseburger. How many calories are in the cheeseburger and the fries?

75. If 12 is added to twice a number, the result is 5 times the number. Find the number.

76. Romualdo bought a 6-month, $10,000 certificate of deposit. At the end of the 6 months, he received $200 simple interest. What rate of interest did the certificate pay?

77. The cost of renting a car (in dollars) is $30 plus $0.25 per mile traveled. If the rental cost for one day amounts to $110, how many miles were traveled?

78. Michael Hebranko decreased his weight by 900 pounds to a manageable 200 with the help of the Richard Simmons Program. How much did he weigh at the beginning of his diet? He has since regained most of the weight, and they had to remove a window to take him to the hospital!

79. A moon rock weighs 30 pounds on the moon. What would be its weight on Earth if objects weigh 6 times as much on Earth as they do on the moon?

Collaborative Learning

Adapted from: The History of Algebra
http://teacherexchange.mde.k12.ms.us/teachnett/history-of-algebra.htm

Form three groups:

GROUP 1. Collect information about al-Khowarizmi.

GROUP 2. Create a timeline to show when the concepts used by al-Khowarizmi were developed in relation to other significant events in history.

GROUP 3. Find examples of the algebraic concepts developed by al-Khowarizmi.

Each group will go on a web quest to find the mathematicians that historians have named as being the *fathers or founders of algebra* and will prepare a written or electronic report to present to the class.

Each group will include: a Surfer, a Writer/Recorder, a Mathematician, and a Reporter.

Surfer—Surf the Internet using search engines. Work with the Recorder to research and record needed information for your topic.

Writer/Recorder—Record information on your topic and the URLs of websites where the information was found. Work with the Surfer and the Reporter to prepare (in hard copy and electronic formats) a report of the findings of your group.

Mathematician—Work with the Surfer and the Recorder to find sites that contain examples of mathematical problems from your assigned topic. Choose two examples from the Internet that you can share with the class; then create a problem in which you can demonstrate the topic for the class.

Reporter—Work with the other members of your group to create a presentation, using computer software or traditional hard copies, which you will present to the class.

Research Questions

1. Find out more about al-Khowarizmi (for example, his exact birthplace and date of birth) and write a few paragraphs about him and his life. Include the titles of the books he wrote and the types of problems appearing in these books.
2. A Latin corruption of *al-Khowarizmi* was interpreted as "the art of computing with Hindu-Arabic numerals." What modern concept is derived from the word, and what does the concept mean in mathematics?
3. Another translation of the word *jabr* is "the setting of broken bones." Write a paragraph detailing how the word reached Spain as *algebrista* and the context in which the word was used.
4. In his book, al-Khowarizmi mentioned six types of equations that can be written using squares, square roots, and constants. What are the six types of equations?
5. Al-Khowarizmi's work was continued in the Islamic world. In the first decade of the eleventh century, a major book dealing with algebra entitled *The Marvelous* was written. Who was the author of this book, what was its original title, and what topics were covered in the book?

Summary Chapter 10

Section	Item	Meaning	Example
10.1	Placeholders (variables or unknowns)	Letters used in place of numbers	In $x + 4 = 7$, x is the placeholder, variable, or unknown.
10.1A	Expression	A collection of numbers and letters connected by operation signs	$xy^2 + 2x + 3$ is an expression.
10.1B	Distributive property	$a(b + c) = ab + ac$	$3(x + 5) = 3x + 15$
10.1C	Factoring	$ax + bx = (a + b)x$	$3x + 6y = 3(x + 2y)$
10.1E	Removing parentheses	$-(a - b) = -a + b$	$-(2x - 5) = -2x + 5$
10.2A	Integer exponents	$x^{-n} = \frac{1}{x^n}$, n an integer, $x \neq 0$.	$3^{-2} = \frac{1}{3^2}$ and $x^{-5} = \frac{1}{x^5}$
10.2B	Product rule of exponents	$x^m \cdot x^n = x^{m+n}$, m and n integers, $x \neq 0$.	$x^5 \cdot x^3 = x^{5+3} = x^8$
10.2C	Quotient rule of exponents	$\frac{x^m}{x^n} = x^{m-n}$, m and n integers, $x \neq 0$.	$\frac{x^7}{x^3} = x^{7-3} = x^4$
10.2D	Power rule of exponents Product to a power rule of exponents	$(x^m)^n = x^{mn}$, m and n integers, $x \neq 0$. $(x^m y^n)^k = x^{mk}y^{nk}$ m, n, k integers, $x, y \neq 0$.	$(x^{-2})^3 = x^{-2 \cdot 3} = x^{-6} = \frac{1}{x^6}$ $(x^3y^{-4})^5 = x^{15}y^{-20} = \frac{x^{15}}{y^{20}}$

(continued)

Section	Item	Meaning	Example
10.2E	Compound interest	$A = P(1 + r)^n$, where A is the amount after n years of a principal P invested at rate r compounded annually.	If $P = \$1000$ is invested for $n = 3$ years at a rate of $r = 6\% = 0.06$, the amount is $A = 1000(1 + 0.06)^3$.
10.3A	Scientific notation	A number is in scientific notation when written as $M \times 10^n$, M between 1 and 10, n an integer.	6.27×10^{-4} is in scientific notation.
10.4	Linear equation	An equation in which the variable occurs only to the first degree.	$x + 7 = 2x - 9$ is a linear equation.
10.4B	Procedure for solving linear equations	1. Simplify the equation if necessary. 2. Add or subtract the same numbers to or from both sides so that one side contains variables only. 3. Add or subtract the same expressions to or from both sides so that the other side contains numbers only. 4. If the coefficient of the variable (the number that multiplies the variable) is not 1, divide both sides by this number. 5. The resulting number is the solution of the equation. Check in the original equation to be sure it is correct.	
10.5A	Procedure for solving word problems	Read the problem. Select the unknown. Translate the problem. Use algebra to solve the equation. Verify your answer.	

› Review Exercises Chapter 10

(If you need help with these exercises, look in the section indicated in brackets.)

1. ‹**10.1A**› *List the terms in the expression.*

a. $6x^3 - 3x^2 + 7x - 7$

b. $5x^3 - 4x^2 + 6x - 6$

c. $4x^3 - 5x^2 + 5x - 5$

d. $3x^3 - 6x^2 + 4x - 4$

e. $2x^3 - 7x^2 + 3x - 3$

2. ‹**10.1B**› *Multiply.*

a. $2(x + 2y)$

b. $3(x + 2y)$

c. $4(x + 2y)$

d. $5(x + 2y)$

e. $6(x + 2y)$

3. ‹**10.1B**› *Multiply.*

a. $-4(x - 2y + z)$

b. $-5(x - 2y + z)$

c. $-6(x - 2y + z)$

d. $-7(x - 2y + z)$

e. $-8(x - 2y + z)$

4. ⟨**10.1C**⟩ *Factor.*

a. $3x - 12y$

b. $3x - 15y$

c. $3x - 18y$

d. $3x - 21y$

e. $3x - 24y$

5. ⟨**10.1C**⟩ *Factor.*

a. $40x - 50y - 60z$

b. $30x - 40y - 50z$

c. $20x - 30y - 40z$

d. $10x - 20y - 30z$

e. $10x - 10y - 20z$

6. ⟨**10.1D**⟩ *Combine like terms (simplify).*

a. $3x + 5y - x + 4y$

b. $4x + 5y - x + 3y$

c. $5x + 5y - x + 2y$

d. $6x + 5y - 2x + 4y$

e. $7x + 5y - 3x + 4y$

7. ⟨**10.1D**⟩ *Combine like terms (simplify).*

a. $0.3x + 0.31y - 0.5x + 0.23y$

b. $0.3x + 0.32y - 0.6x + 0.23y$

c. $0.3x + 0.33y - 0.7x + 0.23y$

d. $0.3x + 0.34y - 0.8x + 0.23y$

e. $0.3x + 0.35y - 0.9x + 0.23y$

8. ⟨**10.1D**⟩ *Combine like terms (simplify).*

a. $\frac{1}{11}x + \frac{9}{13}y + \frac{3}{11}x - \frac{2}{13}y$

b. $\frac{2}{11}x + \frac{8}{13}y + \frac{3}{11}x - \frac{2}{13}y$

c. $\frac{3}{11}x + \frac{7}{13}y + \frac{3}{11}x - \frac{2}{13}y$

d. $\frac{4}{11}x + \frac{6}{13}y + \frac{3}{11}x - \frac{2}{13}y$

e. $\frac{5}{11}x + \frac{5}{13}y + \frac{3}{11}x - \frac{2}{13}y$

9. ⟨**10.1E**⟩ *Simplify.*

a. $(4a + 5b) - (a + 2b)$

b. $(5a + 5b) - (a + 2b)$

c. $(6a + 5b) - (a - 2b)$

d. $(7a + 5b) - (-a + 2b)$

e. $(8a + 5b) - (a - 3b)$

10. ⟨**10.2A**⟩ *Write with positive exponents and simplify.*

a. 2^{-2} **b.** 2^{-3}

c. 2^{-4} **d.** 2^{-5}

e. 2^{-6}

11. ⟨**10.2A**⟩ *Write using negative exponents.*

a. $\frac{1}{2^3}$ **b.** $\frac{1}{2^4}$

c. $\frac{1}{2^5}$ **d.** $\frac{1}{2^6}$

e. $\frac{1}{2^7}$

12. ⟨**10.2B**⟩ *Multiply and simplify.*

a. $x^3 \cdot x^4$ **b.** $x^3 \cdot x^5$

c. $x^3 \cdot x^6$ **d.** $x^3 \cdot x^7$

e. $x^3 \cdot x^8$

13. ⟨**10.2B**⟩ *Multiply and simplify.*

a. $2^3 \cdot 2^{-5}$ **b.** $2^3 \cdot 2^{-6}$

c. $2^3 \cdot 2^{-7}$ **d.** $2^3 \cdot 2^{-8}$

e. $2^3 \cdot 2^{-9}$

14. ⟨**10.2B**⟩ *Multiply and simplify.*

a. $y^3 \cdot y^{-5}$ **b.** $y^3 \cdot y^{-6}$

c. $y^3 \cdot y^{-7}$ **d.** $y^3 \cdot y^{-8}$

e. $y^3 \cdot y^{-9}$

15. ⟨**10.2C**⟩ *Divide and simplify.*

a. $\frac{2^4}{2^{-2}}$ **b.** $\frac{2^5}{2^{-2}}$

c. $\frac{2^6}{2^{-2}}$ **d.** $\frac{2^7}{2^{-2}}$

e. $\frac{2^8}{2^{-2}}$

16. ⟨**10.2C**⟩ *Divide and simplify.*

a. $\frac{x^2}{x^5}$ **b.** $\frac{x^2}{x^6}$

c. $\frac{x^2}{x^7}$ **d.** $\frac{x^2}{x^8}$

e. $\frac{x^2}{x^9}$

17. ⟨**10.2C**⟩ *Divide and simplify.*

a. $\frac{y^{-2}}{y^{-4}}$ **b.** $\frac{y^{-2}}{y^{-5}}$

c. $\frac{y^{-2}}{y^{-6}}$ **d.** $\frac{y^{-2}}{y^{-7}}$

e. $\frac{y^{-2}}{y^{-8}}$

18. ⟨**10.2D**⟩ *Simplify.*

a. $(2^2)^1$ **b.** $(2^2)^2$

c. $(2^2)^3$ **d.** $(2^2)^4$

e. $(2^2)^5$

19. **⟨10.2D⟩** *Simplify.*

a. $(y^4)^{-2}$ **b.** $(y^4)^{-3}$

c. $(y^4)^{-4}$ **d.** $(y^4)^{-5}$

e. $(y^4)^{-6}$

20. **⟨10.2D⟩** *Simplify.*

a. $(z^{-3})^{-2}$ **b.** $(z^{-3})^{-3}$

c. $(z^{-3})^{-4}$ **d.** $(z^{-3})^{-5}$

e. $(z^{-3})^{-6}$

21. **⟨10.2D⟩** *Simplify.*

a. $(x^3y^{-2})^2$ **b.** $(x^3y^{-2})^3$

c. $(x^3y^{-2})^4$ **d.** $(x^3y^{-2})^5$

e. $(x^3y^{-2})^6$

22. **⟨10.2D⟩** *Simplify.*

a. $(x^{-2}y^3)^{-2}$

b. $(x^{-2}y^3)^{-3}$

c. $(x^{-2}y^3)^{-4}$

d. $(x^{-2}y^3)^{-5}$

e. $(x^{-2}y^3)^{-6}$

23. **⟨10.3A⟩** *Write in scientific notation.*

a. 44,000,000

b. 450,000,000

c. 4,600,000

d. 47,000

e. 48,000,000

24. **⟨10.3A⟩** *Write in scientific notation.*

a. 0.0014

b. 0.00015

c. 0.000016

d. 0.0000017

e. 0.00000018

25. **⟨10.3A⟩** *Write in standard notation.*

a. 7.83×10^3

b. 6.83×10^4

c. 5.83×10^5

d. 4.83×10^6

e. 3.83×10^7

26. **⟨10.3A⟩** *Write in standard notation.*

a. 8.4×10^{-2}

b. 7.4×10^{-3}

c. 6.4×10^{-4}

d. 5.4×10^{-5}

e. 4.4×10^{-6}

27. **⟨10.3B⟩** *Perform the indicated operations and write the answer in scientific notation.*

a. $(2 \times 10^2) \times (1.1 \times 10^3)$

b. $(3 \times 10^2) \times (3.1 \times 10^4)$

c. $(4 \times 10^2) \times (3.1 \times 10^5)$

d. $(5 \times 10^2) \times (3.1 \times 10^6)$

e. $(6 \times 10^2) \times (3.1 \times 10^6)$

28. **⟨10.3B⟩** *Perform the indicated operations and write the answer in scientific notation.*

a. $(1.1 \times 10^3) \times (2 \times 10^{-5})$

b. $(1.1 \times 10^4) \times (3 \times 10^{-5})$

c. $(1.1 \times 10^3) \times (4 \times 10^{-6})$

d. $(1.1 \times 10^3) \times (5 \times 10^{-7})$

e. $(1.1 \times 10^4) \times (6 \times 10^{-8})$

29. **⟨10.3B⟩** *Perform the indicated operations and write the answer in scientific notation.*

a. $(1.15 \times 10^{-3}) \div (2.3 \times 10^{-4})$

b. $(1.38 \times 10^{-2}) \div (2.3 \times 10^{-4})$

c. $(1.61 \times 10^{-3}) \div (2.3 \times 10^{-5})$

d. $(1.84 \times 10^{-4}) \div (2.3 \times 10^{-4})$

e. $(2.07 \times 10^{-3}) \div (2.3 \times 10^{-4})$

30. **⟨10.4A⟩** *Solve.*

a. $y + \frac{1}{3} = \frac{1}{2}$

b. $y + \frac{1}{4} = \frac{1}{2}$

c. $y + \frac{1}{5} = \frac{1}{2}$

d. $y + \frac{1}{6} = \frac{1}{2}$

e. $y + \frac{1}{7} = \frac{1}{2}$

31. **⟨10.4A⟩** *Solve.*

a. $\frac{-y}{3} = 2$

b. $\frac{-y}{3} = 3$

c. $\frac{-y}{3} = 4$

d. $\frac{-y}{3} = 5$

e. $\frac{-y}{3} = 6$

32. **⟨10.4A⟩** *Solve.*

a. $-1.2z = 2.4$

b. $-1.2z = 3.6$

c. $-1.2z = 4.8$

d. $-1.2z = 7.2$

e. $-1.2z = 8.4$

33. **⟨10.4B⟩** *Solve.*

a. $-3x + 5 = 10$

b. $-3x + 5 = 13$

c. $-3x + 5 = 15$

d. $-3x + 5 = 18$

e. $-3x + 5 = 21$

34. ⟨**10.4B**⟩ *Solve.*

a. $2x - 6 = 10$

b. $2x - 6 = 12$

c. $2x - 6 = 14$

d. $2x - 6 = 16$

e. $2x - 6 = 18$

35. ⟨**10.4B**⟩ *Solve.*

a. $4x + 5.4 = 2x + 9.6$

b. $5x + 5.4 = 3x + 9.6$

c. $6x + 5.4 = 4x + 9.6$

d. $7x + 5.4 = 5x + 9.6$

e. $8x + 5.4 = 6x + 9.6$

36. ⟨**10.4B**⟩ *Solve.*

a. $36 = 3(x + 1)$

b. $27 = 3(x + 1)$

c. $24 = 3(x + 1)$

d. $21 = 3(x + 1)$

e. $18 = 3(x + 1)$

37. ⟨**10.4B**⟩ *Solve.*

a. $\frac{2}{7}x = -3$

b. $\frac{2}{7}x = -5$

c. $\frac{2}{7}x = -7$

d. $\frac{2}{7}x = -9$

e. $\frac{2}{7}x = -11$

38. ⟨**10.4B**⟩ *Solve.*

a. $\frac{1}{4}x - \frac{3}{8} + \frac{3}{2}x = \frac{1}{4} + 2x$

b. $\frac{1}{4}x - \frac{3}{8} + \frac{5}{2}x = \frac{1}{4} + 3x$

c. $\frac{1}{4}x - \frac{3}{8} + \frac{7}{2}x = \frac{1}{4} + 4x$

d. $\frac{1}{4}x - \frac{3}{8} + \frac{9}{2}x = \frac{1}{4} + 5x$

e. $\frac{1}{4}x - \frac{3}{8} + \frac{11}{2}x = \frac{1}{4} + 6x$

39. ⟨**10.5A**⟩ *The daily cost C of renting a car is $30 plus $0.20 per mile traveled. If m is the number of miles traveled, then C = 30 + 0.20m. Find the number m of miles traveled if a day's rental amounted to:*

a. $70.40 **b.** $70.60

c. $70.80 **d.** $71.20

e. $71.40

40. ⟨**10.5A**⟩ *An investor bought a 6-month, $1000 certificate of deposit. What rate of interest did the certificate pay if at the end of the 6 months the interest the investor received was:*

a. $30 **b.** $40

c. $50 **d.** $60

e. $70

Practice Test Chapter 10

(Answers on page 631)

Visit www.mhhe.com/bello to view helpful videos that provide step-by-step solutions to several of the problems below.

1. List the terms in the expression $6x^3 - 3x^2 + 8x - 9$.

2. Multiply.

a. $3(x + 2y)$

b. $-4(x - 2y + z)$

3. Factor.

a. $4x - 8y$

b. $30x - 40y - 50z$

4. Combine like terms (simplify).

a. $4x + 6y - x + 4y$

b. $\frac{2}{7}x + \frac{4}{5}y + \frac{3}{7}x - \frac{3}{5}y$

5. Simplify: $(4a + 5b) - (a + 2b)$.

6. Write as a fraction with positive exponents and simplify: 3^{-4}.

7. Multiply and simplify.

a. $x^2 \cdot x^6$

b. $4^3 \cdot 4^{-6}$

c. $y^3 \cdot y^{-6}$

8. Divide and simplify.

a. $\frac{2^4}{2^{-2}}$

b. $\frac{x^3}{x^7}$

c. $\frac{y^{-3}}{y^{-6}}$

9. Simplify.

a. $(2^2)^4$

b. $(y^4)^{-4}$

c. $(z^{-3})^{-3}$

10. Simplify.

a. $(x^4y^{-2})^3$

b. $(x^{-3}y^4)^{-3}$

11. Write in scientific notation.

a. 88,000,000

b. 0.0000013

12. Write in standard notation.

a. 6.57×10^4

b. 9.4×10^{-3}

13. Perform the indicated operations and write the answer in scientific notation.

a. $(3 \times 10^2) \times (7.1 \times 10^3)$

b. $(3.1 \times 10^3) \times (3 \times 10^{-6})$

c. $(1.84 \times 10^{-3}) \div (2.3 \times 10^{-4})$

14. Solve.

a. $y + \frac{3}{8} = \frac{3}{4}$

b. $\frac{-y}{2} = 6$

c. $-1.8z = 5.4$

15. Solve.

a. $-4x + 9 = 14$

b. $2x - 6 = 12$

16. Solve.

a. $3x + 4.4 = x + 8.6$

b. $4(x + 1) = 2x + 12$

17. Solve.

a. $12 = 3(x + 1)$ **b.** $\frac{3}{5}x = -5$

18. Solve. $\frac{1}{4}x - \frac{1}{8} + \frac{1}{2}x = \frac{1}{2} + x$

19. The daily cost C of renting a car is \$30 plus \$0.20 per mile traveled. If m is the number of miles traveled, then $C = 30 + 0.20m$. If a person paid \$70.60 for a day's rental, how many miles did this person travel?

20. Toni bought a 6-month, \$1000 certificate of deposit. At the end of the 6 months, she received \$50 simple interest. What rate of interest did the certificate pay?

Answers to Practice Test Chapter 10

Answer	If You Missed	Review		
	Question	Section	Examples	Page
1. $6x^3$; $-3x^2$; $8x$; -9	1	10.1	1	576
2. **a.** $3x + 6y$ **b.** $-4x + 8y - 4z$	2	10.1	2	577
3. **a.** $4(x - 2y)$ **b.** $10(3x - 4y - 5z)$	3	10.1	3	577–578
4. **a.** $3x + 10y$ **b.** $\frac{5}{7}x + \frac{1}{5}y$	4	10.1	4	578
5. $3a + 3b$	5	10.1	5, 6	579–580
6. $\frac{1}{3^4} = \frac{1}{81}$	6	10.2	1	588
7. **a.** x^8 **b.** $4^{-3} = \frac{1}{64}$ **c.** $\frac{1}{y^3}$	7	10.2	3	589
8. **a.** $2^6 = 64$ **b.** $\frac{1}{x^4}$ **c.** y^3	8	10.2	4	590
9. **a.** $2^8 = 256$ **b.** $\frac{1}{y^{16}}$ **c.** z^9	9	10.2	5	591
10. **a.** $\frac{x^{12}}{y^6}$ **b.** $\frac{x^9}{y^{12}}$	10	10.2	6	592
11. **a.** 8.8×10^7 **b.** 1.3×10^{-6}	11	10.3	1	597
12. **a.** 65,700 **b.** 0.0094	12	10.3	2	597
13. **a.** 2.13×10^6 **b.** 9.3×10^{-3} **c.** 8×10^0 or 8	13	10.3	3, 4	598
14. **a.** $y = \frac{3}{8}$ **b.** $y = -12$ **c.** $z = -3$	14	10.4	1, 2	603
15. **a.** $x = -\frac{5}{4}$ **b.** $x = 9$	15	10.4	3, 4	605–606
16. **a.** $x = 2.1$ **b.** $x = 4$	16	10.4	5, 6, 7	606–607
17. **a.** $x = 3$ **b.** $x = -\frac{25}{3}$	17	10.4	5, 6, 7	606–607
18. $x = -\frac{5}{2}$	18	10.4	8	608
19. 203	19	10.5	1–4	612–614
20. 10%	20	10.5	5	614

› Cumulative Review Chapters 1–10

1. Simplify: $49 \div 7 \cdot 7 + 8 - 5$

2. Divide: $80 \div 0.13$ (Round answer to two decimal digits.)

3. Write 23% as a decimal.

4. 90% of 40 is what number?

5. What percent of 8 is 4?

6. 6 is 30% of what number?

7. Find the simple interest earned on \$300 invested at 7.5% for 5 years.

8. The following table shows the distribution of families by income in Portland, Oregon.

Income Level	Percent of Families
\$0–9999	3
10,000–14,999	8
15,000–19,999	26
20,000–24,999	34
25,000–34,999	11
35,000–49,999	9
50,000–79,999	5
80,000–119,000	3
120,000 and over	1

What percent of the families in Portland have incomes of \$120,000 and over?

9. The number of hours required in each discipline of a college core curriculum is represented by the following circle graph. What percent of these hours is in art and math combined?

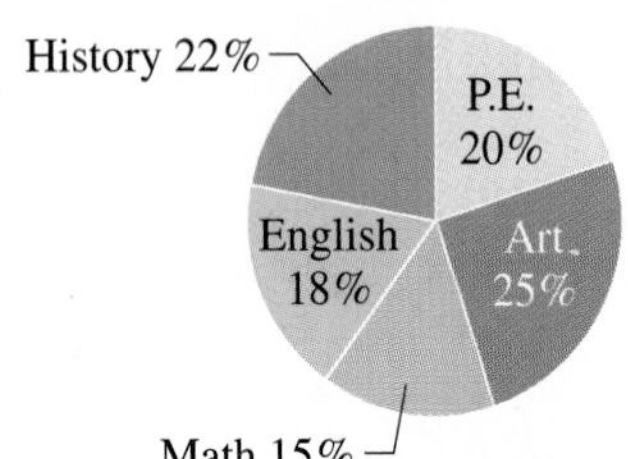

10. What is the mode of the following numbers?
10, 5, 5, 12, 5, 7, 23, 24, 5, 5, 9

11. What is the mean of the following numbers?
2, 16, 1, 28, 1, 30, 1, 25

12. What is the median of the following numbers?
9, 5, 6, 10, 29, 7, 14, 14, 14

13. Convert 4 yards to inches.

14. Convert 17 inches to feet.

15. Convert 17 feet to yards.

16. Convert 7 feet to inches.

17. Convert 240 meters to decimeters.

18. Find the perimeter of a rectangle 4.8 centimeters long by 2.8 centimeters wide.

19. Find the circumference of a circle whose radius is 9 in. (Use 3.14 for π.)

20. Find the area of a triangular piece of cloth that has a 10-centimeter base and is 16 centimeters high.

21. Find the area of a circle with a radius of 3 centimeters. (Use 3.14 for π.)

22. Find the volume of a box 7 centimeters wide, 5 centimeters long, and 3 centimeters high.

23. Find the volume of a sphere with a 12-inch radius. Use 3.14 for π and give the answer to two decimal places.

24. Classify the angle.

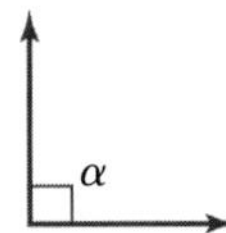

25. If in triangle ABC, $m\angle A = 35°$ and $m\angle B = 25°$, what is $m\angle C$?

26. Find $\sqrt{8100}$.

27. Find the hypotenuse h of the given triangle.

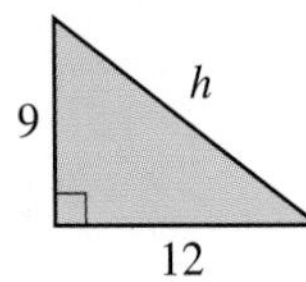

28. Find the additive inverse (opposite) of $-3\frac{4}{5}$.

29. Find: $|-3.5|$

30. Add: $-4.7 + (-6.8)$

31. Add: $\frac{2}{5} + \left(-\frac{1}{8}\right)$

32. Subtract: $-3.8 - (-4.1)$

33. Subtract: $-\frac{2}{5} - \frac{2}{7}$

34. Multiply: $-\frac{7}{8} \cdot \frac{9}{7}$

35. Divide: $-\frac{10}{7} \div \frac{3}{7}$

36. Evaluate: $(-7)^2$

37. Multiply: $5(4x - 8y)$

38. Factor: $60x - 30y - 40z$

39. Simplify: $(2x + y) - (7x - 2y)$

40. Write as a fraction with positive exponents and simplify 9^{-2}.

41. Multiply and simplify: $x^{-2} \cdot x^{-5}$

42. Simplify: $(3x^3y^{-2})^2$

43. Write in scientific notation: 0.000000024

44. Write in standard notation: 2.82×10^7

45. Divide and express the answer in scientific notation: $(2.695 \times 10^{-3}) \div (7.7 \times 10^4)$

46. Solve for x: $\frac{5}{7}x = -35$

47. Solve for x: $9x - 10 = 8$

48. Solve for x: $2x - 4 = x + 3$

49. Solve for x: $16 = 4(x - 9)$

Selected Answers

The brackets preceding answers for the Chapter Review Exercises indicate the Chapter, Section, and Objective for you to review for further study. For example, [3.4C] appearing before answers means those exercises correspond to Chapter 3, Section 4, Objective C.

Chapter 1

Exercises 1.1

1. 2 **3.** 800 **5.** 40 **7.** Tens **9.** Ones (Units) **11.** 30 + 4 **13.** 100 + 8 **15.** 2000 + 500 **17.** 7000 + 40 **19.** 20,000 + 3000 + 10 + 8 **21.** 600,000 + 4000 **23.** 90,000 + 1000 + 300 + 80 + 7 **25.** 60,000 + 8000 + 20 **27.** 80,000 + 80 + 2 **29.** 70,000 + 100 + 90 + 8 **31.** 78 **33.** 308 **35.** 822 **37.** 701 **39.** 3473 **41.** 5250 **43.** 2030 **45.** 8090 **47.** 7001 **49.** 6600 **51.** Fifty-seven **53.** Three thousand four hundred eight **55.** One hundred eighty-one thousand, three hundred sixty-two **57.** Forty-one million, three hundred thousand **59.** One billion, two hundred thirty-one million, three hundred forty-one thousand **61.** 809 **63.** 4897 **65.** 2003 **67.** 2,023,045 **69.** 345,033,894 **71.** One hundred seventy-three thousand, eight hundred eighty **73.** Thirteen million, five hundred thirty-seven thousand **75.** 14,979,000 **77.** Fourteen thousand, eight hundred seventy-two dollars; Sixty-three thousand, six hundred twenty-seven dollars; Thirty-two thousand, forty-four dollars; One hundred thirty-seven thousand, ninety-one dollars **79.** Sixteen thousand, two hundred forty-one dollars; Sixty-nine thousand, four hundred eighty-two dollars; Thirty-four thousand, nine hundred ninety-three dollars; One hundred forty-nine thousand, seven hundred seven dollars **81.** Seventeen thousand, seven hundred thirty-six dollars; Seventy-five thousand, eight hundred seventy-six dollars; Thirty-eight thousand, two hundred thirteen dollars; One hundred sixty-three thousand, four hundred eighty-four dollars **83.** Eight hundred twenty-three thousand, five hundred **85.** Nine hundred fifty-four thousand **87.** Forty-six thousand, three hundred dollars **89.** Thirty-three thousand, six hundred dollars **91.** 5182 kilowatt-hours **93.** 7001 kilowatt-hours **95.** 9799.00 **97.** Twenty-eight thousand, nine hundred **99.** Ninety thousand, eight hundred **103.** 8 **105.** Seven **107.** $4x^2 + 8x + 6$ **109.** whole **111.** natural **113.** numeral **115.** Twelve thousand, eight hundred forty-nine **117.** Hundreds **119.** Three hundred five

Exercises 1.2

1. $<$ **3.** $>$ **5.** $<$ **7.** $>$ **9.** $<$ **11.** 70 **13.** 90 **15.** 100 **17.** 100 **19.** 400 **21.** 1000 **23.** 2000 **25.** 7000 **27.** 10,000 **29.** 9000 **31.** 590; 600; 1000 **33.** 29,450; 29,500; 29,000 **35.** 49,990; 50,000; 50,000 **37.** 259,910; 259,900; 260,000 **39.** 289,000; 289,000; 289,000 **41.** 150 **43.** 1100 **45.** 11,000 **47.** 7,900,000 **49.** $86,000,000 **51. (a)** \$689 $<$ \$968 $<$ \$1019; **(b)** \$1019 $>$ \$968 $>$ \$689; **(c)** Dimension 8400; **(d)** Dimension E310 **53. (a)** $<$; **(b)** $<$; **(c)** $>$ **55.** $23,900 **57.** $349 **59.** $25,676 **63.** left **65.** place **67.** add one **69. (a)** 770; **(b)** 360; **(c)** 900 **71.** $50,000 **73.** Nine thousand, nine hundred ninety-five dollars **75.** Ten thousand, five hundred forty-four dollars **77.** Ten thousand, nine hundred ninety-five dollars

Exercises 1.3

1. 11 **3.** 11 **5.** 16 **7.** 30 **9.** 165 **11.** 129 **13.** 60 **15.** 11 **17.** 34 **19.** 35 **21.** 23 **23.** 81 **25.** 127 **27.** 125 **29.** 132 **31.** 152 **33.** 400 **35.** 6773 **37.** 1813 **39.** 3723 **41.** 4644 **43.** 2340 **45.** 15,190 **47.** 91,275 **49.** 220,810 **51.** 582 **53.** 64 **55.** 195 **57.** 1472 ft **59.** 2280 **61.** 360 ft **63.** 1040 ft **65.** 64 ft **67.** $483 **69.** Family A: \$620; Family B: \$1008 **71.** A = 10 (average); B = 18 (high); C = 6 (below average) **73. (a)** Basic Cow, \$500; Shipping and handling, 36; Extra stomach, 79; Two-tone exterior, 142; Produce storage compartment, 127; Heavy-duty straw chopper, 190; 4 spigot/high output drain system, 149; Automatic fly swatter, 89; Genuine cowhide upholstery, 180; Deluxe dual horns, 59; Automatic fertilizer attachment, 339; 4 × 4 traction drive assembly, 884; Pre-delivery wash and comb (Farmer Prep), 70; Additional Farmer Markup and hay fees, 300; **(b)** \$2843.36; \$2844; **(c)** \$3143.36; **(d)** 4 × 4 traction drive assembly (\$884.16); **(e)** Shipping and handling (\$35.75) **75. (a)** 74 mi; **(b)** 337 mi; **(c)** via Salinas, Paso Robles, and Bakersfield **77.** 482 mi **83. (a)** (101 + 96) + 62; 259 mi; **(b)** 101 + (96 + 62); 259 mi; **(c)** The associative property of addition **85.** $11x + 16$ **87.** $12x^3 + 16x^2 + 11x + 14$ **89.** addends **91.** 0 **93.** Perimeter **95.** 25,549 **97.** 1347 **99.** \$9000; the exact answer is \$8988 **101.** One thousand forty; Two hundred twenty-five; Seventy-two

Exercises 1.4

1. 8 **3.** 7 **5.** 17 **7.** 8 **9.** 34 **11.** 63 **13.** 34 **15.** 15 **17.** 212 **19.** 79 **21.** 407 **23.** 215 **25.** 282 **27.** 179 **29.** 209 **31.** 2621 **33.** 3505 **35.** 4088 **37.** 2893 **39.** 889 **41.** 7246 **43.** 4291 **45.** 5714 **47.** 5880 **49.** 26,431 **51.** 867 ft **53.** 4 **55.** 2019 ft **57.** \$201 **59.** \$1013 **61.** \$12,445 **63. (a)** \$18,975; **(b)** \$3025 **65.** \$51 **67.** \$279 **69.** \$330 **71.** \$4100 **73.** \$16,800 **75.** Non-Hispanic White and White **77.** \$3300 **79.** \$19,044 **81.** 210 **83.** 122 **85.** 79°F **87.** 332°F **89.** 358°F **95.** $2x^2 + 2x + 5$ **97.** $3x^3 + 3x^2 + 2x + 6$ **99.** $a = c + b$ **101.** inverse **103.** 795 **105.** 446 **107.** \$288 **109.** 5745 **111.** 2000 + 300 + 40 + 8

Exercises 1.5

1. (a) 12; **(b)** 21 **3. (a)** 9; **(b)** 8 **5. (a)** 40; **(b)** 72 **7. (a–b)** 0 **9. (a)** 8; **(b)** 4 **11.** 48 **13.** 5 **15.** 81 **17.** 0 **19.** 16 **21.** 90 **23.** 160 **25.** 318 **27.** 816 **29.** 19,456 **31.** 3702 **33.** 563,344 **35.** 2520 **37.** 12,450 **39.** 145,320 **41.** 452,000 **43.** 1,223,100 **45.** 100 **47.** 56,000 **49.** 60,000 **51.** 200 ft **53. (a)** 6,000,000 km; **(b)** 7,500,000 km; **(c)** 9,000,000 km **55.** \$9888 **57. (a)** 260; **(b)** 520 **59.** 64 billion barrels **61.** 300 ft^2 **63.** 8100 ft^2 **65.** 57,600 ft^2 **67.** 2,100,000,000 trees (2 billion, 100 million trees) **69.** 1825 gallons **71.** 1,250,000 gallons **73.** 158 lb **75. (a)** 2250; **(b)** 2550 **81.** $10x + 35$ **83.** $6x^2 + 8x + 2$ **85.** $a \cdot b$, $a \times b$, and $(a)(b)$ **87.** factors **89.** 1 **91.** $a \cdot (b \cdot c)$ **93.** 12,600 **95.** 143,550 **97.** 41,022 **99.** 392 **101.** 30 **103.** 200 + 30 + 4 **105.** 1135 **107.** 2772

Exercises 1.6

1. 6 **3.** 7 **5.** 3 **7.** 0 **9.** 1 **11.** 4 **13.** not defined **15.** 4 **17.** 24 **19.** 10 r 2 **21.** 61 **23.** 631 **25.** 513 r 3 **27.** 24 **29.** 21 **31.** 60 r 5 **33.** 44 **35.** 9 r 76 **37.** 42 r 1 **39.** 59 r 7 **41.** 214 r 25 **43.** 45 r 48 **45.** 87 **47.** 630 **49.** 934 r 466 **51.** 504 **53.** 3 **55.** \$600 **57.** 1160 **59.** 300 **61.** \$114 **63.** \$638 **65.** \$52,500 **67.** \$37,500 **69.** \$25,000 **71.** \$225 **73.** \$25 **79.** $(x + 3)$ r 2 **81.** $(x + 1)$ r 3 **83.** $a = b \times c$

85. inverse **87.** *a* **89. (a)** $48 = 6 \times \square$; 8; **(b)** $37 = 1 \times \square$; 37 **91. (a)** $9 = 9 \times \square$; 1; **(b)** $7 = 1 \times \square$; 7 **93. (a)** 132; **(b)** 192 **95.** < **97.** 305,915

Exercises 1.7

1. prime **3.** 1, 2, 3, 6 **5.** 1, 2, 3, 4, 6, 8, 12, 24 **7.** 1, 5, 25 **9.** prime **11.** 2, 7 **13.** 2, 3 **15.** 29 **17.** 2, 11 **19.** 3, 7 **21.** 2×17 **23.** prime **25.** 2^6 **27.** 7×13 **29.** $2 \times 5 \times 19$ **31.** 4 **33.** 1 **35.** 100 **37.** 128 **39.** 200 **41.** 15 **43.** 2×10^9 **45.** $2^4 = 16$; $2^5 = 32$ **47.** 10^9 **49.** 30,000 **51.** 1,500,000 **53.** 100 **55. (a)** Yes; **(b)** No; **(c)** Yes **57. (a)** Yes; **(b)** No; **(c)** Yes **59. (a)** Yes; **(b)** Yes; **(c)** No; **(d)** Yes **61. (a)** Yes; **(b)** No; **(c)** Yes; **(d)** No **63. (a)** Yes; **(b)** No; **(c)** No; **(d)** Yes **69.** sum **71.** even **73.** 1 **75. (a)** prime; **(b)** composite **77. (a)** $2^4 \times 3^2 \times 5^2$; **(b)** $2^3 \times 3^2 \times 5$ **79.** 100 **81.** $1 \times 10^2 + 3 \times 10 + 8$ **83.** $1 \times 10^3 + 2 \times 10^2 + 8$

Exercises 1.8

1. 26 **3.** 13 **5.** 53 **7.** 5 **9.** 3 **11.** 10 **13.** 21 **15.** 18 **17.** 8 **19.** 10 **21.** 10 **23.** 20 **25.** 27 **27. (a)** $3 \times 10 + 2$; **(b)** $32 **29.** 26% **31.** 45% **33.** 89% **35.** 43% **37.** 20% **39.** 13% **41.** 5 milligrams **43.** 3 tablets every 12 hours **45.** $(8 \div 2)(3 + 3)$ **47.** $(8 \div 2)(9 - 3)$ **49.** 7 **51.** 35 **53.** 2 **55.** 8 **57.** parentheses **59.** multiplication **61.** addition **63.** 16 **65.** 2 **67.** 84 **69.** 28 **71.** 144 **73.** Seven thousand, two hundred

Translate This

1. I **3.** B **5.** C **7.** D **9.** M

Exercises 1.9

1. $m = 8$ **3.** $x = 4$ **5.** $x = 5$ **7.** $x = 6$ **9.** $y = 5$ **11.** $t = 7$ **13.** $z = 48$ **15.** $p = 52$ **17.** $x = 14$ **19.** $m = 13$ **21.** $x = 7$ **23.** $x = 4$ **25.** 633 mi/hr **27.** 37 miles **29.** 15,000 lb **31.** 150 **33.** 280 **35.** Fries 210, cheeseburger 330 **37.** $1572 **39.** Grants $2000, scholarships $1600 **41.** $1000 + 5p$; $223 **43.** $1212 **45.** $450 **47.** 158 **49. (a)** 52,500; **(b)** 500; **(c)** 105 **51.** Tip $16, Meal $80 **53.** 6 hours **57.** 1; yes **59.** solution **61.** Addition **63.** $a \div c = b \div c$; $c \neq 0$ **65.** $x = 9$ **67.** $x = 5$ **69.** $m = 2$ **71.** $x = 15$ **73.** $x = 15$ **75.** 9 r 2

Review Exercises

1. [1.1A, B] **(a)** 100 + 20 + 7; seven; **(b)** 100 + 80 + 9; 80; **(c)** 300 + 80; 300; **(d)** 1000 + 400 + 90; 1000 **(e)** 2000 + 500 + 50 + 9; 2000 **2.** [1.1C] **(a)** 49; **(b)** 586; **(c)** 503; **(d)** 810; **(e)** 1004 **3.** [1.1D] **(a)** Seventy-nine; **(b)** One hundred forty-three; **(c)** One thousand, two hundred forty-nine; **(d)** Five thousand, six hundred fifty-nine; **(e)** Twelve thousand, three hundred forty-seven **4.** [1.1E] **(a)** 26; **(b)** 192; **(c)** 468; **(d)** 1644; **(e)** 42,801 **5.** [1.2A] **(a)** <; **(b)** >; **(c)** <; **(d)** <; **(e)** > **6.** [1.2B] **(a)** 2800; **(b)** 9700; **(c)** 3600; **(d)** 4400; **(e)** 5600 **7.** [1.2C] **(a)** $21,100; **(b)** $27,300; **(c)** $35,500; **(d)** $26,500; **(e)** $23,000 **8.** [1.3A] **(a)** 11,978; **(b)** 4481; **(c)** 9646; **(d)** 13,166; **(e)** 13,249 **9.** [1.3B] **(a)** 7 ft; **(b)** 11 ft; **(c)** 14 ft; **(d)** 12 ft; **(e)** 24 ft **10.** [1.4A] **(a)** 29; **(b)** 17; **(c)** 29; **(d)** 9; **(e)** 49 **11.** [1.4A] **(a)** 187; **(b)** 89; **(c)** 89; **(d)** 178; **(e)** 186 **12.** [1.4B] **(a)** $4534; **(b)** $4625; **(c)** $4414; **(d)** $4727; **(e)** $4638 **13.** [1.5A] **(a)** 1620; **(b)** 1372; **(c)** 1833; **(d)** 1344; **(e)** 4416 **14.** [1.5A] **(a)** 26,568; **(b)** 95,403; **(c)** 225,630; **(d)** 194,733; **(e)** 500,151 **15.** [1.5B] **(a)** 77,220; **(b)** 120,120; **(c)** 178,200; **(d)** 55,800; **(e)** 437,080 **16.** [1.5C] **(a)** $7920; **(b)** $5280; **(c)** $10,560; **(d)** $6600; **(e)** $13,200 **17.** [1.5D] **(a)** 360 in.^{2}; **(b)** 240 in.^{2}; **(c)** 360 in.^{2}; **(d)** 180 in.^{2}; **(e)** 288 in.^{2} **18.** [1.6A] **(a–c)** 0 **19.** [1.6A] **(a–c)** not defined **20.** [1.6A] **(a)** 15; **(b)** 12; **(c)** 15; **(d)** 11; **(e)** 17 **21.** [1.6B] **(a)** 31; **(b)** 42; **(c)** 103; **(d)** 21; **(e)** 65 **22.** [1.6B] **(a)** 46 r 1; **(b)** 42 r 1; **(c)** 25 r 1; **(d)** 37 r 1; **(e)** 48 r 2 **23.** [1.6C] **(a)** $1248; **(b)** $936; **(c)** $468; **(d)** $432; **(e)** $216 **24.** [1.7A] **(a)** prime; **(b)** composite; **(c)** prime; **(d)** composite; **(e)** prime **25.** [1.7B] **(a)** 2, 5; **(b)** 5; **(c)** 3, 5; **(d)** 2; **(e)** 2, 17 **26.** [1.7C] **(a)** $2 \times 5 \times 5$ or 2×5^2; **(b)** 2×17; **(c)** $2 \times 2 \times 19$ or $2^2 \times 19$; **(d)** 3×13; **(e)** $3 \times 3 \times 3 \times 3$ or 3^4 **27.** [1.7D] **(a)** 4; **(b)** 9; **(c)** 125; **(d)** 128; **(e)** 243 **28.** [1.7D] **(a)** 1125; **(b)** 27; **(c)** 225; **(d)** 200; **(e)** 80 **29.** [1.7D] **(a)** 12; **(b)** 50; **(c)** 675; **(d)** 25; **(e)** 1 **30.** [1.8A] **(a)** 54; **(b)** 45; **(c)** 36; **(d)** 27; **(e)** 19 **31.** [1.8A] **(a)** 50; **(b)** 56; **(c)** 62; **(d)** 68; **(e)** 74 **32.** [1.8A] **(a)** 5; **(b)** 2; **(c)** 7; **(d)** 10; **(e)** 17 **33.** [1.8A] **(a)** 29; **(b)** 11; **(c)** 17; **(d)** 9; **(e)** 11 **34.** [1.8B] **(a)** 25; **(b)** 35; **(c)** 45; **(d)** 57; **(e)** 64 **35.** [1.8C] **(a)** $170; **(b)** $230; **(c)** $140; **(d)** $200; **(e)** $260 **36.** [1.9A] **(a)** $x = 12$; **(b)** $x = 11$; **(c)** $x = 10$; **(d)** $x = 9$; **(e)** $x = 8$ **37.** [1.9A] **(a)** $x = 7$; **(b)** $x = 6$; **(c)** $x = 5$; **(d)** $x = 4$; **(e)** $x = 3$ **38.** [1.9A] **(a)** $x = 5$; **(b)** $x = 4$; **(c)** $x = 2$; **(d)** $x = 1$; **(e)** $x = 10$ **39.** [1.9A] **(a)** $x = 7$; **(b)** $x = 6$; **(c)** $x = 5$; **(d)** $x = 4$; **(e)** $x = 3$ **40.** [1.9B] **(a)** $n = 21$; **(b)** $n = 26$; **(c)** $n = 40$; **(d)** $n = 62$; **(e)** $n = 33$ **41.** [1.9B] **(a)** $m = 32$; **(b)** $m = 58$; **(c)** $m = 25$; **(d)** $m = 57$; **(e)** $m = 65$ **42.** [1.9B] **(a)** $m = 15$; **(b)** $m = 13$; **(c)** $m = 18$; **(d)** $m = 22$; **(e)** $m = 29$ **43.** [1.9B] **(a)** $x = 12$; **(b)** $x = 13$; **(c)** $x = 12$; **(d)** $x = 9$; **(e)** $x = 12$ **44.** [1.9B] **(a)** $x = 5$; **(b)** $x = 4$; **(c)** $x = 4$; **(d)** $x = 6$; **(e)** $x = 6$ **45.** [1.9C] **(a)** 90 ft; **(b)** 100 ft; **(c)** 110 ft; **(d)** 85 ft; **(e)** 75 ft

Chapter 2

Exercises 2.1

1. $\frac{1}{2}$ **3.** $\frac{1}{3}$ **5.** $\frac{5}{12}$ **7.** $\frac{1}{4}$ **9.** $\frac{3}{4}$ **11.** $\frac{2}{4}$ **13.** $\frac{3}{4}$ **15.** $\frac{4}{4}$ or 1 **17.** $\frac{2}{3}$ **19.** 1 **21.** proper **23.** proper **25.** proper **27.** proper **29.** proper **31.** $3\frac{1}{10}$ **33.** $1\frac{1}{7}$ **35.** $3\frac{5}{8}$ **37.** $7\frac{6}{9} = 7\frac{2}{3}$ **39.** $10\frac{1}{10}$ **41.** $\frac{36}{7}$ **43.** $\frac{41}{10}$ **45.** $\frac{13}{11}$ **47.** $\frac{83}{10}$ **49.** $\frac{13}{6}$ **51.** $\frac{7}{24}$ **53.** $\frac{7}{16}$ **55.** $\frac{5}{8}$ **57.** $\frac{51}{100}$ **59. (a)** $\frac{60}{60} = 1$; **(b)** $\frac{90}{60} = \frac{3}{2} = 1\frac{1}{2}$; **(c)** $\frac{45}{60} = \frac{3}{4}$; **(d)** $\frac{15}{60} = \frac{1}{4}$ **61.** $\frac{25}{98}$ **63.** $\frac{6}{98}$ or $\frac{3}{49}$ **65.** $\frac{3}{98}$ **67.** $\frac{37}{99}$ **69.** $\frac{2}{99}$ **71.** $\frac{7}{20}$ **73.** $\frac{3}{20}$ **75.** $\frac{1}{10}$ **77.** 0, 1 **79.** $\frac{2}{4}$ or $\frac{1}{2}$, $\frac{2}{4}$ or $\frac{1}{2}$ **81.** 18 **83.** 21 **85.** 350 mi **87.** No. You need $\frac{340}{20} = 17$ gallons and the tank holds 14. **89.** 12 gallons **95.** denominator **97.** greater **99.** undefined **101.** $\frac{a}{b}$ **103.** $\frac{5}{12}$ **105.** $8\frac{1}{3}$ **107.** improper **109.** $2^2 \times 3^2$ **111.** $2^2 \times 7$ **113.** $2^2 \times 3^2 \times 5$

Exercises 2.2

1. 30 **3.** 30 **5.** 45 **7.** 15 **9.** 72 **11.** 4 **13.** 12 **15.** 3 **17.** $\frac{14}{15}$ **19.** $\frac{1}{4}$ **21.** $\frac{7}{3} = 2\frac{1}{3}$ **23.** $\frac{3}{4}$ **25.** $\frac{2}{3}$ **27.** $\frac{3}{14}$ **29.** $\frac{3}{13}$ **31.** $\frac{23}{50}$ **33.** $\frac{1}{6}$ **35.** $\frac{20}{73}$ **37. (a)** $\frac{1}{2}$; **(b)** $\frac{1}{4}$; **(c)** $\frac{1}{13}$ **39.** the first recipe **41.** $\frac{1}{4}$ **43.** $\frac{1}{8}$ **45.** $\frac{1}{8}$ **47.** $\frac{7}{40}$ **49.** $\frac{13}{35}$ **51. (a)** $\frac{5}{258}$; **(b)** $\frac{5}{129}$ **53. (a)** $\frac{9}{43}$; **(b)** $\frac{53}{258}$ **55. (a)** $\frac{8}{129}$; **(b)** $\frac{3}{86}$ **57.** $\frac{5}{6}$ **59. (a)** $\frac{1}{28}$; **(b)** 400 **65.** $\frac{3x}{4}$ **67.** $\frac{2x}{3}$ **69.** equivalent **71.** reduced **73.** $\frac{15}{25}$ **75.** $\frac{3}{25}$ **77.** $\frac{4}{23}$ **79.** $\frac{27}{8}$ **81.** $\frac{79}{10}$ **83.** $\frac{132}{13}$

Exercises 2.3

1. $\frac{21}{32}$ **3.** $\frac{1}{7}$ **5.** $\frac{2}{3}$ **7.** $\frac{6}{5} = 1\frac{1}{5}$ **9.** $\frac{1}{2}$ **11.** $\frac{3}{2} = 1\frac{1}{2}$ **13.** 4 **15.** 2 **17.** 7 **19.** $\frac{21}{2} = 10\frac{1}{2}$ **21.** 13 **23.** 62 **25.** $\frac{1}{9}$ **27.** $\frac{25}{4} = 6\frac{1}{4}$ **29. (a)** $\frac{2}{15}$; **(b)** $\frac{1}{2}$ **31.** $\frac{1}{3}$ **33.** $\frac{2}{21}$ **35.** $\frac{8}{27}$ **37.** $\frac{15}{2} = 7\frac{1}{2}$ **39.** $\frac{2}{15}$ **41.** $\frac{7}{9}$ **43.** $\frac{3}{2} = 1\frac{1}{2}$ **45.** $\frac{8}{5} = 1\frac{3}{5}$ **47.** 1 **49.** $\frac{2}{5}$ **51.** 10 **53.** $\frac{13}{5} = 2\frac{3}{5}$ **55.** $\frac{75}{32} = 2\frac{11}{32}$ **57.** 1 **59.** $\frac{4}{33}$ **61.** $\frac{2}{7}$ square mile **63.** 72 people **65.** 16 days **67.** 8 turns **69.** $\frac{84}{5} = 16\frac{4}{5}$ or 16 vests **71.** 660 **73.** 11 **75.** 13 **77.** $1\frac{1}{8}$ **79.** $1\frac{1}{12}$ **81.** 96 mi **83.** 168 mi **85.** $\frac{20}{3} = 6\frac{2}{3}$ in. **87.** $25\frac{5}{32}$ square inches **89.** $39\frac{3}{16}$ square inches **91.** $43\frac{1}{6}$ square inches **93.** $134\frac{2}{5}$ square inches **95.** $4\frac{1}{2}$ yards **97.** 26 ft **99.** $1\frac{1}{8}$ cm **101.** $3\frac{1}{8}$ inches **107.** $\frac{a \cdot c}{b \cdot d}$ **109.** $\frac{1}{2}$ cup

111. $\frac{15}{4} = 3\frac{3}{4}$ **113.** $\frac{27}{32}$ **115.** 4 **117.** $\frac{140}{9} = 15\frac{5}{9}$ yd^2 **119.** 2^7 **121.** $2^2 \times 3^2 \times 5$ **123.** $2^2 \times 3^2 \times 5^2$

Exercises 2.4

1. 40 **3.** 48 **5.** 18 **7.** 42 **9.** 60 **11.** 6; $\frac{2}{6}, \frac{1}{6}$ **13.** 21; $\frac{1}{21}, \frac{3}{21}$ **15.** 20; $\frac{15}{20}, \frac{2}{20}$ **17.** 24; $\frac{4}{24}, \frac{2}{24}, \frac{1}{24}$ **19.** 40; $\frac{24}{40}, \frac{25}{40}, \frac{14}{40}$ **21.** 72; $\frac{4}{72}, \frac{3}{72}$ **23.** 160; $\frac{5}{160}, \frac{2}{160}$ **25.** 20; $\frac{15}{20}, \frac{6}{20}$ **27.** 24; $\frac{4}{24}, \frac{2}{24}, \frac{1}{24}$ **29.** 40; $\frac{24}{40}, \frac{25}{40}, \frac{14}{40}$ **31.** $\frac{7}{8}$ **33.** $\frac{5}{11}$ **35.** $<$ **37.** $<$ **39.** $<$ **41.** 60 min **43.** 221 years **45.** 12 hours **47.** 20 days **49.** 12 days **51.** It takes 60 years, so in 2060 **53.** 420 years **57.** They are the same. **59.** LCM **61.** LCD **63.** $>$ **65.** 45 **67.** $\frac{15}{35}, \frac{28}{35}$ **69.** 180 **71.** $\frac{40}{72}, \frac{27}{72}$

Exercises 2.5

1. $\frac{2}{3}$ **3.** $\frac{5}{7}$ **5.** $\frac{2}{3}$ **7.** 1 **9.** 2 **11.** $\frac{8}{15}$ **13.** $\frac{2}{3}$ **15.** $\frac{13}{10} = 1\frac{3}{10}$ **17.** $\frac{11}{14}$ **19.** $\frac{7}{8}$ **21.** $\frac{29}{360}$ **23.** $\frac{19}{130}$ **25.** $\frac{23}{360}$ **27.** $\frac{50}{60} = \frac{5}{6}$ **29.** $\frac{316}{126} = \frac{158}{63} = 2\frac{32}{63}$ **31.** $\frac{2}{7}$ **33.** $\frac{2}{3}$ **35.** $\frac{1}{6}$ **37.** $\frac{1}{10}$ **39.** $\frac{3}{40}$ **41.** $\frac{11}{24}$ **43.** $\frac{47}{240}$ **45.** $\frac{5}{9}$ **47.** 1 **49.** $\frac{13}{6} = 2\frac{1}{6}$ **51.** $\frac{37}{32} = 1\frac{5}{32}$ in. **53.** $\frac{1}{8}$ **55.** $\frac{9}{20}$ **57.** $\frac{1}{6}$ **59.** $\frac{3}{4}$ **61.** $\frac{1}{5}$ **63.** $\frac{1}{2}$ **65.** $\frac{7}{10}$ **67.** **(a)** $\frac{1}{2}$; **(b)** $1500 **69.** **(a)** $\frac{9}{20}$; **(b)** $1350 **71.** 4 packages of hot dogs, 5 packages of buns **77.** $\frac{a+b}{c}$ **79.** 90 **81.** $\frac{7}{24}$ **83.** $\frac{67}{120}$ **85.** $\frac{31}{120}$ **87.** $\frac{16}{5}$ **89.** $\frac{55}{8}$ **91.** $1\frac{2}{3}$

Exercises 2.6

1. $4\frac{4}{7}$ **3.** $2\frac{4}{7}$ **5.** $5\frac{1}{2}$ **7.** $4\frac{2}{5}$ **9.** $5\frac{1}{7}$ **11.** $2\frac{53}{60}$ **13.** $4\frac{13}{60}$ **15.** $4\frac{7}{12}$ **17.** $11\frac{16}{63}$ **19.** $12\frac{21}{110}$ **21.** $2\frac{2}{7}$ **23.** $1\frac{2}{3}$ **25.** $1\frac{5}{6}$ **27.** $\frac{7}{10}$ **29.** $\frac{39}{40}$ **31.** $2\frac{11}{24}$ **33.** $\frac{47}{240}$ **35.** 4 **37.** $3\frac{2}{3}$ **39.** $5\frac{5}{12}$ **41.** $12\frac{3}{130}$ **43.** $18\frac{11}{60}$ **45.** 3 **47.** $1\frac{5}{16}$ lb **49.** $3\frac{1}{4}$ cups **51.** $5\frac{3}{4}$ lb **53.** $\frac{3}{10}$ **55.** $7\frac{7}{10}$ hr **57.** $24\frac{13}{20}$ **59.** 52 ft **61.** $3\frac{3}{8} + \frac{1}{4}$; $3\frac{5}{8}$ **63.** $2\frac{2}{3} - 1\frac{5}{8}$; $1\frac{1}{24}$ **65.** $\$3\frac{1}{8}$ **67.** $\$62\frac{5}{8}$ **69.** No **73.** LCD **75.** $2\frac{49}{60}$ **77.** $6\frac{7}{12}$ **79.** $2\frac{31}{180}$ **81.** $\frac{1}{4}$ **83.** $\frac{128}{75}$ or $1\frac{53}{75}$

Exercises 2.7

1. $\frac{13}{60}$ **3.** $\frac{19}{84}$ **5.** 0 **7.** $\frac{13}{30}$ **9.** $\frac{7}{6} = 1\frac{1}{6}$ **11.** $\frac{1}{2}$ **13.** $\frac{7}{36}$ **15.** $\frac{52}{15} = 3\frac{7}{15}$ **17.** $\frac{63}{80}$ **19.** $\frac{17}{20}$ **21.** $\frac{6}{5} = 1\frac{1}{5}$ **23.** $\frac{35}{8} = 4\frac{3}{8}$ **25.** 0 **27.** $2\frac{17}{20}$ in. **29.** $31\frac{11}{30}$ lb **31.** **(a)** $\$149\frac{1}{30}$ million; **(b)** $\$349\frac{8}{15}$ million **33.** $8\frac{8}{15}$ nights **35.** $14\frac{1}{6}$ hr per wk **37.** $20\frac{11}{30}$ hr per wk **39.** Women 18 and over **41.** $5\frac{1}{6}$ lb **43.** $23\frac{7}{16}$ lb **45.** **(a)** 8, 4, 2, 1; **(b)** $A_8 = \frac{15}{4} = 3\frac{3}{4}$; **(c)** $H_8 = \frac{32}{15} = 2\frac{2}{15}$; **(d)** Yes **49.** parentheses **51.** multiplication **53.** addition **55.** $\frac{83}{20} = 4\frac{3}{20}$ **57.** $\frac{25}{36}$ **59.** $\frac{83}{162}$ **61.** $3\frac{7}{8}$ **63.** $x = 6$ **65.** $x = 3$

Exercises 2.8

Translate This

1. I **3.** B **5.** C **7.** D **9.** M

1. $=$ **3.** $\times$ **5.** $+$ **7.** $-$ **9.** $2 \cdot n$ **11.** $5 + n$ **13.** $n - 7$ **15.** $\frac{3}{4} \div n = 5$ **17.** $\frac{1}{2} \cdot 3 \cdot n = 2$ **19.** $\frac{n}{2} - 4 = \frac{3}{2}$ **21.** $m + \frac{1}{8} = \frac{3}{7}$; $m = \frac{17}{56}$ **23.** $p + \frac{2}{5} = 1\frac{3}{4}$; $p = \frac{27}{20} = 1\frac{7}{20}$ **25.** $y - \frac{3}{4} = \frac{4}{5}$; $y = \frac{31}{20} = 1\frac{11}{20}$ **27.** $\frac{u}{6} = 3\frac{1}{2}$; $u = 21$ **29.** $3 \cdot t = 2\frac{1}{5}$; $t = \frac{11}{15}$ **31.** $1\frac{1}{2} \cdot n = 7\frac{1}{2}$; $n = 5$ **33.** $n \cdot 1\frac{2}{3} = 4$; $n = \frac{12}{5} = 2\frac{2}{5}$ **35.** $n \cdot 2\frac{1}{2} = 6\frac{1}{4}$; $n = \frac{5}{2} = 2\frac{1}{2}$ **37.** $1\frac{1}{3} \cdot n = 4\frac{2}{3}$; $n = \frac{7}{2} = 3\frac{1}{2}$ **39.** $1\frac{1}{8} \cdot 2\frac{1}{2} = n$; $n = \frac{45}{16} = 2\frac{13}{16}$ **41.** 30 in. **43.** 1 cup **45.** 24 lb **47.** $\frac{5}{2} = 2\frac{1}{2}$ km **49.** 70¢ **51.** $\frac{13}{25}$ **53.** 77 **55.** 69°F **57.** **(a)** $\frac{1}{6}$; **(b)** $\frac{1}{3}$; **(c)** $\frac{1}{2}$; **(d)** 25 **59.** **(a)** $7\frac{1}{12}$; **(b)** $56\frac{2}{3}$; **(c)** $5\frac{2}{3}$ **61.** $3\frac{1}{4}$ **63.** $118\frac{4}{5}$ lb **65.** $4\frac{1}{2}$ oz **67.** $3\frac{1}{2}$ oz **69.** 5 oz **75.** $a - c = b - c$ **77.** $a \div c = b \div c$ **79.** $2\frac{1}{2}$ **81.** **(a)** $3n = 9$; **(b)** $n - 5 = 2$; **(c)** $n + 8 = 7$ **83.** $9\frac{2}{7}$ **85.** 3 mi **87.** 190 **89.** 6

Review Exercises

1. [2.1B] **(a)** Proper; **(b)** Proper; **(c)** Improper; **(d)** Proper; **(e)** Improper **2.** [2.1C] **(a)** $3\frac{1}{7}$; **(b)** $2\frac{4}{7}$; **(c)** $9\frac{2}{3}$; **(d)** $3\frac{1}{2}$; **(e)** $1\frac{8}{11}$ **3.** [2.1D] **(a)** $\frac{9}{2}$; **(b)** $\frac{28}{9}$; **(c)** $\frac{22}{5}$; **(d)** $\frac{115}{14}$; **(e)** $\frac{63}{8}$ **4.** [2.1E] **(a)** 8; **(b)** 10; **(c)** 4; **(d)** 2; **(e)** 5 **5.** [2.2A] **(a)** 8; **(b)** 15; **(c)** 24; **(d)** 28; **(e)** 18 **6.** [2.2A] **(a)** 7; **(b)** 5; **(c)** 8; **(d)** 8; **(e)** 10 **7.** [2.2B] **(a)** $\frac{1}{2}$; **(b)** $\frac{2}{3}$; **(c)** $\frac{2}{5}$; **(d)** $\frac{2}{7}$; **(e)** $\frac{2}{19}$ **8.** [2.2B] **(a)** 12, $\frac{1}{3}$; **(b)** 10, $\frac{1}{5}$; **(c)** 9, $\frac{2}{5}$; **(d)** 14, $\frac{2}{3}$; **(e)** 17, $\frac{3}{2}$ **9.** [2.3A] **(a)** $\frac{2}{21}$; **(b)** $\frac{2}{9}$; **(c)** $\frac{1}{3}$; **(d)** $\frac{3}{2} = 1\frac{1}{2}$; **(e)** 1 **10.** [2.3B] **(a)** $\frac{38}{21} = 1\frac{17}{21}$; **(b)** 2; **(c)** $\frac{3}{2} = 1\frac{1}{2}$; **(d)** $\frac{81}{40} = 2\frac{1}{40}$; **(e)** 4 **11.** [2.3C] **(a)** $\frac{2}{15}$; **(b)** 1; **(c)** $\frac{2}{3}$; **(d)** $\frac{7}{6} = 1\frac{1}{6}$; **(e)** 2 **12.** [2.3D] **(a)** $\frac{7}{8}$; **(b)** $\frac{7}{16}$; **(c)** $\frac{36}{25} = 1\frac{11}{25}$; **(d)** $\frac{15}{7} = 2\frac{1}{7}$; **(e)** $\frac{1}{2}$ **13.** [2.3E] **(a)** $\frac{45}{16} = 2\frac{13}{16}$; **(b)** $\frac{176}{49} = 3\frac{29}{49}$; **(c)** $\frac{169}{8} = 21\frac{1}{8}$; **(d)** $\frac{3}{2} = 1\frac{1}{2}$; **(e)** $\frac{435}{98} = 4\frac{43}{98}$ **14.** [2.3E] **(a)** $\frac{1}{2}$; **(b)** $\frac{4}{17}$; **(c)** $\frac{3}{16}$; **(d)** $\frac{2}{35}$; **(e)** $\frac{16}{225}$ **15.** [2.3G] **(a)** $15\frac{5}{9}$ square yards; **(b)** $15\frac{3}{4}$ square yards; **(c)** 15 square yards; **(d)** $15\frac{1}{6}$ square yards; **(e)** $24\frac{3}{4}$ square yards **16.** [2.4A] **(a)** 24; **(b)** 30; **(c)** 36; **(d)** 120; **(e)** 540 **17.** [2.4A] **(a)** 33; **(b)** 34; **(c)** 57; **(d)** 40; **(e)** 92 **18.** [2.4B] **(a)** 48, $\frac{28}{48}, \frac{9}{48}$; **(b)** 45, $\frac{6}{45}, \frac{25}{45}$; **(c)** 144, $\frac{45}{144}, \frac{40}{144}$; **(d)** 35, $\frac{15}{35}, \frac{28}{35}$; **(e)** 45, $\frac{25}{45}, \frac{12}{45}$ **19.** [2.4B] **(a)** 12, $\frac{9}{12}, \frac{6}{12}, \frac{10}{12}$; **(b)** 72, $\frac{30}{72}, \frac{8}{72}, \frac{27}{72}$; **(c)** 144, $\frac{117}{144}, \frac{8}{144}, \frac{132}{144}$; **(d)** 120, $\frac{12}{120}, \frac{45}{120}, \frac{10}{120}$; **(e)** 360, $\frac{72}{360}, \frac{160}{360}, \frac{45}{360}$ **20.** [2.4C] **(a)** $>$; **(b)** $>$; **(c)** $>$; **(d)** $<$; **(e)** $>$ **21.** [2.5A] **(a)** $\frac{3}{5}$; **(b)** 1; **(c)** $\frac{4}{7}$; **(d)** $\frac{1}{3}$; **(e)** 8 **22.** [2.5B] **(a)** 6; $\frac{7}{6} = 1\frac{1}{6}$; **(b)** 45; $\frac{14}{45}$; **(c)** 42; $\frac{53}{42} = 1\frac{11}{42}$; **(d)** 60; $\frac{37}{60}$; **(e)** 105; $\frac{51}{105} = \frac{17}{35}$ **23.** [2.5B] **(a)** 12; $\frac{109}{12} = 9\frac{1}{12}$; **(b)** 6; $\frac{31}{6} = 5\frac{1}{6}$; **(c)** 16; $\frac{101}{16} = 6\frac{5}{16}$; **(d)** 9; $\frac{58}{9} = 6\frac{4}{9}$; **(e)** 72; $\frac{233}{72} = 3\frac{17}{72}$ **24.** [2.5D] **(a)** $\frac{67}{84}$; **(b)** $\frac{19}{24}$; **(c)** $\frac{21}{16} = 1\frac{5}{16}$; **(d)** 1; **(e)** $\frac{4}{3} = 1\frac{1}{3}$ **25.** [2.5D] **(a)** $\frac{1}{8}$; **(b)** $\frac{19}{36}$; **(c)** $\frac{13}{48}$; **(d)** $\frac{4}{35}$; **(e)** $\frac{83}{216}$ **26.** [2.5E] **(a)** $\frac{3}{4}$; **(b)** $\frac{19}{50}$; **(c)** $\frac{23}{100}$; **(d)** $\frac{13}{25}$; **(e)** $\frac{3}{25}$ **27.** [2.6B] **(a)** $7\frac{11}{30}$; **(b)** $5\frac{5}{12}$; **(c)** $7\frac{23}{28}$; **(d)** $7\frac{4}{9}$; **(e)** $8\frac{7}{8}$ **28a.** [2.5C] **(a)** $\frac{5}{24}$; **(b)** $\frac{26}{15} = 1\frac{11}{15}$; **(c)** $\frac{13}{15}$; **(d)** $\frac{39}{40}$; **(e)** $\frac{23}{72}$ **28b.** [2.6D] **(a)** $3\frac{209}{360}$; **(b)** $4\frac{28}{45}$; **(c)** $5\frac{37}{72}$; **(d)** $6\frac{17}{36}$; **(e)** $7\frac{37}{72}$ **29.** [2.6E] **(a)** $19\frac{1}{2}$ yards; **(b)** $15\frac{2}{3}$ yards; **(c)** $19\frac{2}{3}$ yards; **(d)** $17\frac{2}{3}$ yards; **(e)** 12 yards **30.** [2.7A] **(a)** $\frac{1}{9}$; **(b)** $\frac{1}{8}$; **(c)** $\frac{1}{9}$; **(d)** $\frac{5}{49}$; **(e)** $\frac{3}{32}$ **31.** [2.7A] **(a–e)** $\frac{1}{3}$ **32.** [2.7A] **(a)** $\frac{89}{96}$; **(b)** $\frac{121}{96} = 1\frac{25}{96}$; **(c)** $\frac{19}{32}$; **(d)** $\frac{2}{3}$; **(e)** $\frac{7}{4} = 1\frac{3}{4}$ **33.** [2.7B] **(a)** $\frac{17}{21}$; **(b)** $\frac{5}{6}$; **(c)** $\frac{13}{15}$; **(d)** $\frac{11}{12}$; **(e)** 1 **34.** [2.7C] **(a)** $4\frac{3}{8}$ lb; **(b)** $5\frac{3}{8}$ lb; **(c)** $6\frac{3}{8}$ lb; **(d)** $7\frac{3}{8}$ lb; **(e)** $8\frac{3}{8}$ lb **35.** [2.8A] **(a)** $n + 8 = 10$; **(b)** $n - 5 = 1$; **(c)** $2n = 12$; **(d)** $\frac{n}{2} = 8$; **(e)** $\frac{n}{7} = 3$ **36.** [2.8B] **(a)** $p + \frac{1}{6} = \frac{1}{3}$; $p = \frac{1}{6}$; **(b)** $q + \frac{1}{5} = \frac{1}{4}$; $q = \frac{1}{20}$; **(c)** $r + \frac{1}{4} = \frac{2}{5}$; $r = \frac{3}{20}$; **(d)** $s + \frac{1}{3} = \frac{5}{6}$; $s = \frac{1}{2}$; **(e)** $t + \frac{1}{2} = \frac{6}{7}$; $t = \frac{5}{14}$ **37.** [2.8B] **(a)** $r - \frac{1}{6} = \frac{2}{7}$; $r = \frac{19}{42}$; **(b)** $s - \frac{1}{5} = \frac{3}{7}$; $s = \frac{22}{35}$; **(c)** $t - \frac{1}{4} = \frac{4}{7}$; $t = \frac{23}{28}$; **(d)** $u - \frac{1}{3} = \frac{5}{7}$; $u = \frac{22}{21} = 1\frac{1}{21}$; **(e)** $v - \frac{1}{2} = \frac{6}{7}$; $v = \frac{19}{14} = 1\frac{5}{14}$ **38.** [2.8B] **(a)** $\frac{v}{3} = \frac{2}{7}$; $v = \frac{6}{7}$; **(b)** $\frac{v}{4} = \frac{3}{7}$; $v = \frac{12}{7} = 1\frac{5}{7}$; **(c)** $\frac{v}{5} = \frac{4}{7}$; $v = \frac{20}{7} = 2\frac{6}{7}$; **(d)** $\frac{v}{6} = \frac{5}{7}$; $v = \frac{30}{7}$; **(e)** $\frac{v}{7} = \frac{6}{7}$; $v = 6$ **39.** [2.8B] **(a)** 16; **(b)** 6; **(c)** 45; **(d)** 49; **(e)** 10 **40.** [2.8C] **(a)** $\frac{77}{100}$; **(b)** $\frac{61}{100}$; **(c)** $\frac{53}{100}$; **(d)** $\frac{9}{20}$; **(e)** $\frac{33}{100}$

Cumulative Review Chapters 1–2

1. $400 + 30 + 8$ **2.** 984 **3.** seventy-four thousand, eight **4.** 6710 **5.** 8600 **6.** 3679 **7.** 154 **8.** 43,703 **9.** $3720 10. 34 r 5 **11.** 2, 3 **12.** $2^2 \times 3^2 \times 5$ **13.** 32 **14.** 40 **15.** 23 **16.** 3 **17.** proper **18.** $5\frac{1}{2}$ **19.** $\frac{9}{4}$ **20.** 14 **21.** 27 **22.** $\frac{5}{6}$ **23.** $<$ **24.** $\frac{19}{6} = 3\frac{1}{6}$ **25.** $\frac{1}{36}$ **26.** $\frac{9}{14}$ **27.** 30; $6\frac{19}{30}$ **28.** $6\frac{16}{63}$ **29.** $z - \frac{6}{7} = \frac{4}{9}$; $z = \frac{82}{63}$ **30.** $5\frac{7}{9}$ **31.** $1.12 **32.** 22 yd **33.** $28\frac{8}{9}$ yd^2 **34.** 80 **35.** 76 **36.** 36; $\frac{28}{36}, \frac{15}{36}$ **37.** 30; $\frac{21}{30}, \frac{25}{30}, \frac{18}{30}$

Chapter 3

Exercises 3.1

1. Three and eight tenths **3.** Thirteen and twelve hundredths
5. One hundred thirty-two and thirty-four hundredths
7. Five and one hundred eighty-three thousandths
9. Two thousand, one hundred seventy-two ten-thousandths
11. $3 + \frac{2}{10} + \frac{1}{100}$ **13.** $40 + 1 + \frac{3}{10} + \frac{8}{100}$
15. $80 + 9 + \frac{1}{10} + \frac{2}{100} + \frac{3}{1000}$
17. $200 + 30 + 8 + \frac{3}{10} + \frac{9}{100} + \frac{2}{1000}$
19. $300 + 1 + \frac{5}{10} + \frac{8}{100} + \frac{7}{1000} + \frac{9}{10{,}000}$ **21.** 0.5 **23.** 1.5 **25.** 0.2

27. 3.1 **29.** 1.8 **31.** 4.9 **33.** 1.9 **35.** 0.3 **37.** 21.23 **39.** 23.33 **41.** 8.95 **43.** $989.07 **45.** 919.154 **47.** 182.103 **49.** 4.077 **51.** 26.85 **53.** 2.38 **55.** 3.024 **57.** 6.844 **59.** 9.0946 **61.** 392.5 mi **63.** $54.58 **65.** 23.61 **67.** 93.5 **69.** 103.4% **71.** $230.50; $82,969.95 **73.** $86.67; $31,195.13 **75.** $30.99; $11,156.11 **77.** $87.50 **79.** $718.70 **81.** 3.7 mi **83.** 1.1 mi **85.** 2.3 mi **87.** 0.2 mi **89.** "And" indicates a decimal point, but this is a whole number **91.** 1, 8 on both cases **93.** $10.6y^2 + 1.2y$ **95.** $0.8y^2 + 0.5y$ **97.** whole, decimal, decimal **99.** decimal **101.** 42.463 **103.** $40 + 1 + \frac{2}{10} + \frac{0}{100} + \frac{8}{1000}$ **105.** 454.83 **107.** 1.6 mi **109.** 230 **111.** 240,000 **113.** 3500

Exercises 3.2

1. 0.35 **3.** 0.64 **5.** 0.00035 **7.** 5.6396 **9.** 95.7 **11.** 0.024605 **13.** 12.90516 **15.** 0.002542 **17.** 423.3 **19.** 1950 **21.** 32,890 **23.** 4.8 **25.** 3.9 **27.** 0.6 **29.** 6.4 **31.** 1700 **33.** 80 **35.** 0.046 **37.** 30 **39.** 100 **41.** 3.2 **43.** 10 **45.** 338.12 **47.** 0.078 **49.** 0.0005 **51.** 0.33 **53.** 0.09 **55.** 0.23 **57.** 0.01 **59.** 0.03 **61.** $35.24 **63.** **(a)** $16.08; **(b)** $482.40 **65.** $54.00 **67.** $58.26 **69.** $380.49 **71.** 0.915 sec **73.** 0.305 sec **75.** 5.55 min **77.** $8.50 **79.** 5.1136 cm **81.** 1129.77 mi **83.** $0.29 or 29¢ **85.** 9.9 **87.** 5.1 **89.** 2.3 **91.** 3.4 **93.** 19¢; $1900 **95.** 69.5 in. **97.** 59.5 in. **99.** 67.4 in. **101.** 73.8 in. **103.** 88.2 in.; table does not apply **109.** right, zeros **111.** left, divisor **113.** 642.86 **115.** 0.0005 **117.** **(a)** 324.23; **(b)** 48,400; **(c)** 32.8 **119.** **(a)** 14.112; **(b)** 75.0924 **121.** 4 **123.** 16

Exercises 3.3

1. 0.5 **3.** 0.6875 **5.** 0.45 **7.** 0.9 **9.** 0.25 **11.** 0.83 **13.** 0.43 **15.** 2.67 **17.** 0.33 **19.** 0.18 **21.** $\frac{4}{5}$ **23.** $\frac{19}{100}$ **25.** $\frac{3}{100}$ **27.** $\frac{31}{10}$ **29.** $\frac{5}{9}$ **31.** $\frac{7}{33}$ **33.** $\frac{1}{9}$ **35.** 0.375 **37.** 25 **39.** 4.8 **41.** 35 **43.** 0.333 **45.** 0.7 **47.** $2.18 **49.** 0.625 **51.** 0.056 **53.** 0.48 **55.** $0.08\overline{3}$ **57.** $40,000 **59.** $30,000 **61.** 0.35 **63.** 0.15 **65.** 0.06 **67.** 0.07 **69.** 0.10 **71.** 0.0625; 0.0625 **73.** 0.09375; 0.0938 **77.** numerator, denominator **79.** numerator **81.** 0.6 **83.** **(a)** $\frac{41}{100}$; **(b)** $\frac{303}{1000}$ **85.** **(a)** $\frac{7}{200}$; **(b)** $\frac{3}{80}$ **87.** **(a)** 0.6; **(b)** 0.225 **89.** $<$ **91.** $<$

Exercises 3.4

1. $66.606 > 66.066 > 66.06$ **3.** $0.5101 > 0.51 > 0.501$ **5.** $9.999 > 9.909 > 9.099$ **7.** $7.430 > 7.403 > 7.043$ **9.** $3.1\overline{4} > 3.\overline{14} > 3.14$ **11.** $5.12 > 5.\overline{1} > 5.1$ **13.** $0.\overline{3} > 0.333 > 0.33$ **15.** $0.\overline{8} > 0.88 > 0.\overline{81}$ **17.** $\frac{1}{9}$ **19.** $\frac{1}{6}$ **21.** $\frac{2}{7}$ **23.** $0.1\overline{4}$ **25.** $0.\overline{9}$ **27.** Constantine > Monjane > Nashnush **29.** 30 **31.** 12.48 **33.** 12.09 **35.** $9.1\overline{6}$ **37.** 3 **39.** 0.495 **41.** 0.33 **43.** 148.3 **45.** 28.68 **47.** $1\frac{1}{6}$ or $\frac{7}{6}$ **49.** $\frac{1}{2}$ **51.** $\frac{7}{36}$ **53.** Copper > Nickel > Cadmium > Brass **55.** $\frac{22}{7}$ **57.** decimal **59.** exponential **61.** 33 **63.** A **65.** $8.015 > 8.01 > 8.005$ **67.** 28.395 **69.** 50 **71.** 18.8

Exercises 3.5

Translate This

1. E **3.** F **5.** H **7.** A **9.** K

1. $x = 1.5$ **3.** $y = 6.5$ **5.** $z = 5.6$ **7.** $z = 16.5$ **9.** $m = 6$ **11.** $m = 0.7$ **13.** $n = 23.8$ **15.** $n = 12.4$ **17.** 2.89 **19.** 99.1°F **21.** 5.2 **23.** $0.5 billion **25.** $36.03 billion **27.** $300 **29.** 1023 **31.** About 42 **33.** 1.9 **35.** 8200 **37.** 172 **39.** $10,000 **41.** **(a)** $T + S = 1062$; **(b)** $T = S + 140$; **(c)** $T = \$461 + \$140 =$ $601 million **43.** **(a)** $T + B = 92$; **(b)** $T = B + 10$; **(c)** $B = 41$ **45.** 55 mph **47.** 3.2 **51.** $a + c = b + c$ **53.** $a \cdot c = b \cdot c$ **55.** 18.4 **57.** 10.1 **59.** $507.16 **61.** $\frac{19}{35}$ **63.** $1\frac{22}{35}$

Review Exercises

1. [3.1A] **(a)** Twenty-three and three hundred eighty-nine thousandths; **(b)** Twenty-two and thirty-four hundredths; **(c)** Twenty-four and five hundred sixty-four thousandths; **(d)** Twenty-seven and eight tenths; **(e)** Twenty-nine and sixty-seven hundredths **2.** [3.1B] **(a)** $30 + 7 + \frac{4}{10}$; **(b)** $50 + 9 + \frac{9}{100}$; **(c)** $100 + 40 + 5 + \frac{3}{100} + \frac{5}{1000}$; **(d)** $100 + 50 + \frac{3}{10} + \frac{9}{1000}$; **(e)** $200 + 30 + 4 + \frac{3}{1000}$ **3.** [3.1C] **(a)** 21.94; **(b)** 25.4257; **(c)** 23.7756; **(d)** 29.452; **(e)** 23.52 **4.** [3.1C] **(a)** 39.36; **(b)** 48.034; **(c)** 27.662; **(d)** 41.12; **(e)** 47.7617 **5.** [3.1D] **(a)** 314.801; **(b)** 323.44; **(c)** 278.275; **(d)** 347.55; **(e)** 23.66 **6.** [3.2A] **(a)** 0.03768; **(b)** 0.3276; **(c)** 3.317175; **(d)** 0.03752; **(e)** 1.026 **7.** [3.2B] **(a)** 370; **(b)** 4.9; **(c)** 2.5; **(d)** 4285; **(e)** 945 **8.** [3.2C] **(a)** 61; **(b)** 63; **(c)** 92; **(d)** 8.07; **(e)** 90.8 **9.** [3.2D] **(a)** 329.7; **(b)** 238.3; **(c)** 887.4; **(d)** 459.4; **(e)** 348.3 **10.** [3.2D] **(a)** 5.33; **(b)** 5.63; **(c)** 6.86; **(d)** 6.46; **(e)** 6.93 **11.** [3.2E] **(a)** 0.0312; **(b)** 0.00418; **(c)** 0.321; **(d)** 8.215; **(e)** 4.723 **12.** [3.2F] **(a)** $0.32 or 32¢; **(b)** $0.34 or 34¢; **(c)** $0.30 or 30¢; **(d)** $0.28 or 28¢; **(e)** $0.27 or 27¢ **13.** [3.3A] **(a)** 0.6; **(b)** 0.9; **(c)** 2.5; **(d)** 0.1875; **(e)** 0.875 **14.** [3.3A] **(a)** $0.\overline{3}$; **(b)** $0.8\overline{3}$; **(c)** $0.\overline{6}$; **(d)** $0.\overline{285714}$; **(e)** $0.\overline{1}$ **15.** [3.3B] **(a)** $\frac{19}{50}$; **(b)** $\frac{41}{100}$; **(c)** $\frac{3}{5}$; **(d)** $\frac{3}{100}$; **(e)** $\frac{333}{1000}$ **16.** [3.3B] **(a)** $\frac{233}{100}$; **(b)** $\frac{347}{100}$; **(c)** $\frac{131}{20}$; **(d)** $\frac{137}{100}$; **(e)** $\frac{1067}{500}$ **17.** [3.3C] **(a)** $\frac{5}{11}$; **(b)** $\frac{8}{99}$; **(c)** $\frac{80}{999}$; **(d)** $\frac{4}{999}$; **(e)** $\frac{11}{999}$ **18.** [3.3D] **(a)** 0.25; **(b)** 0.6; **(c)** 0.5; **(d)** 0.75; **(e)** 0.1875 **19.** [3.3D] **(a)** 0.2; **(b)** 0.25; **(c)** $0.\overline{142857}$; **(d)** 0.3; **(e)** $0.1\overline{6}$ **20.** [3.4A] **(a)** $1.032 > 1.03 > 1.003$; **(b)** $2.032 > 2.03 > 2.003$; **(c)** $3.033 > 3.032 > 3.03$; **(d)** $4.055 > 4.052 > 4.05$; **(e)** $5.033 > 5.03 > 5.003$ **21.** [3.4A] **(a)** $1.21\overline{6} > 1.2\overline{16} > 1.216$; **(b)** $2.33\overline{6} > 2.3\overline{36} > 2.336$; **(c)** $3.21\overline{6} > 3.2\overline{16} > 3.216$; **(d)** $4.5\overline{42} > 4.54\overline{2} > 4.542$; **(e)** $5.12\overline{3} > 5.1\overline{23} > 5.123$ **22.** [3.4B] **(a)** $>$; **(b)** $>$; **(c)** $<$; **(d)** $<$; **(e)** $>$ **23.** [3.4C] **(a)** 29; **(b)** 25; **(c)** 33; **(d)** 39; **(e)** 23 **24.** [3.5A] **(a)** $x = 4.3$; **(b)** $x = 2.3$; **(c)** $x = 0.5$; **(d)** $x = 3.6$; **(e)** $x = 2.7$ **25.** [3.5A] **(a)** $y = 7.3$; **(b)** $y = 7.7$; **(c)** $y = 13.32$; **(d)** $y = 10$; **(e)** $y = 11.5$ **26.** [3.5A] **(a)** $y = 5$; **(b)** $y = 7$; **(c)** $y = 4$; **(d)** $y = 8$; **(e)** $y = 8$ **27.** [3.5A] **(a)** $z = 24.6$; **(b)** $z = 35.7$; **(c)** $z = 10.8$; **(d)** $z = 44.02$; **(e)** $z = 56.7$ **28.** [3.5B] **(a)** $44.23; **(b)** $42.96; **(c)** $47.68; **(d)** $46.78; **(e)** $49.50

Cumulative Review Chapters 1–3

1. 394 **2.** 3210 **3.** 2, 5 **4.** $2^2 \times 3 \times 5$ **5.** 20 **6.** 52 **7.** improper **8.** $5\frac{1}{2}$ **9.** $\frac{43}{8}$ **10.** 18 **11.** 35 **12.** $\frac{31}{12} = 2\frac{7}{12}$ **13.** $\frac{1}{4}$ **14.** $\frac{81}{20} = 4\frac{1}{20}$ **15.** $15\frac{1}{24}$ **16.** $11\frac{7}{12}$ **17.** $z - \frac{6}{7} = \frac{3}{5}; z = \frac{51}{35} = 1\frac{16}{35}$ **18.** 4 **19.** $0.98 or 98¢ **20.** One hundred thirty-five and sixty-four hundredths **21.** $90 + 4 + \frac{4}{10} + \frac{7}{100} + \frac{8}{1000}$ **22.** 56.344 **23.** 228.92 **24.** 11.362 **25.** 1700 **26.** 250 **27.** 76.92 **28.** $0.38 or 38¢ **29.** $0.8\overline{3}$ **30.** $\frac{7}{20}$ **31.** $\frac{26}{33}$ **32.** 0.2 **33.** $\frac{1}{10}$ **34.** $5.31\overline{4} > 5.3\overline{14} > 5.314$ **35.** $<$ **36.** $x = 7.3$ **37.** $y = 8$ **38.** $z = 33.6$

Chapter 4

Exercises 4.1

1. $\frac{3}{8}$ **3.** $\frac{1}{7}$ **5.** $\frac{8}{1}$ **7.** $\frac{11}{3}$ **9.** $\frac{10}{3}$ **11.** $\frac{1}{4} = \frac{5}{20}$ **13.** $\frac{a}{3} = \frac{b}{7}$ **15.** $\frac{a}{6} = \frac{b}{18}$ **17.** $\frac{3}{a} = \frac{12}{b}$ **19.** no **21.** yes **23.** yes **25.** yes **27.** no **29.** $x = 8$ **31.** $x = 20$ **33.** $x = 48$ **35.** $x = 15$ **37.** $x = 25$ **39.** $x = 14$ **41.** $x = 12$ **43.** $x = 9$ **45.** $x = \frac{128}{9}$ **47.** $x = 11$ **49.** $x = 6.125$ **51.** $\frac{2}{26} = \frac{8}{d}; d = 104$ **53.** $\frac{5}{4} = \frac{8}{x}; x = 6.40$ **55.** $\frac{4.38}{4} = \frac{y}{6}; y = 6.57$ **57.** $\frac{9}{3} = \frac{z}{5}; z = 15$ **59.** $\frac{6.95}{5} = \frac{5.56}{x}; x = 4$ **61.** $\frac{17}{50}$ **63.** $\frac{57}{5000}$ **65.** $\frac{205}{87}$ **67.** 50 **69.** $360 **71.** 5760 cm or 57.60 m **73.** 120,000 **75.** **(a)** $\frac{17}{20}$; **(b)** 425; **(c)** 680 **77.** 4000 **79.** 21,250 **81.** **(a)** $\frac{4}{3}$; **(b)** 24 in. **83.** **(a)** $\frac{959}{485}$; **(b)** 1.977 **85.** $\frac{15}{8}$ **87.** $\frac{4}{25}$ **91.** quotient **93.** proportion **95.** $\frac{17}{21}$ **97.** **(a)** $\frac{2}{5} = \frac{4}{10}$; **(b)** $\frac{5}{8} = \frac{15}{x}$ **99.** yes **101.** 9 hours **103.** $\frac{7}{20}$ **105.** $\frac{3}{2}$

Exercises 4.2

1. \$4.75 **3.** 21.5 **5.** 560 mi/hr **7.** 1229.5 mi/hr **9.** 2.5 tortillas/min **11.** 2.63 **13.** 12 **15.** 1.5 in./hr **17.** 740 calories/lb **19.** 30.1 points/game **21.** \$10.38 **23.** \$21.75 **25.** \$25.57 **27.** 200 **29.** 75 **31.** 38¢/oz; 40¢/oz; First **33.** 8¢/oz; 9¢/oz; First **35.** \$1.80/oz; \$2.53/oz; First **37.** **(a)** 6¢; **(b)** 5¢; **(c)** White Magic **39.** **(a)** $\frac{4}{5}$; **(b)** $\frac{80}{100}$; **(c)** 80% **41.** **(a)** 0.83; **(b)** $\frac{83}{100}$; **(c)** 83% **43.** **(a)** 0.667; **(b)** $\frac{66.7}{100}$; **(c)** 66.7%; **(d)** Technically, no! **45.** **(a)** \$34.80; **(b)** \$3.48 **51.** rate **53.** 38.3¢/oz **55.** **(a)** \$254.20; **(b)** \$6.20 **57.** 22 **59.** $x = 6$ **61.** $x = \frac{15}{7}$ **63.** $x = 15$

Exercises 4.3

1. 15 lb **3.** 760 **5.** 8 **7.** 250 **9.** 40 min **11.** 7700 **13.** 24 **15.** 20 **17.** \$5.75 billion **19.** 86,250 **21.** 39 **23.** 28 **25.** 700 million **27.** 360 **29.** 70 mi **31.** \$859.00 **33.** \$1297.80 **35.** 24,500 **37.** No **39.** Translate **41.** 2 teaspoons/day **43.** 200 **45.** 400 **47.** 245.9 **49.** 250

Review Exercises

1. [4.1A] **(a)** $\frac{1}{10}$; **(b)** $\frac{1}{5}$; **(c)** $\frac{3}{10}$; **(d)** $\frac{2}{5}$; **(e)** $\frac{1}{2}$ **2.** [4.1A] **(a)** $\frac{17}{40}$; **(b)** $\frac{13}{40}$; **(c)** $\frac{3}{10}$; **(d)** $\frac{3}{20}$; **(e)** $\frac{1}{25}$ **3.** [4.1A] **(a)** $\frac{22}{7}$; **(b)** $\frac{729}{232}$; **(c)** $\frac{292}{93}$; **(d)** $\frac{732}{233}$; **(e)** $\frac{1468}{467}$ **4.** [4.1A] **(a)** $\frac{29}{21}$; **(b)** $\frac{11}{9}$; **(c)** $\frac{13}{12}$; **(d)** $\frac{12}{13}$; **(e)** $\frac{23}{27}$ **5.** [4.1B] **(a)** $\frac{3}{7} = \frac{6}{x}$; **(b)** $\frac{4}{7} = \frac{12}{x}$; **(c)** $\frac{5}{7} = \frac{x}{21}$; **(d)** $\frac{6}{7} = \frac{33}{x}$; **(e)** $\frac{7}{35} = \frac{5}{x}$ **6.** [4.1C] **(a)** no; **(b)** yes; **(c)** no; **(d)** yes; **(e)** yes **7.** [4.1D] **(a)** $x = 2$; **(b)** $x = 3$; **(c)** $x = 4$; **(d)** $x = 5$; **(e)** $x = 6$ **8.** [4.1D] **(a)** $x = 5$; **(b)** $x = 10$; **(c)** $x = 15$; **(d)** $x = 25$; **(e)** $x = 30$ **9.** [4.1D] **(a)** $x = 18$; **(b)** $x = 3$; **(c)** $x = 2$; **(d)** $x = 1$; **(e)** $x = 6$ **10.** [4.1D] **(a)** $x = 5$; **(b)** $x = 10$; **(c)** $x = 20$; **(d)** $x = 25$; **(e)** $x = 30$ **11.** [4.1E] **(a)** 9; **(b)** 18; **(c)** 36; **(d)** 45; **(e)** 54 **12.** [4.2A] **(a)** \$6; **(b)** \$4; **(c)** \$3; **(d)** \$2; **(e)** \$1.50 **13.** [4.2A] **(a)** 22; **(b)** 21; **(c)** 20; **(d)** 19; **(e)** 18 **14.** [4.2A] **(a)** 250; **(b)** 227; **(c)** 208; **(d)** 200; **(e)** 100 **15.** [4.2B] **(a)** 10¢; **(b)** 15¢; **(c)** 7¢; **(d)** 6¢; **(e)** 5¢ **16.** [4.2B] **(a)** Brand X; **(b)** Brand X; **(c)** Brand X; **(d)** Brand X; **(e)** generic **17.** [4.3A] **(a)** 15 lb; **(b)** $17\frac{1}{2} = 17.5$ lb; **(c)** 18 lb; **(d)** $22\frac{1}{2} = 22.5$ lb; **(e)** 25 lb **18.** [4.3A] **(a)** 1800; **(b)** 2400; **(c)** 3000; **(d)** 3600; **(e)** 4800 **19.** [4.3A] **(a)** 21 oz; **(b)** 28 oz; **(c)** 35 oz; **(d)** 42 oz; **(e)** 49 oz **20.** [4.3A] **(a)** 200; **(b)** 400; **(c)** 350; **(d)** 450; **(e)** 600

Cumulative Review Chapters 1–4

1. 9810 **2.** 2, 7 **3.** 56 **4.** 29 **5.** Improper **6.** $9\frac{3}{4}$ **7.** $\frac{23}{3}$ **8.** $\frac{11}{7}$ **9.** $\frac{1}{49}$ **10.** $\frac{7}{10}$ **11.** $9\frac{4}{15}$ **12.** $3\frac{3}{8}$ **13.** $c - \frac{7}{9} = \frac{1}{2}$; $c = \frac{23}{18}$ **14.** $7\frac{41}{55}$ **15.** Two hundred forty-one and thirty-five hundredths **16.** $40 + 4 + \frac{8}{10} + \frac{7}{100} + \frac{4}{1000}$ **17.** 46.144 **18.** 328.92 **19.** 0.08310 **20.** 500 **21.** 450 **22.** 76.92 **23.** $0.58\overline{3}$ **24.** $\frac{3}{20}$ **25.** $\frac{28}{33}$ **26.** 0.75 **27.** $6.43\overline{5} > 6.4\overline{35} > 6.435$ **28.** > **29.** 4 **30.** 7 **31.** 62.1 **32.** $\frac{21}{50}$ **33.** $\frac{6}{2} = \frac{54}{x}$ **34.** No **35.** $j = \frac{1}{5}$ **36.** $c = 15$ **37.** 100 **38.** 35 mpg **39.** 12¢/oz **40.** 40 **41.** 24 **42.** 800

Chapter 5

Exercises 5.1

1. 0.03 **3.** 0.10 **5.** 3.00 **7.** 0.1225 **9.** 0.115 **11.** 0.003 **13.** 4% **15.** 81.3% **17.** 314% **19.** 100% **21.** 0.2% **23.** $\frac{3}{10}$ **25.** $\frac{3}{50}$ **27.** $\frac{7}{100}$ **29.** $\frac{9}{200}$ **31.** $\frac{1}{75}$ **33.** $\frac{17}{500}$ **35.** $\frac{21}{200}$ **37.** 60% **39.** 50% **41.** $83\frac{1}{3}\%$ **43.** $37\frac{1}{2}\%$ **45.** $133\frac{1}{3}\%$ **47.** 81% **49.** 15% **51.** **(a)** $\frac{41}{100}$; 0.41; **(b)** $\frac{33}{100}$; 0.33; **(c)** $\frac{8}{25}$; 0.32; **(d)** $\frac{31}{100}$; 0.31 **53.** **(a)** $\frac{11}{25}$; 0.44; **(b)** $\frac{1}{5}$; 0.2; **(c)** $\frac{3}{20}$; 0.15 **55.** **(a)** $\frac{7}{10}$; **(b)** 70%; **(c)** 0.70 **57.** **(a)** $\frac{11}{14}$; **(b)** 79%; **(c)** 0.79 **59.** **(a)** $\frac{1}{48}$; **(b)** 2%; **(c)** 0.02 **61.** **(a)** $\frac{17}{960}$; **(b)** 2%; **(c)** 0.018 **63.** **(a)** $\frac{3}{10}$; **(b)** 30%; **(c)** 0.30 **65.** **(a)** $\frac{13}{25}$; **(b)** 0.52 **67.** No; due to rounding **69.** 70% **71.** **(a)** 40%; **(b)** $\frac{2}{5}$ **73.** **(a)** $\frac{49}{100}$; **(b)** 0.49 **75.** 5% **77.** 1% **81.** two, left **83.** **(a)** 100; **(b)** Reduce; **(c)** Omit **85.** **(a)** $\frac{41}{50}$; **(b)** 0.82 **87.** **(a)** $\frac{13}{200}$; **(b)** $\frac{131}{2000}$ **89.** **(a)** 6%; **(b)** 619%; **(c)** 4220% **91.** **(a)** 0.38; **(b)** 0.293 **93.** 29 **95.** $x = 40$ **97.** $x = \frac{1}{3}$ or $0.\overline{3}$

Exercises 5.2

Translate This

1. D **3.** M **5.** C **7.** K **9.** L

1. 32 **3.** 12 **5.** \$3 **7.** 28.8 **9.** 5 **11.** 2.1 **13.** 20 **15.** 10% **17.** 200% **19.** 10% **21.** $12\frac{1}{2}\%$ **23.** 25% **25.** 60 **27.** 200 **29.** $83\frac{1}{3}$ **31.** $16\frac{2}{3}$ **33.** 200 **35.** 400 **37.** 50 **39.** 45 **41.** $16\frac{2}{3}$ **43.** 18.75 **45.** 800 **47.** 3 **49.** 300 **51.** 73 million **53.** 23 **55.** 250 **57.** **(a)** 40%; **(b)** more **59.** 900 **61.** 1000 billion barrels **63.** 1,000,000 short tons **65.** 18 million or 18,000,000 **67.** 26% **69.** 7.68 oz of sugar; 6.56 oz of water **71.** 720 **73.** 710 **75.** \$15 **77.** $7\frac{1}{2}\%$ **79.** Three **81.** base **83.** percent **85.** \$795; \$705 **87.** 50 **89.** 60 **91.** 20 **93.** $n = 16$ **95.** $W = 50$

Exercises 5.3

1. 24 **3.** 30 **5.** \$6 **7.** 27 **9.** 7.5 **11.** 2.1 **13.** 22 **15.** 10% **17.** 400% **19.** 20% **21.** 20% **23.** 25% **25.** 80 **27.** 200 **29.** $83\frac{1}{3}$ **31.** $16\frac{2}{3}$ **33.** 300 **35.** 800 **37.** 150 **39.** 90 **41.** 35 **43.** 18.75 **45.** 1600 **47.** 49 **49.** 300 **51.** 48 **53.** 80 **55.** $h < 30$ in. **57.** $h < 32$ in. **61.** $\frac{\text{Percent}}{100}$ **63.** 32 **65.** 600 **67.** 0.055 **69.** $0.1\overline{6}$

Exercises 5.4

1. \$3640 **3.** \$13.00 **5.** \$660; \$13,860 **7.** 4% **9.** \$24 **11.** \$30 **13.** \$37.50 **15.** \$4.50 **17.** **(a)** \$530; **(b)** \$5300 **19.** \$160 **21.** \$6077.53 **23.** \$60,775.31 **25.** \$510.05 **27.** \$280 **29.** \$626 **31.** \$13.05 **33.** \$414 **35.** \$5 **37.** \$20.08 **39.** \$12.38 **41.** **(a)** \$13.50; **(b)** $2 \times 5.40 + \frac{1}{2} \times 5.40$; **(c)** \$13.50, yes **43.** **(a)** \$3.72; **(b)** $2 \times \$3.72 = \$7.44$; **(c)** $3 \times \$3.72 = \$11.16$ **45.** **(a)** \$4.64; **(b)** \$9.28; **(c)** \$6.96 **47.** \$3050 **49.** \$8620 **51.** \$18,620 **53.** 20% **55.** **(a)** Move the decimal point left one place **57.** principal **59.** year **61.** \$9.60; \$70.40 **63.** \$10,824.32; \$824.32 **65.** \$128.70 **67.** 167% **69.** 56% **71.** 83%

Exercises 5.5

1. 20% **3.** 17% **5.** 1% **7.** 4% **9.** 4% **11.** 14% **13.** 28% **15.** 338% **17.** 158% **19.** **(a)** 30%; **(b)** 390 **21.** **(a)** 243%; **(b)** 343,000 **23.** **(a)** 30%; **(b)** 7800 **25.** **(a)** \$10,150; **(b)** \$7308 **27.** **(a)** 338%; **(b)** 3%; **(c)** 12% **29.** **(a)** 8%; **(b)** 73% **31.** 17% **33.** **(a)** 7%; **(b)** 2003–2004; **(c)** 19% **35.** **(a)** \$723; **(b)** 10% **37.** 128% **39.** \$6800; no **45.** original **47.** 12% **49.** 27% **51.** 11% **53.** \$53.25 **55.** 3.9 **57.** \$1013.38

Exercises 5.6

1. \$90; \$1.35; \$141.35 **3.** \$109.39; \$1.64; \$185.01 **5.** \$303.93; \$4.56; \$557.48 **7.** \$200; \$3; \$453 **9.** **(a)** \$1.28; **(b)** \$236.28; **(c)** \$11.81 **11.** **(a)** \$5.16; **(b)** \$409.16; **(c)** \$20.46 **13.** **(a)** \$1.21; **(b)** \$180.39; **(c)** \$10 **15.** **(a)** \$0.84; **(b)** \$92.73; **(c)** \$10 **17.** **(a)** \$50; \$143.12; **(b)** \$87.50; **(c)** \$6262.80 **19.** **(a)** \$50; **(b)** \$61; **(c)** \$3320 **21.** **(a)** \$100; **(b)** \$122; **(c)** \$6640 **23.** **(a)** \$47,500; **(b)** \$2500; **(c)** \$38.92; \$277.08 **25.** **(a)** \$500; \$99.55; \$99,900.45; **(b)** \$499.50; \$100.05; \$99,800.40 **27.** **(a)** \$875; \$122.95; \$149,877.05; **(b)** \$874.28; \$123.67; \$149,753.38 **29.** \$225,000 **31.** **(a)** \$144,798; **(b)** \$269,798 **33.** \$5.00 **35.** $\frac{1}{12}$ **37.** $\frac{1}{4}$ **39.** loan **41.** **(a)** \$1000; **(b)** \$200 **43.** **(a)** \$150; **(b)** \$2.25; **(c)** \$212.25; **(d)** \$10.61 **45.** 50 **47.** 700

Review Exercises

1. [5.1A] **(a)** 0.39; **(b)** 0.01; **(c)** 0.13; **(d)** 1.01; **(e)** 2.07
2. [5.1A] **(a)** 0.032; **(b)** 0.112; **(c)** 0.714; **(d)** 0.1751; **(e)** 1.425
3. [5.1A] **(a)** 0.0625; **(b)** 0.715; **(c)** 0.05375; **(d)** 0.1725; **(e)** 0.52125
4. [5.1A] **(a)** $0.06\overline{3}$; **(b)** $0.08\overline{6}$; **(c)** $0.011\overline{6}$; **(d)** $0.188\overline{3}$; **(e)** $0.20\overline{1}$
5. [5.1B] **(a)** 1%; **(b)** 7%; **(c)** 17%; **(d)** 91%; **(e)** 83%
6. [5.1B] **(a)** 320%; **(b)** 110%; **(c)** 790%; **(d)** 910%; **(e)** 432%
7. [5.1C] **(a)** $\frac{17}{100}$; **(b)** $\frac{23}{100}$; **(c)** $\frac{51}{100}$; **(d)** $\frac{111}{100}$; **(e)** $\frac{201}{100}$ **8.** [5.1C] **(a)** $\frac{1}{10}$; **(b)** $\frac{2}{5}$; **(c)** $\frac{3}{20}$; **(d)** $\frac{7}{20}$; **(e)** $\frac{21}{50}$ **9.** [5.1C] **(a)** $\frac{1}{6}$; **(b)** $\frac{1}{3}$; **(c)** $\frac{5}{8}$; **(d)** $\frac{5}{6}$; **(e)** $\frac{7}{8}$
10. [5.1D] **(a)** $37\frac{1}{2}$%; **(b)** $62\frac{1}{2}$%; **(c)** $6\frac{1}{4}$%; **(d)** $18\frac{3}{4}$%; **(e)** $31\frac{1}{4}$%
11. [5.2A] **(a)** 18; **(b)** 98; **(c)** 32; **(d)** 27; **(e)** 36.75 **12.** [5.2A] **(a)** 10; **(b)** 24.3; **(c)** 38.75; **(d)** 32; **(e)** 1732.5 **13.** [5.2B] **(a)** 25%; **(b)** 20%; **(c)** $33\frac{1}{3}$%; **(d)** 200%; **(e)** 20% **14.** [5.2B] **(a)** 75%; **(b)** 40%; **(c)** 200%; **(d)** $66\frac{2}{3}$%; **(e)** $16\frac{2}{3}$% **15.** [5.2B] **(a)** 50%; **(b)** $33\frac{1}{3}$%; **(c)** 75%; **(d)** 150%; **(e)** $37\frac{1}{2}$% **16.** [5.2C] **(a)** 60; **(b)** 50; **(c)** 20; **(d)** $33\frac{1}{3}$; **(e)** $66\frac{2}{3}$ **17.** [5.2C] **(a)** 150; **(b)** 300; **(c)** 50; **(d)** 1500; **(e)** 84 **18.** [5.2C] **(a)** 25; **(b)** 10; **(c)** 200; **(d)** 300; **(e)** 250
19. [5.3A] **(a)** $\frac{30}{100} = \frac{\text{Part}}{40}$; 12; **(b)** $\frac{40}{100} = \frac{\text{Part}}{72}$; 28.8; **(c)** $\frac{50}{100} = \frac{\text{Part}}{94}$; 47; **(d)** $\frac{60}{100} = \frac{\text{Part}}{50}$; 30; **(e)** $\frac{70}{100} = \frac{\text{Part}}{70}$; 49 **20.** [5.3B] **(a)** $\frac{\text{Percent}}{100} = \frac{80}{800}$; 10%; **(b)** $\frac{\text{Percent}}{100} = \frac{22}{110}$; 20%; **(c)** $\frac{\text{Percent}}{100} = \frac{28}{70}$; 40%; **(d)** $\frac{\text{Percent}}{100} = \frac{90}{180}$; 50%; **(e)** $\frac{\text{Percent}}{100} = \frac{80}{40}$; 200% **21.** [5.3C] **(a)** $\frac{10}{100} = \frac{20}{\text{Whole}}$; 200; **(b)** $\frac{12}{100} = \frac{30}{\text{Whole}}$; 250; **(c)** $\frac{90}{100} = \frac{45}{\text{Whole}}$; 50; **(d)** $\frac{25}{100} = \frac{50}{\text{Whole}}$; 200; **(e)** $\frac{18}{100} = \frac{63}{\text{Whole}}$; 350
22. [5.4A] **(a)** \$21.20; **(b)** \$52; **(c)** \$18.90; **(d)** \$85.20; **(e)** \$313.50
23. [5.4B] **(a)** \$20; **(b)** \$90; **(c)** \$90; **(d)** \$108; **(e)** \$765
24. [5.4B] **(a)** \$10; **(b)** \$18; **(c)** \$10; **(d)** \$27; **(e)** \$32.50
25. [5.4C] **(a)** \$10,824.32; \$824.32; **(b)** \$11,255.09; \$1255.09; **(c)** \$11,698.59; \$1698.59; **(d)** \$12,155.06; \$2155.06; **(e)** \$12,624.77; \$2624.77 **26.** [5.4D] **(a)** \$8; **(b)** \$15; **(c)** \$1225; **(d)** \$19.50; **(e)** \$38.50 **27.** [5.4E] **(a)** \$21; **(b)** \$20; **(c)** \$40; **(d)** \$81; **(e)** \$459 **28.** [5.5A] **(a)** 50%; **(b)** 55%; **(c)** 60%; **(d)** 65%; **(e)** 70% **29.** [5.5A] **(a)** 1%; **(b)** 3%; **(c)** 4%; **(d)** 5%; **(e)** 6%
30. [5.5A] **(a)** \$56 million; **(b)** \$64 million; **(c)** \$66 million; **(d)** \$80 million; **(e)** \$88 million **31.** [5.6A] **(a)** \$80; \$1.20; \$131.20; **(b)** \$120; \$1.80; \$181.80; **(c)** \$160; \$2.40; \$232.40; **(d)** \$200; \$3; \$303; **(e)** \$300; \$4.50; \$604.50 **32.** [5.6B] **(a)** \$12.50; \$35.78; \$21.88; \$1566; **(b)** \$25; \$71.56; \$43.75; \$3130.80; **(c)** \$37.50; \$107.34; \$65.63; \$4696.80; **(d)** \$50; \$143.12; \$87.50; \$6262.80; **(e)** \$62.50; \$178.90; \$109.38; \$7827.60

Cumulative Review Chapters 1–5

1. 6510 **2.** 40 **3.** improper **4.** $7\frac{3}{4}$ **5.** $\frac{58}{9}$ **6.** $\frac{1}{36}$ **7.** $\frac{6}{35}$
8. $x - \frac{3}{4} = \frac{1}{3}; x = \frac{13}{12}$ **9.** $7\frac{37}{45}$ **10.** Three hundred forty-two and forty-one hundredths **11.** $30 + 4 + \frac{7}{10} + \frac{7}{100} + \frac{3}{1000}$ **12.** 626.92
13. 8.910 **14.** 700 **15.** 749.9 **16.** 76.92 **17.** $\frac{4}{33}$ **18.** 0.25
19. $4.2\overline{93} > 4.29\overline{3} > 4.293$ **20.** $<$ **21.** $x = 3.9$ **22.** $y = 7$
23. $z = 21$ **24.** $\frac{12}{25}$ **25.** $\frac{5}{9} = \frac{40}{x}$ **26.** No **27.** $f = \frac{1}{4}$ **28.** $f = 18$
29. 19 mpg **30.** 12¢/oz **31.** 4 lb **32.** 112 oz **33.** 0.12
34. 0.0725 **35.** 3% **36.** 8 **37.** 36 **38.** 50% **39.** 20
40. \$16.80 **41.** \$68

Chapter 6

Exercises 6.1

1. Wendy's **3.** 69% **5.** 22 **7.** 94 **9.** Most times **11.** Internet **13.** Products, Spiritual **15.** Health **17.** 50 **19.** About 22 **21.** Administrative **23.** Microsoft **25.** 70 million **27.** 35 million **29.** 5 million **31.** 59% **33.** Hearts **35.** 36% **37.** Wendy's **39.** Jack in the Box **43.** table **45.** White **47.** \$181.8 billion **49.** 7 **51.** 3 **53.** 14.5 **55.** 507.5 **57.** 25%

Exercises 6.2

1. **(a)** 10; **(b)** 16; **(c)** Medicine **3.** **(a)** About 4; **(b)** About 5; **(c)** Medicine **5.** College; \$413 **7.** \$294 **9.** **(a)** 59%; **(b)** 20%; **(c)** 145 **11.** **(a)** 20 to 29; **(b)** <1; answers may vary; **(c)** Less than 50 years old; **(d)** 90+; smaller population **13.** **(a)** 39; **(b)** 29; **(c)** 0.20–0.29; 14
15. **(a)**

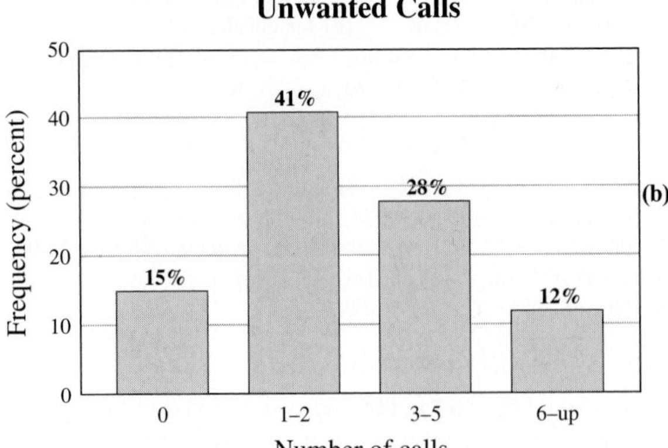

1–2; **(c)** 15
17. **(a)**

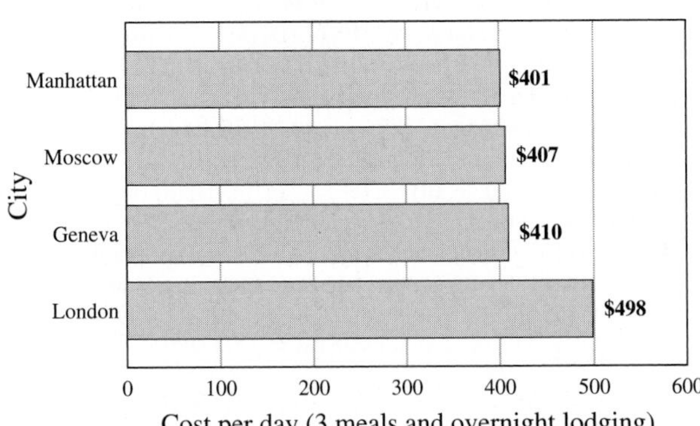

(b) London; **(c)** Manhattan; **(d)** \$97 **19.** **(a)** 3 and 4; **(b)** 5; **(c)** 5 **21.** **(a)** About 47%; **(b)** About 33%; **(c)** About 14%; **(d)** About 20%; **(e)** 1980–1982 **23.** **(a)** 1982; 2001; **(b)** 28%; **(c)** 2001

25.

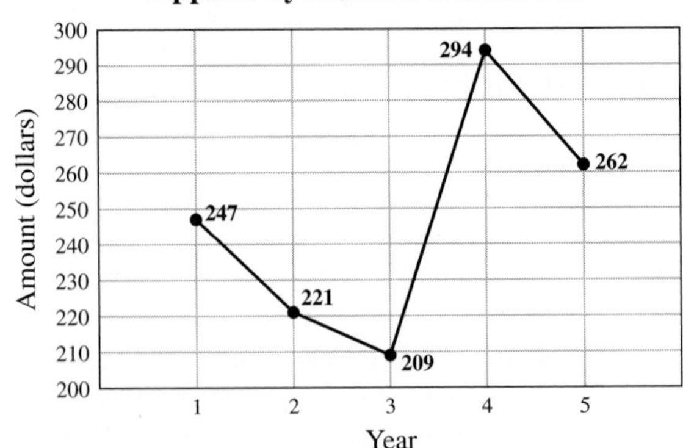

27.

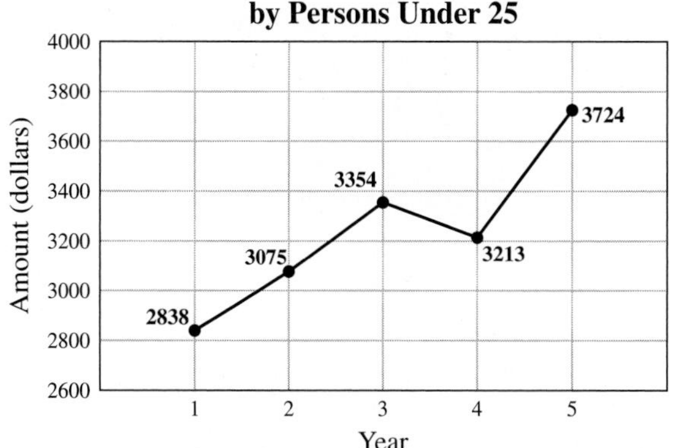

29.

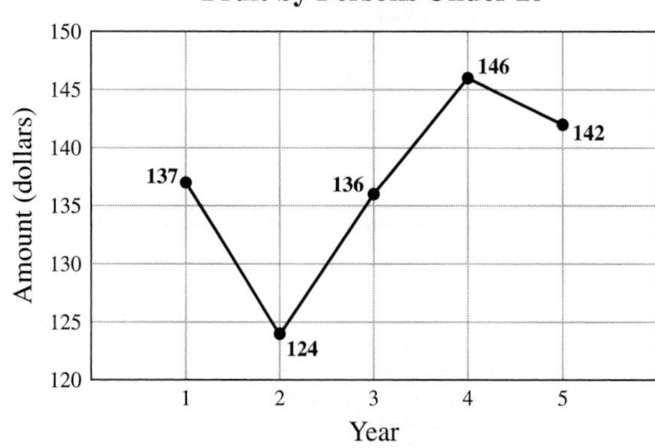

31.

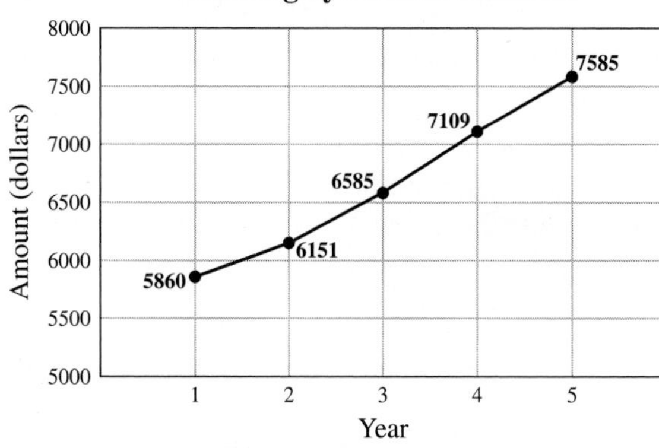

33.

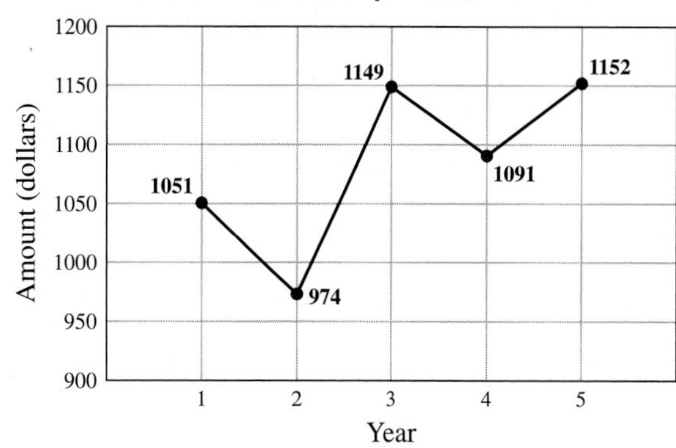

35.

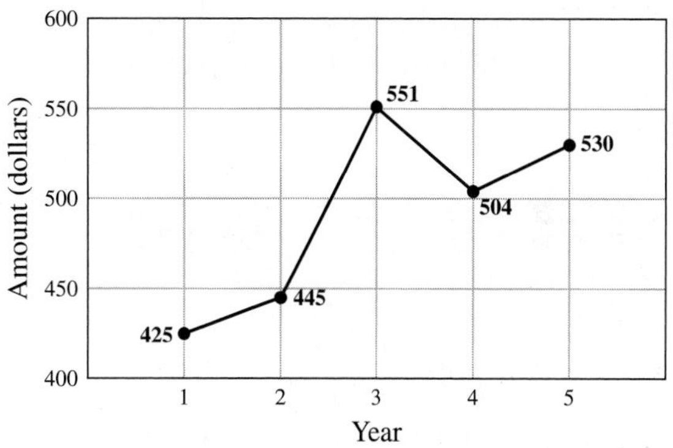

37.

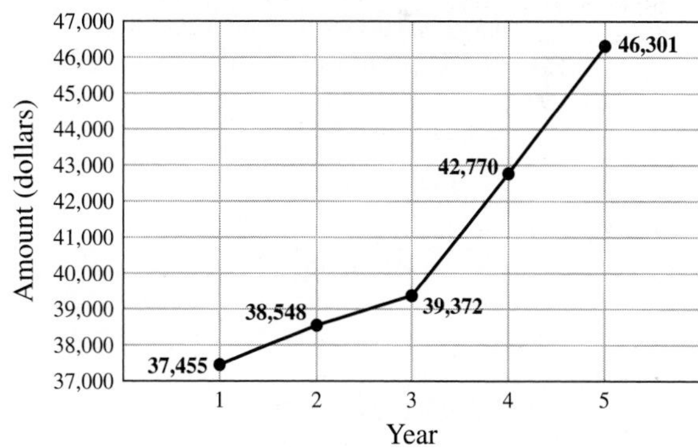

39.

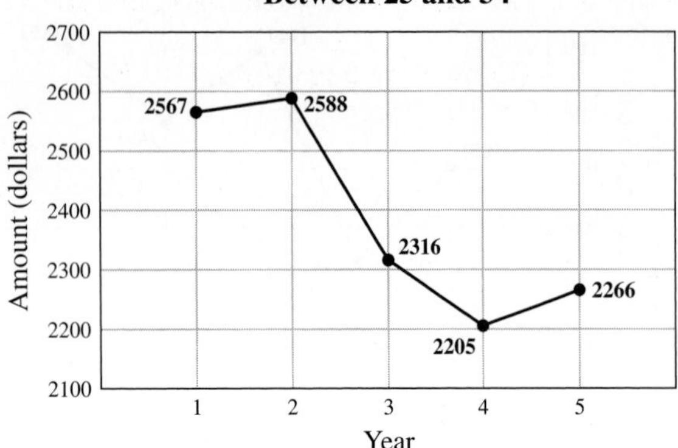

41. 3%; No **45.** proportional

47.

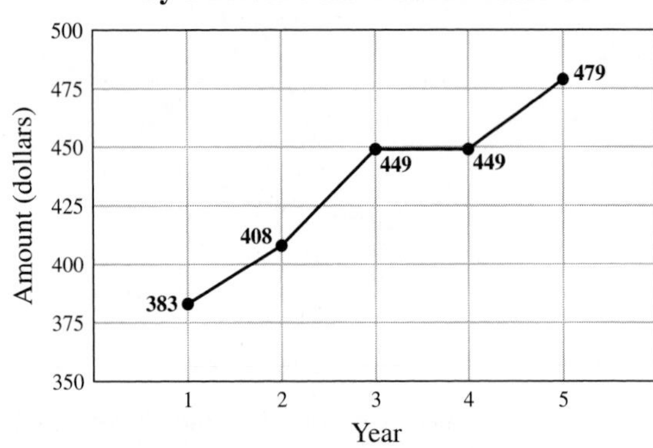

49.

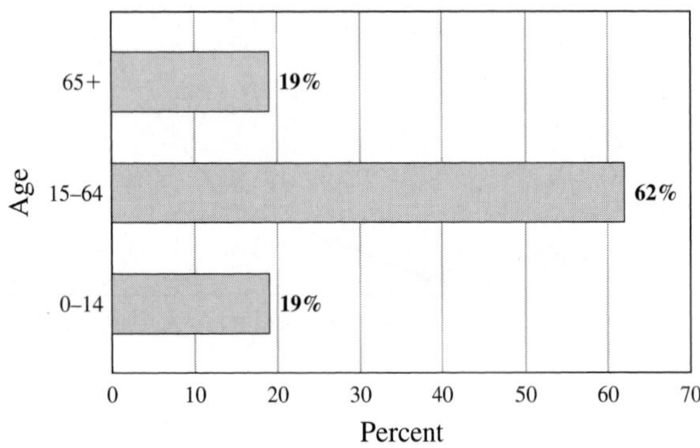

51. **(a)** 47%; **(b)** 57%; **(c)** September 7–10 **53.** 110 **55.** 25

Exercises 6.3

1. **(a)** bus; **(b)** bike; **(c)** 71% **3.** **(a)** Cheddar; **(b)** Swiss; **(c)** Mozzarella **5.** **(a)** Toilet; **(b)** 85 gal; **(c)** Faucet; **(d)** 30 gal **7.** **(a)** Paper; **(b)** Yard trimmings; **(c)** 20 lb; **(d)** 9 lb **9.** **(a)** Oil; **(b)** Nuclear; **(c)** Natural gas **11.** Food **13.** Books

15.

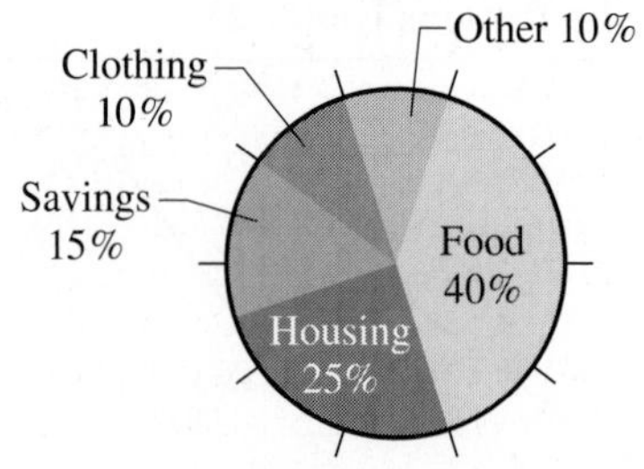

17.

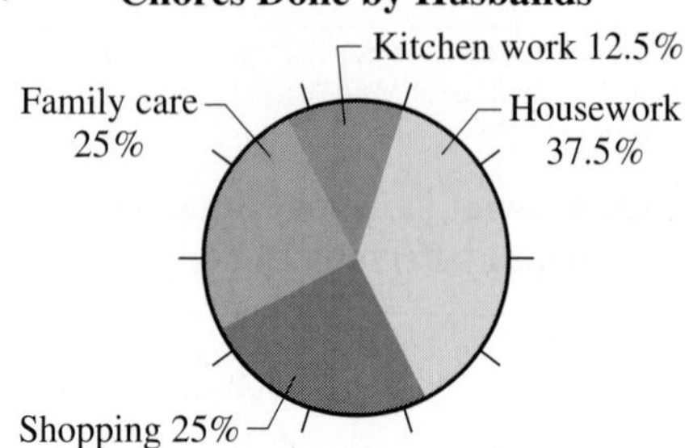

19.

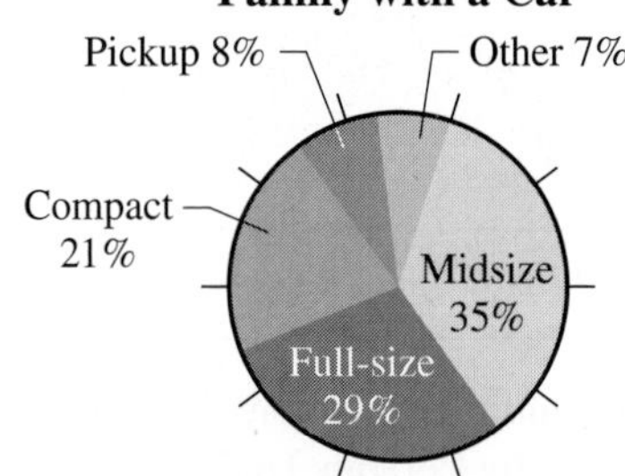

21.

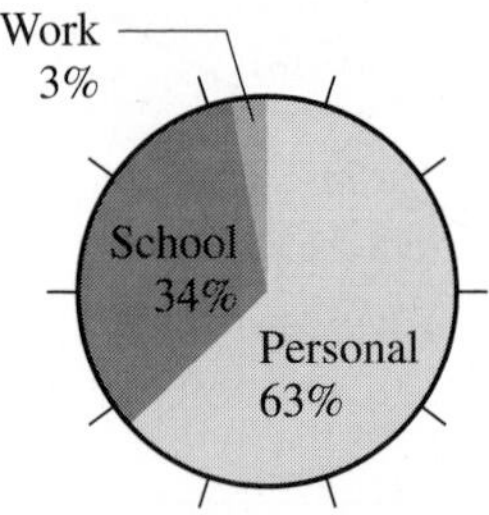

23.

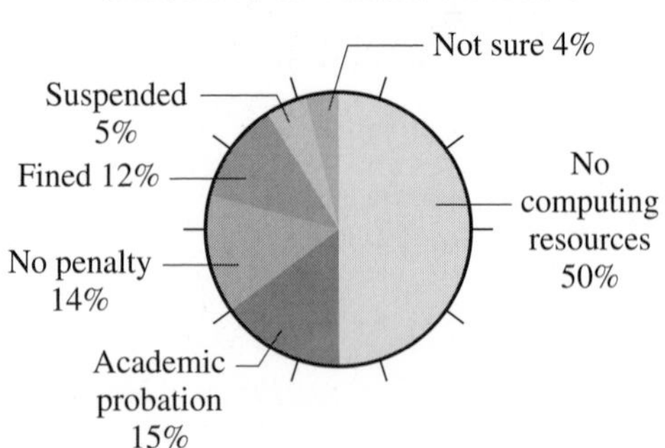

25. 144° **27.** 54° **31.** circle **33.** 34% **35.** 65 **37.** 2.51 **39.** 1.5

Exercises 6.4

1. 9 **3.** $15.1\overline{6}$ **5.** 9 **7.** 12.5 **9.** 5 and 52 **11.** 11 **13. (a)** Mean: \$4.61, median: \$4.41, mode: none; **(b)** no; **(c)** Answers may vary; **(d)** Mean and median **15.** 1.85 **17. (a)** Mean: 43,216.4, median: 40,827, mode: none; **(b)** Median Answers may vary. **19. (a)** Rodriguez \$26,000,000, Jeter \$21,000,000, Giambi \$20,000,000, Bonds \$20,000,000, Bagwell 19,000,000, Mussina \$19,000,000, Ramirez \$18,000,000, Helton \$17,000,000, Pettitte \$16,000,000, Ordonez \$16,000,000; **(b)** Mean: \$19,200,000, median: \$19,000,000, mode: \$20,000,000; \$19,000,000; \$16,000,000; **(c)** All three are good measures because they are close. (Answers will vary.) **21. (a)** Park \$15,000,000, Milton \$10,000,000, Kendall \$12,000,000, Matsui \$8,000,000, Casey \$9,000,000; **(b)** Mean: \$10,800,000, median: \$10,000,000, mode: none; **(c)** Answers may vary. **23.** Mean: \$51,117.50, Median: \$51,785.50 **25. (a)** Mean: 2440.2 (thousand), Median: 1175 (thousand); **(b)** Mean: 1523.6 (thousand), Median: 803 (thousand) **27.** Mode **31.** mean **33.** even, median, average **35.** mode **37.** $22.\overline{2}$ **39.** 28 **41.** Internet **43.** 5 **45.** 2 **47.** 105

Review Exercises

1. [6.1A] **(a)** 42%; **(b)** 39%; **(c)** \$1514.44; **(d)** \$1878.16; **(e)** \$0.01 **2.** [6.1B] **(a)** McDonald's; **(b)** Wendy's; **(c)** 200; **(d)** 550; **(e)** 100 **3.** [6.2A] **(a)** Ages 18–24; **(b)** Ages 25–34; **(c)** 44%; **(d)** 10%; **(e)** 37% **4.** [6.2B]

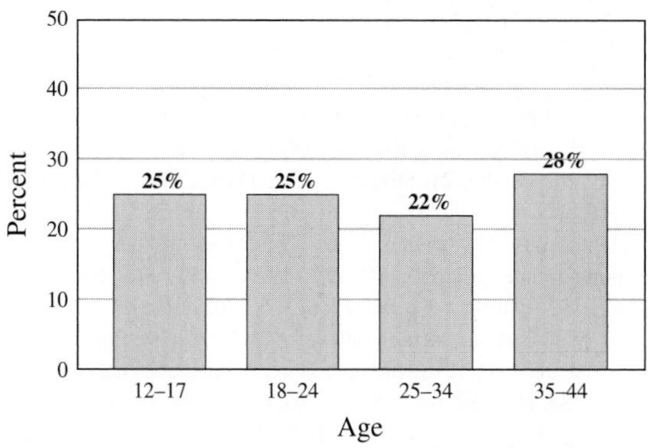

5. [6.2B]

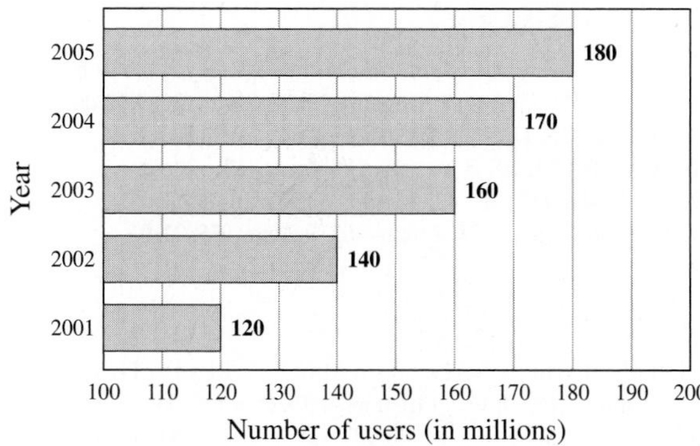

6. [6.2C] **(a)** About 175,000,000; **(b)** About 200,000,000; **(c)** About 250,000,000; **(d)** About 300,000,000; **(e)** About 325,000,000 **7.** [6.2C] **(a)** About 12,500,000; **(b)** About 25,000,000; **(c)** About 50,000,000; **(d)** About 75,000,000; **(e)** About 80,000,000 **8.** [6.2D]

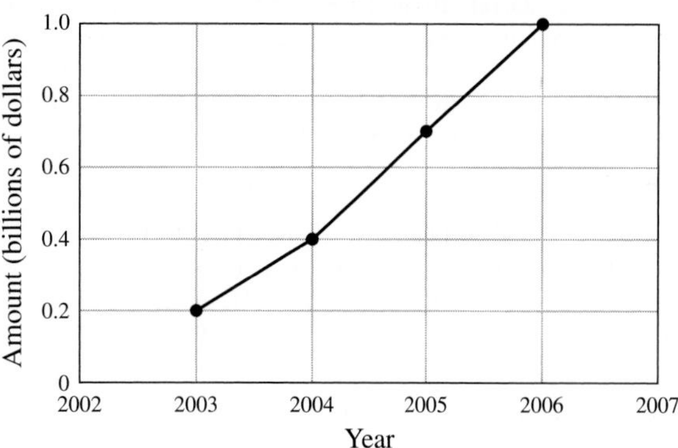

9. [6.3A] **(a)** 13%; **(b)** 42%; **(c)** 22%; **(d)** Home; **(e)** 20% **10.** [6.3A] **(a)** 100; **(b)** 130; **(c)** 130; **(d)** 220; **(e)** 420 **11.** [6.3B]

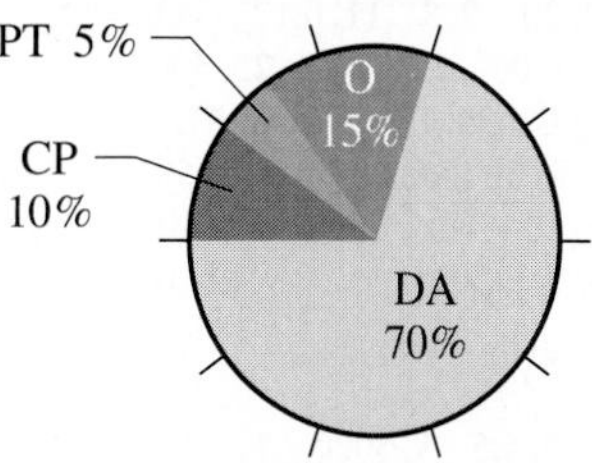

12. [6.4A, B, C] **(a)** 93 cents; **(b)** 99 cents; **(c)** 99 cents; **(d)** 94 cents; **(e)** 99 cents **13.** [6.4A, B, C] **(a)** 15 grams; **(b)** 14 grams; **(c)** None; **(d)** 16 grams; **(e)** 16 grams **14.** [6.4A] 2.8 **15.** [6.4A] 3.0

Cumulative Review Chapters 1–6

1. 6510 **2.** 7 **3.** Proper **4.** $3\frac{1}{8}$ **5.** $\frac{13}{6}$ **6.** $\frac{1}{25}$ **7.** $\frac{9}{5} = 1\frac{4}{5}$ **8.** $z - \frac{3}{4} = \frac{3}{8}, z = \frac{9}{8} = 1\frac{1}{8}$ **9.** $4\frac{25}{32}$ **10.** Three hundred fifty-two and fifty-one hundredths **11.** $60 + 4 + \frac{1}{10} + \frac{7}{100} + \frac{5}{1000}$ **12.** 528.92 **13.** 0.7787 **14.** 600 **15.** 749.9 **16.** 615.38 **17.** $\frac{26}{33}$ **18.** 0.3 **19.** $9.56\overline{8} > 9.5\overline{68} > 9.568$ **20.** $<$ **21.** 4.2 **22.** 9 **23.** 13.2 **24.** $\frac{7}{25}$ **25.** $\frac{4}{7} = \frac{28}{x}$ **26.** Yes **27.** $x = \frac{1}{2}$ **28.** $p = 15$ **29.** 26 mpg **30.** 12¢ **31.** 4 lb **32.** 39 oz **33.** 0.67 **34.** 0.0375 **35.** 9% **36.** 48 **37.** 36 **38.** 50% **39.** 30 **40.** \$18.72 **41.** \$67.50 **42.** 27% **43.** About 26 in. **44.** 7 degrees **45.** 50% **46.** 2 **47.** 9 **48.** 18

Chapter 7

Exercises 7.1

1. 144 **3.** 90 **5.** 84 **7.** $4\frac{1}{6}$ **9.** 7 **11.** $\frac{1}{4}$ **13.** $\frac{3}{4}$ **15.** 10 **17.** $12\frac{1}{3}$ **19.** 1 **21.** 48 **23.** 3 **25.** 5280 **27.** 1760 **29.** 1 **31. (a)** 36 in.; **(b)** yes **33. (a)** 30 in.; **(b)** yes **35.** 4029 yd **37.** 265 in. **39.** 1 mi **41.** 141 in. **43.** 64 in. **45.** 10,560 ft/hr **47. (a)** 2 yd/sec; **(b)** 120 yd **49.** 23,178; the 4.55-mi = 24,024-ft split **51.** 34 yd/sec **53.** $33\frac{1}{3}$ ft/sec **55. (a)** $833\frac{1}{3}$ yd; **(b)** To prove this, you have to know that 1 mile = 1760 yd (see Problem 27) $\frac{60 \text{ miles}}{\text{hour}} = \frac{60(1760 \text{ yd})}{3600 \text{ sec}} = \frac{(176 \text{ yd})}{6 \text{ sec}} \approx \frac{29 \text{ yd}}{\text{sec}}$ **(c)** $\frac{833\frac{1}{3} \text{ yd}}{29 \text{yd/sec}} \approx 29$ sec **57.** 194,040 ft

63. U.S., metric **65.** 78 **67.** 6 **69.** $8\frac{1}{3}$ **71.** $6\frac{1}{4}$ **73.** 4.232 **75.** 8350 **77.** 46.5

Exercises 7.2

1. 5000 **3.** 1.877 **5.** 0.4 **7.** 490 **9.** 1.82 **11.** 2200 **13.** 3000 **15.** 2.358 **17.** 300 **19.** 6.7 **21.** c **23.** a **25.** b **27.** 0.6 **29.** 160 **31.** 5000 steps **33.** **(a)** 110 cm, 320 cm, 150 cm; **(b)** African elephant: 320 cm, chimp: 110 cm, zebra: 150 cm **35.** **(a)** 2.40 m, 0.76 m, and 0.17 m; **(b)** copperhead: 0.76 m, salamander: 0.17 m, Madagascar boa: 2.40 m **37.** **(i)** d; **(ii)** e; **(iii)** b; **(iv)** c; **(v)** a **41.** substitute **43.** 9240 **45.** 0.3 **47.** 2.3 **49.** 182.88 **51.** 80 **53.** 4

Exercises 7.3

1. 121.92 **3.** 27.42 **5.** 144 **7.** 8 **9.** 660 **11.** 6.2 **13.** 89.6 **15.** 9700 **17.** 62 **19.** 36-24-36 **21.** 2.8 in. per century **23.** 2.54 cm **25.** 1.178 or about 1.2 **27.** 24 **29.** larger **31.** smaller **33.** 24.8 **35.** 96 **37.** 2.179 **39.** 731.2 **41.** 240 **43.** 12.4 **45.** 3 **47.** 9.88

Exercises 7.4

1. 3,000,000 **3.** 288 **5.** 3 **7.** 6 **9.** 9680 **11.** 7.41 **13.** 20,000 **15.** 17,139 **17.** 8,373,200 **19.** **(a)** 1500; **(b)** $166\frac{2}{3}$ **21.** 2.87 **23.** 2.96 **25.** 5214 **27.** $14\frac{2}{3}$ **29.** 9000 **31.** 3 **33.** One square meter. **35.** 6 **37.** 432 **39.** 9.88 **41.** 48,400 **43.** $2\frac{1}{12}$

Exercises 7.5

1. 3 **3.** 2 **5.** 2 **7.** 20 **9.** 8 **11.** 0.177 **13.** 3.847 **15.** 2.05 **17.** 5.5 **19.** 60 **21.** 700 **23.** 6000 **25.** 500 **27.** 4000 **29.** 900 **31.** **(a)** 0.2 L; **(b)** 1.8 L **33.** 1.5 **35.** 5 **37.** 10 **39.** 480 **41.** 450 **43.** 15 **45.** 8 **47.** **(a)** 50 fl oz per day; **(b)** $6\frac{1}{4}$ cups per day **49.** **(a)** 3120 mL; **(b)** 104 fl oz **51.** **(a)** 16; **(b)** 3.2; **(c)** 4.6875 **55.** cm **57.** **(a–b)** 750 **59.** 1.25; 240 **61.** 10,000 **63.** 5 **65.** 2.7

Exercises 7.6

1. 48 **3.** 72 **5.** 4 **7.** 4.5 **9.** 8000 **11.** 5000 **13.** 1.5 **15.** 2 **17.** 200,000 **19.** 200 **21.** 200 **23.** 50 **25.** 0.899 **27.** 0.030 **29.** 0.57 **31.** 8.8 **33.** 22 **35.** 0.9 **37.** 100 **39.** 0.45 **41.** 15 **43.** 25 **45.** 45 **47.** 50 **49.** 95 **51.** 1832 **53.** 40 **55.** 72 **57.** 158 **59.** $52 \times 2.2 \approx 114$ **61.** 100 to 200 **63.** 23,949 **65.** 59 **67.** 71 **69.** 83 **71.** 61 **73.** 70 **77.** gram **79.** 104° **81.** 17.6 **83.** 0.384 **85.** 6 **87.** 10 **89.** 130 **91.** 74

Review Exercises

1. [7.1A] **(a)** 72; **(b)** 108; **(c)** 144; **(d)** 216; **(e)** 252 **2.** [7.1A] **(a)** 1; **(b)** 2; **(c)** 3; **(d)** $3\frac{1}{3}$; **(e)** $5\frac{5}{12}$ **3.** [7.1A] **(a)** $\frac{1}{6}$; **(b)** $\frac{1}{4}$; **(c)** $\frac{5}{12}$; **(d)** $\frac{7}{12}$; **(e)** $1\frac{1}{6}$ **4.** [7.1A] **(a)** 2; **(b)** 4; **(c)** 6; **(d)** $6\frac{2}{3}$; **(e)** $9\frac{2}{3}$ **5.** [7.1A] **(a)** 1; **(b)** 2; **(c)** 5; **(d)** 4; **(e)** 3 **6.** [7.2A] **(a)** 2000; **(b)** 7000; **(c)** 4600; **(d)** 450; **(e)** 45,000 **7.** [7.2A] **(a)** 20; **(b)** 30; **(c)** 70; **(d)** 90; **(e)** 100 **8.** [7.2A] **(a)** 1000; **(b)** 3000; **(c)** 3500; **(d)** 4500; **(e)** 6000 **9.** [7.2A] **(a)** 2; **(b)** 3.95; **(c)** 4.05; **(d)** 2.34; **(e)** 4.99 **10.** [7.3A] **(a)** 76.2; **(b)** 101.6; **(c)** 1524; **(d)** 1828.8; **(e)** 2133.6 **11.** [7.3A] **(a)** 91.4; **(b)** 182.8; **(c)** 319.9; **(d)** 411.3; **(e)** 457 **12.** [7.3A] **(a)** 48; **(b)** 80; **(c)** 144; **(d)** 160; **(e)** 400 **13.** [7.3A] **(a)** 24.8; **(b)** 31; **(c)** 37.2; **(d)** 43.4; **(e)** 49.6 **14.** [7.4A] **(a)** 20,000; **(b)** 30,000; **(c)** 40,000; **(d)** 50,000; **(e)** 60,000 **15.** [7.4A] **(a)** 2,000,000; **(b)** 3,000,000; **(c)** 4,000,000; **(d)** 5,000,000; **(e)** 6,000,000 **16.** [7.4B] **(a)** 288; **(b)** 432; **(c)** 576; **(d)** 720; **(e)** 864 **17.** [7.4B] **(a)** 1; **(b)** 2; **(c)** 2.5; **(d)** 3; **(e)** 3.5 **18.** [7.4B] **(a)** 4840 yd²; **(b)** 14,520 yd²; **(c)** 9680 yd²; **(d)** 7260 yd²; **(e)** 19,360 yd² **19.** [7.4C] **(a)** 12.35; **(b)** 9.88; **(c)** 2.47; **(d)** 7.41; **(e)** 4.94 **20.** [7.5A] **(a)** 7; **(b)** 9; **(c)** 10; **(d)** 13; **(e)** 2.2 **21.** [7.5B] **(a)** 4000; **(b)** 7000; **(c)** 9000; **(d)** 2300; **(e)** 5970 **22.** [7.5B] **(a)** 0.452; **(b)** 0.048; **(c)** 0.003; **(d)** 1.657; **(e)** 0.456 **23.** [7.5C] **(a)** 960; **(b)** 1200; **(c)** 1440; **(d)** 1680; **(e)** 2160 **24.** [7.5C] **(a)** 0.5; **(b)** 1.5; **(c)** 2; **(d)** 2.5; **(e)** 3 **25.** [7.5C] **(a)** 3600; **(b)** 5400; **(c)** 7200; **(d)** 9000; **(e)** 10,800 **26.** [7.5C] **(a)** 105; **(b)** 120; **(c)** 50; **(d)** 55; **(e)** 65 **27.** [7.6A] **(a)** 48; **(b)** 64; **(c)** 80; **(d)** 96; **(e)** 112 **28.** [7.6A] **(a)** 1; **(b)** 1.5; **(c)** 2; **(d)** 2.5; **(e)** 3 **29.** [7.6A] **(a)** 4000; **(b)** 6000; **(c)** 8000; **(d)** 10,000; **(e)** 12,000 **30.** [7.6A] **(a)** 1.5; **(b)** 2.5; **(c)** 3.5; **(d)** 4.5; **(e)** 9 **31.** [7.6B] **(a)** 100; **(b)** 300; **(c)** 500; **(d)** 400; **(e)** 200 **32.** [7.6B] **(a)** 0.307; **(b)** 0.040; **(c)** 3.245; **(d)** 0.002; **(e)** 10.342 **33.** [7.6C] **(a)** 2.2; **(b)** 15.4; **(c)** 13.2; **(d)** 8.8; **(e)** 17.6 **34.** [7.6C] **(a)** 0.45; **(b)** 1.35; **(c)** 2.7; **(d)** 1.8; **(e)** 4.5 **35.** [7.6D] **(a)** 0; **(b)** 5; **(c)** 10; **(d)** 15; **(e)** 100 **36.** [7.6D] **(a)** 50; **(b)** 59; **(c)** 68; **(d)** 77; **(e)** 86

Cumulative Review Chapters 1–7

1. 2910 **2.** 6 **3.** $5\frac{1}{6}$ **4.** $\frac{13}{3}$ **5.** 727.92 **6.** 0.08048 **7.** 549.9 **8.** 384.62 **9.** 0.25 **10.** $y = 4$ **11.** $z = 35.1$ **12.** No **13.** $\frac{1}{3}$ **14.** 45 oz **15.** 0.12 **16.** 0.0725 **17.** 20 **18.** 2 **19.** 50% **20.** 20 **21.** $212.50 **22.** Buses and trucks
23.

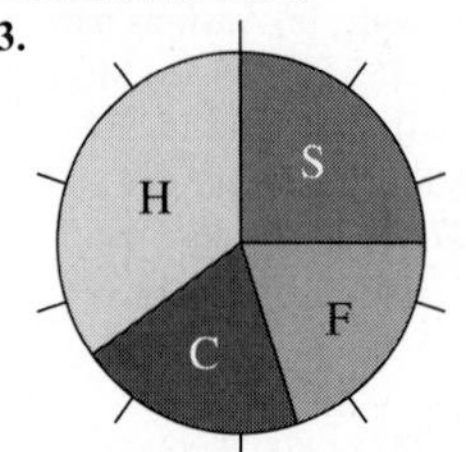

24. 43% **25.** About 34 in. **26.** 5 degrees **27.** 30% **28.** 6 **29.** 8 **30.** 25 **31.** 432 in. **32.** $1\frac{5}{12}$ ft **33.** $15\frac{2}{3}$ yd **34.** 5 mi **35.** 96 in. **36.** 2000 m **37.** 50 m **38.** 1500 dm **39.** 45.7 m **40.** 105.6 km **41.** 62 mi **42.** 19,360 yd² **43.** 14.82 acres

Chapter 8

Exercises 8.1

1. Ray $\overrightarrow{RS}$ **3.** Line segment $\overline{TU}$ **5.** Line $\overleftrightarrow{PQ}$ or $\overleftrightarrow{QP}$ **7.** Line $\overleftrightarrow{CD} \parallel \overleftrightarrow{EF}$ **9.** Line $\overleftrightarrow{EF}$ intersects line $\overleftrightarrow{CD}$ at point G. **11.** $\angle\delta$, $\angle P$, $\angle QPR$ **13.** $\angle BAC$ or $\angle CAB$ **15.** $\angle EAF$ or $\angle FAE$ **17.** Acute **19.** Right **21.** Straight **23.** $\angle DAB$, $\angle DAF$ **25.** $\angle FAB$ **27.** $\angle DAE$ **29.** $\angle CAD$ **31.** $\angle EAB$ **33.** 75° **35.** 55° **37.** Scalene right **39.** Scalene acute **41.** Isosceles acute **43.** Scalene obtuse **45.** 55° **47.** 15° **49.** 25° **51.** 90° **53.** Acute **55.** Acute **57.** Obtuse **59.** Acute **61.** 45° **63.** Obtuse **65.** Right **67.** Isosceles acute **69.** 45° **71.** 138° **73.** Rap **75.** 3 **77.** 90° **79.** 36° **81.** A ray has an endpoint, a line does not have an endpoint. **83.** No **89.** line **91.** $\parallel$ **93.** angle **95.** straight **97.** obtuse **99.** Supplementary **101.** P, Q, M **103.** $\overline{PQ}$, $\overline{ON}$, $\overline{NM}$ **105.** $\overrightarrow{PN}$ **107.** **(a)** Obtuse scalene; **(b)** Right isosceles; **(c)** Acute equilateral; **(d)** Acute isosceles **109.** 145° **111.** $\angle\theta$, $\angle C$, $\angle DCE$ or $\angle ECD$ **113.** 16.6 **115.** 13

Exercises 8.2

1. 20 ft **3.** 8.6 cm **5.** $16\frac{3}{4}$ in. **7.** 21 cm **9.** $17\frac{1}{3}$ yd **11.** 9 ft **13.** $11\frac{11}{12}$ yd **15.** $7\frac{1}{4}$ ft **17.** 78.6 m **19.** 98.6 km **21.** 9.42 cm **23.** 10.99 ft **25.** 13.188 m **27.** 7.536 mi **29.** 1110 m **31.** 4404.8 ft **33.** 102.8 in. **35.** 188.4 cm **37.** 320 yd **39.** 52 yd **41.** 489 mi **43.** 564 mi **47.** circumference **49.** 2512 million km **51.** 21 cm **53.** 9.2 cm **55.** 21 ft **57.** 150 **59.** $\frac{15}{2}$

Exercises 8.3

1. 150 ft² **3.** 6 in.² **5.** 72 cm² **7.** 81 in.² **9.** 81 yd² **11.** 28 cm² **13.** 175 mm² **15.** 10 ft² **17.** 30 in.² **19.** 20 km² **21.** 15 in.² **23.** $13\frac{3}{4}$ cm² **25.** 60 m² **27.** 45 yd² **29.** $11\frac{1}{2}$ in.² **31.** 5625 ft² **33.** 195 ft² **35.** 50.24 in.² **37.** 153.86 cm² **39.** 3.14 m² **41.** 24 ft² **43.** 28 ft² **45.** 30 ft² **47.** 957 cm² **49.** 3.87 cm² **51.** 8100 ft² **53.** 5024 ft² **55.** 21,226,400 ft² **57.** 113.04 m² **59.** 314.00 in.² **61.** 1256.00 ft² **63.** 14.13 ft² **65.** 2289.06 ft²

67. 792 yd² **69.** 6400 m² **71.** 15 in.² **73.** 27.56 in.² **75.** 3 in.² **77.** Two small pizzas **79.** LW **81.** bh **83.** 11,304 ft² **85.** 18 cm² **87.** 100 cm² **89.** 24 m² **91.** $\frac{10}{3}$ **93.** 220

Exercises 8.4

1. 240 cm³ **3.** $128\frac{1}{4}$ in.³ **5.** 7904 in.³ **7.** 440 in.³ **9.** 120 yd³ **11.** 1350 gal **13.** 8 gal **15.** 2512.0 in.³ **17.** 6280.0 cm³ **19.** 31.8 m³ **21.** 3.14 ft³ **23.** 35.33 in.³ **25.** 12,560.0 ft³ **27.** 28.3 in.³; 17 fl oz **29.** 434,073.6 gal **31.** 6,782,400.0 gal **33.** 628.0 in.³ **35.** 52,333.3 ft³ **37.** 0.5 m³ **39.** 76.9 cm³ **41.** 11,134,944 ft³ **43. (a)** 1232 ft³; **(b)** 63 ft³; **(c)** No; 1300 > 1295 **45. (a)** 20,569.1 in.³ ≈ 11.9 ft³; **(b)** 10,284.6 in.³ ≈ 6 ft³ **47. (a)** 17,280 in.³ = 10 ft³; **(b)** 11,520 in.³ ≈ 6.7 ft³ **49.** 14,130,000 ft³ **51.** 0.9 ft³ **53.** 63.6 ft³ **55.** 7.9 buildings **57.** 100.5 in.³ **59.** 25.1 in. **63.** LWH **65.** $\frac{4}{3}\pi r^3$ **67.** 376.8 cm³ **69.** 38.8 in.³ **71.** 25.1 in.³ **73.** 185

Exercises 8.5

1. 10 **3.** 14 **5.** 19 **7.** 20 **9.** 13 **11.** 2.828 **13.** 3.317 **15.** 4.796 **17.** 5.385 **19.** 10.392 **21.** 11.662 **23.** 15 **25.** 20 **27.** 47.170 **29.** 45.122 **31.** 12 in. **33.** 12 in. **35.** 2.236 ft **37.** 4.472 m **39.** 0.533 m **41.** 247.024 ft **43.** 118.811 ft **45.** 17.804 cm **47.** 126 ft **49.** 310 ft **51. (a)** $5\frac{1}{11}$; **(b)** $5\frac{3}{11}$; **(c)** $5\frac{5}{11}$ **55.** $a^2 + b^2 = c^2$ **57.** 8.660 **59.** 7.810 **61.** 4.583 **63.** 13 **65.** 4 **67.** 14

Review Exercises

1. [8.1A] **(a)** A; **(b)** B; **(c)** C; **(d)** D; **(e)** E **2.** [8.1A] **(a)** $\overleftrightarrow{AB}$; **(b)** $\overleftrightarrow{EC}$ **3.** [8.1A] **(a)** $\overline{AB}$; **(b)** $\overline{AD}$; **(c)** $\overline{DB}$; **(d)** $\overline{ED}$; **(e)** $\overline{DC}$ **4.** [8.1A] $\overleftrightarrow{AB}$ and $\overleftrightarrow{EC}$ **5.** [8.1A] **(a)** $\overrightarrow{AD}$; **(b)** $\overrightarrow{BD}$ (Answers may vary.) **6.** [8.1B] **(a)** $\angle\alpha$, $\angle A$, $\angle CAB$ or $\angle BAC$; **(b)** $\angle\delta$, $\angle D$, $\angle EDF$ or $\angle FDE$; **(c)** $\angle\gamma$, $\angle G$, $\angle HGI$ or $\angle IGH$; **(d)** $\angle\xi$, $\angle J$, $\angle KJL$ or $\angle LJK$; **(e)** $\angle\mu$, $\angle M$, $\angle NMO$ or $\angle OMN$ **7.** [8.1C] **(a)** Acute; **(b)** Acute; **(c)** Right; **(d)** Straight; **(e)** Obtuse **8.** [8.1D] **(a)** 80°; **(b)** 75°; **(c)** 70°; **(d)** 60°; **(e)** 10° **9.** [8.1D] **(a)** 170°; **(b)** 165°; **(c)** 160°; **(d)** 150°; **(e)** 100° **10.** [8.1E] **(a)** Acute equilateral; **(b)** Acute isosceles; **(c)** Right isosceles; **(d)** Obtuse scalene; **(e)** Right scalene **11.** [8.1F] **(a)** 110°; **(b)** 90°; **(c)** 70°; **(d)** 50°; **(e)** 30° **12.** [8.2A] **(a)** 16.6 m; **(b)** 14.6 cm; **(c)** 15.2 in.; **(d)** 18.6 yd; **(e)** 20.6 ft **13.** [8.2A] **(a)** 15 m; **(b)** 19 yd; **(c)** 15 ft; **(d)** 17 cm; **(e)** 12 in. **14.** [8.2B] **(a)** 37.68 cm; **(b)** 50.24 cm; **(c)** 62.8 in.; **(d)** 75.36 in.; **(e)** 87.92 ft **15.** [8.3A] **(a)** 30 m²; **(b)** 56 m²; **(c)** 24 in.²; **(d)** 21 in.²; **(e)** 108 in.² **16.** [8.3B] **(a)** 60 in.²; **(b)** 40 in.²; **(c)** 24 cm²; **(d)** 12 m²; **(e)** 4 m² **17.** [8.3C] **(a)** 6 in.²; **(b)** 7.5 cm²; **(c)** 8 m²; **(d)** 35 ft²; **(e)** 64 m² **18.** [8.3D] **(a)** 8 in.²; **(b)** 60 ft²; **(c)** 12 m²; **(d)** 15 cm²; **(e)** 28 m² **19.** [8.3E] **(a)** 153.86 in.²; **(b)** 3.14 cm²; **(c)** 28.26 in.²; **(d)** 12.56 ft²; **(e)** 78.5 yd² **20.** [8.4A] **(a)** 84 cm³; **(b)** 60 cm³; **(c)** 40 cm³; **(d)** 90 cm³; **(e)** 70 cm³ **21.** [8.4B] **(a)** 18.84 in.³; **(b)** 21.98 in.³; **(c)** 25.12 in.³; **(d)** 113.04 in.³; **(e)** 125.60 in.³ **22.** [8.4C] **(a)** 904.32 in.³; **(b)** 1436.03 in.³; **(c)** 2143.57 in.³; **(d)** 3052.08 in.³; **(e)** 4186.67 in.³ **23.** [8.4D] **(a)** 1570 in.³; **(b)** 803.84 in.³; **(c)** 339.12 in.³; **(d)** 100.48 in.³; **(e)** 12.56 in.³ **24.** [8.4E] **(a)** 1,764,000 m³; **(b)** 1,866,240 m³; **(c)** 1,936,000 m³; **(d)** 1,953,640 m³; **(e)** 2,025,000 m³ **25.** [8.5A] **(a)** 5; **(b)** 3; **(c)** 11; **(d)** 15; **(e)** 13 **26.** [8.5A] **(a)** 1.414; **(b)** 2.828; **(c)** 3.464; **(d)** 4.690; **(e)** 4.123 **27.** [8.5B] **(a)** 13 cm; **(b)** 5 cm; **(c)** 15 cm; **(d)** 20 cm; **(e)** 25 cm **28.** [8.5B] **(a)** 3.606 cm; **(b)** 4.472 cm; **(c)** 6.403 cm; **(d)** 6.708 cm; **(e)** 7.071 cm **29.** [8.5B] **(a)** 4; **(b)** 6.928; **(c)** 7.483; **(d)** 8; **(e)** 8.944

Cumulative Review Chapters 1–8

1. 10 **2.** 727.92 **3.** 750 **4.** 307.69 **5.** $y = 6$ **6.** $z = 17.6$ **7.** $j = \frac{1}{5}$ **8.** 0.89 **9.** 45 **10.** 25% **11.** 60 **12.** $78 **13.** Automobiles

14.

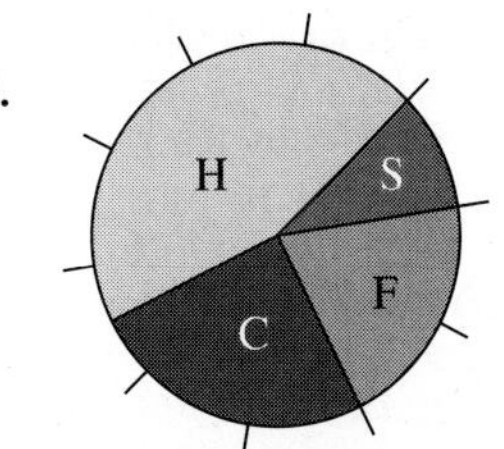

15. 10% **16.** 5 degrees **17.** 40% **18.** 18 **19.** 20 **20.** 21 **21.** 396 in. **22.** $\frac{3}{4}$ ft **23.** $2\frac{1}{3}$ yd **24.** 60 in. **25.** 4000 m **26.** 20 m **27.** 1400 dm **28.** 48,400 yd² **29.** 15.8 in. **30.** 37.68 cm **31.** 120 cm² **32.** 200.96 cm² **33.** 168 in.² **34.** 72 cm³ **35.** 904.32 in.³ **36.** Acute **37.** 125° **38.** 60 **39.** $c = 5$

Chapter 9

Exercises 9.1

1. (a) 0, 1, 3; **(b)** 1, 3 **3.** −5, −2, 0, 1, 3 **5.** Indicates change in population since the year 2000 **7. (a)** +$50 million (or $50 million); **(b)** −$30 million

9. (a) −5 −4 −3 −2 −1 0 1 2 3 4 5

(b) >, > **11.** 0 **13.** 17 **15.** 2 **17.** 11 **19.** 30 **21.** 3 **23.** −1 **25.** 4 **27.** −6 **29.** 0 **31.** 2 **33.** 7 **35.** 5 **37.** −13 **39.** 8 **41.** −11 **43.** −20 **45.** −4 **47.** 12 **49.** 4 **51.** 16,000 ft **53.** 73°C **55.** 3500°C **57.** 145°F **59.** ninth floor **61.** +19; 525 **63.** +53; 559 **65.** 4 **67.** 0 **69.** About $27,000; 36–40 **71.** 46–50 **73.** About −$12,000 **75. (a)** $231 billion; **(b)** −$158 billion; **(c)** $389 billion **77. (a)** −$307 billion; **(b)** −$412 billion; **(c)** $105 billion **79.** 799 **81.** 284 **83.** 2262 **89.** −1, −2, −3, . . . **91.** $b > a$ **93.** distance **95.** add **97. (a)** 8; **(b)** 4; **(c)** 0 **99.** −2 **101. (a)** −12; **(b)** −4; **(c)** 4 **103.** Q1 2003, −2% **105. (a)** +80 points or 80 points; **(b)** −$100,000

107. −5 −4 −3 −2 −1 0 1 2 3 4 5

109. > **111.** 25 **113.** 72

Exercises 9.2

1. 32 **3.** −56 **5.** −10 **7.** 20 **9.** 70 **11.** 5 **13.** −4 **15.** −5 **17.** −10 **19.** −20 **21.** 7 **23.** 14 **25.** 0 **27.** 0 **29.** Not defined **31.** 16 **33.** −25 **35.** −216 **37.** 81 **39.** −125 **41.** 60 **43.** 40 **45.** −30 **47.** 120 **49.** −48 **51.** $2 **53.** 15 gain **55.** 100 gain **57.** $30 **59.** $250 **61.** 5(−4); −20°F **63.** 15(−4); −60°F **65.** 5(−7); −35°C **67.** 70°F −14(4) or 70°F + 14(−4); 14°F **69.** −22°C **71.** −3 **73.** 0 **75.** 0 **79.** positive **81.** 0 **83. (a)** 100; **(b)** −100 **85. (a)** 6; **(b)** −4; **(c)** −3; **(d)** 4 **87. (a)** 60; **(b)** −30 **89.** $\frac{9}{5}$ **91.** $\frac{10}{7}$ **93.** 2.56 **95.** 0.07 or $\frac{7}{100}$ **97.** 0.8 or $\frac{4}{5}$

Exercises 9.3

1. $-\frac{7}{3}$ **3.** 6.4 **5.** $-3\frac{1}{7}$ **7.** $\frac{4}{5}$ **9.** 3.4 **11.** $1\frac{1}{2}$ **13.** −4.7 **15.** −5.4 **17.** −8.6 **19.** $\frac{3}{7}$ **21.** $-\frac{1}{2}$ **23.** $-\frac{1}{12}$ **25.** $\frac{7}{12}$ **27.** $-\frac{13}{21}$ **29.** $-\frac{31}{18}$ **31.** −2.6 **33.** 5.2 **35.** −3.7 **37.** $\frac{4}{7}$ **39.** $-\frac{29}{12}$ **41.** −7.26 **43.** 2.86 **45.** $-\frac{25}{42}$ **47.** $\frac{1}{4}$ **49.** $-\frac{15}{4}$ **51.** $-\frac{21}{20}$ **53.** $\frac{4}{7}$ **55.** $-\frac{5}{7}$ **57.** $-\frac{1}{2}$ or −0.5 **59.** $\frac{1}{6}$

	61. $\sqrt{16}$	**63.** 0	**65.** 3	**67.** $0.\overline{68}$
Natural number	✓		✓	
Whole number	✓	✓	✓	
Integer	✓	✓	✓	
Rational number	✓	✓	✓	✓
Irrational number				
Real number	✓	✓	✓	✓

69. $\subset$, natural number, whole number **71.** $\subset$, integer, rational number **73.** -1.35 **75.** -3.045 **77.** 5.425 **81.** $-a$ **83.** positive **85.** $\frac{b}{a}, a \neq 0$ **87.** $-\frac{13}{10}$ **89.** $1\frac{1}{5}$ **91.** $\frac{6}{17}$ **93.** $7\frac{9}{11}$ **95.** -0.3 **97.** -3.5 **99.** $\frac{1}{30}$ **101.** -9.5 **103.** $-\frac{7}{24}$ **105.** 6.24 **107.** $-\frac{1}{2}$ **109.** $\frac{9}{7}$ **111.** -2 **113.** 0.25 or $\frac{1}{4}$

115.

	(a) 3.2	(b) $\frac{-8}{9}$	(c) 9	(d) $\sqrt{21}$
Natural number			✓	
Whole number			✓	
Integer			✓	
Rational number	✓	✓	✓	
Irrational number				✓
Real number	✓	✓	✓	✓

117. 58 **119.** 4

Exercises 9.4

1. 2 **3.** -4 **5.** -17 **7.** 17 **9.** 59 **11.** 22 **13.** -1 **15.** 1 **17.** 0 **19.** -22 **21.** 1 **23.** -24 **25.** -33 **27.** 11,800 **29.** -1 **31.** -31 **33.** 36 **35.** 87 **37.** **(a)** 144; **(b)** 126 **39.** **(a)** 78.5 in.^{2}; **(b)** 10.2 cents **41.** **(a)** 153.86 in.^{2}; **(b)** 8.1 cents **43.** One possible answer: $3 + 2 \cdot 5 + 6$
45. One possible answer: $1 + (2 - 3 - 4 - 5 + 6 + 7 + 8) \cdot 9$
47. No **51.** exponent **53.** division **55.** subtraction **57.** -12 **59.** -21 **61.** -5 **63.** 119 **65.** 45 **67.** 30

Review Exercises

1. [9.1A] **(a)** $-6, -4, -1$; **(b)** 2, 5; **(c)** 0, 2, 5; **(d)** 2, 5; **(e)** $-6, -4, -1, 0, 2, 5$ **2.** [9.1A] **(a)** Price changes in financial markets; **(b)** Outdoor temperature; **(c)** Highest, lowest points in North America; **(d)** Elevations in New Orleans; **(e)** Estimated population changes from 2000 to 2050 (in millions) **3.** [9.1A] **(a)** 129°F; **(b)** -63°C; **(c)** $-\$53$; **(d)** $-\$8$ trillion; **(e)** $-10{,}897$ m
4. [9.1A]

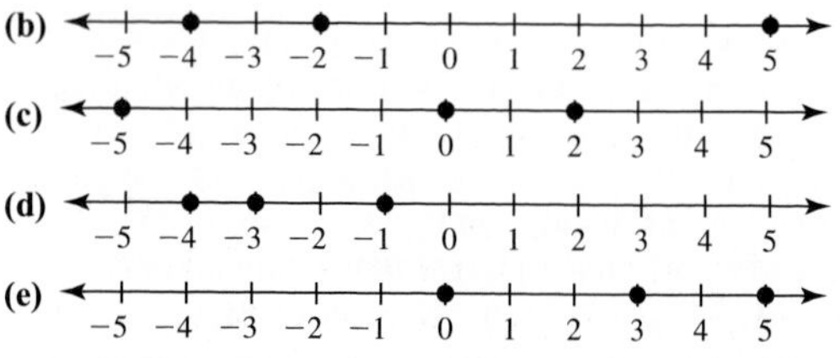

5. [9.1A] **(a)** $<$; **(b)** $>$; **(c)** $<$; **(d)** $<$; **(e)** $>$ **6.** [9.1B] **(a)** -10; **(b)** -11; **(c)** -12; **(d)** -13; **(e)** -14 **7.** [9.1B] **(a)** 9; **(b)** 8; **(c)** 7; **(d)** 6; **(e)** 5 **8.** [9.1C] **(a)** 7; **(b)** 8; **(c)** 9; **(d)** 10; **(e)** 11
9. [9.1E] **(a)** -13; **(b)** -12; **(c)** -11; **(d)** -10; **(e)** -9
10. [9.1E] **(a)** -7; **(b)** -6; **(c)** -5; **(d)** -4; **(e)** -3
11. [9.1E] **(a)** 10; **(b)** 9; **(c)** 8; **(d)** 7; **(e)** 6 **12.** [9.1F] **(a)** -18; **(b)** -19; **(c)** -20; **(d)** -21; **(e)** -22 **13.** [9.1F] **(a)** -8; **(b)** -7; **(c)** -9; **(d)** -6; **(e)** -5 **14.** [9.1F] **(a)** -5; **(b)** -6; **(c)** -7; **(d)** -8; **(e)** -9 **15.** [9.2A] **(a)** -24; **(b)** -30; **(c)** -36; **(d)** -42; **(e)** -48 **16.** [9.2A] **(a)** 40; **(b)** 48; **(c)** 56; **(d)** 64; **(e)** 72
17. [9.2A] **(a)** -25; **(b)** -30; **(c)** -35; **(d)** -40; **(e)** -45
18. [9.2A] **(a)** -29; **(b)** -24; **(c)** -19; **(d)** -14; **(e)** -9
19. [9.2A] **(a)** -6; **(b)** -4; **(c)** -3; **(d)** -2; **(e)** -1
20. [9.2A] **(a)** 3; **(b)** 5; **(c)** 7; **(d)** 9; **(e)** 11 **21.** [9.2B] **(a)** 16; **(b)** 25; **(c)** 36; **(d)** 49; **(e)** 64 **22.** [9.2B] **(a)** -16; **(b)** -25; **(c)** -36; **(d)** -49; **(e)** -64 **23.** [9.3A] **(a)** $-\frac{3}{11}$; **(b)** $-\frac{4}{11}$; **(c)** $-\frac{5}{11}$; **(d)** $-\frac{6}{11}$; **(e)** $-\frac{7}{11}$ **24.** [9.3A] **(a)** 3.4; **(b)** 4.5; **(c)** 5.6; **(d)** 6.7; **(e)** 7.8
25. [9.3A] **(a)** $3\frac{1}{2}$; **(b)** $4\frac{1}{2}$; **(c)** $5\frac{1}{2}$; **(d)** $6\frac{1}{2}$; **(e)** $7\frac{1}{2}$ **26.** [9.3B] **(a)** $\frac{2}{11}$; **(b)** $\frac{3}{11}$; **(c)** $\frac{4}{11}$; **(d)** $\frac{5}{11}$; **(e)** $\frac{6}{11}$ **27.** [9.3B] **(a)** $3\frac{1}{4}$; **(b)** $4\frac{1}{4}$; **(c)** $5\frac{1}{4}$; **(d)** $6\frac{1}{4}$; **(e)** $7\frac{1}{4}$ **28.** [9.3B] **(a)** 5.1; **(b)** 6.2; **(c)** 7.3; **(d)** 8.4; **(e)** 9.5 **29.** [9.3C] **(a)** -5.6; **(b)** -5.5; **(c)** -5.4; **(d)** -5.3; **(e)** -5.2
30. [9.3C] **(a)** -3.1; **(b)** -3.2; **(c)** -3.3; **(d)** -3.4; **(e)** -3.5
31. [9.3C] **(a)** -5.3; **(b)** -5.4; **(c)** -5.5; **(d)** -5.6; **(e)** -5.7
32. [9.3C] **(a)** $\frac{2}{11}$; **(b)** $\frac{3}{11}$; **(c)** $\frac{4}{11}$; **(d)** $\frac{5}{11}$; **(e)** $\frac{6}{11}$ **33.** [9.3C] **(a)** $-\frac{1}{45}$; **(b)** $-\frac{11}{45}$; **(c)** $-\frac{16}{45}$; **(d)** $-\frac{26}{45}$; **(e)** $-\frac{31}{45}$ **34.** [9.3D] **(a)** -2.8; **(b)** -2.7; **(c)** -2.6; **(d)** -2.5; **(e)** -2.4 **35.** [9.3D] **(a)** 4.3; **(b)** 4.4; **(c)** 4.5; **(d)** 4.6; **(e)** 4.7 **36.** [9.3D] **(a)** $\frac{5}{11}$; **(b)** $\frac{6}{11}$; **(c)** $\frac{7}{11}$; **(d)** $\frac{8}{11}$; **(e)** $\frac{9}{11}$
37. [9.3D] **(a)** $-\frac{13}{6}$; **(b)** $-\frac{5}{2}$; **(c)** $-\frac{19}{6}$; **(d)** $-\frac{7}{2}$; **(e)** $-\frac{3}{2}$
38. [9.3E] **(a)** -13.02; **(b)** -13.33; **(c)** -13.64; **(d)** -13.95; **(e)** -14.26 **39.** [9.3E] **(a)** 6.51; **(b)** 6.82; **(c)** 7.13; **(d)** 7.44; **(e)** 7.75 **40.** [9.3E] **(a)** $\frac{4}{9}$; **(b)** $\frac{8}{9}$; **(c)** $\frac{10}{9}$; **(d)** $\frac{14}{9}$; **(e)** $\frac{16}{9}$ **41.** [9.3E] **(a)** $-\frac{5}{7}$; **(b)** $-\frac{15}{14}$; **(c)** $-\frac{10}{7}$; **(d)** $-\frac{15}{7}$; **(e)** $-\frac{20}{7}$ **42.** [9.3F] **(a)** $-\frac{11}{2}$; **(b)** $-\frac{11}{3}$; **(c)** $-\frac{11}{4}$; **(d)** $-\frac{11}{5}$; **(e)** $-\frac{11}{6}$ **43.** [9.3F] **(a)** $-\frac{7}{5}$; **(b)** $-\frac{7}{10}$; **(c)** $-\frac{7}{15}$; **(d)** $-\frac{7}{20}$; **(e)** $-\frac{7}{25}$ **44.** [9.3F] **(a)** 10; **(b)** 15; **(c)** 20; **(d)** 25; **(e)** 30 **45.** [9.3F] **(a)** $-\frac{2}{3}$; **(b)** $-\frac{1}{2}$; **(c)** $-\frac{2}{5}$; **(d)** $-\frac{1}{3}$; **(e)** $-\frac{2}{7}$
46. [9.3F] **(a)** -2; **(b)** -3; **(c)** -4; **(d)** -5; **(e)** -6 **47.** [9.3F] **(a)** $\frac{1}{2}$; **(b)** $\frac{1}{3}$; **(c)** $\frac{1}{4}$; **(d)** $\frac{1}{5}$; **(e)** $\frac{1}{6}$ **48.** [9.3F] **(a)** -2; **(b)** -3; **(c)** -4; **(d)** -5; **(e)** -6
49. [9.3G]

	(a) 3.7	(b) $\sqrt{121}$	(c) $0.\overline{56}$	(d) $-3\frac{1}{5}$	(e) $\sqrt{21}$
Natural number		✓			
Whole number		✓			
Integer		✓			
Rational number	✓	✓	✓	✓	
Irrational number					✓
Real number	✓	✓	✓	✓	✓

50. [9.4A] **(a)** -12; **(b)** -13; **(c)** -14; **(d)** -15; **(e)** -16
51. [9.4A] **(a)** -47; **(b)** -57; **(c)** -77; **(d)** -87; **(e)** -97
52. [9.4A] **(a)** -29; **(b)** -13; **(c)** -5; **(d)** -1; **(e)** 1
53. [9.4B] **(a)** -40; **(b)** -51; **(c)** -49; **(d)** -62; **(e)** -77

Cumulative Review Chapters 1–9

1. 10 **2.** 450 **3.** 692.31 **4.** $y = 8$ **5.** $z = 39.6$ **6.** $c = \frac{1}{4}$ **7.** 0.23 **8.** 32 **9.** 50% **10.** 75 **11.** \$162 **12.** Agriculture
13.

14. 4% **15.** 2 degrees **16.** 47% **17.** 6 **18.** 17 **19.** 4 **20.** 324 in. **21.** $1\frac{1}{3}$ ft **22.** $5\frac{2}{3}$ yd **23.** 72 in. **24.** 1700 dm **25.** 13.4 in. **26.** 50.24 ft **27.** 160 cm^2 **28.** 113.04 cm^2 **29.** 261 in.^{2} **30.** 200 cm^3 **31.** 3052.08 in.^{3} **32.** Obtuse **33.** 125° **34.** 70 **35.** 10 **36.** $4\frac{3}{4}$ **37.** 9.8 **38.** -9.8 **39.** $\frac{2}{35}$ **40.** 4.3 **41.** $-\frac{11}{30}$ **42.** $-\frac{1}{3}$ **43.** $-\frac{9}{5}$ **44.** 81

Chapter 10

Exercises 10.1

1. $2x; -3$ **3.** $-3x^2; 0.5x; -6$ **5.** $\frac{1}{5}x; -\frac{3}{4}y; \frac{1}{8}z$ **7.** $6x + 3y$ **9.** $3a - 3b$ **11.** $-10x + 5y$ **13.** $24x^2 + 16$ **15.** $-12x^2 + 18$ **17.** $6x^2 + 9x + 15$ **19.** $-15x^2 + 10x + 15$ **21.** $0.5x + 0.5y - 1.0$

23. $\frac{6}{5}a - \frac{6}{5}b + 6$ **25.** $-2x + 2y - 8$ **27.** $-0.3x - 0.3y + 1.8$ **29.** $-\frac{5}{2}a + 5b - \frac{5}{2}c + \frac{5}{2}$ **31.** $3(x + 5)$ **33.** $9(y - 2)$ **35.** $-5(y - 4)$ **37.** $-3(x + 9)$ **39.** $b(x - y + z)$ **41.** $10a$ **43.** $6x$ **45.** $10a^2$ **47.** $11x$ **49.** $5y^2$ **51.** $5x + 3y$ **53.** $10x - 3y - 4$ **55.** $2.4a - 3b - 7$ **57.** $\frac{4}{7}a - \frac{4}{5}b$ **59.** $\frac{1}{4}x - \frac{5}{11}b$ **61.** $-6a - 13b$ **63.** $a - 5b$ **65.** $-3b^2 + 5a$ or $5a - 3b^2$ **67.** $3a^2 - 2b$ **69.** $-4b^2 + 8a$ or $8a - 4b^2$ **71.** -2 **73.** -8 **75.** 135 **77.** 8 **79.** 32 **81.** 29.5, 31.2, 32.9, 34.6, 36.3 **83.** 32.82, 33.34, 33.86, 34.38, 34.9 **85.** 716.2, 766.7, 817.2, 867.7, 918.2 **87.** 380.8, 393.4, 406, 418.6, 431.2 **89.** 9.72 million **91.** 10.92 million **93.** 19.5 million **99.** terms **101.** trinomial **103.** $-a + b$ **105.** evaluated **107.** (a) $3x + y$; (b) $-2x + 5y$ **109.** (a) $4a$; (b) $4a + 6b$; (c) $\frac{5}{9}a + \frac{1}{7}b$ **111.** (a) $3a + 6b$; (b) $10a - 15b$; (c) $-4a + 12b - 4c$ **113.** $9 \cdot 25 \cdot 1 = 225$ **115.** $1000 \cdot 1 \cdot 25 \cdot 8 = 200{,}000$

Exercises 10.2

1. (a) $\frac{1}{16}$; (b) $\frac{1}{x^2}$ **3.** (a) $\frac{1}{125}$; (b) $\frac{1}{y^3}$ **5.** (a) $\frac{1}{81}$; (b) $\frac{1}{z^4}$ **7.** (a) 2^{-3}; (b) x^{-3} **9.** (a) 4^{-5}; (b) y^{-5} **11.** (a) 3^{-5}; (b) z^{-5} **13.** (a) 243; (b) x^{13} **15.** (a) 4; (b) y^5 **17.** (a) $\frac{1}{16}$; (b) $\frac{1}{x^4}$ **19.** (a) $\frac{1}{216}$; (b) $\frac{1}{y^7}$ **21.** (a) $\frac{1}{64}$; (b) $\frac{1}{x^{10}}$ **23.** x^2 **25.** y^2 **27.** $\frac{1}{a^5}$ **29.** $\frac{1}{x^2}$ **31.** $\frac{1}{x^2}$ **33.** $\frac{1}{a^5}$ **35.** $b^0 = 1$ **37.** 243 **39.** $\frac{1}{64}$ **41.** $\frac{1}{y^2}$ **43.** x^3 **45.** $\frac{1}{x^2}$ **47.** $\frac{1}{x^7}$ **49.** x^3 **51.** 81 **53.** $\frac{1}{9}$ **55.** 64 **57.** $\frac{1}{9}$ **59.** $\frac{1}{x^9}$ **61.** $\frac{1}{y^6}$ **63.** a^6 **65.** $\frac{x^9}{y^6}$ **67.** $\frac{y^6}{x^4}$ **69.** $\frac{1}{x^9y^6}$ **71.** $\frac{1}{x^{12}y^6}$ **73.** $x^{12}y^{12}$ **75.** \$1259.71 **77.** \$1464.10 **79.** 11,397,000,000 km **81.** \$270,000,000 = \$270 million **83.** \$28,100 **85.** (a) 123,454,321; (b) 12,345,654,321 **87.** (a) $5^2 = 25$; (b) $7^2 = 49$ **93.** 1 **95.** x^{m+n} **97.** $x^{m \cdot n}$ **99.** $P(1 + r)^n$ **101.** (a) 64; (b) $\frac{1}{x^6}$; (c) $\frac{1}{a^{15}}$; (d) b^{10} **103.** (a) $\frac{a^{12}}{b^8}$; (b) $\frac{b^{12}}{a^{12}}$; (c) $\frac{a^6}{b^8}$ **105.** (a) 6^{-3}; (b) c^{-4} **107.** 73.1 **109.** 73,100

Exercises 10.3

1. 6.8×10^7 **3.** 2.93×10^8 **5.** 1.9×10^9 **7.** 2.4×10^{-4} **9.** 2×10^{-9} **11.** 235 **13.** 8,000,000 **15.** 6,800,000,000 **17.** 0.23 **19.** 0.00025 **21.** 1.5×10^{10} **23.** 3.06×10^4 **25.** 1.24×10^{-4} **27.** 2×10^3 **29.** 2.5×10^{-3} **31.** (a) 4.8×10^{10}; (b) 48,000,000,000 **33.** 4.63 **35.** 2×10^7 **37.** 9.66×10^6 = 9,660,000 people **39.** 1.8×10^4 people/km^2 = 18,000 people/km^2 **41.** $5 \times 10^2 = 500$ sec **43.** 2.99792458×10^8 **45.** 3.09×10^{13} **47.** 3.27 **51.** left **53.** 40 **55.** 2.5×10^5 **57.** 4,500,000,000 **59.** $x = -3$ **61.** $x = -\frac{1}{10}$ **63.** $x = -0.5$ **65.** $x = -3$

Exercises 10.4

1. $x = -2$ **3.** $y = -\frac{1}{2}$ **5.** $x = -7$ **7.** $y = \frac{3}{10}$ **9.** $x = 6.1$ **11.** $x = -8$ **13.** $y = \frac{1}{5}$ **15.** $x = -2.3$ **17.** $x = -16$ **19.** $y = -12.4$ **21.** $z = -2$ **23.** $x = -\frac{1}{2}$ or -0.5 **25.** $y = \frac{4}{3}$ or $1.\overline{3}$ **27.** $x = 1.1$ **29.** $y = 1.1$ **31.** $x = \frac{10}{3}$ **33.** $x = \frac{6}{5}$ **35.** $x = 1$ **37.** $y = 2$ **39.** $y = -10$ **41.** $x = \frac{12}{5}$ **43.** $y = -\frac{10}{3}$ **45.** $y = 1$ **47.** $x = 1$ **49.** $x = 0$ **51.** $x = -\frac{20}{3}$ **53.** $y = -\frac{16}{3}$ **55.** $x = \frac{15}{2}$ **57.** $y = \frac{1}{2}$ **59.** $y = \frac{11}{6}$ **61.** (a) $2.8 + I = 3.4$; (b) Subtraction Principle; (c) $I = 0.6$ (billion gallons) **63.** (a) $\frac{1}{50}T = 2.8$; (b) Multiplication Principle; (c) $T = 140$ (billion gallons) **65.** \$110 **67.** \$122 **69.** 70 in. **73.** equation **75.** $a - c = b - c$ **77.** $\frac{a}{c} = \frac{b}{c}, c \neq 0$ **79.** $x = -1$ **81.** $y = -7$ **83.** $z = -3$ **85.** $x = 5$ **87.** $x = 4$ **89.** $x = -9$ **91.** $\frac{2000}{3}$ **93.** 270

Exercises 10.5

Translate This

1. K **3.** A **5.** B **7.** F **9.** J

1. \$18 **3.** 10,207 **5.** 10,196 **7.** \$68 **9.** 10 **11.** 12 **13.** 27 gal **15.** 700 ft **17.** \$36 **19.** 72 **21.** 8% **23.** 8% **25.** 21% **27.** 4 **29.** 9 **31.** 3 hr **33.** 8000 gigs **35.** 500 employees, 351 IT decision-makers **37.** 175 SD, 155 LS, 133 IS **39.** 800 teens, 573 tweens **41.** 104 teens, 69 tweens **43.** (a) BMR = $12W$; (b) 150 lb **45.** (a) $700 = 0.20O$; (b) 3500 calories; (c) 1 lb **47.** (a) $C = 920\ h$; (b) 2760 calories; (c) the 150-lb person **49.** (a) 170 lb; (b) 140 lb; (c) 119 lb **51.** (a) $P = 0.067E$; (b) 10.05 lb; (c) 140 lb **53.** (a) 2170; (b) 2710; (c) 3070 **55.** (a) 3.3 lb; (b) 0.5 lb **57.** (a) 893 million L; (b) 3524 million L; (c) 6155 million L **59.** (a) 310.8 million; (b) 74 yr; (c) 2080 (2006 + 74) **61.** +6 **63.** +8 **69.** Read **71.** Translate **73.** Verify **75.** 4 **77.** 320 mi **79.** 180 lb

Review Exercises

1. [10.1A] (a) $6x^3; -3x^2; 7x; -7$; (b) $5x^3; -4x^2; 6x; -6$; (c) $4x^3; -5x^2; 5x; -5$; (d) $3x^3; -6x^2; 4x; -4$; (e) $2x^3; -7x^2; 3x; -3$ **2.** [10.1B] (a) $2x + 4y$; (b) $3x + 6y$; (c) $4x + 8y$; (d) $5x + 10y$; (e) $6x + 12y$ **3.** [10.1B] (a) $-4x + 8y - 4z$; (b) $-5x + 10y - 5z$; (c) $-6x + 12y - 6z$; (d) $-7x + 14y - 7z$; (e) $-8x + 16y - 8z$ **4.** [10.1C] (a) $3(x - 4y)$; (b) $3(x - 5y)$; (c) $3(x - 6y)$; (d) $3(x - 7y)$; (e) $3(x - 8y)$ **5.** [10.1C] (a) $10(4x - 5y - 6z)$; (b) $10(3x - 4y - 5z)$; (c) $10(2x - 3y - 4z)$; (d) $10(x - 2y - 3z)$; (e) $10(x - y - 2z)$ **6.** [10.1D] (a) $2x + 9y$; (b) $3x + 8y$; (c) $4x + 7y$; (d) $4x + 9y$; (e) $4x + 9y$ **7.** [10.1D] (a) $-0.2x + 0.54y$; (b) $-0.3x + 0.55y$; (c) $-0.4x + 0.56y$; (d) $-0.5x + 0.57y$; (e) $-0.6x + 0.58y$ **8.** [10.1D] (a) $\frac{4}{11}x + \frac{7}{13}y$; (b) $\frac{5}{11}x + \frac{6}{13}y$; (c) $\frac{6}{11}x + \frac{5}{13}y$; (d) $\frac{7}{11}x + \frac{4}{13}y$; (e) $\frac{8}{11}x + \frac{3}{13}y$ **9.** [10.1E] (a) $3a + 3b$; (b) $4a + 3b$; (c) $5a + 7b$; (d) $8a + 3b$; (e) $7a + 8b$ **10.** [10.2A] (a) $\frac{1}{4}$; (b) $\frac{1}{8}$; (c) $\frac{1}{16}$; (d) $\frac{1}{32}$; (e) $\frac{1}{64}$ **11.** [10.2A] (a) 2^{-3}; (b) 2^{-4}; (c) 2^{-5}; (d) 2^{-6}; (e) 2^{-7} **12.** [10.2B] (a) x^7; (b) x^8; (c) x^9; (d) x^{10}; (e) x^{11} **13.** [10.2B] (a) $\frac{1}{4}$; (b) $\frac{1}{8}$; (c) $\frac{1}{16}$; (d) $\frac{1}{32}$; (e) $\frac{1}{64}$ **14.** [10.2B] (a) $\frac{1}{y^2}$; (b) $\frac{1}{y^3}$; (c) $\frac{1}{y^4}$; (d) $\frac{1}{y^5}$; (e) $\frac{1}{y^6}$ **15.** [10.2C] (a) 64; (b) 128; (c) 256; (d) 512; (e) 1024 **16.** [10.2C] (a) $\frac{1}{x^3}$; (b) $\frac{1}{x^4}$; (c) $\frac{1}{x^5}$; (d) $\frac{1}{x^6}$; (e) $\frac{1}{x^7}$ **17.** [10.2C] (a) y^2; (b) y^3; (c) y^4; (d) y^5; (e) y^6 **18.** [10.2D] (a) 4; (b) 16; (c) 64; (d) 256; (e) 1024 **19.** [10.2D] (a) $\frac{1}{y^8}$; (b) $\frac{1}{y^{12}}$; (c) $\frac{1}{y^{16}}$; (d) $\frac{1}{y^{20}}$; (e) $\frac{1}{y^{24}}$ **20.** [10.2D] (a) z^6; (b) z^9; (c) z^{12}; (d) z^{15}; (e) z^{18} **21.** [10.2D] (a) $\frac{x^6}{y^4}$; (b) $\frac{x^9}{y^6}$; (c) $\frac{x^{12}}{y^8}$; (d) $\frac{x^{15}}{y^{10}}$; (e) $\frac{x^{18}}{y^{12}}$ **22.** [10.2D] (a) $\frac{x^4}{y^6}$; (b) $\frac{x^6}{y^9}$; (c) $\frac{x^8}{y^{12}}$; (d) $\frac{x^{10}}{y^{15}}$; (e) $\frac{x^{12}}{y^{18}}$ **23.** [10.3A] (a) 4.4×10^7; (b) 4.5×10^8; (c) 4.6×10^6; (d) 4.7×10^4; (e) 4.8×10^7 **24.** [10.3A] (a) 1.4×10^{-3}; (b) 1.5×10^{-4}; (c) 1.6×10^{-5}; (d) 1.7×10^{-6}; (e) 1.8×10^{-7} **25.** [10.3A] (a) 7830; (b) 68,300; (c) 583,000; (d) 4,830,000; (e) 38,300,000 **26.** [10.3A] (a) 0.084; (b) 0.0074; (c) 0.00064; (d) 0.000054; (e) 0.0000044 **27.** [10.3B] (a) 2.2×10^5; (b) 9.3×10^6; (c) 1.24×10^8; (d) 1.55×10^9; (e) 1.86×10^9 **28.** [10.3B] (a) 2.2×10^{-2}; (b) 3.3×10^{-1}; (c) 4.4×10^{-3}; (d) 5.5×10^{-4}; (e) 6.6×10^{-4} **29.** [10.3B] (a) 5×10^0 or 5; (b) 6×10^1; (c) 7×10^1; (d) 8×10^{-1}; (e) 9×10^0 or 9 **30.** [10.4A] (a) $y = \frac{1}{6}$; (b) $y = \frac{1}{4}$; (c) $y = \frac{3}{10}$; (d) $y = \frac{1}{3}$; (e) $y = \frac{5}{14}$ **31.** [10.4A] (a) $y = -6$; (b) $y = -9$; (c) $y = -12$; (d) $y = -15$; (e) $y = -18$ **32.** [10.4A] (a) $z = -2$; (b) $z = -3$; (c) $z = -4$; (d) $z = -6$; (e) $z = -7$ **33.** [10.4B] (a) $x = -\frac{5}{3}$; (b) $x = -\frac{8}{3}$; (c) $x = -\frac{10}{3}$; (d) $x = -\frac{13}{3}$; (e) $x = -\frac{16}{3}$ **34.** [10.4B] (a) $x = 8$; (b) $x = 9$; (c) $x = 10$; (d) $x = 11$; (e) $x = 12$

35. [10.4B] **(a)** $x = 2.1$; **(b)** $x = 2.1$; **(c)** $x = 2.1$; **(d)** $x = 2.1$; **(e)** $x = 2.1$ **36.** [10.4B] **(a)** $x = 11$; **(b)** $x = 8$; **(c)** $x = 7$; **(d)** $x = 6$; **(e)** $x = 5$ **37.** [10.4B] **(a)** $x = -\frac{21}{2}$; **(b)** $x = -\frac{35}{2}$; **(c)** $x = -\frac{49}{2}$; **(d)** $x = -\frac{63}{2}$; **(e)** $x = -\frac{77}{2}$ **38.** [10.4B] **(a)** $x = -\frac{5}{2}$; **(b)** $x = -\frac{5}{2}$; **(c)** $x = -\frac{5}{2}$; **(d)** $x = -\frac{5}{2}$; **(e)** $x = -\frac{5}{2}$ **39.** [10.5A] **(a)** $m = 202$; **(b)** $m = 203$; **(c)** $m = 204$; **(d)** $m = 206$; **(e)** $m = 207$

40. [10.5A] **(a)** 6%; **(b)** 8%; **(c)** 10%; **(d)** 12%; **(e)** 14%

Cumulative Review Chapters 1–10

1. 52 **2.** 615.38 **3.** 0.23 **4.** 36 **5.** 50% **6.** 20 **7.** $112.50 **8.** 1% **9.** 40% **10.** 5 **11.** 13 **12.** 10 **13.** 144 in. **14.** $1\frac{5}{12}$ ft **15.** $5\frac{2}{3}$ yd **16.** 84 in. **17.** 2400 dm **18.** 15.2 cm **19.** 56.52 in. **20.** 80 cm^2 **21.** 28.26 cm^2 **22.** 105 cm^3 **23.** 7234.56 $in.^3$ **24.** Right **25.** 120° **26.** 90 **27.** 15 **28.** $3\frac{4}{5}$ **29.** 3.5 **30.** -11.5 **31.** $\frac{11}{40}$ **32.** 0.3 **33.** $-\frac{24}{35}$ **34.** $-\frac{9}{8}$ **35.** $-\frac{10}{3}$ **36.** 49 **37.** $20x - 40y$ **38.** $10(6x - 3y - 4z)$ **39.** $-5x + 3y$ **40.** $\frac{1}{9^2} = \frac{1}{81}$ **41.** $\frac{1}{x^7}$ **42.** $\frac{9x^6}{y^4}$ **43.** 2.4×10^{-8} **44.** 28,200,000 **45.** 3.5×10^{-8} **46.** $x = -49$ **47.** $x = 2$ **48.** $x = 7$ **49.** $x = 13$

Photo Credits

Chapter 1

Page **1:** © Anne Gransden/SuperStock, Inc; **5:** © AP Photo/Gail Oskin; **6:** © Brand X Pictures/Creatas/Corbis; **13:** © The McGraw-Hill Companies, Inc./Emily and David Tietz, photographer;**19:** © The McGraw-Hill Companies, Inc./Connie Mueller, photographer; **23:** © Reuters/Corbis; **32(left):** Courtesy NPS/Canyon de Chelly National Monument;, **32(right):** © Hisam F. Ibrahim/Getty Images RF; **48:** © Todd Pearson/The Image Bank/Getty Images; **55:** © Dale Wilson/Masterfile; **56:** Courtesy of The Florida Strawberry Festival ®; **62, 88:** © The McGraw-Hill Companies, Inc./Emily and David Tietz, photographer; **96:** Courtesy Ignacio Bello.

Chapter 2

Page **109:** © Edward Cohen/SuperStock; **110, 114, 116:** © The McGraw-Hill Companies, Inc./Emily and David Tietz, photographer; **118, 120, 129, 136:** Courtesy Ignacio Bello; **139(top):** © Steve Cole/Getty Images/V51, **139(left):** © Ryan McVay/Getty Images; **139(right):** © Nicholas Eveleigh/Superstock; **148(left):** © John A. Rizzo/Getty Images RF; **148(right):** © Image Source/Getty RF; **149(left):** Courtesy USDA ARS; **149(right):** © Spike Mafford/Getty Images RF; **151:** © The McGraw-Hill Companies, Inc./Emiily and David Tietz, photographer; **169:** Courtesy Ignacio Bello; **182:** © The McGraw-Hill Companies, Inc./Emily and David Tietz, photographer.

Chapter 3

Page **199:** © Robert Brenner/PhotoEdit; **200:** Courtesy Jim Goetz; **205, 209, 210, 211(all), 214, 223(both):** Courtesy Ignacio Bello; **239:** © Tony Freeman/Photo Edit.

Chapter 4

Page **255:** © Susan VanEtten/Photo Edit; **256(left):** © Underwood & Underwood/Corbis; **256(right):** © C. Borland/PhotoLink/Getty Images RF; **258:** © The McGraw Hill Companies, Inc./Connie Mueller Photo; **258:** © Royalty Free/Corbis; **259:** © Tom Pantages adaptation/Courtesy BLM; **261:** © Corbis/Sygma; **272, 273:** © Gary Meissner Goldennumber.net; **276:** © Pam Gardner, Frank Lane Picture Agency/Corbis.

Chapter 5

Page **287:** © Alan Schein Photography/Corbis; **293:** Courtesy Ignacio Bello; **314(top):** © Ryan McVay/Getty Images/VOS56; **314(bottom):** © Lake County Museum/Corbis; **317:** © Bill Bachmann/Photo Researchers, Inc.; **325:** Courtesy Honda; **330:** © Alan Schein Photography/Corbis.

Chapter 6

Page **349:** Courtesy National Library; **353(left):** © Ryan McVay/Getty Images/V114; **353(right):** © C. Borland/PhotoLink/Getty Images/V23; **353(lower):** © Steve Cole/Getty Images/SS38; **379:** © David Job/The Image Bank/Getty Images; **386(both), 389(both):** Courtesy Ignacio Bello; **391:** © AP/Wide World Photos/Chris O'Meara.

Chapter 7

Page **409:** © Richard Cummins/Corbis; **410:** © Photodisc/Getty Images; **411:** © Photos at Your Place/http://PhotosatYourPlace.com; **414(right):** © Photodisc/Getty Images; **414(left):** © Flat Earth Images RF; **417:** © Photodisc/Getty Images; **426:** © Chromosohm Media Inc./The Image Works; **430:** Courtesy Michel Verdure/Royal Caribbean International; **431:** © The McGraw-Hill Companies, Inc./Emily and David Tietz, photographer; **442:** © Creatas/PunchStock RF.

Chapter 8

Page **453:** © David Young-Wolff/PhotoEdit; **454:** Mondrian, Piet (1872–1944) II: Blanc et Rouge (white and red), 1937/Composition in Red, Blue, and Yellow, 1937-42. Oil on canvas, $23\frac{3}{4} \times 21\frac{7}{8}$ (60.3 × 55.4 cm). © 2007 Mondrian/Holtzman Trust c/o HCR International, Warrenton VA Digital Image © The Museum of Modern Art /Licensed by SCALA /Art Resource, NY; **456:** © Michael T. Sedam/Corbis; **460, 466(left, right), 467(top left, top right):** © David Young-Wolff/Photo Edit; **467(bottom):** © Michael T. Sedam/Corbis; **479:** © Vol. 24/PhotoDisc/Getty Images. **483(left):** Courtesy of Fermi National Accelerator Lab/U.S. Department of Energy; **483(right):** © San Francisco Chronicle Corbis SABA; **489(left, right), 490(top, bottom):** Courtesy Ignacio Bello; **491(left, right):** Viz-A-Ball is a registered trademark of Brunswick Bowling & Billiards Corporation. U.S. Patent No. 6,524,419 and 6,691,759; **491(bottom):** © Burazin/Masterfile; **492(top):** Courtesy of elope Inc; **492(bottom):** © Royalty Free/Corbis; **493:** Courtesy of EPA; **494(top):** © Photodisc/Getty Images; **494(middle left):** © David Young-Wolff/Photo Edit; **494(middle right):** © Michael Newman/Photo Edit; **494(bottom left):** © Bill Aron/Photo Edit; **494(bottom right):** © Felicia Martinez/Photo Edit; **495(top left):** Courtesy of UltraTech International; **495(top right):** Courtesy Ignacio Bello; **495(bottom):** © Lawrence Lawry/Getty Images/ V56; **496(top left):** Courtesy Ignacio Bello; **496(top right):** © David Young-Wolff/Photo Edit; **496(bottom left, Pyramid of the Sun):** © Neil Beer/PhotoDisc/Getty Images; **496(bottom left, Pyramids at Giza):** © Digital Vision/Getty RF; **496(bottom right):** © Danny Lehman/Corbis RF; **497(left, lower left):** Courtesy Ignacio Bello; **497(bottom left):** © Royalty Free Corbis/V188; **498(top):** © Michael Newman/Photo Edit; **498(bottom):** Courtesy Ignacio Bello; **499:** © David Young-Wolff/PhotoEdit; **502:** © Steve Cole/Getty Images/V51; **504(left, right, bottom):** Courtesy Ignacio Bello; **505:** Courtesy HNBT.

Chapter 9

Page **521:** © Robert George Young/Masterfile; **532:** © F. Vnoucek/Peter Arnold, Inc; **546:** © David Young-Wolff/Photo Edit; **556:** © Rim Light/PhotoLink/Getty Images/V27.

Chapter 10

Page **575:** © Neal & Molly Jansen/Superstock; **596:** Courtesy Ignacio Bello; **602:** © Science Source/Photo Researchers, Inc; **611:** Courtesy of Nasa/James B. Irwin; **618:** © Adam Crowley/PhotoDisc/Getty Images; **619(top):** © TRBfoto/PhotoDisc/Getty Images; **619(bottom):** © Flat Earth Images RF; **620(top, bottom):** © Royalty Free/Corbis; **621(left, right):** © JPL/NASA; **622:** © Mitch Hrdilcka/PhotoDisc/Getty Images.

Index

Index

I

J

K

L

M

N

Index